Chemistry of Environmental Systems

Chemistry of Environmental Systems

Fundamental Principles and Analytical Methods

Jeffrey S. Gaffney
University of Arkansas at Little Rock (retired)
USA

Nancy A. Marley
Argonne National Laboratory (retired)
USA

This edition first published 2020

Registered Offices
John Wiley & Sons, Inc., 111 River Street, Hoboken, NJ 07030, USA
John Wiley & Sons Ltd, The Atrium, Southern Gate, Chichester, West Sussex, PO19 8SQ, UK

Editorial Office
The Atrium, Southern Gate, Chichester,West Sussex, PO19 8SQ, UK

For details of our global editorial offices, customer services, and more information about Wiley products visit us at www.wiley.com.

Wiley also publishes its books in a variety of electronic formats and by print-on-demand. Some content that appears in standard print versions of this book may not be available in other formats.

Library of Congress Cataloging-in-Publication Data

Names: Gaffney, Jeffrey S., 1949- author. | Marley, Nancy A., 1948- author.
Title: Chemistry of environmental systems : fundamental principles and analytical methods / Jeffrey S. Gaffney, Ph.D. (emeritus professor of chemistry, University of Arkansas at Little Rock, Department of Chemistry), Nancy A. Marley, Ph.D. (scientist, Argonne National Laboratory, retired).
Description: First edition. | Hoboken, NJ : Wiley, 2020. | Includes bibliographical references and index. |
Identifiers: LCCN 2019014828 (print) | LCCN 2019017607 (ebook) | ISBN 9781119313632 (Adobe PDF) | ISBN 9781119313588 (ePub) | ISBN 9781119313403 (hardcover)
Subjects: LCSH: Environmental chemistry–Textbooks. | Atmospheric chemistry–Textbooks. | Chemistry–Textbooks.
Classification: LCC TD193 (ebook) | LCC TD193 .G34 2019 (print) | DDC 577/.14–dc23
LC record available at https://lccn.loc.gov/2019014828

Cover Design: Wiley
Cover Image: © anucha sirivisansuwan/Getty Images

Set in 10/12pt WarnockPro by SPi Global, Chennai, India
Printed and bound in Singapore by Markono Print Media Pte Ltd

10 9 8 7 6 5 4 3 2 1

In memory of Dr. James N. Pitts, Jr., Dr. F. Sherwood Rowland, and Dr. Jack G. Calvert who taught us about the importance of the fundamental principles of environmental chemistry

Dr. James N. Pitts, Jr. (January 10, 1921–June 19, 2014)

Dr. F. Sherwood Rowland (June 28, 1927–March 10, 2012)

Dr. Jack G. Calvert (May 9, 1923–June 1, 2016)

Contents

About the Authors

Dr. Jeffrey S. Gaffney obtained his undergraduate and graduate training in physical organic chemistry from the University of California at Riverside, under the guidance of Dr. James N. Pitts, Jr. He was a nationally and internationally known Senior Chemist at three of the U.S. Department of Energy (DOE) National Labs (Brookhaven, Los Alamos, and Argonne) for 31 years before coming to the University of Arkansas at Little Rock (UALR) in 2006 as Chair and Tenured Professor of Chemistry. On July 1, 2016 Dr. Gaffney retired as Emeritus Professor of Chemistry at UALR. Dr. Gaffney has published over 200 peer-reviewed papers, 25 book chapters, and edited two American Chemical Society symposium book volumes. His research spans a wide range of chemistry in all of the basic areas, and he is internationally known for his work in air and water chemistry and global change research. Dr. Gaffney has taught undergraduate and graduate courses in general chemistry, organic chemistry, environmental chemistry, and the history of chemistry. He developed a one-semester course for senior undergraduates and graduate chemistry students entitled "Environmental Chemistry," that he has taught for eight years. This course combines the chemistry of the atmosphere, hydrosphere, geosphere, and biosphere – and the interactions between each of these areas – which is needed to successfully understand the chemistry of the environment as a whole system.

Dr. Nancy A. Marley obtained her B.S. in Chemistry from Jacksonville University and her Ph.D. in Analytical Chemistry and Optical Spectroscopy from Florida State University under the direction of Dr. Thomas J. Vickers. Between her B.S. degree and doctoral degree, Dr. Marley was District Chemist for the Department of Environmental Quality in Florida and also the Lead Chemist for the Childhood Lead Poisoning Program for the Florida Department of Health and Rehabilitative Services. She was a postdoctoral researcher at Los Alamos National Laboratory before joining Argonne National Laboratory, where she worked for 18 years. After retiring from Argonne National Laboratory, she joined the University of Arkansas at Little Rock as an Associate Research Professor. Currently a retired scientist/scholar, she has published over 140 peer-reviewed papers, 14 book chapters, and edited two American Chemical Society symposium book volumes. Her research has focused on the applications of laser Raman, infrared, and UV–visible–near IR spectroscopy to problems in environmental chemistry and geochemistry. She has also developed a number of analytical methods using chemiluminescent reactions and fast gas chromatography in collaboration with Dr. Gaffney.

Both Dr. Gaffney and Dr. Marley worked closely as co-principal investigators with the geochemistry/environmental chemistry/atmospheric science staff at both Los Alamos National Laboratory and Argonne National Laboratory. They also collaborated with geoscientists at the University of Chicago and the University of Illinois at Chicago on projects ranging from atmospheric chemistry to climate change, aqueous transport of radionuclides, and instrument development.

Preface

Chemistry of Environmental Systems is written with the overall concept of teaching the subject in a manner that develops a proactive science, learning from our past experiences and using the knowledge obtained from these experiences to prevent future environmental impacts. This is differentiated from the reactive science that environmental chemistry has been in the past. Thus, environmental chemistry as a subject is not simply the measurement of pollution species in the various environmental phases, but is the understanding of the chemical processes in the natural environment and how these natural chemical processes react when perturbed by either acute or chronic additions of chemical compounds into the whole Earth system.

This book addresses the development of the chemistry of environmental systems as it has changed over the past 50 years. *Chemistry of Environmental Systems* evolved from a senior-level undergraduate/lower-level graduate course in Environmental Chemistry taught by one of the authors (JSG) at the University of Arkansas at Little Rock. This course used two textbooks as resources: *Chemistry of the Upper and Lower Atmosphere*, by Dr. Barbara Finlayson-Pitts and Dr. James N. Pitts, Jr. and the Second Edition of *Chemistry of the Environment*, by Dr. Ronald A. Bailey, Dr. Herbert M. Clark, Dr. James P. Ferris, Dr. Sonja Krause, and Dr. Robert L. Strong. Both of these texts have influenced the writing of this book and while *Chemistry of the Upper and Lower Atmosphere* details atmospheric chemistry as it was known at the date of publication (2000), *Chemistry of the Environment* covers the subject with a more holistic approach, including the environmental problems of the atmosphere, hydrosphere, and geosphere that were clearly identified by the year of publication (2002). While teaching the Environmental Chemistry course, it became increasingly apparent that the impacts of pollution on the chemistry of the environment are global in nature, even though they were originally thought to be local issues. Also, environmental chemical reactions are not restricted to one specific environmental phase. For example, the fundamental reaction mechanisms that are important in gas-phase chemistry of the troposphere and stratosphere are, in many cases, also important mechanisms for aqueous and surface phase reactions.

Major environmental problems, such as urban air pollution, stratospheric ozone depletion, acid rain, and catastrophic releases of pollutants like oil spills, have been well known for decades, if not centuries. The potential impacts of climate change have now brought even more attention to the fact that these environmental problems are not standalone one-time events, but will continue to be major issues as we move forward in the Anthropocene. The impacts of the ever-growing human population on the environment are real and require that we address these issues in a more fundamental way, focusing on the principles of environmental chemistry. This approach is very important in training future scientists in all disciplines to begin to develop critical thinking skills in order to avoid future major environmental impacts and to move toward a sustainable global environment. Indeed, fully understanding environmental impacts requires that students and future environmental researchers have a solid background in all areas of chemistry and are trained to be able to ask the right questions of engineers, biologists, and ecologists, to aid in providing the answers to important questions of energy and technology development in the years to come. These scientists will also need to be able to effectively communicate their findings to public policy makers and leaders to insure that we, as a global community, make the best decisions regarding chemical and energy use to insure that our air and water quality is maintained and to prepare to adapt to the potential impacts of our changing climate and weather patterns, induced by anthropogenic emissions from an increasing world population.

This book is intended not only for use as a textbook, but also as a reference guide to provide a solid background in environmental chemistry and the basic mechanisms involved in the various phases of the environment, which are in constant equilibrium. Some key references used in this text are pre-2000 and are typically not accessed by current scientists, who rely on the Internet as their major source of information. Following in the tradition of the influential environmental texts noted earlier, units of measurement that are still commonly used in the environmental chemistry literature are used in this text but they are defined in terms of the accepted *SI* units to assist the reader in connecting current and past work in environmental chemistry with the most ease. The organization of the book first addresses the chemistry and radiative balance of the atmosphere associated with climate change and air quality. It then continues into the areas of aqueous and heterogeneous chemistry. The linkages between these areas and the biosphere are stressed to indicate the important concept that environmental impacts, in many cases, can be amplified by biospheric or geospheric feedbacks. Key concepts that are fundamental to many areas of chemistry – such as the basic principles and laws of photochemistry, thermodynamics, and kinetics – are reviewed, along with principles of organic and nuclear chemistry related to energy use. Analytical methods that are commonly used in environmental chemistry for both air and water analyses are also included so that the reader will appreciate the relationship between the fundamental principles of chemistry and the methods used to perform analyses on key environmental species.

Throughout the chapters, brief histories of the subject areas are presented to give the reader a perspective of the developing awareness of some environmental problems that we have come to understand today. In many cases these problems actually arose from perceived solutions of another problem. A classic example of this is the development of chlorofluorocarbons (CFCs) as the working fluids in refrigeration by Thomas Midgley. Seen as a major safety improvement over ammonia and butane as working fluids, the potential impacts of CFCs on the protective ozone layer in the stratosphere were not considered until almost 40 years later. The recognition and evaluation of the impacts of the CFCs led to a Nobel Prize in Chemistry to Rowland, Molina, and Crutzen in 1995, as well as to a major change in the global use of chlorinated chemicals (the Montreal Protocol of 1992). These and other examples are used to emphasize a basic principle of environmental chemistry, which is the analysis of "cradle to grave" impacts of chemicals on the environment in order to develop safer chemical products and processes for tomorrow's sustainable economies.

During seven years of teaching a senior undergraduate and first-year graduate one-semester course entitled "Environmental Chemistry," Dr. Gaffney compiled student responses to the material presented in the course. Working with Dr. Marley, they determined that there was a real need for an environmental chemistry textbook that taught a proactive approach to environmental problems, stressed the fundamental principles of chemistry necessary to their understanding, and also included information about the analytical methods used in environmental measurements. This recognition has led to the joint effort by Dr. Gaffney and Dr. Marley to produce *Chemistry of Environmental Systems: Fundamental Principles and Analytical Methods.*

While this text is not meant to be an extensive review of the literature, it is aimed at covering the key aspects and mechanisms of the currently identified environmental issues, which can be used to address both current and future environmental problems in a routine manner. Key fundamental properties of chemical compounds, such as their solubilities and volatilities, along with basic chemical reaction mechanisms, are reviewed and stressed. It is hoped that this textbook will encourage future environmental chemists to learn from past mistakes and use our current knowledge of the fundamental chemistry and physics of molecules and atoms to minimize future environmental impacts. Indeed, future environmental chemists will need to work together with engineers and biotechnologists to develop a safe, sustainable methodology for the global community. As part of that approach, study problems are provided at the end of each chapter, with answers provided in Appendix A. These can be used as class assignments or as individual exercises to reinforce the reader's comprehension of the material.

Acknowledgments

Chemistry of Environmental Systems was conceived from a one-semester course that Dr. Gaffney developed and taught during his tenure as Chair and Professor of Chemistry at the University of Arkansas at Little Rock. During the evolution of this course, it was clear that available textbooks in this area were either outdated or did not cover the breadth of the chemistry encountered in environmental chemistry. While writing this textbook, a number of questions arose concerning how best to present the material in the context of the various chemical disciplines that are fundamental to environmental chemistry (i.e. analytical, biochemical, inorganic, organic, and physical chemistry). The authors wish to acknowledge all of the faculty members in the Department of Chemistry at the College of Arts, Letters, and Science, University of Arkansas at Little Rock for their assistance and helpful discussions in the various chemical disciplines that helped make this a better book. We also wish to thank the upper-level undergraduate and graduate students who took CHEM 4352/5352 Environmental Chemistry over the years for their useful feedback and encouragement, which ultimately affirmed that we were on the right track with *Chemistry of Environmental Systems*.

Supplementary Material

To access supplementary materials for this book please use the download links shown below.

http://booksupport.wiley.com

Please enter the book title, author name or ISBN to access this material.

Here you will find valuable material designed to enhance your learning including:

- Figure PPTs

1

Introduction to Environmental Chemistry

1.1 What is Environmental Chemistry?

Environmental chemistry is traditionally defined as "the study of the sources, reactions, transport, effects and fates of chemical species, in water, soil, and air environments, and the effect of human activity on these" (Simple English Wikipedia, n.d.). This definition implies a major focus on the measurement of pollutants in the various environments. However, a more precise definition would be the study of chemistry in natural systems and how this chemistry changes when perturbed by anthropogenic activities and/or the release of chemicals into the environment which changes their natural background levels. Environmental chemistry is not simply the measurement of air, water, or soil pollutants. It is a multidisciplinary subject that requires the environmental chemist to have a solid background in all areas of chemistry, including analytical, physical, organic, inorganic, and simple biochemistry. It is also a discipline that requires studies across boundaries between the different fields of chemistry, physics, biology, ecology, meteorology, and others involved in the various environmental systems, which include the atmosphere, hydrosphere, geosphere, and biosphere. Understanding how these different systems interact allows the environmental chemist to give an accurate assessment of the impacts of chemical species on the environment as a whole.

Thus, the scope of environmental chemistry covers all of the Earth's systems and their interactions with each other. Future environmental chemists need to recognize these interactions and include them when predicting how energy and chemical usage will affect each of the various systems. Traditionally, environmental chemistry is taught in a compartmentalized manner, treating the chemistries of the air, water, soil, and biota separately. Although this is done in an effort to simplify the treatment of these complicated systems, it does not address the interactions between the different systems or the connections between the physics and chemistry of each system. Therefore, a major focus of this text will be on the connections between the various environmental compartments, with special attention given to the fundamental chemistry that is common to all of the various systems.

Environmental chemistry should not be confused with "green chemistry." The goal of green chemistry is the development of chemical processes that use smaller amounts of safer chemicals with less energy use in order to lower their environmental impacts. The goal of environmental chemistry is the understanding of the chemical reactions and processes that control the environmental systems and how these are impacted by the addition of anthropogenic chemicals. This approach allows environmental chemistry to be proactive instead of reactive. In the past, environmental chemistry has had a more reactive approach, identifying problems after the occurrence of a tragedy or a clearly obvious impact on the environment with loss of life, damage to plants and animals, or radical changes in environmental conditions – such as the loss of visibility in air and water systems or the depletion of the stratospheric ozone layer. Some of

Chemistry of Environmental Systems: Fundamental Principles and Analytical Methods, First Edition.
Jeffrey S. Gaffney and Nancy A. Marley.

these past incidents will be examined in this text as examples of events that have taught us the potential impacts of anthropogenic pollution on the environment and the importance of understanding how the different environmental systems are linked. The lessons learned from these events have helped us to understand how the whole Earth system works and how the separate environmental systems interact. So, this book will stress how the various chemistries of the natural systems in the atmosphere, geosphere, hydrosphere, and biosphere interact with each other, sometimes enhancing environmental impacts and sometimes mitigating those impacts.

1.2 Anthropogenic Pollution

Anthropogenic pollution is defined as the introduction of harmful substances or the creation of harmful impacts in the environment that are directly tied to man's activities, including: agriculture, industry, and energy production and use. It also includes the release and deposition into the environment of waste materials that end up in landfill or are incinerated. Anthropogenic pollution can be described as being intentional or nonintentional, with intentional pollution in many cases being tied to warfare. Our knowledge of the chemistry of the Earth's environmental systems is based on the studies of these anthropogenic impacts on our air and water resources due to the release of chemicals that are not natural to the environment or are emitted at much higher levels than natural.

The potential impacts on air and water quality from anthropogenic pollution have been recognized for a long time. The intentional pollution of water supplies with poisons or waste was used in ancient times, during long-term sieges of major cities such as the siege of Tortona, Italy in 1155 led by Frederick Barbarossa (Bradbury, 1992). It has been documented as early as 1000 BCE with the ancient Chinese putting arsenic in the water supplies of their enemies (Kroll, 2006). Unintentional pollution was recognized to correlate with higher human population densities in cities. It was strongly linked to the observation of polluted air and water, which not only impact the environmental systems, but also human health and longevity. As early as the twelfth century, Moses Maimonides (Goodhill, 1971; Finlayson-Pitts and Pitts, 2000), a well-known philosopher and physician shown in Figure 1.1, noted in his writings about the city of Cairo that:

> Comparing the air of cities to the air of deserts and arid lands is like comparing waters that are befouled and turbid to waters that are fine and pure. In the city, because of the height of its buildings, the narrowness of its streets, and all that pours forth from its inhabitants and their superfluities… the air becomes stagnant, turbid, thick, misty, and foggy… If there is no choice in this matter, for we have grown up in the cities and have become accustomed to them, you should… select from the cities one of open horizons… endeavor at least to dwell at the outskirts of the city… If the air is altered ever so slightly, the state of the Psychic Spirit will be altered perceptibly. Therefore you find many men in whom you can notice defects in the actions of the psyche with the spoilage of the air, namely, that they develop dullness of understanding, failure of intelligence and defect of memory…

This quote from Maimonides predates the Industrial Revolution, when the invention of the steam engine and enhanced farming techniques led to rapidly increasing populations. It clearly notes that, in past as well as present urban settings, anthropogenic pollution is strongly tied to population densities. The increase in population in centralized areas requires increases in energy and water usage, along with the increased transportation necessary to bring the required

Figure 1.1 Moses Maimonides, a twelfth-century philosopher and physician who reported on the degradation of water and air quality in Cairo. Source: U.S. National Library of Medicine Digital Collection.

goods to and from the urban centers. Indeed, the remarks made by Maimonides hit on a number of the present impacts on the chemistry of natural systems, such as the degradation of air and water quality leading to an unpleasant and unhealthy environment.

Another example of the early recognition of health issues caused by exposure to environmental contaminants is the linking of cancer in young male chimney sweeps to environmental exposure by Sir Percival Pott, an English surgeon shown in Figure 1.2. In 1775 Pott attributed the high incidence of scrotal cancer in young chimney sweeps to the exposure and inhalation of chimney soot (Dobson, 1972). This was the first documented identification of the connection of an environmental occupational exposure to an unintentional health effect. It also led to one of the first child labor laws based on an environmental exposure impact on children's health called the Chimney Sweepers Act of 1788 (Hayes, 2008).

Occupational exposure to chemicals and the associated health risks are a significant part of the impacts of anthropogenic pollution. The identification of mercury as a toxic metal with adverse health effects, including dementia, originated with the use of mercuric nitrate in the production of felt for hats in the eighteenth and nineteenth centuries. The term "mad as a hatter" is now a colloquialism meaning crazy, but its first use referred to hatters who developed a mental disorder called mad hatter syndrome after being exposed to mercury during their work. One particular city that had a thriving hat industry was Danbury, CT, which had so many cases of mercury poisoning that the condition was referred to as the "Danbury Shakes" and led to the banning of mercury compounds in the city in 1941 (New England Historical Society, 2016). Other occupations that are well known for impacts from chemical and environmental exposures include: coal miners, asbestos workers, industrial chemists, and refinery workers. These well-known occupational exposure problems led to the establishment of the Occupational and Safety Administration (OSHA) in the United States and similar agencies throughout the world. These agencies have responsibility for establishing regulations detailing occupational exposure limits and monitoring requirements for hazards to workers in the workplace.

Figure 1.2 Sir Percival Pott, an English surgeon who found the association between soot exposure and cancer in young male chimney sweeps in 1775. Source: U.S. National Library of Medicine Digital Collection.

One of the major changes in our view of the impacts of anthropogenic pollution on the environment is the recognition that man occupies a planet that is not limitless in its resources, especially as the human population increases and the demands on those resources increases. Originally recognized as local problems, we have come to realize that pollutants can have a wide variety of impacts and that the various chemicals and materials released into the environment can have effects on more than one environmental system, depending on their chemical and physical properties. An understanding of these properties and the chemical lifecycles of pollutant species is an important key to our developing environmentally sustainable practices with minimal environmental impacts. Thus, this text will focus on the connections between the chemical reactions of both natural and anthropogenic species in the different environmental systems.

1.3 A Planet at Risk

Our planet is a precious resource on which we have evolved. In its orbit around the Sun, the Earth is at the right distance and has all of the necessary components to capture the Sun's energy in the proper amount to allow life to develop and flourish for billions of years. As a relative latecomer to the planet, the genus *Homo* has occupied it for approximately 2.8 million years. Our species, *Homo sapiens*, appeared about 200 000 years ago and modern man developed only about 50 000 years ago (Stringer, 2002). While only on the planet for a relatively short time, our ability to pass on ideas and concepts from generation to generation, along with the discovery of fossil fuels as an energy source, led to the Industrial Revolution and the dramatic increase in population over the past 200 years.

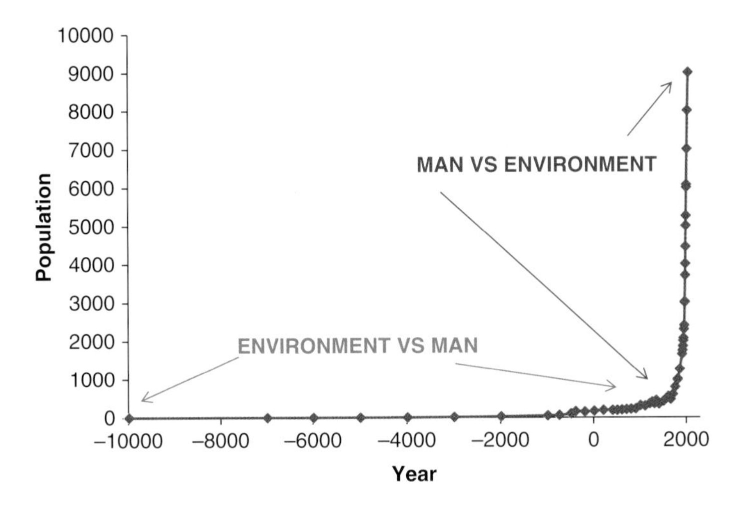

Figure 1.3 Global human population over the last 12 000 years showing the very fast population increase since the beginning of the Industrial Revolution from 1760 to 1840. Source: Data from El T, Wikimedia Commons.

In the early years of the Industrial Revolution, we were able to use dilution as a means of reducing pollution levels. Hence the slogan: "the solution to pollution is dilution." Since the human population has now grown to the extent that it inhabits most of all of the land masses and tremendous amounts of pollutants in the form of gases, liquids, and solids are released by man's activities into our environment on a daily basis worldwide, we can no longer rely on dilution to adequately reduce the impacts of these pollutants on the environment. Pollutant emissions impact local and regional air quality and pollute our surface and groundwater supplies, as well as our soils, vegetation, and food supplies. The emissions of pollutant species, including both gases and aerosols that affect the radiative balance of the atmosphere, are leading to impacts that can change our weather and climate. Changing climates are particularly important when one realizes that they are linked to sea-level rise, storm frequency and severity, droughts, flooding, and the spread of tropical diseases into regions where they were not observed before. We are truly all aboard "Spaceship Earth," where we have limited air and water resources, as well as energy options, and we can no longer ignore the effects that our activities have on the environment.

Figure 1.3 shows the human population growth as a function of recent time. Up until the late 1800s, the development of our society and its survival was strongly determined by environmental factors and the local production of the necessary food and supplies. This period can be considered to be an era of "environment vs man," since our technology had not advanced to a point where we could control or harness our environment and we lived at the mercy of environmental conditions. After the Industrial Revolution, in about 1760–1840, the ability to grow and store more food, build stronger buildings capable of protecting us from the weather, and transport materials globally led to an incredibly rapid increase in population and the beginning of a "man vs environment" era, where we began to control the impacts of the environment on society. However, we also began to change the environment, both intentionally and unintentionally, as we focused on the growth and development of human society.

The impacts of anthropogenic pollution can now be seen on the entire planet, and they affect all of our environmental systems, including the atmosphere, hydrosphere, geosphere, and biosphere. The presence of *H. sapiens* has had such a large impact on the Earth that many geologists now consider the most recent geological period, covering about 8000 BCE to the present, to be the Anthropocene; the period during which human activity has been the dominant influence on the climate and environment. The impacts of our species on the environment are both short and long term, and we must recognize these effects and begin to minimize and, if possible, reverse these trends through a thorough understanding of the chemistry and physics of the environmental systems. We have put increasing demand on our resources, especially water, fuel, and food, and continuing our current practices and population growth rates will clearly result in worldwide shortages of these necessities and a reduction in the living standards of future generations. Other, more severe, consequences include regional-scale war, severe population decline, and a planet that is not as habitable for man, animals, and plants. Our planet is at risk, and the recognition and understanding of the chemistry of environmental systems on all scales is needed to determine the impacts of environmental pollution and how to reverse them. This will require that future chemists and environmental scientists have an understanding of how the chemistry and physics are linked to the biological systems in order to make decisions regarding possible strategies for managing our rapidly decreasing global resources.

1.4 Energy, Water, and Population Connections

Let us look more closely at the dates when the human population reached each milestone of an added billion people on the planet. Table 1.1 shows the approximate years at which the human population increased by one billion. While it took us over 100 years to reach the second billion, it took only 32 years to reach the third billion, and we are now increasing our global population by about one billion every 12–13 years. This means that the human population increases by a factor of 6–7 billion in an average lifespan of 80 years. This rapidly increasing population requires a corresponding increase in food and potable water to be able to survive. Considering that the resources on the planet are finite, this rate of population increase cannot be sustained indefinitely.

The very rapid rate of human population increase in recent years has primarily been due to our ability to access stored fossil energy sources. Most of our advances that increase lifespan and decrease the death rate – including industrial and agricultural improvements, medical advances, and transportation and trade developments – are linked to the increase in the availability of low-cost fossil energy sources. In addition, fossil fuels are not only used as our primary source of energy, but also serve as raw materials for the chemical production of plastics and other commonly used polymers, as well as for the manufacture of synthetic drugs and

Table 1.1 The increase in global population in milestones of billions and the approximate time it took to reach each milestone.

Population	Year reached	Years to milestone
1×10^9	1804	—
2×10^9	1927	123
3×10^9	1959	32
4×10^9	1974	25
5×10^9	1987	13
6×10^9	1999	12
7×10^9	2012	13
8×10^9	2025	13
9×10^9	2040	15

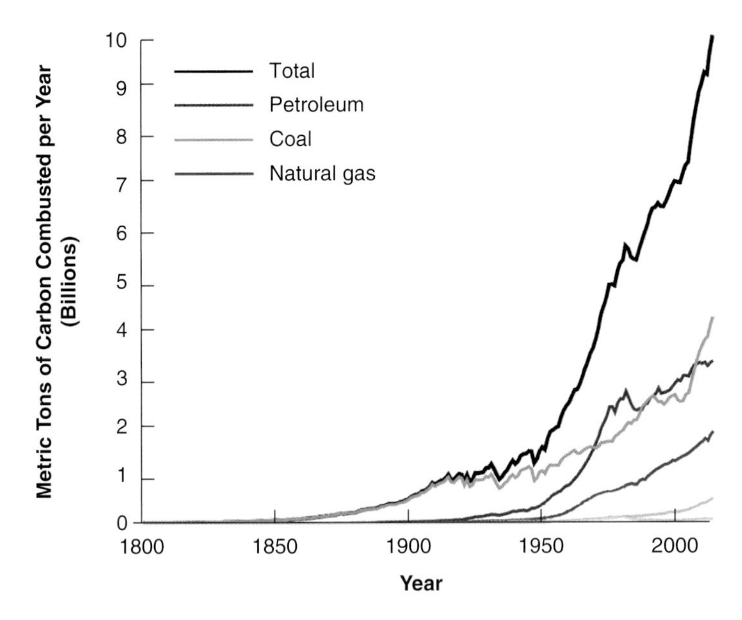

Figure 1.4 Global use of fossil fuel in metric tons of carbon combusted as a function of time from 1800 to 2007. Source: Adapted from Bramfab, Wikimedia Commons.

pharmaceuticals. The trend in fossil fuel use from 1800 to 2007 is shown in Figure 1.4 as the amount of carbon combusted per year. The total amount of fossil energy used in the form of petroleum, coal, and natural gas follows a similar trend as that of the population increase for this same time period (Marland et al., 2007).

Fossil fuels are not renewable, and their supply is also not infinite. The Earth's fossil fuel resources took hundreds of millions of years to be produced. Beginning with the fixation of carbon dioxide by ancient plants, the organic carbon in the dead plants was slowly deposited as kerogen sediments, which were buried deeper and deeper by sedimentation processes and finally transformed into large amounts of fossil carbonaceous materials in the form of coals, petroleum, and natural gas now found in the geosphere. The earliest known deposits of fossil fuels were first formed during the Cambrian Period, about 500 million years ago. More recent deposits were formed in the Pliocene Period, about five million years ago. It is obvious that the process that creates fossil fuels does not occur fast enough to replenish the rapidly decreasing supplies.

The use of fossil fuels as an energy source in the late nineteenth century and continuing into the twentieth and twenty-first centuries has allowed for the rapid development of our species, but it is also accompanied by a huge change in the biosphere and the surface of the geosphere, as we converted forests and grasslands to farming operations worldwide to feed the rapidly growing population. Ecosystems are also destroyed during the mining of coal, and during drilling for oil and natural gas. Oil spills alone have devastated ecosystems, especially in coastal areas. Add to this the damage incurred by the emission of pollutants during the combustion of the fuels, which contributes to photochemical smog and climate change, and the result is that the planet has dramatically changed due to the widespread use of fossil energy since the late nineteenth century.

While, as shown in Figure 1.5, there are renewable energy sources that have also increased during this period – including nuclear, solar, wind, biofuels, geothermal, and hydroelectric – fossil fuels are currently the dominant source of energy in the world. The total world energy

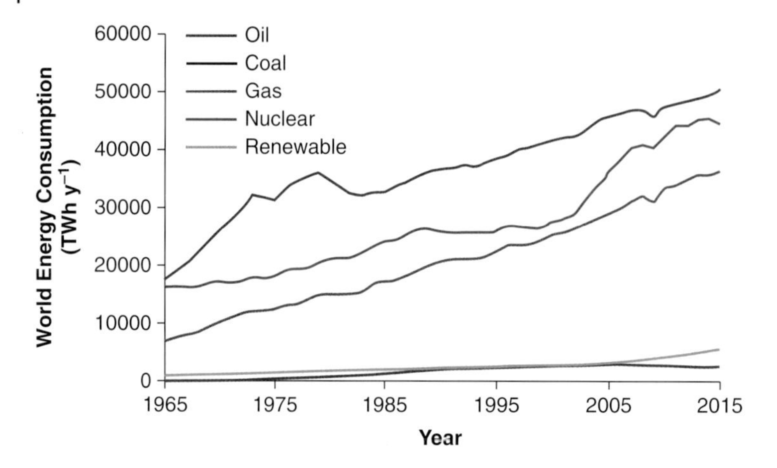

Figure 1.5 World energy consumption in terawatt hours per year (TWh yr^{-1}) over the last 50 years. Source: Adapted from Martinburo, Wikimedia Commons.

demand is about 8.8×10^{17} Wh yr^{-1}, with oil, coal, and natural gas supplying about 92% of that demand. Renewable energy sources supply 6% of this demand, and nuclear energy supplies only 2%. The rapid use of the limited fossil fuel supplies has led to projections that within the next few decades, an energy crisis is likely to occur. While we have begun to search for and increase the use of renewable energy sources, these sources still have a long way to go to reach the huge energy demand of our world's populations. In order to avoid this projected energy crisis, renewable energy sources must become an increasingly larger fraction of our energy demand. Not only must they keep up with the increasing population growth, but they must begin to replace fossil fuel if we are to meet future energy needs.

The increased use of carbonaceous fossil fuels has led to associated increases in ozone-producing gases formed during combustion, as well as increases in the atmospheric levels of carbon dioxide, one of the major greenhouse gases. The associated increases in agricultural crop production, particularly rice, as well as increases in animal husbandry, have also led to higher levels of other greenhouse gases, such as methane and nitrous oxide. Not surprisingly, the atmospheric levels of the greenhouse gases have also followed a similar increasing trend with increasing human population and increasing fossil energy use, as shown in Figure 1.6.

While methane and nitrous oxide have major roles as greenhouse gases in the lower part of our atmosphere, they also play an important role in stratospheric chemistry tied to ozone depletion, which will be discussed in detail in Chapters 2, 4, 11, and 13. The important thing to note here is that many of the chemical pollutant species, like methane and nitrous oxide, have more than one impact on the environment. They can impact chemical and physical processes that can lead to air pollution, and they can also impact climate change by acting as greenhouse gases. In addition, they can be transported into the stratosphere where they can increase stratospheric ozone depletion. Thus, in order to fully understand the environmental chemistry of any molecule or material introduced into the environment, they must be examined with regard to the potential for multiple environmental impacts.

The use of higher amounts of fertilizers and pesticides for crops and antibiotics in animal husbandry is well known to lead to water pollution problems. Also, air pollutants such as nitrogen oxides and sulfur dioxide can undergo chemical transformation, transport, and wet deposition, contributing to significant water quality degradation as well as ecosystem impacts. Enhanced use of water for agriculture and for industry has consequences in terms of water quality and supply. While the Earth is the water planet, most of that water is salt water and not usable for

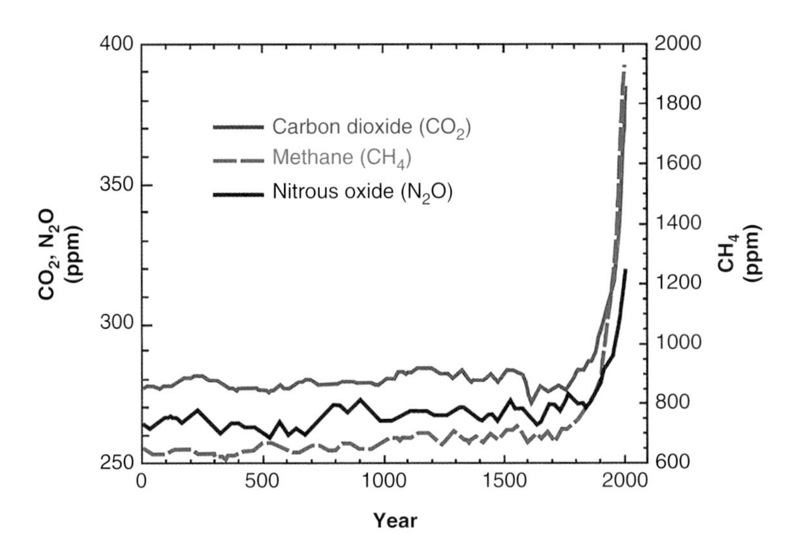

Figure 1.6 Atmospheric concentrations (ppm) of three important greenhouse gases: CO_2, CH_4, and N_2O over time. Source: U.S. National Climate Assessment, 2014.

growing terrestrial plants and crops. Agricultural demand for fresh water has led to overuse of surface waters and depletion of groundwater supplies in most populated areas. These agricultural practices have caused available freshwater supplies to be further limited due to the addition of pollutants, causing them to become no longer useful as a freshwater resource without water processing.

A further complication of the increasing human population is that it is not evenly distributed over the globe. We have developed concentrated areas of population in urban centers worldwide, which have their own set of environmental problems, particularly in the areas of increased water and air pollution. In 1800, only 3% of the population lived in cities. As of 2014, the number of humans living in urban centers has risen to 53% (United Nations, 2014). These large urban centers are known as megacities, large metropolitan areas with a population of 10 million or greater (Lewis, 2007). They can consist of a single city or two or more metropolitan areas that function as a single city. The first city to reach a population of 10 million was New York City in the United States in 1950. As of 2015, there are 35 megacities worldwide. Tokyo, Japan is number one, while New York City has dropped to eighth. A map of the world's population density is shown in Figure 1.7 for the year 1994. It is clear that the majority of the human population lives in the northern hemisphere, and Asia, Africa, and South America are areas where these urban centers are rapidly growing.

This change in population density is projected to continue, with urban populations increasing and rural populations declining. In addition, the population of these urban megacities has also been increasing along with the increase in overall global population. All of the large urban centers suffer from similar issues of high air pollution levels, water shortages, high energy demand, and waste-handling problems. Consequently, the megacities are areas where environmental pollution and energy, food, and water shortages are most likely to occur in the future. The megacities also act as very large pollution sources, which can impact regional and global air and water quality. These increasing problems in the megacities worldwide demonstrate the need to develop clean and safe mass-transit systems, efficient sanitation systems, and the wide use of renewable-energy sources that can sustain the needs of industry as well as urban households and recreation facilities. Balancing the increasing energy, food,

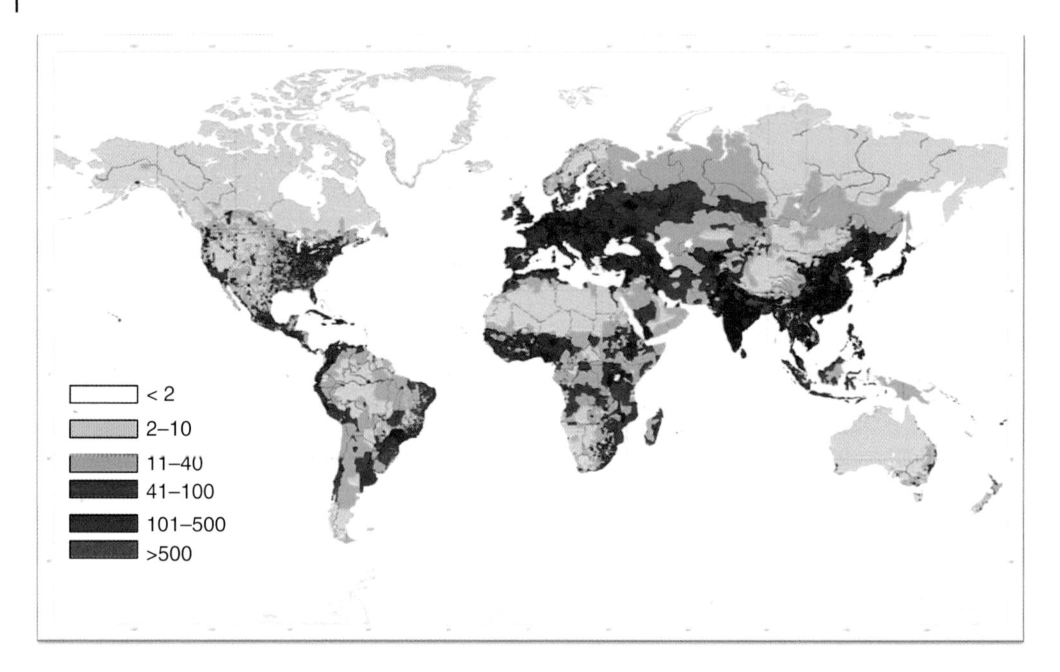

Figure 1.7 World population density in people per square kilometer as of 1994. Source: U.S. Department of Agriculture.

and water demands against chemical pollution in the megacity centers is one of the major problems we face in the coming decades.

While we have recognized that the population–energy–food web has primarily evolved in areas where increasing population requires more energy, water, and food leading to more pollution and a degradation of air and water quality, we have not yet been successful in putting better practices and technologies in place to insure a sustainable system in the future. With a rapidly increasing population and rapidly depleting limited resources, we may soon be faced with a dire situation globally with no good solutions available. There is clearly a need for future scientists to be trained to think proactively in order to address these problems before they become critical, and to be able to communicate their ideas and findings in a timely fashion so that national and international policies can be put in place to develop a sustainable situation for humankind and the planet. As noted earlier, the chemical and physical properties of pollutant species and their potential impacts on environmental systems need to be addressed based on our best understanding of all of the Earth's systems. These impact assessments need to be strongly based on the fundamental principles of environmental chemistry and physics. While we may not be able to find the perfect solution for these very large and complex problems linked to increasing human populations, we do need to determine the best solutions that will minimize the ever-growing impacts.

1.5 The Need to Understand Environmental Problems

So, it is clear that the human population is dramatically increasing and is expected to continue to increase. More human beings means an increased demand for energy, water, food, and other resources needed to sustain our species, and this leads to increasing levels of pollution in the air, fresh and salt waters, soils, subsurface, and biosphere. Past environmental problems, when the world populations were smaller, occurred primarily in urban centers. London's use of coal in the early seventeenth and eighteenth centuries allowed it to grow, but also produced significant levels of air pollution, which was dominated by soot and sulfur dioxide emissions resulting in reduced visibility. The term "smog" was first used in London because these air pollution events looked like heavy smoky fogs (smoke + fog = smog). The high-density population and traffic in Los Angeles, CA in the 1950s led to the formation of "photochemical smog," which was determined to be due to the emissions of nitrogen oxides and unburned volatile organic hydrocarbons from the internal combustion engine reacting to form ozone. Details on these air pollution issues will be discussed further in Chapter 5. These observations led to research which developed a fundamental understanding of the chemistry involved in the processes. This understanding has been used to develop control strategies to reduce the levels of these pollutants in many urban areas. While these problems are still affecting all major cities, we have begun to reduce the levels of key air pollutants, such as ozone and carbon monoxide, in many of the megacity environments.

The need to fully understand the chemistry of any material added to our environment is emphasized by cases where a chemical species that was developed to solve a specific problem was found later to lead to more severe environmental and health impacts and to the eventual regulation or banning of the materials. One example of this is the use of chlorofluorocarbons (CFCs), whose chemistry will be described in detail in Chapter 4. One of the first commercial CFCs was dichlorodifluoromethane (CCl_2F_2), with the trademark name Freon. It was developed by Thomas Midgely, Jr. and Albert Henne, who worked for General Motors Research in the late 1920s. The compound was specifically developed as an operating fluid to replace the toxic and flammable gases of ammonia, sulfur dioxide, propane, and chloroform (chloromethane) that were being used in refrigeration at that time. Midgely, shown in Figure 1.8, received awards for the development of Freon, as it was thought to be environmentally benign due to its low reactivity.

At the time, we did not understand the chemistry of the stratosphere, nor did we understand atmospheric chemical transport. A better understanding of these atmospheric processes would soon lead to the discovery that Freon and other CFCs would build up in the lower atmosphere, be transported to the upper atmosphere, and lead to the potential catalytic destruction of the stratospheric ozone that was shielding all life from harmful ultraviolet (UV) radiation. Later work also discovered that these same CFCs were very strong infrared-radiation absorbers, allowing them to act as potent greenhouse gases. These problems made the CFCs

Figure 1.8 Thomas Midgely, Jr., a chemist for General Motors Research who developed the first chlorofluorocarbon, trademarked as Freon, in the late 1920s. Source: Author unknown, Wikimedia Commons.

environmentally unsound for long-term usage, and led to their worldwide ban by the Montreal Protocol. Thus, a better understanding of the chemistry and physics of the atmosphere led to the determination that these apparently safe compounds had other significant impacts on the global systems that were unforeseen at the time.

Another example is the use of alkyl lead products, tetramethyllead ($(CH_3)_4Pb$) and tetraethyllead ($(CH_3CH_2)_4Pb$), chemical additives that were added to gasoline to improve motor vehicle operation. These fuel additives were developed to reduce knocking in motor vehicles by enhancing and stabilizing the fuel energy content. At the time, little attention was paid to the impacts of the alkyl lead combustion products. We now know that combustion of these additives leads to the formation of lead-containing aerosol particles emitted to the atmosphere in the vehicle's exhaust. The emission of these lead aerosols results in elevated lead levels in the nearby environment approaching harmful levels, especially for young children. This led to their eventual elimination and ban as a fuel additive, followed by the development of unleaded gasolines now used primarily throughout the United States and the world. These new unleaded gasolines are more refined with a higher octane content, which solves the engine-knocking problem while avoiding the toxic lead combustion products. Interestingly, both the alkyl lead gasoline additives and the CFCs were developed by Thomas Midgely, Jr.

The increasing need for food production led to the development and widespread use of agricultural pesticides to prevent crop losses. One of the most well known of these pesticides is 1,1′-(2,2,2-trichloroethane-1,1-diyl)bis(4-chlorobenzene) or dichlorodiphenyltrichloroethane (DDT), shown in Figure 1.9. DDT was first synthesized in the late nineteenth century, but its toxicity to insects and, in particular, disease-carrying mosquitos was first discovered by chemist Paul Hermann Müller in 1939, who won the Nobel Prize in Medicine for his work. Later, it was discovered that DDT was bioaccumulated in the food chain and caused serious problems with bird reproduction. This problem was publicized by marine biologist Rachel Carson in her book *Silent Spring* (Carson, 1962), which led to an environmental movement and the eventual ban of the use of DDT except in the most serious situations. Since that time,

Figure 1.9 Chemist Hermann Müller (left) who discovered the toxicity of DDT to insects, particularly disease-carrying mosquitos, and marine biologist Rachel Carson (right) whose book *Silent Spring* described the effects of DDT on bird reproduction. Source: (left) Author unknown; (right) U.S. Fish and Wildlife Service, Wikimedia Commons.

chemists involved in the development of pesticides are required to investigate their toxicity and chemical reactivity in a broad range of environmental systems to insure that the compounds do not lead to similar problems.

Once it was recognized that the emission of chemicals into our air, water, and soils could lead to significant short- and long-term impacts that were unintentional, regulations and control strategies were initiated to attempt to limit the impacts. However, in many cases control strategies were implemented before a thorough understanding of the chemical and physical impacts of the chemicals were well understood with regard to their transport, transformation, and removal processes on the various scales involved in the environmental systems. An example of a pollution control strategy that solved a local problem but led to a regional one was the use of tall stacks for the release of combustion gases from power plants and industrial sites. These tall stacks were designed to release pollutant gases at a higher altitude in the atmosphere. The gases then entered the atmosphere's mixing layer, resulting in the dilution of the gases by mixing them together with uncontaminated air. This also kept them above ground level, where they would not be a direct risk to local populations. Although the release of sulfur dioxide from coal-fired power plants at mixing layer height reduced the immediate damage to the local area, this high-altitude release allowed the sulfur dioxide and associated air pollutants to be transported long distances from the source. During the transport time, the sulfur dioxide was converted to sulfuric acid, which was eventually deposited on the ground in rain, leading to "acid rain" on a regional scale. We now recognize that the solution to pollution is not dilution and, in most cases, we work to either trap pollutants before release into the atmosphere or to convert any toxic or environmentally damaging pollution to as inert a form as possible before release in order to minimize the environmental impacts.

These examples and others, which we will discuss in the following chapters, clearly indicate why a thorough understanding of environmental problems and how their chemistry is linked to all the environmental systems is essential to preventing additional, otherwise unforeseen impacts. Our past problems have given us the opportunity to learn from our mistakes and we are now at a point in time when the chemistry involved in these environmental processes is fairly well understood, allowing for the prediction of possible negative feedbacks of new technologies, materials, and chemicals before they happen in order to propose more effective mitigation or remediation strategies. The past limitations of our understanding of the environmental chemistry of the various environmental systems led to an inability to consider the "cradle to grave" consequences of the use of chemicals and their release to the environment. However, our current understanding of the processes has progressed far enough to allow us to predict impacts and work toward using environmental chemistry to limit these impacts as much as possible as we examine the overall lifecycles of chemicals in the environment. The fundamental principles of environmental chemistry require that we examine all of the potential systems where the chemicals can have impacts, so that we do not repeat our mistakes from the past and do not create additional unforeseen problems. Thus, we can now work toward a proactive approach to environmental chemistry that considers all of the Earth's environmental systems.

1.6 Atmosphere–Hydrosphere–Geosphere–Biosphere Linkages

The Earth's systems are linked through physical, chemical, and biological processes. Each of the environmental systems consists of the three basic chemical phases of matter: gases, liquids, and solids. However, while chemists normally study homogeneous, well-mixed systems, these are very rare in the natural systems. Environmental chemistry is not the chemistry of well-mixed beakers, nor does it involve reactions of large amounts of chemical species. Rather, it is the chemistry of open systems that include transport and transformation of often trace chemical species over a range of scales and time frames. But although environmental systems are very large open systems, they interact with each other in an attempt to reach equilibrium states. In many cases it may take long time periods for the systems to reach equilibrium, but we can still use the principles of chemical kinetics and equilibrium to make approximations that can be very useful to understand their behavior. For example, the reactive lifetimes of chemicals in the atmosphere and hydrosphere, and/or their transport properties, can be evaluated using vapor pressures and solubility constants.

Water is one of the key linking species between all of the environmental systems, promoting the transport of chemical species between them. Our atmosphere consists of both unreactive gases, such as nitrogen and argon, as well as reactive gases, such as oxygen, carbon dioxide, and trace gases. Some gases have a low water solubility and others have a high water solubility, which allows them to interact with the hydrosphere, including fresh and salt surface waters, clouds, and precipitation in the form of rain, snow, hail, and sleet. Clouds themselves are not pure water but are made up of large aqueous droplets that can contain trace particles as well as soluble chemicals. These water droplets can act as important chemical reactors due to water's ability to act as a universal solvent. If the water droplets become large and numerous enough, clouds can produce a wide variety of precipitation, recycling the water and its chemicals back to the land in the form of fresh water. These surface waters can act to solubilize compounds from the geosphere and ultimately transport them to lakes and seas by way of rivers and other hydrological systems. These hydrological systems are strongly linked to the geosphere and biosphere, and the feedback between them is key in determining many factors relating to the currently recognized problems of climate change and air and water quality degradation.

The natural systems can interact with anthropogenic pollution, leading to synergistic effects that sometimes have positive outcomes and sometimes have very negative outcomes. This leads to a complex chemistry, which impacts many if not all of the natural systems. An example of this type of complexity is the release of mercury vapor (Hg°) into the environment from the combustion of coal or smelter operations, as demonstrated in Figure 1.10 (Gaffney and Marley, 2014).

Once released into the lower atmosphere as a gas, mercury can equilibrate with cloud water due to its small water solubility. The cloud water can contain significant amounts of hydrogen peroxide, which is formed as a trace gas in the atmosphere from free radical chemical reactions that lead to the formation of the hydroperoxyl radical (HO_2) followed by the reaction of HO_2 with oxygen. Although the concentrations of hydrogen peroxide in air are small (low ppb), the cloudwater concentrations can be around a hundred micromolar, since hydrogen peroxide is extremely water soluble. This hydrogen peroxide concentration can effectively oxidize the Hg° to HgO. That same cloud water can contain organic acids and compounds containing hydroxyl groups, formed from the oxidation of anthropogenic and natural organic compounds. These oxidized organics can act as chelating agents for the oxidized forms of elemental mercury. Once deposited to the surface in rain water, this organically complex mercury is more bioavailable in the hydrosphere and geosphere and thus can accumulate in organisms in the biosphere. Once the mercury is in the biosphere, it can be converted into organomercury compounds that have much higher toxicity than elemental mercury.

The transformation and stabilities of these mercury compounds are very dependent on the chemistry of the environmental systems. They are more stable in reducing environments, such as in lake sediments or anoxic lake or ocean waters, where they can accumulate. But they can be recycled back to elemental mercury and revolatilized back into the air if they are introduced into an oxidizing environment. Once it enters the environment, the mercury levels eventually increase in all of the environmental systems due to its complex environmental chemical behavior. It does not exist simply in one environmental compartment, but is in equilibrium with all of the various compartments: atmosphere, hydrosphere, geosphere, and biosphere.

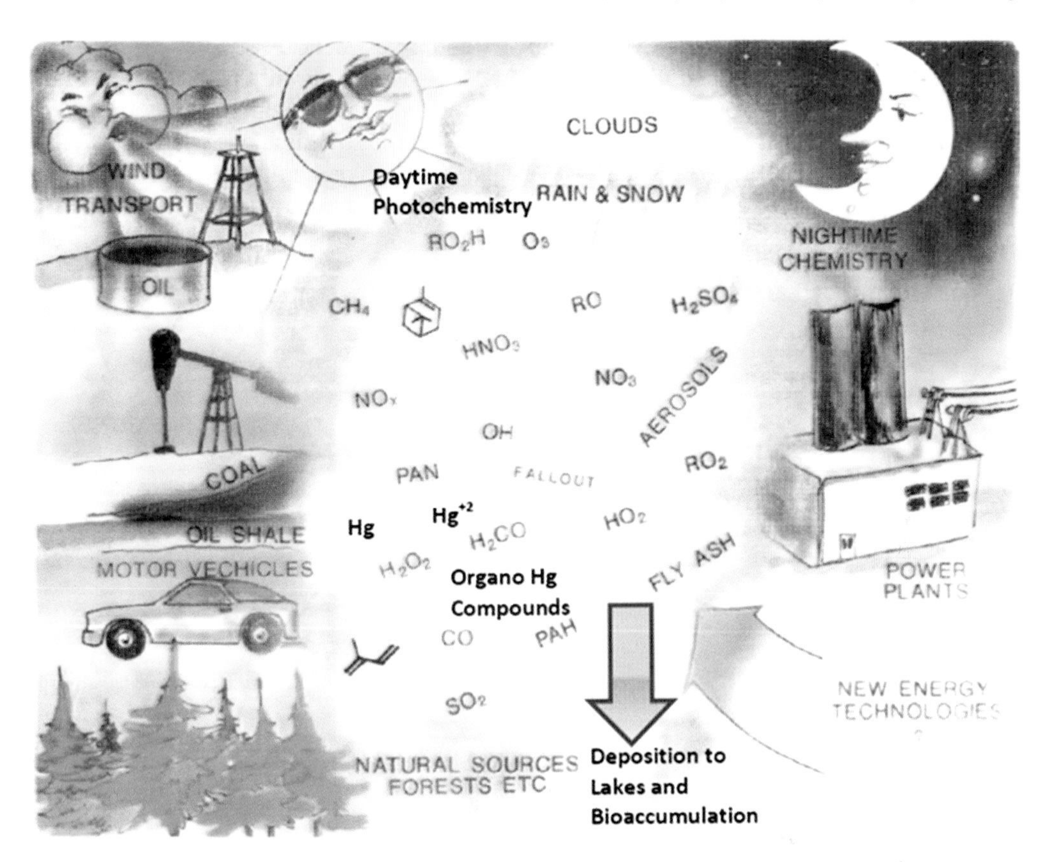

Figure 1.10 The complex environmental chemistry of mercury in the atmosphere and its connection to agricultural and biomass burning. Elemental mercury (Hg°) can react with H_2O_2 to form Hg^{2+}. Mercury's complexation with organomercury compounds in clouds and wet aerosols leads to increasing deposition in lakes, and subsequent bioaccumulation in fish through the food chain.

As we develop new energy technologies, we introduce similar chemicals in the environment and they will also undergo transport, transformation, and deposition, and will become incorporated into the atmosphere–hydrosphere–geosphere–biosphere system, dependent upon the chemistry of each species. New materials, such as novel polymers or nanomaterials, will also undergo similar chemical processes. The fundamental properties of these materials – including photochemical stability, susceptibility to key oxidants in the atmosphere and hydrosphere, ability to be reduced in anoxic environments, solubility and volatility, as well as organic or inorganic complexation – need to be known in order to evaluate their environmental fate and impacts. These are some of the types of chemical reactions and properties that we will be examining in detail in the following chapters.

As we begin to examine the environmental chemistries of each of the environmental systems, it is important to remember that these systems are strongly connected, with the atmosphere being the most mobile system, followed by the hydrosphere. There is truly one atmosphere across the globe, even though it is not well mixed on short time scales. This leads to observations of clear skies, strong storms and flooding, as well as droughts, occurring over the same areas as a function of time. But these short-term physical processes, which we call weather, are linked to climate over longer time scales and both are linked to the chemistry and physics of the atmosphere and hydrosphere, and are strongly moderated by the geosphere and biosphere. Thus, these environmental systems are all linked and these linkages are at the heart of their chemistry. A review of our current understanding of the fundamental chemical and physical principles of these systems will be accomplished by starting with the atmosphere and moving down and into the hydrosphere, geosphere, and biosphere, noting the important connections between them along the way. The goal of environmental chemistry is to make use of this knowledge to predict the behavior of natural and polluted systems, and to minimize the environmental impacts of anthropogenic activities in order to allow for the development of a sustainable, sound, and safe Earth system for future generations.

References

Bradbury, J. (1992). *The Medieval Siege*. Woodbridge, UK: Boydell Press.

Carson, R. (1962). *Silent Spring*. Boston, MA: Houghton Mifflin.

Dobson, J. (1972). Percival Pott. *Ann. R. Coll. Surg. Engl.* 50: 54–65.

Environmental Chemistry n.d. *Simple English Wikipedia*. Available online at: https://simple.wikipedia.org/wiki/Environmental_chemistry (accessed September 7, 2017).

Finlayson-Pitts, B.J. and Pitts, J.N. Jr., (2000). *Chemistry of the Upper and Lower Atmosphere: Theory, Experiments, and Applications*. San Diego, CA: Academic Press.

Gaffney, J.S. and Marley, N.A. (2014). In-depth review of atmospheric mercury: Sources, transformations, and potential sinks. *Energy Emissions Control Technol.* 2: 1–21.

Goodhill, V. (1971). Maimonides – modern medical relevance. *Trans. Am. Acad. Ophthalmol. Otolaryngol.* 75: 463–491.

Hayes, A.W. (ed.) (2008). *Principles and Methods of Toxicology*, 5e. New York: Informa Health Care.

Kroll, D.J. (2006). *Securing our Water Supply: Protecting a Vulnerable Resource*. Tulsa, OK: PennWell.

Lewis, M. (2007). *Megacities of the Future*. Jersey City, NJ: Forbes Available online at: https://www.forbes.com/2007/06/11/megacities-population-urbanization-biz-cx_21cities_ml_0611megacities.html (accessed September 15, 2007).

Marland, G., Boden, T.A., and Andres, R.J. (2007). Global, regional, and national CO_2 emissions. In: *A Compendium of Data on Global Change*. Oak Ridge, TN: Carbon Dioxide Information Analysis Center, Oak Ridge National Laboratory, US Department of Energy.

New England Historical Society (2016). *The Mad Hatters of Danbury Connecticut*. Arlington, VA: NEHS Available online at: http://www.newenglandhistoricalsociety.com/mad-hatters-danbury-conn (accessed September 8, 2017).

Stringer, C. (2002). Modern human origins: Progress and prospects. *Philos. Trans. R. Soc. London, Ser. B* 357: 563–379.

United Nations (2014). *2014 Revision of World Urbanization Prospects*. New York: UN Department of Economic and Social Affairs, Population Division. Available online at: http://www.un.org/en/development/desa/news/population/world-urbanization-prospects.2014.html (accessed September 15, 2017).

Study Problems

1.1 What is the traditional definition of environmental chemistry?

1.2 What are the four major environmental systems?

1.3 What is green chemistry? How does it differ from environmental chemistry?

1.4 What areas of chemistry are important to the study of environmental chemistry?

1.5 What is anthropogenic pollution? How are the two types of anthropogenic pollution described in terms of intent?

1.6 The recognition of occupational exposure problems led to the establishment of what administration in the United States?

1.7 What organic compound, trademarked as "Freon," was developed for use in refrigeration units as an alternative to ammonia, propane, sulfur dioxide, and chloroform?

1.8 What two previously unforeseen impacts do we now know Freon to have on the environment?

1.9 As the population has changed over the past 50 000 years, how can we describe the time period before the Industrial Revolution and the time period between the Industrial Revolution and the present in terms of the relationship between man and the environment?

1.10 The most recent geological era, beginning about 8000 BCE, is considered by most geologists to be called by what name? Why?

1.11 How long does it currently take to increase the Earth's human population by a billion?

1.12 What are cities with more than 10 million inhabitants called?

1.13 How do the population trends compare to our use of fossil fuels?

1.14 How are fossil fuels classified when compared to other types of fuels with respect to the amount and type of energy source?

1.15 What is the chemical name of the pesticide DDT?

1.16 Why was DDT banned for use in pest control except in the most serious situations?

1.17 What was the name of the book written by Rachel Carson that described the problems associated with the use of the pesticide DDT?

1.18 What were the two alkyl lead compounds that were used as additives in gasoline to reduce engine knocking?

1.19 One of the most stable and insoluble forms of mercury is cinnabar. What is its chemical formula?

1.20 What is the lowest level of the atmosphere called?

1.21 Which of the four environmental systems is the most mobile?

1.22 What is the goal of environmental chemistry?

2

Atmospheric Composition and Basic Physics

Of all our planetary resources, the atmosphere is the one most taken for granted. The air we breathe and that supports plant and animal life is crucial to the existence of all life. The present-day atmosphere is primarily made up of three major gases: nitrogen (N_2) at 78%, oxygen (O_2) at 20.9%, and argon (Ar) at 0.9%, with a number of other trace gases making up the remaining 0.2%. These other trace gases include water vapor (H_2O), carbon dioxide (CO_2), methane (CH_4), nitrous oxide (N_2O), the chlorofluorocarbons (CFCs), ozone (O_3), and many others. Water vapor is the one atmospheric gas that is highly variable. It is also the one that is most mobile, since it can condense to form clouds and precipitate back to the ground in the form of solid or liquid water. In this chapter we will examine the current composition of our atmosphere and how it evolved from the primordial to the present state. We will also discuss the circulation patterns of the atmosphere and how they relate to planetary physics and rotation, as well as how they can be affected by variations in terrain. Finally, we will discuss how atmospheric circulation relates to the radiative balance of the atmosphere and, in turn, how this relates to climate forcing.

The concepts covered in this review will be important to the chemistry that we will examine in the chapters that follow, since the chemistry is affected by atmospheric transport and lifetime as well as by the available solar energy. We will also find that the composition and chemistry of the air are not only important in the atmosphere, but have important effects on the hydrosphere and soil environments. The changing physics and chemistry of the atmosphere affect the chemistry and biochemistry in both fresh and salt waters, as well as in soils, and therefore have important impacts on the biosphere.

2.1 Evolution of the Atmosphere

The Earth's atmosphere is unique in our solar system as it contains significant amounts of the reactive gas oxygen. The observation of oxygen in the atmosphere is an indication of how important plant life is on the planet, since the source of this oxygen is photosynthesis. Without life, that oxygen would rapidly be depleted over time, forming oxidized forms of carbon and creating an atmosphere that would be more like our neighboring planet, Venus. Our atmosphere was not always oxygen rich. Indeed, it originally had little or no oxygen needed for life as we know it, and evolved over time to be what it is today.

Our Earth is approximately 4.5 billion years old, based on radionuclide dating of meteorites found on the planet (Dalrymple, 2001). Meteorites are believed to be formed from the earliest materials in our solar system, at the time when the planets were initially formed, and the meteoritic data has been used to establish the start of our planetary geological clock. The early Earth

Chemistry of Environmental Systems: Fundamental Principles and Analytical Methods, First Edition.
Jeffrey S. Gaffney and Nancy A. Marley.

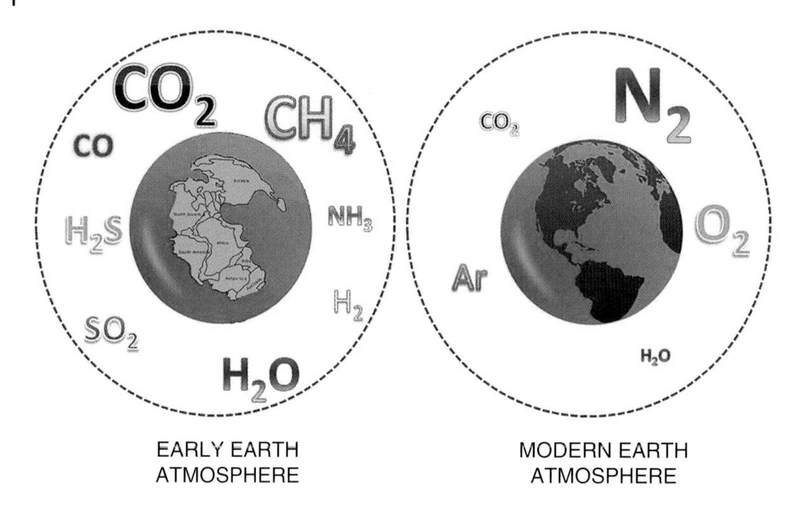

Figure 2.1 Comparison of the Earth's early atmosphere to the modern Earth atmosphere. The size of the chemical symbols for each of the atmospheric gases is relatively proportional to its atmospheric concentration. Source: Kieff, Wikimedia Commons.

is thought to have formed from a mixture of hot gases and solids that became a molten surface. These hot gases and solids gradually cooled with little or no atmosphere at all. Once formed, the planet began to outgas materials from the mantle and crust into the atmosphere, with the lighter gases being lost to space over time and the heavier gases being held in the atmosphere by the planet's gravity. After approximately half a billion years, the earth cooled sufficiently to allow water to condense on the surface. The water would evaporate, returning to the atmosphere, and recondense in the form of precipitation. This hydrological cycle led to dissolution of water-soluble salts in the rocks, resulting in the formation of our oceans.

The atmosphere is thought to have evolved with time as gases were emitted from the planet's volcanic activity. These gases were likely similar to those emitted from volcanos today and primarily contained hydrogen sulfide (H_2S), carbon dioxide (CO_2), methane (CH_4), and water (H_2O), as shown in Figure 2.1. Argon was also emitted into the atmosphere from the Earth's surface, from the radioactive decay of potassium-40, which has a radioactive half-life of 1.25 billion years.

Argon has three stable isotopes, ^{36}Ar, ^{38}Ar, and ^{40}Ar. While the argon in our atmosphere is 99.6% ^{40}Ar, the argon in the Sun is mostly ^{36}Ar with about 15% ^{38}Ar, and this is believed to be the isotopic composition of the original argon produced when the solar system was formed (Porcelli and Turekian, 2011). The isotopic signature of the Earth's atmospheric argon (^{40}Ar) clearly shows that its source is the radioactive decay of ^{40}K and its half-life of over a billion years is consistent with the age of the Earth being more than 4 billion years. It is interesting to note that the atmosphere of Mars has a similar atmospheric abundance of ^{40}Ar, with a smaller amount of ^{36}Ar than that of the Earth. This is due to Mars's lower gravity, causing a loss of the lighter isotope. The presence of argon in the atmosphere of both planets is clearly due to the radioactive decay of ^{40}K over a long period of time.

There are two proposed sources for the large amount of nitrogen gas (N_2) in our atmosphere. One is that plate tectonics caused the movement of rocks containing ammonium to become heated, resulting in the decomposition of the ammonium in the subsurface, producing nitrogen gas which was then released to the atmosphere over time during volcanic activity (Mikheil and Sverjensky, 2014). This is based on a comparison of the nitrogen levels in the atmosphere of our neighboring planets, Venus and Mars, which have much lower levels of nitrogen than in the Earth's atmosphere. The other proposed source of nitrogen is the addition of ammonia directly to the Earth's atmosphere, either from an interaction with Jupiter's ammonia ice clouds or from comets striking the Earth during the planet's formation. Over time, this ammonia could react in the atmosphere to form nitrogen gas. This theory is consistent with a comparison between nitrogen isotopic abundances of meteorites and the Earth's atmospheric nitrogen (Harries et al., 2015). Interestingly enough, ammonia currently exists in the atmosphere at very trace levels, and is the only basic gas in our modern atmosphere. It is currently uncertain which of these proposed mechanisms was the most dominant source of nitrogen gas in the Earth's atmosphere.

The primary gases in the Earth's early atmosphere – methane, carbon dioxide, and water – kept the planet from freezing, since they are greenhouse gases and at that time the Sun was producing only about 70% of the energy that it does today. There was little or no molecular oxygen in the atmosphere at this time. All of the oxygen in the atmosphere was bound up in water or carbon dioxide. During this period, called the Archean Eon, the first life on the planet was thought to be anaerobic bacteria, single-celled organisms that were able to use sulfur and other elements as energy sources on which they could survive. At the end of this period, about 2.7 billion years ago, cyanobacteria evolved in the early oceans. These bacteria used sunlight, carbon dioxide, and water to produce organic compounds, while releasing oxygen into the atmosphere in the process called photosynthesis. These microorganisms were the major source of oxygen in the planet's early atmosphere. Once oxygen levels began to increase, life on the planet evolved rapidly with increasing plant life, which allowed for increased conversion of carbon dioxide and water to oxygen. This led to the oxygen levels that we have in the modern atmosphere, which is approximately 21%. The production of this oxygen led to further evolution of the atmosphere due to photochemical reactions in the presence of ultraviolet (UV) radiation from our Sun.

While photosynthesis is responsible for the increased oxygen levels as plant populations increased both on land and in the oceans, plant respiration is also an important source of carbon dioxide. In the early atmosphere, with high levels of carbon dioxide, significant amounts of water-soluble carbon dioxide would have interacted with the oceans. The equilibrium of carbon dioxide with water leads to carbonic acid, which is in equilibrium with bicarbonate and carbonate. This carbon dioxide dissolution in water initiated the carbonate cycle, which formed insoluble inorganic compounds such as calcium and magnesium carbonates. The carbonate cycle also acted to reduce the carbon dioxide levels in the atmosphere, while the less reactive gases such as nitrogen and argon continued to build up to the current levels. While the increasing oxygen content led to animal life on the planet, the increasing carbonate levels in the oceans led to shellfish, which produced calcium carbonate shells. Limestone deposits formed over hundreds of millions of years from shellfish and other carbonate-fixing ocean life account for most of the sequestered carbon on the planet.

The production of oxygen in our atmosphere also has an important outcome with regard to protecting organisms from damaging UV radiation from the Sun. As most of the early life was in the oceans, it was protected from harmful UV radiation due to absorption by water. At some depth the UV radiation is reduced sufficiently that plant and animal life can be maintained. But the development of life in terrestrial areas required significant oxygen levels to be formed in order to screen the harmful UV radiation. The increased levels of oxygen also produced ozone in the upper atmosphere that would further protect life in terrestrial environments from UV radiation. Thus, animal life – including man – evolved along with our atmosphere to its current state.

2.2 Structure and Composition of the Modern Atmosphere

Some of the first investigations of the modern atmosphere were made by the early chemists who founded the principles of modern chemistry, and who were also interested in determining the composition and behavior of gases in the air that they breathed. Their work resulted in the development of the ideal gas law ($PV = nRT$), which described the relationship between the pressure of the gas measured in atmospheres (P) and the volume measured in liters (V), to the amount measured in moles (n) and the temperature measured in degrees Kelvin (T). The proportionality constant for this relationship (R) is called the ideal gas constant. These early chemists included: Joseph Black, who discovered carbon dioxide; Robert Boyle, who determined the relationship between the pressure and volume of gases; Henry Cavendish, who discovered the gases of hydrogen and argon; John Dalton, who determined the law of partial pressures, made early meteorological measurements of the atmosphere, and first proposed the concept of atoms forming into molecules; Antoine Lavoisier, who used mass balance to follow chemical reactions; Joseph Priestly, who discovered 10 atmospheric gases, including O_2, N_2, NO, CO, NO_2, and SO_2, as well as four other trace gases; and Joseph Louis Gay-Lussac, who determined the relationship between the pressure and temperature of gases. Gay-Lussac conducted some of the first experiments using hot air balloons to determine the composition of air as a function of altitude. Gay-Lussac, accompanied by physicist Jean-Baptiste Biot, obtained samples of air in evacuated flasks as they rose to an altitude of 4000 m (see Figure 2.2). They found that the gases of N_2 and O_2 remained in the same proportion (4 to 1) throughout the ascent, even as the overall pressure dropped, thus determining that the lower atmosphere was well mixed.

Early meteorologists measured the temperatures and pressure of the atmosphere as a function of altitude, initially by climbing mountains and later by using hydrogen or helium-filled balloons as with Gay-Lussac and Jean-Baptiste Biot. We are now able to take measurements at much higher altitudes and with finer resolution, which has led to the determination of our

Figure 2.2 Chemist Gay-Lussac and physicist Jean-Baptiste Biot ascend to an altitude of 4000 m to collect samples of air in 1804. Source: Author unknown, Wikimedia Commons.

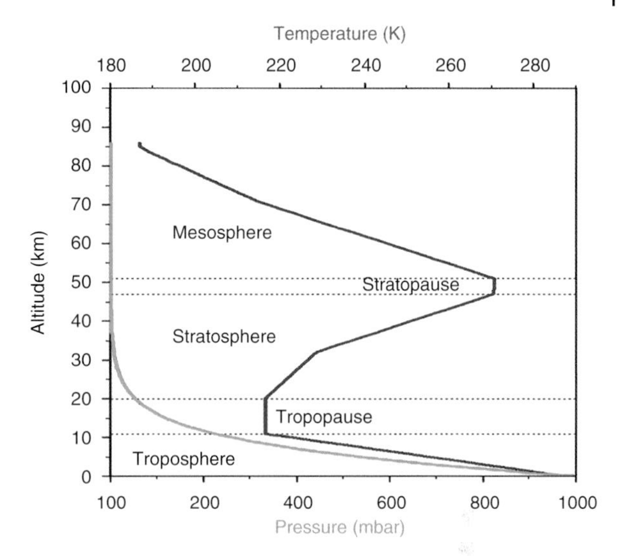

Figure 2.3 Atmospheric pressure and temperature as a function of altitude. The temperature profile defines the layers of the atmosphere as the troposphere, tropopause, stratosphere, stratopause, and mesosphere.

atmosphere's physical structure from ground level to space. Figure 2.3 shows a diagram of the change in atmospheric pressure and temperature as a function of altitude up to a height of 85 km. Note that the pressure simply drops off with increasing altitude, steadily decreasing until you reach the vacuum of outer space. But the change in temperature with increasing altitude shows a much different profile, and it is this profile that defines the different regions of the atmosphere.

The lowest layer of the atmosphere, the one closest to the surface of the Earth, is the troposphere. It is the region where we live and where weather and climate occur. The temperature in this layer decreases steadily up to an altitude of about 10 km. So the temperature of the air at sea level is higher than that of the air at higher altitudes. The warmer air rises and mixes with the air above it, while the colder air sinks and mixes into the air below it. This causes the atmosphere in the troposphere to become well mixed, as observed by Gay-Lussac and Jean-Baptiste Biot during their hot air balloon experiment. While the ratio of nitrogen to oxygen remains the same throughout the troposphere, the concentrations of both gases drop as the overall pressure drops according to the ideal gas law. As you climb a mountain, you will notice that the temperature gets cooler and you may have difficulty breathing as you ascend due to the drop in both temperature and pressure with altitude in the troposphere. If you were to climb Mt. Everest, you would likely need special equipment to survive the ascent because the concentration of oxygen decreases sufficiently that humans who have not become acclimatized to low oxygen levels require a breathing device.

The temperature in the troposphere begins to level off at higher altitudes and finally reaches a region of constant temperature called the tropopause at an altitude of about 10–21 km. Because of this constant temperature, there is very little mixing of the atmosphere in the tropopause. The atmospheric layer above the tropopause is the stratosphere. The intensity of UV radiation from the Sun is higher in the stratosphere, which causes the photolysis of oxygen, forming ozone. Both molecular oxygen and ozone in the stratosphere act to absorb most of the incoming UV radiation from the Sun, protecting the troposphere from its harmful effects. This absorption of UV radiation in the stratosphere causes an increase in temperature with altitude. So the stratosphere has warmer air at higher altitudes and colder air at lower altitudes, resulting in much less turbulent mixing than occurs in the troposphere and creating a generally calmer atmospheric region. Because of this, the lower stratosphere is where most airlines like to fly, as the air convection is minimal, creating a smoother ride.

Above the stratosphere, at an altitude of about 48–52 km, is another region where the temperature of the atmosphere is fairly constant, called the stratopause. Above the stratopause is the mesosphere, where the temperature again decreases with altitude, creating atmospheric mixing similar to that in the troposphere. At an altitude of about 85 km, the temperature again becomes fairly constant at the mesopause. Above the mesopause the temperature increases rapidly at very low pressure, at altitudes from 100 to 700 km. This atmospheric region is called the thermosphere, also known as the ionosphere. It is made up of very high-energy molecules as well as high-energy radiation and high-energy atomic particles (protons, electrons, and neutrons) from the Sun. The high-energy molecules undergo dissociation to atoms after collision with the high-energy atomic particles. This dissociation is followed by chemiluminescent recombination reactions, which result in the emission of visible light. These chemiluminescent reactions are especially prevalent over the planet's polar regions, where the charged high-energy atomic particles (electrons and protons) are attracted to the Earth's magnetic poles. The visible light emission in the polar regions is called the "northern lights" or the Aurora Borealis in the northern polar region, and the "southern lights" or the Aurora Australis in the southern polar region.

At the top of the thermosphere the temperature is again constant with increasing altitude in the thermopause. The thermopause is followed by the exosphere, which lies from 700 to 10 000 km. The exosphere contains light atoms and molecules such as helium, hydrogen, carbon dioxide, and atomic oxygen, along with smaller concentrations of heavier species. The pressure in the exosphere is extremely low, since this region of the atmosphere is essentially mixing with outer space. Most of our low Earth-orbiting satellites are located at altitudes of 160–2000 km in the thermosphere and the exosphere. Medium-orbiting satellites are located outside of the atmosphere at about 20 000 km, while geostationary satellites are located at a further distance of approximately 36 000 km. For the purposes of environmental chemistry, we will focus primarily on the lower portions of our atmosphere, the troposphere and stratosphere, as this is where most of the reactive gases are located and where almost all of the relevant chemistry occurs that affects the hydrosphere, geosphere, and biosphere.

In atmospheric science, the composition of the atmosphere is typically described in terms of mixing ratios. Mixing ratios are defined in meteorology as a unitless ratio of the mass of one gas species in the mixture to the mass of the total mixture, sometimes called the mass ratio. For atmospheric trace gases and substances such as aerosols (fine solid particulates or aqueous droplets), this is approximated by using the mass of the substance measured relative to the total mass of air analyzed. The only exception in meteorology is for measurements of water vapor, where the mass of the water is measured relative to the mass of dry air. In atmospheric chemistry, mixing ratios are defined as the ratio of the amount of one chemical species to the amount of the entire air sample. But since the amount of water vapor in an air sample varies with time, it is usually excluded from the determination of the total amount. This is indicated by describing the mixing ratio as determined "in dry air." The expression of mixing ratios can be obtained by several methods, depending on the species. Some important ways of describing mixing ratios in atmospheric chemistry are listed in Table 2.1. Since the accepted unit for the amount of a substance in chemistry is the mole, the mixing ratio of a gas species in an air sample can easily be described as a mole ratio, known in chemistry as a "mole fraction," which is equal to the ratio of the number of moles of the gas species to the sum of the number of moles of all the gases in the air sample (usually excluding water vapor). The mole fraction is signified by the Greek letter χ. But since the mole is related to the number of molecules of a gas and also to the volume of the gas by the ideal gas law, mixing ratios can also be described in terms of the number of molecules or the volume of a gas.

The mixing ratios of the atmospheric gases that are present in larger concentrations – nitrogen, oxygen, and argon – are usually expressed as a percent. This can be either a mole

Table 2.1 Methods used to express mixing ratios of different gas and aerosol species in the atmosphere.

Expression	Symbol	Definition	Species
Mole fraction	χ	moles of gas per moles of air	trace gas
Mole percent	% (χ)	moles of gas per moles of air × 100	major gas
Volume percent	% (v/v_T)	m^3 of gas per m^3 of air × 100	major gas
Part per million	ppm	m^3 per 10^6 m^3 of air	trace gas
Part per billion	ppb	m^3 per 10^9 m^3 of air	trace gas
Part per trillion	ppt	m^3 per 10^{12} m^3 of air	trace gas
Micrograms per cubic meter	$\mu g\ m^{-3}$	μg of aerosol per m^3 of air	solid or liquid aerosol

percent ($\chi \times 100$) or a volume percent. A volume percent (v/v) is equal to the ratio of the volume of the gas species (v) to the volume of the total air sample (v_T) measured in cubic meters and multiplied by 100 ($v/v_T \times 100$). Trace gases are often expressed as parts per million (ppm), parts per billion (ppb), or parts per trillion (ppt). An air sample that contains 1 m^3 of a gas species in 1×10^6 m^3 of air is equal to 1 ppm, 1 m^3 of a gas species in 1×10^9 m^3 of air is equal to 1 ppb, and 1 m^3 of a gas species in 1×10^{12} m^3 of air is equal to 1 ppt. The mixing ratios of liquid or solid aerosols in air are expressed as the weight of aerosol in micrograms (μg) per unit volume of gas in cubic meters.

A typical composition of dry air at sea level under ambient conditions is given in Table 2.2. These mixing ratios are given at standard ambient temperature and pressure (SATP). Atmospheric conditions of SATP are a pressure of 1 atm and a temperature of 298 K (25 °C) and are not the same as the atmospheric conditions of standard temperature and pressure (STP), which are a pressure of 1 atm and a temperature of 273 K (0 °C). Since most atmospheric chemistry measurements are taken at ambient conditions, they are normally reported using SATP and not

Table 2.2 Typical gas concentrations in dry air at standard ambient temperature and pressure assuming ideal gas behavior.

Gas	Chemical formula	Mole percent	Mixing ratio (ppb)	Reference
Nitrogen	N_2	78.08	780 800 000	Atmosphere of Earth (n.d.)
Oxygen	O_2	20.95	209 500 000	Atmosphere of Earth (n.d.)
Argon	Ar	0.934	9 340 000	Atmosphere of Earth (n.d.)
Carbon dioxide	CO_2	0.0407	407 000	U.S. Department of Commerce (2017)
Neon	Ne	0.001818	18 180	Atmosphere of Earth (n.d.)
Helium	He	0.000524	5240	Atmosphere of Earth (n.d.)
Methane	CH_4	0.0001834	1834	Blasing (2016)
Nitrous oxide	N_2O	0.0000328	328	Blasing (2016)
Ozone (troposphere)	O_3	0.0000337	337	Blasing (2016)
CFC-12	CCl_2F_2	0.0000000516	0.516	Blasing (2016)
HCFC-22	$CHClF_2$	0.0000000233	0.233	Blasing (2016)
Carbon tetrachloride	CCl_4	0.0000000082	0.082	Blasing (2016)
Carbon tetrafluoride	CF_4	0.0000000080	0.080	Aoki and Makide (2004)

STP. The relative number of moles of the trace gas species per cubic centimeter of air, as well as the number of molecules per cubic centimeter of air, can then be calculated at STP by using the ideal gas law.

The three major gases in the dry atmosphere are nitrogen at 78.08%, oxygen at 20.95%, and argon at 0.934%. All other gases are at concentrations of less than 0.1%. However, one of the most important gases in the atmosphere not listed in Table 2.2 is water vapor. Liquid water on the surface of the Earth evaporates from lakes, oceans, rivers, soils, and plant leaves. This water vapor enters the lower atmosphere at amounts directly proportional to the temperature of the air, with warmer air causing more water to evaporate. The water content in the air is reported as relative humidity, which is defined as the percent of the amount needed for saturation at the same air temperature. Dry air has a low percent relative humidity and moister air has a higher percent relative humidity at the same temperature.

The water vapor content of air also has a direct effect on atmospheric pressure. The molecular weight of water is approximately 18 g mol^{-1}. This can be compared to N_2 at 28 g mol^{-1}, O_2 at 32 g mol^{-1}, and Ar at 40 g mol^{-1}. So, water is less dense than any of the other major atmospheric gases. Warmer air normally contains more water vapor due to the promotion of evaporation, and so is less dense and lighter than the cooler air above it. Low-pressure systems in meteorology always consist of warm humid air, while high-pressure systems consist of cold dry air. When warm moist air (low-pressure systems) interacts with cold dry air (high-pressure systems), the differences in pressure and temperature between the systems produce winds. The larger the differences in temperature and pressure, the stronger the winds and the more severe weather they produce. Variations in the temperature and water vapor content of air parcels at different altitudes can lead to winds occurring in different directions. This creates wind shear aloft and, in some cases, the formation of circular wind patterns that can create tornados, which are associated with large amounts of damage due to the high wind velocities in the vortex.

As warm air rises in the troposphere it begins to cool and the water vapor it contains begins to condense and form clouds. But a small amount of aerosol is required as a substrate in order for this process to occur. These aerosols, called cloud condensation nuclei (CCN), can be solids, liquids, or mixtures of both. The air in the upper troposphere and tropopause is very cold and dry, because the condensation in the lower altitudes has removed the water vapor. Although vertical mixing from the lower troposphere into the stratosphere can be caused by some violent events, such as volcanic eruptions or very strong thunderstorms, it is extremely difficult to carry water vapor into the stratosphere directly. But water vapor can be formed in the stratosphere by the chemical oxidation of hydrogen and methane. These highly volatile gases have sufficiently long atmospheric lifetimes that they can eventually mix from the troposphere into the stratosphere by strong thunderstorms. Once in the stratosphere, they undergo rapid oxidation, due to the high UV intensity, allowing them to be an important source of water vapor for the formation of stratospheric clouds.

The concentrations of trace gases and aerosols in the troposphere and stratosphere depend on their chemical and physical properties, as well as their sources and sinks. The atmospheric lifetimes of these gases and aerosols are affected by their transport and transformation mechanisms, which can occur on local, regional, and global scales, depending on the chemical species. The atmospheric lifetimes of many of the trace gases are determined by their reactivity with hydroxyl radical (OH). The atmospheric OH concentration can vary from remote to rural to urban locations, and is dependent on a number of factors that will be detailed in Chapters 4 and 5. In general, the lifetimes of the reactive gases can be a few minutes to several hours for natural hydrocarbons, such as the sesquiterpenes and monoterpenes emitted from pine forests, to several months for molecules such as carbon monoxide, and several years

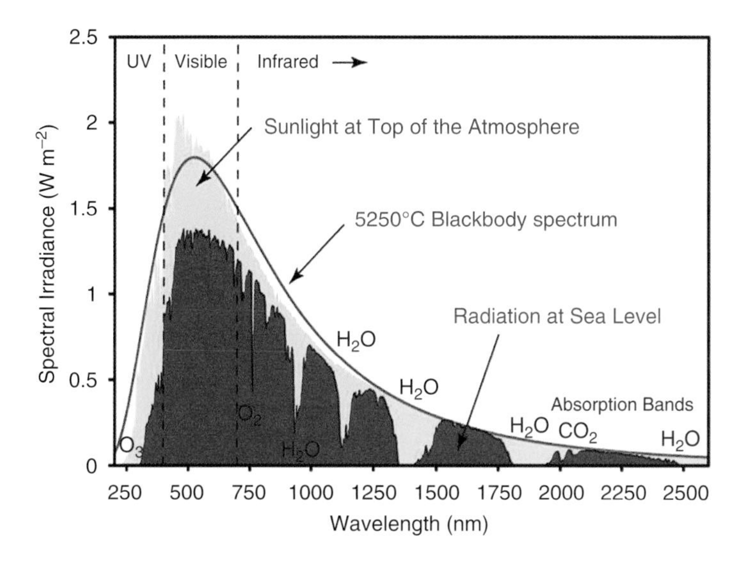

Figure 2.4 The solar radiation spectrum comparing the blackbody emission at 5250 °C (black line), the radiation at the top of the atmosphere (yellow area), and the radiation at sea level (red area). Source: Robert A. Rhode, Wikimedia Commons.

and decades for molecules like methane and the hydrochlorofluorocarbons (HCFCs) used as replacements for CFCs.

The atmospheric lifetimes of both trace gases and aerosols also depend on the solar light intensity, which drives the heating of the Earth's surface and the water evaporation from its oceans and lakes. As a source of UV light, solar light intensity also drives photochemical reactions in both the stratosphere and troposphere, including the formation of the OH radical. The solar irradiance at the top of the atmosphere is shown in Figure 2.4 (yellow), in watts per square meter. Also shown is the solar irradiance at sea level after it has been modified by the Earth's atmosphere (red). The Sun can be viewed as a blackbody, a theoretically ideal radiator and absorber of energy at all electromagnetic wavelengths, at a temperature of approximately 5250 °C. The radiation emitted by this blackbody source is shown in Figure 2.4 as a black line. Notice that the peak intensity of the radiation is in the visible spectral region, at a wavelength of approximately 500 nm. A significant amount of solar radiation in the UV region of 200–300 nm is removed by absorption due to ozone and oxygen in the stratosphere and mesosphere. The major absorbers above 1000 nm are water vapor and carbon dioxide.

2.3 Atmospheric Circulation

As noted in Section 2.2, the mixing of the troposphere is created by changes in temperature, pressure, and relative humidity. The vertical mixing of air masses is due to the rising of less dense warmer air and the sinking of more dense colder air. Changes in pressure (due to temperature and relative humidity changes) are what cause winds to occur, which blow from high-pressure to low-pressure areas. The Earth's orbit (distance from the Sun), shape, rotation, and angle of rotation play an important role in the balance of temperature and pressure that gives rise to winds and atmospheric circulations.

The intensity of the incoming radiation from the Sun is greater at the equator than at either higher or lower latitudes due to the curvature of the planet, as shown in Figure 2.5. This is

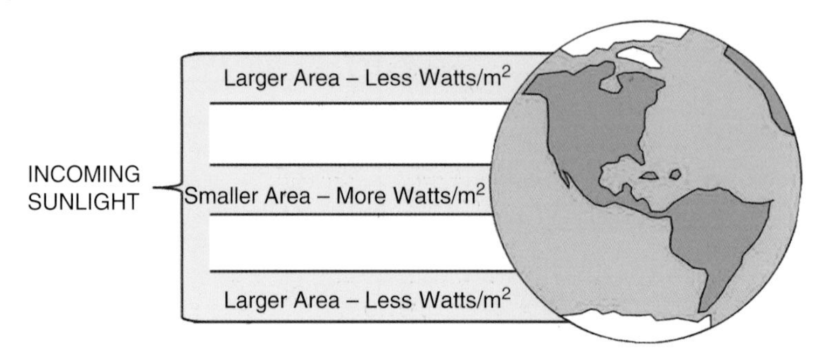

Figure 2.5 The curvature of the Earth leads to the incoming radiation from the Sun being spread out over a larger area, resulting in cooler regions in the northern and southern temperate latitudes.

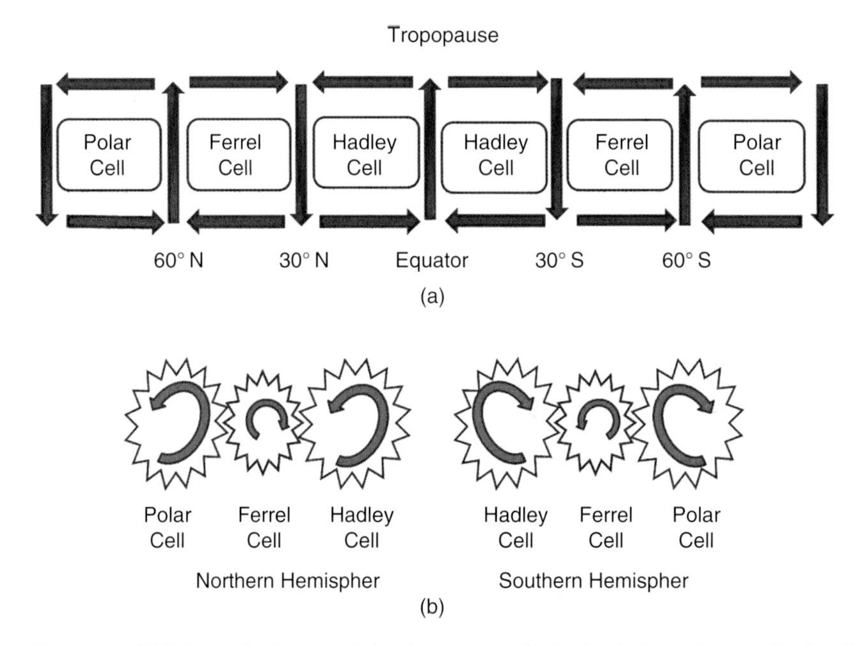

Figure 2.6 (A) Schematic diagram of the three atmospheric circulation cells – the Hadley Cell, the Ferrel Cell, and the Polar Cell – located above and below the equator. (B) Movement of the secondary Ferrel Cells driven by the presence of the Hadley Cells and the Polar Cells on either side.

essentially because the light travels in straight lines from the Sun and when it hits the center of the sphere at the equator, more energy is deposited per unit area (watts m^{-2}). At higher or lower latitudes, the curvature of the Earth leads to the same amount of sunlight hitting a larger surface area, which results in less energy per unit area in the northern and southern latitudes. So, the more north or more south you go, the less solar energy is received per unit of surface area and the colder the temperatures are relative to the equatorial regions.

The large-scale circulation patterns in the atmosphere consist of three cells, shown in Figure 2.6(A), that are basically driven by the surface temperature and the humidity of the air. These three cells make up a global atmospheric circulation pattern, which results in the transport of heat from the equatorial to the polar regions. Hadley Cells are named for the eighteenth-century lawyer and amateur meteorologist George Hadley, who studied equatorial wind patterns in an attempt to explain the mechanism that sustained the "trade winds." In the Hadley Cells, warm moist air rises near the equator until it reaches the tropopause. This creates an area of low pressure at the surface, which then causes air at the surface to move in to replace the rising air. When the rising air reaches the tropopause, it moves toward the poles where it cools and becomes denser. The denser air then descends at about 30° latitude and travels back toward the equator along the surface, replacing the air that rose from the equatorial zone.

A similar cell exists over the polar regions above 60° latitude, called the Polar Cell. This cell circulates cold, polar air toward the equator. The cold air over the poles sinks and flows over the surface toward the warmer regions at 60° latitude, where it rises and flows back toward the poles. The atmospheric circulation in the mid-latitudes is driven by Ferrel Cells, named after American meteorologist William Ferrel, who first proposed them. In the Ferrel Cells air sinks at 30° and flows toward the poles near the surface, where it rises. It then flows toward the equator at higher altitudes. This movement is the reverse of the air flow in the Hadley or Polar Cells. The Ferrel Cell is actually a secondary circulation pattern driven by the presence of the Hadley and Polar Cells on either side of it. It occurs as a result of eddy circulations between the high and low-pressure areas in the Hadley and Polar Cells. The Ferrel Cell has neither a strong source of heat nor a strong heat sink to drive its convection. It therefore acts like a secondary gear driven by the motions of the primary circulation patterns in the Hadley and Polar Cells, as shown in Figure 2.6(B).

Although the major driving force behind the circulation of the atmosphere is the uneven distribution of solar heating across the Earth, the Earth's rotation also plays an important role in atmospheric circulation. As the Earth is a spinning sphere, the rotating motion creates a Coriolis force on the atmosphere. The Coriolis effect is the generation of a force on a rotating object which acts perpendicular to the direction of motion or the axis of rotation. This force generated by the rotating Earth acts to deflect the motion of the atmosphere to the right in the Northern Hemisphere (NH) and to the left in the Southern Hemisphere (SH). An air mass on the Earth moving at a given speed v and subject only to the Coriolis force will travel in a circular motion called an "inertial circle," as shown in Figure 2.7. Since the Coriolis force is directed at right angles to the direction of rotation, the air mass will move with a constant speed around a circle whose radius is determined by the Coriolis parameter, which is dependent on the latitude:

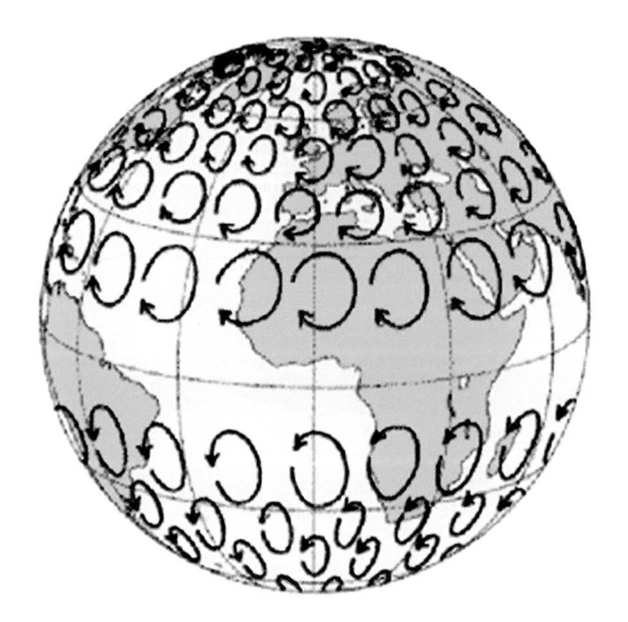

Figure 2.7 Inertial circles of air masses caused by the Coriolis effect created by the Earth's rotation. Source: Anders Persson, Wikimedia Commons.

These inertial circles are clockwise in the NH (where air movement is deflected to the right) and counterclockwise in the SH (where air movement is deflected to the left).

George Hadley was the first to include the effects of the rotation of the Earth in his explanation of the movement of air masses in the tropics. In the Hadley Cell the air mass at the surface, which flows toward the equator, is turned toward the west by the Coriolis effect, generating the trade winds, also called the tropical easterlies. So the trade winds have a sustained westerly flow instead of flowing directly north to south as predicted by temperature and pressure patterns. These prevailing winds were the reason that most European sailing ships reached North America. The trade winds travel from the northeast to the southwest in the NH and from the southeast to the northwest in the SH. In Ferrel Cells, the polar flowing surface air is deflected to the east by the Coriolis effect, resulting in westerly winds. Winds are named by the direction from which they blow, so the Ferrel Cell westerlies travel from the southwest to the northeast in the NH and from the northwest to the southeast in the SH. Ferrel's model was the first to explain these westerly winds between latitudes 35° and 60° in both hemispheres. In the Polar Cells, the surface winds above 60° latitude are prevailing easterlies similar to those seen in the Hadley Cells flowing from northeast to southwest in the NH and southeast to northwest in the SH.

These major wind circulation patterns, which are driven by solar heating at the surface, the rotation of the planet, and cooling aloft, are large-scale phenomena that are used to explain the observations of the trade winds, the tropical rain belts (warm and wet), and subtropical deserts (cold and dry). Convergence zones, which are regions between the cells where the winds travel in opposite directions, can give a spin to the atmosphere, leading to cyclonic storms. Where the cells converge, there are regions of either high or low pressure. If the column of air above the convergence zone is dry and cold (denser), a high-pressure (**H**) region results, while if the air above the convergence zone is wet and warm a low-pressure (**L**) region results, as shown in Figure 2.8.

At the equator, the convergence between the northern and southern Hadley Cells results in a low-pressure (**L**) region. In this convergence region the northeast trade winds above the equator and the southeast trade winds below the equator converge, creating an area of low winds

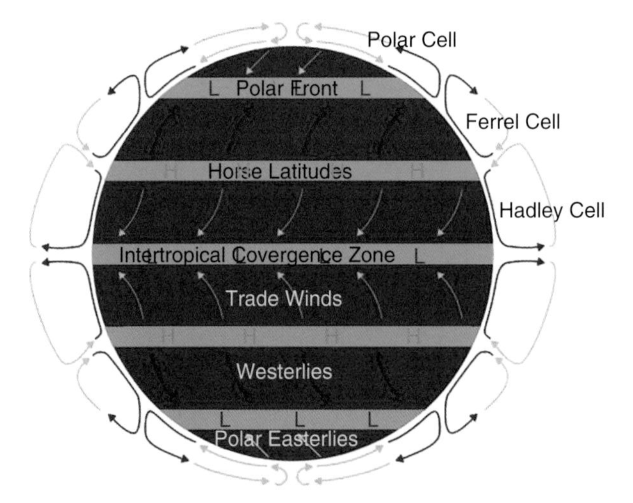

Figure 2.8 The major wind patterns due to the Hadley, Ferrel, and Polar Cell circulations. Low-pressure (L) or high-pressure (H) areas form at the convergence between the cells, depending on the position of the warm and cold portions of the circulation. The convergence between the Hadley Cells at the equator is called the Intertropical Convergence Zone and is characterized by very low winds called the doldrums. Source: DWindrim, Wikimedia Commons.

known as the Intertropical Convergence Zone (ITCZ), also called the doldrums by sailors. The area between the Hadley and the Ferrel Cells in the subtropical latitudes is an area of high pressure (H), called the Horse Latitudes, where precipitation is suppressed. This subtropical high-pressure area is again dominated by lower wind speeds, which leads to calm winds and seas. The Horse Latitudes then converge into the Ferrel Cell regions, where the winds are westerly. In the United States it is this convergence zone that causes weather patterns normally to travel from the west coast to the east coast. Another low-pressure (L) area is found at the convergence of the Ferrel and Polar Cells creating the Polar Front. There is a sharp temperature gradient at this convergence zone since the two air masses in the Ferrel and Polar Cells are at very different temperatures. The Polar Front occurs as a result of the cold polar air in the Polar Cell meeting the warm mid-latitude air in the Ferrel Cell. It is a stationary front since the two air masses are not moving against each other.

At the top of the convergence zones between the Hadley and Ferrel Cells and the Ferrel and Polar Cells near the tropopause are areas of high-altitude winds that are known as the jet streams. These jet streams are areas of very strong winds ranging from 120 to 250 mph. Often, airlines try to fly with the jet stream because their flight times can be affected dramatically by either flying with the flow or against the flow of the jet stream, resulting in significant fuel and time cost savings. Jet streams are caused by the large temperature differences between the upper air masses in the neighboring cells, resulting in a very large increase in horizontal wind speed with altitude. Figure 2.9(A) shows the formation of the Subtropical Jet between the Hadley and Ferrel Cells and the Polar Jet between the Ferrel and Polar Cells. The mixing of cold dry air in the Polar Cell and warm wet air in the Ferrel Cell leads to strong storms in the Polar Jet region, around 45–60° in latitude. The Subtropical Jet lies above the subtropical high-pressure belt and is generally associated with fair weather. The Polar Jet is more widely known because of its importance in predicting weather patterns in the northern latitudes.

Figure 2.9(B) shows how the jet streams move across the globe. The Polar and Subtropical Jets are both westerly, and both jets move north and south with the seasons as the horizontal

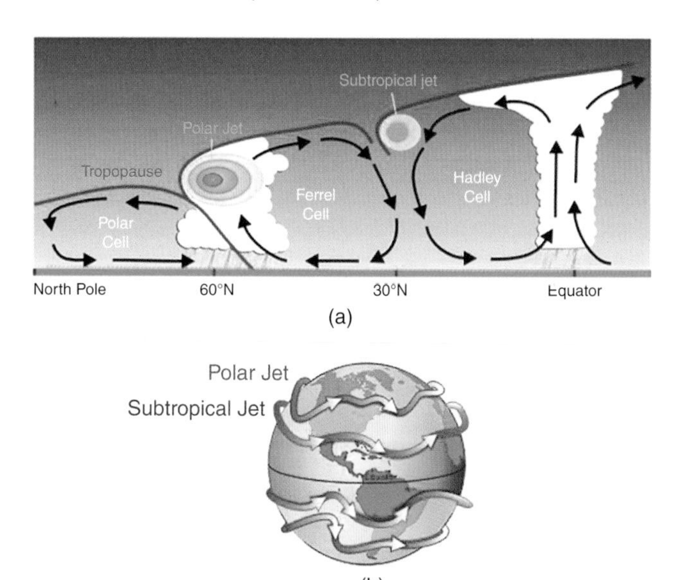

Figure 2.9 (A) Creation of the Polar and Subtropical Jets. (B) Movement of the Polar and Subtropical Jets around the globe. Source: (A) Lyndon State College Meteorology; (B) National Weather Service Jet Stream, Wikimedia Commons.

temperature fields across the globe shift with the areas of strongest sunlight. The paths of the jet streams typically have a meandering pattern caused by the changes in the Coriolis effect with altitude. Jet streams can also split in two when they encounter an upper-level low that diverts a portion of the jet stream. As the northern Polar Jet dips southward creating a trough, cold air from the polar region moves with it. Any region located within the trough will experience colder temperatures. Similarly, as the jet moves northward creating a ridge, warmer air from the equatorial regions is able to move toward the poles and areas located within these ridges will experience warmer temperatures. So, the position and curvature of the polar jet streams are a great indicator of prevailing weather experienced at the Earth's surface.

So far, we have examined large-scale atmospheric circulation patterns. However, regional factors are also very important in determining atmospheric mixing and precipitation patterns over short distances. An important example of smaller-scale factors that influence the distribution of precipitation is the effects of mountainous terrain located near oceans or large lakes, as shown in Figure 2.10. Warm moist air moving over the body of water is drawn toward the higher altitudes of the mountains. This effect, which forces an air mass from a low altitude to a higher altitude due to a rising terrain, is called an orographic lift and the side of the mountains with prevailing winds flowing toward them is called the windward side. At the higher altitudes the air is cooled and the water vapor it carries condenses and precipitates in the form of rain or snow. Having lost its moisture, the drier air then crosses the tops of the mountains creating a drier side on the leeward, or downwind side, of the mountain range. This drier side is called a "rain shadow," which is an area where no rain falls due to the prevailing winds having lost their moisture before crossing the mountain range. This terrain effect has significant impacts on the deposition of water-soluble air pollutants as the pollutants dissolved in the rain are deposited only on the windward side of the mountain range. This affects both water sheds and ecosystems in mountainous regions. It also leads to the formation of deserts on the leeward side of mountain ranges, which has implications for climate.

Another example of a terrain effect also associated with mountains or hills and their accompanying valleys that commonly occurs at night is shown in Figure 2.11. During the evening hours the sides of hills or mountains cool and downslope winds carry colder air from aloft into the valleys. At the same time the valleys that were heated during the day give off heat, warming the air directly above the valley. The warmer air from the valley rises and the colder

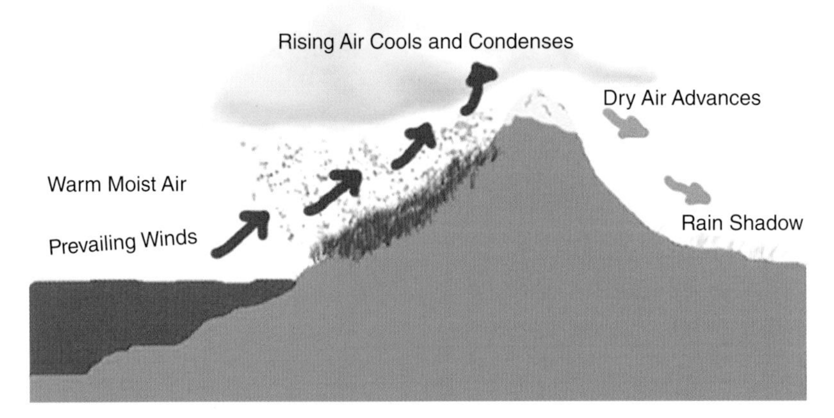

Figure 2.10 The effect of mountainous terrain on precipitation distribution. The orographic lifting of the air over a mountain leads to precipitation on the windward side and a rain shadow on the leeward side of the mountain. Source: Bariot, Wikimedia Commons.

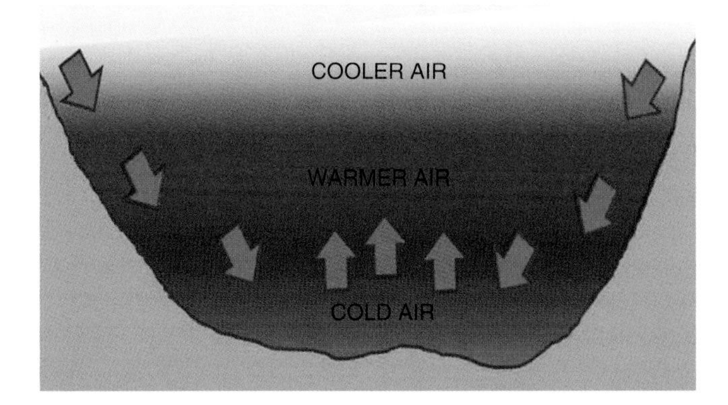

Figure 2.11 The formation of a nocturnal thermal inversion in a valley. Cold air flows down the slopes while warmer air in the valley rises above the colder layer. The presence of warmer air above cold air creates a thermal inversion which traps air in the valley.

air from the mountain sides or hill sides displaces the warm air with cold air. This results in an increase in temperature with height, the inverse of the normal condition in the troposphere. This condition is called a thermal inversion or temperature inversion. It creates a stable atmospheric condition that stops atmospheric convection, which promotes atmospheric mixing. This thermal inversion leads to a thermal capping of the air in the valley during the night and early morning periods. Later in the day the valley will absorb radiation from the Sun, heating the air, breaking the thermal inversion, and again promoting mixing. This type of inversion can also occur in areas next to oceans when a warm, moist (less dense) air mass moves over a cooler, drier (denser) air mass during the movements of warm fronts. In many cases, when the air is humid, the inversion leads to low-lying clouds or fog. These thermal inversions can also cause significant air-pollution events to occur, as emitting air pollutants into the capped cold air leads to higher concentrations at ground level until the thermal inversion can be broken by solar heating.

Normally, the warmer air near the ground mixes with the colder air aloft. The height to which this mixing occurs in one hour or less is called the mixing height. Warm air from the surface will rise to over 1500 m before cooling to a temperature where it does not rise at a significant rate. This height is called the boundary layer and represents the height in the troposphere where atmospheric mixing is reduced. Terrain-induced thermal inversions usually extend to a height of a few hundred meters, which is significantly less than the usual 1500 m of the boundary layer. Thermal inversions result in a lowering of the mixing height, limiting the amount of air that can mix with or dilute pollutants released in the inversion zone. So, the release of the pollutants into a smaller mixing height can lead to elevated pollutant concentrations since they can no longer mix with the upper air. To overcome the potential for releasing emissions into a nocturnal thermal inversion, industries have used tall stacks to release pollutants into the atmosphere above the potential height of thermal inversions. These tall stacks then reduce the high concentrations that would occur if emitted at ground level into the cold air pool caused by the thermal inversion. While this solution lowers the air pollutants locally, the release of air pollutants at a higher altitude allows them to be transported long distances, leading to pollution problems downwind of the stack. We will discuss this further in Chapter 5. Figure 2.12 shows an example of tall industrial smoke stacks at a coal-fired power plant (Cumberland Power Plant, Cumberland City, TN).

The basic physics driving the circulation of the atmosphere is the same as that driving the circulation of large bodies of water, which will be discussed in detail in Chapter 8. Sunlight causes

Figure 2.12 An example of tall industrial smoke stacks used to release pollutant emissions above the thermal boundary layer. Source: Steven Greenwood, Wikimedia Commons.

warming of the upper portions of the water surface, which does not mix well with the colder water below. This leads to a thermal profile that is the opposite of that in the troposphere. The heated surface waters promote the evaporation of water from lakes and ocean surfaces, ultimately leading to cloud formation in the atmosphere. The evaporation in the thermally isolated surface ocean waters leads to increased salt content, which causes the surface waters to become denser, causing them to sink and mix with the water below. So, both the atmosphere and large water bodies are fluids whose mixing is governed by heat and density differences. The atmosphere and hydrosphere are closely linked systems that are both tied to sunlight and the energy balance.

2.4 Energy Balance

We will cover the topic of the Earth's energy balance in detail as it relates to weather and climate change in Chapter 11. For now, we will briefly review the fundamental aspects of energy balance, also called radiative balance. The Earth's energy budget is made up of the balance between the energy the Earth receives from the Sun and the energy the Earth radiates back into space after being absorbed and distributed throughout the four Earth systems: the atmosphere, the hydrosphere (including water, water vapor, and ice), the geosphere (Earth's crust), and the biosphere (all living organisms). Radiative forcing is the change in this energy balance when one of the variables is changed while all others are held constant. It is typically determined either at the tropopause or more commonly at the top of the atmosphere, in units of watts per square meter of the Earth's surface. When the Earth receives more incoming energy from sunlight than it

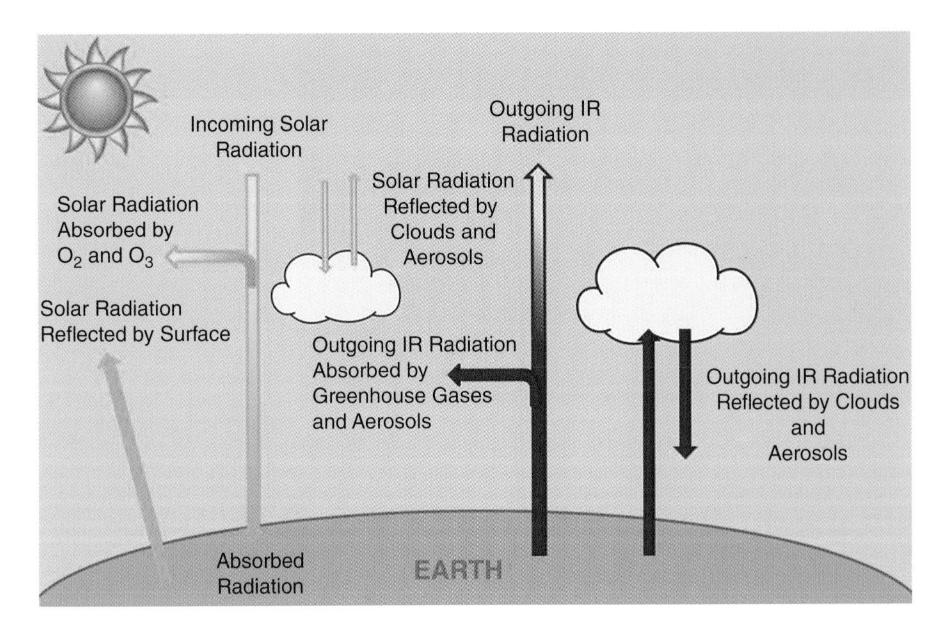

Figure 2.13 The major process involved in the energy balance of the Earth.

radiates to space, the radiative forcing is positive. This net gain of energy in the Earth's systems will cause warming. In contrast, if the Earth loses more energy to space than it receives from the Sun, the radiative forcing is negative and the result is cooling.

Figure 2.13 shows a diagram of some of the major processes involved in the Earth's energy balance. The incoming solar radiation enters the Earth's upper atmosphere where the UV portion is absorbed by oxygen and ozone. The solar radiation not absorbed in the upper atmosphere can be reflected back to space by clouds or atmospheric aerosols. The portion that is not absorbed or reflected in the atmosphere is either absorbed or reflected by the surface of the Earth. The solar light absorbed by the surface heats it up, converting the solar light to infrared (IR) radiation, which is released into the atmosphere. This outgoing IR radiation can be absorbed by gases in the atmosphere called greenhouse gasses. It can also be absorbed by IR-absorbing aerosols in the atmosphere, such as carbonaceous soot, or it can be reflected back toward the surface by clouds.

The amount of energy received from the Sun depends on: changes in the Sun's radiation output, changes in the Earth's orbit, and changes in the Earth's axis of rotation. The amount of energy released back to space is determined by the difference between the amount absorbed and the amount reflected by the Earth's surface and atmosphere. This includes the effects of surface albedo, clouds and aerosols, greenhouse gases, and surface absorption.

2.4.1 Milankovitch Cycles

The variation of solar energy impacting the surface of the Earth due to the angle of the Earth's axis has been discussed in Section 2.3. The variation of energy received by the Sun is also controlled by variations in the Earth's rotation and orbit. The Earth orbits our Sun every year, but this orbit is not always the same. There are variations in the orbit that are due to gravitational interactions with other planets (particularly Jupiter and Saturn) that are also orbiting the Sun. These variations are called Milankovitch cycles after their discoverer, Milutin Milankovitch. They show a periodicity due to the effects on the orbital eccentricity (shape), axial obliquity

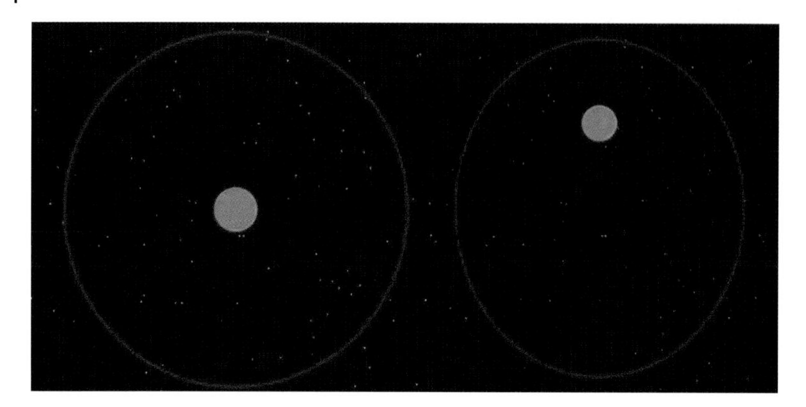

Figure 2.14 The comparison of a planetary orbit eccentricity of 0 (left) and 0.5 (right). Source: NASA, Wikimedia Commons.

(tilt), and axial precession (axis orientation) of the Earth relative to the Sun. The eccentricity of a planetary orbit around the Sun relates to the orbital pattern or shape. It is defined at values from 0 to 1, where 0 is a perfect circle and 1 is a parabolic escape orbit. The closer the eccentricity is to 0, the closer the orbit is to a perfect circle around the central object. A comparison of orbit eccentricity for a value of 0 and a value of 0.5 is shown in Figure 2.14. The Earth's eccentricity value has been very close to a circular orbit and over its recent past has not varied significantly from 0 with a range of 6×10^{-5} to 0.07 (slightly elliptical) and a geometric mean of 0.02 (Laskar et al., 2011). The Earth's orbit eccentricity changes on approximately a 100 000-year cycle, a long-term effect compared to human history. Currently the eccentricity is at a value of 0.0017 (Berger and Loutre, 1991).

The axial obliquity, or tilt of the Earth's rotational axis shown in Figure 2.15(A), also changes slightly over time with a range from 22.1° to 24.5° over a cycle of 40 000 years. We are currently at a midpoint of about 23.5° (Seidelmann, 1992). The axial precession is a description of the movement of the axis as the Earth rotates. As the Earth makes a complete rotation once a day

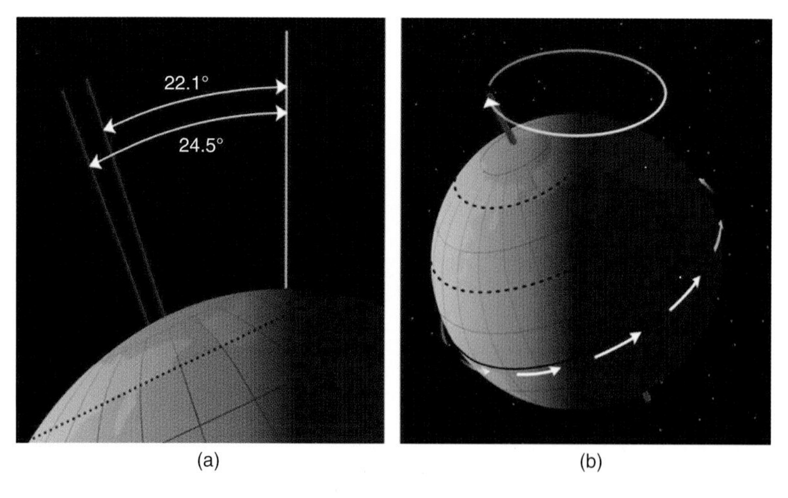

Figure 2.15 (A) The angle of the Earth's axial tilt, which varies between 22.1° and 24.5°, over a cycle of about 41 000 years. (B) Precessional movement of the Earth over a period of about 26 000 years. As the Earth rotates (white arrows) once a day around its rotational axis (red), the axis itself rotates slowly (white circle). Source: NASA Mysid, Wikimedia Commons.

around its rotational axis, the axis itself rotates slowly in a motion similar to a wobbling top, as shown in Figure 2.15(B), completing a rotation in about 26 000 years. This axial precession is caused by gravitational interactions of the Sun and Moon and is measured relative to the position of the stars. The Earth is not a perfect sphere but is an ellipsoid with a diameter at the equator about 43 km larger than that at the poles. Because of the axial tilt, the gravitational pull on the half of the equatorial bulge that is closer to the Sun and Moon is stronger, creating a small torque on the Earth. It is this gravitational force on the equatorial bulge that results in the precession. If the Earth were a perfect sphere, there would be no precession. This effect causes some differences in the amount of energy received from the Sun between the NH and the SH, and currently impacts the SH more than the NH (Berger, 1976).

The variations in incoming solar energy reaching the Earth due to these orbital and axial processes over past and current time periods are shown in Figure 2.16. Note that these effects are all on long time scales. The solar energy on the day of the summer solstice, 2000 CE at 65 N latitude (blue dot in Figure 2.16) is 480 W m^{-2}. Over the next 50 000 years this value is predicted to vary from 500 to 470 W m^{-2}. Current models indicate that about 6000 years ago we began a cooling trend due to these cycles which is expected to last for another 23 000 years (Imbrie and Imbrie, 1980). This predicted cooling trend is based only on the Earth's orbit and rotation cycles, and represents only the solar energy received at the top of the atmosphere. This does not include any impacts that the atmosphere has on the incoming solar energy before reaching the surface of the planet. Some of these factors, which can increase or decrease the amount of energy at the surface, include surface albedo, volcanic activity, changes in land and ice mass, aerosol and cloud formation, and anthropogenic impacts. These anthropogenic impacts, which include the addition of greenhouse gases and aerosols into the atmosphere, will be discussed briefly here and in more detail in Chapter 11.

Another important factor influencing the amount of solar energy reaching the Earth is the variability of the Sun's radiation output, which is dependent on solar magnetic and sunspot activity. This cycle occurs approximately every 11 years. The period of increased magnetic and sunspot activity is called the "solar maximum." Even though sunspots are colder areas, solar irradiance is actually higher during a period of sunspots due to the higher-temperature areas surrounding them. The solar maximum is followed by a quiet period called the "solar minimum." The variation in sunspot activity since 1985 is shown in Figure 2.17 (in red). The impact of the sunspot activity on the Earth's global temperature variation is expected to be 0.1 K at the solar minimum and 0.3 K at the solar maximum (Camp and Tung, 2007).

The changes in the solar energy output cause impacts in space as well as in the Earth's atmosphere and surface. In general, sunspot activity is directly correlated with total solar irradiance (TSI), as shown in Figure 2.17. The TSI is the amount of solar radiative energy incident on the

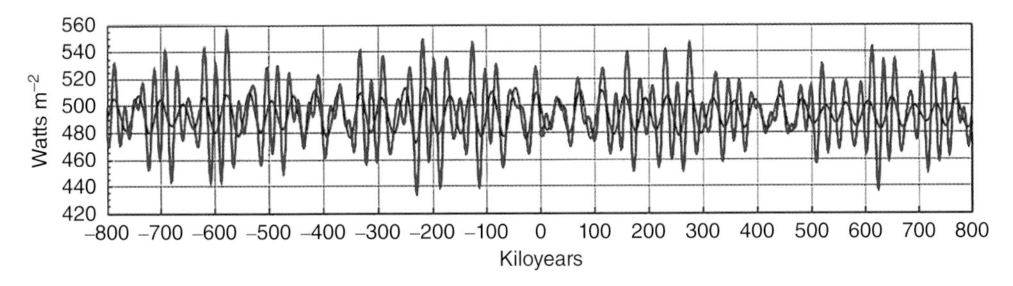

Figure 2.16 Past and future variations of solar radiation at the top of the atmosphere on the day of the summer solstice at 65 N latitude. The red curve is based on the predicted values for eccentricity. The green curve is based on an eccentricity of 0. The blue dot is the conditions for the year 2000. Source: Incredio, Wikimedia Commons.

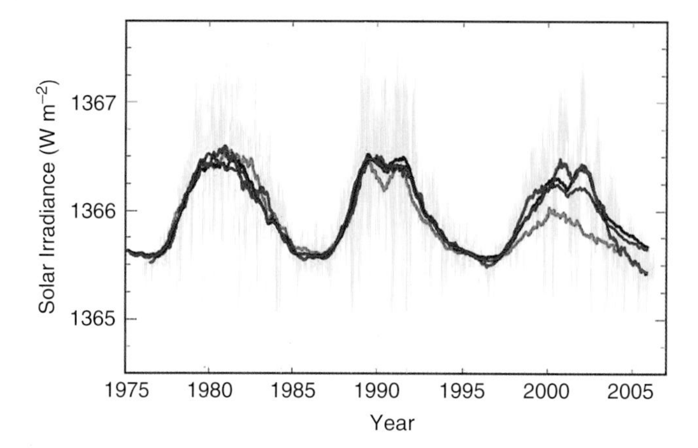

Figure 2.17 Solar activity cycles occurring from 1975 to 2005 as seen in daily total solar irradiance (yellow), TSI yearly average (red), sunspot number (blue), 10.7 cm radio flux (purple), and solar flare index (green). The vertical scales for all quantities except TSI have been adjusted to permit their plotting on the same vertical axis. Source: Robert A. Rhode, Wikimedia Commons.

top of the Earth's atmosphere. It is also highly correlated with extreme ultraviolet (EUV) or X-ray emissions in the 10–125 nm region. The EUV can affect the upper stratosphere and is an ionizing radiation that is of concern for astronauts and materials in space. Its overall impact on the atmosphere is not well understood and is an area of ongoing study, particularly as it relates to stratospheric chemistry and the stratospheric ozone cycles, which will be examined in Chapter 4.

2.4.2 Planetary Albedo

Once the total solar irradiance at the top of the atmosphere is determined, the next important factor in determining the energy balance, or radiative balance, of the Earth is how much radiation is reflected back from the surface of the planet into the atmosphere and finally into space. This is called the planetary albedo. The albedo is that fraction of the radiation that is reflected by the different surfaces of the Earth's crust. So, it is determined as the ratio of the reflected light to the incident light as:

$$A = \frac{\text{reflected light}}{\text{incident light}}$$

The albedo has values between 0 and 1, where a perfectly reflective white surface would have an albedo of 1, while a totally absorbing black surface would have an albedo of 0. Some examples of albedos from different surfaces are given in Table 2.3 for visible light in the wavelength region of 400–800 nm.

It is clear from Table 2.3 that most of the surface of the planet has fairly low albedos, with the exception of frozen areas and deserts. The average albedo of the Earth is usually taken to be around 30%, which also includes the albedo of atmospheric clouds. So, about 70% of the incoming radiation from the Sun in the UV and visible region is absorbed by the surface of the Earth. That energy heats the surface and is converted into IR radiation, which is re-emitted from the heated surfaces. Figure 2.18 shows the IR emission from the surface of the Earth approximated as a blackbody at a temperature of 300 K. This IR emission peaks at about 10 μm.

Clouds and wet aerosols are also included in the planetary albedo, since both can act to reflect and scatter the incoming sunlight. Light scattering causes the direct sunlight to be scattered in

Table 2.3 The albedo of surfaces in the visible-light wavelength range (400–800 nm).

Surface	Albedo	Reference
Asphalt	0.04–0.16	Pomerantz et al. (2000)
Boreal forest – grass	0.2	Betts and Ball (1997)
Boreal forest – snow-covered grass	0.75	Betts and Ball (1997)
Boreal forest – conifers	0.083	Betts and Ball (1997)
Boreal forest – snow-covered grass	0.13	Betts and Ball (1997)
Sahara desert	0.2–0.50	Pinty and Ramond (1987)
Snow	0.54–0.8	McKenzie et al. (1998)
Open ocean	0.05–0.25	Jin et al. (2004)
Lakes	0.08–0.17	Maykut and Church (1973)
Tundra	0.1	Maykut and Church (1973)
Frozen tundra – snow-covered	0.75	Maykut and Church (1973)
Soil	0.1–0.5	Houldcroft et al. (2009)

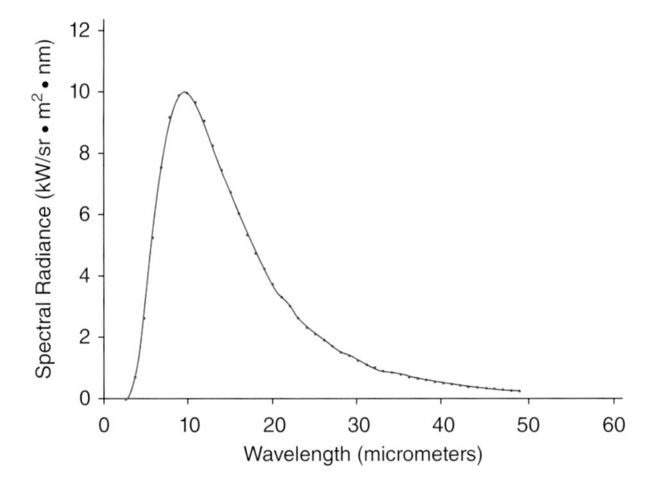

Figure 2.18 The emission of IR radiation from the Earth approximated as a blackbody at 300 K.

many directions, creating a diffuse radiation. When clouds are thick enough, they reflect the incoming sunlight giving a white color to the cloud and any white surface has an albedo of 1. As the cloud passes over, you can see its shadow on the ground, similar to the shadow that a tree or building casts. It is cooler in the cloud's shadow because the direct sunlight does not hit the surface under the cloud. Only the less intense diffuse radiation created by light scattering can hit the surface under the cloud shadow. This leads to less light absorption and less heating of the surface. So, cloudy days are generally cooler than sunny days. When the cloud is very thick, there is less direct sunlight that penetrates to the lower portion of the cloud, making it appear darker. These darker clouds can have albedos less than 1 due to absorption of light by the water droplets. We know these dark clouds as storm clouds, since they are typically associated with large rain events. Figure 2.19 shows a photograph of some typical clouds and their varying appearance from white to gray, which is an indication of the cloud optical thickness and liquid water content.

Figure 2.19 Shadows cast by clouds and trees on the campus of the University of Arkansas at Little Rock. The clouds in the sky are also casting shadows onto the adjacent clouds, showing how they act to scatter the incoming sunlight.

2.4.3 Greenhouse Gases

The greenhouse gases get their name from the so-called "greenhouse effect." Greenhouses are buildings with glass walls and ceilings that are used to grow plants year-round. The sunlight coming through the glass hits the plants and other surfaces in the room, where the sunlight is absorbed. The heated surfaces emit IR radiation similar to the heated Earth surfaces (Figure 2.18). This IR radiation is trapped in the room by the air and IR-reflective glass, causing the room to heat up to a higher temperature than the outside. The same effect happens in a closed automobile, where sunlight comes in through the windows causing the inside of the vehicle to heat up. Greenhouse gases are those atmospheric gases that can absorb the IR radiation that is emitted from the Earth's surface after it has absorbed the incoming solar radiation. This traps the IR radiation in the atmosphere, preventing it from exiting into space.

The Earth's atmosphere plays a very important role in trapping the energy emitted from the heated surface. To fully understand this impact, we can compare Earth with our nearest neighboring planets – Venus and Mars – as shown in Table 2.4. While Venus and Earth are about the same size, Mars is only about half the size of Earth, with a gravity that is only about 40% of the gravity of either Venus or Earth. So, over time Mars has lost most of its atmosphere to outer space. The atmospheres of these three planets are quite different, as Venus has a very heavy atmosphere composed of 96% carbon dioxide and 3.5% nitrogen with a pressure of about 90 atm at the surface, while Earth's atmosphere is 98% nitrogen and only 4% carbon dioxide, with a pressure of 1 atm at the surface. Although the composition of the atmosphere of Mars is more similar to that of Venus than Earth (96% carbon dioxide and 1.9% nitrogen), it is very thin with a surface pressure of only 0.006 atm, 0.007% of that of Venus and 0.6% of that of Earth.

Table 2.4 The comparison of the atmospheric composition and pressure with surface temperature predicted with no atmosphere and observed with the atmosphere for Venus, Earth, and Mars.

Gas	Venus	Earth	Mars
Nitrogen (%)	3.5	98	1.9
Carbon dioxide (%)	96	4	96
Oxygen (%)	0	20.9	Tr
Argon (%)	Tr	1	1.9
Surface pressure (atm)	90	1	0.006
Gravity (g)	0.9	1	0.4
Temperature (K) with atmosphere	750	288	226
Temperature (K) without atmosphere	227	255	216

Tr = Trace.

The atmosphere of Venus has 24 times the amount of carbon dioxide as that of Earth. Since carbon dioxide is a very important greenhouse gas, with the ability to trap a significant amount of heat, the surface of Venus is very hot (750 K). Conversely, Mars, with very little atmosphere, is very cold (226 K), while the Earth's atmosphere, with a smaller amount of carbon dioxide (4%), is at an intermediate temperature (288 K). As described in Section 2.2, pressure and temperature decrease with altitude in Earth's troposphere. On Venus, the pressure and temperature also decrease with altitude. At an altitude of about 50 km, the pressure is about 1 atm at a temperature of about 295 K, more similar to Earth's surface conditions. Indeed, some NASA scientists have proposed that this would be a suitable place for humans to establish "floating" cities. Since nitrogen and oxygen are lighter gases than carbon dioxide, these floating cities could use air in large balloons to lift the cities to the pleasant 50 km altitude. This would also be a source of breathing air for the inhabitants (Landis, 2003).

Figure 2.20(A) shows the relationship between the incoming solar radiation and the outgoing IR radiation emitted by the Earth's surface. Incoming solar radiation between 100 and 400 nm is the UV spectral region. This region is divided into three wavelength ranges: UVA (315–400 nm), UVB (280–315 nm), and UVC (100–280 nm). While the UVC is strongly absorbed by oxygen and ozone in the stratosphere, the UVB is absorbed partially by ozone with a small amount reaching the surface and UVA is nearly 100% transmitted to the surface. Light in this region contains enough energy to break the chemical bonds in some molecules in the troposphere. Overall, the incoming solar radiation is 70–75% transmitted by the atmosphere, with the majority of the absorption occurring in the stratosphere in the UV region between 0.2 μm (200 nm) and 0.28 μm (280 nm) (Figure 2.20(C)). Figure 2.20(B) shows that most of the incoming light in the 0.4–0.8 μm (400–800 nm) range reaches the surface. It is interesting to note that this is the "visible region" of the electromagnetic spectrum, where our vision developed. This is consistent with light of this wavelength range being dominant at the Earth's surface.

The blackbody radiation curves shown in Figure 2.20(A) (right) are the outgoing radiation for three temperatures: 210 K (black line), 260 K (blue line), and 310 K (purple line). The outgoing IR radiation is only 15–30% transmitted through the atmosphere to space, because the greenhouse gases in the lower atmosphere absorb the radiation that is emitted from the planet's surface. Figure 2.20(C) shows the major absorbing gases in the atmosphere – water vapor, carbon dioxide, oxygen, ozone, methane, and nitrous oxide – along with the wavelength region of their absorption. Oxygen and ozone absorb mostly in the UV and are responsible for the

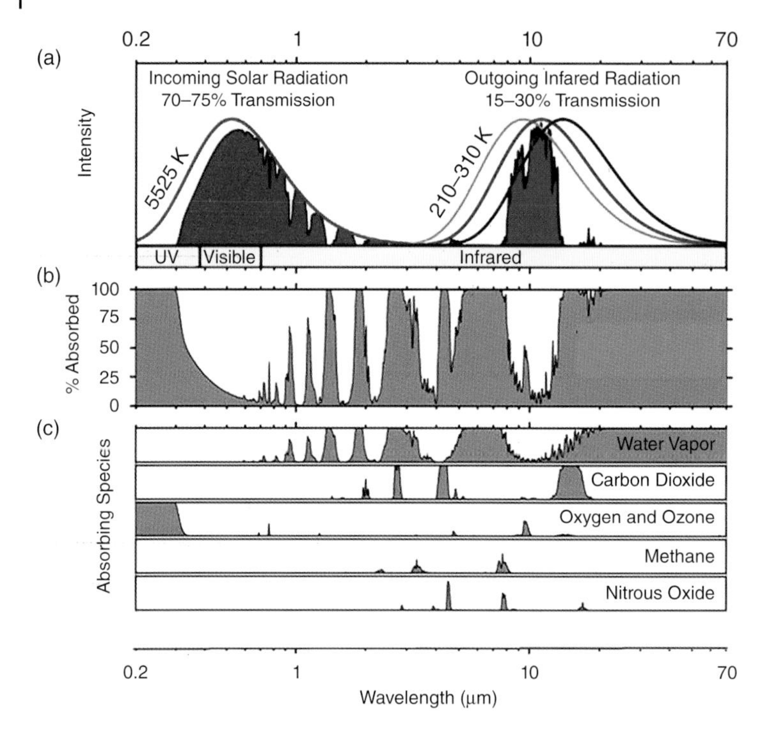

Figure 2.20 (A) The intensity of the incoming solar radiation (left) and the outgoing IR radiation (right). (B) The percent of the incoming solar radiation and the outgoing IR radiation that is absorbed by the atmosphere. (C) The species responsible for the atmospheric absorption of both incoming and outgoing radiation. Source: Adapted from Robert A. Rhode, Wikimedia Commons.

major absorption of the incoming solar radiation. The other four absorbing gases, shown in Figure 2.20(C), absorb in the IR and are responsible for the absorption of the outgoing IR radiation. Although the outgoing IR radiation is not well transmitted through the atmosphere due to absorption by these major greenhouse gases, there is a region between 8 and 13 μm (shown in Figure 2.20(B)) where a significant amount of the outgoing IR radiation is not absorbed. This region is called the window region, as it occurs at wavelengths where the major greenhouse species do not absorb. The window region is very important when considering the impacts of other trace greenhouse species, since any absorption of the outgoing IR radiation in this region will lead to an increased warming effect on the atmosphere.

Carbon dioxide, methane, and nitrous oxide are all naturally occurring greenhouse gases. However, the increase in their concentration due to man's activities will cause an increase in the natural greenhouse effect. Carbon dioxide is increasing due to fossil fuel combustion and methane and nitrous oxide are increasing due to agricultural activities. The increased warming caused by the increases in these greenhouse gases results in increased surface water evaporation and thus increases in atmospheric water vapor. Since water vapor is also a strong greenhouse gas, as shown in Figure 2.20(C), this increase in atmospheric water vapor also enhances radiative forcing and further warming. There are also numerous greenhouse species besides water vapor, carbon dioxide, methane, and nitrous oxide. Indeed, any trace gas species that can absorb in the IR window region will act as a greenhouse species. These include all of the chlorofluorocarbons as well as the hydrochlorofluorocarbons that are currently being used in refrigeration units.

The greenhouse gas that has received the most attention is carbon dioxide. This gas occurs naturally and plays an important role in maintaining the oxygen content of the atmosphere and

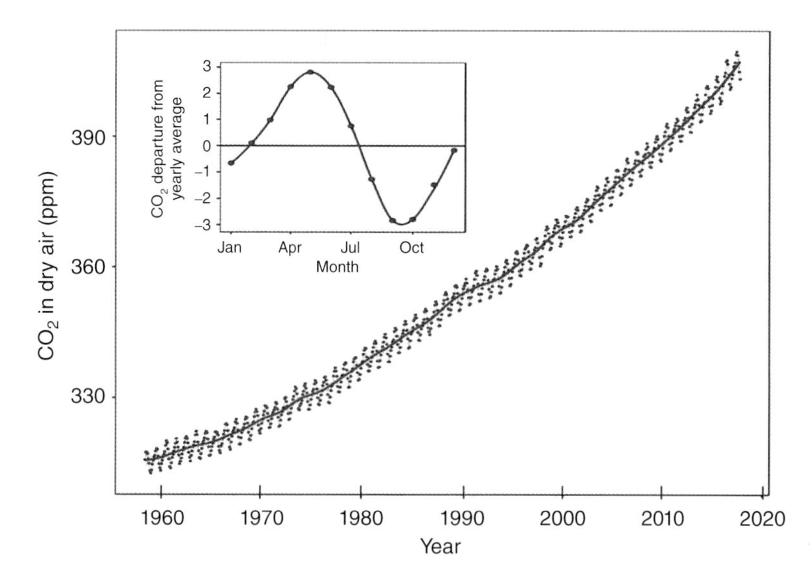

Figure 2.21 The mean monthly carbon dioxide concentrations (ppm in dry air) at Mauna Loa, Hawaii from 1958 to 2017. The insert at the upper left demonstrates the seasonal variation plotted as the departure from the yearly average. Source: Delorme, Wikimedia Commons.

is well recognized as being connected to planetary temperature. Its concentration has varied significantly over geological time scales from about 200 ppm in the Quaternary Period (last 2 million years) to over 7000 ppm in the Cambrian Period (500 million years ago). During the past few hundred thousand years, concentrations of carbon dioxide are believed to have been stable at about 280 ppm. This is based on measurements of trapped carbon dioxide in ice cores (Fischer et al., 1999). The first careful measurements of carbon dioxide concentration in the troposphere began in 1958 at a sampling site on the top of Mauna Loa, Hawaii, an area not impacted by local sources. These measurements have continued to the present day and show that the level of carbon dioxide in the atmosphere is steadily increasing and is now much higher than 280 ppm (Keeling, 1960; Keeling et al., 2005). These continuous measurements, shown in Figure 2.21, are now known as the "Keeling curve," named in honor of David Keeling who pioneered the measurements at the site.

The latest measurements made at the Mauna Loa site have recorded a yearly average carbon dioxide concentration of 405 ppm in 2016. This is an increase of 125 ppm since the Industrial Revolution. The measurements fluctuate from −3 to +3 ppm around the yearly average, as shown in the insert to Figure 2.21 (upper left). This monthly variation occurs due to the biosphere's influence on the carbon dioxide concentrations. During the spring and summer months the carbon dioxide is taken up by forests and agricultural activity in the NH, so the atmospheric carbon dioxide concentrations drop. During the fall and winter months the crops are harvested and deciduous forests go dormant. Although dormant, they still respire carbon dioxide and so the CO_2 values rise.

2.4.4 Aerosols

Atmospheric aerosols are particulates suspended in the air with diameters between 0.002 and 100 μm. They can be solid, liquid, or a mixture of both. Aerosols are another important part of the atmospheric system, which can play an important role in the energy balance as well as the chemistry of both the troposphere and the stratosphere. Solid aerosols can be hydrophilic or hydrophobic, with most of the hydrophobic aerosols consisting of large complex organic mixtures, including carbonaceous soot. Carbonaceous soot is produced from both natural fires and anthropogenic combustion, such as agricultural burning, trash burning, wild fires, and diesel engines. Other aerosol components include sea salt and wind-blown soil and dust. Aerosols are classified as primary and secondary. Primary aerosols are those that are emitted directly into the atmosphere, while secondary aerosols are those that are formed in the atmosphere from the oxidation of gases. Most secondary aerosols are highly water soluble and so are incorporated rapidly into the liquid phase of clouds and wet aerosols. The chemistry and physics of these aerosols will be detailed in Chapter 6.

The lifetimes of aerosols in the atmosphere are dependent on their size and chemical composition. Very large aerosols (>10 μm) have lifetimes of a few minutes, depending on the height at which they are released into the atmosphere, where small aerosols (0.1 to 1.0 μm) have lifetimes from 1 to 100 days depending on their ability to condense water (Gaffney and Marley, 2005). The atmospheric lifetimes of aerosols are affected by their transport and transformation mechanisms, which can occur on local, regional, and global scales, depending on the chemical species. For example, volcanic eruptions are known to emit a large amount of particulate matter in the form of volcanic dust, as well as significant amounts of sulfur dioxide gas (SO_2). Some of these volcanic events are large enough to produce particulate matter in the stratosphere at around 18 km that are predominantly made up of aqueous acidic sulfate aerosols. The existence of this stratospheric particulate layer was originally discovered by Christian Junge in 1960, while investigating the effects of above-ground nuclear weapons testing on the atmosphere. This stratospheric aerosol layer, called the Junge layer, can last for about 3 years. This is the amount of time required for the aerosols at this altitude to travel from the lower latitudes to the polar regions.

Atmospheric aerosols can have both direct and indirect effects on the radiative balance of the Earth. They have a direct effect as they can scatter and absorb radiation directly. White aerosols, such as sulfate, are very efficient light-scattering species. Aerosols that absorb IR radiation, such as carbonaceous soot or water-soluble compounds like sulfates, nitrates, and oxidized organics, can act as greenhouse species trapping the IR radiation in the lower troposphere (Gaffney and Marley, 1998). Atmospheric aerosols can also have an indirect effect on radiative balance, since they function as CCN which lead to the formation of clouds and clouds are important in scattering both incoming and outgoing radiation. Also, when submicron aerosols are in high concentrations, they can form haze layers. The presence of an aerosol haze can lead to the reduction of cloud formation. Cloud formation is inhibited by the presence of too many small aerosols, which all compete for the available atmospheric water vapor needed for the growth of cloud droplets.

2.5 Global Climate Models

Global climate models, also called general circulation models (GCMs), are a type of climate model that employs a mathematical description of the general circulation of the Earth's atmosphere and oceans. These models use Navier–Stokes equations on a rotating sphere to describe the motion of the fluids, with a coupling of the atmosphere and ocean systems. They are designed to represent mathematically the physical processes in the atmosphere, ocean, cryosphere, and land surface. They also include the thermodynamics of the various energy balance terms described in Section 2.4. These GCMs are the basis for computer programs used to simulate the radiative forcing of the Earth's energy balance. They are currently the most advanced tools available for simulating the response of the global climate system to future increases in greenhouse gas concentrations.

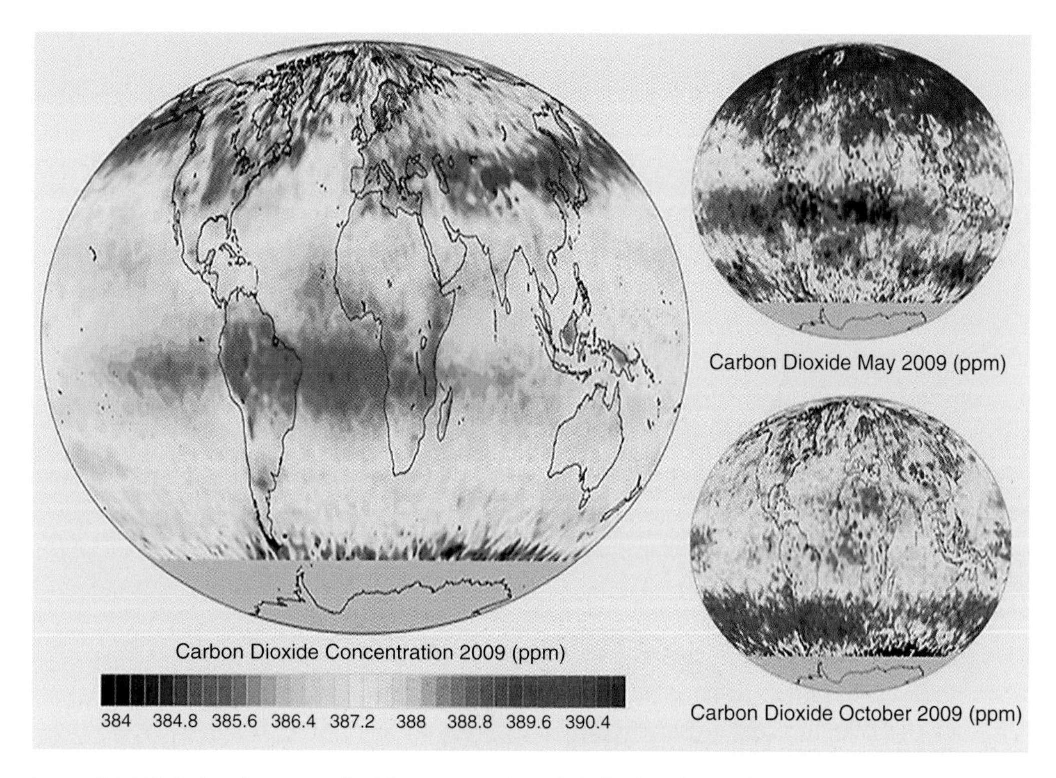

Figure 2.22 Variations in carbon dioxide concentrations globally show higher levels of CO_2 in the Northern Hemisphere than in the Southern Hemisphere and a strong seasonal variation caused by the biosphere. Source: Giorgiogp2, Wikimedia Commons.

The full description of the atmosphere in a GCM must include the chemistry and physics of each greenhouse gas, their spatial and temporal variations, and the possible feedbacks between each of the Earth's systems. For example, Figure 2.22 shows that CO_2 is not evenly distributed across the globe. This is also true of other greenhouse gases. Although the concentration of CO_2 and other greenhouse gases is generally higher in the NH due to the higher concentration of anthropogenic sources, the SH also has a not insignificant seasonal component of CO_2. Remembering that the seasons alternate in the NH and SH, with summer in the NH being winter in the SH, there is a seasonal variation in both hemispheres. These spatial and temporal differences in greenhouse gas concentrations lead to hemispheric and regional differences in the composition of the atmosphere, as well as in the energy balance. This requires that the GCMs take into account the sources, sinks, and transport mechanisms of each of the climate-relevant gases, including both the natural and the anthropogenic sources.

About 71% of the Earth's surface is covered by oceans. The major ocean currents are driven by winds and evaporation, leading to increasing salt content and ocean density. Similar to the general circulation patterns in the atmosphere, our oceans have circulation patterns that are driven by the same fundamental principles: warm vs cold and lower density vs higher density. Higher-density waters occur primarily in equatorial regions where more energy is absorbed by the water, causing more evaporation. These warmer ocean currents act to warm the air above them, resulting in a coupling of the ocean–atmosphere systems. Figure 2.23 shows the thermohaline circulation of warm (red) and cold (blue) currents in the oceans. Note that the warmer currents located near the coast of Great Britain have a significant influence on the regional energy balance of the area, leading to a much milder climate than that of Nova Scotia, which

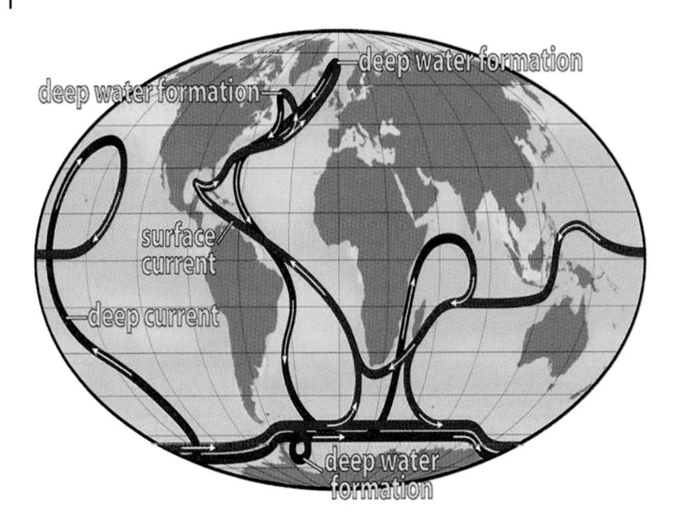

Figure 2.23 The path of the deep-water (blue) and surface-water (red) ocean thermohaline circulation. Source: Robert Simmon, Wikimedia Commons.

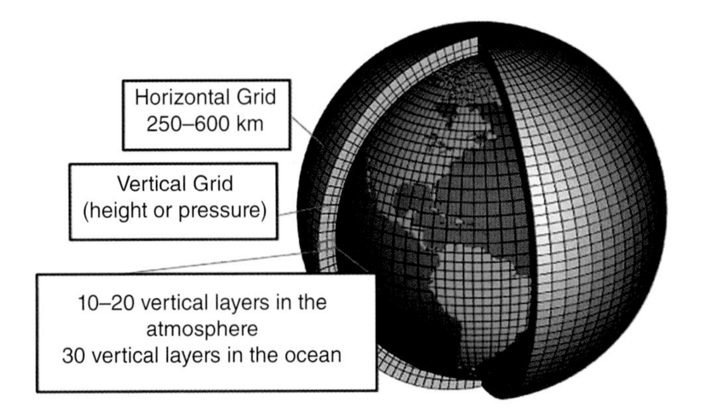

Figure 2.24 A typical grid scale used in global climate models. Source: Adapted from NOAA, Wikimedia Commons.

is near the colder current. Similarly, the annual variations in the Pacific warm waters of the equatorial region, or the so-called "Pacific warm pool," regulate the weather conditions called the El Niño and La Niña. These have significant impacts on annual weather patterns due to their effects on the jet stream circulation patterns. These atmosphere–ocean feedbacks are the reason that current GCMs are focused on coupling the circulation and energy balance of the oceans with those of the atmosphere (NASA, 2016).

The GCMs make use of a three-dimensional grid that is placed over the surface of the globe to model the various components of the Earth's climate system. A diagram of a typical GCM grid is shown in Figure 2.24. The models are limited in the number and size of the grids used due to computational limitations for the necessary large data sets. These models were extremely limited by computing power for many years, but recent advances in computational capacity and large data-set manipulation techniques have improved their resolution significantly. The grids currently have a horizontal resolution of between 250 and 600 km, with 10 to 20 vertical layers in the atmosphere and sometimes as many as 30 layers in the oceans (IPCC, 2013). So, their resolution is still coarse relative to the scale of some climate impacts, such as those related to clouds and aerosols.

Major problems associated with the GCMs are primarily related to the scale of the grid compared to the scale of many important processes in the energy balance. This requires that the processes have to be averaged, with mean values placed in the larger grid cells. Regional models and other submodeling practices are being implemented for use with the GCMs to improve the accuracy in their predictions for both short and long time scales. One particular area of importance is the effects of clouds, as they typically occur on much smaller scales than the GCM grid cells and have significant impacts on weather and climate that need to be better determined. Other sources of uncertainty in the GCMs include the treatment of various feedback mechanisms, such as the increase in water vapor from atmospheric warming.

References

Aoki, N. and Makide, Y. (2004). Precise determination of the atmospheric CF_4 concentration by using natural Kr in the atmosphere as an internal reference in the preconcentration/GC/MS analysis. *Chem. Lett.* 33: 1634–1635.

Atmosphere of Earth n.d. *Wikipedia*. Available online at: https://en.wikipedia.org/wiki/Atmosphere_of_Earth (accessed October 14, 2017).

Berger, A.L. (1976). Obliquity & precession for the last 5 million years. *Astron. Astrophys.* 51: 127–125.

Berger, A. and Loutre, M.F. (1991). Insolation values for the climate of the last 10 million years. *Quat. Sci. Rev.* 10: 297–317.

Betts, A.K. and Ball, J.H. (1997). Albedo over the boreal forest. *J. Geophys. Res.* 102: 28901–28909.

Blasing, TJ 2016 *Current greenhouse gas concentrations*. Carbon Dioxide Information Analysis Center (CDIAC). Available online at: https://en.wikipedia.org/wiki/Atmosphere_of_Earth (accessed October 14, 2017).

Camp, C.D. and Tung, K.K. (2007). Surface warming by the solar cycle as revealed by the composite mean difference projection. *Geophys. Res. Lett.* 34: L14703. https://doi.org/10.1029/2007GL030207.

Dalrymple, G.B. (2001). The age of the Earth in the twentieth century: A problem (mostly) solved. *Geol. Soc. Spec. Publ.* 190: 205–221.

Fischer, H., Wahlen, M., Smith, J. et al. (1999). Ice core records of atmospheric CO_2 around the last three glacial terminations. *Science* 283: 1712–1714.

Gaffney, J.S. and Marley, N.A. (1998). Uncertainties in climate change predictions: Aerosol effects. *Atmos. Environ.* 32: 2873–2874.

Gaffney, J.S. and Marley, N.A. (2005). The importance of the chemical and physical properties of aerosols in determining their transport and residence times in the troposphere. In: *Urban Aerosols and their Impacts: Lessons Learned from the World Trade Center Tragedy* (ed. J.S. Gaffney and N.A. Marley). Washington, D.C.: American Chemical Society.

Harries, D., Hoppe, P., and Langenhorst, F. (2015). Reactive ammonia in the solar protoplanetary disk and the origin of the Earth's nitrogen. *Nat. Geosci.* 8: 97–101.

Houldcroft, C.J., Grey, W.M.F., Barnsley, M. et al. (2009). New vegetation albedo parameters and global fields of soil background albedo derived from MODIS for use in a climate model. *J. Hydrometeorol.* 10: 183–198.

Imbrie, J. and Imbrie, J.Z. (1980). Modeling the climatic response to orbital variations. *Science* 207: 943–953.

IPCC 2013 *What is a GCM?* Intergovernment Panel on Climate Change, Data Distribution Center. Available online at: http://www.ipcc-data.org/guidelines/pages/gcm_guide.html (accessed January 3, 2018).

Jin, Z., Charlock, T.P., Smith, W.L., and Rutledge, K. (2004). A parameterization of ocean surface albedo. *Geophys. Res. Lett.* 31: 2230–2234.

Keeling, C.D. (1960). The concentration and isotopic abundances of carbon dioxide in the atmosphere. *Tellus* 12: 200–203.

Keeling, C.D., Piper, S.C., Bacastow, R.B. et al. (2005). Atmospheric CO_2 and $^{13}CO_2$ exchange with the terrestrial biosphere and oceans from 1978 to 2000: Observations and carbon cycle implications. In: *A History of Atmospheric CO_2 and its Effects on Plants, Animals, and Ecosystems* (ed. J.R. Ehleringer, T.E. Cerling and M.D. Dearing). New York: Springer-Verlag.

Landis, GA 2003 Colonization of Venus, in Proceedings of the American Institute of Physics, Conference on Human Space Exploration, Space Technology & Applications, Albuquerque, NM, vol. 654, pp. 1193–1198.

Laskar, J., Fieng, A., Gastineau, M., and Manche, H. (2011). La2010: A new orbital solution for the long-term motion of the Earth. *Astron. Astrophys.* 532 (A89): 1–15.

Maykut, G.A. and Church, P.E. (1973). Radiation climate of Barrow, Alaska, 1962–66. *J. Appl. Meteorol.* 12: 620–628.

McKenzie, R.L., Paulin, K.J., and Madronich, S. (1998). Effects of snow cover on UV irradiance and surface albedo: A case study. *J. Geophys. Res.* 103: 28785–28792.

Mikheil, S. and Sverjensky, D.A. (2014). Nitrogen speciation in upper mantle fluids and the origin of Earth's nitrogen rich atmosphere. *Nat. Geosci.* 7: 816–819.

NASA 2016 *Global climate modeling*. Available online at https://www.giss.nasa.gov/projects/gcm (accessed January 7, 2018).

Pinty, B. and Ramond, D. (1987). A method for the estimate of the broadband directional surface albedo from a geostationary satellite. *J. Climate Appl. Meteorol.* 26: 1709–1722.

Pomerantz, M, Pon, B, Akbari, H & Chang, S-C 2000 *The effect of pavement temperatures on air temperatures in large cities*. Technical Report No. LBNL-43442, Environmental Energy Technology Division, Earnest Orlando Lawrence Berkeley National Laboratory, Berkeley, CA, USA, April 2000, 24 pp.

Porcelli, D. and Turekian, K.K. (2011). The history of planetary outgassing as recorded by noble gases. In: *Geochemistry of Earth Systems: From the Treatise on Geochemistry*, 1e (ed. H.D. Holland and K.K. Turekian). London: Elsevier.

Seidelmann, P.K. (ed.) (1992). *Explanatory Supplement to the Astronomical Almanac*, 733. Herndon, VA: University Science Books.

U.S. Department of Commerce 2017 *Trends in atmospheric carbon dioxide*. Available online at https://www.esrl.noaa.gov/gmd/ccgg/trends/index.html#mlo_growth (accessed October 14, 2017).

Study Problems

2.1 What is the major inert (or noble) gas that is in the Earth's atmosphere? What is its source?

2.2 What are the two major gases in the atmosphere? Of these, which is the most reactive and which environmental system is its source?

2.3 What two carbon-containing gases are thought to be the major gases in the very early atmosphere of the Earth?

2.4 If the current population of the Earth is 7.5 billion people, how many people would be considered one part per billion? One part per million?

2.5 Name the three lower portions of the atmosphere as defined by changes in the temperature profile in order of increasing altitude.

2.6 Why is the troposphere well mixed?

2.7 The three planets Venus, Earth, and Mars have very different surface temperatures. What is the main gas that acts to maintain the heat in each of these atmospheric systems? Why is Mars colder than Venus even though the atmospheric composition is similar?

2.8 Why is the Earth's south polar region very cold during December even though it receives a lot of solar radiation for almost 24 hours a day during the month?

2.9 What is a temperature inversion in the atmosphere? Is this a normal condition in the troposphere?

2.10 Why would power plants and industrial sites located in a valley use tall stacks to release pollutants?

2.11 What are the three major atmospheric circulation patterns that start at the equator and move toward the poles?

2.12 What is the most dominant greenhouse gas on Earth? What is the second? Which of these gases is being increased by the burning of fossil fuels?

2.13 What is the name given to the plot of recent high-precision measurements of carbon dioxide as a function of time made at the Mauna Loa Observatory?

2.14 What name is given to the long-term changes in solar radiation due to astronomical variance of the Earth's orbit?

2.15 What causes the temperature in the stratosphere to increase above the tropopause?

2.16 What role do clouds play in the Earth's energy balance?

2.17 What causes the temperature differences to be larger from day to night in deserts than in grasslands and forests?

2.18 What causes the seasonal variation of carbon dioxide concentrations at locations in the Northern Hemisphere to be larger than those in the Southern Hemisphere in remote locations?

2.19 What is the name of the large circulation pattern in the oceans that is caused by changes in density?

2.20 If high soot levels were to be deposited on top of snow and ice in Greenland, what effect would this have on the energy balance? On the thermohaline circulation?

3

The Fundamentals of Photochemistry

As discussed in Chapter 2, the energy balance of the Earth is primarily dependent on the amount of solar radiation entering the atmosphere and reaching the surface of the planet. Since oxygen and ozone are important in absorbing the harmful UVC and UVB radiation in the stratosphere, most of the solar radiation reaching the troposphere is in the UVA and visible regions, with a small amount of UVB. While a significant amount of the light energy in this region is absorbed by surface materials, leading to heating of the Earth and lower atmosphere, there are also important light-driven chemical reactions that occur in the atmosphere, in surface waters, and on the surfaces of the terrestrial ecosystems. These reactions are known as photochemical reactions, chemical reactions initiated by the absorption of energy in the form of light. If the light is of sufficient energy, this absorption can cause molecular bonds to break and, since most chemical bonds are fairly strong, the majority of the important photochemical reactions occur after absorption of UV radiation. This chapter reviews the fundamentals of photochemistry and the basic laws that govern photochemical reactions. Also discussed is the difference between photochemical processes and thermal processes.

3.1 Light and Photochemistry

Photochemistry is the study of chemical reactions and physical behavior of molecules caused by the absorption of visible and/or UV light. Photochemical reactions involve the interaction of a photon, or quantized amount of light energy, with a molecule and its subsequent effects on the chemistry and physical properties of the molecule. A photochemical reaction can be considered to be a "bimolecular" process, where the photon is one reactant and the absorbing molecule is the other reactant (Calvert and Pitts, 1966). Indeed, the name "photochemistry" indicates that a photon is required for the chemistry to occur. In order to understand photochemical reactions, it is necessary to review the properties of light and, in particular, sunlight when considering the photochemistry of environmental systems.

Recall that light is a form of energy and that sunlight, or solar energy, can cause heating of a surface. This becomes very apparent when a magnifying glass is used to focus sunlight. When the sunlight is focused on an absorbing piece of paper, the heat evolved can rapidly cause the paper to reach its kindling temperature and ignite. This simple experiment shows that light is a form of radiant energy that can be converted into heat when it is absorbed by a surface. Light can also interact with molecules to cause the electrons to move from the ground state to an electronically excited state. It is these electronic transitions that can cause electrons to move

Chemistry of Environmental Systems: Fundamental Principles and Analytical Methods, First Edition.
Jeffrey S. Gaffney and Nancy A. Marley.

Figure 3.1 Portrait of Ole Christensen Rømer, Danish astronomer who first showed that the speed of light was finite and constant in a vacuum by timing the eclipses of Jupiter's moon, Io. Source: Jacob Coning, Wikimedia Commons.

from bonding orbitals into antibonding orbitals, causing the chemical bonds in the molecule to break.

One of the most important fundamental properties of light is its velocity. The first successful measurement of the velocity of light was done by the Danish astronomer Ole Rømer, shown in Figure 3.1. Working at the Royal Observatory in Paris in 1676, Rømer demonstrated that the speed of light was finite. He did this by timing eclipses of Io, a moon of Jupiter. He concluded that light had a finite velocity of approximately 2.2×10^8 m s^{-1}. While his measurement is actually about 74% of the actual speed of light, which has been more accurately determined to be $2.997\,924\,58 \times 10^8$ m s^{-1}, his experiment established that light moved at a constant speed in the vacuum of space (MacKay and Oldford, 2000).

Further study of the theory of light followed Rømer's measurements of its velocity, leading to Einstein's famous paper on the electrodynamics of moving bodies. This paper was published as part of his work on the theory of relativity, where he hypothesized that the speed of light is independent of the motion of its source in a vacuum (Einstein, 1905a). The work established that the frequency of light emitted from a moving body is affected by the speed and direction of the moving body relative to an observer, but it does not affect the velocity of the emitted light. This is known as the Doppler effect, defined as an apparent change in the frequency of an electromagnetic wave, such as light or sound, that occurs when the source is in motion relative to an observer. The frequency of the light emitted from the source increases when the source is moving toward the observer and decreases when it is moving away from the observer. This effect is used by astronomers to determine if stars are moving toward or away from the Earth, by examining the shifts in the frequency of their emitted light. The Doppler effect is most commonly observed as the change in pitch of the whistle of a train as it passes by. The pitch is higher as the train approaches you, decreases as it becomes closer, and becomes lower as it moves away from you.

The Doppler effect occurs because light behaves as an electromagnetic wave. But, in some cases, light also behaves as a particle. This seemingly inconsistent behavior of light is called the wave–particle duality. Albert Einstein and Leopold Infeld (1938, p. 263) described this observed duality as follows:

> It seems as though we must use sometimes the one theory and sometimes the other, while at times we may use either. We are faced with a new kind of difficulty. We have two contradictory pictures of reality; separately neither of them fully explains the phenomena of light, but together they do.

So, light possesses both wave properties and particle properties. The wave properties are more apparent over relatively large time scales and large distances, while the particle properties are more apparent over smaller time scales and smaller distances. Let us first examine the wave properties of light.

James Clerk Maxwell presented the now classical theory of electromagnetic radiation (Maxwell, 1865). Electromagnetic radiation is defined as a form of energy that is produced by oscillating electric and magnetic waves, as shown in Figure 3.2. Maxwell described electromagnetic radiation as a transverse plane wave that is associated with an electric field vector (**E**) and a magnetic field vector (**B**), which are aligned in planes that are at right angles to each other. Both **E** and **B** travel perpendicular to the direction of the wave propagation. The variation of the field strength as a function of time (t) and distance (x) along the wave propagation axis is given by two sine functions:

$$\mathbf{E}_y = A \sin 2\pi \left(\frac{x}{\lambda} - \nu t \right)$$

$$\mathbf{B}_z = \sqrt{\varepsilon/\mu}\, A \sin 2\pi \left(\frac{x}{\lambda} - \nu t \right)$$

where $\mathbf{E}_y$ is the electric field vector in the xy plane increasing along the y axis, and $\mathbf{B}_z$ is the magnetic field vector in the xz plane increasing along the z axis. The amplitude of the wave is given by A and the intensity of the wave is proportional to the square of the amplitude. The other important values in these equations are the dielectric constant (ε), the magnetic permeability of the medium in which the wave is moving (μ), the wavelength (λ), and the frequency (ν) of the wave. In a vacuum, both ε and μ are equal to 1.

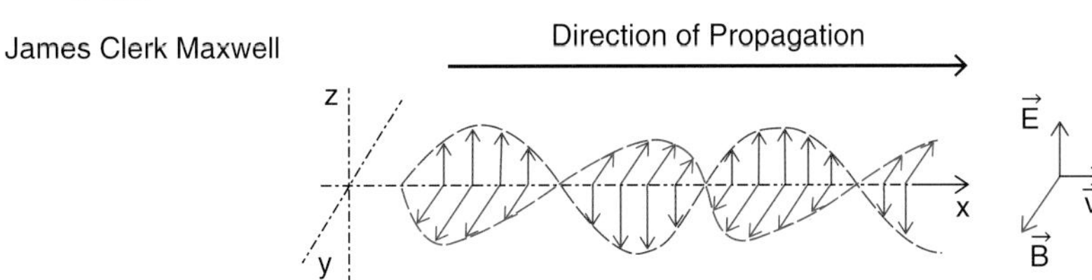

Figure 3.2 Portrait of James Clerk Maxwell and a graphical representation of his classical theory for the motion of electromagnetic radiation. The electric field vector (**E**) is at right angles to the magnetic field vector (**B**) and the wave moves along the axis in the x direction. Source: Photo by George J. Stodart, Wikimedia Commons; figure by SuperManu, Wikimedia Commons.

An electromagnetic wave consists of successive troughs and crests, as shown in Figure 3.2. The distance between two adjacent crests is the wavelength, which is typically given in nanometers. The number of crests that move past a given point in a given unit of time is the frequency, which is typically given in Hertz or s^{-1}. The wavelength and frequency of electromagnetic radiation in a vacuum are related to each other by the velocity of the radiation (c):

$$\nu = c/\lambda$$

The frequency of the electromagnetic radiation is not dependent on the medium through which the electromagnetic wave moves. As an electromagnetic wave crosses a boundary between different media, its velocity changes but its frequency remains constant. The velocity of the electromagnetic radiation in a nonconducting medium (c') is dependent on the dielectric constant and magnetic permeability of the medium:

$$c' = \frac{c}{\sqrt{\varepsilon\mu}}$$

Classical electromagnetic theory for the behavior of light as a wave can be used to understand the reflection and refraction of light, as shown in Figure 3.3. Reflection is the change in direction of an electromagnetic wave at an interface between two media with different densities, so that the wave returns to the medium from which it originated. The angle of reflection (θ_2) is equal to the angle of the incidence (θ_1). Refraction is the change in direction of an electromagnetic wave when traveling from one medium to another of different density. The angle of refraction is given by Snell's Law, named after Dutch astronomer Willebrod Snellius, which states that the ratio of the sines of the angle of incidence ($\sin\theta_1$) and the angle of refraction ($\sin\theta_2$) is equal to the ratios of the speed of light in the two media:

$$\frac{\sin\theta_1}{\sin\theta_2} = \frac{c'_1}{c'_2}$$

Since the refractive index is defined as the speed of light in a vacuum (c) divided by the speed of light in the medium ($n = c/c'$), the ratio of the sines is also equal to the reciprocal of the ratio of the refractive indices of the two media:

$$\frac{\sin\theta_1}{\sin\theta_2} = \frac{n_2}{n_1}$$

where n_1 is the refractive index of the first medium and n_2 is the refractive index of the second medium.

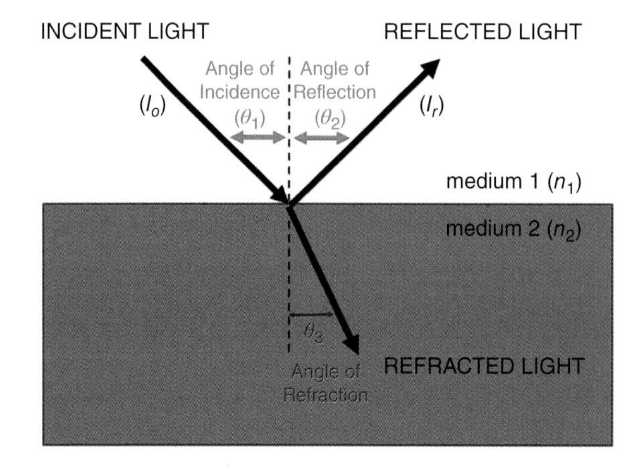

Figure 3.3 Reflection and refraction of a beam of light traveling from medium 1 with refractive index n_1 to medium 2 with refractive index n_2. The angle of incidence is θ_1, the angle of reflection is θ_2, and the angle of refraction is θ_3.

Unless the surface of the second medium is totally reflective, some of the light will be reflected and some will be refracted, continuing to travel through the second medium. French civil engineer and physicist Augustin-Jean Fresnel derived an equation experimentally that describes the ratio of the intensity of reflected light (I_r) to the intensity of incident light (I_o) for a perpendicular beam of unpolarized light:

$$\frac{I_r}{I_o} = \left(\frac{n_1 - n_2}{n_1 + n_2}\right)^2$$

Both Snell's Law and the Fresnel equation are very useful in evaluating the loss of incoming sunlight due to reflection by surfaces, as in albedo effects in the environment.

When a beam of white light passes from air into a material having a refractive index that varies with wavelength, the refraction angle (θ_3 in Figure 3.3) varies with wavelength. This results in dispersion of the light, where light of different wavelengths is refracted at different angles. So, different colors of light are bent at different angles at the interface of two media, resulting in separation of the light into colors. For the case of sunlight striking the surface of water, a common occurrence in the environment, the refractive index of water (n_2) varies with wavelength in the visible spectrum, as shown in Table 3.1, and the refractive index of air is approximately 1.0003. So, the angle of refraction for an incident angle of 45° varies from 31.90° in the red to 32.12° in the violet. The long-wavelength light (red) is refracted less than the short-wavelength light (violet). This is the basis of the formation of a rainbow after a rain event. After the rain clouds dissipate, the incoming sunlight strikes the surface of water droplets still in the atmosphere, causing the separation of the white sunlight into its component wavelengths. Although the wavelength separation is not large, at long distances from the interface the colors we know as a rainbow (red, orange, yellow, green, blue, and violet) can be seen in the sky. This process is also the basis of the function of a prism to separate the wavelengths of light. The materials used for prisms (glass, quartz, plastic, fluorite) have much higher wavelength-dependent refractive indices, so the wavelength separation can be seen at short distances.

The other two common properties of light that are related to its wave behavior are interference and diffraction. Interference is the phenomenon where two waves of the same frequency superimpose to form a new wave with a higher or lower amplitude. If the crest of one wave meets the crest of the second wave at the same point, the amplitude of the new wave is equal to the sum of the amplitudes of the two waves. This is called constructive interference. But, if a crest of the first wave meets a trough of the second wave at the same point, the amplitude of the new wave is equal to the difference in the amplitudes of the two waves. This is called destructive interference. An example of constructive and destructive interference between two waves of the same frequency and amplitude is shown in Figure 3.4. In this case, constructive

Table 3.1 The variation of the average refractive index of water with wavelength for the visible wavelength region of the light spectrum.

Color	Wavelength (nm)	Refractive index	Angle of refraction (degrees)
Red	620–750	1.3385	31.90
Orange	590–620	1.3375	31.93
Yellow	570–590	1.3345	32.01
Green	495–570	1.3328	32.03
Blue	450–495	1.3317	32.08
Violet	380–450	1.3305	32.12

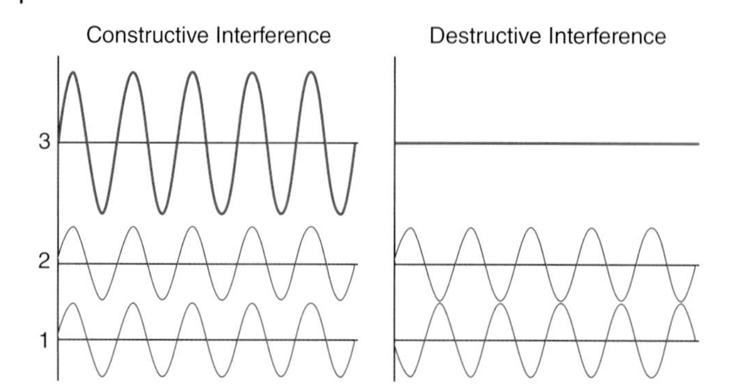

Figure 3.4 Constructive and destructive interference between light waves 1 and 2 with the same frequency and amplitude giving the resulting waves 3. Source: Haade, Wikimedia Commons.

interference results in a wave with twice the amplitude and destructive interference results in zero amplitude. Constructive and destructive interference are described as a phase difference between the two waves, typically expressed in time or radians. Two waves that have no phase difference are said to be in phase, whereas two waves that have the same frequency and different phases are said to be out of phase.

Diffraction is defined as the bending of light around the corners of an obstacle or surface opening. The amount of bending depends on the relative size of the wavelength of the light to the size of the opening or obstacle. If the opening is much larger than the wavelength, the bending will be very small and if the wavelength is close or equal to the wavelength, the bending will be very large. This results in a separation of the wavelengths of light. Diffraction can be described as constructive and destructive interference between the waves making up the light after it is bent, which is observed by the presence of a pattern of closely spaced dark and light bands, called a diffraction pattern, at the edge of a shadow. In the atmosphere, sunlight can be bent around atmospheric particles or water droplets in clouds. This diffracted light can produce fringes of light and dark or colored bands, such as seen around the edges of clouds or the crown of light surrounding the Sun or Moon, called a corona. Diffraction is also the basis for the function of a "diffraction grating" used to separate the wavelengths of light. A diffraction grating consists of a large number of evenly spaced parallel groves or ridges in a transparent or reflecting plate. As the light waves bend around the ridges, they interact with each other leading to constructive and destructive interference, depending on the wavelength of the light.

While the wave properties of light are adequate to explain light scattering, the particle properties of light are necessary to explain light absorption and emission by molecules. The absorption of electromagnetic radiation is defined as the process where the energy of the radiation is taken up by matter, typically the electrons of an atom. When the wave model of electromagnetic radiation was used to explain the intensity of the radiation given off by a blackbody, it could not fit the observations at short wavelengths. This prompted Max Planck (1901) to propose the theory that electromagnetic radiation could only be emitted in quantized form. In other words, it could only be a multiple of an elementary unit: $E = h\nu$ So, the light emitted by a blackbody must be of specific value and exist as particles called photons that have a specific energy associated with their wavelength or frequency:

$$E = h\nu = \frac{hc}{\lambda}$$

where E is the energy of the photon, ν is the frequency, λ is the wavelength, and h is a constant, now called Planck's constant, which is equal to 6.626×10^{-34} m^2 kg s^{-1}. The quantized units of energy are represented by $h\nu$ and a photon of frequency ν will have its own specific and unique energy. The total energy at that frequency is then equal to $h\nu$ multiplied by the number of photons at that frequency.

Other experiments also showed that light can behave as particles. It is well known that when short-wavelength light impacts a metal surface, electrons are emitted from the metal. This is known as the photoelectric effect. The number of electrons ejected from the metal surface is directly proportional to the intensity of the light, while the kinetic energy of the electrons is related to the frequency of the light. This observation cannot be explained by the classical wave description of electromagnetic radiation, but it was adequately explained by Einstein using Planck's concept of light behaving as quantized particles (Einstein, 1905b). Einstein considered that in order to release an electron from the metal, the light absorbed by the metal had to exceed a threshold frequency or energy ($h\nu$). By relating the energy required to eject an electron to the frequency of the light and not to the intensity of the light, he was able to predict the photoelectric behavior observed. Also, the Compton effect, which found that X-rays scattered off a surface resulted in an increase in the wavelength of the X-ray, could only be explained if the X-rays were behaving as particles. Some of the energy of the photon is transferred to the electron, resulting in a decrease in the energy and an increase in the wavelength of the photon. Again, the concept of light behaving as individual amounts of energy, or photons containing that energy, explained a phenomenon that could not be explained by the classical wave description of light. All of these experiments showed that light has particle properties and follows physical laws that are consistent with collisions of particles with matter. This is called the quantum theory of light. It is this quantum property of light that is found to best describe the processes of photochemical reactions.

3.2 The Laws of Photochemistry

Photochemistry assumes that the quantum theory of light, which describes light as being composed of photons that have discrete energy values, best explains the chemical changes induced when light interacts with molecules. A photon is a packet or quantum of energy that has no mass and travels at the speed of light. When a photochemical process happens, the photochemical reaction is written as the reactant interacting with a photon ($h\nu$) to form products. This usually occurs by a decomposition reaction, where a bond(s) is broken in the reactant molecule after the absorption of light. Photochemists refer to the yield of a photochemical reaction as the quantum yield (Φ), defined as the ratio of the number of molecules formed per unit time divided by the number of photons (quanta) of light absorbed by the reactant per unit time. The photons absorbed represent a specific amount of energy defined by $h\nu$. Since $h\nu$ is also equal to hc/λ, the energy absorbed can be calculated from the frequency or the reciprocal of the wavelength multiplied by Planck's constant (h) and the speed of light (c).

When studying photochemical reactions, the apparent size of Φ can vary significantly from very small values to very large values, depending on the type of reaction and the chemical reaction mechanism. For example, in the case of a chain reaction the first step might be the photolysis of molecular chlorine (Cl_2) to form two chlorine atoms (2Cl), which initiate the chain reaction. In this case one photon leads to two Cl atoms that can reform after the formation of products. Since the chain reaction reforms the reactant without requiring a second photon, the apparent quantum yield Φ can be as high as 10^6 or more before the radical chain reaction is terminated. The concept of the quantum yield is a simple one, but it can become complicated depending on the reaction mechanism. This difficulty led to the development of some simple laws that are used in photochemistry to better define the photochemical process.

The first law of photochemistry is called the Principle of Photochemical Activation, also called the Grotthus–Draper Law after two chemists who formulated it in the early nineteenth century. The law states that:

> Only the light which is absorbed by a molecule can be effective in producing a photochemical change in the molecule.

This law may seem very simple, but it is important to remember that a molecule being photolyzed must absorb the light in order for a reaction to happen. This means that we need to know the absorption spectrum of the molecules being studied as well as the wavelength of the light being absorbed.

The second law of photochemistry is the Photochemical Equivalence Law or the Stark–Einstein Law, named after the German physicists Johannes Stark and Albert Einstein who formulated it in the early twentieth century. The law states that:

> The absorption of light by a molecule is a one-quantum process, so that the quantum yields of the primary process must be unity.

This law deals with the primary photochemical process, which is the light-induced reaction. But the primary process is usually followed by secondary processes, which are normal chemical reactions that do not require light absorption. The inclusion of these secondary processes results in an apparent quantum yield that is greater than one. After the absorption of light, the primary processes for the electronically excited molecule can include simple fluorescence or phosphorescence, collisional deactivation, bond breaking, and isomerization reactions. The Stark–Einstein Law states that all of the reaction pathways available to the excited molecule must add up to one. So, for x number of processes:

$$\sum \Phi x = 1$$

The Stark–Einstein Law can be restated as: for every mole of reactant, an equivalent mole of photons must be absorbed. Then the total energy required for 1 mol of reactant is

$$E = N_A h\nu$$

where N_A is Avogadro's number.

The Stark–Einstein Law works primarily due to the fact that the light absorption and subsequent reactions all happen on fast time scales, so the opportunity for a second photon to be involved in the reaction is essentially zero. With the advent of high-intensity lasers, the potential for two photon processes can now be realized in the laboratory. But, for the case of environmental photochemistry with the light source being the Sun, the law still holds true. The Grotthus–Draper Law and the Stark–Einstein Law both form the basis for photochemical studies, along with the familiar Beer–Lambert Law, sometimes simply referred to as Beer's Law.

The Beer–Lambert Law deals with the absorption of a monochromatic source of light and relates the intensity of the light absorbed to the concentration of the absorbing species. If a monochromatic light of intensity I_o passes through an absorbing sample, the intensity will be reduced to I as it exits the sample, as shown in Figure 3.5. The ratio of these two intensities is called the transmittance of the sample, which is related to the absorbance of light (A) by the absorbing species:

$$A = \ln\left(\frac{I_o}{I}\right)$$

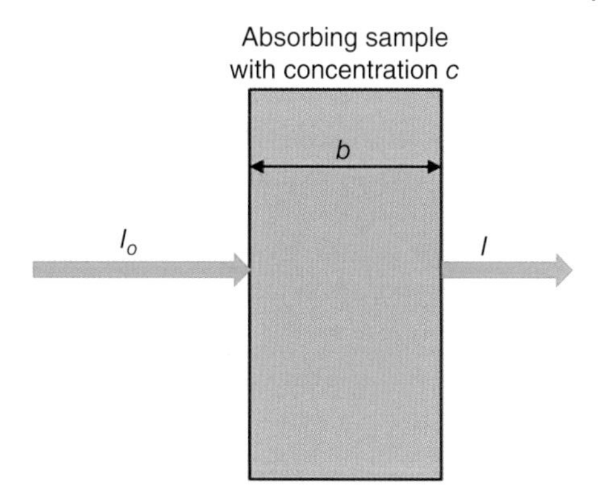

Figure 3.5 The absorption of monochromatic light of intensity I_o by an absorbing sample of concentration c and path length b.

Beer's Law relates the absorbance of light by an absorbing species to the concentration (c):

$$A = \varepsilon bc$$

or

$$\frac{I}{I_o} = e^{-\varepsilon Cb}$$

where b (cm) is the path length through the sample, c (mol l^{-1}) is the concentration of the absorbing species, and ε (l mol^{-1} cm^{-1}) is the molar extinction coefficient of the absorbing species at that specific wavelength.

If there is more than one absorbing species in the sample, the Beer–Lambert Law requires that each of the molar extinctions ($\varepsilon_1, \varepsilon_2, \varepsilon_3, \ldots$) and their concentrations ($c_1, c_2, c_3, \ldots$) be included to determine the total absorption or transmission of the light:

$$A = \sum_{i=1}^{N} A_i = \sum_{i=1}^{N} \varepsilon_i \int_0^b c_i\,(z)\,\mathrm{d}z$$

or

$$\frac{I}{I_o} = e^{-\sum_{i=1}^{N} \varepsilon_i \int_0^b c_i\,(z)\mathrm{d}z}$$

The Beer–Lambert Law can also be applied to the transmittance of sunlight through the atmosphere. So, the available sunlight incoming to the surface of the Earth is decreased by absorption due to absorbing molecules in the atmosphere above the surface. The attenuation is dependent on the molecular species and their concentration in the atmosphere. This is an important aspect of how photochemical processes play a role in our environment, particularly in the stratospheric chemistry of ozone, which will be discussed in detail in Chapter 4.

3.3 Thermochemical and Photochemical Processes

Chemical reactions that occur in the environment can be driven by either thermochemical (also called thermal) or photochemical processes. Thermochemical reactions involve the absorption or emission of heat, while photochemical reactions involve the absorption of light. One of the main differences between these two types of chemical reactions is that thermal reactions involve reactants in their ground electronic states, while photochemical reactions involve reactants in their electronically excited states. However, the thermal reactions do involve molecules in higher vibrational, rotational, and translational states, because they require molecules of higher kinetic energies for the reaction to occur.

3.3.1 Activation Energy

In order for a thermochemical reaction to occur, the reactant molecules must have the minimum amount of kinetic energy necessary and must collide with the appropriate orientation for the formation of products. According to the Maxwell–Boltzmann Law, there is a distribution of kinetic energies of the molecules in a system, as shown in Figure 3.6. The shape of this distribution depends on the temperature of the system. Some of the molecules at the upper end of the distribution will have the minimum energy needed for the chemical reaction to occur, called the activation energy (E_a). The higher the temperature of the system, the larger the number of molecules that will have the minimum energy required for reaction and thus the higher the rate of reaction. So, the rate of thermochemical reaction depends on the statistical thermal energy distribution of the molecules in the system, called the Maxwell–Boltzmann distribution.

The energy required for a photochemical reaction to occur, called the photochemical activation energy (E_{pa}), is equal to the energy needed to promote the electrons from the ground-state orbitals (S_o) into the excited-state orbitals (S_a^*), as shown in Figure 3.7. If the excited state orbital is an antibonding orbital, this leads to breaking of the bonds in the molecule followed by the formation of new products. The photochemical activation energy is only dependent on the energy of the incoming photon, which is directly related to the frequency or wavelength of light:

$$E = h\nu = hc/\lambda$$

While the rate of a thermochemical reaction is determined by the temperature of the system, the rate of a photochemical reaction is determined by the intensity of the light at the proper frequency or the number of photons that have sufficient energy to promote the photochemical reaction. The amount of photons in the photochemical reaction is measured in moles, the same as the amount of reactant molecules. The unit for 1 mol of photons is called an einstein. The einstein is therefore equivalent to an energy unit which is dependent on the frequency or

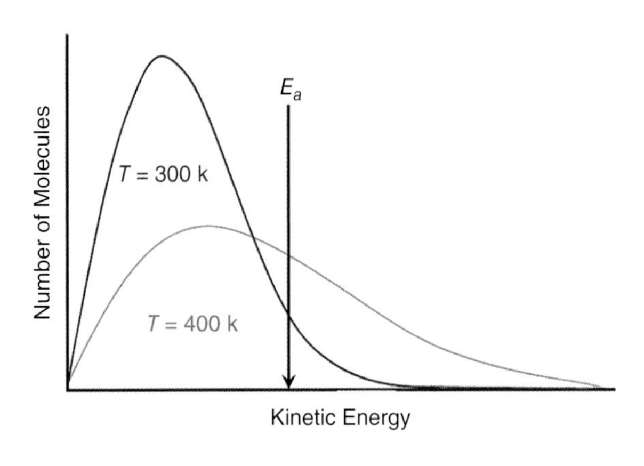

Figure 3.6 The kinetic energy distribution of a system, known as a Maxwell–Boltzmann distribution, for temperatures of 300 and 400 K. The molecules achieving the minimum amount of energy required for a thermochemical reaction are those with kinetic energy at or above the activation energy (E_a).

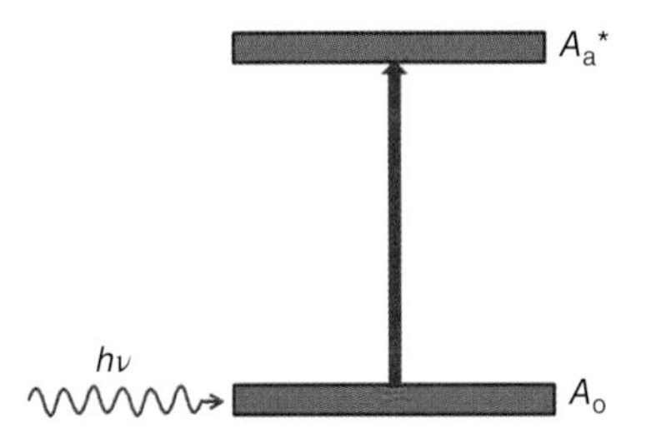

Figure 3.7 Energy-level diagram for the absorption of light of frequency ν, which results in the promotion of an electron from the molecular ground-state bonding orbital A_o to the excited-state antibonding orbital A_a^*.

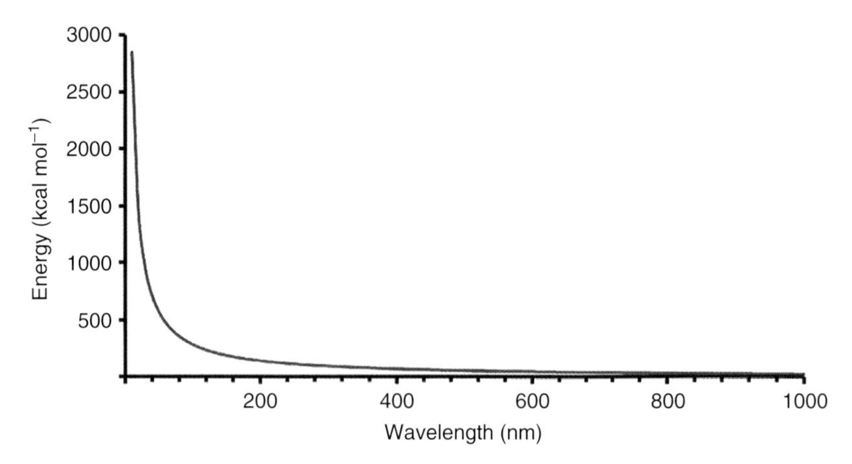

Figure 3.8 The energy (kcal) of one mole of photons at the top of the atmosphere as a function of wavelength.

wavelength of the photons:

$$E = \frac{Nhc}{\lambda}$$

where N is Avogadro's number (6.022×10^{23} molecules mol^{-1}), h is Planck's constant (1.58×10^{-34} cal s^{-1}), c is the speed of light (3.0×10^{17} nm s^{-1}), and λ is the wavelength (nm).

The amount of energy per mole of photons at the top of the atmosphere as a function of wavelength is shown in Figure 3.8. This energy received from the Sun at the top of the atmosphere is between the wavelengths of 10 and 1000 nm and includes the very high-energy photons in the UV wavelength range of 10–280 nm, with energies of 2850–100 kcal mol^{-1}. The high-energy photons below 200 nm are removed in the upper layers of the atmosphere due to absorption by molecular and atomic nitrogen and oxygen. This leaves wavelengths above 200 nm, which are most important to the photochemistry of the stratosphere and troposphere. The wavelengths of light entering the stratosphere are shown in Figure 3.9(A), which includes the region of 200–1000 nm with energies of 140–30 kcal mol^{-1}. The wavelengths of light entering the troposphere shown in Figure 3.9(B) are in the region of 300–1000 nm, with energies of 95–30 kcal mol^{-1}. The photochemical activation energy for the decomposition of molecular oxygen is 118 kcal mol^{-1} at a wavelength of 245 nm. This photochemical reaction can only occur in the stratosphere, since the energies of the photons reaching the troposphere are not of high enough energy. The photochemical reactions occurring in the troposphere are those requiring energies less than 95 kcal mol^{-1}.

The energy available in a solar photon, expressed as the wavelength, is compared with the bond-dissociation energy for some key photochemical species in Table 3.2. Since the bond-dissociation energy is the minimum energy required to break the bond, it is also equal to the photochemical activation energy for the photochemical reaction involving the bond. A bond can only be broken in a photochemical reaction if the molecule absorbs light of the wavelength listed.

Ozone is formed in the atmosphere by the reaction of molecular oxygen with atomic oxygen. Clearly, the data from Table 3.2 show that the photolysis of molecular oxygen to form the oxygen atoms that are needed to form ozone requires UVB light shorter than 245 nm in wavelength. This UV light is not available in the troposphere, and so ozone can only be formed from oxygen photolysis in the stratosphere. However, nitrogen dioxide can undergo photochemical reaction to form NO and an oxygen atom at wavelengths around or below 390 nm, which is available in

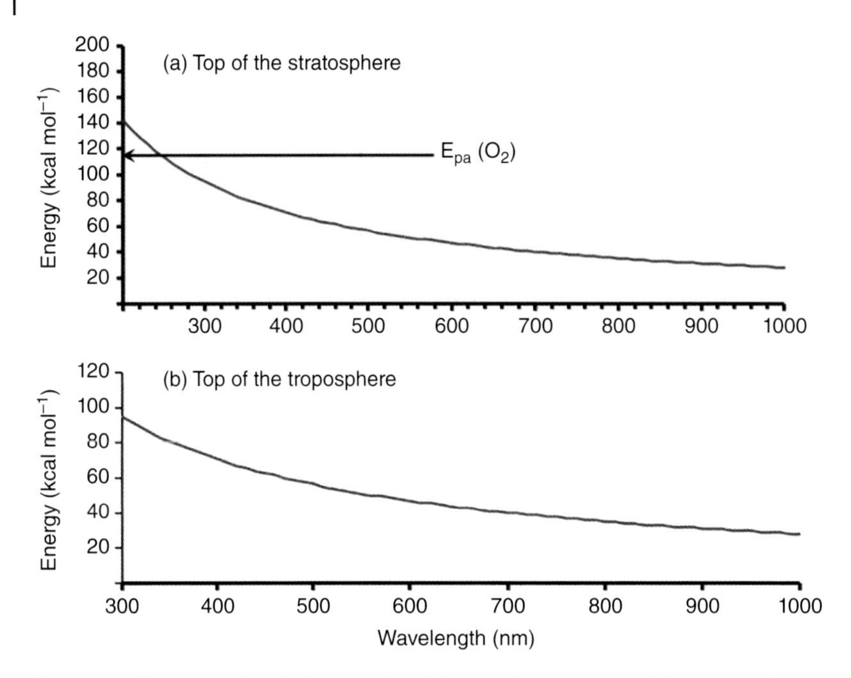

Figure 3.9 The energy (kcal) of one mole of photons (a) at the top of the stratosphere and (b) at the top of the troposphere as a function of wavelength.

Table 3.2 The bond dissociation energies (kcal mol^{-1}) compared to the wavelength of light needed to supply the photochemical activation energies for some important photochemical species.

Species	Bond broken	Dissociation energy (kcal mol^{-1})	Wavelength (nm)	Reference
Molecular oxygen	O=O	118	245	Calvert and Pitts (1966)
Chlorofluoromethane	CF_3—Cl	89	320	Luo (2003)
Formaldehyde	HCO—H	88	325	Calvert and Pitts (1966)
Nitrogen dioxide	ON—O	73	390	Calvert and Pitts (1966)
Nitrate radical	NO_2—O	50	560	Davis et al. (1993)
Nitrous acid	ON—OH	49	580	Calvert and Pitts (1966)

the troposphere. So, nitrogen dioxide is the tropospheric species responsible for the formation of tropospheric ozone in smog chemistry, as will be discussed in detail in Chapter 5. Other species listed in Table 3.2 are important in photochemical reactions that initiate chain reactions, which will be covered in Chapters 4 and 5.

3.3.2 Kinetics

In order to better appreciate the differences between thermochemical and photochemical reactions, it is necessary to briefly review the principles of chemical kinetics. The kinetics of thermochemical reactions involves molecular collisions between the reactants that are in higher vibrational and rotational states. The kinetics is usually quite complex, involving more than one reacting species, and can usually take place by way of a series of steps. Recall from general chemistry that the chemical equation for a general reaction involving two reactants A and B forming two products C and D is written as follows:

$$aA + bB \rightarrow cC + dD$$

where the letters a, b, c, and d represent the stoichiometry of the reaction. The rate of disappearance of the reactants, or the rate of appearance of the products with time, is written as

$$\text{rate} = -\frac{1}{a}\frac{\mathrm{d}[A]}{\mathrm{d}t} = -\frac{1}{b}\frac{\mathrm{d}[B]}{\mathrm{d}t} = \frac{1}{c}\frac{\mathrm{d}[C]}{\mathrm{d}t} = \frac{1}{d}\frac{\mathrm{d}[D]}{\mathrm{d}t}$$

where $[A]$, $[B]$, $[C]$, and $[D]$ are the concentrations (mol l^{-1}) of the reactants A and B and the products C and D. In a thermal reaction, the rate of the reaction is determined by measuring the change in concentration of one of the products or reactants with time, corrected using the stoichiometric coefficients. For reactions in the gas phase, the reaction conditions must be such that the volume does not change.

When determining the rate of a multistep thermochemical reaction, it is commonly treated as a single-step process with the slowest step being the rate-determining step. Thermal reactions are collisional reactions with reactants going to products by way of a transition state. The number of the reactants involved in the transition state is known as the molecularity of the reaction, and the kinetics of the reaction can be described in terms of this molecularity. A unimolecular reaction is one where the transition state involves only one reactant molecule, a biomolecular reaction is one where the transition state involves two molecules, and a termolecular reaction is one where the transition state involves three molecules. Tetramolecular reactions (four-molecule transition states) are very seldom seen except in solution, where the solvent molecules are involved in the transition state.

When writing an equation for the rate of a reaction, it is common to consider the order of the reaction, which is the number of reactant molecules involved in the reaction process. For example, let us consider the equilibrium reaction between the hydroperoxyl radical (HO_2) and nitrogen dioxide (NO_2) to form peroxynitric acid (HO_2NO_2). This reaction is very temperature dependent, favoring the formation of products at temperatures at or below 273 K. Because of this the reaction rate has to be studied as a function of temperature. Consider the forward reaction for the reaction of HO_2 with NO_2 at 298 K:

$$HO_2 + NO_2 \xrightarrow{k_1} HOONO_2$$

The rate law for this reaction is written as

$$\text{rate of reaction} = k_1[HO_2][NO_2]$$

where k_1 is the rate constant for the reaction at 298 K. This transition state involves only the two reactants (HO_2 and NO_2), so it is bimolecular. However, it involves only one molecule of HO_2 and one molecule of NO_2, so it is first order in HO_2 and first order in NO_2. The overall order of the reaction is equal to the sum of the orders for each reactant. So, this reaction is second order overall, since it is first order in each of the reactants. The rate constant will have units of $\text{l mol}^{-1}\,\text{s}^{-1}$, with the concentration units for the two reactants in mol l^{-1}, this gives a rate of formation for the product in units of $\text{mol l}^{-1}\,\text{s}^{-1}$. The reverse reaction involves only the decomposition of $HOONO_2$ to HO_2 and NO_2. Thus, the back reaction is unimolecular, first order in the reactant $HOONO_2$, and first order overall. The rate law for the reverse reaction would have the form

$$\text{rate of reaction} = k_2[HOONO_2]$$

Table 3.3 A summary of the rate laws for unimolecular, bimolecular, and termolecular reactions.

Molecularity	Rate law	Order in *A*	Order in *B*	Order in *C*	Order overall	Units of *k*
Unimolecular	rate = $k[A]$	first	—	—	first	s^{-1}
Bimolecular	rate = $k[A]^2$	second	—	—	second	$l\,mol^{-1}\,s^{-1}$
Bimolecular	rate = $k[A][B]$	first	first	—	second	$l\,mol^{-1}\,s^{-1}$
Termolecular	rate = $k[A]^3$	third	—	—	third	$l^2\,mol^{-2}\,s^{-1}$
Termolecular	rate = $k[A]^2[B]$	second	first	—	third	$l^2\,mol^{-2}\,s^{-1}$
Termolecular	rate = $k[A][B][C]$	first	first	first	third	$l^2\,mol^{-2}\,s^{-1}$

where the rate constant k_2 will have units of s^{-1} in order for the reaction rate to be in units of $mol\,l^{-1}\,s^{-1}$.

Many of the gas-phase thermal reactions that we will be discussing in Chapters 4 and 5 follow similar simple kinetics. These can be generalized using *A*, *B*, and *C* to represent reactants. The rate laws, reaction order, and units for the rate constants for each of the commonly observed gas-phase reactions are summarized in Table 3.3.

So, a unimolecular reaction only involves one reactant *A* going to products with the rate law

$$\text{rate} = k[A]$$

This unimolecular reaction is first order overall and first order in the reactant *A*. The units for the rate constant are s^{-1}. A biomolecular reaction involves two reactants, which need to collide in order to form products. This can be two of the same reactants $(A + A)$ or two different reactants $(A + B)$. The rate laws for these two possible bimolecular reactions are

$$\text{rate} = k[A]^2$$

$$\text{rate} = k[A][B]$$

Both of these reactions are second order overall. The first reaction that involves only one reactant (A) is second order in that reactant. The second reaction that involves two reactants $(A + B)$ is first order in each of the reactants. In this case the rate constant is in units of $l\,mol^{-1}\,s^{-1}$. A termolecular reaction requires three molecules to collide together to form products. This can be three molecules of the same reactant $(A + A + A)$, two molecules of one reactant with one molecule of a second reactant $(2A + B)$, or one molecule of three different reactants $(A + B + C)$. The rate laws of each of these reactions are of the form

$$\text{rate} = k[A]^3$$

$$\text{rate} = k[A]^2[B]$$

$$\text{rate} = k[A][B][C]$$

These reactions are all third order overall. The order of each of the reactants is equal to the power that the concentration is raised to in the rate law. So, the reaction involving only one reactant (A) is third order in that reactant. The reaction that involves two molecules of one reactant (A) and one molecule of another reactant (B) is second order in the first reactant and first order in the second reactant. The reaction involving three different reactants $(A, B, \text{and } C)$ is first order in each reactant. The rate constant units for all possible third-order reactions are $l^2\,mol^{-2}\,s^{-1}$.

There are several thermal atmospheric reactions that include a nonreactive species called a third body, which is represented in the chemical equation by the letter M:

$$A + B + M \rightarrow AB + M$$

Here M can be any nonreactive gas species such as nitrogen (N_2), oxygen (O_2), or argon (Ar). These reactions involve two reactive species (A and B) that react to yield a thermally excited product (AB^*). Collision of the product with the third body removes the excess energy, stabilizing the product (AB). Without the interaction with the third body, the product would quickly decompose back into the reactants A and B:

$$AB^* + M \rightarrow AB$$

or

$$AB^* \rightarrow A + B$$

The rate law for the third body reaction is

$$\text{rate} = k[A][B][M]$$

Third-body reactions are termolecular as all three species are involved in the transition state. They are also first order in each of the three reactants (A, B, and M). Although the third-body reaction is third order overall, in cases where the atmospheric pressure is constant and the concentration of M does not change, the reaction can be considered to be second order overall, also called pseudo second order. In this case, since $[M]$ is a constant, the rate law can be simplified by combining k with $[M]$ to give the second-order rate constant k' ($k' = k[M]$). In cases where the atmospheric pressure changes, as with changing altitude, $[M]$ is not constant and the reaction rate is represented by the third-order rate equation. Because of this, third-body reactions are considered to be pressure-dependent reactions.

The rate laws listed in Table 3.3 are in the form of differential equations where the rate of reaction is the change in the concentration of reactants with respect to time ($d[A]/dt$). These rate laws, known as the differential rate laws, describe how the rate of the reaction varies with the reactant concentrations. The integrated rate laws, obtained by integrating the differential rate laws, express the reaction rate as a function of the initial reactant concentrations ($[A]_o$) and the concentrations after an amount of time has passed ($[A]_t$). These integrated forms of the rate laws are listed in Table 3.4. They are given for the simple cases of one reactant (A, $2A$, or $3A$). The rate laws for two reactants ($A + B$ and $2A + B$) or three reactants ($A + B + C$) can be simplified to the case for one reactant if the stoichiometries are all 1 : 1 ($[A] = [B]$ or $[A] = [B] = [C]$).

The half-life of a chemical reactant is defined as the amount of time required for the reactant concentration to fall to half (50%) of its original value. It is most often used for first-order reactions, such as a radioactive decay process, which will be discussed in Chapter 12. The half-life

Table 3.4 Rate parameters for unimolecular, bimolecular, and termolecular reactions.

	Unimolecular	**Bimolecular**	**Termolecular**
Differential rate law	$\text{rate} = k[A]$	$\text{rate} = k[A]^2$	$\text{rate} = k[A]^3$
Integrated rate law	$[A]_t = [A]_o\, e^{-kt}$	$\frac{1}{[A]_t} = \frac{1}{[A]_o} + kt$	$\frac{1}{[A]_t^2} = \frac{1}{[A]_o^2} + 2kt$
Half-life	$0.693/k$	$1/(k[A]_0)$	$3/(2[A]_0^2)$

($t_{1/2}$) of the reactant species can be predicted from the integrated rate law equations by substituting $1/2[A_o]$ for $[A]_t$ and $t_{1/2}$ for t then solving for $t_{1/2}$:

$$\frac{[A]_{1/2}}{[A]_o} = \frac{1}{2} = e^{-kt_{1/2}}$$

$$\ln 0.5 = -kt_{1/2}$$

$$t_{1/2} = \frac{0.693}{k}$$

So, the half-life of a first-order reaction is a constant and does not depend on the initial reactant concentration ($[A]_o$). This means that the time required for 1 mol of a reactant to be reduced to 0.5 mol is the same time required for 100 mol to be reduced to 50 mol. Also, the rate constant and the half-life of a first-order reaction are inversely related by the value 0.693 (ln 0.5). Since the units for the rate constant for a first-order reaction are s^{-1}, the half-life is given in seconds (s).

Because thermal reactions are temperature dependent, the rate constants must be determined experimentally for different temperatures to determine the temperature effects. The rate of reaction is dependent on the number of molecular collisions, which are dependent on the kinetic energy of the system, and the molecular orientation during collisions. Svante Arrhenius in 1889 developed an empirical relationship between the temperature and the observed reaction rate constant based on the activation energy of a thermal reaction. This relationship, called the Arrhenius equation, is:

$$k = Ae^{-Ea/RT}$$

where k is the reaction rate constant, E_a is the activation energy, R is the ideal gas constant (8.314×10^{-3} kJ K^{-1} mol^{-1}), and T is the temperature in Kelvin. The parameter A is called the frequency factor, which is related to the number of collisions with the correct orientation for reaction. The factor $e^{-Ea/RT}$ is the fraction of molecules with the minimum amount of energy required for reaction. While the frequency factor has a very small temperature dependence, the exponential factor is strongly temperature dependent. An alternate form of the Arrhenius equation:

$$\ln k = \ln A - \frac{E_a}{RT}$$

shows that the relationship between ln k and $1/T$ is a straight line, where the y intercept is ln A, and the slope of the line is $-E_a/R$. This form of the Arrhenius equation is used to predict the effect of a change in temperature on the rate constant, and therefore on the rate of the reaction.

Since Arrhenius developed this empirical relationship there have been two other theoretical approaches presented, which have led to similar equations. The first was developed around 1916 by Max Trautz and William Lewis based on molecular collision theory. Collision theory states that in order for molecules to react they must collide with each other and the collisions must be energetic enough to be able to break and reform molecular bonds. Also, the reactant atoms must be oriented in such a way that they can easily reform bonds to yield the products. A collision between molecules that results in a chemical reaction is called an effective collision. So, according to collision theory, the rate of a chemical reaction is equal to the frequency of the effective collisions.

Trautz and Lewis proposed that two reactant molecules must come within the combined radius of each other along a reaction coordinate axis in order to result in an effective collision. The necessary energy required for reaction is the energy of collision (E_c). The rate of the reaction (k) is proportional to the collisional frequency (Z), the concentration of the reactants [A] and [B] (mol l^{-1}), and a steric factor (ρ):

$$k = \rho Z[A][B]e^{-E_c/RT}$$

where T is temperature in K, and R is the ideal gas constant. The form of this equation is very similar to the Arrhenius equation, where the term $\rho Z[A][B]$ is equivalent to the frequency factor (A) and E_c is equivalent to E_a.

The Trautz–Lewis equation is useful for evaluating reaction rates for very fast reactions, since it allows for the maximum number of collisions to be evaluated at any temperature. The activation energy for some of the important atmospheric radical reactions is very low and, in some cases, essentially zero. So, it is important that they be examined at a temperature where there is a reaction "speed limit," which represents the fastest speed at which a reaction can proceed. This reaction "speed limit" is the rate of collision at the reaction temperature. This will be important as we examine other empirical techniques for predicting the reaction rates of radicals and atoms with organic molecules in the atmosphere, as the predicted reaction rates cannot exceed these collisional limits.

The second theoretical approach was first proposed by Henry Eyring in the 1930s based on transition state theory. Transition state theory states that the rates of reaction are determined by a quasi-equilibrium between the reactants and the transition state, which is then converted into products. The transition state activation energy is included as the thermodynamic state function; the Gibbs free energy of activation ($\Delta G‡$), proposed by Rene Marcelin. Erying's equation takes the form:

$$k = \frac{BT}{h}e^{-\Delta G‡/RT}$$

where k is the bimolecular rate constant, B is Boltzmann's constant, h is Planck's constant, T is temperature in K, and R is the ideal gas constant. This equation also has a very similar form to the Arrhenius equation, where BT/h represents the frequency factor (A), except that it now includes a specified temperature dependence.

Recall that the Gibbs free energy function $\Delta G‡$ also has a temperature dependence in that it is related to enthalpy ($\Delta H‡$) and entropy ($\Delta S‡$) by the equation:

$$\Delta G‡ = \Delta H‡ - T\Delta S‡$$

So, the Eyring equation for the reaction rate constant can be rewritten as:

$$k = \frac{BT}{h}e^{\Delta S‡/R}e^{-\Delta H‡/RT}$$

This form is useful in evaluating the thermodynamic properties of the transition state by comparison with the experimentally determined E_a values. This temperature dependence becomes important for aqueous-phase thermochemical reactions. The activation energies for many aqueous-phase reactions, such as nucleophilic substitution in chlorinated organics, are on the order of 12 kcal mol^{-1}. With this activation energy, the reaction rates will change by a factor of two for every 10 K change in temperature. So, these reactions will slow down in colder climates such as in the Arctic and Antarctic, and will speed up in areas with higher temperatures such as in the tropical regions. For most of the gas-phase atmospheric reactions involving radicals, such as the hydroxyl radical (OH) with organics, the activation energies are usually less than a few kcal mol^{-1}, so the temperature dependence is not a significant factor in the reaction rates.

The primary process in a photochemical reaction is the absorption of a photon ($h\nu$) by the reactant (A), leading to the electronically excited state (A*):

$$A + h\nu \rightarrow A^*$$

The rate of loss of reactant A is given by:

$$\text{rate} = \frac{d[A]}{dt} = -k_p[A]$$

where k_p is the photolysis rate constant which includes the absorption cross-section of A (σ) in cm^2 $molecule^{-1}$, the quantum yield of the photolysis reaction (Φ), and the spherically integrated actinic flux (F) in photons cm^{-2} s^{-1} (Finlayson-Pitts and Pitts, 2000). So, the photolysis rate constant is expressed as:

$$k_p = \int_{\lambda_1}^{\lambda_2} \Phi \sigma F \, d\lambda$$

The absorption cross-section (σ) is derived from the Beer–Lambert Law discussed in Section 3.2. It can be measured experimentally using a spectrometer, but the form used to calculate photolysis rates must be expressed as the natural logarithm (ln) not the base-10 logarithm (log), which most spectrometers use.

The spherically integrated actinic flux is the total amount of light available to the absorbing molecule and includes both direct and scattered sunlight. The actinic flux can be obtained directly by measuring the photolysis rate of a well-characterized species with known values for the absorption cross-section and the quantum yield, such as NO_2. This method is called chemical actinometry. It can also be obtained from the measured solar light intensity. The diffuse and direct solar light intensity is commonly measured using a multi-filter rotating shadow band radiometer (MFRSR), as shown in Figure 3.10. The MFRSR measures the diffuse and total components of the solar irradiance at wavelengths of 415, 500, 615, 673, 870, and 940 nm. There is also a UV-sensitive MFRSR that measures the irradiance at 300, 305, 311, 317, 325, 332, and 368 nm. The diffuse component is obtained when the shadow band moves to cast a shadow over the detector sensor (Figure 3.10, position A), removing the direct component. The total amount is measured when the shadow band moves out of the view of the detector sensor (Figure 3.10, position B). The direct component is calculated from the difference between the total and diffuse measurements.

While the actinic flux is the spherically integrated photon flux incident at a point, the irradiance is the photon flux crossing a surface. The direct irradiance component (E_0) is related to the direct actinic flux (F_0) as:

$$E_0 = \mu_0 F_0$$

where μ_0 is the cosine of the solar zenith angle (Kylling et al., 2003). The diffuse component of the actinic flux is more difficult to calculate exactly from diffuse irradiance. It depends on the surface albedo, aerosol loading, and cloud cover. However, the albedo effects are not as important for wavelengths in the photochemically active UV range, unless there is snow cover.

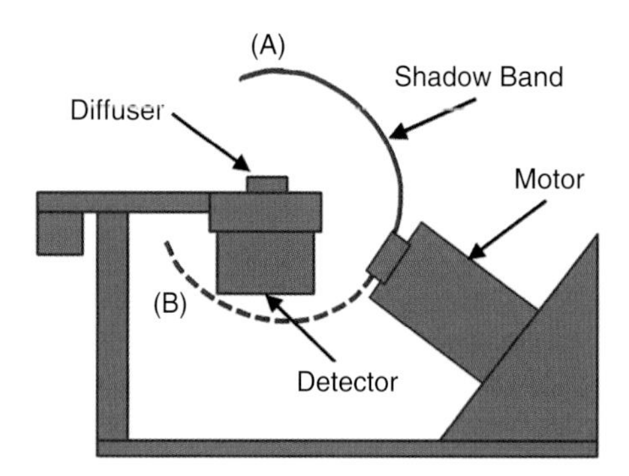

Figure 3.10 A schematic diagram of the MFRSR. The shadow band (red) casts a shadow over the detector when in position (A), giving measurements of diffuse irradiance. It then moves out of the detector view in position (B), giving measurements of total irradiance.

So, the cosine-corrected total irradiance is a very good estimate of the total actinic flux. If more accurate measurements are required, several computer algorithms are available that correct for aerosol and cloud effects on total solar irradiance.

3.4 Photochemical Deactivation Processes

After the molecule or atom absorbs a photon of light and has acquired the additional energy of the absorbed photon, it will undergo either a chemical reaction or a physical process in order to reduce the extra energy and return to the ground state. The most important chemical reactions that an excited molecule AB^* can undergo are shown in Figure 3.11. These are dissociation, ionization, and molecular rearrangement including isomerization. The reaction pathway is determined by the electronic bonding and energy levels of the electronically excited molecule. The most important photochemical reaction in the atmosphere is dissociation, where the electronic excitation results in breaking of the bonds between the atoms of the molecule:

$$AB + h\nu \rightarrow AB^* \rightarrow A + B$$

Ionization is not generally important in the lower atmosphere, since ionization potentials for most atmospheric molecules are greater than 9 eV, which would require high-energy (short-wavelength) light that is not available in the stratosphere and troposphere.

The extra energy of an electronically excited molecule can also be reduced through a physical deactivation process. In these processes the excited molecule emits the extra energy in the form of light or heat to return to the ground state without undergoing a chemical reaction. The two most common methods of deactivation by light emission are fluorescence and phosphorescence, as shown in Figure 3.12. Fluorescence is the emission of light of longer wavelength than that of the absorbed photon after a very short time interval (10^{-9}–10^{-7} s), called the fluorescence lifetime. The emitted light is shifted to longer wavelengths because the electronically excited molecule is also in a vibrationally excited state. It first loses this extra vibrational energy in the form of heat before returning to the ground state by the emission of light. The energy of this fluorescence light ($h\nu_2$) is equal to the light absorbed ($h\nu_1$) less the energy lost in the vibrational deactivation (E_ν):

$$h\nu_2 = h\nu_1 - E_\nu$$

The difference between fluorescence and phosphorescence is that the electronic transition in fluorescence takes place between two electronic energy levels with the same electron spin,

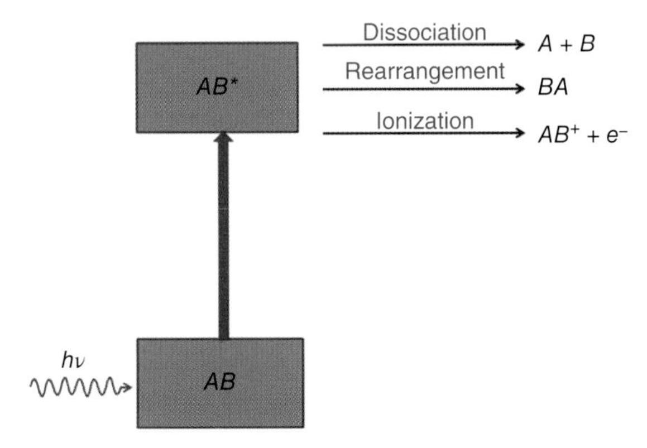

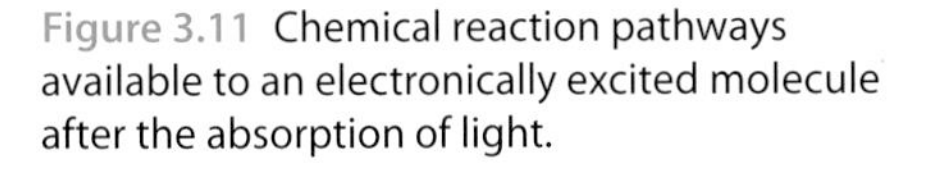
Figure 3.11 Chemical reaction pathways available to an electronically excited molecule after the absorption of light.

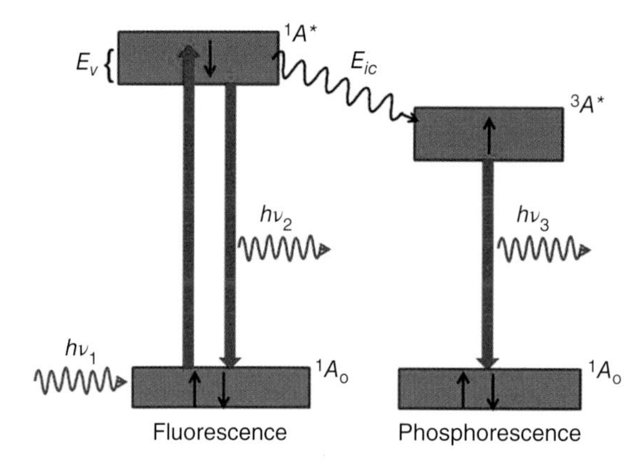

Figure 3.12 A comparison of the fluorescence and phosphorescence processes. An electron is promoted from the singlet ground state (^1A_o) to the excited singlet state ($^1A^*$) by absorption of a photon ($h\nu_1$). After losing excess vibrational energy (E_v), the electron returns to the ground singlet state by emission of light of a longer wavelength ($h\nu_2$) in fluorescence. The electron can also change spin, resulting in a triplet excited state ($^3A^*$) returning to the singlet ground state with the emission of light of longer wavelength ($h\nu_3$), in phosphorescence.

called singlet states (^1A and $^1A^*$). A singlet state is a molecular electronic state in which all the electron spins are paired and so the total electron spin in the molecule is equal to zero ($-\frac{1}{2} + \frac{1}{2} = 0$). The electronic transition in phosphorescence involves two electronic energy levels of different electron spins, a singlet state and a triplet state ($^1A^*$ and $^3A^*$). A triplet state is a molecular electronic state in which the total electron spin of the molecule is equal to one ($\frac{1}{2} + \frac{1}{2} = 1$). So, there must be two unpaired electrons in the triplet state. During phosphorescence the electron in the excited singlet ($^1A^*$) changes spin, resulting in an excited triplet state. The transition between the excited singlet state and the excited triplet state is called intersystem crossing. In the process of intersystem crossing, the molecule loses energy (E_{ic}) and so the excited triplet state is of lower energy than the excited singlet state. The energy lost in intersystem crossing to the triplet state results in the light emitted during phosphorescence being at longer wavelengths than the light originally absorbed by the molecule:

$$h\nu_3 = h\nu_1 - E_{ic}$$

In order to return to the singlet ground state ($^1A_0$), the electron must change spin a second time, resulting in a longer lifetime of the excited state (10^{-3}–10^2 s) than observed with florescence. So, there is a much longer time between the absorption of light and phosphorescence than there is between the absorption of light and fluorescence.

Other photophysical processes that return the electronically excited molecule to the ground state are listed in Table 3.5, along with their lifetimes. These deactivation processes include energy transfer, collisional deactivation, and internal conversion. In energy transfer the excited state of one molecule ($^1A^*$) is deactivated to the ground state (^1A_o) by

Table 3.5 Time scales for photochemical absorption and decay processes.

Process	Mechanism	Lifetime (s)	Emission
Absorption	$^1A_o + h\nu \rightarrow {}^1A^*$	10^{-15}	—
Fluorescence	$^1A^* \rightarrow {}^1A_o + h\nu$	10^{-9} to 10^{-7}	light
Phosphorescence	$^3A^* \rightarrow {}^1A_o + h\nu$	10^{-3} to 10^2	light
Energy transfer	$^1A^* + {}^1B_o \rightarrow {}^1A_o + {}^1B^*$	10^{-12}	—
Collisional deactivation	$^1A^* + M \rightarrow {}^1A_o + M$	10^{-14} to 10^{-11}	heat
Internal conversion	$^1A^* \rightarrow {}^1A_o$	10^{-12} to 10^{-10}	heat

transferring its energy to a second molecule (^1B_o), which is then raised to a higher-energy state ($^1B^*$) as

$$^1A^* + {}^1B_o \rightarrow {}^1A_o + {}^1B^*$$

The new excited molecule $^1B^*$ can then undergo its own deactivation process by chemical reaction, emission of light, or emission of heat.

Both collisional deactivation and internal conversion are sometimes called "radiationless de-excitation" processes, because they involve the deactivation of the electronically excited state without the emission of a photon. Both processes transfer the electronic excitation energy into vibrational excitation energy. Collisional deactivation involves the transfer of the excitation energy to other molecules in the system through energetic collisions. The energy is then converted to increased vibrational energy throughout the system, which is observed as an increase in temperature. In internal conversion the excitation energy is transferred to vibrational modes of the same molecule. Energy transfer, collisional deactivation, and internal conversion are all very fast processes compared to fluorescence and phosphorescence. They effectively reduce the excited state lifetime and so prevent the molecule from undergoing any of the slower processes. Any process which decreases the radiative deactivation of a molecule is called quenching.

The quantum yield of a photolysis process is the quantitative representation of the relative efficiency of the process. It is equal to the number of molecules undergoing a specific process (photochemical or photophysical) divided by the number of photons absorbed:

$$\Phi = \frac{\text{number of molecules undergoing process } i}{\text{number of photons absorbed}}$$

Most electronically excited molecules can undergo more than one deactivation process. As stated by the second law of photochemistry discussed in Section 3.2, the sum of all the quantum yields for all the photochemical and photophysical processes that a molecular species can undergo after electronic excitation must be equal to one:

$$\sum(\Phi_f + \Phi_p + \Phi_{ET} + \Phi_{Coll} + \Phi_{IC} + \Phi_{CD} + \Phi_{CR} + \Phi_{CI}) = 1.0$$

where Φ_f is the fluorescence quantum yield, Φ_p is the phosphorescence quantum yield, Φ_{ET} is the energy transfer quantum yield, Φ_{Coll} is the collisional deactivation quantum yield, Φ_{IC} is the internal conversion quantum yield, Φ_{CD} is the chemical dissociation quantum yield, Φ_{CR} is the chemical rearrangement quantum yield, and Φ_{CI} is the chemical ionization quantum yield.

This relationship deals with the primary photochemical and photophysical processes. The primary chemical processes can often be followed by secondary chemical reactions that do not require light absorption. The inclusion of these secondary processes results in apparent quantum yields greater than one. It is important to recognize that many of the chemical reactions in the environment, particularly those in the atmosphere and in surface waters, can be initiated by a photochemical reaction. However, they can continue to react by way of the more common thermal reactions. Most of these secondary processes are free radical chain reactions involving oxidation. One example of a photo-initiated chain reaction is the photolysis of the CFCs by UVB and UVC in the stratosphere to form chlorine atoms (Cl). The chlorine atoms then undergo chain reactions with ozone, leading to catalytic ozone depletion, which will be discussed in detail in Chapter 4. Another example is the photolysis of ozone in the troposphere by UVB, which leads to the formation of an electronically excited oxygen atom. The electronically excited oxygen atom then reacts with water vapor to produce two hydroxyl radicals (OH). The OH in turn can react with numerous organic compounds to produce radical oxidation chain reactions. These reactions are important in the lower atmosphere and in surface waters, and will be covered in Chapters 5 and 8.

References

Calvert, J.G. and Pitts, J.N. Jr. (1966). *Photochemistry*. New York: Wiley.

Davis, H.F., Bongsoo, K., Johnston, H.S., and Lee, Y.T. (1993). Dissociation energy and photochemistry of NO_3. *J. Phys. Chem.* 97: 2172–2180.

Einstein, A. (1905a). On the electrodynamics of moving bodies. *Ann. Phys.* 17: 891–922.

Einstein, A. (1905b). On a heuristic viewpoint concerning the production and transformation of light. *Ann. Phys.* 17: 132–148.

Einstein, A. and Infeld, L. (1938). *The Evolution of Physics from Early Concepts to Relativity and Quanta*. New York: Simon & Schuster.

Finlayson-Pitts, B.J. and Pitts, J.N. Jr. (2000). *Chemistry of the Upper and Lower Atmosphere: Theory, Experiments, and Applications*. San Diego, CA: Academic Press.

Kylling, A., Webb, A.R., Bais, A.F. et al. (2003). Actinic flux determination from measurements of irradiance. *J. Geophys. Res.* 108: 4524–4531.

Luo, Y.-R. (2003). *Handbook of Bond Dissociation Energies in Organic Molecules*, 142–143. Boca Raton, FL: CRC Press.

MacKay, R.J. and Oldford, R.W. (2000). Scientific method, statistical method and the speed of light. *Stat. Sci.* 15: 254–278.

Maxwell, J.C. (1865). A dynamical theory of the electromagnetic field. *Philos. Trans. R. Soc. London* 155: 459–512.

Planck, M. (1901). On the law of the energy distribution in the normal spectrum. *Ann. Phys.* 4: 553–564.

Further Reading

Evans, R.C., Douglas, P., and Burrows, H.D. (eds.) (2013). *Applied Photochemistry*. Berlin: Springer-Verlag.

Wayne, C.E. and Wayne, R.P. (1996). *Photochemistry*. Oxford: Oxford University Press.

Study Problems

3.1 What theory best explains the reflection and refraction of light?

3.2 What law describes the change in direction of an electromagnetic wave when traveling from one medium to another of different density?

3.3 What is the experimentally derived equation that describes the ratio of the reflected light intensity (I_r) to the intensity of incident light (I_o) for a perpendicular beam of unpolarized light?

3.4 What are the two types of light interference?

3.5 (a) What is the speed of light in a vacuum in meters per second ($m\,s^{-1}$)?
(b) In kilometers per second ($km\,s^{-1}$)?

3.6 (a) When light crosses a boundary between two different media as an electromagnetic wave, what property changes?
(b) What property remains the same?

3.7 When light travels from air into water, which wavelengths are refracted to a lesser degree?

3.8 What is the first law of photochemistry?

3.9 What is the second law of photochemistry?

3.10 What effect does the inclusion of secondary processes have on the calculation of the photochemical quantum yield?

3.11 What law relates the intensity of the light absorbed by a molecule to the concentration of the absorbing species?

3.12 What are the two most important differences between thermochemical and photochemical processes?

3.13 What is one mole of photons called?

3.14 (a) How much energy is there in a mole of photons with a wavelength of 260 nm?
(b) 200 nm?

3.15 (a) What is the shortest wavelength of light entering the stratosphere?
(b) The troposphere?

3.16 What is the overall order of reactions with the following rate laws:
(a) rate = $k[A]^2[B]$?
(b) rate = $k[A]$?
(c) rate = $k[A][B][C]$?
(d) rate = $k[A][B]$?

3.17 What is the order of reactant A in each of the rate laws in Problem 3.16?

3.18 What is the function of the third body in a third-body reaction?

3.19 (a) What is the overall molecularity of a third-body reaction involving two reactants?
(b) What is the order of the reaction at constant pressure?

3.20 (a) What is the Arrhenius equation?
(b) What is the parameter A in the Arrhenius equation related to?

3.21 What is the relationship between the initial reactant concentration and the half-life for a first-order reaction?

3.22 (a) What are the three possible photochemical pathways for an electronically excited molecule?
(b) Which is the most important pathway for the lower atmosphere?

3.23 (a) What are the physical deactivation pathways that involve the emission of light?
(b) What is the difference between these pathways?

3.24 (a) What are the physical deactivation pathways that involve the emission of heat?
(b) What is the difference between these pathways?

4
Chemistry of the Stratosphere

The photochemistry in the thermosphere, mesosphere, and upper stratosphere involves very high-energy radiation. Molecular oxygen (O_2) is the principle absorbing molecule in the vacuum UV (10–200 nm). It strongly absorbs this very short-wavelength (high-energy) radiation entering the upper stratosphere, undergoing dissociation to produce oxygen atoms. One of the most important reactions in the stratosphere is the reaction of these oxygen atoms with molecular oxygen to form ozone in the third body reaction:

$$O + O_2 + M \rightarrow O_3 + M$$

A primary topic in this chapter involves this production of ozone in the stratosphere and the absorption of light by both molecular oxygen and ozone, which protects the surface of the planet from harmful UV radiation.

Although molecular nitrogen (N_2) is prevalent in the stratosphere, it has a bond energy of 226 kcal mol^{-1}, which would require a photon with a minimum wavelength of approximately 130 nm to promote bond dissociation. Indeed, molecular nitrogen has one of the strongest bonds known for a diatomic molecule. This is one of the reasons why it remains the dominant gas in our atmosphere. Nitrogen molecules can undergo photoionization to N_2^+ at wavelengths below 100 nm, but this can only occur above the stratosphere where photochemical reactions with N_2^+ contribute to the aurora phenomena at the poles. Molecular nitrogen is only important in the stratosphere as a third body in the kinetics of ozone formation. Above the stratosphere, ionized nitrogen can react with oxygen to produce the reactive nitrogen compound nitric oxide (NO). But NO is only produced in the stratosphere from two other sources: from nitrous oxide (N_2O), which is mixed into the stratosphere from the troposphere, and from direct emission by high-flying commercial and military aircraft.

4.1 Structure and Composition of the Stratosphere

The stratosphere was named in 1900 by the French scientist Teisserienc de Bort, who discovered it using an instrumented balloon. With this equipment he was able to obtain a thermal profile of the atmosphere as a function of altitude. What he observed was the expected drop in temperature from the surface up to the tropopause. But above this altitude he found a thermally stratified layer that was warmer as the altitude increased, as shown in Figure 4.1. The name stratosphere in French means literally "sphere of layers." The region appears layered because the temperature profile in the stratosphere is stable and so mixing is much faster horizontally than it is vertically. Conversely, in the troposphere, where the warmer air is below the colder air, vertical mixing is predominant because the warm air rises and cools while the cooler air sinks. This means that the stratosphere has little vertical wind turbulence, unlike in the troposphere.

Chemistry of Environmental Systems: Fundamental Principles and Analytical Methods, First Edition.
Jeffrey S. Gaffney and Nancy A. Marley.

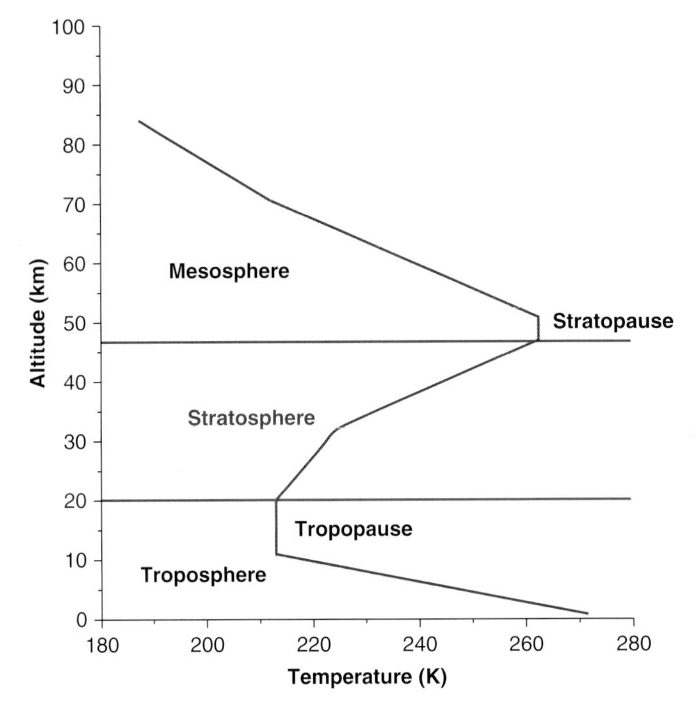

Figure 4.1 The temperature profile of the stratosphere.

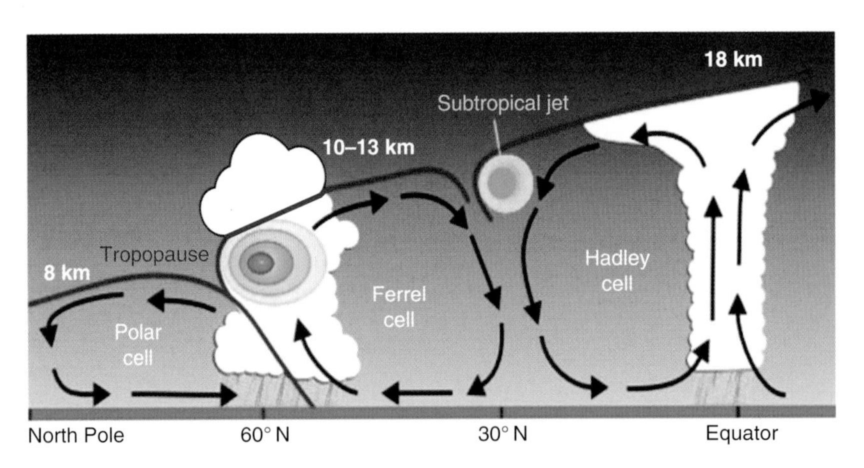

Figure 4.2 The height of the tropopause (red) above sea level varies from 8 km at the poles to 18 km at the equator. Source: Adapted from Lyndon State College Meteorology, Wikimedia Commons.

The stratosphere is located between the tropopause and the stratopause. It is not at a uniform height across the globe, as the spinning of the Earth causes the air to bulge at the equator and thin out at the poles. The height of the lower stratosphere is dependent on the height of the tropopause. The height variation of the tropopause from the equator to the poles is shown in Figure 4.2. The stratospheric layer closest to the surface of the Earth, just above the tropopause, is about 18 km above sea level at the equator. In the mid-latitudes it is between 10 and 13 km, while at the poles it is lowered further to 8 km. The top of the stratosphere, at the stratopause, is about 50 km above sea level. It varies somewhat depending on the extent of vertical mixing events, and ranges from about 45 to 58 km with some seasonal variation due to changes in the solar heating of the troposphere and stratosphere.

Since pressure drops continuously with altitude from the Earth's surface to outer space, most of the atmosphere is located in the troposphere while the stratosphere contains only about 20% of the total mass of the atmosphere. The stratosphere is a fairly cold region with temperatures ranging from 215 K at the lower altitudes to 270 K at the upper altitudes. Because of this, molecular water cannot make its way up into the stratosphere easily, as it will freeze and fall as precipitation long before it reaches the stratosphere under most circumstances. So, the stratosphere is a very dry region with very small amounts of water vapor present. This means that it contains very few clouds, except for polar stratospheric clouds, which can form in the lower stratosphere at altitudes of 15–25 km in winter when temperatures fall below −78 °C.

The overall circulation patterns in the stratosphere are called the Brewer–Dobson stratospheric circulation, as shown in Figure 4.3. This large cell circulation consists of an upwelling of air in the tropics, a horizontal movement at high altitudes toward the mid-latitudes, and a downwelling of air near the poles. The circulation is caused by atmospheric wave motion produced by the rotation of the planet. This wave motion, called Rossby waves, is responsible for the upwelling in the tropics which initiates the circulation. The Brewer–Dobson circulation was originally proposed independently by Alan Brewer and Gordon Dobson to explain why the stratospheric ozone levels are higher in the polar regions than in the tropical regions, even though most of the ozone is produced in the tropics. The proposed circulation was later confirmed by studies of stratospheric transport of radioactive fallout during atmospheric nuclear testing in the late 1950s (Sheppard, 1963).

Localized vertical mixing can also occur in the stratosphere through upward mixing from the troposphere due to episodic thunderstorms that can break through the tropopause and bring in warmer air, as shown in Figure 4.2. After entering the stratosphere, this warmer tropospheric air rises, causing some vertical mixing. These storm events are local and occur sporadically around the globe. Their occurrence leads to a low-frequency oscillation of the stratosphere in both the vertical and horizontal directions, which can affect the upper-level winds in the regions

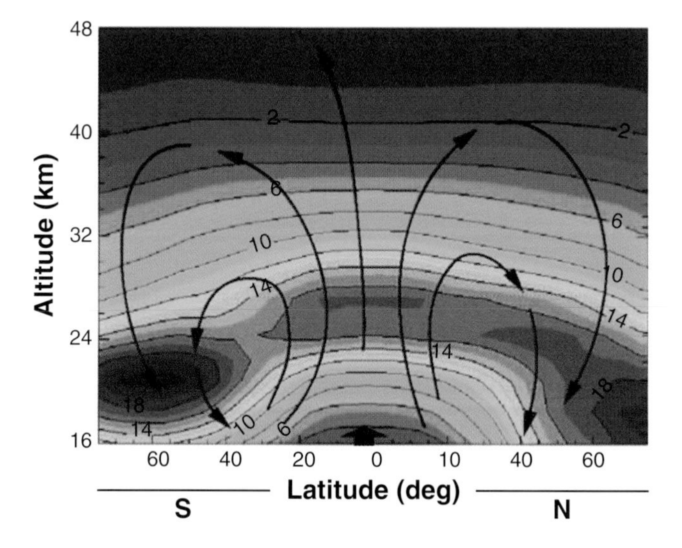

Figure 4.3 A contour plot of the stratosphere showing the Brewer–Dobson circulation (black arrows) which begins with an upwelling of stratospheric air in the tropics followed by horizontal movement toward the poles and a downwelling at high latitudes. The ozone layer is shown as the red, orange, and yellow bands, with darker reds representing the highest ozone concentrations. Source: NASA, Wikimedia Commons.

surrounding the thunderstorms. Due to the predominance of horizontal transport and the lack of vertical mixing, substances that are transported into the stratosphere from the troposphere can remain there for long periods of time. This can happen during large volcanic eruptions or major meteorite impacts, which can project aerosol particles up into the stratosphere where they remain for months or years. Thunderstorms can also transport nonreactive pollutants, such as CFCs, from the troposphere into the stratosphere.

The increase in temperature with altitude in the stratosphere is due to a series of exothermic reactions, beginning with the absorption of UVC radiation by molecular oxygen at wavelengths below 240 nm, resulting in its dissociation to oxygen atoms. The oxygen atoms produced react with molecular oxygen to form ozone, as shown above. This exothermic reaction involves the formation of a chemical bond with the release of heat and requires a third-body collision to stabilize the product. In addition, the ozone that is formed absorbs UV radiation from 200 to 300 nm with an absorbance maximum at 250 nm. The electronically excited state that is formed can re-emit the energy into the stratosphere through collisional deactivation or it can dissociate back into molecular oxygen and an oxygen atom. All three chemical reactions involved in this cycle are exothermic. The chemical heat that is released causes the temperature to rise in the stratospheric layers where the solar UV energy is being absorbed. Since the UV solar light intensity increases with altitude, so the temperature increases with altitude in the stratosphere.

4.2 The Ozone Layer

About 90% of the ozone in the atmosphere is found in the stratosphere. The largest concentration of ozone in the stratosphere, called the ozone layer, lies between 20 and 30 km in altitude, falling to lower altitudes near the poles due to the Brewer–Dobson circulation, as shown in Figure 4.3. The ozone layer contains about 10 ppm of ozone, while the naturally formed ozone levels in the troposphere, not including anthropogenically produced ozone, are around 10 ppb. The ozone layer absorbs 97–99% of the solar UV radiation from 200 to 315 nm. These wavelengths are harmful to life because they can be absorbed by the nucleic acid in cells, causing mutations resulting in cancer. It is especially harmful to humans because we don't have the protective covering of fur or feathers that other animals do, but it can also cause harm to some crops and marine life.

The stratospheric ozone layer was first discovered by the French scientists Charles Fabry and Henri Buissonin in 1913, when studying the incoming UV radiation from the Sun. The solar radiation measured at ground level matched that expected from a blackbody at a temperature of 5500–6000 K, except that the radiation below 310 nm was missing. They concluded that the radiation in the lower end of the spectrum was being absorbed by some molecular species in the upper atmosphere. It was known that molecular oxygen absorbed UV radiation strongly in the region from about 120 to 240 nm, as shown in Figure 4.4(A), but this did not explain the missing wavelengths above 240 nm.

Ozone was discovered around 1850 and determined to exist in the atmosphere, but the measurement of its absorption profile was difficult due to the very low intensity of the source light at these lower wavelengths. In 1880 Fabry designed a double spectrograph which eliminated scattered light from longer wavelengths and overcame the problem. This allowed him and his partner Buissonin to accurately determine the absorption coefficients of ozone from 200 to 320 nm, as shown in Figure 4.4(B). They compared these measurements to the measurements of solar radiation below 310 nm and concluded that the total amount of ozone in the atmosphere was equivalent to an ozone layer at SATP that would be approximately 5 mm in thickness. In addition, the absorption of UV light by both molecular oxygen and ozone together accounted for the missing UV wavelengths in the solar radiation measured at ground level.

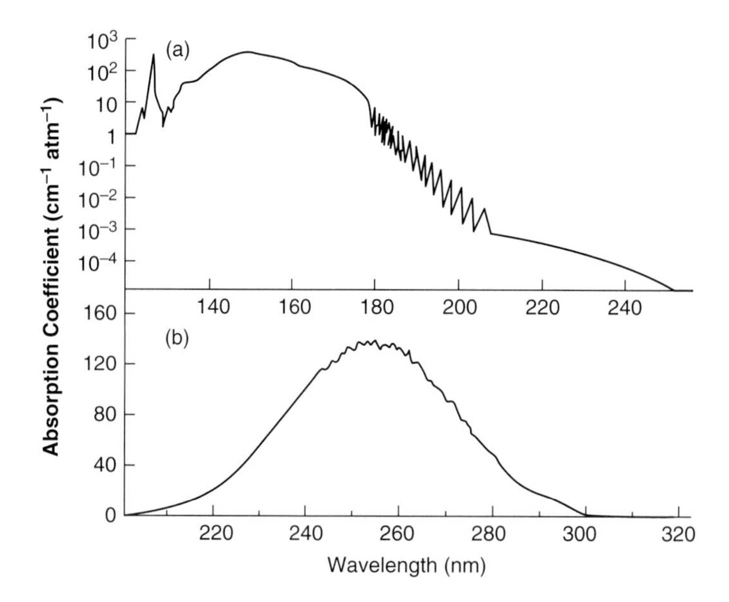

Figure 4.4 The absorption coefficient of (A) molecular oxygen and (B) ozone as a function of wavelength. The sharp bands in the spectrum of molecular oxygen between 176 and 193 nm are known as the Schumann–Runge bands, caused by rotationally excited states of the molecule.

Ozone is a strong absorber of UV radiation in the wavelength range of 220–300 nm. Its absorption band lies outside the absorption range of molecular oxygen, which is from 120 to 240 nm, and so it absorbs radiation in the wavelength range where O_2 no longer has strong absorption. Figure 4.5 shows the effects of the stratosphere on the incoming solar UV radiation. Most of the UVC (100–280 nm) is absorbed by molecular oxygen in the upper layers of the stratosphere. The remaining UVC is absorbed in the ozone layer. The UVB (280–315 nm) is drastically reduced by both O_2 and O_3 absorption, with a small amount reaching the troposphere. So, between O_2 and O_3, the harmful UVC and most of the UVB radiation is absorbed

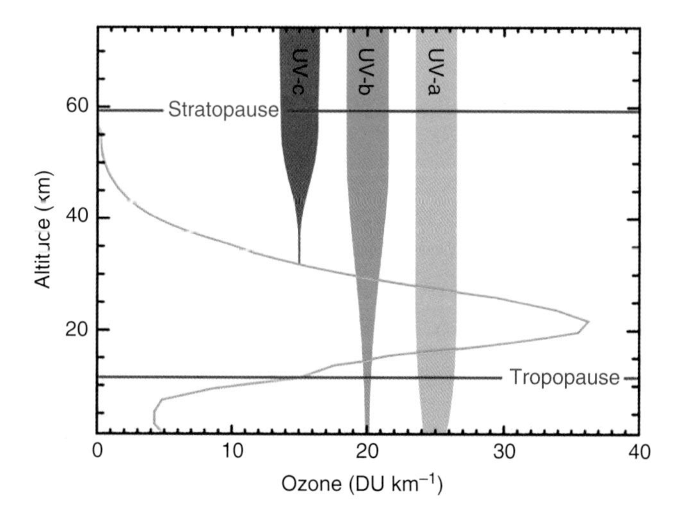

Figure 4.5 Relative ozone concentrations as a function of altitude in the stratosphere. Essentially all UVC (100–280 nm) and most UVB (280–315 nm) is absorbed by molecular oxygen in the range of 120–240 nm, or by ozone in the range of 220–300 nm. The UVA (315–400 nm) is not absorbed in the stratosphere. Source: NASA, Wikimedia Commons.

in the stratosphere, preventing it from reaching the troposphere. The UVA (315–400 nm) is not affected by either O_2 or O_3 absorption. So, some of the UVB and most of the UVA radiation does enter the troposphere and this plays an important role in the photochemistry below the stratosphere, which will be covered in Chapter 5.

The absorption spectrum of ozone, shown in Figure 4.4(B), was used in 1924 by Gordon Dobson to obtain the amount of total column ozone in the atmosphere. He realized that he could use the measurement of the UVA intensity, which was not absorbed appreciably in the atmosphere, as an estimate of the solar light intensity at the top of the atmosphere (I_o). He used the ratio of the intensity of a selected wavelength in the UVA (325 nm) to the intensity of a nearby wavelength in the UVB (305 nm) to determine the absorption of solar light by ozone in the atmosphere from the definition of light transmittance:

$$A = \ln \frac{I}{I_o} = e^{abc}$$

where I is the intensity of solar light at ground level measured at 305 nm, a is the absorptivity of ozone at 305 nm, and I_o is the solar light intensity at the top of the atmosphere at a wavelength of 305 nm, which is estimated by the measurement of solar light intensity at ground level at 325 nm. The path length of the solar light through the atmosphere (b), multiplied by the concentration of ozone in the atmosphere (c), gives the amount of ozone in the atmosphere in Dobson units, named in honor of Gordon Dobson. The Dobson unit (DU) is defined as the thickness (in units of 10 μm) of a layer of pure ozone at STP that would give the same absorbance as that measured in the total atmospheric column. The amount of atmospheric ozone is still measured in Dobson units, equivalent to 0.4462 mmol m^{-2} and 2.69×10^{20} molecules m^{-2}. Normal values for the total atmospheric column of ozone are about 300 DU (3 mm).

4.3 Ozone Formation in the Stratosphere

The ozone in the stratosphere is formed naturally by chemical reactions involving solar UV light and molecular oxygen, which makes up 21% of the atmosphere. Most of the ozone in the stratosphere is formed over the equator, where the level of solar UV radiation is highest. The Brewer–Dobson circulation then transports it toward the poles. So, the amount of stratospheric ozone above a location on the Earth varies naturally with latitude and season.

4.3.1 The Chapman Cycle

In 1930 Sydney Chapman, shown in Figure 4.6, proposed a mechanism for the formation of ozone in the stratosphere, now called the Chapman Cycle (Chapman, 1930). This mechanism involves four reaction steps: (1) initiation, (2) ozone formation, (3) ozone photolysis, and (4) termination. The initiation step is the creation of O from O_2 photolysis after absorption of UVC and UVB radiation (< 242 nm). Ozone is formed in the second step by the three-body reaction of O with O_2 to form O_3. Ozone is then photolyzed after absorption of UVB radiation (240–310 nm) reforming O_2 and O, which can again react to reform O_3 (step 2). The termination step removes O_3 by reaction with O to form two O_2.

The Chapman Cycle

1. Initiation: $O_2 + h\nu_{(<242\,nm)} \rightarrow 2O$
2. Ozone formation: $O + O_2 + M \rightarrow O_3 + M$
3. Ozone photolysis: $O_3 + h\nu_{(240-310\,nm)} \rightarrow O_2 + O$
4. Termination: $O_3 + O \rightarrow 2O_2$

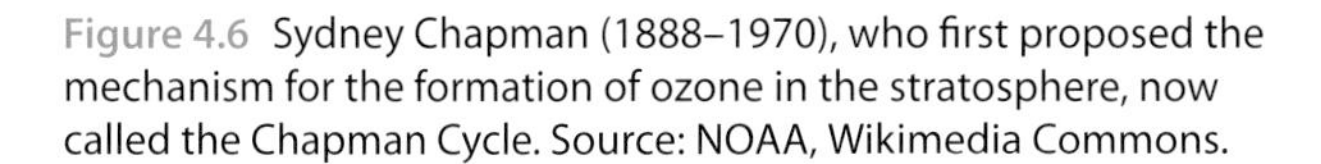

Figure 4.6 Sydney Chapman (1888–1970), who first proposed the mechanism for the formation of ozone in the stratosphere, now called the Chapman Cycle. Source: NOAA, Wikimedia Commons.

The formation of O_3 in step 2 is a bond-forming process that releases heat into the stratosphere causing the observed warming. This reaction is a termolecular process, which is dependent on a third body (M), and so the rate of reaction is proportional to atmospheric pressure. At higher altitudes the atmospheric pressure and concentration of M decrease, causing the rate of the formation of O_3 to decrease. At some point, ozone is no longer formed and no more heat is released into the stratosphere. At the stratopause the pressure and temperature are too low to support the third-body reaction, preventing further ozone formation. The temperature remains constant with altitude to the top of the stratopause. At the top of the stratopause the normal lowering of temperature with increasing altitude resumes, as shown in Figure 4.1.

The Chapman Cycle predicted the observed temperature profile and the formation of ozone in the stratosphere fairly well, but it overpredicted the actual ozone concentrations. This is because it only addresses the formation of ozone and does not address the natural chemical cycles that are responsible for the loss of ozone. In order to adequately predict stratospheric ozone levels, other important chemical reaction cycles, which control the atmospheric concentrations of ozone, need to be addressed. While ozone formation involves the ground-state oxygen atom, the two major ozone loss mechanisms are initiated by an oxygen atom in an excited state. In order to understand these chemical cycles, the differences between the ground state and the excited state of the oxygen atom must first be understood.

4.3.2 Term Symbols

In step 2 of the Chapman Cycle, O_3 can photolyze to give three different electronic states for the oxygen atom. The O atom has electronic configuration [He]$2s^2 2p^4$. The four valence electrons in the three $2p$ electronic orbitals ($2p_x$, $2p_y$, $2p_z$) can have three possible electronic configurations, shown in Figure 4.7. These three electron configurations differ in the placement and spin of the electrons in the p orbitals. Each of the configurations is designated by a term symbol shown in red in Figure 4.7. These term symbols are an abbreviated description of the total angular momentum quantum numbers of the atom in each electronic configuration. It is important to understand the meaning of the term symbols for the electronic states of the oxygen atom in order to fully understand the photochemistry of ozone.

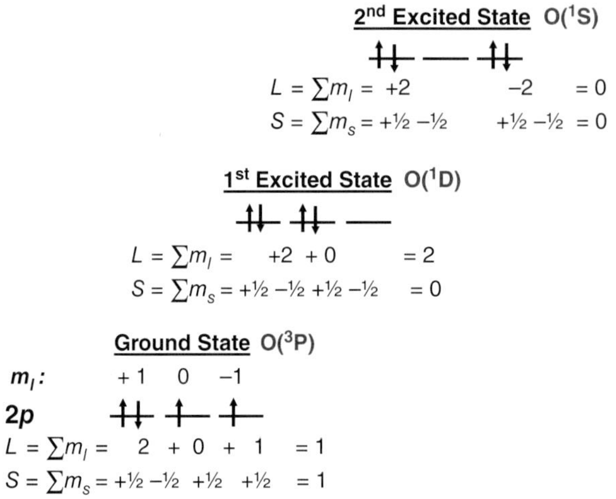

2nd Excited State O(^{1}S)

⇅ — ⇅

$L = \sum m_l = +2 \quad -2 \quad = 0$

$S = \sum m_s = +½ −½ \quad +½ −½ = 0$

1st Excited State O(^{1}D)

⇅ ⇅ —

$L = \sum m_l = +2 + 0 \quad = 2$

$S = \sum m_s = +½ −½ +½ −½ \quad = 0$

Ground State O(^{3}P)

m_l: + 1 0 −1

2p ⇅ ↑ ↑

$L = \sum m_l = 2 + 0 + 1 \quad = 1$

$S = \sum m_s = +½ −½ +½ +½ \quad = 1$

Figure 4.7 The three possible electronic configurations for atomic oxygen and the determination of their term symbols (shown in red).

The relative energies of the three electronic configurations are determined by the multiplicity of each configuration, which is equal to $2S + 1$where S is the total spin angular momentum. The total spin angular momentum is equal to the sum of the spin angular momentum quantum numbers ($\sum m_s$) for all valence electrons, which have the allowed values of +½ (up) and −½ (down). The determinations of the total spin angular momentum quantum number for each of the three electronic configurations are shown in Figure 4.7. The electronic configuration that has two unpaired electrons with the same spin ($S = +½ - ½ + ½ + ½ = 1$) has a multiplicity of $(2 \times 1) + 1 = 3$. The other two possible electronic configurations, with all electrons paired, have a spin angular momentum of 0 and a multiplicity of $(2 \times 0) + 1 = 1$. The terms used for these multiplicities are triplet (3) and singlet (1) states. According to Hund's rules for atomic energy levels:

The electronic configuration with the maximum multiplicity has the lowest energy and is the ground state of the atom.

Since the highest multiplicity of all the possible electronic configurations of the oxygen atom is 3, the triplet state is the ground state. The other two electronic configurations are both singlets with higher energies than the ground state, and so they are excited states. The multiplicity of the electronic configuration is designated in the term symbol by the numerical exponent to the left of the letter in parentheses.

The relative energies of the excited states with the same multiplicity are determined by the differences in the total orbital angular momentum quantum number (L). The excited state with the largest value of L has the lowest energy and is the first excited state, while the excited state with the lowest value of L has a higher energy and so is the second excited state. The total orbital angular momentum quantum number is equal to the sum of the magnetic angular momentum quantum numbers (m_l) of each of the valence electrons. The magnetic angular momentum quantum number distinguishes between the orbitals in a subshell and has values of $-l$ to $+l$ including 0, where l is the azimuthal quantum number assigned to the electron subshell. The l value for the p subshell is equal to 1, and so the values of m_l are −1, 0, and +1. The total angular momentum values for the three possible electronic states of the oxygen atom are 1, 2, and 0, as shown in Figure 4.7. The electronic configuration determined to be the ground state by its multiplicity has a total orbital angular momentum quantum number of 1. The two excited states have L values of 0 and 2. So, the electronic configuration with an L value of 2 is the first excited

Table 4.1 The term symbol letters used for values of the total angular momentum quantum numbers (L) for a multi-electron atom.

L value	Term symbol letter
0	S
1	P
2	D
3	F

state and the electronic configuration with an L value of 0 is the second excited state. These L values also determine the letter designations used in the term symbols according to Table 4.1.

The term symbols are written in the form

$$Z(^{2S+1}L)$$

where Z is the atomic symbol for the element. So, according to Table 4.1, the letter used in the term symbol for the ground state of oxygen is P, the letter used for the first excited state is D, and the letter used for the second excited state is S. Accordingly, the ground state for the O atom is denoted as $O(^3P)$, the first excited state is $O(^1D)$ and the second excited state and highest-energy state is $O(^1S)$.

4.3.3 The HO_x and NO_x Cycles

The two major chemical reaction cycles responsible for the difference between the observed ozone concentrations and those predicted from the Chapman Cycle are the HO_x Cycle (OH and HO_2) and the NO_x Cycle (NO and NO_2). Both of these reaction cycles are chain mechanisms that are initiated by reaction with $O(^1D)$. The dominant product of the photolysis of O_2 at wavelengths below 240 nm in the stratosphere is $O(^3P)$, but the photolysis of O_3 in the stratosphere produces oxygen atoms in both the $O(^3P)$ and $O(^1D)$ electronic states. The $O(^1D)$ excited state can react with water vapor to form two hydroxyl radicals (OH), initiating the HO_x Cycle. It can also react with N_2O to form two NO molecules, initiating the NO_x Cycle. Both these chemical reaction cycles consume O_3 and are thus considered to be ozone depletion cycles.

The HO_x Cycle involves reactions of both OH and HO_2 with O_3 as follows:

The HO_x Cycle

1. Initiation: $H_2O + h\nu \rightarrow H + OH$
   $O(^1D) + H_2O \rightarrow 2OH$
2. Ozone depletion: $OH + O_3 \rightarrow HO_2 + O_2$
   $HO_2 + O \rightarrow OH + O_2$
3. Termination: $OH + HO_2 \rightarrow H_2O + O_2$

It was first suggested in 1950 that the production of OH in the stratosphere was due to the photolysis of water vapor (Bates and Nicolet, 1950). Water vapor absorbs solar light strongly in the UVC and weakly in the UVB. The resulting photolysis of water leads to the production of a hydrogen atom (H) and OH. This is one of the initiating steps of the HO_x Cycle. The other initiating reaction is $O(^1D)$ with H_2O to produce two OH radicals. The OH from either initiating reaction then reacts with O_3 to form HO_2 and molecular oxygen. Then the HO_2 reacts with O to regenerate OH. The net reaction for the two ozone depletion reactions is:

$$\text{Net ozone depletion}: O_3 + O \rightarrow 2O_2$$

The two ozone depletion reactions consume O_3 while regenerating OH and HO_2. So, these two species, known as HO_x, act as a catalyst for O_3 loss. The production of one OH molecule by the initiation reaction results in the loss of a large number of O_3 molecules because the HO_x is regenerated by the ozone depletion reactions until they become involved in a termination reaction. Termination of the cycle requires loss of HO_x by a reaction such as the reaction of OH and HO_2 to form water and O_2.

The HO_x Cycle depends on the presence of H_2O in the stratosphere. Since the stratosphere is very cold, it is not easy for H_2O to reach the stratosphere directly. Under most conditions, it will form precipitation and return to the hydrosphere before reaching the stratosphere. The only exceptions are large storms that sporadically break through the tropopause or explosively erupting volcanoes, which occur very infrequently. So, the main sources of water vapor in the stratosphere are indirect sources that involve reactions of methane and hydrogen gases transported from the troposphere. These gases can slowly mix into the lower stratosphere with a half-life of about 8–10 years. Once in the stratosphere they are rapidly oxidized by O and OH to form water. Since the source of H_2O in the stratosphere is linked to the troposphere, the HO_x Cycle primarily impacts the lower portion of the stratosphere.

After the accurate determination of the rate constants for the HO_x Cycle reactions in the 1950s and 1960s it was determined that, although it represented a significant O_3 sink in the lower stratosphere, it still did not account for the entire difference between the ozone formation predicted by the Chapman Cycle and the direct observations of ozone concentrations in the stratosphere. So, a second ozone depletion cycle must naturally exist. Interest in the NO_x Cycle began in the 1960s, with a proposal to use supersonic transport (SST) to reduce flying times across the world. Since these aircraft would fly in the upper stratosphere, their emission of nitric oxide (NO), produced by the high-temperature oxidation of N_2 in the aircraft engine, could react with O_3. It was then proposed that nitrous oxide (N_2O) could be a natural source of NO in the stratosphere, initiating the NO_x ozone depletion cycle (Crutzen, 1970). Nitrous oxide is essentially an inert gas in the troposphere that, like methane, is also primarily produced from anaerobic bacteria in soils and wetland environments. Because it has a very long lifetime in the troposphere, it can slowly mix into the stratosphere where it can be converted to NO.

The reaction of N_2O with $O(^1D)$ is the initiation step in the NO_x Cycle, producing the reactive species NO. The NO_x Cycle involves NO and NO_2 in the following reactions.

The NO_x Cycle

1. Initiation: $O(^1D) + N_2O \rightarrow 2NO$
2. Ozone depletion: $NO + O_3 \rightarrow NO_2 + O_2$
   $NO_2 + O \rightarrow NO + O_2$
3. Termination: $OH + NO_2 + M \rightarrow HNO_3 + M$
   $NO_2 + O_3 \rightarrow NO_3 + O_2 \text{(night)}$
   $NO_3 + NO_2 + M \rightarrow N_2O_5 + M \text{ (night)}$

The ozone depletion reaction is NO with O_3 to produce NO_2 and O_2. The NO_2 product then reacts with oxygen to regenerate NO, thus continuing the ozone depletion cycle. The net reaction for the two ozone depletion reactions is:

$$\text{Net ozone depletion}: O_3 + O \rightarrow 2O_2$$

The two ozone depletion reactions consume O_3 while regenerating NO and NO_2. So, the two NO_x species (NO and NO_2) act as the catalysts for the O_3 loss. As with the HO_x Cycle, the production of one NO molecule in the initiation reaction results in the loss of a large number of O_3 molecules due to the regeneration of NO, which restarts the cycle. The termination reactions consume NO_2 without regenerating either of the NO_x species. The reaction of OH and

NO_2 to form nitric acid (HNO_3) in a third-body reaction occurs during the day because the formation of OH by photolysis of water vapor requires sunlight. There are two major termination reactions that occur during the night: (1) NO_2 reacts with O_3 to give the nitrate radical (NO_3) and (2) NO_3 reacts with NO_2 to form dinitrogen pentoxide (N_2O_5) in a third-body reaction. As soon as there is sunlight, N_2O_5 will photolyze rapidly to NO_3 and NO_2, and NO_3 will photolyze back to NO_2 and O. Also, HNO_3, produced during the day, slowly photolyzes back to OH and NO_2. Since N_2O_5, NO_3, and HNO_3 can all photolyze to give back NO_2, they act as reservoirs for NO_x. These and other reservoir species are collectively called NO_y. They can only be completely removed from the NO_x Cycle when they are slowly mixed into the troposphere with stratospheric lifetimes of weeks to years.

By including the NO_x and HO_x ozone depletion cycles with the Chapman ozone production cycle, the overall stratospheric ozone concentration profile could be closely reproduced. This coupling of the Chapman Cycle with the HO_x and NO_x Cycles, along with the Brewer–Dobson stratospheric circulation from the equator to the polar regions, provides a mechanism for determining the natural ozone concentrations in the stratospheric ozone layer.

4.4 Ozone Depletion

The catalytic ozone destruction cycles, including the HO_x and NO_x Cycles, are of the general form:

$$O_3 + X \rightarrow XO + O_2$$

$$O + XO \rightarrow X + 2O_2$$

where X is the chain reaction catalyst, commonly formed by photolysis, and XO reacts with O to regenerate X. The net reaction for each cycle is:

$$O_3 + O \rightarrow 2O_2$$

In the HO_x Cycle, X is OH and XO is HO_2 while in the NO_x Cycle, NO is X and NO_2 is XO. There are many chemical cycles that follow this general reaction scheme and they are dominant processes in stratospheric chemistry that contribute to ozone loss. While the HO_x and NO_x Cycles are natural ozone depletion cycles, other cycles exist that are initiated by anthropogenic species. Some highly stable, long-lived, anthropogenic compounds are capable of mixing into the stratosphere where they can be photolyzed by UVC to generate ozone-depleting species in this general catalytic reaction cycle.

4.4.1 Chlorofluorocarbons

The CFCs are fully halogenated alkanes (methane, ethane, and propane) that contain chlorine, fluorine, and carbon. They were originally released under the trademark Freon. A related set of compounds have one or more hydrogens along with chlorine, fluorine, and carbon and are called hydrochlorofluorocarbons (HCFCs). All of these compounds are manmade and are not found in the natural environment. As discussed in Chapter 1, the CFCs were first developed by General Motors researchers led by Thomas Midgely Jr., to be used as operating fluids in refrigeration. They were to replace other operating fluids in use at the time, which were either flammable or toxic. The CFCs became popular because they were inflammable, chemically stable, and of low to moderate toxicity. The first of the CFCs to be used in refrigeration was dichlorodifluoromethane (CF_2Cl_2). After producing this first compound many others followed, and soon the use of the CFCs (Freons) included many diverse applications such as aerosol spray can propellants, organic solvents, insulating foam, and packaging blowing agents.

The industry used a naming method for the CFCs and HCFCs that is still used today. This naming method uses three numbers to define the chemical structure of the compound. The first number is the number of carbon atoms minus one. The second number is the number of hydrogen atoms plus one. Finally, the third number is the number of fluorine atoms. The remaining atoms are the number of chlorine atoms. These three numbers follow the three-letter designation for chlorofluorocarbons as:

$$\mathrm{CFC} - (\mathrm{C} - 1)(\mathrm{H} + 1)\mathrm{F}$$

For example, dichlorodifluoromethane (CCl_2F_2) has one carbon. Normally, the first number would be the number of carbons minus one, except $1 - 1 = 0$ and zeros are not listed in the number designation. So, the number designation for a substituted methane has only two numbers. In this case the first number is the number of hydrogens plus 1, or $(0 + 1 = 1)$. The second number is the number of fluorine atoms (2). So, the industrial name for dichlorodifluoromethane is CFC-12, also called Freon-12. When determining the chemical formula from an industrial name, if the industrial name contains only two numbers it will be a methane derivative with only one carbon. For example, CFC-21 has one carbon $(1 - 1 = 0)$, one hydrogen $(1 + 1 = 2)$, and one fluorine. It must have two chlorines since there are four bonds to carbon in methane. The chemical formula is CCl_2HF, with the chemical name of dichlorofluoromethane. Table 4.2 shows some of the most common CFC and HCFC chemical formulas, chemical and industrial names, and their primary uses.

These CFCs and HCFCs do not absorb solar light in the UVB and UVA, so they do not photolyze in the troposphere. They are also very inert to oxidation by the OH radical, which is the main path for removal of most photolysis-resistant compounds in the troposphere. Because they do not react with OH, all of the CFCs and HCFCs have very long tropospheric lifetimes. In many cases the lifetimes can reach thousands of years for CFCs and tens of years for HCFCs. So, when they are released into the atmosphere, their concentrations continue to build up in the troposphere. This allows them to mix into the stratosphere where they can either photolyze or react with $O(^1D)$ atoms. The stratospheric lifetimes of the CFCs are in the range of 40–100 years. The main process for their loss is photolysis in the wavelength range of 185–235 nm for the CFCs and 225–280 nm for the HCFCs.

In 1970, Paul Crutzen demonstrated the potential for long-lived tropospheric gases to mix into the stratosphere (Crutzen, 1970). Using this information, Mario Molina and F. Sherwood Rowland proposed that the widespread use of CFCs was causing their build-up in

Table 4.2 Some common CFCs and HCFCs with their chemical structures, chemical and industrial names, and common uses.

Chemical formula	Chemical name	Industrial name	Common use
CCl_3F	trichlorofluoromethane	CFC-11	refrigerant/propellant
CCl_2F_2	dichlorodifluoromethane	CFC-12	refrigerant/propellant
$C_2Cl_3F_3$	trichlorotrifluoroethane	CFC-113	solvent
$C_2Cl_2F_4$	dichlorotertfluoroethane	CFC-114	refrigerant
C_2ClF_5	chloropentafluoroethane	CFC-115	refrigerant
$CHCl_2F$	dichlorofluoromethane	HCFC-21	refrigerant/propellant
$CHClF_2$	chlorodifluoromethane	HCFC-22	air conditioning
CH_2ClF	chlorofluoromethane	HCFC-31	refrigerant

the troposphere and, due to their long tropospheric lifetimes, they could be transported into the stratosphere. Once in the stratosphere the CFCs would undergo photolytic dissociation to release chlorine atoms (Cl), which could react with stratospheric ozone in a catalytic chain reaction. In their 1974 paper in *Nature* (Molina and Rowland, 1974), they proposed the following Cl atom-initiated ozone depletion cycle:

The Cl/ClO Cycle

1. Initiation: $CFC + h\nu \rightarrow Cl + products$
2. Ozone depletion: $Cl + O_3 \rightarrow ClO + O_2$
   $ClO + O \rightarrow Cl + O_2$
3. Termination: $Cl + CH_4 \rightarrow HCl + CH_3$
   $ClO + NO_2 \rightarrow ClONO_2$

The Cl atoms formed from the photolysis of the CFC by UVC in the upper stratosphere would react rapidly with O_3 to produce chlorine monoxide (ClO). The ClO in turn would react with an O to reform Cl and O_2. The net reaction for the two ozone depletion reactions is:

$$\text{Net ozone depletion}: O_3 + O \rightarrow 2O_2$$

This leads to a catalytic chain reaction where the Cl atoms act as the catalyst to deplete the amount of ozone in the stratosphere. As with the HO_x and NO_x Cycles, the production of one Cl in the photolytic initiation reaction results in the loss of a large number of O_3 molecules due to the regeneration of Cl, which restarts the cycle. Estimates of the chain length of these reactions were about 100 000 molecules of O_3 lost per Cl atom released from CFC photolysis. The potential chain termination reactions are Cl reacting with CH_4 to form hydrochloric acid (HCl) or the reaction of ClO with NO_2 to form chlorine nitrate ($ClONO_2$). As it turns out, these two possible termination reactions form Cl reservoir species similar to those in the NO_x Cycle. The HCl can react with OH to form water and Cl with a HCl lifetime of a few weeks. The $ClONO_2$ can undergo slow photolysis to form Cl and NO_3 with a $ClONO_2$ lifetime of about a day. These two reformation reactions are sometimes referred to as the ClO_x Cycle, with the two reservoir species referred to as ClO_y as with NO_x and NO_y.

The concern raised by Molina and Rowland was that the increasing use of the chlorine-containing CFCs and HCFCs would lead to higher tropospheric concentrations and so to higher amounts reaching the stratosphere. This would cause enhanced ozone depletion in the stratospheric ozone layer, which would result in a decrease in the layer's UVB filtering capacity. This means that there would be an increase in the UVB radiation entering the troposphere. This increase in UVB could lead to a number of unfortunate consequences, including enhancing the chemistry of the troposphere and increasing exposure to higher levels of harmful UV radiation. Some of the potential impacts to the biosphere from the enhanced UV radiation include increased cases of sunburn, skin cancers, and eye damage such as cataracts, as well as damage to ocean phytoplankton, corals, and terrestrial and aquatic plants.

When this hypothesis was proposed it initially met with a great deal of skepticism, primarily by the industrial chemistry community, which had developed these compounds as environmentally safe and nontoxic materials. There were also questions regarding the stratospheric chemistry, as it had not been well studied and scientists were unsure if there might be missing reactions that would limit the ozone-depleting capacity of chlorine-containing compounds. There are natural chlorine and bromine compounds, such as methyl chloride (CH_3Cl) and methyl bromide (CH_3Br), which are known to be produced from biological decay processes. These natural chlorine-containing compounds do provide a natural level of chlorine and bromine atoms in the troposphere and stratosphere. However, they are more reactive to OH oxidation and so their lifetimes in the troposphere are sufficiently short that they only contribute a small amount of background chlorine and bromine in the stratosphere. Since the CFCs and HCFCS have much longer lifetimes, they produce additional chlorine and bromine in the stratosphere over that which exists naturally. So, their effects needed to be understood.

At the time, photochemistry and free radical reactions had not been studied under stratospheric conditions. So, a number of laboratory studies were undertaken to examine the photolysis cross-sections of the CFCs and HCFCs and to determine the temperature-dependent rate constants for the reactions proposed by Molina and Rowland. The laboratory studies continued to substantiate their hypothesis and finally, in 1985, it was reported that measurements of total atmospheric ozone column densities determined that there was significant ozone depletion observed during the spring months over Antarctica (Farman et al., 1985). This observation of stratospheric ozone depletion was called the "ozone hole."

4.4.2 The "Ozone Hole"

An atmospheric circulation pattern, called the circumpolar vortex, is an upper-level low-pressure area that rotates counterclockwise around the North Pole and clockwise around the South Pole. Both vortexes strengthen in the winter and weaken in the summer, because of their dependence on the temperature difference between the equator and the polar regions. Due to the very cold air at the South Pole, the southern polar vortex is stronger and lasts longer than the Arctic vortex. During the long winter night (March through September), the Antarctic area receives no sunlight at all, preventing the formation of oxygen atoms necessary to initiate the Chapman Cycle and produce stratospheric ozone. In addition, the polar vortex isolates the air, preventing it from mixing with mid-latitude stratospheric air containing ozone. During the summer, the levels of ozone in the south polar stratosphere are higher than the levels at the equator due to the Brewer–Dobson circulation. But during the winter months, lower levels of ozone occur naturally due to the lack of sunlight and the isolated air mass.

During the spring months (September through October) the Sun begins to rise, initiating photochemistry, but it is very low in the sky, reaching a maximum of 23.5° only in midsummer (December). The amount of daylight increases from 1.1 hours in September to 12.6 hours in October. But since temperatures are still very low, with average temperatures of −59 °C in September to −52 °C in October, there is still little mixing with mid-latitude stratospheric air during the spring. Normal ozone levels during October at the South Pole range from 300 to 325 DU. But measurements of stratospheric ozone levels during the spring month of October in Antarctica since about 1975 have shown levels even lower than normal. These low ozone levels have been called the "ozone hole," because it appears as a hole in colored representations. But it is important to recognize that this is not an actual hole, as if ozone did not exist in this region, instead it is a reduction in the normal ozone levels in these areas of the stratosphere. It was originally considered to be the area where ozone levels dropped below 220 DU.

In 1985 the British Antarctic Survey was the first to report total column ozone measurements that showed a decline over the Antarctic (Farman et al., 1985; Shanklin, 2017). Monthly averages were obtained continuously from 1956 at the Halley Station in Antarctica for the spring months of September through October. The measurements indicated that there was an apparent decline in the concentration of ozone outside of the normal variability that began around 1975 and continued to decrease. At the same time, the U.S. National Oceanic and Atmospheric Administration (NOAA) measured total column ozone using a Dobson spectrometer at the Amundsen Station, Antarctica (NOAA, n.d.). The October monthly average ozone values obtained from 1956 to 1985 at the Halley Station by the British Antarctic Survey are compared to measurements made at the Amundsen Station in Figure 4.8. These lower ozone values were interpreted as being consistent with the Molina and Rowland hypothesis.

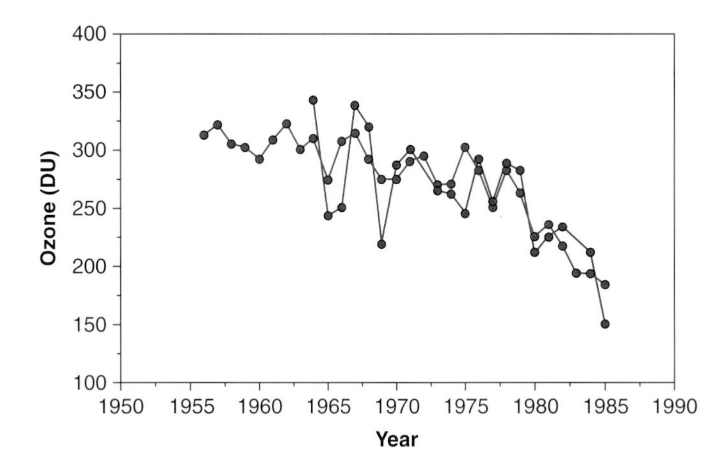

Figure 4.8 Mean monthly total column ozone measured in Dobson units in the month of October at Halley Station, Antarctica by the British Antarctic Survey (red) and at the Amundsen Station, Antarctica by the National Oceanic and Atmospheric Administration (blue).

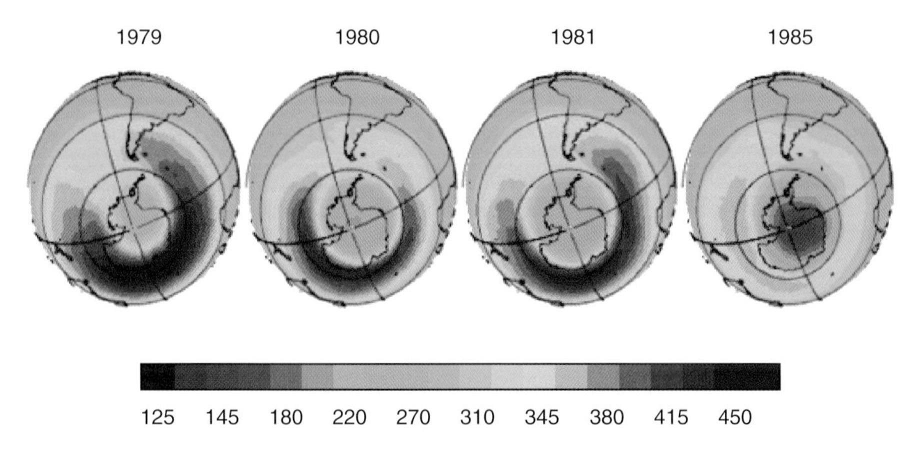

Figure 4.9 Ozone measurements over the South Pole obtained from the Total Ozone Mapping spectrometer (TOMS) on the Nimbus-7 satellite by the U.S. National Aeronautics and Space Administration (NASA). Results are given in Dobson units. Source: NASA, Earth Observatory.

In 1970 the Total Ozone Mapping spectrometer (TOMS) was launched aboard the Nimbus-7 satellite by the U.S. National Aeronautics and Space Administration (NASA). In 1978 the TOMS began to also record low ozone values over Antarctica. But the TOMS data analysis software was originally set to remove data below 180 DU because it was thought to be out of the range of possibility and thus due to instrumental error. Upon re-examination of the raw data, NASA scientists concluded that the originally excluded data were real and were due to a reduction in ozone over Antarctica in the spring months. Figure 4.9 shows a color representation of the data obtained from the TOMS on the Nimbus satellite. In 1979, the lowest ozone levels observed over Antarctica were recorded as 255 DU, in 1980 these levels had dropped to 200 DU. In 1981 the ozone levels were approximately the same at 220 DU, but in 1985 further decrease in the ozone over Antarctica was dramatic, reaching levels of 140 DU, the lowest ever measured. While the British Antarctic Survey scientists were the first to report the ozone decline in the literature, measurements by scientists from the NOAA and NASA in the United States

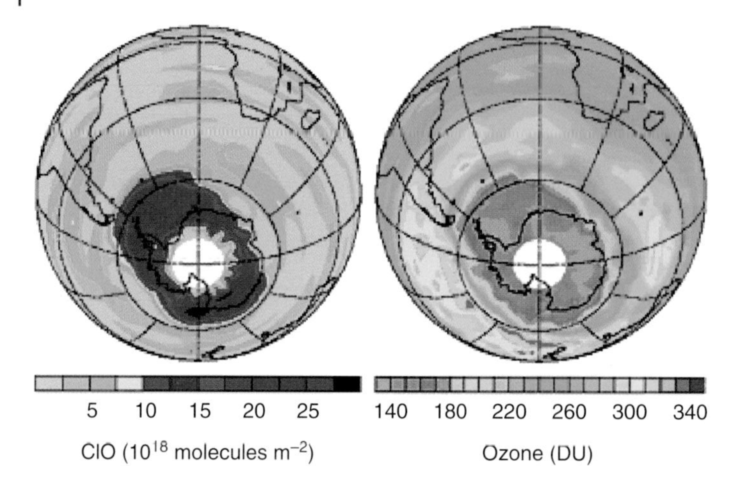

Figure 4.10 A comparison of ClO (right) and O_3 (left) measurements by the NASA satellite on August 30, 1996. Source: NASA, MLS.

confirmed that ozone depletion was occurring in Antarctica, with the largest losses in the springtime.

Although these dramatically lower ozone values over the southern polar regions were interpreted as being consistent with the Molina and Rowland hypothesis, confirmation of the Cl/ClO reaction mechanism was not made until 1990. At this time, instruments were deployed on the ground and in aircraft to examine the stratospheric ClO concentrations, an important intermediate in the ozone depletion reactions of the Cl/ClO Cycle. The Upper Atmosphere Research Satellite (UARS) was launched in 1990, which could measure both ClO and O_3 concentrations simultaneously over global scales. Figure 4.10 shows a comparison of the O_3 and ClO concentrations measured by the UARS instrumentation in August 1996. This provided the first maps showing the full spatial extent of enhanced ClO concentrations and its coincidence with depleted ozone. The white area in Figure 4.10 is the area where no measurements were made. The fact that ClO was found in parts per billion (ppb) concentrations in the stratosphere over Antarctica, and that it was anti-correlated with O_3 concentrations, confirmed that the chain reaction mechanism involving Cl was the cause of the O_3 depletion in this area.

The confirmation of the chlorine free radical reaction mechanism and the recognition of the potential damage caused by the stratospheric ozone depletion from the use of CFCs led to the passing of the Montreal Protocol in 1987. In this agreement, the countries of the world began to phase out the use of CFCs to protect the Earth's ozone layer. It is interesting to note that CFCs are also very strong greenhouse gases, so reducing their release into the atmosphere also had an important impact in the area of climate change. For their work, Drs. Rowland and Molina shared the Nobel Prize in Chemistry in 1995 with Dr. Paul Crutzen, who predicted that a number of trace gases could be mixed into the stratosphere from the troposphere through strong thunderstorm activity and that this mixing could affect stratospheric ozone chemistry.

4.4.3 Altitude Dependence

The reductions in the stratospheric ozone observed in the early 1980s over Antarctica were startling, as they were much larger than expected from the predicted impacts of CFCs on the ozone cycles. They were also much larger in the Antarctic than in the Arctic. This concern led to high-altitude balloon launches in Antarctica, which carried ozone instruments called ozonesondes, to obtain ozone concentration profiles as a function of altitude (Angell

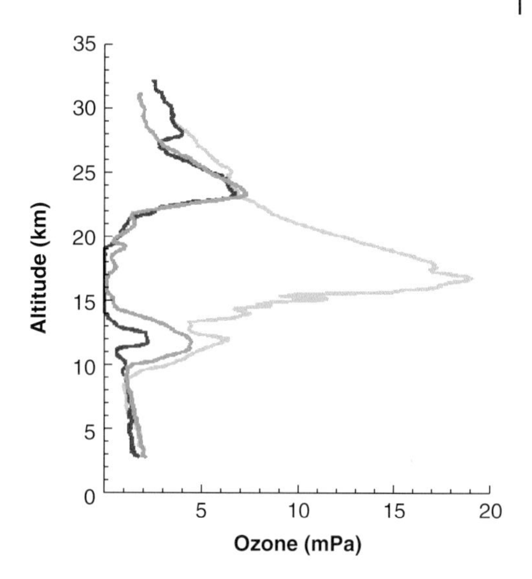

Figure 4.11 Ozone profiles in partial pressure (mPa) as a function of altitude measured by ozonesondes at Amundson Station, South Pole on August 23, 1993 (blue), October 12, 1993 (red), and October 5, 1995 (green). Source: NOAA, National Weather Service, Climate Prediction Center.

et al., 1995). Examples of the ozone profiles obtained during the polar winter night (August 23, 1993) and during the early polar spring (October 12, 1993 and October 5, 1995) are shown in Figure 4.11. The August profile represents the lowest natural ozone concentration, since it was obtained in the winter when there is no solar light to initiate ozone formation. The October profiles show nearly complete destruction of ozone from 15 to 20 km, with a gradual rise to reach the levels seen in August at an altitude of 20–25 km. These measurements showed that the ozone depletion was primarily confined to the lower stratosphere. This observation was of great importance, since the photolysis of the CFCs should occur primarily in the upper stratosphere where the UVC intensity is higher, as shown in Figure 4.5. So, the gas-phase Cl/ClO Cycle should have resulted in the reduction in ozone occurring at higher altitudes. The large destruction of ozone in the lower stratosphere pointed to some other mechanism affecting the Cl/ClO Cycle.

There are three reaction mechanisms that contribute to the higher than expected ozone depletion rates in the lower stratosphere: (1) a gas-phase reaction involving the dimerization of ClO during the polar night, (2) a heterogeneous reaction that results in the formation of Cl_2, and (3) the sequestration of HNO_3 in polar stratospheric clouds (PSCs). In the gas-phase reaction, it was determined that the reactive ClO ozone depletion intermediate could form a dimer during the very cold polar night, sequestering the ClO until polar sunrise in the spring (Molina and Molina, 1987). During the cold polar night, two ClO radicals will react in a third-body reaction to form the chlorine peroxide dimer (ClOOCl). Chlorine peroxide absorbs in the UVB and UVA with a maximum absorbance in the UVB at 254 nm. So, it will act as a nocturnal chlorine reservoir species, which will rapidly photolyze back to Cl and O_2 in the spring daylight hours by a two-step process:

$$\text{Dimer formation: ClO} + \text{ClO} + \text{M} \rightarrow \text{ClOOCl} + \text{M (night)}$$

$$\text{Photolysis: ClOOCl} + h\nu \rightarrow \text{Cl} + \text{ClOO (day)}$$

$$\text{ClOO} + \text{M} \rightarrow \text{Cl} + \text{O}_2 + \text{M}$$

The first Cl is lost by photolysis to form ClOO. The ClOO radical is only slightly stable and quickly decomposes to Cl and O_2 by way of a third-body reaction. This releases two Cl atoms in the polar spring, which initiate the Cl/ClO ozone depletion cycle.

Table 4.3 The different types of polar stratospheric clouds and the chemical composition and properties of the particulates.

Type	Chemical composition	Shape	Properties
Ia	$HNO_3 \bullet 3H_2O$	non-spherical	large
Ib	H_2SO_4, HNO_3, H_2O	spherical	small, super-cooled
Ic	H_2O, HNO_3	non-spherical	metastable
II	H_2O ice	non-spherical	—

Since both the dimerization reaction and the decomposition of ClOO are third-body reactions, they are pressure dependent and so they more readily occur in the lower stratosphere where the pressure is higher. Also, the dimerization reaction is second order in ClO. The rate of the dimer formation proceeds as the square of the ClO concentration:

$$k = [\text{ClO}]^2$$

Since the concentrations of ClO in the Antarctic are high, the ozone depletion mechanism of ClOOCl could potentially account for about 60% of the ozone loss in the spring.

The heterogeneous mechanism responsible for the enhanced ozone depletion in the lower stratosphere involves reactions of Cl reservoir species on the surfaces of frozen particulates in PSCs, which result in the enhancement of the catalytic ozone depletion reactions in the polar spring (Solomon et al., 1986). Polar stratospheric clouds require temperatures of 195 K to form in the stratosphere. The lower stratosphere in the region of the South Pole is typically at a temperature of 185 K during the dark winter months. The North Pole only periodically reaches 195 K, so PSCs are not common over the north polar region. The reason for this temperature difference between the poles is due to Antarctica being at higher altitudes than the Arctic, which is mainly composed of sea ice. The PSCs form primarily at an altitude of 10–20 km, where water vapor can enter from the troposphere. This is the same region where the enhanced loss of stratospheric ozone was observed over Antarctica.

There are four types of PSCs that are classified according to the chemical composition of their particles, given in Table 4.3. Type Ia particles consist of nitric acid trihydrate (NAT) particles that are not spherical in shape. Type Ib particles consist of a super-cooled mixture of sulfuric acid (H_2SO_4), nitric acid (HNO_3), and water and are spherical in shape. Type Ic particles are composed of a metastable mixture of water and HNO_3 and are very small and nonspherical. Type II particles consist of water ice particles only. Type I (Ia, Ib, and Ic) PSCs form at temperatures around 195 K, while type II PSCs require temperatures of 188 K. Since type II PSCs form at colder temperatures, they are normally not seen in the Arctic and are more common in the colder Antarctica. All types of PSCs are formed in the lower portion of the stratosphere, in the region of 10–20 km.

In the south polar winter, the frozen particles in the PSCs provide surfaces for heterogeneous chemical reactions and the south polar vortex acts to contain the chemical species in the region where the PSCs are prevalent. The very slow gas-phase reactions of the reservoir species to regenerate reactive chlorine can go very quickly as a heterogeneous reaction. The Cl reservoir species HCl and $ClONO_2$ or HOCl can be trapped on the particles during the winter months, where surface reactions convert them into Cl_2 in the reaction:

$$\text{HCl(s)} + \text{ClONO}_2\text{(s)} \rightarrow \text{Cl}_2\text{(g)} + \text{HNO}_3\text{(s)}$$

$$\text{HCl(s)} + \text{HOCl} \rightarrow \text{Cl}_2\text{(g)} + \text{H}_2\text{O(s)}$$

The Cl_2 can then rapidly photolyze with the spring sunrise to produce two Cl free radicals initiating the ozone depletion cycle.

In addition, the type I PSCs act to sequester any available HNO_3, a reservoir species for NO_2, within their frozen particles. If left in the gas phase, HNO_3 will slowly photolyze back to OH and NO_2, as in the NO_x Cycle, during polar spring. The NO_2 can then react with ClO in the Cl/ClO termination reaction, interrupting the Cl/ClO ozone depletion cycle. Since the HNO_3 is taken up in the type I PSCs, it cannot regenerate NO_2 and so cannot participate in the termination reaction. Therefore, one Cl atom can result in more ozone depletion reactions than if the HNO_3 were available in the gas phase.

Submicron aerosols can also act to enhance ozone depletion in the lower stratosphere in the same manner as the PSCs. There has always been a small amount of sulfate aerosol in the lower stratosphere, called the Junge layer. This aerosol layer was named after Christian Junge, who discovered the fine aerosol layer in 1960 while studying the transport of fallout from above-ground nuclear tests. This layer of aerosols is located just above the tropopause, extending to an altitude of 30 km. The particles in the layer are composed of sulfuric acid (H_2SO_4), formed from sulfur dioxide (SO_2) and water. Sulfur dioxide is too reactive in the troposphere to reach the stratosphere under normal conditions, since it is rapidly oxidized to H_2SO_4, which is highly water soluble and quickly removed in precipitation. However, SO_2 can be injected directly into the stratosphere by explosive volcanic eruptions (Portmann et al., 1996). It can also be produced directly in the stratosphere by photolysis of carbonyl sulfide (COS) followed by reaction with O atoms. Carbonyl sulfide is the most abundant sulfur compound naturally present in the troposphere. It is emitted from oceans, volcanoes, and deep-sea vents. Because of its long tropospheric lifetime, most of the COS in the troposphere reaches the stratosphere. Both SO_2 and COS are oxidized to SO_4^{2-} through heterogeneous reactions on the surface of the aerosols present in the Junge layer. The surfaces of the H_2SO_4/H_2O aerosols can also sequester reservoir species and promote heterogeneous reactions that generate Cl_2 in the same way as the PSCs. The Cl_2 is then photolyzed with the return of daylight, resulting in enhanced ozone depletion in the lower stratosphere during the polar spring.

4.4.4 Ozone-Depleting Substances

The most well known of the ozone-depleting substances are the CFCs. However, stratospheric ozone depletion can also be caused by trace gases containing nitrogen, hydrogen, bromine, and chlorine that are long-lived enough in the troposphere to become mixed into the stratosphere. The contributions from the major gas species in the Antarctic stratosphere are shown in Figure 4.12. These include CFC-12, CFC-11, CFC-113, methyl chloride (CH_3Cl), methyl bromide (CH_3Br), carbon tetrachloride (CCl_4), trichloroethane or methylchloroform (CH_3CCl_3), the HCFCs, and the halons. The HCFCs include HCFC-22, HCFC-141, and HCFC-142. The halons include halon-1211 (bromochlorodifluoromethane, CF_2ClBr), halon-1301 (bromotrifluoromethane, $CBrF_3$), and halon-2402 (dibromotetrafluoroethane, $C_2Br_2F_4$).

As discussed in Section 4.4.1, the CFCs were used as refrigerants, aerosol propellants, foam blowing agents and solvents until their ban by the Montreal Protocol in 1987. The HCFCs are now being used as replacements for the CFCs. Carbon tetrachloride is most often used as a dry-cleaning agent, although its use is also being phased out in favor of liquid CO_2. Methyl chloroform is used as a solvent and the halons are used in fire extinguishers. Methyl chloride and methyl bromide are unique among the ozone-depleting gases because they have substantial natural sources. Both CH_3Cl and CH_3Br are produced in large quantities by plants in salt marshes. It is estimated that this source may produce roughly 10% of the total fluxes of atmospheric CH_3Br and CH_3Cl (Rhew et al., 2000). Methyl chloride was once used as a refrigerant, until it was replaced by the CFCs. It is produced naturally in the oceans by phytoplankton. It

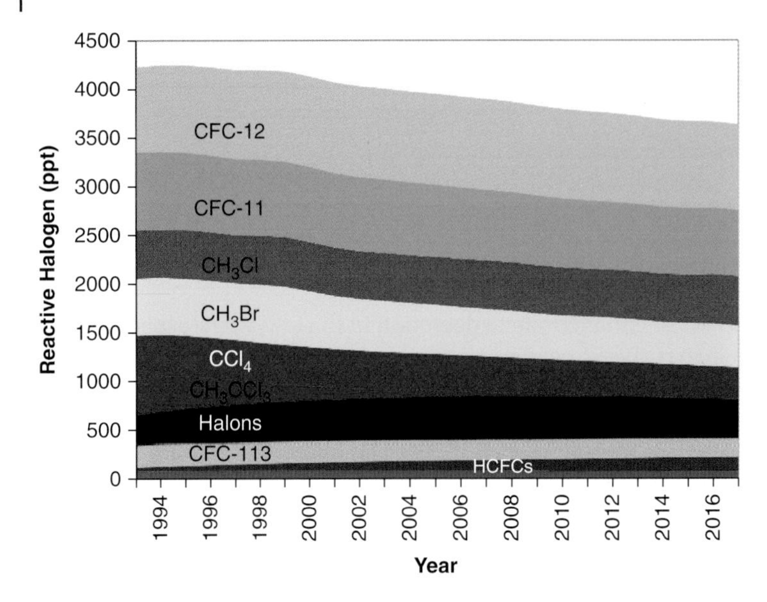

Figure 4.12 The contribution of the long-lived chlorine and bromine-containing species to ozone destruction in the Antarctic stratosphere from 1993 to 2017. The HCFCs include HCFC-22, HCFC-141, and HCFC-142. The halons include halon-1211, halon-1301, and halon-2402. Source: Adapted from NOAA, Earth Systems Laboratory.

also occurs in emissions from biomass burning. Methyl bromide is used as a fumigant and soil sterilizer in agriculture. However, its use is being restricted by the U.S. Environmental Protection Agency due to its ozone depletion potential. It is produced naturally by marine organisms, which are estimated to produce 56 000 tons annually (Gribble, 1999).

The bromine-containing gases such as methylbromide and the halons, which have similar structures to the CFCs, are photolyzed in the stratosphere to produce bromine atoms (Br), which have a catalytic ozone depletion cycle similar to the Cl/ClO Cycle.

The Br/BrO Cycle

1. Initiation: $CH_3Br + h\nu \rightarrow Br + CH_3$
2. Ozone depletion: $Br + O_3 \rightarrow BrO + O_2$
   $BrO + O \rightarrow Br + O_2$
3. Termination: $BrO + NO_2 \rightarrow BrONO_2$
4. Net ozone depletion: $O_3 + O \rightarrow 2O_2$

But Br is almost 60 times more effective than Cl in destroying ozone. This is because $BrONO_2$ formed in the termination reaction is much less stable than $ClONO_2$ and can easily photolyze to form either BrO or Br:

$$BrONO_2 + h\nu \rightarrow BrO + NO_2$$
$$BrONO_2 + h\nu \rightarrow Br + NO_3$$

In addition, the formation of HBr in the analogous termination reaction:

$$Br + CH_4 \rightarrow HBr + CH_3$$

is thermodynamically unfavorable since it is an endothermic reaction, unlike the formation of HCl. The BrO chain carrier species can also react with ClO to give both Br and Cl atoms. This results in the enhanced ozone depletion cycle:

$$BrO + ClO \rightarrow Br + Cl + O_2$$
$$Cl + O_3 \rightarrow ClO + O_2$$
$$Br + O_3 \rightarrow BrO + O_2$$
$$\text{Net ozone depletion}: 2O_3 \rightarrow 3O_2$$

This coupled reaction cycle is much more effective in destroying ozone than the Cl/ClO Cycle alone, since it regenerates both Cl and Br reactive species. The Br catalytic reactions are estimated to contribute 30% of the ozone destruction in the polar regions.

4.5 Summary

The key to stratospheric chemistry is the catalytic formation and destruction of ozone. The important reactions involved in this formation–destruction cycle are outlined in Figure 4.13. The ozone formation reactions are outlined in the Chapman Cycle (Figure 4.13, black). The natural destruction cycles are the HO_x and NO_x Cycles (Figure 4.13, green). Other ozone destruction cycles are caused by long-lived manmade gases transporting into the stratosphere from the troposphere. The most important of these involve the halogenated compounds (Figure 4.13, red). These chlorine and bromine compounds undergo photolysis in the stratosphere to give chlorine and bromine atoms, which are the main ozone-destroying species. The important reservoir species (Figure 4.13, blue) tie up reactive Cl and Br and thus remove them from their role in the catalytic destruction of ozone. But, under the right conditions, they can release the free Br and Cl, regenerating the ozone depletion cycle. This can be accomplished by slow photolysis, oxidation, or heterogeneous reaction on PSCs (Figure 4.13, blue arrow). The naturally occurring Junge layer of fine aerosol particles, which consist of sulfuric acid and

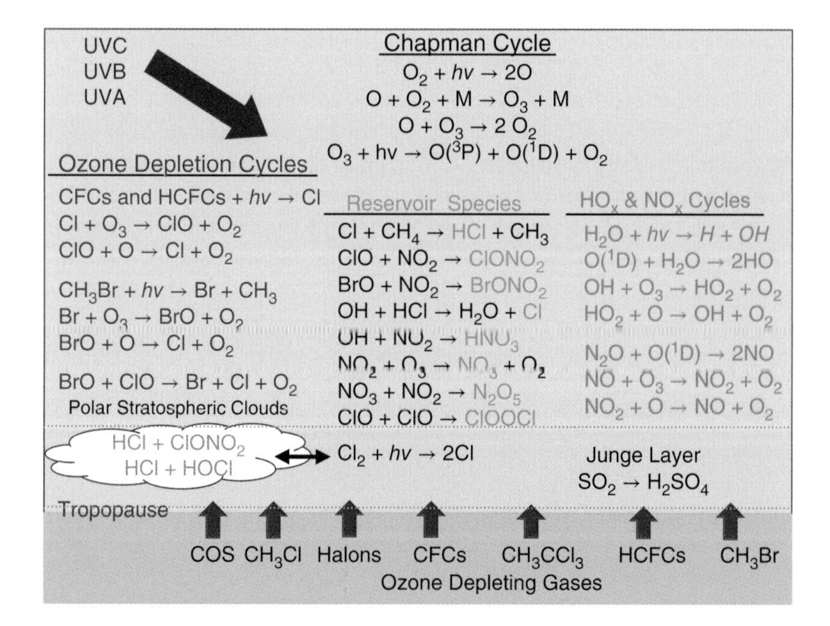

Figure 4.13 Summary of the important reactions involved in both ozone formation (Chapman Cycle) and ozone depletion in the stratosphere.

water, is formed by heterogeneous oxidation of SO_2 (Figure 4.13, gray). The ozone-depleting gases are shown coming from the troposphere and moving into the stratosphere (Figure 4.13, red arrows), where they can undergo the reactions listed, including interactions of reservoir species with the PSCs and the Junge layer of stratospheric aerosols.

These many reactions lead to the enhanced depletion of ozone in the lower to middle stratosphere, primarily over the polar regions. This occurs to a larger extent over Antarctica due to the strong polar vortex and colder temperatures. To complicate the reaction systems, the different reaction cycles are also linked to each other. For example, when the NO_x concentration increases, the ozone depletion from the Cl/ClO Cycle decreases due to the increased formation of the reservoir species $ClONO_2$. Also, when the HO_x concentration increases, the ozone depletion from the NO_x Cycle decreases due to the increased formation of the reservoir species HNO_3. At the same time the increase in HO_x increases the ozone depletion from the Cl/ClO Cycle by regenerating Cl from the reservoir species HCl. In addition, both the Cl/ClO and Br/BrO Cycles strengthen each other due to the coupled reaction that releases both of the ozone-depleting species, Br and Cl.

The confirmation that chlorine and bromine-containing compounds were causing the depletion of stratospheric ozone led the countries of the world to agree to curtail the production and use of these materials worldwide. On September 16, 1987 a treaty was signed, called the Montreal Protocol on Substances that Deplete the Ozone Layer (or simply the Montreal Protocol), which went into effect two years later. All of the countries of the world signed this legally binding international treaty, which is considered to be one of the most successful international environmental treaties ever. It is an example of how the global community can reach a positive agreement to ensure that the Earth's environment is protected. Since signing this agreement, the levels of CFCs have decreased and the stratospheric ozone levels have begun to recover. As noted earlier, Mario Molina, F. Sherwood Rowland, and Paul Crutzen were awarded the Nobel Prize in Chemistry in 1995 for their pioneering work on the stratospheric chemistry of N_2O and the CFCs that could cause ozone depletion.

The measurements in Figure 4.12 show that the atmospheric concentrations of nearly all the ozone-depleting gases restricted by the Montreal Protocol were decreasing in the atmosphere by 2017. The most rapid decline is seen in CH_3Cl, CH_3Br, CH_3CCl_3, and CCl_4 due to their rapid phase-out coupled with their shorter atmospheric lifetimes compared to the CFCs. The halogenated hydrocarbons CH_3Cl, CH_3Br, and CH_3CCl_3 show the largest loss because of their higher reactivity with OH radicals in the troposphere. This is due to their each having three hydrogens that can react with OH by abstraction. The atmospheric concentration of CH_3Br has declined most of all the ozone-depleting substances since 1998, despite its large natural source.

The decline of the two major CFCs (CFC-11 and CFC-12) is less dramatic, because of continuing emissions and their extremely long atmospheric lifetimes (50–100 years). The Montreal Protocol did not address existing sources of the ozone-depleting substances such as stockpiles or discarded products and equipment. These continuing emissions of CFC-11, CFC-12, and CFC-113 are likely due to slow leakage from discarded refrigeration products. The increase in the halons, shown in Figure 4.12, is also likely due to substantial banks of fire-extinguishing equipment that gradually release halon to the atmosphere. So, the atmospheric abundance of halons is expected to remain high well into the twenty-first century, because of its long lifetime (65 years) and continued release. The increase in the HCFCs is due to their use as replacements for CFCs. Unlike the CFCs, the HCFCs can react with OH in the troposphere by abstraction of their hydrogen, resulting in shorter lifetimes. Although concentrations of HCFCs continue to increase in the background atmosphere and a phase-out of production is not scheduled until 2030, their current rates of increase have been reduced and they contribute relatively little ($< 5\%$) to the atmospheric levels of the reactive halogens. The total amount of

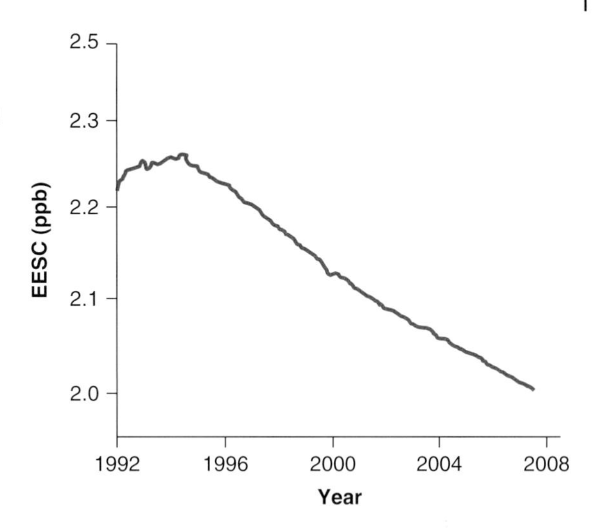

Figure 4.14 The equivalent effective stratospheric chlorine (EESC) in parts per billion (ppb) calculated from the tropospheric concentrations of ozone-depleting halogenated compounds, tropospheric lifetimes, and ozone-depleting potentials. Source: NOAA, Wikimedia Commons.

HCFCs in the troposphere is still much lower than that of CFC-12 and CFC-22 by factors of 10 to 20 or more.

The equivalent effective stratospheric chlorine (EESC) provides an estimate of the total effective amount of ozone-depleting halogens in the stratosphere. The EESC is calculated from the amounts of ozone-depleting compounds in the troposphere, their ozone depletion potentials (ODPs), and their transport times from the troposphere to the stratosphere. The ODP is the relative amount of degradation that a chemical species can cause to the stratospheric ozone layer. Since CFC-11 has the maximum ozone depletion potential of all the CFCs due to the presence of three chlorine atoms in the molecule, its ODP is fixed at 1.0. So, CFC-22, with only one chlorine atom, has an ODP of 0.5. Figure 4.14 shows EESC from 1992 to 2008. The EESC peaked in the mid-1990s and is now dropping due to the curtailing of the use of chlorine and bromine-containing gases in industrial, agricultural, and residential use. The decrease in the effective chlorine content in the stratosphere is about 12% over a 10-year period and appears to be fairly constant. If this pattern continues, the stratospheric ozone layer may recover to 1980s levels in the mid-latitudes by about 2050 and in the Antarctic region by about 2070.

There are a couple of issues that may affect the predicted recovery of the stratospheric ozone layer. First, there is concern that the ozone layer and stratospheric chemistry could be altered by unforeseen natural volcanic events, which inject significant amounts of water and sulfate aerosol into the stratosphere. The injection of SO_2 from volcanic events acts to enhance the levels of the sulfuric acid aerosol layer (Junge layer), which can remove the reservoir species HNO_3 and $ClONO_2$, enhancing ozone depletion in the lower stratosphere. The effects of these events last about 3–4 years, the removal time for the HNO_3 and H_2SO_4 by stratospheric transport to the polar regions followed by washout. Another issue that is becoming of more concern is the increasing levels of N_2O and CH_4 in the troposphere. Both N_2O and CH_4 are emitted by anaerobic bacteria in wetlands and soils. Nitrous oxide is converted to NO in the stratosphere by reaction with $O(^1D)$ (Figure 4.13, green), which then reacts with O_3 in the NO_x Cycle. Methane is the primary source of water vapor in the stratosphere, by the reaction:

$$CH_4 + OH \rightarrow CH_3 + H_2O$$

which is necessary to the HO_x Cycle as well as for the production of sulfate aerosols and PSCs.

Although both N_2O and CH_4 are naturally produced, changing agricultural practices have led to rising levels of these gases in the troposphere. The increase of N_2O and CH_4 in the troposphere in recent years has been primarily linked to fertilization practices in rice cultivation. Organic fertilizer use in the wetland production of rice leads to anaerobic bacteria, which increases the production of CH_4. The use of inorganic nitrogen fertilizer in wetland rice production has led to the enhanced emission of N_2O into the troposphere. In addition, any increases in the emission rates of other naturally produced ozone-depleting gases, such as CH_3Br and CH_3Cl, due to changes in their sources, will also add to ozone depletion. The increased levels of these gases in the troposphere is an indication of the importance of how changes in the biosphere could affect the emission sources and may have feedback impacts on the stratospheric chemistry. This is one reason why the continued monitoring of these trace gases with time is very important.

The final issue to be addressed is the effect of climate change on the stratosphere. It is generally agreed that the increasing levels of greenhouse gases will lead to warming of the troposphere. The heat radiated from the surface of the Earth is trapped in the troposphere by the greenhouse gases. The excess heat trapped in the troposphere cannot enter the stratosphere, resulting in cooling. Although ozone is a greenhouse gas and does account for some heating of the stratosphere, it is small compared to the chemical heat released during its formation. So, the reduction in stratospheric ozone will also add to the cooling. If the stratosphere cools, then stratospheric clouds may form outside of the polar regions, leading to more heterogeneous ozone depletion reactions over a larger scale. Also, climate change may lead to more and stronger storm events due to more energy being trapped in the troposphere. This will enhance the mixing of ozone-depleting gases from the troposphere into the stratosphere. This rate of mixing controls the stratospheric levels of key species such as N_2O, CH_4, CFCs, HCFCs, COS, as well as CH_3Cl and CH_3Br. These potential climate connections show that the stratospheric chemistry and the stratospheric ozone layer that protects the planet from UV radiation are strongly tied to the dynamics of the troposphere.

References

Angell, JK, Gelman, ME, Hoffman, D, Long, CS, Miller, AJ, Nagatani, RM, Oltmans, S & Planet, WG 1995 *Southern hemisphere winter summary*. National Weather Service, Climate Prediction Center, National Atmospheric and Oceanic Administration. Available online at http://www.cpc.ncep.noaa.gov/products/stratosphere/winter_bulletins/sh_95.

Bates, D.R. and Nicolet, M. (1950). The photochemistry of water vapor. *J. Geophys. Res.* 55: 301.

Chapman, S. (1930). A theory of upper atmospheric ozone. *Mem. Royal Meteorol. Soc.* 3: 103–125.

Crutzen, P.J. (1970). The influence of nitrogen oxides on the atmospheric content. *Quart. J. Royal Meteorol. Soc.* 96: 320–325.

Farman, J.C., Gardner, B.G., and Shanklin, J.D. (1985). Large losses of total ozone in Antarctica reveal seasonal ClO_x/NO_x interaction. *Nature* 315: 207–210.

Gribble, G.W. (1999). The diversity of naturally occurring organobromine compounds. *Chem. Soc. Rev.* 28: 335–336.

Molina, L.T. and Molina, M.H. (1987). Production of chlorine oxide (Cl_2O_2) from the self-reaction of the chlorine oxide (ClO) radical. *J. Phys. Chem.* 91: 433–436.

Molina, M.J. and Rowland, F.S. (1974). Stratospheric sink for chlorofluorocarbons: Chlorine atom-catalysed destruction of ozone. *Nature* 249: 810–812.

NOAA n.d. *Archived data*. National Oceanic and Atmospheric Administration, Boulder, CO. Available online at ftp://aftp.cmdl.noaa.gov/data/ozwv/Dobson/WinDobson/South_Pole_AMS/dobson_to111.txt (accessed February 28, 2018).

Portmann, R.W., Solomon, S., Garcia, R.R. et al. (1996). Role of aerosol variations in anthropogenic ozone depletion in polar regions. *J. Geophys. Res.* 101: 22,991–23,006.

Rhew, R.C., Miller, B.R., and Weiss, R.F. (2000). Natural methylbromide and methyl chloride emissions from coastal salt marshes. *Nature* 403: 492–495.

Shanklin, J. (2017). *Provisional monthly mean ozone values for Halley between 1956 and 1985*. UK: Polar Data Centre, Natural Environment Research Council https://doi.org/10.5285/d16ac510-3b63-41c0-a6b6-a85bc2ca4d7e.

Sheppard, P.A. (1963). Atmospheric tracers and the study of the general circulation of the atmosphere. *Rep. Prog. Phys.* 26: 213–267.

Solomon, S., Garcia, R.R., Rowland, F.S., and Wuebbles, D.J. (1986). On the depletion of Antarctic ozone. *Nature* 321: 755–758.

Further Reading

Finlayson-Pitts, B.J. and Pitts, J.N. Jr., (2000). *Chemistry of the Upper and Lower Atmosphere: Theory, Experiments, and Applications*. San Diego, CA: Academic Press.

Polvani, LM, Sobel, AH & Waugh, DW 2010 *The Stratosphere: Dynamics, Transport, and Chemistry*. Geophysical Monograph Series, vol. 190. Wiley, New York.

Study Problems

4.1 The stratosphere lies between what two layers of the atmosphere?

4.2 How much of the mass of the atmosphere is in the stratosphere?

4.3 (a) What is the approximate altitude of the top of the stratosphere?
(b) What is the approximate altitude of the bottom of the stratosphere near the equator?
(c) At middle latitudes?
(d) Over the polar regions?

4.4 Why is vertical mixing slower in the stratosphere than in the troposphere?

4.5 What is responsible for the increase in temperature with altitude in the stratosphere?

4.6 (a) What two gases are responsible for most of the absorption of the UVC and UVB radiation in the stratosphere?
(b) Which of the two gases absorbs primarily in the UVC?
(c) Which one absorbs primarily in the UVB?

4.7 The first ozone spectrometer measured the solar light intensities at 305 nm and 325 nm.
(a) How was the total column ozone determined from these two wavelengths?
(b) What were the units used and how are they defined?

4.8 What is the source of the oxygen atoms in the stratosphere?

4.9 What is the name given to the stratospheric ozone formation cycle first proposed in 1930?

4.10 (a) How do the predicted stratospheric ozone concentrations obtained from the ozone formation cycle compare to directly measured values?
(b) What is the reason for this?

4.11 What two chemical cycles are natural ozone loss mechanisms?

4.12 What is the term symbol of the electronically excited oxygen atom species that initiates both natural ozone loss cycles?

4.13 (a) What is the term symbol of the ground state of the O atom?
(b) What is the term symbol of the first excited state of the O atom?

4.14 (a) What chemical species reacts with ozone in the HO_x Cycle?
(b) What chemical reaction regenerates this ozone-reacting species?

4.15 (a) What chemical species reacts with ozone in the NO_x Cycle?
(b) What chemical reaction regenerates this ozone-reacting species?

4.16 What reaction between the HO_x Cycle and the NO_x Cycle produces an acidic and highly water-soluble reservoir species?

4.17 What is the net ozone depletion reaction for both the HO_x and NO_x Cycles?

4.18 In the Molina and Rowland hypothesis, what was the initiation reaction of their suggested ozone depletion cycle?

4.19 What are the products of the reaction of Cl with ozone?

4.20 (a) What is the reservoir species produced from interaction of the NO_x Cycle and the Cl/ClO Cycle in the stratosphere?
(b) What is the chemical reaction that forms this species?

4.21 (a) Why are ozone concentrations lower at the equator than the middle latitudes and the polar regions?
(b) What process is responsible for this difference?

4.22 (a) Ground-based observations of stratospheric ozone depletion were first observed over what continent?
(b) What were the two main reasons that the major depletions occurred at this location?

4.23 (a) Why are polar stratospheric clouds seen commonly over the South Pole and not over the North Pole?
(b) What two mechanisms are responsible for the enhanced ozone depletion due to polar stratospheric clouds?

4.24 What were three major uses of CFCs that led to their build-up in the troposphere?

4.25 Why do HCFCs have a lower impact on the stratosphere than CFCs?

4.26 What is the name of the global treaty that agreed to reduce and ultimately cease the use of CFCs?

4.27 (a) What is the name of the layer of sulfate aerosols that occurs naturally in the lower stratosphere?
(b) What are the two major sources of the sulfate in this layer?

4.28 What is the chemical formula for: (a) CFC-11? (b) CFC-12?

4.29 What would be the industrial name for:
(a) $C_2F_3Cl_3$?
(b) $C_2F_4Cl_2$?

4.30 (a) What is the ozone-depleting species that results from the photolysis of a halon?
(b) What is the product of the reaction of this species with ozone?

4.31 Why is the reaction of BrO with ClO especially effective in amplifying ozone depletion?

4.32 Why is the reaction $ClO + ClO + M \rightarrow ClOOCl + M$ only important at night?

4.33 What are the products of the reaction of Cl + water?

4.34 (a) Why have methane and nitrous oxide levels been increasing in the troposphere, leading to their increased impacts in the stratosphere?
(b) How does this impact the stratosphere?

4.35 (a) What is the equivalent effective stratospheric chlorine (EESC) content?
(b) What three parameters is it calculated from?

4.36 What has been the major reason for the EESC decline since 1992?

4.37 (a) What is the expected change in temperature in the stratosphere relative to the troposphere if greenhouse gases increase?
(b) How will this affect the formation of polar stratospheric clouds?

4.38 (a) Why is vertical transport of tropospheric water vapor not an important source of water vapor in the stratosphere?
(b) What is the major source of stratospheric water vapor?

4.39 (a) How many types of polar stratospheric clouds have been identified?
(b) Which types are not found in the Arctic?

4.40 (a) What chemical species does all type I polar stratospheric clouds contain?
(b) What species does type II polar stratospheric clouds contain?

4.41 (a) What organic gas is produced primarily from anaerobic bacteria and is the primary source of water vapor in the stratosphere?
(b) What inorganic gas is produced primarily from anaerobic bacteria and is the major source of NO in the stratosphere?

5

Chemistry of the Troposphere

The chemistry of the troposphere is essentially oxidative chemistry. The oxidation processes in the troposphere are important because the troposphere contains most of the mass of the atmosphere and the trace gases are generally emitted either naturally or anthropogenically from the surface of the Earth. Many of these important trace gases are removed from the atmosphere by oxidation in the troposphere before they can reach the stratosphere. Some gases, such as the CFCs, reach the stratosphere because they are not easily oxidized in the troposphere. The chemically important oxidants in the troposphere are OH, NO_3, O_3, and H_2O_2. The OH is by far the most effective oxidant in the day-time troposphere but since it is formed by the photolysis of O_3, it is not important at night. The nitrate radical is the most important oxidant at night. It is not important during the day because it is easily photolyzed. Hydrogen peroxide is important to oxidation processes in cloud water.

Most of the ozone in the atmosphere (90%) resides in the stratosphere and, while it is occasionally mixed from the stratosphere into the upper troposphere, the ozone concentrations in the troposphere are typically much lower than the average 600 ppb found in the stratosphere. So, background ozone concentrations in the troposphere are about 10–20 times lower than those in the stratosphere. While the source of ozone in the stratosphere is the reaction of O atoms with O_2, the major source of ozone in the troposphere is the photochemical reactions of NO_x, CO, and volatile organic carbon compounds (VOCs). Also, while ozone in the stratosphere can be important to protect the biosphere from harmful radiation, in the troposphere ozone is a strong oxidant and is a concern for human health effects: reduction of lung function, increase in asthma, and possible lung damage after prolonged exposure. It can also cause damage to trees and other plants, reducing forest growth and crop yields.

5.1 Structure and Composition of the Troposphere

The troposphere was named in 1914 by the French meteorologist Leon Phillippe Teisserenc de Bort. The name means literally "turning sphere," emphasizing the importance of turbulent mixing to the dynamics and properties of the troposphere. The troposphere is the lowest and densest layer of the atmosphere. It contains approximately 75% total mass of the atmosphere and 99% of water vapor and aerosols. This means that it also contains almost all of the clouds and all of the precipitation in the atmosphere. So, it is responsible for the cycling of water in the hydrosphere.

The basic structure of the troposphere is shown in Figure 5.1. The height of the troposphere is determined by the height of the tropopause, which varies with latitude from about 16 km in the tropics to 8 km in the polar regions. The lowest part of the troposphere is the planetary boundary layer (PBL), also called the atmospheric boundary layer. The PBL is the layer of air

Chemistry of Environmental Systems: Fundamental Principles and Analytical Methods, First Edition.
Jeffrey S. Gaffney and Nancy A. Marley.

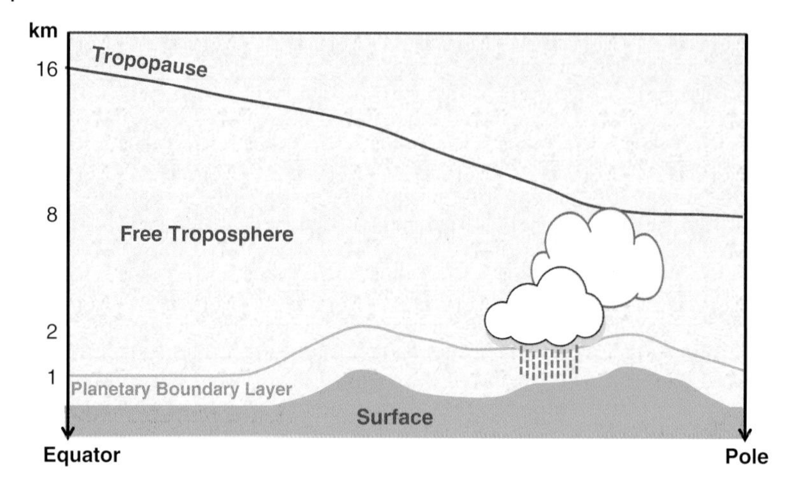

Figure 5.1 The vertical structure of the troposphere. The lowest level of the troposphere is the planetary boundary layer (PBL), where the properties and dynamics of the air are directly influenced by the Earth's surface.

whose behavior is directly influenced by its contact with the planetary surface. The depth of the PBL depends on the surface temperature and the terrain. It can be as deep as 4 km over desert areas where thermal mixing is strong, and less than 1 km over the oceans. The depth of the PBL is also not constant. Over most land areas during the day, the PBL normally extends to an altitude of 1–2 km. However, at night and in the colder season, the PBL tends to be shallower. This is due to lack of sunlight heating the ground, resulting in a reduction in air rising from the surface. The PBL expands during the day and in warm seasons due to increased surface heating, resulting in stronger convection.

The behavior and dynamics of the PBL are characterized by strong vertical mixing and rapid fluctuations in wind speed, wind direction, temperature, and relative humidity. These fluctuations are caused by rapid changes in solar heating and cooling of the surface, as well as variations in terrain. The winds in the PBL are directly influenced by its contact with the Earth's surface. The wind speeds are weaker close to the surface due to surface friction and increase with increasing height above ground level. The surface friction also causes winds in the PBL to blow toward low-pressure areas.

Above the PBL is the "free troposphere," where the wind is driven by pressure gradients and by the Coriolis force, discussed in Section 2.3. Turbulence in the free troposphere is intermittent and typically caused by the formation of convective clouds. The division between the PBL and the free troposphere is often marked with a temperature inversion or a difference in air masses, with different wind speeds, moisture content, and/or wind direction. The occurrence of temperature inversions traps air within the PBL, inhibiting convection between the PBL and the free troposphere. There can also be orographically induced temperature inversions within the PBL, as shown in Figure 2.11, which can trap air in localized areas.

The temperature of the free troposphere decreases with height at about 6.5 K km^{-1}, with temperatures at the top of the troposphere typically close to 220 K. This is because the surface of the Earth absorbs most of the energy from the Sun. The heated ground then radiates heat energy to warm the PBL. The temperature variation within the PBL depends on the type of air mass just above the PBL, the surface albedo, surface thermal characteristics, amount of solar radiation at the surface, and variation of solar radiation with height. The variation in temperature in the free troposphere is due to decreasing pressure with altitude. Since the temperature of a gas is directly proportional to the pressure (Gay-Lussac's Law), the air at higher altitudes (lower pressure) is cooler than the air at lower altitudes (higher pressure).

The pressure of the atmosphere is at a maximum at sea level and decreases with altitude. This is because the pressure is equal to the weight of the air above any point. The variations in pressure within the troposphere are due to variations in air density, which is dependent on water vapor content. Water vapor is released into the PBL by evaporation from the oceans, lakes, and other parts of the hydrosphere, as well as evapotranspiration from plants. This water vapor is a major contributor to the dynamics of the troposphere. Since water is lighter than the major atmospheric gases, changes in water vapor concentrations produce changes in density of the air, which produce changes in pressure. Air masses with higher levels of water vapor are less dense than dry air masses. The water vapor that is released from the surface rises in the troposphere until it reaches areas of lower temperature, where it condenses onto nucleation particles producing clouds. Clouds result in precipitation in the form of rain, snow, and hail. These precipitation events are usually accompanied with a low atmospheric pressure, since air masses that contain more water vapor are less dense. The changes in pressure due to changes in water vapor content are the major driving factors for large-scale circulation patterns that generally move the tropospheric air from the equator to the North and South Poles, as described in Section 2.3. On smaller scales, the warm, wet, less dense air masses rise while the dry, colder, denser air masses sink. The differences in pressure caused by differences in water vapor content cause winds that lead to regional weather patterns in both the PBL and the free troposphere.

Dry air in the troposphere is made up of four major gases: N_2 (78%), O_2 (21%), Ar (0.9%), and CO_2 (0.04%), along with trace gases (see Table 2.2). The concentrations of these major gases are uniform throughout the troposphere. However, the amount of water vapor throughout the troposphere is variable. Wet air contains about 1% water vapor at sea level and 0.4% over the entire troposphere. The trace gases in the troposphere include chemical species that have both short and long lifetimes and can come from both natural and anthropogenic sources. There are natural background levels of most all of the trace gases that we normally consider to be air pollutants. Similar to the naturally occurring CH_3Cl and CH_3Br that release a natural background of Cl and Br into the stratosphere, there are natural sources of SO_2, NO, and the VOCs in the troposphere. Our air is considered to be polluted when these natural background concentrations are exceeded by a significant amount from anthropogenic emissions, to the extent that they can produce harmful effects to the biosphere.

5.2 History of Smog

As discussed briefly in Chapter 1, air pollution has been known to be a serious problem since the twelfth century, particularly in urban regions where large populations increase energy use as well as industrial and agricultural activities. These increases in energy use and industrial activities led to the release of large amounts of pollutant gases and aerosols into the local PBL. At first the air pollution was tied to the combustion of wood and coal used for domestic heating and cooking. The combustion of coal released the sulfur contained in the fuel as SO_2. This SO_2 was accompanied by the release of soot from wood burning and small black particles, called fly ash, from coal combustion. The SO_2 was rapidly oxidized to H_2SO_4, which is very hygroscopic and forms an acidic aerosol quickly in humid air, resulting in a type of irritating fog. The combination of soot, fly ash, and wet acidic aerosols was considered to be a mixture of smoke and fog, which was called "smog."

The most prevalent and well-documented early episodes of smog were seen in London. The emissions from high-sulfur coal used as a fuel in fireplaces and industries in the highly populated city was made worse by the local weather, which created thermal inversions in the early mornings. The exposure to fly ash particulates from the uncontrolled burning of coal, combined with irritation from the SO_2 and H_2SO_4 aerosols, was known to be a health hazard as early as the seventeenth century. In 1661, just after the Restoration, King Charles II of England assigned to John Evelyn the task of investigating the London smog. Evelyn describes the air pollution in a pamphlet published with his findings, entitled *Fumifugium* (Evelyn, 1661, p. 6):

> For when in all other places the Aer is most Serene and Pure, it is here Ecclipsed with such a Cloud of Sulphure, as the Sun it self, which gives day to all the World besides, is hardly able to penetrate and impart it here; and the weary Traveller, at many Miles distance, sooner smells, then sees the City to which he repairs. This is that pernicious Smoake which fullyes all her Glory, superinducing a sooty Crust or furr upon all that it lights, spoyling the moveables, tarnishing the Plate, Gildings and Furniture, and corroding the very Iron-bars and hardest stones with those piercing and acrimonious Spirits which accompany its Sulphure; and executing more in one year, then expos'd to the pure Aer of the Country it could effect in some hundreds… It is this horrid Smoake which obscures our Churches, and makes our Palaces look old, which fouls our Clothes, and corrupts the Waters, so as the very Rain, and refreshing Dews which fall in the several Seasons, precipitate this impure vapour, which, with its black and tenacious quality, spots and contaminates whatsoever is expos'd to it.

He continues to describe its health effects on the inhabitants of London (Evelyn, 1661, p. 12):

> But in the mean time being thus incorporated with the very Aer, which ministers to the necessary respiration of our Lungs, the Inhabitants of London, and such as frequent it, find it in all their Expectorations; the Spittle, and other excrements which proceed from them, being for the most part of a blackish and fuliginous Colour: Besides this acrimonious Soot produces another sad effect, by rendring the people obnoxious to Inflammations, and comes (in time) to exulcerate the Lungs, which is a mischief so incurable, that it carries away multitudes by Languishing and deep Consumptions, as the Bills of Mortality do Weekly inform us.

The connection between the burning of coal in large amounts in a confined urban area and the degradation of air quality led Evelyn to suggest moving the primary polluting industries to the outskirts of the city. He also recommended the creation of clear areas or parks to improve the air quality. These are similar to the suggestions that Moses Maimonides made in the twelfth century, as described in Chapter 1.

The air pollution problem from coal burning in London continued until a particularly serious event occurred from December 5 to December 9, 1952 called the "Great London Smog" or the "Great Smoke of 1952." During this period, the cold windless weather conditions, along with the emissions from coal combustion in London, trapped very high levels of SO_2 and acidic aerosols in the lower PBL. The dramatic visibility loss due to the smoke and aerosol formation can be seen in Figure 5.2, which compares images of Nelson's Column in London on December 1952 and on an undisclosed date in 2012. Many people were hospitalized during this episode, which was initially credited with causing the deaths of approximately 4000 citizens and causing health effects to over 100 000 Londoners. Later, revised estimates of the impact of this air pollution event have the death toll as high as 12 000. This led to the passing of the Clean Air Act by the British Parliament in 1956. The Clean Air Act put regulations in place to lower the particulate and sulfur dioxide emissions. It included the assignment of smoke control regions where only

Figure 5.2 Photograph of Nelson's Column in London during a London smog event in December 1952 (left) and in 2012 (right). Source: N T Stobbs, Wikimedia Commons.

smokeless fuels such as natural gas could be burned, required the use of low-sulfur coal, and encouraged the use of electricity for home heating and cooking. The power plants that produced the electricity were relocated outside urban areas and the height of the stacks was increased to allow emissions to mix more efficiently with the free troposphere. Note that these solutions were essentially very similar to those proposed centuries before by John Evelyn. However, it took a catastrophic event for regulations and emission controls to finally be implemented.

Similar events also occurred in the United States due to the emission of smoke, CO, and SO_2 from coal and wood combustion. One event occurred in Donora, PA from October 27 to October 31, 1948. The town of 14 000 inhabitants was impacted by hydrofluoric acid (HF) and SO_2 emissions from local steel plants, railroad yards, and zinc works. This episode was accompanied by a thermal inversion, which capped the PBL, trapping the emissions near the ground, leading to very high levels of the toxic gases accompanied by loss of visibility. About a third to a half of the town's residents were affected during the pollution episode, including the death of 20 people. Another 50 deaths occurred within a month from exposure to the killer smog. Other similar types of pollution events were seen in St. Louis in the late 1930s, due to the domestic and industrial use of coal. This type of smog was always characterized by SO_2 emissions and heavy particulate levels. It was accompanied by high loss of visibility, particularly during weather conditions that reduced vertical mixing.

In the 1950s, New York City also began to observe significant smog events, similar to those that occurred in London. Although New York City is located on a coastal island and winds usually clear the region daily, a number of cold weather events occurred that produced thermal inversions, causing a stagnation of the air. The thermal inversions trapped New York's high levels of pollutant emissions in the PBL, leading to two deadly smog events. The first event in 1953 and the second in 1963 resulted in an estimated 200 to 400 deaths each. About 10% of the population showed respiratory distress, with symptoms of coughing and tearing eyes. Figure 5.3 shows a photograph taken on November 20, 1953 from the Empire State Building looking toward the Chrysler Building. The layer of smog and loss of visibility that occurred in

Figure 5.3 The New York City smog event on November 20, 1953. Source: Photo taken from the Empire State Building looking toward the Chrysler Building (Walter Albertin, Library of Congress photograph collection).

New York City during these smog events can easily be seen in the figure. These smog events, which were caused by high emissions of SO_2 and heavy particulates, accompanied by high loss of visibility and lung distress, came to be called "London Smog."

Around this same time, a new type of smog was seen in Los Angeles, CA. Los Angeles and the south coastal region of California are located in a basin between the Pacific Ocean and the Peninsular mountain ranges, which surround it on three sides. This basin geography routinely forms orographic thermal inversions due to the nocturnal sea breeze, which is cooler than the surrounding air mass, filling the basin at night. These regular nocturnal thermal inversions trap and concentrate pollutants in the basin. The region is also semi-arid with little cloud cover, resulting in high-intensity solar radiation in the PBL, which increases the potential for air pollution photochemistry to occur. Figure 5.4 shows the Los Angeles basin as seen from Hollywood Hills in late afternoon, 1955. The low-lying layer of smog with accompanying visibility loss can clearly be seen, trapped in the basin.

The first major smog event in Los Angeles occurred on July 26, 1943. The concentrated pollution emissions and photochemical reaction products trapped in the basin caused severe eye irritation and upper respiratory distress among residents. The reactions were so severe that, since this occurred during World War II, the event was originally thought to be a chemical warfare attack. Although Los Angeles did not use either coal or wood fuel to any extent, since the event was connected to high visibility loss it was believed at the time that smoke was the primary cause. This led to the State of California passing the Air Pollution Control Act in 1947, which

Figure 5.4 The Los Angeles basin from Hollywood Hills (Griffith Observatory at left) in late afternoon 1955 showing the low-lying photochemical haze filling the basin. Source: Diliff, Wikimedia Commons.

required California counties to establish Air Pollution Control Districts for smoke abatement. It was not until 1952 that a natural products chemist, Arie Haagen-Smit from the California Institute of Technology, determined that the air pollution in the Los Angeles basin was not caused by smoke.

As an organic chemist, Haagen-Smit recognized that ozone was produced in the smog, along with oxidized organics in the form of aldehydes and ketones. He performed photochemical experiments by placing the exhaust from motor vehicles in sunlight, which demonstrated that ozone and organic aerosols were being formed in a new type of smog, unlike the London smog. Downwind of Los Angeles, toward the mountain ranges, crops began to show significant damage that could not be attributed to SO_2 exposure. Plant damage studies were conducted at the University of California at Riverside (Figure 5.5, right), which identified ozone as responsible for a portion of the plant damage. This helped to confirm that ozone was being produced in the smog. This new type of smog is now called "photochemical smog." Although Arie Haagen-Smit did not understand the mechanism of the reaction, he was correct in concluding that the NO_x and VOC emissions from motor vehicles, oil refineries, and industries were causing the problems. The expansion of freeways and increase in motor vehicle use (Figure 5.5, left) was a major factor in the photochemical smog formation.

This event and others like it led to the passing of the Air Pollution Control Act of 1955 by the U.S. Congress, which recognized the air pollution problems but left the solutions up to the state and local community governments. The Air Pollution Control Act was succeeded by the Clean Air Act of 1963, which began to establish controls on air pollution and placed the responsibility for implementing these controls in the U.S. Public Health Service. The U.S. Public Health Service began funding research into the monitoring and control of air pollutants and by 1965 the Clean Air Act was amended with the Motor Vehicle Air Pollution Control Act, which began to require emission controls on vehicles, starting in 1968. In 1967, the United States again amended the Clean Air Act to include the determination of the effects of air pollution transport. To accomplish this, the Clean Air Act included support for ambient monitoring of air pollutants

Figure 5.5 Traffic at Venice Boulevard and La Cienega Boulevard in Los Angeles, CA, 1953 (left) and the California Air Resources Board (CARB) Laboratory located on the University of California at Riverside campus (right). Source: left, Los Angeles Daily News, Wikimedia Commons; right, National Archives Archeological Site, Wikimedia Commons.

and the measurement of pollution emissions from stationary sources. In 1970, the Clean Air Act had its most significant amendment, which established the U.S. Environmental Protection Agency (EPA). The EPA was given responsibility for establishing regulations to protect air and water quality, consolidating environmental regulations into one agency and expanding federal and state enforcement authority. In 1990 the act was further amended to address stratospheric ozone depletion, acidic rain, and air toxics.

5.3 The Clean Air Act

The Clean Air Act of 1970 is a U.S. federal law. It authorized the development of comprehensive federal and state regulations to limit emissions from both stationary and mobile sources. It is one of the first environmental laws of the United States and is one of the most comprehensive air pollution laws in the world. The Clean Air Act is administered by the EPA in coordination with state governments. It required the EPA to identify pollutants and to set standards for their ambient levels in order to maintain the public health and protect the environment. These standards were chosen after review of the available chemical data and the known health impacts of the identified pollutants. Based on this data, the EPA established the National Air Quality Standards for selected pollutants and then enforced regulations on pollutant emitters in order to keep the ambient levels below these standards. It is the EPA's responsibility to re-examine and modify the standards as better data is obtained in order to protect the health and welfare of the citizens of the United States.

5.3.1 Criteria Pollutants

The criteria air pollutants were the first set of pollutants recognized by the EPA as needing air quality standards on a national level. They are the only air pollutants that have national air quality standards which define the allowable concentrations of these substances in ambient air.

The EPA has identified six criteria air pollutants: NO_2, SO_2, ground-level O_3, CO, particulate matter (PM), and lead. Five of the criteria pollutants – NO_2, SO_2, CO, PM, and lead – are primary pollutants because they are emitted directly from the source into the troposphere. Ozone, however, is a secondary pollutant because it is produced from chemical reactions of primary pollutants in the troposphere. The criteria pollutant PM can be either primary or secondary.

The emissions of the primary pollutants are from both stationary and mobile sources. The stationary sources are primarily power plants, industrial activities such as steel mills and smelters, mining operations, and agriculture. The mobile sources are motor vehicles, aircraft, and trains, which all use fossil fuels. Most of the lead emissions were from motor vehicles due to the use of tetraethyl- and tetramethyl-lead as fuel additives to enhance octane levels and vehicle performance. The use of alkyl-lead fuel additives was developed by Thomas Midgely, as discussed in Section 1.5. Since the fuel also contained alkyl bromides and chlorides, the lead was emitted from the engines as a fine aerosol of lead bromide, lead bromochloride, and lead chloride. The high toxicity of the atmospheric lead and the standards set by the EPA required that motor vehicles use unleaded fuels in order to reduce lead emissions into the air. Although leaded fuel is no longer sold in the United States for motor vehicles, lead is still considered a criteria pollutant since elevated lead emissions can still be a problem from smelters, ore and metal processing, waste incinerators, and lead acid battery processing. Leaded fuel is also still used in piston engine aircraft.

Once the criteria pollutants were identified, standards were set to limit their levels in ambient air. There are two types of standards: primary and secondary. The primary standards are developed from health exposure studies and are intended to protect the health of sensitive populations such as asthmatics, children, and the elderly. The secondary standards are developed based on environmental exposure studies and are intended to prevent environmental degradation, including visibility loss and/or damage to crops, vegetation, buildings, and animals. The primary standards are targeted at short-term acute exposure levels, while the secondary standards are targeted at more long-term chronic exposure levels.

Table 5.1 gives the current primary and secondary standards for the six criteria pollutants. Periodically, the standards are reviewed and may be revised depending on the most recent health recommendations. The units of measurement for the standards are parts per million (ppm) by volume and parts per billion (ppb) by volume for gases and micrograms per cubic meter of air ($\mu g\, m^{-3}$) for PM. The PM standards are measured by mass and are divided into two size ranges. The first is PM less than 10 μm in aerodynamic diameter (PM_{10}) and the second is PM less than 2.5 μm in aerodynamic diameter ($PM_{2.5}$). These PM standards are based on total weight and not on the composition of the aerosol. Because of this, it does not consider the relative toxicity of the PM. It also does not distinguish between primary or secondary PM.

The criteria pollutants listed in Table 5.1 are all compounds that were easily measured in the 1970s using wet chemical or gravimetric methods. In some cases, the standard was based on the limit of detection for the analytical method used at the time. These original analytical methods that were used to generate the first air quality standards are now considered to be the EPA primary reference methods. In order for other, more modern methods to be used they must first be standardized using these primary reference methods. This requirement will be discussed in detail in Chapter 7 under the analytical methods used in air analysis.

The criteria air pollutants receive more attention and are monitored more often, with better time and geographical resolution than the non-criteria pollutants. Other important pollutants that threaten our health and environment, while receiving attention, have not been adopted as criteria pollutants, but there are many other toxic substances that have been identified as hazardous air pollutants by the EPA. They continue to be considered for standards and in some cases have specific laws that limit their use. However, they have not yet been considered for

Table 5.1 National air quality standards for the six criteria pollutants.

Pollutant	Type	Concentration	Unit	Averaging time	Exceedance limit
CO	Primary	35	ppm	1 h	1 yr^{-1}
CO	Primary	9	ppm	8 h	1 yr^{-1}
NO_2	Primary	100	ppb	1 h	3 yr average
NO_2	Primary and secondary	53	ppb	1 yr	1 yr^{-1}
SO_2	Primary	75	ppb	1 h	3 yr average
SO_2	Secondary	500	ppb	3 h	1 yr^{-1}
O_3	Primary and secondary	70	ppb	8 h	3 yr average
Lead	Primary and secondary	0.15	$\mu g\ m^{-3}$	3 mo	none
PM_{10}	Primary and secondary	150	$\mu g\ m^{-3}$	24 h	1 in 3 yr
$PM_{2.5}$	Primary	12.0	$\mu g\ m^{-3}$	1 yr	3 yr average
$PM_{2.5}$	Secondary	15.0	$\mu g\ m^{-3}$	1 yr	3 yr average
$PM_{2.5}$	Primary and secondary	35	$\mu g\ m^{-3}$	24 h	3 yr average

criteria pollutant status. Elemental mercury is an example of this type of non-criteria pollutant. Also, in response to climate change issues, such as global warming, the EPA has considered adding some of the key greenhouse gases, including CO_2, CH_4, N_2O, and chlorine-containing compounds as criteria pollutants. This recommendation is still under review and has not been implemented.

5.3.2 Non-Criteria Pollutants

Non-criteria pollutants are those that have been identified as posing a health hazard but are not listed as criteria pollutants. They are called hazardous air pollutants by the EPA and are listed as air toxics. They originally included 190 species, listed in Appendix B, which were under some form of regulation by the EPA. These compounds were chosen because they were known to have human health effects or effects on biota. The air toxics included 173 organic compounds including pesticides, benzene, and other aromatic compounds, chlorinated and brominated organics, aldehydes, and acetates. Also included are 13 elements (11 metals, phosphorous, and chlorine), radionuclides including radon, and fibrous materials (asbestos and mineral fibers). Listed separately is "coke oven emissions." This is because coke oven workers were found to have a high incidence of cancer. Although the carcinogen in the emissions has not been identified, the mixture of emissions is considered to be carcinogenic.

In the list of air toxics (Appendix B), the compounds that cause eye irritation and tearing are called lachrymators. Other species, referred to as irritants, cause skin irritation or upper respiratory stress. Many of the hazardous air pollutants are known to be toxic, highly toxic, and/or carcinogenic. Compounds that cause genetic mutations are listed as mutagens. Other compounds, known or suspected to cause birth defects, are teratogens. The term "toxic" refers to humans and animals only. Those that are especially toxic to aquatic organisms are noted separately. Five compounds have been removed from the original list of air toxics: methyl ethyl ketone (2005), ethylene glycol monobutyl ether (2004), caprolactam (1996), hydrogen sulfide (1991), and surfactant alcohol ethoxylates (2000). *N*-propylbromide has been under consideration for addition to the list of hazardous pollutants since December 28, 2016 but no decision has been made. The emissions of these air toxics are to be controlled by what is determined to be the current maximum achievable technology at the time for each species.

5.4 Formation of Ozone in the Troposphere

The compounds identified as air toxics were all considered to be primary air pollutants with direct emission sources. Some species, which were thought to be primary pollutants, such as the aldehydes, also have secondary sources. But the secondary air chemistry that produces toxic substances was not really known or understood at the time the list was created. This atmospheric chemistry of these species was tied to the oxidative production of ozone in the troposphere. The mechanism of ozone formation in the troposphere was a bit of a mystery, since the photolysis of molecular oxygen is not possible in the troposphere and the mixing of ozone from the stratosphere would not lead to high levels over an urban area. Although Arie Haagen-Smit had shown that automobile emissions of NO_x and VOCs would produce ozone when irradiated in sunlight, he did not understand the mechanisms that drove the chemistry.

Recall that ozone is produced in the stratosphere by the photolysis of O_2 with UVC radiation. This produces two O atoms that can combine with O_2 in the presence of a third body (M) to form O_3. It was understood that the formation of O_3 from O_2 and an O atom was the reaction that produced O_3 in the troposphere. However, the problem was determining how the O atom was produced, since UVC does not penetrate into the troposphere, as shown in Figure 4.5. The O atom could be produced from the photolysis of a primary pollutant containing oxygen, such as SO_2, NO, and NO_2. But SO_2 and NO also require UVC radiation to produce an O atom. So, the most likely candidate for production of O atoms in the troposphere was NO_2. Frances Blacet at the University of California at Los Angeles found that the photolysis of NO_2 has a quantum yield of 0.7–0.9 for wavelengths between 375 and 425 nm and quantum yields very close to one at shorter wavelengths (Finlayson-Pitts and Pitts, 1986). So, NO_2 can be photolyzed by UVB and UVA in the troposphere to form NO and an O atom, which can then produce ozone.

The NO_2/O_3 Cycle

$$NO_2 + h\nu \xrightarrow{k_p} NO + O$$

$$O + O_2 + M \xrightarrow{k_1} O_3 + M$$

$$NO + O_3 \xrightarrow{k_2} NO_2 + O_2$$

This photochemistry is similar to that in the stratosphere: NO_2 photolyzes to form an O atom, which reacts with O_2 in a third-body reaction to form O_3. The NO can then react with O_3 to reform NO_2. So, the reaction with NO reforms NO_2 while destroying O_3. The net O_3 production from this cycle depends on the kinetics of the reactions.

5.4.1 The Photostationary State

The reaction of O + O_2 is much faster than the photolysis of NO_2, so the production of ozone is limited by the photolysis of NO_2. The reaction of O with O_2 is so rapid that the $[O_3]$ is in steady state determined by the ratio of $[NO_2]$ to [NO], given by

$$[O_3] = \frac{k_p[NO_2]}{k_2[NO]}$$

where k_p is the photolysis rate constant, as defined in Section 3.3.2. It is the product of the quantum yield of the photolysis reaction (Φ), the absorption cross-sections for NO_2 in the region of 280–425 nm, and the spherically integrated actinic flux (F). This relationship is a photostationary state equation, a steady state reached by a photochemical reaction in which the rates of formation and reaction of a species are equal. The photostationary state for the formation of ozone from NO_2 was first described by Philip Leighton and is sometimes referred to as the Leighton relationship for predicting ozone levels (Leighton, 1961).

It is important to note that the photostationary state equation for the formation of O_3 predicts that

$$[O_3] \propto \frac{[NO_2]}{[NO]}$$

So, the ozone concentration should remain constant and is dependent on the ratio of the NO_2 to NO concentrations. It also predicts that ozone formation will only occur if $[NO_2] > [NO]$ and, since O is produced by photolysis of NO_2, at night the ozone will return to the steady-state value that is dependent on the ratio of the NO_x species. But in the Los Angeles area, ozone was found to build up over many days and the emissions of NO and NO_2 were primarily from motor vehicles with NO greater than 90% of NO_x emissions. So, in order to account for the O_3 levels observed in the Los Angeles area, there must be another mechanism for oxidizing NO to NO_2.

If NO is converted to NO_2 by way of the thermal reaction

$$2NO + O_2 \rightarrow 2NO_2$$

the reaction rate follows the square of the NO concentration:

$$\text{rate} = k[NO]^2[O_2]$$

So, the thermal reaction is fast at high [NO]. For example, at [NO] = 100 Torr at atmospheric pressure (760 Torr), the reaction reaches 85% conversion in 15 s. But at very low [NO], the reaction rate is very slow. For example, at [NO] = 100 ppb the reaction reaches 85% conversion in 226 days. At [NO] levels typically seen in the lower troposphere, the thermal conversion of NO to NO_2 cannot generate enough NO_2 to form the observed ozone levels in Los Angeles by the NO_2/O_3 Cycle. This meant that there must be another oxidant that is produced in the atmosphere which can convert NO to NO_2 when the ratio of $[NO_2]/[NO]$ is much less than one.

5.4.2 The Hydroxyl Radical

As discussed in Section 4.3.3, the OH radical can be produced in the stratosphere by the reaction of $O(^1D)$ with H_2O. This same reaction can occur in the troposphere, since $O(^1D)$ is produced from the photolysis of O_3 in the UVB. In the stratosphere OH has limited reactions, since there are few unoxidized species with which it can react. This is not the case in the troposphere. In fact, tropospheric chemistry is oxidative chemistry and OH is the main oxidizing agent for these oxidation reactions in the troposphere.

In 1969, Julian Heicklen proposed that OH could react with carbon monoxide (CO) to form the hydroperoxyl radical (HO_2). Carbon monoxide was being emitted in large concentrations from mobile vehicles but was not building up to high levels in the urban atmosphere. So, there must be a loss mechanism for CO in the troposphere. What Heicklen and others proposed was that OH could be generated by the same reaction as in the stratosphere. It could then react with CO, methane (CH_4), and other hydrocarbons, initiating chain reactions that would produce the hydroperoxyl radical (HO_2), as well as alkyl peroxy radicals (RO_2) and acyl peroxy radicals (RCO_3). Under high NO conditions these peroxy radicals could react with NO to form NO_2, which could then photolyze to form NO and an O atom. The O atom would then form ozone in the same reaction that occurs in the stratosphere. The first simple scheme for these reaction cycles involves the oxidation of CO as follows.

The OH/CO Cycle

1. Initiation: $O_3 + h\nu \rightarrow O(^1D) + O_2$
   $O(^1D) + H_2O \rightarrow 2\mathbf{OH}$
2. Peroxy radical formation: $\mathbf{OH} + CO \rightarrow CO_2 + H$
   $H + O_2 \rightarrow \mathbf{HO_2}$
   $\mathbf{HO_2} + NO \rightarrow \mathbf{NO_2} + OH$
3. Ozone formation: $\mathbf{NO_2} + h\nu \rightarrow O + NO$
   $O + O_2 + M \rightarrow O_3 + M$

The reaction of OH (bold) with CO produces carbon dioxide (CO_2) and a hydrogen atom (H). In the atmosphere, the H atom reacts rapidly with the abundant O_2 to form HO_2 (red). It is the HO_2 radical that reacts with the NO to form NO_2 (blue), and it is the photolysis of NO_2 in the far UVB and UVA region that yields the O atom needed to form O_3 in the troposphere. This simple set of reactions provides the necessary sink for CO, while at the same time providing the oxidant needed to convert NO to NO_2 without using O_3 as the oxidant for that conversion.

5.4.3 Hydroxyl Radical Abstraction Reactions

The OH/CO Cycle is just one of the many reaction cycles of OH with molecules that can undergo oxidation. The OH can undergo two types of reactions with organic compounds, abstraction and addition, both leading to the formation of peroxy radicals. The reaction of OH with CH_4 is the simplest example of an abstraction reaction which occurs with many saturated organic molecules.

OH/CH_4 Abstraction Cycle

$$\mathbf{OH} + CH_4 \rightarrow CH_3 + H_2O$$
$$CH_3 + O_2 \rightarrow \mathbf{CH_3O_2}$$
$$\mathbf{CH_3O_2} + NO \rightarrow \mathbf{NO_2} + CH_3O$$
$$CH_3O + O_2 \rightarrow CH_2O + \mathbf{HO_2}$$
$$CH_2O + \mathbf{OH} \rightarrow CHO + H_2O$$
$$CHO + O_2 \rightarrow CO + \mathbf{HO_2}$$
$$CH_2O + h\nu \rightarrow H_2 + CO\ (55\%)$$
$$\rightarrow H + CHO\ (45\%)$$
$$H + O_2 \rightarrow \mathbf{HO_2}$$
$$\mathbf{HO_2} + NO \rightarrow \mathbf{NO_2} + \mathbf{OH}$$

This OH abstraction reaction cycle occurs with all types of alkane compounds. The abstraction of methane by OH produces methyl peroxy radicals (CH_3O_2), which can oxidize NO to NO_2 in the same way as HO_2. This produces a methoxy radical (CH_3O), which loses an H atom by reaction with O_2 to form HO_2 and CH_2O. Formaldehyde and many of the aldehydes can undergo photolysis in the UVA. In the case of formaldehyde there are two pathways, the formation of H_2 and CO and the formation of the formyl radical (HCO) and an H atom. The HCO and H atoms both react further to form HO_2 radicals, which can then react with NO to form NO_2. This abstraction reaction cycle forms a chain reaction mechanism that converts NO to

Table 5.2 The OH radical abstraction reaction rate constants (k_{OH}) in cm^3 molecule^{-1} s^{-1} for some selected organic compounds at 298 K and 1 atm pressure.

Compound	Chemical formula	k_{OH}	Relative rate	Half-life
Methane	CH_4	6.9×10^{-15}	1	4.6 yr
Ethane	CH_3CH_3	2.6×10^{-13}	38	1.5 mo
Propane	$CH_3CH_2CH_3$	1.2×10^{-12}	176	9.7 d
Butane	$CH_3CH_2CH_2CH_3$	2.5×10^{-12}	370	4.6 d
2-Methylpropane	$CH_3CH(CH_2)CH_3$	2.3×10^{-12}	340	5.0 d
Pentane	$CH_3CH_2CH_2CH_2CH_3$	4.0×10^{-12}	577	3.0 d
2,2-Dimethylpropane	$CH_3C(CH_3)_2CH_3$	8.5×10^{-13}	124	13.8 d
2,3-Dimethlybutane	$CH_3CH(CH_3)CH(CH_3)CH_3$	6.0×10^{-12}	873	1.96 d
Cyclopentane	C_5H_{10}	5.1×10^{-12}	741	2.3 d
Cyclohexane	C_6H_{12}	7.5×10^{-12}	1092	1.6 d
Formaldehyde	CH_2O	9.2×10^{-12}	1341	1.3 d
Acetaldehyde	CH_3CHO	1.6×10^{-11}	2332	17.8 h
Methanol	CH_3OH	9.3×10^{-13}	136	12.6 d
Ethanol	CH_3CH_2OH	3.3×10^{-12}	481	3.6 d
Peroxyacetyl nitrate	$CH_3COO_2NO_2$	1.1×10^{-13}	16	3.5 mo

Relative rates are compared to the rate of reaction of methane. The half-life is estimated using a steady-state concentration for $[OH] = 1 \times 10^6$ molecule cm^{-3}.

NO_2. So, in the abstraction cycle of CH_4, four molecules of HO_2 and one molecule of CH_3O_2 are formed for every one molecule of CH_4. This means the five molecules of NO_2 are formed per molecule of CH_4 oxidized.

So, the production of ozone in the troposphere depends on the production of peroxy radicals (RO_2 and HO_2) and their ability to oxidize NO to form NO_2. The peroxy radical formation depends on the reaction of OH with the VOCs and CO. In addition, the reactivity of the VOCs with OH determines the rate of ozone production. The abstraction rate constants for some important organic molecules with OH are given in Table 5.2 (Atkinson et al., 1997). The k_{OH} value for the abstraction reactions ranges over four orders of magnitude. Methane is the slowest to react (6.9×10^{-15}) and therefore its lifetime in the atmosphere is much longer than the other molecules ($t_{1/2} = 4.6$ years). This is one of the reasons that methane has an atmospheric concentration in the ppm range, while other organic compounds have atmospheric concentrations in the ppb range. It is also the reason that methane can last long enough in the troposphere to reach the stratosphere, where it is important as a source of water vapor. The relative reaction rates listed in Table 5.2 are the comparison of the reaction rate of each compound to that of methane (rate of compound/rate of methane) and represents how much faster each compound reacts with OH. The half-life for each compound is determined assuming a bimolecular reaction according to Table 3.4 ($t_{1/2} = 1/k_{OH}[OH]_o$), with $[OH]_o = 1 \times 10^6$ molecule cm^{-3}.

The half-lives for the abstraction reactions are shorter for organics that have more available hydrogens. The OH radical reacts with C–H bonds in the order of bond strengths. So, for the alkanes the rate of reaction is in order of tertiary > secondary > primary > methane, which is the order of the bond energies: C—H = 96, HC—H = 98, H_2C—H = 101, H_3C—H = 105 kcal mol^{-1}. The aldehydes react faster than the alkanes, since the aldehyde C–H bond strength is only about 89 kcal mol^{-1}. The OH abstraction rates for the linear alkanes as a function of the number of

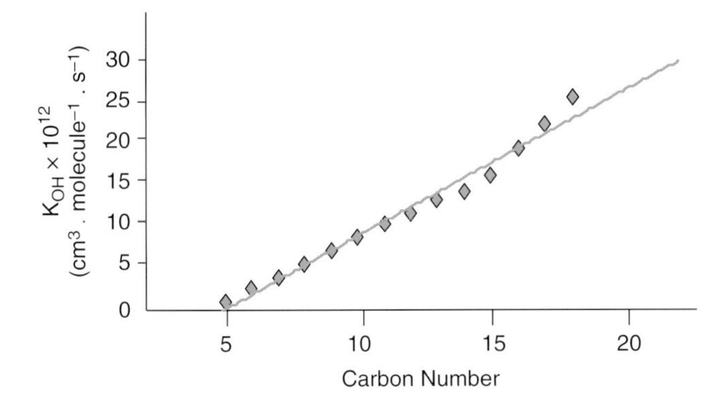

Figure 5.6 The OH abstraction rate constants for a series of straight-chain alkanes ($r^2 = 0.98$).

Table 5.3 Partial k_{OH} abstraction rate constants (cm^3 $molecule^{-1}$ s^{-1}) at 300 K for the three types of hydrocarbon hydrogens.

Type of hydrogen	Partial k_{OH}
Primary	6.5×10^{-14}
Secondary	5.8×10^{-13}
Tertiary	2.1×10^{-12}

carbon atoms are shown in Figure 5.6. This relationship is near linear, with a regression coefficient (r^2) of 0.98.

The k_{OH} value reported for an organic molecule is an average of the total number of C–H bond types in a molecule. However, the analysis of this systematic increase in reactivity with increasing carbon number led to the determination of a partial k_{OH} rate constant for each type of C–H bond (Darnall et al., 1978). These partial rate constants, listed in Table 5.3, allow for the calculation of the total k_{OH} for an alkane with a chemical formula of three carbon atoms or greater. For example, propane has six primary hydrogens and two secondary hydrogens. So, the predicted k_{OH} value is $(6)(6.5 \times 10^{-14}) + (2)(5.8 \times 10^{-13}) = 1.6 \times 10^{-12}$ cm^3 $molecule^{-1}$ s^{-1}. Pentane, with six primary hydrogens and six secondary hydrogens, has a predicted k_{OH} of 3.9×10^{-12} cm^3 $molecule^{-1}$ s^{-1}. The predicted values are in very good agreement with the experimentally determined values in Table 5.2, which have an experimental error of about 10%. This approach has been extended to allow for the estimation of abstraction reactions for more complicated hydrocarbons that contain other functional groups (Atkinson, 1986).

The OH abstraction reactions of the large hydrocarbons also produce large alkyl peroxy radicals (RO_2), which react with NO to form NO_2 and the alkoxy radical (RO), as shown in Figure 5.7 for hexane. However, unlike with the smaller alkanes, the larger alkoxy radicals can internally isomerize by intramolecular hydrogen abstraction, which involves a six-member cyclic intermediate including the oxygen atom and the hydrogen on the fifth carbon, as shown in reaction 3 in Figure 5.7. The hydrogen is then transferred from the fifth carbon to the oxygen atom. Because of the formation of the six-member cyclic intermediate, this intramolecular isomerization is called a 1,6 H shift. This leads to a hydroxyl functional group on the second carbon atom and a radical at the fifth carbon atom. This product can then add O_2 to form a hydroxy-substituted peroxy radical ($OHRO_2$), which will react with NO to form

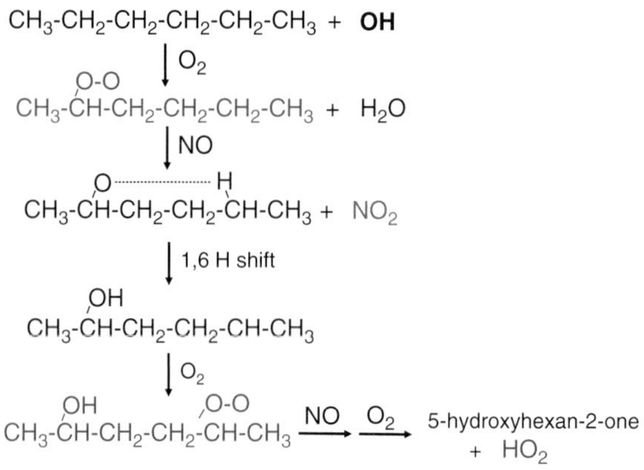

CH_3-CH_2-CH_2-CH_2-CH_2-CH_3 + **OH**

↓ O_2

O-O
CH_3-CH-CH_2-CH_2-CH_2-CH_3 + H_2O

↓ NO

O ········· H
CH_3-CH-CH_2-CH_2-CH-CH_3 + NO_2

↓ 1,6 H shift

OH
CH_3-CH-CH_2-CH_2-CH-CH_3

↓ O_2

OH O-O
CH_3-CH-CH_2-CH_2-CH-CH_3 $\xrightarrow{NO}$ $\xrightarrow{O_2}$ 5-hydroxyhexan-2-one + HO_2

Figure 5.7 The OH abstraction reaction with hexane showing the 1,6 H-shift of the alkoxy radical that ultimately produces the hydroxy ketone, 5-hydroxyhexan-2-one.

a hydroxy-substituted alkoxy radical (OHRO). The alkoxy radical then reacts with O_2 to give HO_2 and a hydroxy ketone (5-hydroxyhexan-2-one).

In the OH abstraction of pentane, the product would be a hydroxy aldehyde since the six-member cyclic intermediate would include the terminal carbon. The larger the alkane, the more of the hyroxy ketone is formed. These products tend to be water soluble and contribute to secondary aerosol formation, which will be discussed in Chapter 6.

5.4.4 Hydroxyl Radical Addition Reactions

The OH radical is isoelectronic with the fluorine atom ($2s^2 2p^5$), which means that it is very electrophilic and undergoes reactions that allow it to add an electron. In abstraction reactions, it gains the electron by removing a hydrogen atom from the reactant along with its electrons. The other important reaction, which allows OH radicals to gain an electron, is the addition reaction to a double bond in an alkene so as to form a hydroxy alkyl radical:

$$\mathrm{OH} + {>}\mathrm{C{=}C}{<} \rightarrow {>}\mathrm{C(OH)C}{<}$$

In the addition reaction, the electron is added to the OH by bonding with a π electron in the double bond of the alkene. Although OH is not highly selective in most of its reactions, it does show a preference in addition reactions to follow Markovnikov addition rules. That is, the stability of the radical formed from the addition of OH follows the order tertiary > secondary > primary, with the more stable radical being the major product. For example, in the case of propene the secondary radical is formed 65% of the time while the primary radical is formed 35% of the time:

$$\mathrm{OH} + \mathrm{CH_2{=}CH_2CH_3} \rightarrow \mathrm{CHOH{-}CH_2CH_3} \quad 65\%$$

$$\mathrm{OH} + \mathrm{CH_2{=}CH_2CH_3} \rightarrow \mathrm{CH_2{-}CHOHCH_3} \quad 35\%$$

The OH addition mechanism for the reaction of OH with ethene is as follows.

OH/CH_2CH_2 Addition Cycle

$$\mathbf{OH} + \mathrm{CH_2{=}CH_2} \rightarrow \mathrm{CH_2OHCH_2}$$

$$\mathrm{CH_2OHCH_2} + \mathrm{O_2} \rightarrow \mathrm{CH_2OHCH_2O_2}$$

$$\mathrm{CH_2OHCH_2O_2} + \mathrm{NO} \rightarrow \mathrm{CH_2OHCH_2O} + \mathrm{NO_2}$$

$$\mathrm{CH_2OHCH_2O} \rightarrow \mathrm{CH_2OH} + \mathrm{CH_2O}$$

$$\mathrm{CH_2OHCH_2O} + \mathrm{O_2} \rightarrow \mathrm{CH_2OHCHO} + \mathrm{HO_2}$$

$$CH_2OH + O_2 \rightarrow HO_2 + CH_2O$$
$$HO_2 + NO \rightarrow NO_2 + \mathbf{OH}$$

The addition of OH to the double bond followed by reaction with O_2 forms the peroxy radical ($CH_2OHCH_2O_2$), which can oxidize NO to NO_2 in the same way as in the abstraction reactions. Since ethene is a small molecule, the peroxy radical can also undergo decomposition to form a CH_2OH radical and formaldehyde or it can react with O_2 to form glycolaldehyde (CH_2OHCHO) and HO_2. The CH_2OH radical can also react with O_2 to form HO_2 and formaldehyde.

The reaction rates of the OH addition reactions are faster than the reaction rates of the abstraction reactions. This is to be expected, since the addition involves breaking of the weaker π bond while abstraction requires breaking of the stronger σ bond. Since addition reactions are faster than abstraction reactions, the alkenes are much more effective in producing peroxy radicals in the atmosphere than the alkanes. The alkenes also have much shorter lifetimes, with half-lives on the order of hours compared to days for the alkanes. So, the concentrations of the alkenes in the troposphere are very low. In most cases, the addition reactions are only effective near the alkene source region, due to their high reactivity and short atmospheric lifetimes. The OH addition reaction rate constants (k_{OH}) and their relative reactivities compared to methane are given in Table 5.4 for some atmospherically important compounds (Atkinson et al., 1997).

The last five compounds listed in Table 5.4 are some important naturally occurring reactive atmospheric organic hydrocarbons: isoprene (2-methyl-1,3-butadiene), α-pinene (2,6,6-trimethylbicyclo[3.1.1]hept-2-ene), β-pinene (6,6-dimethyl-2-methylenebicyclo[3.1.1] heptane), limonene (1-methyl-4-(prop-1-en-2-yl)cyclohexene), and terpinolene (1-methyl-4-propan-2-ylidenecyclohexene). The chemical structures of these natural organic hydrocarbons are shown in Figure 5.8. These are very reactive compounds, with half-lives on the order of hours. Isoprene is a hemiterpene with the molecular formula C_5H_8. It is released into the atmosphere from broadleaf trees and shrubs. This natural emission of isoprene accounts for about one-third of all the VOCs released into the atmosphere, especially during warm weather. Limonene, terpinoline, and the pinenes (α-pinene and β-pinene) are cyclic monoterpenes with molecular formula $C_{10}H_{16}$. They are all made up of two isoprene units

Table 5.4 The OH radical addition reaction rate constants (k_{OH}) in cm^3 $molecule^{-1}$ s^{-1} for some selected organic compounds at 298 K and 1 atm pressure.

Compound	Chemical formula	k_{OH}	Relative rate	Half-life
Ethene	$CH_2{=}CH_2$	8.0×10^{-12}	1 242	1.4 d
Propene	$CH_2{=}CHCH_3$	2.6×10^{-11}	3 834	10.7 h
1-Butene	$CH_2{=}CHCH_2CH_3$	3.1×10^{-11}	4 577	8.9 h
Cyclohexene	C_6H_{12}	6.8×10^{-11}	9 869	4.1 h
1,3-Butadiene	$CH_2{=}CHCH{=}CH_2$	6.7×10^{-11}	9 709	4.2 h
Isoprene	$CH_2{=}C(CH_3)CH{=}CH_2$	1.0×10^{-10}	14 723	2.8 h
Limonene	$C_{10}H_{16}$	1.7×10^{-10}	24 927	1.7 h
α-Pinene	$C_{10}H_{16}$	5.4×10^{-11}	7 828	5.2 h
β-Pinene	$C_{10}H_{16}$	7.9×10^{-11}	11 502	3.5 h
Terpinolene	$C_{10}H_{16}$	2.3×10^{-10}	32 799	1.2 h

Relative rates are compared to the rate of reaction of methane. The half-life is estimated using a steady-state concentration for $[OH] = 1 \times 10^6$ molecule cm^{-3}.

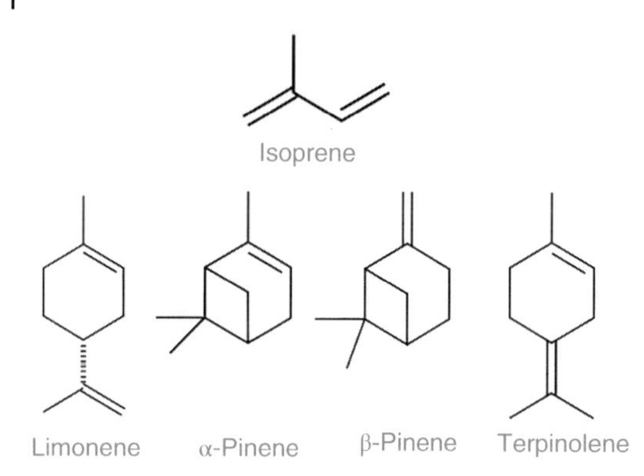

Figure 5.8 The chemical structures of the naturally emitted terpenes: isoprene, a hemiterpene, and the four monoterpenes, limonene, α-pinene, β-pinene, and terpinoline. Source: Fvasconcellos, Benjah-bmm27, J. delanoy, Jan Herold, and Ed, Wikimedia Commons.

and have the same carbon framework but they differ in the position of the C=C bond. The monoterpenes are also emitted in substantial amounts by vegetation, with emission rates dependent on temperature and light intensity. Limonene is found principally in citrus. It is used in large quantities as degreasing and household cleaning products. Both α- and β-pinene are structural isomers of the bicyclic monoterpene pinene, with α-pinene being the most abundant in nature. They are emitted to the atmosphere mainly by conifers.

The addition reaction rate of aromatic hydrocarbons with the OH radical is slower than the addition reaction rates of the alkenes. This is consistent with the added stability that resonance gives to the aromatic ring. The OH addition reaction rates for some of the common simple aromatic hydrocarbons found in the troposphere are listed in Table 5.5 (Atkinson, 1994). The OH addition reaction rates with the aromatic hydrocarbons are similar to the OH abstraction rates for some of the larger alkanes with half-lives of about a day. The OH addition reaction to the aromatic ring proceeds by a more complicated mechanism than that seen for the reaction with the alkenes. The reaction mechanism of OH addition with the simple aromatic compounds, such as toluene, xylene, or ethyl benzene, is believed to occur by OH addition to the ring followed by the reaction with O_2 to give a peroxy radical. This adduct can further react with NO to give NO_2 and an alkoxy radical, which can undergo ring opening of the aromatic compounds. This

Table 5.5 The OH radical addition reaction rate constants (k_{OH}) in cm^3 $molecule^{-1}$ s^{-1} for some selected aromatic hydrocarbons at 298 K and 1 atm pressure.

Compound	k_{OH}	Half-life
Benzene	1.2×10^{-12}	9.3 d
Toluene	6.0×10^{-12}	1.9 d
Ethylbenzene	7.1×10^{-12}	1.6 d
o-Xylene	1.4×10^{-11}	0.8 d
m-Xylene	2.4×10^{-11}	11 h
p-Xylene	1.4×10^{-11}	0.8 d
Propylbenzene	6.0×10^{-12}	1.9 d
1,2,3-Trimethylbenzene	3.3×10^{-11}	8 h
1,3,5-Trimethylbenzene	5.8×10^{-11}	4.5 h

The half-life is estimated using a steady-state concentration for $[OH] = 1 \times 10^6$ molecule cm^{-3}.

produces reactive dienes and conjugated aldehydes, which can also undergo addition reactions with OH producing peroxy radicals.

5.5 Nitrate Radical and Ozone

The other important oxidants in the troposphere that affect the concentrations of VOCs are O_3 and NO_3. The NO_3 radical is formed from the reaction of NO_2 with O_3. It is only important at night since it photolyzes rapidly in sunlight to give O and NO_2. The NO_3 radical is also very electronegative, but it is less reactive than OH since it rarely reacts by abstraction. However, it does undergo addition reactions with alkenes to form a nitrate adduct with a mechanism similar to OH addition. For example, the addition of NO_3 to ethene is as follows.

NO_3/CH_2CH_2 Addition Cycle

$$\mathbf{NO_3} + CH_2{=}CH_2 \rightarrow CH_2NO_3CH_2$$
$$CH_2NO_3CH_2 + O_2 \rightarrow CH_2NO_3CH_2O_2$$
$$CH_2NO_3CH_2O_2 + NO \rightarrow CH_2NO_3CH_2O + NO_2$$
$$CH_2NO_3CH_2O \rightarrow CH_2NO_3 + CH_2O$$
$$CH_2NO_3CH_2O + O_2 \rightarrow CH_2NO_3CHO + HO_2$$
$$HO_2 + NO \rightarrow NO_2 + OH$$

The reaction scheme is the same as for OH additions, with the NO_3 adding to the double bond followed by reaction with O_2 to form the peroxy radical. This reaction mechanism leads to the formation of oxidized organic nitrates instead of the oxidized alcohols formed with OH addition.

The NO_3 radical can also react with NO_2 to form N_2O_5:

$$NO_3 + NO_2 \leftrightarrow N_2O_5$$

Since the reaction is in equilibrium, the formation of N_2O_5 is favored under high-NO_2 conditions. The formation of N_2O_5 is also only important at night, since the NO_3 rapidly photolyzes, shifting the equilibrium to the left during the day. Although the primary formation of atmospheric HNO_3 is from the reaction of OH with NO_2 in the gas phase, N_2O_5 can react with water in aerosols and clouds to form aqueous HNO_3. Both $HNO_3(g)$ and $HNO_3(aq)$ can react with NH_3 to form ammonium nitrate NH_4NO_3 in the solid or liquid phase. Since the formation of N_2O_5 is only important at night, the formation of $HNO_3(aq)$ is also only important during the night time.

Ozone is also an electrophilic oxidant. It is more stable than either OH or NO_3, and so is only important in addition reactions with unsaturated hydrocarbons. The important unsaturated hydrocarbons that undergo ozone addition are simple alkenes, dienes, and trienes, and the naturally occurring unsaturated hydrocarbons, such as isoprene, the pinenes, monoterpenes, and sesquiterpenes, which consist of three isoprene units with molecular formula $C_{15}H_{24}$. The addition of ozone to the alkene forms an intermediate adduct called a molozonide, as shown in Figure 5.9. The molozonide is not stable, since it contains three oxygen atoms bonded to each other across the two carbon atoms that contained the π bond. The molozonide rapidly decomposes to an aldehyde or ketone and a zwitterion or a diradical called the Criegee biradical. The Criegee biradical can decompose to yield the OH radical and other products, or it can react with water to form the organic acid (RCO_2H). It can also react with SO_2, NO, NO_2, CO, RCHO, or $R_2C{=}O$ giving stable products. However, the reaction with water usually dominates,

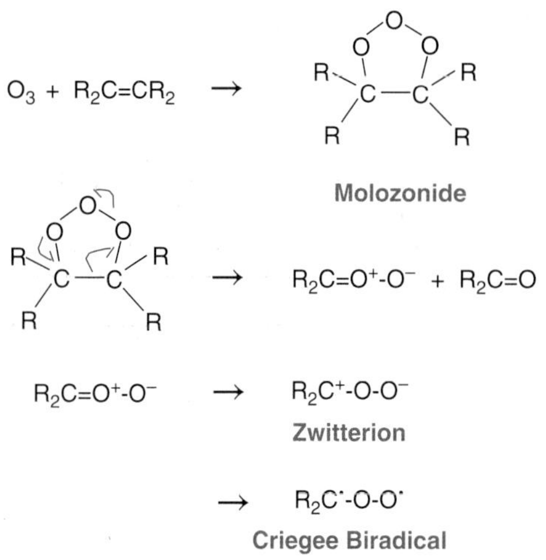

$$O_3 + R_2C{=}CR_2 \rightarrow \text{Molozonide}$$

$$\text{Molozonide} \rightarrow R_2C{=}O^+\text{-}O^- + R_2C{=}O$$

$$R_2C{=}O^+\text{-}O^- \rightarrow R_2C^+\text{-}O\text{-}O^- \ (\text{Zwitterion})$$

$$\rightarrow R_2C^{\cdot}\text{-}O\text{-}O^{\cdot} \ (\text{Criegee Biradical})$$

Figure 5.9 Ozone addition to an alkene forms the molozonide intermediate, which forms an aldehyde or ketone and a zwitterion or Criegee biradical.

since it is in much higher concentrations than these other reactants except in very polluted urban atmospheres.

The addition reaction of ozone with alkenes is important as a source of night-time OH radicals from the decomposition of the Criegee biradical. The amount of OH production from these reactions varies from about 10% for the smaller alkenes to over 80% for some of the larger alkenes and monoterpenes (Paulson and Orlando, 1996). This reaction is particularly important over agricultural or forested areas, where the emission of naturally reactive hydrocarbons is high. This addition of ozone to the natural hydrocarbons emitted from vegetation allows for the generation of HO_x chemistry in areas with low NO_x levels.

5.6 The Peroxyacyl Nitrates

The OH abstraction reaction also occurs with hydrocarbon compounds other than alkanes, such as the aldehydes. The OH abstraction reaction with acetaldehyde (CH_3CHO) is as follows.

OH/CH_3CHO Abstraction Cycle

$$\mathbf{OH} + CH_3CHO \rightarrow CH_3CO + H_2O$$

$$CH_3CO + O_2 \rightarrow \mathbf{CH_3COO_2}$$

$$\mathbf{CH_3COO_2} + NO \rightarrow NO_2 + \mathbf{CH_3CO_2}$$

$$CH_3CO_2 \rightarrow CH_3 + CO_2$$

$$CH_3 + O_2 \rightarrow CH_3O_2$$

$$CH_3O_2 + NO \rightarrow NO_2 + CH_3O$$

$$CH_3O + O_2 \rightarrow CH_2O + \mathbf{HO_2}$$

$$CH_2O + h\nu \rightarrow H_2 + CO\ (55\%)$$

$$\rightarrow H + CHO\ (45\%)$$

$$H + O_2 \rightarrow \mathbf{HO_2}$$

$$HO_2 + NO \rightarrow NO_2 + \mathbf{OH}$$

$$CH_3COO_2 + NO_2 \leftrightarrow CH_3COO_2NO_2$$

This aldehyde is oxidized by OH to produce the peroxy radicals H_3COO_2, CH_3CO_2, CH_3O_2, and HO_2, which can all oxidize NO to NO_2. The last step in the CH_3CHO abstraction reaction cycle is the reaction of the peroxyacetyl radical (CH_3COO_2) with NO_2 to form the stable compound peroxyacetyl nitrate ($CH_3COO_2NO_2$), called PAN. PAN is in equilibrium with CH_3COO_2 and NO_2. This equilibrium has a strong temperature dependence, with PAN being more stable at colder temperatures.

PAN is a potent lachrymator and is highly toxic to plants. Although O_3 is also toxic to plants, PAN shows different plant damage and, in fact, it was initially discovered due to its plant toxicity. As ozone and smog levels increased in the Los Angeles basin, farmers noticed an increase in crop damage. In order to determine what was responsible for the damage, researchers at the University of California at Riverside exposed plants to the known pollutants SO_2, O_3, and NO_2, and found that the effects observed on a number of plant species could not be reproduced. They concluded that the damage must be caused by an as yet unknown compound, which they called "compound X." At the same time, scientists studying the photochemistry of NO_x and the VOCs using long-path infrared spectroscopy found an unusual compound with a strong carbonyl band at a higher frequency than previously observed. After isolating this compound, it was confirmed that "compound X" was PAN and that it was responsible for the plant damage and eye irritation reported in the region.

PAN is always found at higher levels in urban air pollution. It is not very reactive with OH and only undergoes photolysis slowly at high altitudes. The temperature-dependent equilibrium with NO_2 favors PAN at lower temperatures, while higher temperatures favor the dissociation of PAN, reforming NO_2 and CH_3COO_2. The OH reaction with NO_2 to form HNO_3 was discussed in Section 4.3.3 as an important removal process for NO_2 in the stratosphere. This reaction is also important in the troposphere, and since HNO_3 is highly water soluble and easily removed from the gas phase in clouds and precipitation. This means that NO_2 has a short lifetime in the troposphere. But as PAN is stable at lower temperatures, it can be transported long distances in cooler air masses and can thus act as an efficient transport mechanism for NO_2 to areas where emissions are low.

The higher analogs of PAN normally observed in the troposphere at measurable levels are peroxypropionyl nitrate (PPN), peroxybutyl nitrate (PBN), peroxyisobutyl nitrate (PiBN), and peroxymethacryloyl nitrate (MPAN), with the chemical structures shown in Figure 5.10. This family of peroxy nitrates is called collectively the peroxyacyl nitrates ($RCOO_2NO_2$). PAN is by far the most abundant of the peroxyacyl nitrates and is commonly observed in polluted urban atmospheres as well as on regional and global scales. In highly polluted environments the higher analogs – PPN, PBN, and PiBN – have also been reported. PAN and MPAN are commonly observed in forested regions as the MPAN is likely formed from the oxidation of isoprene.

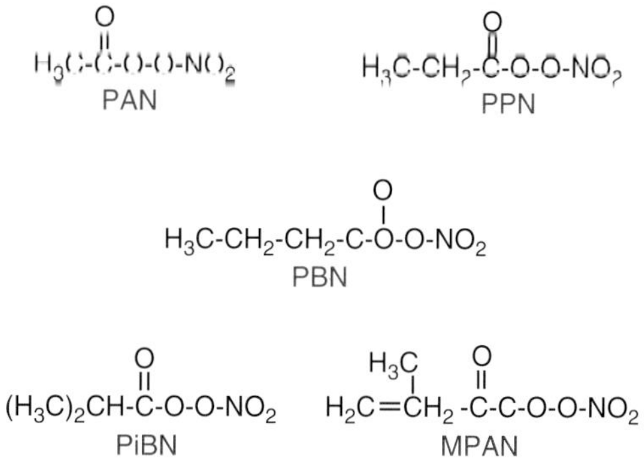

Figure 5.10 The chemical structures of the five peroxyacyl nitrates ($RCOO_2NO_2$) that have been measured in the troposphere: peroxyacetyl nitrate (PAN), peroxypropionyl nitrate (PPN), peroxybutyl nitrate (PBN), peroxyisobutyl nitrate (PiBN), and peroxymethacroyl nitrate (MPAN).

$H_3C\text{-}C(=O)\text{-}O\text{-}O\text{-}NO_2$ PAN

$H_3C\text{-}CH_2\text{-}C(=O)\text{-}O\text{-}O\text{-}NO_2$ PPN

$H_3C\text{-}CH_2\text{-}CH_2\text{-}C(=O)\text{-}O\text{-}O\text{-}NO_2$ PBN

$(H_3C)_2CH\text{-}C(=O)\text{-}O\text{-}O\text{-}NO_2$ PiBN

$H_2C{=}CH_2(CH_3)\text{-}C(=O)\text{-}O\text{-}O\text{-}NO_2$ MPAN

All of the PAN analogs are lachrymators, as well as mutagenic, carcinogenic, and potent plant toxins. They are all also in equilibrium with NO_2 and the peroxyacyl radical ($RCOO_2$), with the equilibrium favoring $RCOO_2NO_2$ at colder temperatures and the dissociation to NO_2 at warmer temperatures. The transport of each of these species at colder temperatures aloft is a source of NO_2 to remote locations and is a mechanism for polluted environments to have an impact on the regional and global-scale formation of ozone.

A slow unimolecular decomposition of PAN has been observed, which leads to the formation of methyl nitrate (CH_3NO_3) and CO_2 (Senum et al., 1986). The organic nitrates can undergo abstraction reactions with OH, which can then lead to the production of further nitrate-substituted peroxy radicals. For smaller nitrates, the abstraction reaction rates with OH are similar to that for PAN given in Table 5.2 (3.5 months). Larger alkyl nitrates will have reaction rates similar to their alkane parent molecule (2–14 days). At colder temperatures (<273 K) the reaction of HO_2 with NO_2 will produce peroxynitric acid (HO_2NO_2), and the reaction of RO_2 with NO_2 forms the peroxynitrates (RO_2NO_2). Like PAN, these species are in equilibrium with the peroxy radicals and NO_2, and will rapidly decompose at warmer temperatures. So, they are also NO_2 reservoir species and, like PAN, can act to transport NO_2 over longer distances than expected.

5.7 Troposphere–Biosphere Interactions

The exchange of gases between the biosphere and the troposphere and the atmospheric chemistry that results is important in determining both air quality and climate forcing on regional and global scales. The emissions of natural VOCs, as well as the general health of the biosphere, are affected by the trace gases and weather conditions in the troposphere. In turn, the levels of trace gases and aerosols in the troposphere are affected by the VOC emissions from the biosphere. Also, both biospheric emissions and tropospheric chemistry are affected by climate change, while climate change itself is determined by the composition of the troposphere. This interplay between the biosphere and the troposphere results in their acting as a dynamically coupled system with both positive and negative feedbacks.

The chemistry in urban atmospheres is driven by NO_x emissions in the presence of reactive VOCs and CO, which are oxidized by the OH radical to form peroxy radicals. This chemistry results in enhanced formation of O_3 in urban areas. But this same chemistry can occur in rural and remote atmospheres where anthropogenic NO_x emissions are lower. A large amount of NO in urban areas is produced by the combustion of fossil fuels, particularly from internal combustion engines in motor vehicles. However, NO can also be formed in rural areas by lightning and is a significant natural source of NO_x in the regional troposphere. In rural areas, NO_x can also be produced by the decomposition of nitrate fertilizers and from biomass and agricultural burning. In addition, while VOCs are sourced from anthropogenic combustion in urban areas, they are also produced in biomass burning as well as by emissions from natural biota, including animals, plants, and microbes. Table 5.6 lists estimates of the annual global emissions from both anthropogenic and natural sources of NO_x and the VOCs (Mueller, 1992).

While the largest emissions of NO_x are from anthropogenic combustion and biomass burning, the largest VOC emissions are from the terrestrial biosphere. The values listed in Table 5.6 are global estimates. But it is important to realize that the source emissions of these important trace gases are quite variable in space and time over the globe. For example, the natural VOC emissions from deserts are much lower than those from dense tropical forests, because of the differences in plant types and densities. In addition, the emissions of VOCs from vegetation are also dependent on temperature and plant biochemistry. In the Northern Hemisphere the

Table 5.6 Global anthropogenic and biogenic emission estimates for NO_2 and VOCs ($Tg\ yr^{-1}$).

Species	Anthropogenic	Biomass burning	Terrestrial	Oceanic	Total
NO_x	72	18	22	0.01	122
VOCs	98	51	500	30	679

deciduous forest emissions are higher in the spring-time and summer, and fall to very low levels during the fall and winter when the trees are dormant and the temperatures are much colder. The natural production of NO by lightning is dependent on the formation of strong storms. Regions where there are warmer air masses and thunderstorms are more common, such as in the lower temperate and tropical regions, will have more lightning strikes and larger natural NO emissions than polar regions where strong thunderstorms are less likely because of the colder air masses.

The transport of NO_2 into remote regions is enhanced by the reaction with OH to form the highly water-soluble HNO_3. The HNO_3 is removed from the gas phase onto wet aerosol surfaces and into clouds, as well as being deposited on wet terrestrial surfaces. So, the gas-phase formation of HNO_3 results in an increase in the acidic content of rain water, which enhances the deposition of nitrate into the biosphere in the form of "acid rain." Another acidic species, sulfuric acid (H_2SO_4), is produced by the reaction of H_2O_2 with SO_2 in wet aerosols and clouds. The H_2O_2 is primarily produced from the reaction of HO_2 with itself in areas where NO is low. This gas-phase reaction of HO_2 with HO_2 to form H_2O_2 has a rate law that is dependent on the square of the HO_2 concentration. So the rate of H_2O_2 formation is proportional to $[HO_2]^2$:

$$2HO_2 \rightarrow H_2O_2 + O_2$$
$$\text{rate} = k[HO_2]^2$$

The reactive natural hydrocarbon isoprene (2-methyl-1,3-butadiene) undergoes addition by both OH and O_3 to give CH_2O as a major product in forested regions with low NO. This is particularly important in deciduous forests (hardwoods such as oaks and maples), which are high isoprene emitters. The CH_2O will react with OH or undergo photolysis to give HO_2 as a product, such as in the OH/CH_4 Abstraction Cycle discussed in Section 5.4.3. So, higher levels of CH_2O result in higher levels of HO_2 and, if NO is low, lead to a higher production of H_2O_2. This ultimately leads to higher H_2SO_4, which forms sulfate aerosols. The HO_2 can react with CH_3COO_2 to produce higher levels of peracetic acid (CH_3CO_3H). In general, HO_2 can react with RO_2 species to increase the levels of organic peroxides (ROOH) in these forested regions.

This interaction of anthropogenic SO_2 and NO_2 with the natural isoprene chemistry leads to seasonal changes in air quality. During the months when the deciduous trees are active, the conversion of SO_2 by aqueous H_2O_2 is fast and leads to significant formation of sulfate aerosol haze, seen over places like the Great Smoky Mountains during the spring and summer months. This also increases the acidic deposition during these warmer seasons, which contributes to lake acidification. When the weather is colder and the deciduous trees are dormant, the overall VOC emissions are reduced. This is because the only trees active in the colder months are the conifers that primarily emit monoterpenes, which are larger and less volatile during the colder months. This reduction in VOC emissions leads to a reduction in sulfate formation and deposition, along with less regional haze and less acidification.

The reactivity of the naturally emitted VOCs varies significantly. Methane has the lowest reactivity with OH and thus the longest lifetime in the troposphere. The oxidation of methane gives a constant background of formaldehyde in the remote lower troposphere of about 250–300 ppt (Jones et al., 2009). Much higher values of formaldehyde have been observed in forested regions from the oxidation of isoprene and the monoterpenes. Values in these areas have been reported to reach a daily maximum as high as 20 ppb, with typical average highs between 4 and 12 ppb (Choi et al., 2010). So, the reactivity of the VOCs is particularly important with regard to both urban and regional air quality over land. The VOC emissions over the open ocean are dramatically lower. In those areas the only real source of hydrocarbons is methane that has been transported from terrestrial regions due to its longer atmospheric lifetime.

Ozone and PAN are known to interact with biota, causing visible damage at high levels and a slowing of the biochemistry at lower levels. This is mostly due to the oxidation of chlorophyll and a lowering of the overall health of trees and plants. This was observed in the San Bernardino Mountains in California by Dr. Paul Miller of the U.S. Department of Agriculture Forestry Service, Pacific Southwest Research Station (Miller et al., 1996). He found that ozone and other oxidants, which were transported into the forests from the Los Angeles air shed, caused damage to the trees, particularly pine trees. The oxidant-stressed pines were seen to contain fewer needles and showed yellowing of the needles, indicating chlorophyll loss, as shown in Figure 5.11. In some cases, the pine needle clusters were also shorter in length than those in healthy trees. These weakened trees were then susceptible to insect infestation such as bark beetles. The oxidant-stressed conifers could not maintain their sap pressure, which would normally force the beetles out of the trees. So, the death of trees originally attributed to insect infestation was primarily caused by stress from ozone and PAN exposure.

Plants are also known to release different hydrocarbons when they undergo drought stress. For example, poplars have been known to release CH_3CHO as well as isoprene when undergoing heat and drought stress (Werner et al., 2016). The higher levels of CH_3CHO emitted in the presence of NO_2 will lead to the formation of PAN and O_3, as discussed in Section 5.4.3, which can enhance stress to the biosphere. The increase in drought, as well as the increased formation of PAN and O_3, is linked to global climate change. For example, regional O_3 production is enhanced by the interactions of NO_x with natural hydrocarbons released from the forests. The O_3 is not only a reactive gas, which can cause oxidative damage to the biosphere, it is also an important greenhouse gas that contributes to warming in the troposphere.

The important greenhouse gas CO_2 is taken up by the biosphere. This is one potential feedback that is hoped to slow the warming effects of greenhouse gases on the troposphere. This

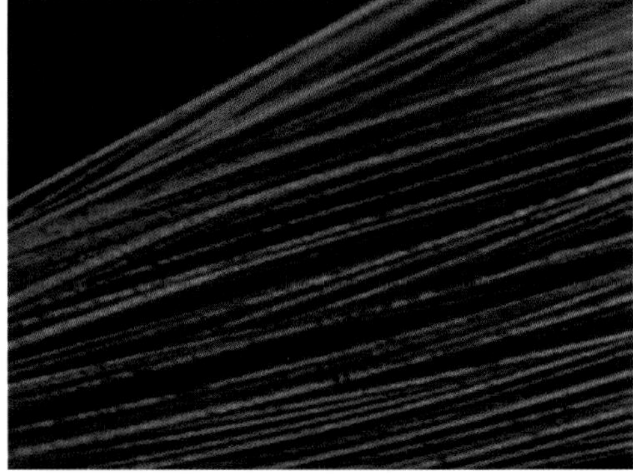

Figure 5.11 Comparison of pine needles from a healthy tree (left) and an oxidant-stressed tree (right). The chlorotic mottle (yellowing) of the pine needles is typical of oxidant damage due to exposure to the atmospheric oxidants ozone and PAN. Source: U.S. Department of Agriculture, Forest Service, Pacific Southwest Research Station.

uptake of CO_2 by plants is seasonal, as shown by the oscillation of the atmospheric CO_2 concentrations in the Northern Hemisphere in Figure 2.21 (Section 2.4.3). These seasonal changes in CO_2 uptake by plants and forested regions are a measure of their biochemical activity, which is also correlated with their emission of VOCs. With the advent of potential changing climates, which are predicted to result in increases in drought and higher temperatures, increased tropospheric chemistry may also result in higher atmospheric ozone levels. Under these combined stresses, the plants' ability to uptake CO_2 could be impaired, even with increasing atmospheric CO_2 levels.

While the oceans are not a significant source of VOCs on a mass basis, they do emit some important compounds that enhance plant and animal life in the terrestrial biosphere. A good example of this is methyl iodide (CH_3I). Marine algae and bacteria have the ability to methylate iodine (Amachi et al., 2001). Once methylated, it is volatile and is released into the troposphere, where it can be transported over long distances before it is oxidized. Methyl iodide is oxidized in the troposphere to the water-soluble iodate, which is then removed from the gas phase in precipitation. This allows iodine, which is in high abundance in the marine environment, to be transported and then deposited onto the iodine-poor terrestrial environment. This oceanic CH_3I is a major source of iodine for terrestrial regions, where the plant and animal life require it.

There are also linkages between sea-salt aerosols and gas-phase tropospheric chemistry (Finlayson-Pitts and Pitts, 2000; Simpson et al., 2015). Chlorine atoms are formed in the troposphere from reactions of N_2O_5 with NaCl in the sea-salt aerosol. This reaction produces $ClNO_2$, which can rapidly photolyze to yield Cl and NO_2. Once formed, Cl reacts with alkanes by abstraction, leading to peroxy radicals and HCl. They also react by addition to alkenes, leading to a chloroperoxy radical. The abstraction reactions of Cl with alkanes are very fast. But since Cl abstraction produces HCl, which is very water soluble, the Cl abstraction chemistry tends to be limited to areas near the source of Cl atoms. Also, since N_2O_5 is formed at night from the reaction of NO_3 with NO_2, the production of Cl requires a significant level of NO_2. So, both abstraction and addition reactions with Cl would be most important near polluted coastal cities.

References

Amachi, S., Kamagata, Y., Kanagawa, T., and Muramatsu, Y. (2001). Bacteria mediate methylation of iodine in marine and terrestrial environments. *Appl. Environ. Microbiol.* 67: 718–722.

Atkinson, R. (1986). Kinetics and mechanisms of the gas phase reactions of the hydroxyl radical with organic compounds under atmospheric conditions. *Chem. Rev.* 86: 69–201.

Atkinson, R. (1994). Gas-phase tropospheric chemistry of organic compounds. *J. Phys. Chem. Ref. Data*, Monograph No. 2 11–216.

Atkinson, R., Baulch, D.L., Cox, R.A. et al. (1997). Evaluated kinetic and photochemical data for atmospheric chemistry: Supplement VI. IUPAC Subcommittee on Gas Kinetic Data Evaluation for Atmospheric Chemistry. *J. Phys. Chem. Ref. Data* 26: 1329–1499.

Choi, W., Faloona, I.C., Bovier-Brown, N.C. et al. (2010). Observations of elevated formaldehyde over a forest canopy suggest missing sources from rapid oxidation of arboreal hydrocarbons. *Atmos. Chem. Phys.* 10: 8761–8781.

Darnall, K.R., Atkinson, R., and Pitts, J.N. Jr. (1978). Rate constants for the reaction of the OH radical with selected alkanes at 300 K. *J. Phys. Chem.* 82: 1581–1584.

Evelyn, J. (1661). *Fumifugium, or, the Inconvenience of the Aer and Smoak of London Dissipated Together with Some Remedies Humbly Proposed by J. E. Esq. to his Sacred Majestie, and to the Parliament Now Assembled*. London: Gabriel Bedel and Thomas Collins Shop.

Finlayson-Pitts, B.J. and Pitts, J.N. Jr. (1986). *Atmospheric Chemistry: Fundamentals and Experimental Techniques*. New York: Wiley.
Finlayson-Pitts, B.J. and Pitts, J.N. Jr. (2000). *Chemistry of the Upper and Lower Atmosphere: Theory, Experiments, and Applications*. San Diego, CA: Academic Press.
Jones, N.B., Riedel, K., Allan, W. et al. (2009). Long-term tropospheric formaldehyde concentrations deduced from ground-based Fourier transform solar infrared measurements. *Atmos. Chem. Phys.* 9: 7131–7142.
Leighton, P.A. (1961). *Photochemistry of Air Pollution*. New York: Academic Press.
Miller, P.R., Stolte, K.W., Duriscoe, D.M., and Pronos, J. (1996). *Evaluating Ozone Air Pollution Effects on Pines in the Western United States*, General Technical Report PSW–GTR–155. Albany, CA: U.S. Department of Agriculture, Forest Service, Pacific Southwest Research Station.
Mueller, J.F. (1992). Geographical distribution and seasonal variation of surface emissions and deposition velocities of atmospheric trace gases. *J. Geophys. Res.* 97: 3787–3804.
Paulson, S.E. and Orlando, J.J. (1996). The reaction of ozone with alkenes: An important source of HO_x in the boundary layer. *Geophys. Res. Lett.* 23: 3727–3730.
Senum, G.I., Fajer, R., and Gaffney, J.S. (1986). Fourier transform infrared spectroscopic study of the thermal stability of peroxyacetyl nitrate. *J. Phys. Chem.* 90: 152–156.
Simpson, W.R., Brown, S.S., Saiz-Lopez, A. et al. (2015). Tropospheric halogen chemistry: Sources, cycling, and impacts. *Chem. Rev.* 115: 4035–4062.
Werner, J., Vanzo, E., Li, Z. et al. (2016). Effects of heat and drought on post illumination bursts of volatile organic compounds in isoprene-emitting and non-emitting poplar. *Plant Cell Environ.* 39: 1204–1215.

Further Reading

Atkinson, R. (1986). Kinetics and mechanisms of the gas-phase reactions of the hydroxyl radical with organic compounds under atmospheric conditions. *Chem. Rev.* 86: 69–201.
Gaffney, JS & Levine, SZ 1979 Predicting gas-phase organic molecule reaction rates using linear free-energy correlations: $O(^3P)$ and OH addition and abstraction reactions. *Int. J. Chem. Kinet.*, vol. 11, pp. 1197–1209.
Gaffney, JS, Fajer, R & Senum, GI 1984 An improved procedure for high purity gaseous peroxyacetyl nitrate production: Use of heavy lipid solvents, *Atmos. Environ.*, vol. 18, pp. 215–218.

Study Problems

5.1 The chemistry of the troposphere is essentially what kind of chemistry?

5.2 (a) What is the most important oxidant in the day-time troposphere?
(b) The night-time troposphere?

5.3 (a) What is the lowest layer of the troposphere?
(b) What defines the behavior of this layer?

5.4 When is our air considered to be polluted?

5.5 What is the major source of the OH radical in the troposphere?

5.6 In the seventeenth century the combustion of coal was observed to be a major problem in what city in England?

5.7 (a) From what two words was the term "smog" derived?
(b) What two pollutant species created this smog?

5.8 What two major emissions always characterized coal burning in cities?

5.9 What was the new type of smog that occurred in Los Angeles, CA not associated with coal burning?

5.10 Arie Haagen-Smit discovered what two emissions from motor vehicles were causing the smog in Los Angeles?

5.11 (a) In 1970, the Clean Air Act was amended to establish what agency in the U.S. Federal Government?
(b) What was that agency's mandate?

5.12 (a) How many criteria pollutants were identified?
(b) What are they?

5.13 What is the difference between primary pollutants and secondary pollutants?

5.14 (a) Which of the criteria pollutants are primary pollutants?
(b) Which are secondary pollutants?

5.15 (a) What are the non-criteria pollutants that are known to be hazardous by the EPA listed as?
(b) How many were originally on the list?
(c) How many have been removed from the original list?
(d) How many have been added?

5.16 What is the source of the O atoms in the troposphere?

5.17 (a) What is a photostationary state?
(b) According to the photostationary state for the formation of ozone from NO_2, what must be true for ozone to form?

5.18 The OH radical reacts with alkanes by what type of reaction mechanism?

5.19 List the following in order of increasing rate of reaction with OH: RCH_3, CH_2O, CH_4, and R_2CH_2.

5.20 (a) What is the product of the OH reaction with an alkane followed by reaction with O_2 in the troposphere?
(b) This product reacts with NO to form what two products?

5.21 What alkane is the slowest to react with OH?

5.22 (a) When NO reacts with larger organic peroxy radicals, what type of intermediate is formed?
(b) This intermediate is followed by what type of internal isomerization?
(c) After isomerization of the intermediate followed by reaction with O_2, NO, and O_2, what water-soluble product can be formed?

5.23 (a) Calculate the rate constant for the OH abstraction reaction with cyclopentane using the partial k_{OH} abstraction rate constants for alkanes from Table 5.3.
(b) How does this compare with the measured value in Table 5.2?

5.24 (a) Calculate the rate constant for the OH abstraction reaction with pentane using the partial k_{OH} abstraction rate constants for alkanes from Table 5.3.
(b) How does this compare to the measured value in Table 5.2?

5.25 The reaction of OH with alkenes proceeds by what type of mechanism?

5.26 Which reaction proceeds by a faster rate: OH addition or OH abstraction?

5.27 After the OH radical reacts with an alkene, the addition of O_2 forms what kind of organic radical?

5.28 (a) How is the nitrate radical formed?
(b) Why is it only important at night?

5.29 (a) The NO_3 radical can react with alkenes by what type of reaction mechanism?
(b) The NO_3/alkene reaction cycle forms what type of product?

5.30 (a) The NO_3 radical can react to form N_2O_5 by an equilibrium reaction with what compound?
(b) Under what conditions is the formation of N_2O_5 favored?

5.31 (a) The reaction of ozone with VOCs is only important with what kind of hydrocarbon?
(b) By what mechanism?

5.32 (a) The reaction of ozone with alkenes forms what intermediate?
(b) What is the final product?

5.33 (a) The reaction of OH with CH_3CHO, followed by the addition of O_2, forms what radical?
(b) This radical can react with NO_2 to form what compound?
(c) This equilibrium reaction favors the product under what conditions?

5.34 (a) How many higher analogs of PAN are normally observed in the troposphere at measurable levels?
(b) What is the general name for this group of compounds and what is their general formula?
(c) They are in equilibrium with what two species?

5.35 (a) On a global level, what is the most important source of NO_x?
(b) Of the VOCs?

5.36 (a) What is the most stable hydrocarbon in the troposphere?
(b) Why is it the most stable?

5.37 What reaction forms HNO_3 in remote regions?

5.38 (a) What forms H_2SO_4 in remote regions?
(b) What is the source of the two reactants?

5.39 What organic halide compound is released from the ocean?

5.40 (a) The reaction of N_2O_5 with NaCl in the sea-salt aerosol forms what compound?
(b) This compound photolyzes to give what products?

5.41 (a) Cl atoms react with alkanes to give what products?
(b) Cl atoms react with alkenes to give what product?
(c) These reactions are most important in what areas? Why?

5.42 (a) What is the rate law for the reaction of two hydroperoxyl radicals (HO_2) reacting to form hydrogen peroxide and oxygen?
(b) What reaction is likely to compete with this reaction in a polluted environment?

6

Aerosols and Cloud Chemistry

Atmospheric aerosols consist of liquids or solids that have aerodynamic diameters between 2 nm and about 100 μm. They are commonly referred to as particulate matter (PM). Aerosols are considered to be a significant health risk, since they can be inhaled and the smaller diameter aerosols are able to reach the bronchial tubes and the lungs. This can cause acute tissue damage or long-term chronic effects depending on their size, shape, and chemical composition. Aerosols are usually considered to be a primary pollutant, since they are normally emitted directly into the atmosphere from a source. These aerosol primary pollutants are called primary aerosols. But aerosols can also be secondary pollutants when they are produced by chemical reactions in the troposphere. These aerosols are referred to as secondary aerosols. The tropospheric chemical reactions that form secondary aerosols often result in the production of an aqueous or a hygroscopic product. Once in the atmosphere they can either add mass by condensation of water or lose mass by evaporation.

Aerosols play an important role in the chemistry and physics of the troposphere, as well as in the formation of clouds. Most all aerosols have some water associated with them, except for those that are hydrophobic in nature, such as aerosols coated with nonpolar organic oils. So, they can provide sites for heterogeneous chemical reactions. The small aerosols that can take up water to assist in the formation of clouds are called cloud condensation nuclei (CCN). Indeed, the presence of CCN is required for the formation of most clouds in the troposphere. Aerosols and clouds play important roles in weather and climate due to their ability to absorb and scatter both incoming and outgoing radiation. They also form an important part of the hydrological cycle, returning evaporated water back to the surface through precipitation.

6.1 Aerosol Size Distributions

Aerosols exist in the troposphere over a wide range of sizes and chemical compositions. The first measurements of atmospheric aerosols were accomplished by using filter samplers to separate the aerosol particulate matter from the gases in the air stream. This sampling method used paper, glass, or quartz fiber filters to collect all the particulates in the air from 0.1 to about 50 μm in size. These first measurements were called total suspended particulates (TSPs) and measured the mass concentration of particulate matter in air. Using this method it was found that there was a significant amount of mass in the TSP, especially in the polluted urban atmospheres. Researchers then began to examine the sizes of the aerosols by using physical size separation methods to determine mass, surface area, and volume of the aerosols in different size ranges. Light-scattering techniques were used to determine the number of particles in each size range. These measurement methods will be discussed in more detail in Chapter 7.

Chemistry of Environmental Systems: Fundamental Principles and Analytical Methods, First Edition.
Jeffrey S. Gaffney and Nancy A. Marley.

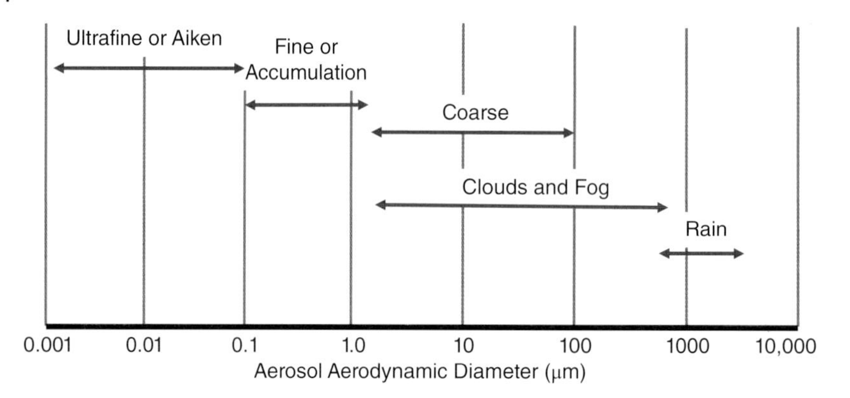

Figure 6.1 Typical size ranges for the three aerosol size modes, clouds, and rain.

Atmospheric aerosols have been found to exist in size ranges covering four orders of magnitude from about 2 nm to 100 μm. They fall into three categories or "modes," as shown in Figure 6.1. These aerosol modes are coarse, fine (also called the accumulation mode), and ultrafine (also called the Aitken mode). The smallest mode of aerosols is the ultrafine or Aitken mode, which are in the size range of 2–100 nm. Because of their very small size, the ultrafine aerosols behave as a gas with very low settling velocities (10^{-5} to 10^{-4} cm s^{-1}) and diffusion coefficients similar to the gases (10^{-2} to 10^{-5} cm^2 s^{-1}). So, they can readily collide and stick to the surfaces of other aerosols, causing them to grow in size and move into the larger size modes. This growth can also happen if the aerosols are very hygroscopic, such as ultrafine sulfuric acid aerosols. These hygroscopic aerosols can take up water vapor onto their surfaces, growing rapidly to larger sizes. The next largest mode is called the fine or accumulation mode. The accumulation-mode aerosols range in size from 0.1 to 2 μm with diffusion coefficients of about 10^{-7} to 10^{-8} cm^2 s^{-1}. Although these aerosols are small, they coagulate too slowly to reach the next larger mode. So, they have a relatively long lifetime in the troposphere, accounting for most of the atmospheric visibility degradation. Aerosols larger than the fine mode are the coarse-mode aerosols, which range in size from 2 to 100 μm. The coarse-mode aerosols have higher settling velocities ($> 1.0 \times 10^{-2}$ cm s^{-1}) and very low diffusion coefficients, so they are rapidly removed from the atmosphere by sedimentation. Since the ultrafine-mode aerosols grow rapidly to reach the next size range and the coarse-mode aerosols rapidly settle out because of their large size, the fine-mode aerosols have longer lifetimes in the atmosphere than aerosols in the other two modes.

Clouds and fogs are formed from atmospheric aerosols (CCN) that have condensed sufficient water vapor to reach sizes in the range of 2–800 μm, as shown in Figure 6.1. At this size the droplets have sufficient mass to fall out rapidly as rain. As the rain falls, droplets can coalesce to form larger rain droplets, but the size of the falling droplets is limited to a maximum of about 5000 μm (5 mm). This limitation is due to frictional interaction between the larger droplets traveling at higher settling velocities and the air flow surrounding them. This friction causes break-up of the larger droplets into smaller droplets, thus limiting the size that the droplets can maintain. Fogs commonly produce liquid precipitation in the form of drizzle. Drizzle is a light precipitation consisting of liquid water droplets smaller than about 0.5 mm. It occurs when the humidity reaches 100% and the small fog droplets begin to coalesce into larger droplets.

There are several methods of determining the amount or concentration of aerosols in the atmosphere. The method most often used for samples that contain a wide range of aerosol sizes, such as TSP or PM_{10}, is to determine the mass of the particles in the sample by weighing the entire collected sample. The mass concentration is defined as the mass of particulate matter per unit volume of air collected (units of μg m^{-3}). But the mass of the aerosol, which depends on the density of the particulates that make up the aerosol mixture, is not a good measure of the number or type of aerosol particles in a sample. Mass measurements are the standard method of determining aerosol amount as regulated by the EPA. But these air quality standards do not take into account the number of aerosols in each aerosol mode or their chemical composition, which can greatly affect the health, climate, and air quality impacts of the aerosols.

Since the sizes of atmospheric aerosols vary widely, aerosol size distributions are used to describe the variation in the sizes of the particles in the atmosphere. The most common method of determining the aerosol size distribution is to divide the complete size range of the aerosols into intervals and determine the amount of particles in each size range interval. Atmospheric aerosols that are composed primarily of solids are usually not spherical and so their size is described in terms of an aerodynamic diameter (D_a), also called the effective aerodynamic diameter. The aerodynamic diameter of an irregularly shaped particle is defined as the diameter of a spherical particle with a density of 1 g cm^{-3} that has the same settling velocity as the irregularly shaped particle. The aerosols are collected by fractionating them into a number of size ranges and the amount of aerosols in each range is measured as the number of particles. The number concentration is defined as the number of particles per unit volume of air collected (units of number m^{-3}). Also, by assuming a spherical shape to the particles, the surface area and the volume of the aerosols in each size range can be calculated from the aerosol aerodynamic diameter, which is defined by the sampling method. The aerosol sampling methods and their relationship to aerosol aerodynamic diameter will be discussed in Chapter 7.

The particle number, surface area, and particle volume size distributions are shown in Figure 6.2, plotted as a normal distribution on a logarithmic scale (lognormal distribution).

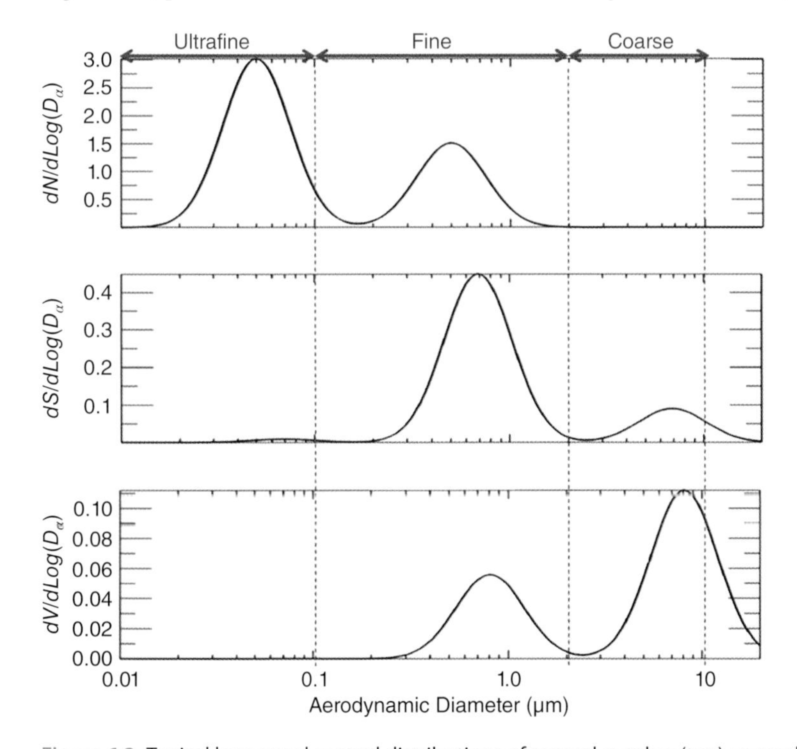

Figure 6.2 Typical lognormal aerosol distributions of aerosol number (top), aerosol surface area (middle), and aerosol volume (bottom) as a function of aerodynamic diameter. The three aerosol modes are shown in red at the top. Source: Adapted from Niall Robinson, Wikimedia Commons.

Traditionally, each of the variables is displayed as normalized to the log of the aerodynamic diameter ($dN/d\log D_a$, $dA/d\log D_a$, $dV/d\log D_a$) in order to compensate for the differences in the width of the aerosol size ranges used to determine the distribution. Typically, the number of aerosol particles is highest in the ultrafine mode, while the aerosol volume and mass are highest in the coarse mode. The aerosol surface area is usually dominated by the fine mode but varies depending on the source of the aerosols.

These aerosol size distributions depend on the emission sources, atmospheric transport times, possible deposition processes, and heterogeneous chemistry. Urban aerosols, consisting primarily of a mixture of primary anthropogenic emissions, have the highest number of aerosols in the ultrafine mode. Aerosols with the highest surface area are usually in the fine mode and those with the largest volume will be divided between the fine and coarse modes. In contrast, remote continental aerosols consisting of natural primary and secondary aerosols have large numbers of aerosols in both the ultrafine and fine modes. Because there is little anthropogenic influence on the remote continental aerosols, the highest surface area and volume are found in the fine mode.

As discussed in Section 5.3.1, the EPA standards for particulate matter are based on the mass of particulate matter less than 10 μm (PM_{10}) and particulate matter less than 2.5 μm ($PM_{2.5}$). The PM_{10} standard typically includes all of the ultrafine and fine aerosols and about two-thirds of the coarse-mode aerosols based on the aerosol volume distributions (Figure 6.2). The $PM_{2.5}$ standard typically includes all of the ultrafine and fine-mode aerosols and excludes the coarse mode. The information most important for aerosol health and climate effects is aerosol chemical composition, surface area, and number. This has led to the proposal that the EPA establishes a $PM_{1.0}$ standard (aerosols < 1.0 μm) in addition to their PM_{10} and $PM_{2.5}$ standards. A $PM_{1.0}$ measurement would be based on optical measurements and would measure the ultrafine and fine aerosols according to their number densities instead of their masses. Although the fine and ultrafine aerosols are included in the $PM_{2.5}$ measurements, the use of mass determinations greatly underestimates the number of aerosols in the range of 0.01–1.0 μm, as can be seen by comparing the aerosol number (Figure 6.2, top) with the aerosol volume (Figure 6.2, bottom) for aerosols less than 2.5 μm. These ultrafine and fine-mode aerosols have significant health and climate impacts and the key properties that are important in determining these impacts include aerosol number, size, bulk and surface chemical composition, physical properties, and morphology. This type of information cannot be obtained from simple mass measurements.

6.2 Aerosol Sources and Sinks

Atmospheric aerosols can be emitted into the atmosphere directly from a source (primary aerosols) or formed in the atmosphere from chemical reactions of gaseous precursors (secondary aerosols). Table 6.1 summarizes the sources, type, and removal processes of the aerosols in the fine and coarse modes. These include primary anthropogenic, primary natural, secondary anthropogenic, and secondary natural aerosol sources. Important primary anthropogenic aerosols include soot from combustion of diesel fuel or biomass and industrial mineral dust. Primary natural aerosols are typically produced by wind action on the ground (mineral) or ocean surface (sea salt). They can also be produced by volcanic activity (dust). Secondary aerosols are produced by chemical reactions, known as gas-to-particle conversion, involving anthropogenic or natural primary emissions of SO_2, NO_x, VOCs, and larger, less volatile organics (organic carbon, OC).

Table 6.1 The sources and removal processes of the important atmospheric aerosol species.

Type	Source	Composition	Mode	Removal process
Anthropogenic primary	combustion (diesel)	soot	fine	oxidation/wet deposition
Anthropogenic primary	biomass burning	soot	fine	oxidation/wet deposition
Anthropogenic primary	industrial dust	mineral	coarse	dry deposition
Anthropogenic secondary	coal burning (SO_2)	sulfates	fine	wet deposition
Anthropogenic secondary	combustion (VOCs)	organic	fine	wet deposition
Anthropogenic secondary	unburned fuel	organic	fine	oxidation/wet deposition
Anthropogenic secondary	combustion (NO_x)	nitrates	coarse	dry deposition
Natural primary	surface abrasion	mineral	coarse	dry deposition
Natural primary	ocean	sea salt	coarse	wet/dry deposition
Natural primary	volcanos	dust	coarse	dry deposition
Natural primary	biogenic (large OC)	organic	coarse	dry deposition
Natural secondary	biogenic (SO_2)	sulfates	fine	wet deposition
Natural secondary	volcanos (SO_2)	sulfates	fine	wet deposition
Natural secondary	biogenic (VOCs)	organic	fine	wet deposition
Natural secondary	lightning (NO_x)	nitrates	coarse	dry deposition

The ultrafine-mode aerosols are produced by high-temperature combustion and secondary aerosol formation processes. The combustion aerosols are found near the combustion sources. Because of the high number concentrations near their source, these ultrafine aerosols coagulate rapidly, giving them lifetimes of only minutes. The ultrafine secondary aerosols are produced by gas-to-particle conversion reactions of SO_2 and VOCs. These newly formed aerosols are very hygroscopic and begin to condense water shortly after formation, growing rapidly in size. So, both primary and secondary ultrafine aerosols grow rapidly to the fine aerosol mode. Because of this they are not included separately in Table 6.1. However, their sources are included in the fine-mode aerosols.

The fine aerosol mode includes combustion aerosols, secondary anthropogenic and natural aerosols, and coagulated ultrafine aerosols. Particles in this mode are small but they coagulate too slowly to reach the coarse mode. So, they have a relatively long lifetime in the atmosphere and can be transported long distances, especially for the non-hygroscopic aerosols that cannot condense water. The coarse-mode aerosols consist of wind-blown dust, large salt particles from sea spray, and mechanically generated anthropogenic particles such as those from agriculture and surface mining. Because of their large size, the coarse particles can readily settle out or impact on surfaces, so their lifetime in the atmosphere is typically a few hours. Fine aerosols are removed from the atmosphere primarily by incorporation into clouds, followed by wet deposition (rainout) since they are not large enough for dry deposition and do not readily coagulate into larger sizes. Fine-mode aerosols that are not hygroscopic, such as combustion soot, must first be oxidized by heterogeneous chemical reactions in order to be efficiently removed by wet deposition. This leads to their long lifetimes in the troposphere compared to the more hygroscopic fine-mode aerosols, such as sulfate. Removal of coarse aerosols is mainly by dry deposition, since their settling velocities are high. The very hygroscopic coarse aerosols, such as sea salt, can also condense water and be removed by wet deposition if the humidity is high enough where they are formed.

6.2.1 Primary Aerosol Emissions

The natural primary aerosols range in size from ultrafine to coarse mode. The fine and ultrafine modes include emissions from wildfires and fine volcanic dust. The coarse mode is generally produced from wind at speeds high enough to loft sand and dust particles greater than 2 μm in size. These high-speed winds are also capable of carrying sea-salt spray from breaking waves near the shore to areas farther inland. The biologically sourced coarse-mode particles – spores, pollen, and bacteria – are also windborne. Although the coarse-mode aerosols are rapidly removed by sedimentation because of their high settling velocities, they can be resuspended if the wind speeds are high enough (>20 mph). The volumes of coarse-mode aerosols are sufficiently large for them to interact with surface winds and undergo re-entrainment back into the atmosphere. In some cases, the prevailing wind speeds are fast enough to cause dust storms that loft the primary aerosols very high into the troposphere, allowing them to be carried long distances. These types of events typically occur in dry areas, such as in the western Saharan Desert in Africa and China. Mineral aerosols from African dust events have been detected as far away as southern Florida, and dust aerosols from western China have been detected in western North America.

The primary anthropogenic coarse-mode aerosols can be produced by mechanical grinding in industrial processes, and from tilling dry soils in agricultural activities. These dusts are typically derived from natural materials and are difficult to differentiate from naturally generated aerosols by simple examination. In these cases, a detailed elemental analysis is usually required to separate the natural from the anthropogenic sources. Most of the primary anthropogenic aerosols are fine-mode aerosols produced from fossil fuel combustion. The most important of these is carbonaceous soot produced from incomplete combustion. Carbonaceous soot is a mixture of black carbonaceous particulate matter, PAHs, and other organics. Although a major source of carbonaceous soot is diesel engines, it is also produced from the incineration of trash and agricultural waste burning. Uncontrolled agricultural burning of plant debris after harvesting is becoming increasingly more important, particularly as controls are placed on aerosol emissions from fossil fuel combustion. The composition of the agricultural aerosol emissions is dependent on the source of the biomass, combustion temperature, and burning conditions. Winter-time wood smoke from fireplaces is also an important source of carbonaceous soot in many areas. This is especially important for residential wood burning in valleys, since the smoke can accumulate during nocturnal weather inversions leading to high concentrations in the lower boundary layer. In the past, the use of smudge pots to warm the air around citrus groves during freezing events was also a major source of primary carbonaceous soot aerosols. However, this practice has been ceased because it is a known health hazard.

Primary carbonaceous aerosols, such as carbonaceous soot, are commonly called black carbon (BC), because they are all highly absorbing and therefore are black in color. Since the carbonaceous aerosols contain primarily sp^2 hybridized carbon atoms, they have also been called elemental carbon or graphitic carbon. But it should be noted that the black carbon aerosols contain a significant amount of hydrogen and so do not have the same thermal properties as graphitic carbon or elemental carbon. The carbonaceous soot aerosols are complex mixtures with highly variable chemical compositions, which depend on the combustion conditions. The name "black carbon" is used as a generic term to describe the black carbonaceous aerosol mixtures in much the same way that the name "petroleum" is used to describe oil mixtures or the name "coal" is used for the many solid fossil fuels with different compositions.

The one thing that all BC aerosols have in common is that they are all produced from incomplete combustion. They are also all associated with PAHs produced during combustion, which typically condense on the surfaces of the BC particles after formation. The fine-mode BC particulates have a high surface area, as shown in Figure 6.2. This surface acts to collect the PAHs and other high molecular weight organics, causing the aerosols to be initially hydrophobic. Both black carbon particulates and the PAHs are produced in significant quantities in diesel engines when they are operated under fuel-rich conditions. Fuel-rich conditions produce more power but burn cooler than fuel-lean conditions. The cooler combustion temperature also results in incomplete combustion of the diesel fuel, producing more BC particles, called diesel soot. Diesel soot particles are typically in the size region of 0.1–0.3 μm at the exit of the exhaust system, so they make up a significant amount (5–10%) of the fine-mode aerosols. Black carbon aerosols are also often associated with primary organic carbon (OC) emissions produced during low-temperature combustion processes, including agricultural burning.

Another primary anthropogenic aerosol produced by fossil fuel combustion is fly ash, which is usually a coarse-mode aerosol. Fly ash is an inorganic material left over from the high-temperature combustion of coal in power plants. The emission of fly ash from power plants was originally completely uncontrolled and so was a significant part of the urban aerosol problem. It is now more commonly removed from power plant emissions by using electrostatic precipitators or bag filters. The fly ash is then combined with bottom ash, which is a heavy granular residue that is left in the furnace after coal combustion. This ash material is stored at power plants in the form of mounds, or is placed into landfill. Fly ash, shown in the electron micrograph in Figure 6.3, varies significantly with the type of coal that is burned. While the fly ash is primarily a mixture of silicon dioxide, aluminum oxide, and calcium oxide, it can contain trace levels of many toxic elements in concentrations of up to hundreds of ppm. These trace elements include arsenic, beryllium, boron, cadmium, chromium, cobalt, lead, manganese, mercury, molybdenum, selenium, strontium, thallium, thorium, vanadium, and uranium, along with low concentrations of PAHs. While fly ash is now primarily collected and stored, it is still an environmental issue with regard to re-entrainment by high winds from the storage areas.

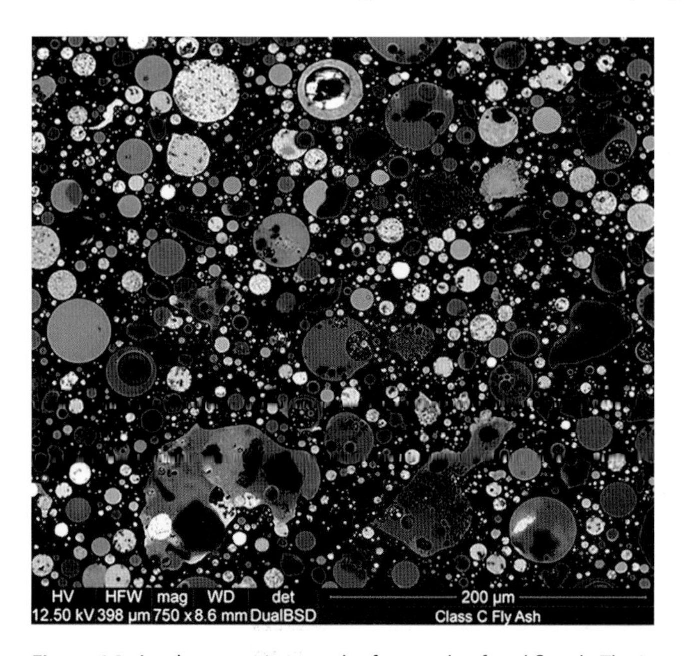

Figure 6.3 An electron micrograph of a sample of coal fly ash. The image was obtained from backscattered electrons and shows the differences in atomic density by variations in gray scale. Source: wabeggs, Wikimedia Commons.

6.2.2 Secondary Aerosol Formation

Secondary aerosols are formed from both gas-phase and aqueous-phase oxidation reactions of primary gas pollutants. Many of these primary gases are produced from fossil fuel or biomass combustion processes. But some are produced from natural emission sources such as vegetation, wetlands, ocean phytoplankton, or volcanos. In most cases the oxidation reactions form a species that has a very low vapor pressure, which condenses to form the aerosol. The one major exception to the oxidation process of secondary aerosol formation is the acid–base reaction of ammonia with HNO_3 and H_2SO_4 to form ammonium salts. These products are very hygroscopic, so they can grow in size by condensation of water vapor.

The reaction of NO_2 with an OH radical leads to the production of gas-phase HNO_3. Since HNO_3 is highly water soluble, it can be taken up in wet aerosols or cloud droplets where it dissociates, forming the aqueous H^+ ion and adding to the acidity of the wet aerosols or cloud water:

$$OH + NO_2 \rightarrow HNO_3(g)$$

$$HNO_3(aq) \rightarrow H^+(aq) + NO_3^-(aq)$$

But gas-phase HNO_3 can also react with one of the few basic gases emitted from the Earth's surface, ammonia (NH_3). In areas of high NH_3 emissions, such as near feedlots or agricultural lands using nitrate fertilizer, the HNO_3 can react with NH_3 to form the fine white ammonium nitrate aerosol:

$$HNO_3(g) + NH_3(g) \rightarrow NH_4NO_3(s)$$

The hygroscopic NH_4NO_3 can then take up water, forming a wet aerosol droplet, or it can be taken up into cloud water.

The principal sulfur compound that is emitted into the atmosphere from fossil fuel combustion is SO_2. Sulfur dioxide can also undergo gas-phase oxidation by OH in the troposphere. However, this reaction with SO_2 is a much slower reaction than with NO_2, and it is believed to involve an adduct ($HOSO_2$) which reacts with O_2 to form SO_3. The SO_3 then reacts rapidly with water to form H_2SO_4:

$$OH + SO_2 \rightarrow HOSO_2$$

$$HOSO_2 + O_2 \rightarrow HO_2 + SO_3$$

$$SO_3 + H_2O \rightarrow H_2SO_4$$

The more important oxidation path for SO_2 involves the reaction with H_2O_2 on wet aerosol surfaces or in cloud droplets. As discussed in Section 5.7, H_2O_2 is formed from the reaction of two HO_2 radicals. Since H_2O_2 is very soluble in water (see Section 6.5), it reacts rapidly with SO_2 on wet aerosol surfaces to form H_2SO_4, which rapidly dissociates as

$$2HO_2(g) \rightarrow H_2O_2(g) + O_2(g)$$

$$H_2O_2(g) + \text{wet aerosol} \rightarrow H_2O_2(aq)$$

$$H_2O_2(aq) + SO_2(aq) \rightarrow H_2SO_4(aq)$$

$$H_2SO_4(aq) \rightarrow 2H^+(aq) + SO_4^{2-}(aq)$$

The presence of HNO_3 and H_2SO_4 in cloud water increases the acidity, decreasing the pH of both the cloud water and the precipitation that forms from the clouds. The natural pH of cloud water is due to the presence of dissolved CO_2, by the following equilibrium reactions:

$$CO_2(g) + H_2O(l) \rightleftarrows H_2CO_3(aq)$$

$$H_2CO_3(aq) \rightleftarrows H^+(aq) + HCO_3^-(aq)$$

$$HCO_3^-(aq) \rightleftarrows H^+(aq) + CO_3^{2-}(aq)$$

So, CO_2 is also an acidic gas and the natural acidity of cloud water or rain water, without any input from HNO_3 or H_2SO_4, is determined by the CO_2 levels in the atmosphere and any CO_3^{2-} present from dust aerosols. The normal pH of clouds and rain from the CO_2/CO_3^{2-} equilibrium reactions is about 5.6. So, values lower than this are usually due to contributions from the strong acids, HNO_3 and H_2SO_4. But some other strong acids, such as HCl, as well as some weak acids, such as acetic acid (H_3CCOOH), can also act to increase cloud water acidity.

As discussed in Sections 5.4.3 and 5.4.4, OH radicals can react with primary organic emissions containing more than four or five carbon atoms to form organic compounds containing –OH and/or –COOH functional groups. These oxidized organics are very water soluble and can easily be taken up into wet aerosols. The five carbon species isoprene and the 10 carbon monoterpenes can react with OH, O_3, and NO_3 to give oxidized products that form secondary organic aerosols (SOAs). At the same time, these reactions also yield a significant amount of formaldehyde, which can photolyze and/or react with OH to give the HO_2 radical. Under low NO conditions, the HO_2 radical will form H_2O_2 and, in the presence of SO_2 and wet aerosols, will form acidic inorganic sulfate ($2H^+ + SO_4^{2-}$). So, in forested regions during the spring and summer months, the emissions from the vegetation can act to increase the SOAs as well as the secondary sulfate aerosols. The organic aerosols formed from these oxidation reactions are fine-mode aerosols and, due to their small size, will scatter the shorter-wavelength light (blue) more efficiently than the longer-wavelength light (red). This light scattering by the secondary fine aerosols gives rise to the blue haze seen over some mountain ranges, leading to their given names: Blue Ridge Mountains and Great Smoky Mountains in the eastern United States, as well as the Blue Mountains in Australia. Figure 6.4 shows an example of the blue haze that can be

Figure 6.4 The Blue Ridge Mountains as seen from Lynchburg, VA showing the blue haze covering the forested regions of the mountains in the spring and summer months. Source: Billy Hathorn, Wikimedia Commons.

seen in the Blue Ridge Mountains in the spring and summer months, which is caused by the fine-mode SOA light scattering.

In the spring and summer months the visibility in the Blue Ridge Mountains decreases due to the formation of acidic sulfate aerosols from the long-range transport of SO_2 emissions from coal-fired power plants. The conversion of the SO_2 to acidic sulfate occurs rapidly over these forested regions, since a significant amount of H_2O_2 is produced from the oxidation of the naturally emitted terpenes. The acidic sulfate aerosols take up water vapor and grow to larger sizes (about 500 nm), where they create a white haze, reducing the visibility significantly. During the fall and winter months, when the deciduous trees are dormant and lower temperatures reduce the natural organic emissions from the conifers, there is a reduction in the H_2O_2 production and a reduction in the sulfate aerosol formation. So, the SO_2 concentrations are higher and the sulfate aerosol concentrations are lower in the winter months. Inorganic nitrate levels can also be affected during the fall and winter months, as the conversion of NO_2 to HNO_3 depends on the atmospheric OH levels, which are formed by photolysis. Although in most cases OH formation is slower during the winter months due to the lower light intensity, in snow-covered areas the light is reflected back into the troposphere by the white snow doubling the photochemical path length and compensating for the lower winter-time light intensity. This causes the HNO_3 rates of formation in snow-covered winter months to be similar to those observed in the summer months.

The relative amounts of primary and secondary aerosols therefore vary with time and season due to the temporal and spatial variability of the natural and anthropogenic emission rates, photochemical reaction rates, and atmospheric conditions such as humidity and temperature. While most of the recorded data for atmospheric aerosols is based on mass, consistent with the EPA standards, there have been attempts to obtain the chemical composition of the aerosols in the $PM_{2.5}$ fraction. Annual averages of the major chemical species in the $PM_{2.5}$ aerosols in the United States from 2000 to 2005 are shown in Figure 6.5, along with winter and summer averages (Bell et al., 2007). The values in Figure 6.5 are given as a percentage by mass of the total mass of $PM_{2.5}$ aerosol. The major aerosol components are NH_4^+ (10–11%), NO_3^- (5–21%), SO_4^{2-} (19–31%), and OC (28–29%). The category labeled as "Other" (14–20%) usually contains more complex inorganic substances such as wind-blown dust and soil along with combustion-derived inorganic oxides that are similar to fly ash. The minor components are BC (4–5%), silicon (1%), and sodium (1%).

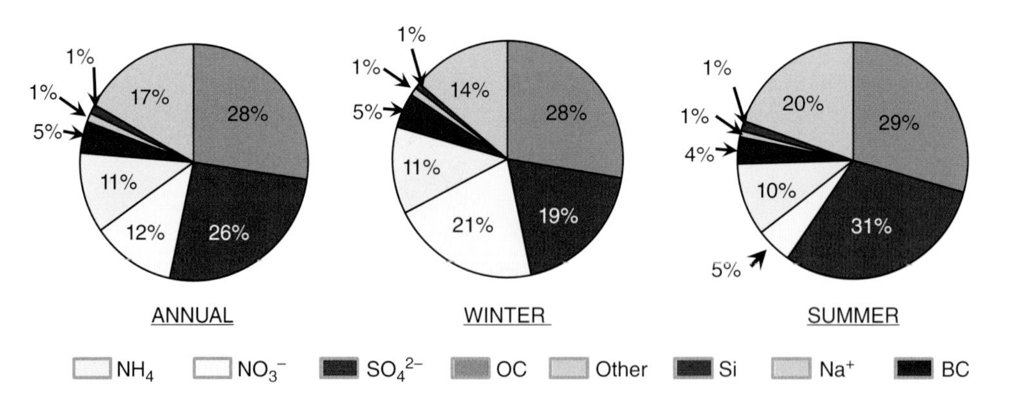

Figure 6.5 Annual average $PM_{2.5}$ aerosol chemical composition in the United States for the period between 2000 and 2005.

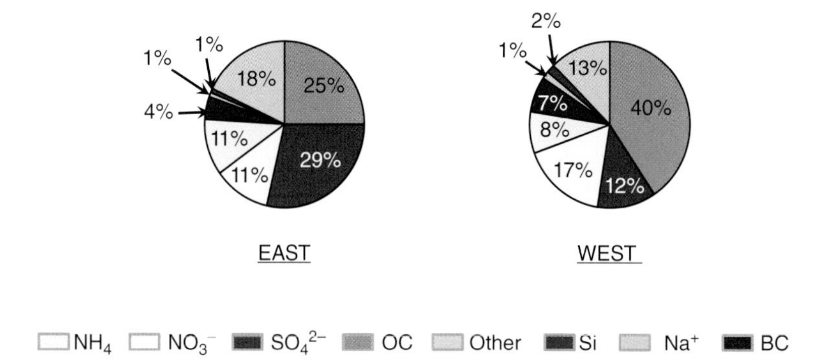

Figure 6.6 Average annual $PM_{2.5}$ aerosol chemical composition in the eastern United States and the western United States.

The SO_4^{2-} is higher in the summer (31%) than in the winter (19%). This is consistent with the oxidation reactions of the higher biogenic organic emissions in the summer and the accompanied increase in H_2O_2 from these reactions, which increases the aqueous reaction of H_2O_2 with SO_2 forming the increased sulfate aerosols. The OC levels are also higher in the summer (29%) than in the winter (28%), although it does not have as large a difference as the SO_4^{2-}. The OC component consists of both SOAs and primary OC, and also has both a biogenic and an anthropogenic component. In the winter, as the biogenic component decreases, the anthropogenic component increases as fireplace and wood stove use increases. This is also reflected in the increase in the primary BC and secondary NO_3^- formation during the winter.

The differences in the $PM_{2.5}$ aerosol composition for the eastern United States and the western United States for the period 2000–2005 is shown in Figure 6.6 (Bell et al., 2007). There is a large difference in the SO_4^{2-} content of the $PM_{2.5}$ aerosols in the eastern (29%) and western (12%) United States. This difference reflects the widespread use of coal in the eastern United States for power generation. Coal-fired power plants are not the main source of energy in the western United States. Most of the SO_2 emissions in the western states are from diesel engines, which are also a source of BC. So, the higher density of diesel vehicles results in higher BC levels in the western United States (7%) than in the eastern United States (4%). The OC content is much higher in the west (40%) than in the east (25%), as are the fine aerosol NO_3^- levels (17% and 11%) and primary BC (7% and 4%). This is most likely due to higher anthropogenic VOCs and NO_x in the higher populated areas of the west, with very high traffic volumes.

6.2.3 Wet Deposition and Henry's Law

Wet deposition is driven by the uptake of water-soluble species into wet aerosols where they can undergo heterogeneous chemical reactions as well as growth into cloud droplets. The fine-mode aerosols are removed from the atmosphere primarily by condensation of water and incorporation into clouds followed by rainout. The rate at which this occurs is dependent on the water solubility of the aerosol surfaces. The more polar compounds have higher water solubility, the nonpolar compounds have low water solubility, and compounds with an intermediate polarity have moderate water solubility. The oxidation of organics to form products with —OH or —COOH functional groups increases their water solubility, and the solubility increases as the number of these groups increases in the molecule. For example, the solubility of 1,2-ethanediol is about four orders of magnitude higher than that of methanol. Further oxidation of larger organic molecules leads to their having significantly increased water solubility, and so they are primarily found in the wet aerosols. This results in the accumulation of these oxidized secondary organic compounds in cloud water as well as in precipitation.

Freshly formed carbonaceous soot or BC species have an oily organic surface. These organics are not water soluble, so the particles will not be likely to uptake water on their surfaces. They have a low hygroscopicity, which is the ability of a molecule to take up water. However, eventually the organics coating the soot surfaces will undergo oxidation by OH radical, giving the particles oxidized functional groups such as —OH and —COOH on their surfaces. This provides the organics with a higher hygroscopicity and allows the particles to take up water, which leads to growth of the soot particles and eventually to their uptake into clouds where they can be removed by rainout. The lifetime of BC particles in the atmosphere will depend on the kind of organic coatings that are on their surfaces. Most of the organic compounds on the surfaces of BC from fossil fuel combustion in diesel engines will be large PAHs that do not react rapidly with the OH radical. But BC particles formed from biomass burning will likely have a significant amount of large alkanes or compounds containing alkene functional groups. These compounds will react faster with OH than the aromatic PAHs. So it is anticipated that the atmospheric lifetimes of BC aerosols will vary depending on their source, the temperature of the combustion processes, and the fuel type, which determines the types of organic surfaces on the BC aerosols.

The heterogeneous reactions that form SOAs depend on the uptake of the gas-phase reactants into the aqueous phase of wet aerosols. The uptake of inorganic and organic gases into wet aerosols and/or clouds can be approximated using Henry's Law, which considers the gas phase and aqueous phase of any species X to be in equilibrium:

$$X(g) \rightleftarrows X(aq)$$

with the equilibrium constant called the Henry's Law constant (H_x). The units of the Henry's Law constant are dependent on how the equilibrium expression is written, since the equilibrium can be viewed either as the gas traveling from the water into the air ($X(aq) \rightarrow X(g)$) or from the air into the water ($X(g) \rightarrow X(aq)$). These two cases result in different Henry's Law constants:

$$X(g) \rightarrow X(aq) \text{ where } H_x = [X(aq)]/[X(g)]$$

$$X(aq) \rightarrow X(g) \text{ where } H_x = [X(g)]/[X(aq)]$$

Since the important process in the atmosphere is the uptake of a gas-phase species into an aqueous environment, the Henry's Law constant important to the atmosphere is described as the solubility of a gas (X) in water:

$$H_x = [X(aq)]/P_x$$

where P_x is the partial pressure of X in atmospheres and the concentration of X in water is given in units of mol l^{-1} or molar (M). This gives the H_x in units of M atm^{-1}. For the opposite case of a water pollutant leaving the aqueous system to the atmosphere, H_x would be described as

$$H_x = P_x/[X(aq)]$$

and the units of H_x would be atm M^{-1}. So, care must be taken when using the values for Henry's law constants to predict the uptake of a gas species onto wet aerosols or into clouds in order to make sure that the units of H_x are correct for the atmospheric application. In addition, Henry's law constants are also temperature dependent and they can be enhanced in cases where chemical reactions are involved in the uptake of the gas species. The H_x values discussed here are given for 298 K and take into account any enhanced uptake due to chemical reactions. Table 6.2 shows

Table 6.2 Henry's Law constants (H_x) at 298 K for some common inorganic and organic gases of importance in environmental chemistry.

Gas	$H_x = [X]/P_x$ ($M\,atm^{-1}$)	$H_x = P_x/[X]$ ($atm\,M^{-1}$)	P_x (atm)	Concentration (M)
O_2	1.3×10^{-3}	7.7×10^{2}	0.21	2.6×10^{-4}
CO_2	3.4×10^{-2}	2.9×10^{1}	4.0×10^{-4}	1.4×10^{-5}
O_3	1.0×10^{-2}	1.0×10^{2}	6.0×10^{-8}	6.0×10^{-10}
SO_2	1.3	7.7×10^{-1}	1.0×10^{-8}	1.3×10^{-8}
H_2O_2	1.0×10^{5}	1.0×10^{-5}	1×10^{-9}	1.0×10^{-4}
NO	2.0×10^{-3}	5.0×10^{2}	1×10^{-9}	2.0×10^{-12}
NO_2	1.0×10^{-2}	1.0×10^{2}	1.0×10^{-8}	1.0×10^{-10}
HNO_3	2.0×10^{5}	5.0×10^{-6}	1.0×10^{-8}	2.0×10^{-3}
NH_3	6.0×10^{1}	1.7×10^{-2}	1.0×10^{-8}	6.0×10^{-7}
H_2CO	6.3×10^{3}	1.6×10^{-4}	1.0×10^{-8}	6.3×10^{-5}
CH_4	1.3×10^{-3}	7.7×10^{2}	2×10^{-6}	1.5×10^{-3}
C_2H_6	2.0×10^{-3}	5.0×10^{2}	1.0×10^{-8}	2.0×10^{-5}
C_2H_4	4.8×10^{-3}	2.1×10^{2}	1.0×10^{-9}	4.8×10^{-6}
CH_3OH	2.2×10^{2}	4.5×10^{-3}	1.0×10^{-9}	2.2×10^{-7}
CH_3COOH	1.0×10^{4}	1.0×10^{-4}	1.0×10^{-9}	1.0×10^{-5}
CH_2OHCH_2OH	4.0×10^{6}	2.5×10^{-7}	1.0×10^{-9}	4.0×10^{-3}

some of the Henry's Law constants for some common inorganic and organic gases of importance in environmental chemistry (Sander, 2015). An expanded list of Henry's Law constants for environmentally important species is given in Appendix C.

Henry's Law constants are very useful in determining the partitioning of chemicals in air and water. For the case of wet aerosols, they are important in determining the concentration levels for gas species in the wet aerosols and thus the concentrations that are available for heterogeneous reactions. The solubility of inorganic and organic species determines their potential for aerosol formation and growth by water condensation, as well as their removal by wet depositional processes, which determine their atmospheric lifetimes. In general, compounds that have a high hygroscopicity will have larger values for Henry's Law constants (units of $M\,atm^{-1}$) ($H_x = [X]/P_x$). For the case of volatile pollutant species in surface waters, the Henry's Law constants are a great aid in determining the volatility and concentration of compounds in the water. This application of Henry's Law will be discussed in Chapter 8.

6.2.4 Dry Deposition

Atmospheric aerosols undergo both dry and wet deposition. In dry deposition, the rate at which the aerosols are deposited on the surface is described as the deposition velocity, which is equal to the rate of deposition. This is typically described as a flux, which is the amount of material moving from the air to a surface, including either the oceans or terrestrial surfaces. The flux (F) is described as the product of the deposition velocity (v) and the concentration (c):

$$F = vc$$

Since the ultrafine aerosol particles behave as gases, they will undergo Brownian motion and diffusion, which allows them to coagulate to form aerosols in the fine mode (0.1–1.0 μm).

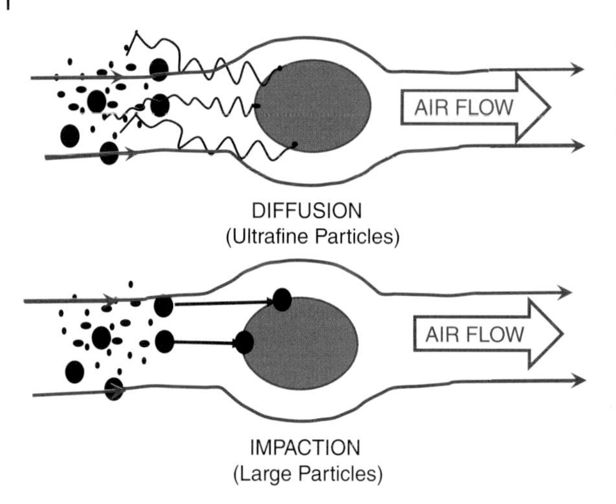

Figure 6.7 A schematic diagram of the removal of ultrafine particles by diffusion to a spherical surface and the removal of large particles by impaction to a spherical surface. The air flow around the surface is shown in red.

The fine-mode aerosols have slow diffusion rates and slow gravitational settling, and so they have the longest atmospheric lifetimes. Aerosols larger than this have significant loss due to gravitational settling and so have shorter lifetimes, which decrease as the aerosol size increases.

Dry deposition is the loss of the aerosol to surfaces, including other aerosols and large droplets. The larger aerosols are then removed by gravitational settling, which depends on the aerosol aerodynamic diameters. Figure 6.7 shows a schematic diagram of aerosol removal by diffusion (Figure 6.7, top) and by impaction onto droplets or larger aerosol surfaces (Figure 6.7, bottom). These mechanisms of dry deposition lead to the loss of the ultrafine aerosol through diffusion and the loss of the larger aerosol particles through impaction. The intermediate fine-mode aerosols have both low diffusion rates and low impaction rates and so they tend to move around the impaction surface with the air flow, resulting in little or no impaction occurring. This is the reason why the fine-mode aerosols tend to accumulate in the atmosphere, hence the name "accumulation mode" given to the fine-mode aerosols.

Diffusional removal is calculated using Fick's first and second laws. The loss by diffusion for both gases and ultrafine aerosol particles that behave as gases is determined by a diffusion depositional flux (J_{diff}), which can be calculated using the following equation:

$$J_{diff} = n\sqrt{D_p/\pi t}$$

where J_{diff} is the depositional flux in number of particles per second, t is the time of diffusion, n is the particle number density at $t = 0$, and D_p is the particle diffusion coefficient. Impaction rates are determined using the Stokes number (Stk). The Stokes number is a unitless number that characterizes the behavior of particles suspended in a flowing fluid. It is defined as the ratio of the characteristic stopping time (t), also known as the relaxation time, of the particle to the characteristic flow time of the fluid around the surface of an object, or:

$$Stk = \frac{tv}{d}$$

where the relaxation time of the particle (t) is the time constant of the exponential decay of the particle velocity due to drag from the fluid, v is the velocity of the fluid flow, and d is the diameter of the object. The movement of a particle with a low Stokes number (< 1) follows the fluid flow and will not impact on the surface of the object, while a particle with a large Stokes number (> 1) is controlled by its inertia and continues along its initial path, impacting on the surface of the object.

The gravitational settling of aerosols is determined using Stokes' Law, which describes the force of viscosity on a small sphere moving through a fluid. The gravitational settling velocity (w_g) is given by Stokes' Law as:

$$w_g = \frac{2(\rho_p - \rho_f)gr^2}{9\mu}$$

where ρ_p and ρ_f are the densities of the particle and the fluid, g is the acceleration due to gravity, r is the particle radius, and μ is the viscosity of the fluid in which the particle is settling. For the case of a dilute suspension of aerosols in air, $\rho_g = 1.225\ \text{kg m}^{-3}$ (1 atm, 288.15 K) and $\mu = 18.27\ \mu\text{Pa s}$ (291.15 K).

Table 6.3 gives gravitational settling velocities as a function of the aerosol diameters and diffusion rates (units of number of particles $\text{cm}^{-2}\ \text{s}^{-1}$). These values assume that air is the settling fluid, the particles have a density of one, and they are striking a horizontal surface. The ratio of the gravitational settling loss (G_{loss}) to the diffusional loss (D_{loss}) is also given for comparison of the loss mechanisms. The gravitational and settling loss are also shown in Figure 6.8 as a function of the log of the aerodynamic particle diameter ($\log(D_a)$). This clearly shows that the rate of aerosol loss is driven by the size of the particles. Particle loss by gravitational settling increases with the size of the particle and dominates the process for particles greater than

Table 6.3 Gravitational settling velocities (G_{loss}) and diffusion rates (D_{loss}) for particles as a function of aerodynamic diameter assuming a particle density of 1 striking a horizontal surface.

Particle diameter (μm)	Gravitational loss (number $\text{cm}^{-2}\ \text{s}^{-1}$)	Diffusional loss (number $\text{cm}^{-2}\ \text{s}^{-1}$)	Ratio of G/D
1×10^{-3}	6.5×10^{-7}	2.5×10^{-2}	2.6×10^{-5}
1×10^{-2}	6.7×10^{-6}	2.6×10^{-3}	2.6×10^{-3}
1×10^{-1}	8.5×10^{-5}	2.9×10^{-4}	1.9×10^{-1}
1	3.5×10^{-3}	5.9×10^{-5}	5.9×10^{1}
1×10^{1}	3.1×10^{-1}	1.7×10^{-5}	1.8×10^{4}
1×10^{2}	2.5×10^{1}	5.5×10^{-6}	4.5×10^{6}

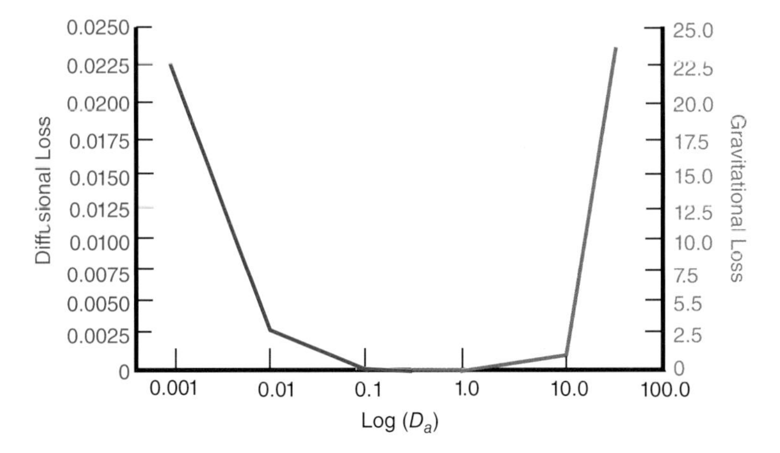

Figure 6.8 Diffusional loss (red) and gravitational loss (blue) (in units of number $\text{cm}^{-2}\ \text{s}^{-1}$) for the unit density particles as a function of aerodynamic diameter (D_a).

10 μm in diameter. The diffusional loss decreases with the size of the particle and is the major mechanism for loss of particles less than 0.01 μm. In addition, the particle size range between 0.01 and 1.0 μm is the most stable region for aerosol particles with regard to both gravitational and diffusional losses (Finlayson-Pitts and Pitts, 2000).

Gravitational settling can also describe the growth and rainout of cloud droplets. As the CCN condense water and become large cloud droplets, they can rapidly gravitationally settle resulting in loss of the aerosols. They can also evaporate water to reform the aerosols, leading to an increase in the aerosol concentrations. These processes are typically referred to as in-cloud scavenging or in-cloud processing of aerosols. Aerosol loss also occurs below the clouds when precipitation falls and the larger rain droplets impact the larger aerosols, causing entrainment into the precipitation. This process is also an impaction-removal mechanism, which again favors the preferential removal of particles greater than 1 μm in size. The exception to this is the removal of an aerosol particle by falling snow. The formation of the snowflakes gives rise to a large surface area (large d) for both impaction and diffusion mechanisms, resulting in the physical removal of fine and ultrafine-mode aerosols. So, while rainfall is not an effective loss mechanism for nonpolar aerosols such as carbonaceous soot or BC because of their low water solubility, they are found to be removed efficiently by snowfall through a diffusion or impaction mechanism. This can also occur during the formation of hail in thunderstorms. The hailstones can impact aerosols as they fall or are carried aloft by updrafts in the complex thunderstorm meteorology. This impaction mechanism can also act to remove non-hygroscopic fine-mode aerosols.

6.3 Aerosol Lifetimes

The lifetimes of atmospheric aerosols are determined by their loss rates, which are dependent on their size and chemical properties. The larger particles (>10 μm) typically have lifetimes of a few minutes or less as they rapidly settle to surfaces. The fine-mode aerosols have longer lifetimes, which range from a few days to months, depending on their ability to uptake water and grow to a size where impaction or gravitational loss can lead to their removal. While the principles controlling aerosol removal rates from the atmosphere are well known, the rate at which these mechanisms occur is not as well known because the direct observations of aerosol lifetimes and removal rates are difficult. The accidental release of the radioactive isotopes ^{137}Cs and ^{133}Xe during nuclear accidents, such as the one at the Fukushima Dai-Ichi nuclear power plant, have been used as tracers to determine aerosol lifetimes (Kristiansen et al., 2016). These studies concluded that the models used to predict aerosol removal rates resulted in aerosol lifetimes that were too short. This is most likely due to the models inaccurately predicting the time that is required for a fine-mode aerosol to grow to a size where it can be removed by rainout or gravitational settling.

The natural radionuclides ^{7}Be and ^{210}Pb have been widely used as tracers in environmental applications, such as soil erosion and transport processes in sediments, surface waters, and ground waters. These isotopes have also been used to determine the atmospheric aerosol lifetimes and removal rates. The formation mechanisms and lifecycles of both ^{7}Be and ^{210}Pb are shown in Figure 6.9. The ^{210}Pb is produced from the radioactive decay of ^{222}Rn gas, which is produced by a series of radioactive decay steps of ^{238}U, commonly found in soils and rocks. Since ^{222}Rn is a gas, it permeates through the subsurface until it is released into the lower troposphere. It then decays through another series of short-lived species to form ^{210}Pb. After formation, the ^{210}Pb acts as an ultrafine aerosol particle and rapidly diffuses to attach to fine-mode aerosols in the lower troposphere. The naturally occurring ^{7}Be is produced in the

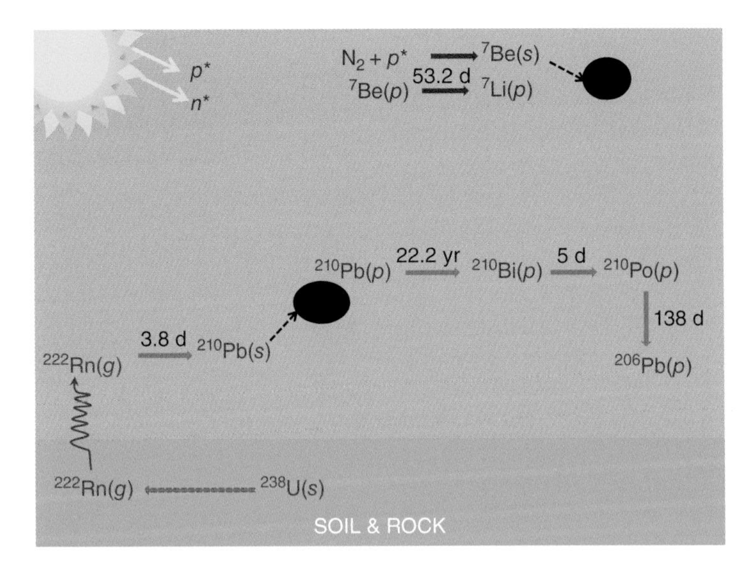

Figure 6.9 The formation of ^{210}Pb in the lower troposphere from the decay of ^{222}Rn and the formation of ^{7}Be in the upper troposphere from cosmogenic processes. Once formed, both radionuclides attach themselves to fine-mode aerosols (black). The $^{210}Pb(p)$ attached to the aerosol particle (*p*) then decays to $^{210}Bi(p)$ and $^{210}Po(p)$ with half-lives of 5 and 138 days.

upper troposphere and lower stratosphere by cosmogenic processes involving the bombardment of nitrogen and oxygen molecules with high-energy particles (excited neutrons and protons) from the Sun. It also rapidly attaches to fine-mode aerosols in the upper troposphere. So, the natural radioisotopes ^{7}Be and ^{210}Pb act as radioactive source labels for the fine-mode aerosols. The aerosols in the lower troposphere are labeled with ^{210}Pb, which has a half-life of 22.2 years, and the aerosols in the upper troposphere are labeled with ^{7}Be, which has a half-life of 53.2 days. The lifetimes of these fine-mode aerosols have been approximated by determining the amounts of ^{7}Be and ^{210}Pb in both wet and dry-deposited aerosols. The comparison of the production rates of the radionuclides with their deposition rates in both wet and dry aerosols gives an estimate of the major loss mechanisms as well as the relative time required for aerosol deposition.

More direct measurements of aerosol lifetimes have also been made by using ^{210}Pb and its daughters, ^{210}Bi and ^{210}Po, which have half-lives of 5 and 138 days, as a radioisotope clock method. Assuming that the only source of these two radionuclides in fine mode aerosols is the ^{210}Pb produced from the decay of ^{222}Rn, the ratios of the activity of the daughter (^{210}Bi or ^{210}Po) to the parent (^{210}Pb) can be used to estimate the atmospheric ages of the fine-mode aerosols (Marley et al., 2000). Aerosol ages determined by this method fall in the range of 5–10 days at urban sampling sites and in the range of 40–80 days in remote areas where background aerosols predominate. At urban sites, most of the mass of the aerosols is introduced into the atmosphere on a daily basis, and therefore should have an age of about 1 day. The longer ages obtained from the ^{210}Pb measurements (5–10 days) in urban areas result from the influence of the older background aerosols on the age estimation. The result of 5–10 days for urban aerosols can be obtained if a portion of the aerosol is from background sources. For example, if we assume that the background aerosol has a typical age of 50 days and there is about 10% of the aerosol loading at the urban sites from this background aerosol, then the "apparent age" of an urban aerosol will be 1 day plus 10% of 50 days, giving an observed age of 6 days, which is the value observed in a typical urban measurement.

The lifetimes of fine-mode aerosols are also dependent on altitude. Lifetimes of fine-mode aerosols in the lower 1 km of the troposphere are estimated to be from 1 to 10 days, where lifetimes in the middle troposphere are estimated to be about 25–40 days. At the tropopause, aerosol lifetimes are estimated to be 100–200 days, which is a similar lifetime to stratospheric aerosols. These much longer lifetimes at very high altitudes are primarily due to a lack of water vapor in these cold and dry atmospheric regions. This lack of water vapor results in very slow condensation processes, even for the most hygroscopic aerosols. Similarly, there are also latitudinal effects for aerosol lifetimes, which are also dependent on the availability of water vapor for condensational growth and loss by cloud processing. At latitudes where it is very cold and dry, the aerosols can transport longer distances before they are removed from the atmosphere. This is particularly important for the transport of aerosol-borne pollutants to the polar regions followed by dry deposition.

With average lifetimes of 10 days or more in the lower to middle troposphere, aerosols can be transported long distances. In a 10-day time frame air parcels can travel over continents and oceans. For example, it is well documented that aerosols have traveled from the Saharan Desert across the Atlantic Ocean to reach Bermuda and eastern North America during dust storm events. Aerosols have also been tracked from China across the Pacific Ocean to western North America. The distances traveled by the aerosols are correlated with both the aerosol hygroscopic nature and the occurrence of rain events during their transport. With the advent of satellite and International Space Station imagery, the long-range transport of aerosols from wind-blown dust events and from large biomass fires has clearly been observed, with the smoke and dust being carried thousands of kilometers. An example of the long-range transport of aerosols from a natural volcanic source, Mt. Etna, is shown in Figure 6.10.

Figure 6.10 The eruption aerosol plume from Mt. Etna on January 2, 2002, which was observed tracking thousands of kilometers from the International Space Station. Source: NASA, International Space Station Crew.

6.4 Determination of Aerosol Sources

Determining the sources of aerosols is important in developing a control strategy to minimize their impact. The determination of the sources of inorganic aerosols has principally been accomplished by measurements of elemental composition and particle morphology. Both major and trace element signatures have been used to determine sources of the inorganic fraction of primary particulate matter such as fly ash and industrial or mineral dusts. For example, fly ash can be identified as round particles containing mostly Al and Si with trace elements of As, Ba, Cd, Sr, and V. Mineral dust particles are mostly aluminum silicates or $CaCO_3$ with complex aggregated shapes. The presence of sulfate and nitrate identifies particulates as secondary aerosols with sources being the precursor gases: SO_2 and NO_x. The anthropogenic combustion sources of these precursor gases are well known and are currently regulated by the EPA, which is leading to the reduction of both the levels of the precursor gases and the secondary aerosols formed from them.

The determination of the sources of organic or carbonaceous aerosols is not as straightforward, but due to their increasing importance and their potential toxicity, the ability to obtain information concerning the various sources of organic aerosols has been pursued using a number of methods. The direct measurement of organic source signature compounds can identify a limited number of sources. For example, species known to be formed during high-temperature combustion, such as the PAHs, can indicate fossil fuel combustion sources. The major problem with using signature compounds to identify organic aerosol sources is that, due to fine-mode aerosol coagulation, many sources may contribute to the aerosol particles. Also, although it can sometimes identify potential source contributions, it cannot usually give quantitative information about these contributions. These problems with identifying the sources of both primary and secondary organic or carbonaceous aerosols have led to the use of carbon isotopic signatures to attempt to differentiate between fossil fuel combustion and biogenically sourced organic aerosols. There are three carbon isotopes found in the environment: stable ^{12}C (99%), stable ^{13}C (1%), and radioactive ^{14}C (trace amounts). The amount of ^{14}C and the $^{13}C/^{12}C$ stable isotopic ratios can be used to determine the amount of aerosol carbon from fossil fuel combustion, biomass burning, and biogenic organic emissions.

The use of carbon isotopes to determine the biogenic source of aerosol carbon is based on the differences in the biochemical pathways in which carbon dioxide is taken up into different types of plants. Prehistoric plants as well as most present-day plants take up carbon dioxide from the air through their stomata and produce the plant's organic matter by way of the Calvin–Benson biochemical cycle, shown schematically in Figure 6.11 (top). The actual photosynthetic process is quite complicated and involves the interaction of ribulose biphosphate (RuBP) with carbon dioxide and nicotinamide adenine dinucleotide phosphate (NADPH), water, and sunlight. It is commonly referred to as the C-3 cycle, since it involves the incorporation of three CO_2 molecules into a triose monosaccharide (glyceraldehyde), which is a three-carbon compound with chemical formula $CH_2OHCHOHCHO$ ($C_3H_6O_3$). The glyceraldehyde is then used to produce glucose and plant organic matter by other biochemical processes. In the formation of glyceraldehyde, oxygen is released to the atmosphere through plant respiration. This process of plant respiration is the source of oxygen in the atmosphere.

The C-3 cycle was the only photosynthetic process used in plants up until about 35 million years ago, when plant biochemistry evolved to produce the Hatch–Slack biochemical cycle, shown schematically in Figure 6.11 (bottom). The Hatch–Slack cycle is known as the C-4 cycle because it fixes one CO_2 molecule with the three-carbon compound phosphoenol pyruvate

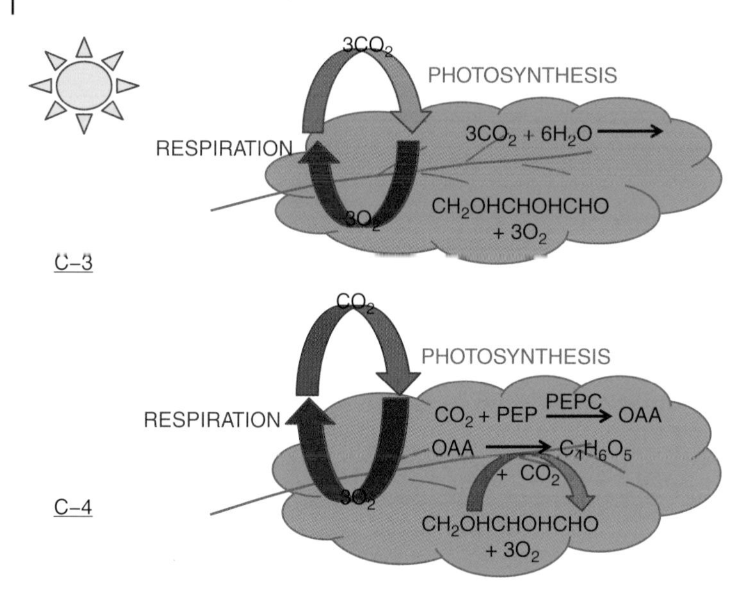

Figure 6.11 A schematic representation of the Calvin–Benson C-3 photochemical cycle, which incorporates three CO_2 molecules into a triose monosaccharide (glyceraldehyde) and the Hatch–Slack C-4 photochemical cycle, which fixes one CO_2 molecule with a three-carbon compound (PEP) in an enzyme-catalyzed reaction to form a four-carbon product (OOA).

(PEP) to form a four-carbon product: oxaloacetic acid (OAA). The C-4 plants have a unique leaf structure called the Kranz leaf anatomy, which contains a complex enzyme called phosphoenolpyruvate carboxylase (PEPC). The PEPC enzyme catalyzes the reaction of PEP with CO_2 to produce the four-carbon diacid (OAA). In the C-4 cycle the OAA is then converted to malic acid ($C_4H_6O_5$), releasing CO_2 inside the leaf. The C-4 biosynthetic pathway then follows the same path as the C-3 cycle using the CO_2 released by the conversion of OAA to form glyceraldehyde.

The C-4 cycle uses less water than the C-3 cycle because the tighter leaf structure of the Kranz anatomy reduces evapotranspiration. This is an evolutionary result of the C-4 plants, which grew in open grasslands where sunlight was abundant but water was less available, developing the Hatch–Slack photosynthetic cycle about six to seven million years ago. It is important to note that the new photosynthetic pathway was not available to the plants that produced the Earth's fossil fuels, which lived hundreds of millions of years ago. So, all plants that produced the fossil fuels used the C-3 biochemical cycle. Today, the C-4 plants make up only about 5% of the current biomass. Some major C-4 plants are corn and sugar cane, which are extremely important in agriculture. This is especially the case in North and South America, where corn is the basis for many food products and for animal feeds. Corn and sugar cane are also the major starting materials for biomass-derived ethanol used for gasohol and other biofuels.

The kinetic isotope effect can lead to the favoring of one isotope over another in a chemical reaction. This is also true for plant biochemical reactions. The effect can be measured as the ratio of the rate constant of the reaction with the light isotope (k_L) to the rate constant of the reaction with the heavy isotope (k_H):

$$\text{kinetic isotope effect} = \frac{k_L}{k_H}$$

The C-4 cycle is more efficient at trapping CO_2 and using it in the formation of the plant biomass than the C-3 cycle. Because of this, the C-4 cycle does not cause a large kinetic isotope effect. Conversely, the C-3 cycle preferentially uses CO_2 containing the lighter carbon isotope (^{12}C), which is the most abundant isotope. So, plant organic matter that is produced by the C-3 cycle will be depleted in CO_2 containing the heavier carbon isotopes ^{14}C and ^{13}C compared to that produced by the C-4 cycle. So, C-4 plant matter will have a larger amount of the heavier carbon isotopes than C-3 plant matter. This variation in carbon isotopic content of plant biomass can be used to determine sources of carbonaceous or organic aerosols formed from different combustion processes.

Radiocarbon (^{14}C) is formed primarily in the upper troposphere and lower stratosphere by high-energy particles from the Sun, primarily high-energy neutrons (n^*), colliding with molecular nitrogen to form ^{14}C and a proton (p):

$$^{14}N + n^* \rightarrow {}^{14}C + p$$

The ^{14}C formed has a half-life of 5700 years. Once formed, the ^{14}C is oxidized to CO_2 and is taken up into plants by photosynthesis, and absorbed into animals who consume the plant material. So, this natural radiocarbon is present in all living organisms that continue to replace their carbon content. Once the organism dies and the carbon is no longer in equilibrium with the atmospheric source through the biochemical pathways, the ^{14}C in the organism will begin to decay. This decay is the basis for age dating of carbonaceous materials. With a half-life of 5700 years, age dating is accurate for ages from about 600–700 years up to about 10 half-lives, which is 57 000 years.

Fossil fuels were formed from decaying plant and animal matter over hundreds of millions of years, and essentially contain no ^{14}C. So, the measurement of the ^{14}C content in atmospheric aerosols can be used to determine the amount of fossil and/or biomass sources of the aerosol carbon. Modern advances in accelerator mass spectrometry (AMS) require only about 20–100 mg of carbon for a ^{14}C determination. This high sensitivity of the AMS method allows for the amount of ^{14}C to be determined on size-fractionated fine-mode aerosols. The aerosol samples are collected on carbon-free quartz filters, which are completely combusted to CO_2 and then converted to graphite for the AMS measurement of ^{14}C content in the aerosol carbon by mass. Corrections are made for any isotopic fractionation that may have occurred during the sample preparation using the $^{13}C/^{12}C$ ratio. Also, corrections are made for sample contamination from ^{14}C bomb carbon in the atmosphere that is derived from above-ground nuclear bomb testing in the 1950s and early 1960s. These corrections to the ^{14}C measurements are discussed in more detail in Chapters 7 and 12.

The results of the ^{14}C determinations are commonly reported as the fraction of modern carbon (fM). This gives a direct mass amount of the aerosol that is sourced from modern biomass. So, aerosols that are totally produced from biomass burning or SOAs formed from plant-derived terpene compounds will have 1.0 fM, while fossil fuel-derived carbonaceous aerosols will have 0.0 fM. Some values of fM from the AMS ^{14}C measurements on PM_{10} and $PM_{2.5}$ aerosol samples collected at 13 different sites are compared in Table 6.4 (Marley et al., 2009). The values range from 0.95 in Launceston, Australia to 0.27 in Denver, CO with a median value of 0.55. This data indicates that a substantial fraction of the carbonaceous aerosols in many locations is from biogenic sources, including the formation of SOAs from the oxidation of terpenes, agricultural burning, and/or wild fires. The fossil fuel component of the aerosols is determined from the fraction of modern carbon and reported as the percentage fossil by mass of aerosol carbon. The aerosol fossil fuel content ranged from 73% in Denver, CO to 5% in Launceston, Australia. Studies in Launceston have estimated that about 73% of the particulate matter is caused by wood smoke in the winter months, while only about 8% is from motor vehicle emissions.

Table 6.4 The ^{14}C content of PM_{10} and $PM_{2.5}$ aerosol samples collected at a number of sites from 1982 to 2006 in winter and summer. Results are reported as the fraction of modern carbon (fM).

Site	Aerosol size	Season	Year	fM	% Fossil
Los Angeles, CA	PM_{10}	summer	1982	0.31	69
Los Angeles, CA	PM_{10}	winter	1982	0.41	59
Long Beach, CA	PM_{10}	summer	1982	0.49	51
Long Beach, CA	PM_{10}	winter	1982	0.47	53
Denver, CO	$PM_{2.5}$	summer	1996	0.44	56
Denver, CO	$PM_{2.5}$	winter	1996–1997	0.27	73
Nashville, TN	$PM_{2.5}$	summer	1999	0.69	31
Houston, TX	$PM_{2.5}$	summer	2000	0.54	46
Tampa, FL	$PM_{2.5}$	summer	2002	0.75	25
Zurich, Ch	$PM_{2.5}$	summer	2002	0.63	37
Mexico City, Mx	$PM_{2.5}$	winter	2003	0.70	30
Launceston, Au	PM_{10}	winter	2003–2004	0.95	5
Seattle, WA	$PM_{2.5}$	summer	2004	0.55	45
Seattle, WA	$PM_{2.5}$	winter	2004–2005	0.58	42
Tokyo, Jp	$PM_{2.5}$	summer	2004	0.38	62
Tokyo, Jp	$PM_{2.5}$	winter	2004–2005	0.47	53
Phoenix, AZ	$PM_{2.5}$	summer	2005	0.60	40
Phoenix, AZ	$PM_{2.5}$	winter	2005–2006	0.53	47
Mexico City, Mx	$PM_{2.5}$	winter	2006	0.60	40
Tecamac, Mx	$PM_{2.5}$	winter	2006	0.76	24

Stable carbon isotopic ratios $^{13}C/^{12}C$ are determined by comparison of the sample with the ^{13}C content of a standard limestone from the Peedee belemnite (PDB) limestone formation. The results are reported as a parts per thousand (‰) difference from the standard in del units ($\delta^{13}C$(‰)) using the following formula:

$$\delta^{13}\text{C (‰)} = \left(\frac{\frac{^{13}\text{C}}{^{12}\text{C}}\text{ Sample} - \frac{^{13}\text{C}}{^{12}\text{C}}\text{ Standard}}{\frac{^{13}\text{C}}{^{12}\text{C}}\text{ Standard}} \right) \times 1000\ \delta(\text{‰})$$

Since the PDB standard is enriched in ^{13}C relative to most other carbonaceous materials, the values obtained for $\delta^{13}C$ (‰) on aerosol samples are negative compared to the standard value. This means that the less negative $\delta^{13}C$ (‰) values are those with a higher ^{13}C content. Table 6.5 gives the $\delta^{13}C$ (‰) values for some important organic samples representative of carbonaceous aerosol source materials. The selected materials include C-3 plants (oats, wheat, and rice), C-4 plants (corn, sorghum, sugar cane), and coal, derived from C-3 plants. The values obtained for atmospheric CO_2 and oxalic acid, used as a standard for ^{14}C measurements, are also given for comparison. Both corn and sugar cane, which are C-4 plants used in making ethanol for gasohol, have values of −10 to −12 $\delta^{13}C$ (‰). This is much higher than the values for the C-3 plants (−21 to −25 $\delta^{13}C$ (‰)), which include the fossil fuel sources. Since gasohol is a blended mixture of ethanol and gasoline, the $\delta^{13}C$ (‰) of the aerosols produced from the combustion of

Table 6.5 $^{13}C/^{12}C$ isotopic ratios ($\delta^{13}C$) in parts per thousand (‰) obtained on selected samples representative of carbonaceous aerosol source materials compared to the PDB standard limestone.

Material	Type	Mean	Range
PDB limestone	^{13}C standard	0	0
Atmospheric CO_2	gas	−8	−6 to −10
Oxalic acid	^{14}C standard	−19	−17 to −21
Wood/charcoal	biogenic	−24	−22 to −26
Tree leaves	biogenic	−27	−25 to −29
C-3 plants	biogenic	−23	−21 to −25
C-4 plants	biogenic	−11	−10 to −12
Coal	fossil	−23	−21 to −25

gasohol would depend on the original percentage of the fuel mixture. The $\delta^{13}C$ (‰) for E10 (10% ethanol) would be around −21.8, while the $\delta^{13}C$ (‰) for E85 (85% ethanol) would be −12.8.

When the ^{14}C fM and $\delta^{13}C$ (‰) values are used together, they can be very useful in differentiating between fossil fuel-derived carbonaceous aerosols (C-3 plant-derived) and those formed from grass corn, and sugar cane burning, which are all C-4 plants. An example of this is shown in Figure 6.12, which correlates ^{14}C fM and $\delta^{13}C$ (‰) measurements of fine-mode aerosol samples obtained during the same time periods at two sites that were 30 km apart in central Mexico. The aerosols at the first site (Figure 6.12, red) were collected in downtown Mexico City while those collected at the second site (Figure 6.12, green) were from a rural region outside of the city. While the urban site was heavily impacted by fossil fuel combustion, the rural site was occasionally impacted by agricultural and biomass burning events. The carbonaceous aerosols from both sites indicated biomass sources at both sites, as shown by the fM values greater than 0.4 in all the samples. However, the aerosols at the rural site had a higher biomass signature

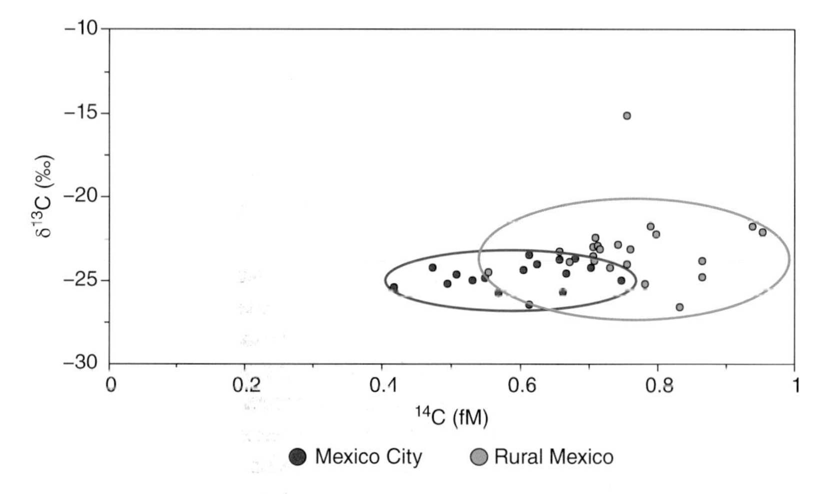

Figure 6.12 The ratio of $^{13}C/^{12}C$ isotopes ($\delta^{13}C$ (‰)) as a function of ^{14}C (fM) for fine-mode aerosol samples taken at two sites in and near Mexico City in March 2006. The samples in red were taken at a site in Mexico City and those shown in green were taken at a rural site 30 km outside the city. The green outlier at $\delta^{13}C$ (‰) = −15 was due to a local grass fire (C-4).

(fM > 0.55–0.95) than the urban site (fM = 0.4–0.75). This was expected due to the larger number of fossil fuel sources, such as diesel trucks and buses, in the city. During one event, the rural site was heavily impacted by a small nearby grass fire. This event resulted in a $\delta^{13}C$ (‰) value of −15 due to much of the aerosol carbon coming from the combustion of the C-4 grasses. So, the identity and amount of the impact of the local biomass combustion source was confirmed by the carbon isotope determinations at these sites. This shows the power of carbon isotope measurements in evaluating sources of carbonaceous aerosols in different environments.

6.5 Aerosol Health Effects

Atmospheric aerosols have been identified as an air pollution health hazard by the EPA since the 1950s, when they were originally classified as TSPs. The EPA standards for TSPs were set in 1971, and in 1987 the standards were revised to include PM_{10}. They were revised again in 1997 to include $PM_{2.5}$. These changes in air quality standards for atmospheric aerosols to include the measurement of the fine-mode aerosols separately reflect our changing knowledge of how particles impact human health. Most of this new information on the health effects of fine aerosols came from studies of the health effects of cigarette smoking, which involves direct inhalation of fine smoke particles derived from the combustion of organic materials. Because of this, there was a growing recognition of the importance of the fine-mode aerosols, since they have the ability to deposit deeper into the respiratory system.

The impact of aerosols after inhalation depends on their size and their ability to take up water and grow to larger sizes. Figure 6.13 shows a diagram of the human respiratory system and the size ranges of the aerosols that are removed from the air flow by impaction, gravitational sedimentation, and diffusion. The larger-size aerosols are removed by impaction in the pharynx (5–10 μm), trachea (3–5 μm), and primary bronchi (2–3 μm). The fine-mode aerosols (1–2 μm) are removed by sedimentation in the bronchioles, and the ultrafine-mode aerosols (1–0.1 μm) are removed by diffusion in the alveoli. Since the human breath is about 5–6% water vapor, hygroscopic aerosols, which have a high water affinity, will grow to a sufficient size rapidly enough that they will be removed by impaction in the nasal and upper respiratory areas. Acidic aerosols are rapidly neutralized by exhaled ammonia, which also affects the impacts of aerosols on tissue surfaces. The fine-mode hydrophobic aerosols, such as carbonaceous soot, can be carried deeper into the lungs, where they are deposited in the alveoli by diffusion. Asbestos, an inorganic hydrophobic aerosol, consists of particles with long spear shapes on the order

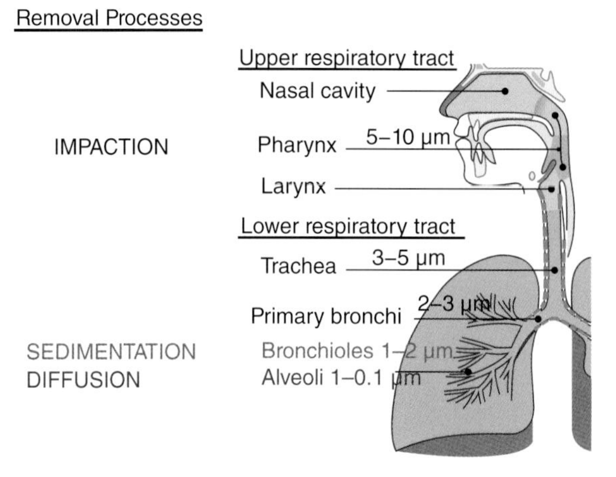

Figure 6.13 Aerosol removal processes from the air flow in the human respiratory system. Source: Adapted from Lord Akryl Jmarchn, Wikimedia Commons.

of 3–10 μm. This size particle would normally be trapped in the upper respiratory system by impaction, but because of their spear shapes they have an effective aerodynamic diameter of a much smaller particle. The removal mechanisms for aerosols in the human respiratory system assume diameters of aerosols with a spherical geometry. So, aerosol materials that are long and narrow in shape have much smaller effective aerosol diameters than their length implies. This causes them to be deposited deeper into the respiratory system than expected, leading to the potential for more damage. For this reason, asbestos fibers and other types of industrial materials of similar sizes and shapes, such as carbon nanotubes, are of major concern as a respiratory health hazard and care must be taken when working with these substances.

When hydrophobic aerosols are inhaled into the lungs, the body does two things to remove them. First, it attempts to convert the aerosols to a more hygroscopic form through oxidative biochemical reactions which produce hydrogen peroxide. Once oxidized, the particles are taken up into the body fluids and into the blood stream, eventually being cleared out through the urinary tract. Second, the body uses macrophages (white blood cells) to seek out and ingest the hydrophobic material producing phlegm, which is usually coughed up from the lungs as sputum. This clearing mechanism is called phagocytosis, the process by which a cell engulfs a solid particle to form an internal compartment known as a phagosome.

If the material is deep in the lungs the macrophages can carry the material into the lymphatic system where it can be carried into the bronchial and carinal lymph nodes. This is the clearing mechanism for carbonaceous BC particles, which are commonly found in these lymph nodes and are known as black pulmonary pigment (Slatkin et al., 1978). Crystalline aerosol species, such as asbestos, are highly toxic to macrophages and lead to their death. In some cases, the length of the particle causes the macrophage wall to break when the macrophage attempts phagocytosis. The breaking of the macrophage cell wall releases biochemicals from the macrophage into the lungs, which can cause damage to the surrounding lung cells, resulting in inflammation and scarring of the tissues.

Once in the respiratory system the aerosols can cause two types of effects, acute and chronic. Acute effects are short-term effects caused by the initial inhalation of the aerosols. This initial inhalation starts the respiratory clearing mechanisms. In some cases, this response can act to amplify any pre-existing respiratory problems, such as asthma or emphysema. In any case, inflammation can occur and difficulty in breathing can be experienced, depending on the aerosol concentration, size, and chemistry. In some severe cases, such as smoke inhalation during a fire, the exposure can lead to pulmonary failure and death. However, in most cases the acute effects subside after the exposure ceases.

A number of hydrophobic aerosol species can also cause chronic health effects, including functional loss and scarring of the lungs. In some cases, the aerosols deposit potential carcinogenic materials, which can be activated through biochemical oxidation reactions. An example of this is the PAH benzo(a)pyrene (BaP), which is associated with carbon soot. BaP is oxidized in the body by biochemical enzymatic reactions in an attempt to solubilize it. These biochemical reactions, shown in Figure 6.14, produce the very active benzo(a)pyrene diol epoxide metabolites, which can bind to DNA and lead to the production of cancerous cells. Similar biochemical reactions of other PAHs and nitro-PAHs present in carbonaceous aerosols also lead to DNA active metabolites that can act as carcinogenic or mutagenic agents. Asbestos exposure has also been linked to mesothelioma, a type of lung cancer. Most of our knowledge of the links between exposure to specific chemical species in aerosols and adverse health effects has come from occupational studies, including asbestos workers, coal miners, and machinists working on metals such as beryllium. Studies of cigarette smoking have also been a major source of information that has linked the aerosol exposure of individuals to PAHs, and other organic aerosol species that have been shown to increase the risk of cancer. All of these studies show

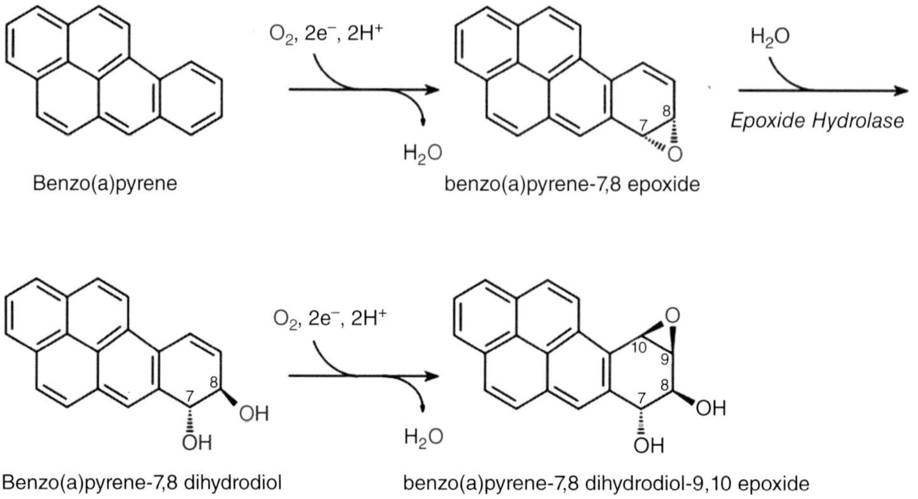

Figure 6.14 Biochemical enzymatic oxidation of benzo(a)pyrene to form the benzo(a)pyrene diol epoxide metabolite that can strongly bind to DNA molecules. Source: Adapted from Elleska, Wikimedia Commons.

that aerosols can act as effective carriers of toxic substances into the lungs, and for these reasons aerosols are and will continue to be a health concern.

The exposure to aerosol particles can also act in a synergetic way to amplify the effects of exposure to gases that are also present in the air. A synergism is when an interaction of two or more substances produces a combined effect greater than the sum of their separate effects. For example, the air pollutant SO_2 will react rapidly with H_2O_2 in aqueous systems to form H_2SO_4, a very strong acid. When exposed to soot from coal fires, the respiratory system's response is to attempt to clear the insoluble particle by oxidation. This is done by producing the oxidizing agent H_2O_2 biochemically deep in the respiratory system. This H_2O_2 can then react with the co-inhaled SO_2 gas to produce H_2SO_4 deep in the lung, causing damage. If the H_2O_2 is instead inhaled as a primary or secondary atmospheric aerosol, it will grow to a size that cannot reach the deep lung because of its high hygroscopicity. It is then removed by impaction in the nasal and pharynx region. This biochemical synergism of SO_2 and carbon soot aerosols leads to an explanation for the large number of people who were affected by the co-exposure to aerosols and SO_2 in the pollution episodes in London and Donora. Such synergistic interactions of particles causing a biochemical response that amplifies the damage caused by a pollutant gas present in the air at the same time demonstrates the need to understand the mechanisms by which organisms respond to multiple air pollutant stresses of all types.

6.6 Aerosol Visibility and Climate Effects

Light scattering by aerosols in the lower troposphere can lead to losses in visibility. Visibility reduction is an important impact of atmospheric aerosols on the environment and is one of the most readily apparent effects. Our ability to see clearly over long distances can be reduced by increased concentrations of light-absorbing gases and aerosols in the atmosphere. Light scattering causes objects to appear less distinct and light absorption can change the perceived colors of objects, as well as the apparent color of the atmosphere. Indeed, the presence of soot and other aerosols, along with NO in the air, tends to give the air pollution haze a reddish-brown color. This effect has given rise to the term "brown cloud" to describe urban air pollution.

The scattering of light by aerosols, especially fine-mode aerosols, can give rise to changes in the observed contrast of objects in the distance. This effect is due to their ability to deflect the light reflected off the object either toward the line of sight, making them appear brighter, or away from the line of sight, making them appear darker. While both light absorption and scattering affect visibility, scattering by aerosols is the dominant factor in visibility reduction. Since aerosols are able to be transported long distances, urban aerosol pollution has led to increases in regional visibility reduction. This regional haze over the western United States is due primarily to anthropogenic emissions, but it also has contributions from natural wildfires. There has been particular concern about visibility reduction and pollution impacts on the environment in the national parks and undeveloped federal lands in the United States. This concern led to the passing of the Wilderness Act in 1964. Under the Wilderness Act, the EPA has regulated anthropogenic aerosol emissions, including the aerosol precursor gases NO_x and SO_2, in order to reduce the levels of regional aerosol haze and improve visibility on federal lands.

Aerosols also have important direct effects on the radiative balance of the atmosphere, as discussed in Section 2.4.4. They do this by either scattering or absorbing the radiation from the Sun. The scattering of light by a particle or gas maintains the total amount of energy in the incident light but, in most cases, alters the direction the light travels. The absorption of light by a particle or gas actually removes energy from the incident light and converts it to another form. Light absorption by dark aerosols such as BC converts energy in the incident light to heat energy, warming the atmosphere. They can also absorb infrared radiation given off from the heated ground, trapping the heat in the lower troposphere in a similar manner as the greenhouse gases. Light scattering by white aerosols such as sulfate can cool the atmosphere by scattering the incoming solar radiation back to space, reducing the light reaching the ground. The total attenuation of light in the atmosphere by aerosols and gases is given by the sum of light scattering and light absorption. It is expressed using an application of the Beer–Lambert Law discussed in Section 3.2:

$$\frac{I}{I_o} = e^{-lb_{ext}}$$

where I is the light intensity at the surface of the Earth, I_o is the light intensity entering the atmosphere, l is the path length of the light through the atmosphere in meters, and b_{ext} is the extinction coefficient of all absorbing and scattering species in the atmosphere in units of m^{-1}. The overall extinction coefficient (b_{ext}) includes both gases and aerosols:

$$b_{ext} = b_g + b_a$$

where b_g is the extinction coefficient for the gases and b_a is the extinction coefficient for the aerosols. The light extinction due to the presence of aerosols in the atmosphere has two components, light scattering and light absorption:

$$b_a = b_{abs} + b_{scat}$$

where b_{abs} is the absorption coefficient and b_{scat} is the scattering coefficient for the aerosols.

The light-scattering coefficients (b_{scat}) and absorption coefficients (b_{abs}) of an aerosol are derived from the complex refractive index (m) of the material that makes up the aerosol particle. The complex refractive index as a function of wavelength is defined as

$$m(\lambda) = n(\lambda) - ik(\lambda)$$

where n is the refractive index and k is the absorption index for the aerosol. The absorption index is related to the aerosol absorption coefficient (b_{abs}) as

$$k = \frac{b_{abs}\lambda}{4\pi}$$

Table 6.6 Values for the refractive index (n) and the absorption index (k) for some typical aerosol materials in the visible ($\lambda = 550$ nm) and the infrared ($\lambda = 3000$ cm^{-1}) wavelength regions.

Aerosol material	n (visible)	k (visible)	n (infrared)	k (infrared)
Water	1.33	0	1.45	0.04
NaCl	1.34	0	—	—
$(NH_4)_2SO_4$	1.53	0	—	—
H_2SO_4	1.44	0	1.43	0.07
NH_4NO_3	1.56	0	1.36	0.08
$CaCO_3$	1.59	0	—	—
SiO_2	1.48	0	—	—
Fe	1.51	1.63	—	—
Fe_3O_4	2.58	0.58	—	—
Fe_2O_3	2.6	1.0	—	—
Diesel soot	1.68	0.56	1.6	0.7
Organic carbon	1.53	0.05	—	—

So the complex refractive index, which is dependent on the chemical make-up of the aerosol, determines the magnitude of the scattering and absorption coefficients. Some selected values for the refractive index and absorption index of materials commonly found in atmospheric aerosols are listed in Table 6.6 for 550 nm and 3000 cm^{-1} (Marley and Gaffney, 2005).

The species listed as organic carbon (OC) is determined by thermal methods and is usually made up of complex organic mixtures. These can contain carbonyl functional groups, PAHs, and organic nitrites and nitrates that absorb radiation in both the visible and infrared wavelength regions. Studies of the aerosols in Grand Canyon National Park have estimated that OC and BC contributions account for about 80–90% of the observed visible light absorption (Malm et al., 1996). The inorganic sulfate, nitrate, and liquid water in wet aerosols also absorb in the infrared region and act as greenhouse species, along with the carbonaceous soot.

The total light absorption by aerosols (A) is dependent on the concentration of each of the absorbing species in the aerosols (c_i), their absorption coefficients (b_{abs}), and the path length of light (l) through the absorbing aerosol according to Beer's Law:

$$A = \sum_{i=1}^{N} A_i = \sum_{i=1}^{N} \varepsilon_i \int_0^b c_i(l)\, dl$$

This assumes that the path length is the same for all the absorbing species. While this is true for the light-absorbing gases that are well mixed in the atmosphere, the path length of light through an absorbing aerosol can be affected by its mixing state. The aerosol mixing states, shown in Figure 6.15, describe how the aerosol components are present in the atmosphere. Atmospheric aerosols commonly contain mixtures of different species, including inorganic compounds and both aqueous and non-aqueous organic compounds. These aerosols are considered to be externally mixed if each of these aerosol components exists in the atmosphere on separate particles, as shown in Figure 6.15(A). They are considered to be internally mixed if they exist combined together on the same particle. In addition, an internally mixed aerosol can be either heterogeneous (Figure 6.15(B)) or homogeneous (Figure 6.15(C)). For example, a wet aerosol that contains both polar and nonpolar organic compounds can exist either in a well-mixed state (homogeneous) or in a micelle-type formation where the nonpolar species

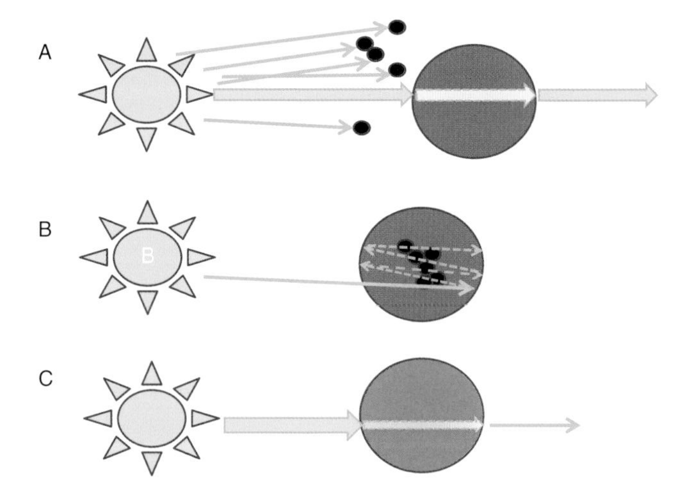

Figure 6.15 The effect of the mixing states (A) externally mixed, (B) heterogeneous internally mixed, and (C) homogeneous internally mixed on light path length for light-absorbing aerosols.

exists on the outside of the wet aerosol particle (heterogeneous). Internally mixed aerosols that contain only very hygroscopic aerosol species, such as aqueous ammonium sulfate or ammonium nitrate, are normally homogeneously mixed.

The aerosol mixing state affects the ability of the aerosol to condense water, gas-phase pollutants, or other particulate matter and can therefore affect aerosol growth. But it also impacts the aerosol light absorption by altering the path length of the light interacting with the absorbing substance. In Figure 6.15(A), the absorbing carbonaceous aerosol is shown as externally mixed fine-mode soot particles with a non-absorbing wet aerosol (blue sphere). Since the BC aerosols are highly absorbing, the light cannot penetrate the particle and so the light absorption occurs at the surface of the particles, resulting in an extremely small path length. In Figure 6.15(B), the BC particles have become oxidized and coagulate with the wet aerosol to form a heterogeneous internally mixed particle. In this aerosol, the carbonaceous particles form a core that is surrounded by the non-absorbing aqueous solution. In this heterogeneous internally mixed state, the incoming light is internally reflected in the liquid outer shell, interacting with the particle surface multiple times, doubling the path length of the single particle shown in Figure 6.15(A) with each reflection. This results in an increase in the path length of the light interacting with the particle and an increase in light absorption by the aerosol. In Figure 6.15(C), the absorbing aerosol species is water soluble, such as a polar organic compound, and is evenly distributed throughout the aerosol, resulting in the largest light path length of the three examples. If the absorbing species in each case has very similar absorption coefficients (b_{abs}), the light absorption by each of the aerosol mixing states in Figure 6.15 is directly proportional to the light path length and would increase in the order $A < B < C$.

As shown in Table 6.6, black carbonaceous soot aerosols are strong absorbers of both longwave and shortwave radiation. However, the warming effects from combustion-sourced aerosols are expected to be important in urban areas as well as in biomass burning event plumes, where the concentrations are highest. The BC from combustion sources can absorb as much as 25% of the incoming solar radiation in urban areas. The heating of the lower troposphere due to the absorption of light by significant amounts of anthropogenic and biomass burning BC has been proposed as a mechanism for the evaporation of clouds over the Indian Ocean and the Amazon forest. This cloud evaporation leads to the loss of large cloud droplets and the formation of haze, with a concurrent reduction in precipitation.

Figure 6.16 Wildfires seen from aloft produce high concentrations of aerosols in a brown haze, which results in a reduction of cloud formation in the area of the fires.

The atmospheric heating effect of BC and light-absorbing organic aerosols on cloud formation has been termed the aerosol "semi-direct" effect. This effect is demonstrated in the aerial photograph of a wildfire shown in Figure 6.16. The wildfire brown haze in the distance is produced from the high concentration of BC in the smoke particles. This high concentration of light-absorbing carbonaceous aerosols heats the air, preventing cloud formation. In the haze region of Figure 6.16 there are no clouds, while clouds are present in the air below the fire and beyond the haze.

Light scattering by aerosols is dependent on the size of the particles relative to the wavelength of light. The different aerosol light-scattering mechanisms are divided into three regimes depending on aerosol size: geometric, Rayleigh, and Mie. Light scattering by particles that are much larger than the wavelengths of light (>10 μm) is governed by geometric optics, which involves reflection from the particle surface, refraction through the particle interior, and diffraction from interaction with the particle edges. The light that is scattered by geometric processes leaves the particle in various directions and with different amplitudes and phases, depending on which process dominates the scattering. Ultrafine particles less than 30 nm scatter light by the same process that gases scatter light, called Rayleigh scattering. The light-scattering coefficient for the ultrafine particles varies as the sixth power of the aerosol aerodynamic diameter (D_a) and the fourth power of the frequency of the incident light (ν^4 or λ^{-4}):

$$b_{scat} = \frac{2\pi^5}{3}\frac{D_a{}^6}{\lambda^4}\left(\frac{n^2-1}{n^2+2}\right)^2$$

where n is the refractive index of the particle.

Aerosols that are approximately the same size as the wavelength of light (0.7–0.3 μm) will scatter it more efficiently and in a more complex manner, known as Mie scattering. Mie scattering is not strongly wavelength dependent. However, unlike Rayleigh scattering, which is isotropic, Mie scattering has a strong component in the forward direction that is larger for larger particles. This is what causes the white glare observed from mist and fog. This forward scattering of UV light toward the surface by fine-mode aerosols can actually lead to enhanced photochemistry and higher ozone formation due to increases in UV radiation in the lower troposphere (Marley et al., 2009).

The determination of the Mie scattering coefficient for an aerosol is based on the calculation of the electric and magnetic fields inside and outside the particle, and varies as the square of the aerosol aerodynamic diameter:

$$b_{scat} = N \frac{\pi D_a^2}{4} Q_{scat}$$

where N is the number concentration of the particles with aerodynamic diameter D_a. The term Q_{scat} is the Mie scattering efficiency, which is a complex function of the refractive index of the particle and the aerosol size parameter α:

$$\alpha = \frac{\pi D_a}{\lambda}$$

which is basically the ratio of the aerosol aerodynamic diameter and the wavelength of light.

The scattering coefficient (b_{scat}) is shown in Figure 6.17 as a function of aerosol aerodynamic diameter for aerosols ranging in size from 0.01 to 100 μm (Sinclair, 1950). The b_{scat} in Figure 6.17 was calculated using Mie theory for incident light of 550 nm and assuming an aerosol refractive index of 1.5. As shown in Figure 6.17, the light scattering of visible radiation is dominated by the fine-mode aerosols in the region of 0.1–1.0 μm, with a small contribution from the aerosols in the 1.0–10.0 μm size range. Since the fine-mode aerosols will have longer atmospheric lifetimes, they have the largest contribution to visible light scattering over regional areas, causing regional haze. Light scattering by fine-mode aerosols will also affect the radiative balance in the Earth's atmosphere by directing some of the incoming solar radiation back to space, leading to cooling of the surface. However, since most of the light scattering by the aerosols in the Mie scattering range will be in the forward direction, most of the backscattered light is due to larger fine-mode aerosols or hygroscopic aerosols that grow to larger sizes, such as sulfate aerosols. In fact, the cooling effects from solar light scattering by sulfate aerosols in the Northern Hemisphere have been estimated to be similar in magnitude to the warming effect from CO_2.

Hygroscopic aerosols, particularly sulfate, act as CCN condensing water and growing in size to form large droplets. If enough CCN is available, these large droplets come together to form clouds. The hygroscopicity of the CCN aerosol is an important controlling factor in cloud

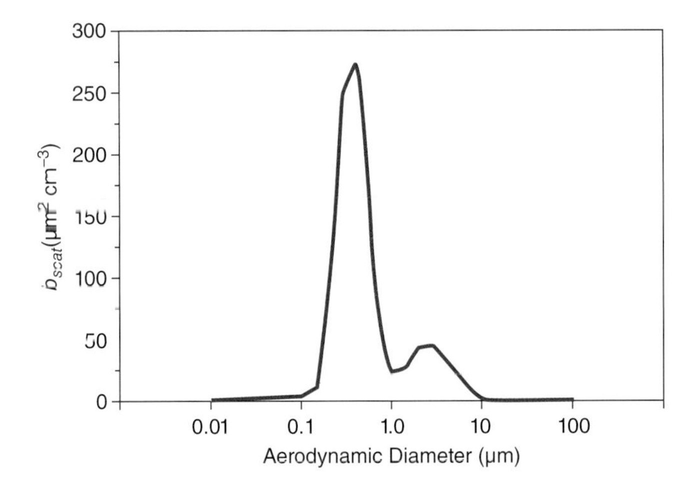

Figure 6.17 Aerosol light-scattering coefficients (b_{scat}) as a function of aerosol aerodynamic diameter for incident light with a wavelength of 550 nm and an aerosol refractive index of 1.5.

formation and cloud evolution. Increases in CCN concentrations result in an increase in cloud formation. This relationship between the increase in CCN and cloud formation is known as the indirect effect. Light scattering from cloud droplets, which are larger in size, is driven by geometric optics and is primarily due to reflection form the droplet surface. This reflection becomes more intense as the numbers of cloud droplets increases. Clouds with a high density of droplets appear dark from the ground as they scatter most of the incoming light from the top of the cloud. This also causes them to appear white from outer space, since most of the incoming light is scattered back to space. Since the light above a dense cloud does not penetrate very far into the cloud, these dense clouds cast shadows on the Earth's surface because of the reduction of light reaching the ground, leading to a cooling effect. The impact of this aerosol indirect effect depends on the aerosol size, chemistry, and concentration, which affect the concentration and size of cloud droplets.

If there is a large concentration of hygroscopic aerosols, they can all uptake water efficiently, resulting in a depletion of the available water vapor. This depletion of available water due to it being spread over the large number of hygroscopic aerosols results in a decrease or cessation of the particle growth rate and leads to aerosols that do not grow above 0.5–1.0 μm in size. Since the size of the aerosol droplets is stable, they are unlikely to grow into cloud droplets. So, the large number of small aerosols form hazes instead of clouds. The cloud formation rate is diminished and the frequency of precipitation events decreased.

6.7 Aqueous Chemistry

The fairly low molecular weight water-soluble secondary organic compounds are taken up into wet aerosols and into cloud droplets, which may have become acidified by the heterogeneous formation of HNO_3 and H_2SO_4. These low molecular weight organics in the wet aerosols can absorb light, leading to photochemical processes, which can produce OH and HO_2 radicals in the presence of dissolved oxygen. This can lead to further oxidation of the organics present in the aerosols, and in many cases these reactions produce carbonyl and hydroxyl compounds. In the wet acidic aerosol or cloud droplets, the oxidized compounds that contain –COOH and –OH functional groups can undergo acid-catalyzed condensation reactions to give ester linkages, leading to higher molecular weight organics. The OC resulting from these condensation reactions have molecular weights in the region of 300–700 Da and have chemical characteristics similar to natural organics found in surface waters, called humic and fulvic acids. So, they have been given the name "humic-like substances" or HULIS.

The humic substances are formed primarily from the oxidation of leaf litter and organic matter in soils and have molecular weights in the range of 500–3000 Da for the fulvic acids and 3000–500 000 Da for the humic acids. In addition, the fulvic acids have been determined to be composed primarily of simple alkyl structures with –COOH and –OH functional groups, while the humic acids have a higher aromatic content. The molecular weights and compositions determined for this high molecular weight OC in wet aerosols and cloud water seem to be more in line with the surface water fulvic acids than the humic acids (Gaffney et al., 1996). Based on their properties, they would be more aptly referred to as "fulvic-like substances" or FULIS. In any case, these complex organic mixtures contain primarily hydroxyl, keto, and carboxylic acid functional groups, which make them excellent complexing agents for any soluble inorganics in the aerosol or cloud water, in a similar manner to which the humic and fulvic acids in natural waters complex metals and other inorganics. For example, the HULIS have been proposed to play a role in the mercury cycle by complexing mercury ions in wet aerosols and clouds, leading to enhanced deposition of mercury during rain events (Gaffney and Marley, 2014).

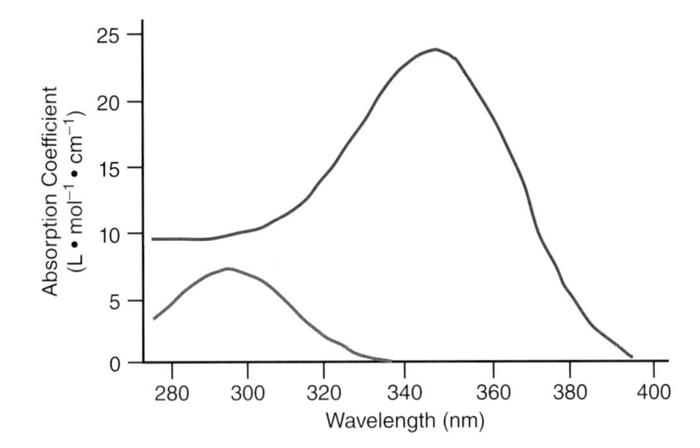

Figure 6.18 The absorption coefficient (b_{abs}) as a function of wavelength for the nitrate ion (green) and the nitrite ion (red).

Both NO_3^- and NO_2^- absorb light in the UVA and UVB wavelength region, as shown in Figure 6.18 (Gaffney et al., 1992; Zuo and Deng, 1998). This light absorption promotes aqueous photochemistry. The photolysis of NO_3^- in wet aerosols and in cloud droplets produces OH radical and NO_2, which is an important process to the recycling of NO_x back into the atmosphere. The NO_2 formed from the photolysis of NO_3^- can be released back into the air or it can be photolyzed to produce NO and an O atom in aqueous solution. Nitrite ion can also undergo aqueous photochemistry to produce NO and radical species. Both NO and NO_2 are not very water soluble, as shown in Table 6.1. So, they are eventually released back into the atmosphere where they play an important role in ozone formation chemistry. These inorganic photochemical processes can also occur on wet surfaces and in surface waters that contain NO_3^- or NO_2^-. This recycling of NO_3^- to NO_x is important in agricultural areas where nitrate fertilizer is used.

The chemistry and physics of aerosols in the atmosphere continues to be an area that needs further study of their complex chemical reactions. The understanding of how the aerosol chemistry is connected to their physical removal processes and radiative impacts also needs to be improved. Indeed, the radiative impacts of aerosols on weather and climate have been identified by the Intergovernmental Panel on Climate Change (IPCC) as one of the areas of least confidence but potentially largest significant impact on climate. The impact of aerosols on radiative forcing will be examined again as it relates to climate in Chapter 11.

References

Bell, M., Dominici, F., Ebisu, K. et al. (2007). Spatial and temporal variation in $PM_{2.5}$ chemical composition in the United States for health effects studies. *Environ. Health Perspect.* 115: 989–995.

Finlayson-Pitts, B.J. and Pitts, J.N. Jr., (2000). *Chemistry of the Upper and Lower Atmosphere: Theory, Experiments, and Applications*, 349–435. San Diego, CA: Academic Press.

Gaffney, J.S. and Marley, N.A. (2014). In-depth review of atmospheric mercury: Sources, transformations, and potential sinks. *Energy Emission Control Technol.* 2: 1–21.

Gaffney, J.S., Marley, N.A., and Cunningham, M.M. (1992). Measurement of the absorption constants for nitrate in water between 270–335 nm. *Environ. Sci. Technol.* 26: 207–209.

Gaffney, J.S., Marley, N.A., and Clark, S.B. (1996). Humic/fulvic acids and organic colloidal materials in the environment. In: *Humic and Fulvic Acids, Isolation, Structure, and*

Environmental Role (ed. J.S. Gaffney, N.A. Marley and S.B. Clark), 2–16. Washington, D.C.: American Chemical Society.

Kristiansen, N.I., Stohl, A., Olivie, D.J.L. et al. (2016). Evaluation of observed and modelled aerosol lifetimes using radioactive tracers of opportunity and an ensemble of 19 global models. *Atmos. Chem. Phys.* 16: 3525–3561.

Malm, W.C., Molenar, J.V., Eldred, R.A., and Sisler, J.F. (1996). Examining the relationship among atmospheric aerosols and light scattering and extinction in the Grand Canyon area. *J. Geophys. Res.* 101: 19251–19265.

Marley, N.A. and Gaffney, J.S. (2005). Introduction to urban aerosols and their impacts. In: *Urban Aerosols and their Impacts: Lessons Learned from the World Trade Center Tragedy* (ed. J.S. Gaffney and N.A. Marley), 2–22. Washington, D.C.: American Chemical Society.

Marley, N.A., Gaffney, J.S., Cunningham, M.M. et al. (2000). Measurement of ^{210}Pb, ^{210}Po, and ^{210}Bi in size fractionated atmospheric aerosols: An estimate of fine aerosol residence times. *Aerosol Sci. Technol.* 32: 569–583.

Marley, N.A., Gaffney, J.S., Castro, T. et al. (2009). Measurements of aerosol absorption and scattering in Mexico City during the MILAGRO field campaign: A comparison of results from the T0 and T1 sites. *Atmos. Chem. Phys.* 9: 189–206.

Sander, R. (2015). Compilation of Henry's law constants (version 4.0) for water. *Atmos. Chem. Phys.* 15: 4399–4981.

Sinclair, D. (1950). Optical properties of aerosols. In: *National Defense Research Committee Handbook on Aerosols*, 64–116. Washington, D.C.: U.S. Atomic Energy Commission.

Slatkin, D.N., Friedman, L., Irsa, A.P., and Gaffney, J.S. (1978). The $^{13}C/^{12}C$ ratio in black pulmonary pigment: A mass spectrometric study. *Human Pathol.* 9: 259–267.

Zuo, Y. and Deng, Y. (1998). The near-UV absorption constants for nitrite ion in aqueous solution. *Chemosphere* 36: 181–188.

Further Reading

Finlayson-Pitts, B.J. and Pitts, J.N. Jr., (1986). *Atmospheric Chemistry: Fundamentals and Experimental Techniques*. New York: Wiley.

Hansen, J. and Nazerenko, L. (2004). Soot climate forcing due to snow and ice albedos. *Proc. Natl. Acad. Sci. U.S.A.* 101: 423–428.

Marley, N.A., Gaffney, J.S., Tackett, M.J. et al. (2009). The impact of biogenic carbon sources on aerosol absorption in Mexico City. *Atmos. Chem. Phys.* 9: 1537–1549.

Study Problems

6.1 What is the size range of atmospheric aerosols?

6.2 (a) What are primary aerosols?
(b) What are secondary aerosols?

6.3 Give the type of aerosol (primary or secondary) for each of the following:
(a) wind-blown dust;
(b) carbonaceous soot from biomass burning;
(c) diesel soot;
(d) sulfate produced from oxidation of SO_2 in the atmosphere;
(e) ammonium nitrate formed from the reaction of ammonia with nitric acid in the gas phase.

6.4 (a) What was the first aerosol measurement technique that used filters to collect all the particulates in the air from 0.1 μm to about 50 μm in size?
(b) What did this method measure?

6.5 What are the names of the aerosol modes that fall in the following size ranges:
(a) 2 nm to 100 nm?
(b) 100 nm to 2 μm?
(c) 2 μm to 100 μm?

6.6 The small aerosols that can take up water to assist in the formation of clouds are called what?

6.7 (a) Which aerosol mode has the highest number of particles?
(b) Which aerosol mode has the most surface area?
(c) Which aerosol mode contains most of the volume?

6.8 What aerosol modes are included in the $PM_{2.5}$ measurement?

6.9 Polycyclic aromatic hydrocarbons (PAHs) formed during incomplete combustion are usually found coating what other type of aerosol also formed during incomplete combustion?

6.10 What coarse-mode aerosol is formed during coal burning?

6.11 What important oxidant reacts with SO_2 in aqueous solution to form sulfuric acid?

6.12 What is the oxidative radical species that converts NO_2 to HNO_3 in the gas phase?

6.13 What white solid aerosol is formed from the reaction of $NH_3(g)$ and $HNO_3(g)$?

6.14 (a) What are the two major inorganic species found in wet aerosols that are derived from secondary oxidation of primary gases produced from combustion?
(b) What are these primary combustion gases?

6.15 (a) The natural pH of cloud water is determined by what dissolving gas?
(b) This gas is in equilibrium with what three species in cloud water?

6.16 The Henry's Law constant for $HNO_3(g)$ is 2×10^5 M atm^{-1}. What would be the concentration of NO_3^- in cloud water if the concentration of the gaseous nitric acid is 1 ppb?

6.17 What are the units for the Henry's Law constants determined from the following equations:
(a) $H_x = [X]/P_x$?
(b) $H_x = P_x/[X]$?

6.18 (a) What is hygroscopicity?
(b) How can it be determined from the Henry's Law constants?

6.19 Are polar compounds or nonpolar compounds expected to have larger Henry's Law constants (in units of M atm^{-1})?

6.20 (a) Ultrafine-mode aerosols coagulate by what process?
(b) Coarse-mode aerosols are deposited by what process?

6.21 What are the two types of aerosol deposition?

6.22 What are the two dry deposition loss mechanisms that determine the lifetimes of the aerosols as a function of their aerodynamic diameter?

6.23 Which aerosol size mode is least affected by diffusion and gravitational settling?

6.24 What is the range of lifetimes for fine-mode aerosols in:
(a) the lower 1 km of the troposphere?
(b) the middle troposphere?
(c) the top of the troposphere?

6.25 What types of plants use the following photosynthetic cycles:
(a) the Calvin–Benson cycle?
(b) the Hatch–Slack cycle?

6.26 (a) Which of the two plant types has more ^{13}C?
(b) Why?

6.27 What two isotopic determinations are useful in differentiating aerosol sources from fossil fuel versus biomass combustion?

6.28 What two important C-4 plants are used to make gasohol?

6.29 Why are there higher levels of ^{14}C in biogenic materials formed in the decade between 1950 and 1960?

6.30 What are the two mechanisms that the body uses to clear hydrophobic aerosols from the respiratory system?

6.31 (a) What is synergism?
(b) Give an example of a synergism between a gas and an aerosol causing unexpected health impacts.

6.32 The oxidation of benzo(a)pyrene in the respiratory system leads to what important derivative that is an active carcinogen?

6.33 What effect causes urban pollution to appear brown, sometimes called the "brown cloud"?

6.34 (a) What are the two processes by which aerosols can directly affect the radiative balance of the Earth?
(b) What effect does each process have?

6.35 (a) The light-scattering coefficients (b_{scat}) and absorption coefficients (b_{abs}) of an aerosol are derived from what complex function?
(b) Write the expression and define its terms.

6.36 (a) What are the three aerosol mixing states?

6.37 If an aerosol is made up of an absorbing carbonaceous aerosol and a non-absorbing wet aerosol:
(a) which mixing state results in the highest light absorption?
(b) which mixing state results in the lowest light absorption?

6.38 (a) What are the three different aerosol light-scattering mechanisms?
(b) Which mechanism is important for fine-mode aerosols?
(c) Coarse-mode aerosols?
(d) Ultrafine-mode aerosols?

6.39 Which light-scattering mechanism has a higher wavelength dependence?

6.40 Light scattering of visible radiation is dominated by what mode of aerosols?

6.41 The relationship between the increase in CCN aerosols and cloud formation is known as what effect?

6.42 The atmospheric heating effect of BC and light-absorbing organic aerosols on cloud formation has been called what effect?

6.43 What is usually formed when there is:
(a) a moderate number of CCN present in a humid atmosphere?
(b) a very high number of hygroscopic aerosols present in a moderately humid atmosphere?

6.44 Light scattering by cloud droplets occurs by what scattering mechanism?

6.45 Black carbon in the troposphere can produce warming by the absorption of what type of radiation?

6.46 What two inorganic species present in cloud water can photolyze to generate OH and HO_2 radicals in aqueous solution?

7

Analytical Methods for Air Analysis

Our understanding of the chemistry of the atmosphere is based on analytical methods that can indicate the presence and amounts of an atmospheric species as a function of space and time. The measurement of the key species in the atmosphere is needed to determine chemical reaction mechanisms and lifetimes of atmospheric gases and aerosols. All measurements are subject to uncertainties caused by errors inherent to the methods of sample collection and measurement. These measurement uncertainties are described by the accuracy and precision of the method. The accuracy of a measurement is how close the result can come to the true value. Any deviations from the true value are caused by systematic errors in the measurement method, which are constant and always of the same sign. They result in a constant offset (positive or negative) from the true value. The accuracy of a measurement is determined by a calibration procedure that involves the measurement of a standard sample with a certified true value. Since systematic errors are constant and always present, any systematic error can be corrected after calibration by electronically or manually subtracting or adding the offset value from the measurement.

The precision of a measurement, often called reproducibility, is the degree to which repeated measurements give the same value whether or not that value is the true value. The precision of a measurement is an indicator of the scatter in a series of measured values made on the same sample. A lack of precision is caused by random errors, both positive and negative, in the measurement and/or sampling methods. They arise from unpredictable fluctuations such as changes in instrumental or environmental conditions or contamination during sampling. The precision of a measurement is determined by repeating the measurement several times on the same sample or on duplicate samples collected in the field and recording the variation in values. The measurement precision is reported by using the average of a series of measurements and including the magnitude of the variation in the measurements. For example, the range of mass measurements 24.5, 24.4, 24.6, 24.5 g would be reported as 24.5 ± 0.1 g.

Other important parameters for the evaluation of the performance of measurement methods are detection limits, sensitivity, resolution, and response time. These parameters are commonly applied only to the measurement method and not to the sampling method. The detection limit of a measurement method is the lowest amount of a chemical species that can be detected by the method. It is usually determined as a signal that is three times the standard deviation of the signal obtained on a sample blank. The sensitivity of a measurement method is the ratio of the change in the measurement signal to the change in the concentration or amount of the species being measured. On a plot of measurement signal (y) as a function of the concentration of the chemical substance (x), which is called the analytical curve of the method, the sensitivity is the slope of the linear portion of the curve ($\Delta y/\Delta x$). The resolution of a measurement method is the ability to detect small changes in the measurement signal or in the concentrations of the species being measured. The response time of a measurement method is

Chemistry of Environmental Systems: Fundamental Principles and Analytical Methods, First Edition.
Jeffrey S. Gaffney and Nancy A. Marley.

the length of time that it takes to make a measurement after the sample has been introduced into the measurement system.

Many of the atmospheric species exist at very low concentrations, which are usually in the range of ppm to ppt in air. Since these low concentrations would be near the detection limits for many measurement methods, good precision of the measurements is sometimes difficult to obtain. In many cases, in order to obtain sufficiently low detection limits, expensive specialized equipment is required for both the collection of the sample and the analysis of the species. Low-cost analytical methods, such as electrochemical sensors or test strips, inevitably have problems with interferences from other chemical species present in the air and have difficulties providing the necessary accuracy and precision for measurement of the low-level concentrations in the atmosphere. An example of this is electrochemical sensors that are used for CO analysis in homes. Methods that are moderate in cost may have high accuracy but would have low sensitivity, high detection limits, and long response times. An example of such a method would be gas analysis using a gas chromatograph equipped with a thermal conductivity detector (GC–TCD). High-cost instruments can show high accuracy, high sensitivity, low detection limits, and short response times, but they are usually large and heavy, require stable power sources, and can only be operated by highly trained personnel. The environmental chemist is faced with these trade-offs when determining the best analytical method to be used for determination of the atmospheric species of interest.

The U.S. Environmental Protection Agency (EPA) maintains a list of approved standard methods for the measurement of pollutant species. The approved methods for the criteria pollutants are published in the Federal Register and are called Federal Reference Methods (FRMs). In pollution monitoring applications, the FRM must be used unless it is not available or not convenient to use. In these cases, an alternative method designated a Federal Equivalent Method (FEM) can be used with the stipulation that the FRM must be used for calibration purposes. The approved methods that must be used for the measurement of the air toxics are listed on the EPA website. These restrictions do not apply for research applications where the range of possible measurement methods is much larger. While the number of chemical measurement methods for atmospheric species is very large, some of the most commonly used methods, both past and present, are presented here with special attention given to the EPA FRMs, FEMs, and approved methods.

7.1 Sampling Methods

Before any atmospheric analysis can be undertaken, the species of interest must first be collected from the air. The sampling procedure should not affect the concentrations of the species and not introduce any contaminants that can act as an interference in the analytical method of measurement. In most cases, the samples should be processed as soon as possible after sample collection to maintain sample integrity. This is important with any analytical measurement technique, especially for reactive species, since the longer the sample is kept before the analysis is performed, the more likely that the sample will be lost to oxidation reactions or to adsorption onto equipment or container walls. These types of losses lead to negative random errors, which can be evaluated by collecting sample standards in the field of similar concentrations expected in the samples. Any possibility of sample contamination during the sampling procedure can also be evaluated by collecting field blanks, such as running purged air through the sampling procedure. It is also recommended to obtain sample duplicates in the field, which can be compared to duplicates of the same sample in the laboratory in order to evaluate the precision of the sampling method versus the precision of the measurement method. Since environmental conditions may change rapidly with time, these field duplicates must be collected at the same time and in the same area.

In the past, because the detection limits of the analytical methods were low, the analyte (species being measured) had to be concentrated in order to collect a sufficient amount of sample for analysis. Analytical methods such as colorimetry, simple chromatography, or electrochemistry used wet scrubbers, impingers (bubblers), or water misters to concentrate the analyte in an aqueous solution during sample collection. Many EPA sample collection methods use impingers, which are specially designed bubble tubes used for collecting airborne chemicals into a liquid medium. An example of a generalized sampling train used in the EPA standard methods is shown in Figure 7.1. This standard sampling method uses a series of impingers to concentrate the sample in order to obtain sufficient sample for measurement. The air is sampled from either stack emissions or from the ambient air using a heated inlet system to prevent water condensation. The first impinger, with a short tube and no reagent, is used to condense water vapor from the air sample. The next four impingers are filled with a chemical reagent solution designed to trap the analyte. The last impinger, called a Mae West impinger because of its shape, is filled with silica gel and functions as a drying tube to remove any water transferred from the reagent solutions. All the impingers are placed in an ice bath to minimize the solution evaporation during the sampling period caused by the rapid bubbling in the impingers. The total amount of air sampled is measured by a dry gas meter after the air exits the impinger line.

Other sample collection methods include chemically impregnated filters or coated wire meshes, which react with the analyte to concentrate it on the collection substrate. The sample is then extracted from the substrate by washing with an appropriate solvent. The air can also be passed through an inert adsorbing material, such as a polar or nonpolar resin to concentrate the analyte, which is then separated from the adsorbing material by solvent extraction. A more

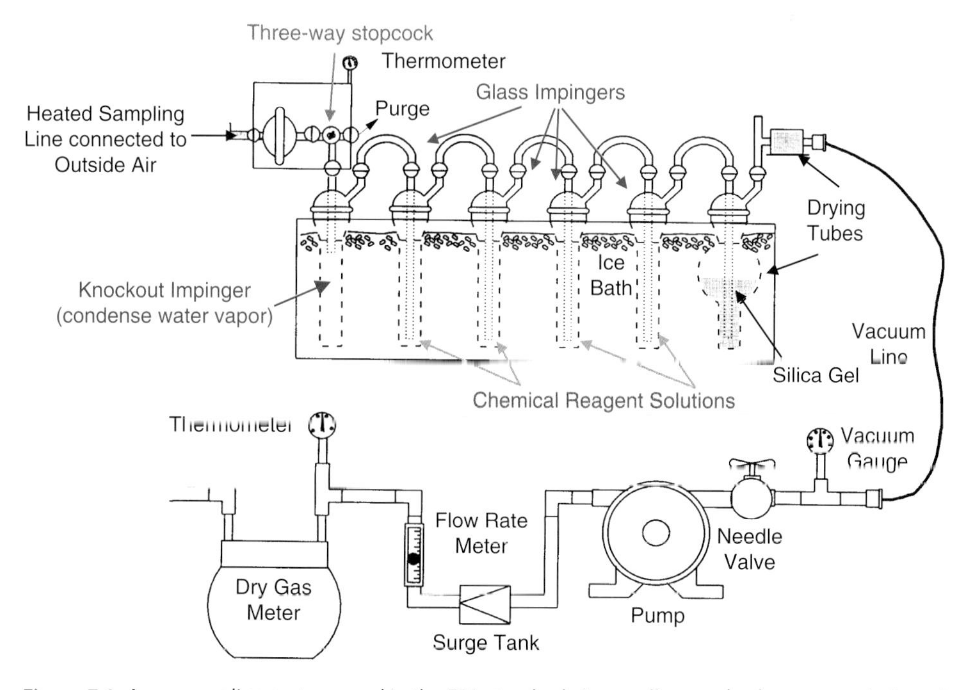

Figure 7.1 A gas sampling system used in the EPA standard air sampling methods composed of a series of glass impingers containing chemical reagent solutions to collect air pollutant gases. Source: Adapted from EPA, SW-846 Test Method 0051.

modern approach has been to cryogenically condense the gases with liquid argon. At the temperature of liquid argon (−186 °C), the atmospheric trace gases are condensed into a small volume without collecting oxygen (bp of −216.8 °C) or nitrogen (bp of −196 °C). All of these sampling methods that are designed to concentrate the analyte during the sampling procedure require sampling times of 30 minutes or longer.

Physical losses of the analyte can occur from absorption on the surfaces of the sampling apparatus unless special precautions are taken. The sample collection of reactive gases, such as acids or oxidants, requires inert surfaces to be used throughout the sampling system to prevent such losses. Sampling systems for collection of oxidants require inlets made from or coated with Teflon (an inert polyperfluorocarbon). In some cases, the tendency for reactive gases to undergo chemical reactions on surfaces has been used as an advantage in sampling procedures. The chemical reaction of the analyte on a surface can be used to remove it from the gas stream. For example, HNO_3 is also measured by the same measurement method as NO and NO_2. Since nylon filters will react with HNO_3, removing it from the gas stream, they are used to obtain the NO_x measurement. Then, the difference between the instrumental signal with and without the nylon filter in place is used to determine the HNO_3 concentration by difference.

These sample concentration methods can suffer from random errors caused by chemical reactions with the trapping reagent from species other than the analyte, which decreases the effective reproducibility of the measurement method. As organic analytes become concentrated in the solution or substrate, they become more susceptible to chemical oxidation and acid catalyzed reactions. Both positive and negative errors can occur from these reactions, depending on the nature of the analyte. For example, the oxidation of an alcohol in the sampling solution leads to a negative error if the alcohol is the analyte. But since the products of the oxidation reaction are aldehydes and ketones, it will produce a positive error if the analyte is an aldehyde. When an absorption, filtration, or chemical derivatization method is used for sample collection in the field, blank measurements should be taken on the absorbing solutions or substrate materials, which have been handled in exactly the same way as the samples without actually pulling air through the solution. This procedure will indicate the presence of any unknown or unexpected contamination during the field operations.

Air samples can be taken either actively or passively. Active sampling uses an air sampling pump to pull the air through the sample collection device. In contrast, passive sampling does not require active air movement from a pump. Passive sampling is dependent on molecular diffusion of the analyte through a diffusive surface onto an adsorbent. Active sampling is most commonly used in air sampling of gases and aerosols. It is accomplished by generating a partial pressure differential between the sampling intake and the outside air. This can be achieved by a variety of mechanical pumps, including advanced vacuum pumps used for mass spectrometers, rotary pumping systems used for high-volume (HI-vol) filter samples, and simple diaphragm pumps used for pulling air into concentrator devices. There are a number of passive sampling methods that have been used for air pollutants. Gaseous diffusion devices using a chemical absorbent are frequently used as a passive sampling device for personal air monitoring. These small samplers are made of chemically treated materials packed into a badge that collect samples over a period of days to weeks. Since the passive monitors require very long times to collect sufficient sample for analysis, they also have high potential for contamination, depending on the nature of the analyte. These passive monitoring devices are sometimes used in industrial environments where workers are exposed to high levels of the analyte. Under these conditions, the time needed to obtain a measurable sample is far reduced. Another example of a passive sampling device uses a series of inert plastic strings to collect air pollutants dissolved in fogs. The prevailing winds carry the fog toward the strings and as the fog intercepts the strings, the impacted droplets flow down the strings into sample collection bottles.

It is important to design the active sampling system so that the pump chosen does not contaminate the samples. For example, oil that is used in some rotary pumps can be exhausted from the pump into the air and, if the pump is near the sample inlet, the oil can be taken up with the air as part of the sample. Another example of this type of contamination occurs in the use of high-speed electrical pumps in high-volume aerosol filter sampling. These pumps use carbon brushes to make electrical contact with the pump rotor. The brushes wear down with time due to friction against the rotor and in the process can release small amounts of carbon particulates into the air. This carbon particulate matter can contaminate the aerosol samples if the pump is not exhausted far enough away from the sample intake.

For aerosol sampling, the filter materials used in active sampling must be carefully chosen when chemical speciation is to be done. The use of a filter material that is low in inorganic content, such as Teflon, is suitable for inorganic speciation. However, quartz-fiber filters, which do not use organic binders, are required for low-level carbon analysis. In all cases, when using any type of concentrating filter materials, the material itself must be analyzed to determine the background content of the analyte. Also, the potential for reactions on the filter media needs to be considered. Chemical reactions can occur in the sample after collection on the filter media as the air containing reactive species is pulled through the filter. If aerosol mass is to be measured, chemical reactions can produce positive or negative errors. In some cases, these types of artifacts can be reduced by using gas denuders, which use a reactive solution to remove acidic gases and oxidants before they reach the aerosol collection device.

7.2 Gas Species Measurement Methods

All analytical methods used for the measurement of gas-phase atmospheric species after the sample has been collected require the use of calibration standards. Gas standards are usually made by gas dilution with very clean air, called "zero air," as the diluent gas. Zero air is a synthetic mixture of ultrahigh purity (UHP) O_2 (22% v/v) and N_2 (78% v/v). Pure analyte standard gases can be diluted in zero air, using calibrated flow measurement devices, and pressurized into lecture bottles or larger compressed gas cylinders for use as calibration standards, depending on the chemical stability of the analyte. Standards for low-volatility organic compounds can be obtained using permeation tubes. The permeation tube is a small tubular device sealed with a gas-permeable membrane. The liquid analyte is sealed inside the tube, which is placed inside a constant-temperature oven. A very small stable flow (ng min^{-1}) of the analyte vapor is emitted through the tube wall and is carried from the oven by a carrier gas, usually zero air. The rate of permeation from the tube is dependent on the temperature of the oven and the zero air gas flow. It can be determined by measuring the rate of the tube's weight loss. The analyte gas from the permeation tube can be used directly for high concentration measurements or it can be further diluted with zero air. Samples of most gases can be produced at concentrations as low as 10 ppb with about 10% error using this method.

7.2.1 The Oxidants: Ozone, Hydroxyl Radical, Peroxyacyl Nitrates, Peroxides, and Peracids

Originally, the method used as the FRM for the measurement of atmospheric ozone was based on the oxidation of a buffered solution of potassium iodide (KI). Air passes through the KI solution using an impinger sampling train similar to that shown in Figure 7.1. The KI solution is oxidized to iodine (I_2) and the concentration of ozone is determined using colorimetric or electrochemical analysis of the solution. But while the method was originally designed for the measurement of atmospheric ozone, the KI will be oxidized by all atmospheric oxidants. So the method actually measures total atmospheric oxidants, including ozone, peroxyacyl nitrates, peroxides, and NO_2. Because of these issues, the KI method is no longer designated as the FRM for the measurement of ozone.

Other methods used for the determination of atmospheric ozone were based on the chemiluminescent reactions of ozone with NO and ethene (C_2H_4). Chemiluminescence is the emission of light by an electronically excited molecule created as the product of a chemical reaction. In both of these methods the air is sampled actively by pulling it into a partially evacuated dark cell where either NO or C_2H_4 is added as the reactant. The reaction of ozone with NO produces electronically excited NO_2^*, which emits light at wavelengths >650 nm and peaks in the near infrared (NIR). The reaction of ozone with C_2H_4 produces electronically excited formaldehyde (CH_2O^*), which emits light with a maximum at 420 nm. The C_2H_4 method has lower detection limits than the NO method, because the photomultiplier detectors used in these instruments have a lower sensitivity and a better response at the shorter wavelengths (sensitivity at 420 nm > sensitivity at 650 nm). Because of the lower detection limits and higher sensitivity, the reaction of C_2H_4 with ozone was adopted as the FRM for atmospheric ozone determinations after the KI method was disapproved. However, the reaction with NO is an approved FEM. Both these methods used pure reactant gases (NO and C_2H_4), which were vented into the atmosphere after passing through the chemiluminescent chamber. This presented problems that arise from inadequate handling of the exhaust from the instruments. Because of these issues, there are currently no commercial C_2H_4–ozone chemiluminescent instruments available on the market and, although the C_2H_4–ozone chemiluminescent method is still listed as the FRM, secondary FEMs are now routinely used for the determination of ozone.

The most commonly used FEM for ozone is based on the use of long-path UV absorption. Figure 7.2 shows a diagram of a typical UV absorption instrument for the determination of atmospheric ozone. The system uses a low-pressure mercury lamp as the source of UV radiation. The maximum emission from this lamp is at 254 nm, where the O_3 absorption is at maximum (λ_{max}). This provides high sensitivity and low detection limits for the method. Although aromatic hydrocarbons and elemental mercury absorb in the same wavelength region as O_3, the use of a sample blank can correct for these interferences. This sample blank is created by passing the sample air through an ozone scrubber, commonly a tube filled with screens coated with MnO_2, which decomposes the ozone on the surface of the mesh. The sample air and scrubbed air are passed alternately through the long-path absorption cell using a solenoid

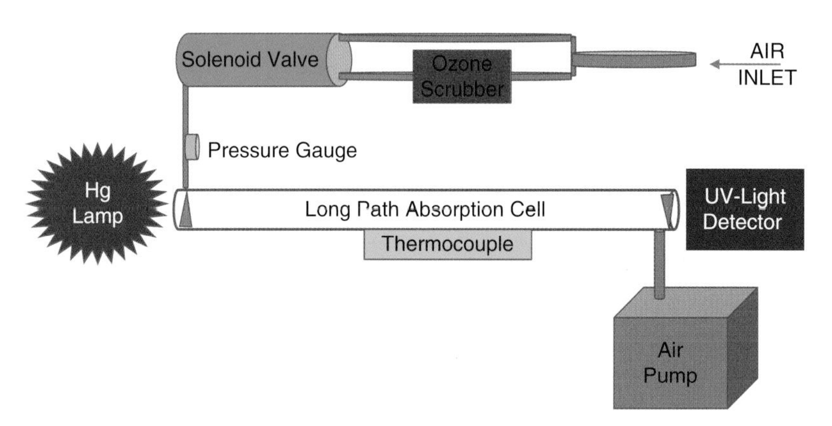

Figure 7.2 A schematic diagram of a typical UV absorption instrument for the measurement of ozone in the atmosphere.

valve. The light absorption of the sample air is then compared with the light absorption by the sample blank, giving the ozone absorption by difference.

The long path lengths, which typically range from 0.1 to 1.0 m, are created using angled mirrors at the end of the absorption tube. The mirrors extend the light path beyond the length of the tube by creating multiple reflections of light. This increased path length increases the sample absorption according to Beer's Law. The air temperature and pressure in the cell are measured by a thermocouple and a diaphragm pressure gauge, so that the ozone measurement can be corrected to standard temperature and pressure (STP). Ozone calibration standards for this method are produced by the photolysis of O_2 in air using a UV lamp or coronal discharge. The use of the UV lamp is preferred as it does not generate NO_x as a side product. The amount of ozone in the standard that is produced in this manner is usually determined by optical spectroscopy. It can also be determined by gas-phase titration of O_3 with an NO gas standard by the reaction

$$O_3 + NO \rightarrow O_2 + NO_2$$

However, the titration method is less accurate than the spectroscopy method, because of the difficulties involved with the formulation of an accurate NO standard at ppb concentration levels.

Other methods used for the measurement of atmospheric ozone include open-path optical methods based on UV, visible, or IR absorption. In open-path spectroscopic methods, light of the appropriate wavelength is transmitted from the source through the open atmosphere to the detector. The path length used in Beer's Law is the distance that the light travels between the light source and the detector. This can be accomplished if the source and the detector are in different locations, with the light traveling between them. Alternately, the source and detector can be at the same location and the light returned back along the incident path to the detector by a reflecting mirror, called a retroreflector. The most commonly used open-path methods for ozone are those that operate in the visible and UV wavelength regions, and are based on differential optical absorption spectroscopy (DOAS). These instruments can use either the Sun or the Moon as the light source in a similar manner as the Dobson meter, discussed in Section 4.2. The difference between the DOAS systems and the way the Dobson meter functions is that DOAS uses a polynomial fit to subtract the broad background from the narrow absorption bands of ozone, instead of comparing two closely related wavelengths. The use of this background subtraction procedure gives them their name: differential absorption spectroscopy. The path length of the light through the atmosphere in these instruments is dependent on the solar or lunar angle to the detection system, and is usually on the order of about 15 km. However, some DOAS systems use a high-intensity light as the source that is directed across an open path to a retroreflector. The retroreflector returns the light to the spectrometer, which then uses a high-resolution monochromator to disperse the light and obtain the optical spectrum. The placement of the spectrometer and the retroreflector results in a fixed path length, which is typically about a kilometer. Since the path length is directly proportional to sample absorption, the shorter path length of these instruments results in lower detection limits. However, this design is sufficient for gases such as ozone that have strong absorption bands.

Open-path Fourier transform infrared spectroscopy (OP-FTIR) can also be used to determine ozone as well as other atmospheric trace gases that absorb in the IR. OP-FTIR is the most versatile of the open-path methods because it can measure many species at the same time, with detection limits in the low ppb. In OP-FTIR the instrument first uses an interferometer, which modulates the IR light beam by using an internal moving mirror. The modulated light beam then passes from the instrument through an open path to a retroreflector that reflects the light back to the instrument and onto a detector. The optical spectrum is produced from the modulated

IR beam using a Fourier transform. The detection limits of OP-FTIR are limited by the concentrations of water vapor and CO_2 in the atmosphere, since both of these species have broad absorption bands in the IR region that interfere with the ozone measurements. For this reason, the OP-FTIR methods require high-resolution instrumentation (typically $0.5\ cm^{-1}$), which can separate the narrow ozone bands from the broader background signals of water and CO_2.

All open-path methods, including DOAS and OP-FTIR, measure an average concentration over the entire path length that the light travels. Comparison studies of the open-path methods with stationary UV ozone monitors have shown good agreement when the air parcel is well mixed. However, variances are seen between the measurements made at a single site and the averages obtained from the open-path methods when the air is not well mixed. There can be high or low concentration variations that are recorded by the UV single-site instruments which are not observed by the open-path measurements because of their averaging function. These differences are likely due to either the local loss of O_3 by reaction with NO or the local production of O_3 due to photochemical reactions with reactive hydrocarbons.

As discussed in Section 4.2, total column ozone measurements are routinely made using a Dobson meter. But since most of the ozone exists in the stratosphere, the ozone levels obtained by the Dobson meter are mostly due to stratospheric ozone. Ozone measurements as a function of altitude are determined using an ozonesonde, which is a small lightweight instrument designed to be used on a balloon platform in conjunction with a radiosonde. The radiosonde measures various atmospheric parameters, including altitude, pressure, temperature, and relative humidity, and transmits the results by radio to a ground receiver. The ozonesonde is based on the ozone–KI reaction because this setup, shown in Figure 7.3, is small and lightweight enough to be used on the balloon platform. The ozone–KI reaction is measured electrochemically and the signal is transmitted continually by the radiosonde to the ground receiver during the balloon ascent, along with the results from the meteorological measurements.

The electrochemical cell used to measure the ozone concentration is an electrochemical concentration cell (ECC), a form of galvanic cell where both half cells have the same composition with differing concentrations. The ECC in the ozonesonde consists of two half cells containing buffered KI solutions connected by a salt bridge. Air is drawn into the cell through Teflon tubing by a Teflon diaphragm pump, which is operated by a standard 9 V battery.

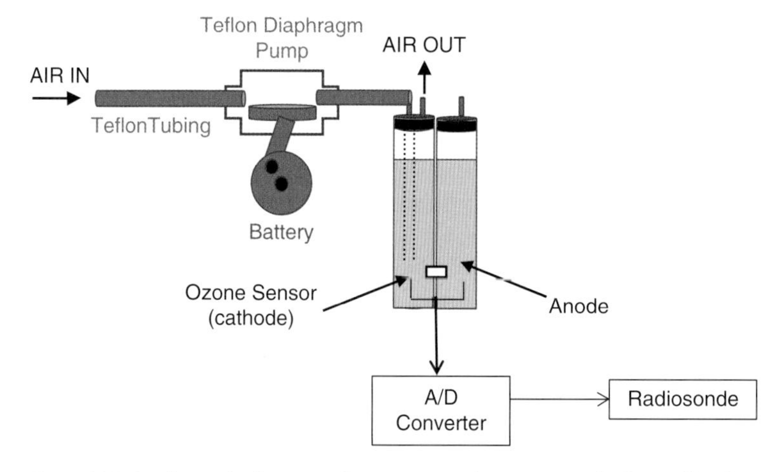

Figure 7.3 A schematic diagram of an ozonesonde constructed of a small Teflon pump, Teflon tubing, and a KI electrochemical concentration cell (ECC). The electrochemical signal produced from the reaction of O_3 with I^- is converted to a digital signal and sent to the radiosonde for transmission to a ground station.

The air containing trace amounts of ozone is pumped into the cathode of the ECC and the ozone in the air reacts with I^- to form I_2 by the reaction

$$2KI + O_3 + H_2O \rightarrow I_2 + O_2 + 2KOH$$

This decreases the concentration of I^- in the cathode solution and causes electrons to flow from the anode to the cathode in the external circuit. This flow of electrons is due to the driving force of the two solutions trying to regain equilibrium. Two electrons are exchanged between the anode and cathode for each O_3 molecule that reacts with two I^- ions to form I_2. This causes the reverse reaction in the cathode, reforming two I^- ions:

$$I_2 + 2e^- \rightarrow 2I^-$$

Thus, the flow of electrons through the external circuit is directly proportional to the partial pressure of ozone in the sampled air (P_{O_3}) measured in nanobars:

$$P_{O_3} = 4.307 \times 10^{-3}(I - I_{bg})Tt$$

where T is the pump temperature (K), t is the pump flow rate (s/100 ml), I is the cell current (μA), and I_{bg} is the cell background current in the absence of ozone.

The atmospheric concentrations of OH radical have routinely been calculated indirectly from the emission rates and atmospheric concentrations of compounds that are known to be removed from the atmosphere primarily by reaction with OH, such as methyl chloroform (CH_3CCl_3). But these mass balance approaches for the determination of OH concentrations give a globally averaged concentration value, which does not indicate the concentration variability from site to site. There have also been questions regarding the applicability of globally averaged values to any specific determination of atmospheric oxidation mechanisms. There are two methods that have been used to measure the concentrations of OH directly. These are DOAS and laser-induced fluorescence (LIF). Since OH is a very small species with only one bond, it has a very simple but characteristic absorption spectrum, which includes two groups of three narrow bands at about 308 nm (308.0 and 308.15 nm). DOAS has been used to determine the vertical column abundances of OH by measuring the absorbance at these wavelengths using the Sun as a source. It has also been used to determine tropospheric concentrations using a retroreflector to return the source light back to the detector. The advantage of the DOAS method is that the absorption coefficient for OH is well known, so OH concentrations can be determined directly from Beer's Law. Also, since the location of the absorption bands is very well known and they have very small bandwidths, any interfering absorption bands can easily be removed by background subtraction.

LIF uses laser light from a tunable dye laser at 616 nm, which is converted to light at 308 nm by a laser doubling crystal. The laser light is directed through the air sample contained in a multi-pass cell with "White cell" optics. A White cell is constructed using three concave mirrors that reflect the light beam between them. The number of passes the light makes through the cell can be changed by making rotational adjustments to the first and third mirrors. So, the path length through the cell can be increased many times without changing the volume of the cell. The laser light passing through the White cell excites the OH radical to its first electronically excited singlet state ($^1A^*$). The excited singlet state decays back to the ground state by the emission of light at 308 nm, which is measured at 90° to the laser beam. A background measurement is made by tuning the laser to a wavelength that is not absorbed by OH. LIF is inherently more sensitive than DOAS because of the high-fluorescence quantum yields. However, the background is higher due to the presence of scattered laser light at the same wavelength as the fluorescence. The major disadvantage of LIF is the need to run field calibrations, and the generation of an OH calibration standard under field conditions is difficult.

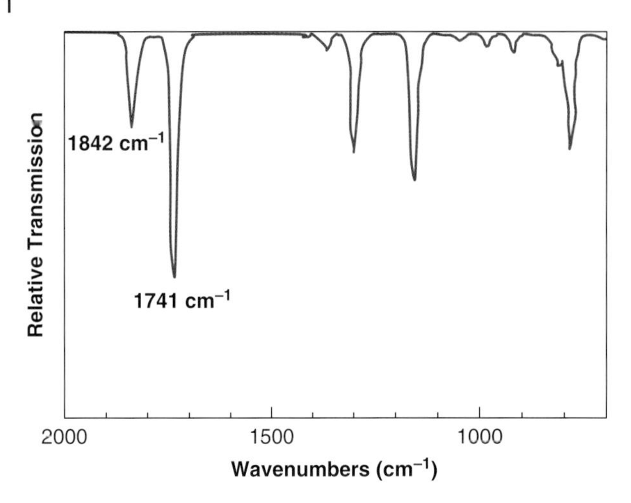

Figure 7.4 Infrared absorption spectrum of PAN ($CH_3C{=}O{-}O_2NO_2$). The characteristic absorption bands due to the C=O stretching vibrations of an anhydride are shown at 1842 and 1741 cm^{-1}.

Peroxy acetyl nitrate was initially measured by open-path IR spectroscopy, which was used to first characterize organic photochemical smog. Perooxyacetyl nitrate (PAN) has a very characteristic IR spectrum, which is shown in Figure 7.4. It has two unusual C=O stretching frequencies at 1842 and 1741 cm^{-1}, which are similar to the frequencies of the carbonyl group in an organic anhydride. Only acid halides and anhydrides show absorption frequencies in the region above 1800 cm^{-1}. The organic acids have C=O stretches at 1700–1720 cm^{-1}. In fact, all the PANs are essentially mixed anhydrides of the organic peracid with HNO_3. In fact, PAN has been produced in the laboratory as the mixed anhydride by the dehydration reaction of peracetic acid ($CH_3C{=}O{-}OOH$) in the presence of HNO_3.

The aqueous reaction of PAN with hydroxide ion was also used to determine its concentration. Although PAN is not highly water soluble ($H_x = 4\,M\,atm^{-1}$) and it does not react with water at low pH, it was found to hydrolyze rapidly under strong basic conditions (pH = 12) by the reaction:

$$CH_3COO_2NO_2 + OH^- \rightarrow CH_3COO^- + H_2O + O_2 + NO_2^-$$

This reaction allowed PAN concentrations to be determined by collecting the samples using an impinger sampling train with a basic reagent solution. The NO_2^- concentrations were then determined using UV spectroscopy, or both NO_2^- and CH_3COO^- concentrations could be measured using ion chromatography.

Since PAN and its analogs contain five highly electronegative oxygen atoms, they have a high electron capture cross-section. Gas chromatography with electron capture detection (GC–ECD), coupled to an automated gas sampling loop, has been used successfully for decades to measure PAN in the atmosphere with detection limits of about 10 ppt. A schematic diagram of a GC–ECD detection system for PAN is shown in Figure 7.5. Since PAN is thermally unstable, the injection port and column are not heated as in traditional gas chromatography (GC). In addition, traditional chromatography uses long columns to increase separation of the gases, but the GC PAN analyzer uses a short column in order to minimize the decomposition of the PAN on column. Originally, the columns were constructed of 0.125 in. diameter Teflon tubing packed with a nonpolar stationary phase such as DB-1 to separate PAN from O_2, chlorofluorocarbons (CFCs), and organic nitrates, which all have a high electron capture cross-section. More recently, the packed column has been replaced with an open tubular glass

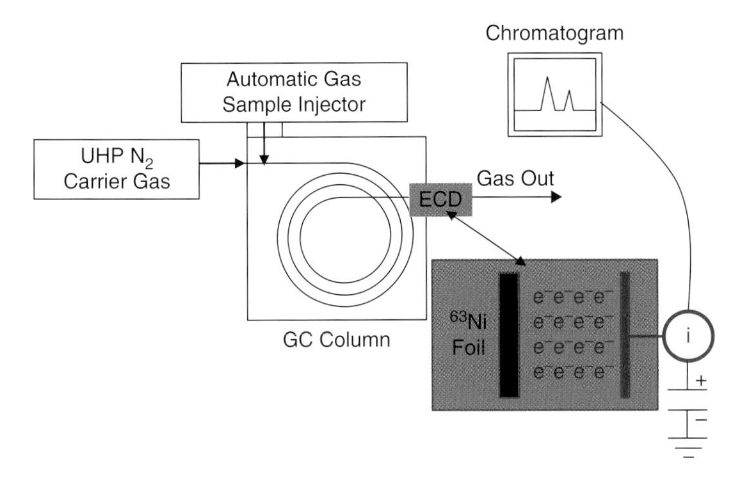

Figure 7.5 A schematic diagram of a gas chromatograph with an electron capture detector (GC–ECD). The ECD contains a ^{63}Ni foil that serves as a source of beta particles (β^-), providing a constant electron current to a positively charged plate.

capillary column with an internal coating of DB-1. The analysis time for this system is usually about 10–15 minutes, because of the time it takes for water vapor to clear off the column. The polar water molecule does not interact with the nonpolar stationary phase and so travels through the column unaffected. However, because of the highly electronegative oxygen atom, water does have an ECD signal. This results in a very broad elevated background signal. So, it takes about 10–15 minutes for the GC–ECD signal to return to baseline before another sample can be injected onto the column.

The electron capture detector (ECD) is a simple device originally invented by James Lovelock for the measurement of oxygen on other planets, as an indicator of the presence of life. The ECD contains a sealed ^{63}Ni radioactive source, which is a beta emitter. The beta particles are fast-moving electrons that are released from the ^{63}Ni into the carrier gas stream as it exits the column, generating a steady electrical current. When an electronegative gas exits from the column, it captures these electrons, causing a drop in the current. This drop in the current is measured by an electrometer, an instrument that measures an electrical charge or electrical potential without drawing current from the circuit. The signal from the electrometer is inversely proportional to the concentration of the electronegative species. The detection limits of the ECD are in the low ppt range for many compounds that contain highly electronegative atoms, such as oxygen and the halogens.

A typical GC–ECD analysis of PAN (30 ppb), peroxypropionyl nitrate (PPN) (2.5 ppb), and peroxybutyl nitrate (PBN) (0.8 ppb) in the air of Mexico City on February 21, 1997 is shown in Figure 7.6. The detection limits of this method for the peroxyacyl nitrates are typically about 10 ppt. The first large peak in the chromatogram at 0.454 minutes is due to O_2, followed by PAN at 3.384 minutes, PPN at 3.919 minutes, and PBN at 5.903 minutes. The peak at 2.350 minutes is due to methyl nitrate (CH_3ONO_2). The time it takes for PAN to pass through the chromatography column from the injection port to the detector, called the retention time, is about 3–4 minutes in this system, depending on the instrumental conditions. But in the sample shown in Figure 7.6, the baseline does not recover until about 14 minutes, when the water finally clears the column. So, the time between analyses depends on the relative humidity. In addition, the PAN retention time can also vary depending on the relative humidity. High relative humidity can interfere with the ability of the stationary support (DB-1) to interact with the PAN molecule as it passes through the column, leading to shorter retention times and broader peaks. This variation

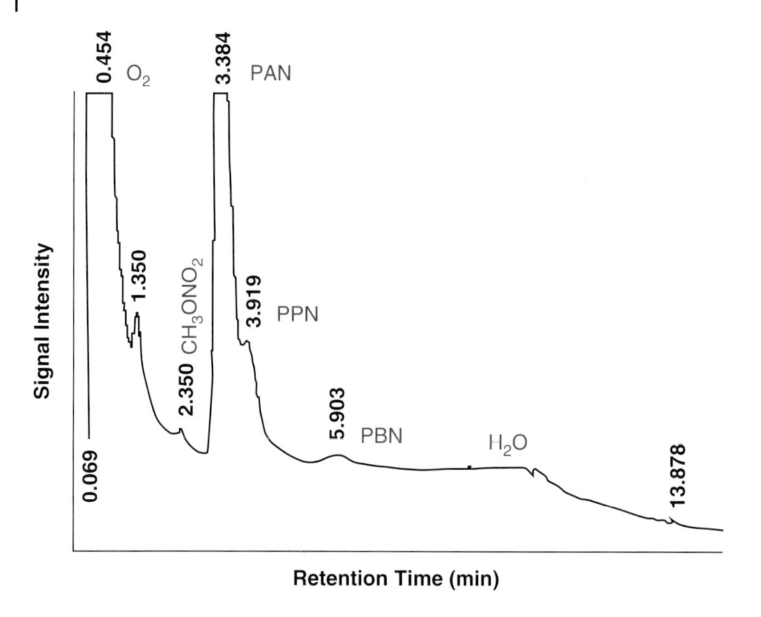

Figure 7.6 A GC–ECD chromatogram of PAN, PPN, and PBN in an air sample taken in Mexico City on February 21, 1997.

in retention times between analyses can make automatic peak location and integration difficult. This is a concern for the use of this method for automated field analysis.

The use of the ^{63}Ni radioactive beta source in the ECD has led to issues regarding special shipping requirements for radioactive materials when operating the system in the field. Also, even though the source is a sealed low-level beta source, it is required to inspect it annually for any evidence of radioactivity leakage, as with all radioactive sources. Also, because the ECD is so sensitive to O_2, the presence of oxygen in the carrier gas would mask the PAN peak. So, the method requires UHP N_2 as a carrier gas along with an oxygen scrubber in order to reduce oxygen interferences and insure a high sensitivity for PAN. The use of a compressed gas cylinder for the carrier gas is also a safety limitation when operating the instrument in an aircraft or remotely at a ground-based site. These operating requirements, along with the required 15-minute analysis times due to water vapor interferences, limit the usefulness of the GC–ECD method for aircraft applications and have led to the development of an alternative detector for GC PAN analysis. This alternative GC detector is based on luminol chemiluminescence (Marley et al., 2004).

Luminol is known to react with a number of oxidants, including H_2O_2, NO_2, PAN, organic peroxides, and organic peracids in a basic solution (pH = 12), which results in chemiluminescence at 425 nm by the processes shown in Figure 7.7. The luminol is activated by reaction with OH^- to produce two resonant forms of a dianion. The dianion then reacts with an oxidant, releasing N_2 and forming another dianion in an excited triplet state ($^3A^*$), which decays to the excited singlet state ($^1A^*$) by intersystem crossing. The excited singlet state then decays to the ground state ($^1A_0$) by the emission of light at 425 nm.

A schematic diagram of the GC–luminol system is shown in Figure 7.8. The air sample is delivered to a capillary column coated with DB-1 by a 5 cc sample loop connected to a six-port switching valve. The carrier gas for the GC can be supplied by a lecture bottle of helium or air that has been scrubbed by a charcoal filter to remove NO_2 and PAN. A peristaltic pump slowly delivers the luminol to a fabric wick in a dark reaction cell and the used luminol flows out of the cell to be collected in a waste reservoir. The gas from the GC column flows across the

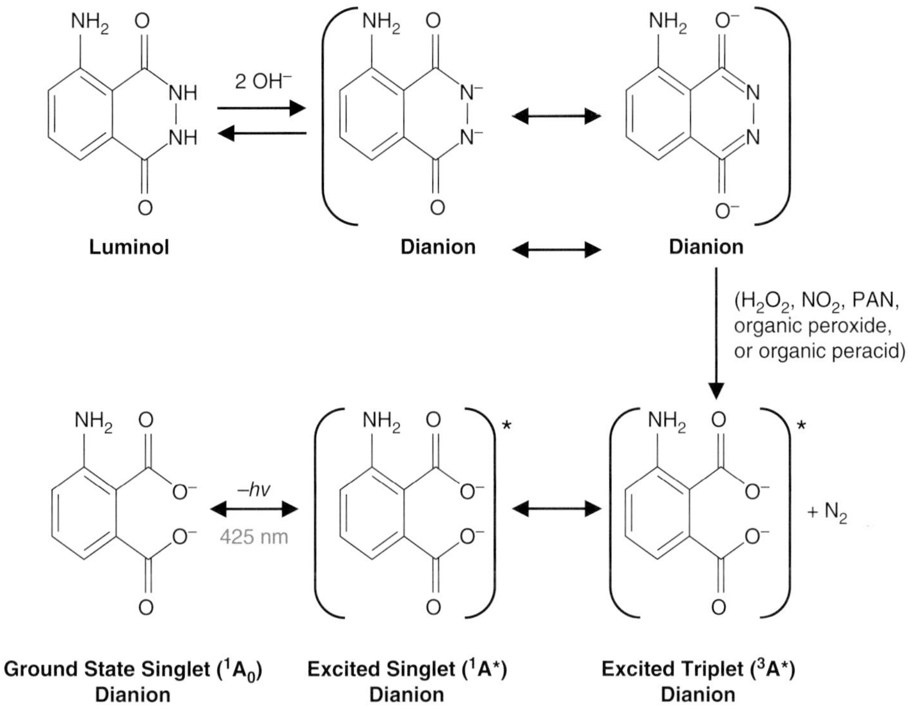

Luminol $\underset{}{\overset{2\ OH^-}{\rightleftharpoons}}$ Dianion ⟷ Dianion

(H_2O_2, NO_2, PAN, organic peroxide, or organic peracid)

Ground State Singlet ($^1A_0$) Dianion $\overset{-h\nu}{\underset{425\ nm}{\longleftrightarrow}}$ Excited Singlet ($^1A^*$) Dianion ⟷ Excited Triplet ($^3A^*$) Dianion + N_2

Figure 7.7 The luminol chemical reaction with organic oxidants to produce chemiluminescence with maximum light emission at 425 nm. Source: Adapted from Deglr6328, Wikimedia Commons.

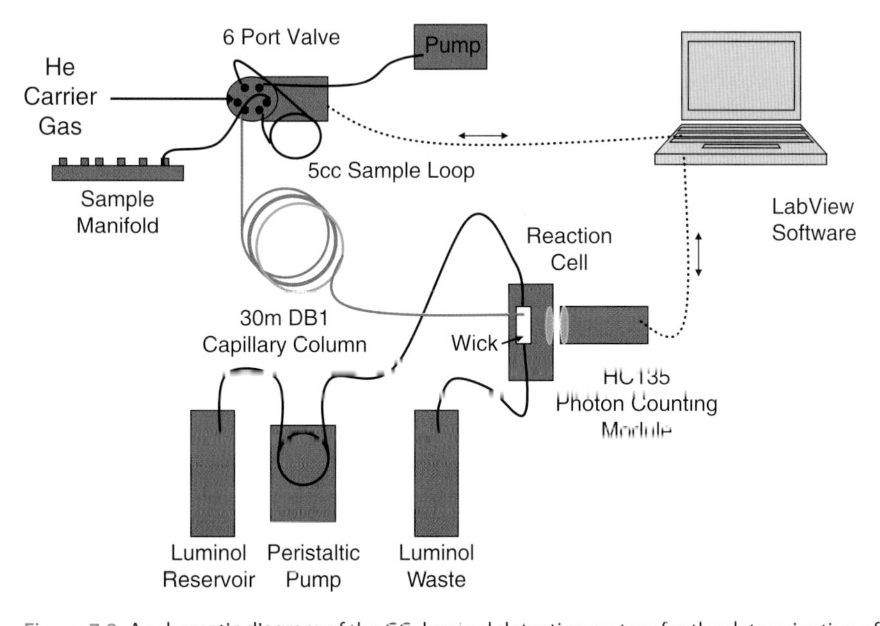

Figure 7.8 A schematic diagram of the GC–luminol detection system for the determination of PAN and NO_2. The chemiluminescent reaction occurs as a gas–liquid surface reaction on a fabric wick.

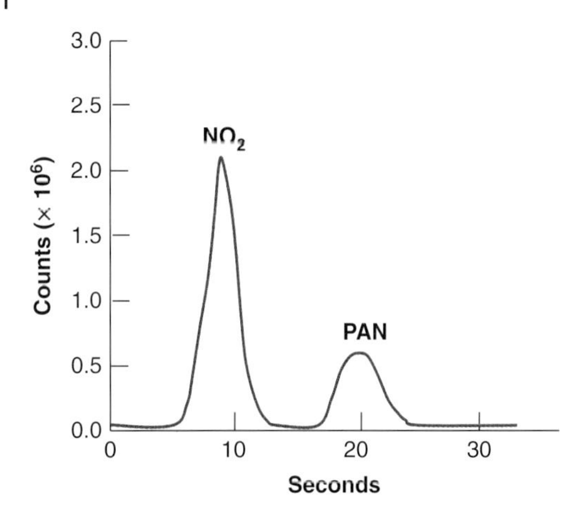

Figure 7.9 A GC–luminol chromatogram of 0.4 ppb PAN and 8 ppb NO_2.

wick, where NO_2 or PAN reacts with the luminol in a gas–liquid surface reaction to produce the chemiluminescence. The light produced by the luminol reaction is measured by a photomultiplier tube (PMT) connected to a photon counting module. Each peak in the resulting chromatogram is integrated using LabVIEW software on a laptop computer. The software also automatically operates the sample valve, coordinating the sample injection sequence with the end of the measurement of the previous sample.

An example of the output of the GC–luminol system for one analysis is shown in Figure 7.9. Because neither water vapor nor oxygen give an observable luminol signal, the instrument has a much faster sample cycle (<30 s) than the GC–ECD system, with similar detection limits (about 10 ppt). This method has been used successfully for aircraft and ground-based measurements of NO_2 and PAN and has been validated by comparison with NO_2 measurements made with DOAS in the field (Gaffney et al., 1999; Marley et al., 2004).

Originally, PAN calibration standards were synthesized in the laboratory by the photolysis of ethyl nitrite (CH_3CH_2ONO) in air. This synthetic procedure was developed by Edgar Stephens at the Statewide Air Pollution Research Center at the University of California, Riverside. The photolysis proceeded by the following reactions:

$$CH_3CH_2ONO + h\nu \rightarrow CH_3CH_2O + NO$$
$$CH_3CH_2O + O_2 \rightarrow HO_2 + CH_3CHO$$
$$HO_2 + NO \rightarrow OH + NO_2$$
$$CH_3CHO + OH \rightarrow CH_3CO + H_2O$$
$$CH_3CO + O_2 \rightarrow CH_3COO_2$$
$$CH_3COO_2 + NO_2 \rightarrow CH_3COO_2NO_2$$

Since the photolysis forms a number of side products, GC was required to purify the gas-phase PAN standard, which was then trapped and pressurized in an evacuated gas container. The formation and purification of PAN by this method was confirmed from the IR spectrum. Due to the thermal instability of PAN, the pressurized gas container needed to be stored in a cold room to prevent catastrophic decomposition. In fact, the use of this collection procedure led to an explosion due to PAN apparently condensing in the metal gas pressure gauge used in the collection of the gas product (Stephens et al., 1969).

This safety issue, along with the need for specialized photolysis equipment for the production of PAN standards by the photolytic method, led to the development of alternative synthetic methods for PAN and its higher analogs (Nielsen et al., 1982; Gaffney et al., 1984). This alternative method for PAN synthesis uses peracetic acid ($CH_3C{=}OOOH$) and HNO_3 with H_2SO_4 as a dehydrating agent to produce PAN as the mixed anhydride by the reaction:

$$CH_3C = OOOH + HNO_3 \rightarrow CH_3C = OO_2NO_2 + H_2O$$

The peracetic acid is first synthesized by the reaction of H_2O_2 with acetic anhydride, producing CH_3COOH as a byproduct. The PAN synthesis is then carried out in a standard Pyrex Erlenmeyer flask in an ice bath at 0 °C by the slow addition of the HNO_3 into an acidified solution of the peracid. Since PAN is very soluble in nonpolar solvents, the PAN was extracted from the reaction solution into heptane. However, this procedure required chromatography to separate the PAN from the heptane (Nielsen et al., 1982). A revised procedure was developed that extracts the PAN into *n*-tridecane (Gaffney et al., 1984). After extraction, the solution is washed with ice water to separate the PAN–tridecane from the water-soluble CH_3COOH, HNO_3, and H_2SO_4. The PAN–tridecane solution is then stored frozen at −10 °C, below the freezing point of *n*-tridecane (−5 °C). The higher peroxyacyl nitrates are synthesized by the same procedure using the appropriate organic peracid ($RC{=}OO_2H$). The synthesis of PAN and its analogs by this method produces very high purity standards (99%). In addition, the *n*-tridecane has a very high flashpoint, so the potential for explosion is greatly lowered. This synthetic method is simple, low cost and safe, and has become the preferred method for preparing PAN calibration standards.

Luminol chemiluminescence has also been used for the measurement of H_2O_2 in the air using a copper ion catalyst and adjusting the conditions of the luminol solution. This system uses a continuous water extraction process to scrub the H_2O_2 from the air and concentrate it in the liquid phase (Kok et al., 1978). However, under these conditions the chemiluminescence is also produced by organic peroxides and peracids, so the method is not specific to H_2O_2. Another frequently used method of measuring H_2O_2 in air involves a chemical reaction that produces a fluorescent product (Lazrus et al., 1985). This procedure also uses continuous water extraction of the air sample, followed by reaction of $H_2O_2(aq)$ with *p*-hydroxyphenylacetic acid (PHPAA) in the presence of a peroxidase enzyme, which produces the PHPAA dimer:

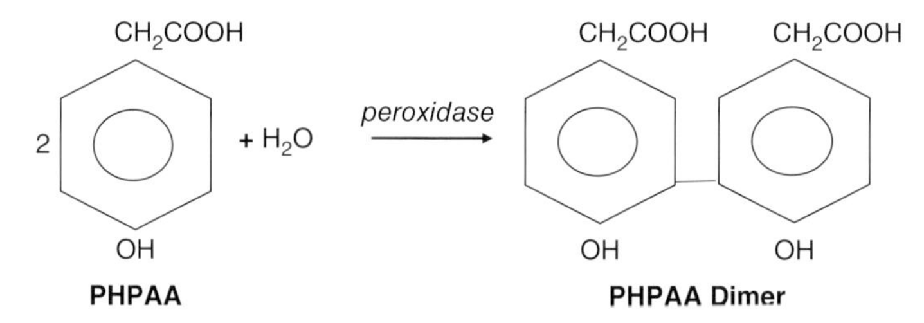

The PHPAA dimer emits fluorescent light at 405 nm after excitation at 320 nm, which is directly proportional to the H_2O_2 concentration. This method also responds to organic peroxides, so the signal is also equal to the total peroxide concentration in the atmosphere.

Either method can be coupled to high-pressure liquid chromatography (HPLC) with a C-18 column to separate the organic peroxides and peracids from the H_2O_2. However, the use of aqueous scrubbers for either method is susceptible to both positive and negative sampling artifacts. Aqueous reaction with O_3 in the sampling air has been shown to give high results for H_2O_2 concentrations. The H_2O_2 is also very susceptible to losses on surfaces in the sample collection system, resulting in lower than expected concentrations of H_2O_2. While H_2O_2 is very soluble in water, the solubility of the organic peroxides and peracids is variable and so their collection efficiencies are also variable. The Henry's Law constants for the organic peroxides and peracids can be used to correct for the differences in collection efficiencies, but their detection limits would be high. There are currently no adequate direct measurement techniques for the determination of H_2O_2 or the organic peroxides and peracids that have sufficient detection limits for atmospheric concentrations. But these same methods are expected to have adequate detection limits with less interference for the determination of peroxides in cloud water and precipitation when used as HPLC detection methods.

7.2.2 The Oxides: Nitric Oxide, Nitrogen Dioxide, Nitric Acid, Carbon Monoxide, Carbon Dioxide, Sulfur Dioxide, and Nitrous Oxide

7.2.2.1 Nitric Oxide, Nitrogen Dioxide, and Nitric Acid

The first measurements of NO and NO_2 were made using colorimetric analysis methods. An early method collected the samples by impingers filled with a reagent solution of $KMnO_4$, which is oxidized by either NO or NO_2 to MnO_2:

$$NO(g) + MnO_4^-(aq) \rightarrow NO_3^-(aq) + MnO_2(s)$$

$$3NO_2(g) + MnO_4^-(aq) + H_2O(l) \rightarrow 3NO_3^-(aq) + MnO_2(s) + 2H^+(aq)$$

The MnO_4^- ion is an intense purple color with a broad absorption band between 460 and 580 nm, while the product MnO_2 is insoluble in water. So, the decrease in the light absorption at 520 nm (λ_{max}) is directly proportional to the concentration of NO_x in the atmosphere. The use of this method was abandoned because of its susceptibility to interferences from other atmospheric species that can also oxidize the MnO_4^- ion. Another colorimetric method more commonly used for the measurement of NO and NO_2 was the Griess–Saltzman method. In this method samples are collected by impingers filled with the Griess–Saltzman absorbing solution, which contains sulfanilic acid ($SO_3HC_6H_4NH_2$) and N-(1-napthy1)-ethylenediamine dihydrochloride. The NO_2 reacts with the sulfanilic acid to form a diazonium salt ($SO_3HC_6H_4PhN_2^+$), which then couples with N-(1-napthy1)-ethylenediamine dihydrochloride to form a deeply colored azo dye. The absorbance of this solution is measured at 550 nm to determine the NO_2 concentration. The method is specific for NO_2. The total NO_x concentration is determined by first oxidizing the sample, converting the NO to NO_2. The NO is then determined by difference.

The FRM for NO and NO_2 measurements in air was originally the Jacobs–Hochheiser method, which is a variation of the Griess–Saltzman method. In the Jacobs–Hochheiser method, the air is passed through an aqueous sodium hydroxide solution to convert the NO_2 to NO_2^-. Since SO_2 is an interference, it is removed from the sample by pretreatment with H_2O_2 and acidification. The rest of the procedure is the same as for the Griess–Saltzman method, except that sulfanilamide ($SO_2NH_2C_6H_4NH_2$) is used instead of sulfanilic acid. The Jacobs–Hochheiser method was listed as the FRM because it allowed for longer sample collection times, as well as delay times between sampling and analysis, than the Griess–Saltzman method. Neither method had sufficient detection limits for short sampling times. The detection limits for the Griess–Saltzman and Jacobs–Hochheiser methods were 20 ppb. They also suffered from a number of interferences.

Currently, the FRM for the determination of NO and NO_2 is based on the chemiluminescent reaction of NO with O_3:

$$NO + O_3 \rightarrow NO_2^* + O_2$$

Ozone reacts with NO to give NO_2 in an excited state. The NO_2^* decays to the ground state by the emission of a broad-wavelength light in the region of 590–3000 nm with a maximum in the NIR at 1270 nm. This same reaction is used to detect atmospheric NO_2 after thermally decomposing it to NO on a heated catalytic surface. The observed chemiluminescence is then due to the sum of atmospheric NO and NO_2 or NO_x. The atmospheric NO_2 concentration is

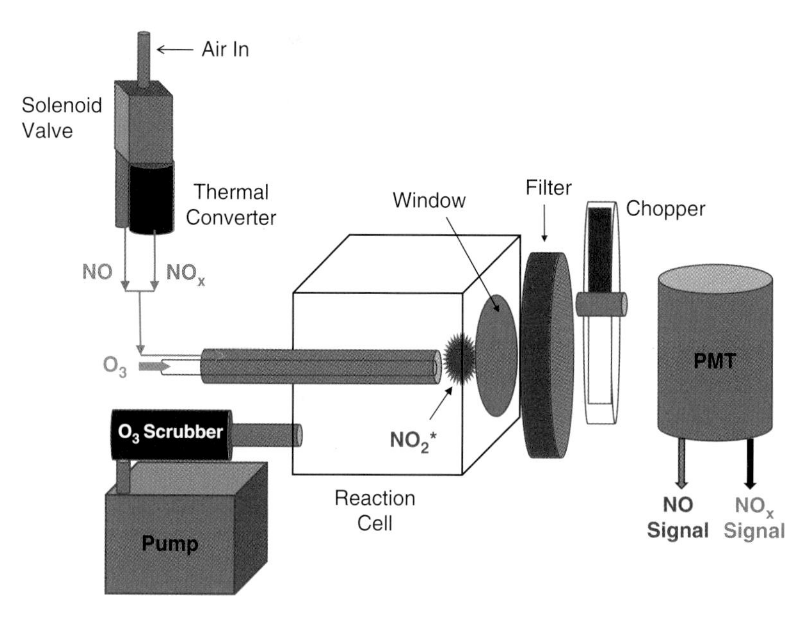

Figure 7.10 Schematic diagram of a chemiluminescence NO and NO_2 instrument.

determined by the difference between the air sample with and without thermal decomposition. Because the instrument responds to total NO_x, it is commonly called a "NO_x Box." A schematic diagram of the NO_x Box is shown in Figure 7.10. Ambient air containing NO is mixed with reagent O_3 produced from a corona discharge, which produces O_3 concentrations in the low percent range. Mixing of the O_3 with the air sample occurs in front of an optical window in a dark reaction cell. The reaction cell is kept at low pressure to minimize collisional quenching of the NO_2^* and maximize the chemiluminescent signal. The light emitted by NO_2^* passes through a red transmission cut-off filter, which transmits light at wavelengths >600 nm. This is done to remove interferences from the chemiluminescent reactions of O_3 with alkenes at wavelengths of 400–600 nm and with reduced sulfur compounds at 340 nm.

The instrument has two channels of operation, the NO channel and the NO_x channel. During NO channel operation, the sample air is passed directly into the reaction cell. During NO_x channel operation, the air sample is first passed through a catalyst heated to 300–400 °C to convert all of the NO_2 to NO, after which it is passed into the reaction cell. However, other atmospheric nitrogen species – including peroxyacyl nitrates, HNO_3, HONO, N_2O_5, HO_2NO_2, and organic nitrates – are also converted to NO in the thermal catalytic converter. The sum of all the atmospheric nitrogen-containing compounds that the NO_x Box responds to ($NO_x + HNO_3 + HONO + N_2O_5 + HO_2NO_2$ + organic nitrates + peroxyacyl nitrates) has now been called NO_y. In order to avoid these many interferences that occur in the NO_x channel, instruments have been developed that replace the thermal catalytic converter with a photolytic converter that uses a UV LED radiation to convert NO_2 to NO by the photolysis reaction discussed in Section 5.4:

$$NO_2 + h\nu \rightarrow NO + O(^3P)$$

One fundamental problem with ozone chemiluminescent detection of NO and NO_2 is the wavelength of the light produced by the NO_2^*. The currently available PMTs are not very sensitive in this wavelength range. Although there are special red-sensitive PMTs available that extend the normal detection cutoff range from 650 to 800 nm, the detection limits for

NO and NO_2 using these detectors are usually around 1 ppb. The luminol chemiluminescent method discussed in Section 7.2.1 for the measurement of PAN has shown much higher sensitivity since the light emission is at 440 nm, which is very easy to detect with low-cost PMTs. The major interfering species with the luminol method when it is used alone is PAN. However, this is easily overcome by using the chemiluminescent reaction as a GC detector. The detection limit for NO_2 with this GC–luminol system is <10 ppt. Other methods currently being explored use dual-wavelength quantum cascade lasers to measure the IR absorption of NO and NO_2 with detection limits of 0.5 ppb for NO and 1.5 ppb for NO_2 (Jagerska et al., 2015). Laser cavity ring-down spectroscopy, an intra-cavity laser absorption technique, has also been used for atmospheric NO_2 measurements (Hargrove et al., 2006). Most all of these optical methods require filtering of the air sample to remove particulates which cause interference from light scattering. In many cases drying the air prior to measurement is also necessary, since water vapor can adsorb on the optics. This also causes light scattering, which adds to the stray light background of the spectrometer, lowering the detection limits of the methods.

7.2.2.2 Nitric Acid, Carbon Monoxide, Carbon Dioxide, Sulfur Dioxide, and Nitrous Oxide

Samples for gas-phase HNO_3 can be collected with impingers filled with water, since it is highly water soluble. The aqueous NO_3^- concentrations are then determined using ion chromatography. Since particulate NH_4NO_3 is also collected in the water, the air samples must first be filtered to remove any particles. However, the sample filtration usually leads to loss of HNO_3 on the filter material. Sampling lines used for sample collection of HNO_3 must also be kept short and heated to minimize the loss of the water-soluble gas to wet surfaces. These interferences can be avoided by the use of diffusion denuders, which collect the gas-phase HNO_3 on the walls of a flow-through reactor while the particulate NH_4NO_3 passes through unaffected. The flow reactors are usually cylindrical or disk-shaped tubes, which maximize the surface area of the reactor. The inner walls of the reactor can also be coated with chemicals that aid in the collection of HNO_3. Some denuder coatings that have been used are $NaCO_3$, $NaHCO_3$, and H_2WO_4. HNO_3 has also been determined by OP–FTIR absorption spectroscopy, which avoids the problems with surface losses. In addition, HNO_3 can be measured with the NO_x Box using a thermal catalytic converter, as discussed in Section 7.1. HNO_3 is removed from the air sample by the use of a nylon filter, and the atmospheric HNO_3 concentration is determined by comparing the signal from the NO_x channel with and without a nylon filter in place.

The atmospheric oxides CO, CO_2, SO_2, and N_2O have characteristic IR absorptions, which are commonly used for their measurements using open-path methods. Diatomic molecules such as CO have only one normal mode of vibration, which involves the stretching of the bond. In addition, since molecules in the gas phase are free to rotate about the molecular axis, their vibrational spectra have rotational components. These are due to the transitions between sequences of rotational levels associated with both the vibrational ground state and the vibrational excited state, called rotational–vibrational transitions. For simple molecules such as CO, these rotational–vibrational transitions can often be resolved into narrow bands or lines at high spectral resolution. Each of these lines represents a transition from one rotational level in the ground vibrational state to one rotational level in the excited vibrational state. For example, the IR absorption spectrum of CO, shown in Figure 7.11, shows both the high-resolution and low-resolution spectra. The low-resolution spectrum obtained at 4 cm^{-1} appears as two broad bands centered at 2140 cm^{-1}, while the high-resolution spectrum obtained at 0.5 cm^{-1} shows the very narrow rotational–vibrational absorption lines which make up these two bands. The narrow absorption lines are due to the transitions that involve changes in both vibrational and rotational states of the CO molecule. These characteristic IR absorption bands of CO have

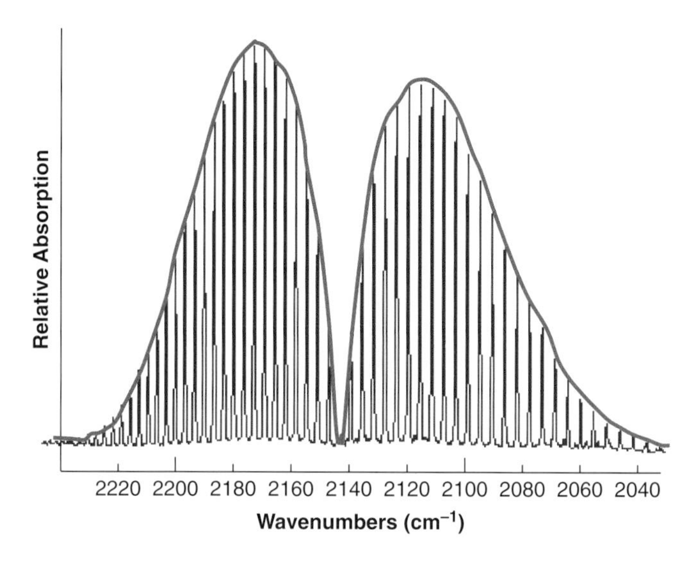

Figure 7.11 The low-resolution spectrum obtained at 4 cm^{-1} (blue) and the high-resolution spectrum obtained at 0.5 cm^{-1} (red) for CO showing the vibrational–rotational lines at high resolution. Source: Adapted from Ian13, Wikimedia Commons.

strong absorption coefficients that can be used for quantitative determinations according to Beer's Law. Also, since the bands are in a wavelength region where there is little interference from other IR-absorbing gases, such as CO_2 or H_2O, either low-resolution or high-resolution spectra can be used for CO measurements at ppb concentrations.

Although either OP–FTIR or nondispersive infrared (NDIR) spectroscopy can be used to obtain accurate measurements with low detection limits for CO, NDIR is the FRM for the measurement of CO concentrations in air. The NDIR system, shown in Figure 7.12, is nondispersive since it does not use a light-dispersing element, such as a grating or prism, to separate the IR source radiation into its component wavelengths. Instead, it uses a bandpass filter to isolate the IR radiation in the region of 2000–2200 cm^{-1} (4.5–5.0 μm). The main components of a CO NDIR system are the IR source, the sample absorption cell, an IR bandpass filter, and an IR detector. The sample air is drawn into the sample absorption cell, which has White cell optics, by a sample pump. The IR source radiation is reflected by the three concave mirrors in the sample absorption cell to give a typical path length of 1 m. The IR beam exits the sample absorption

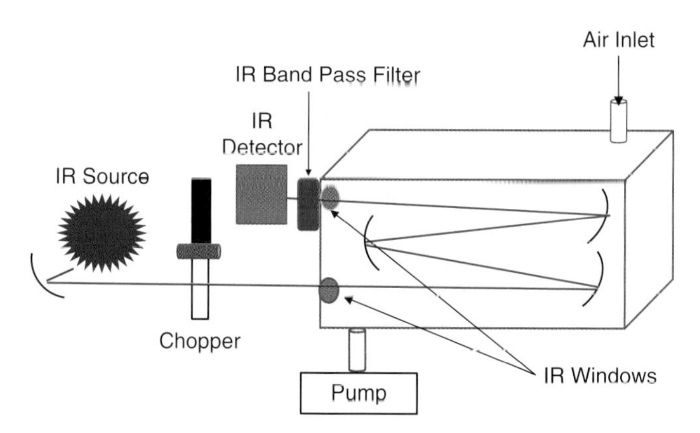

Figure 7.12 A schematic diagram of a nondispersive infrared (NDIR) CO measurement system.

Table 7.1 The IR wavelengths (μm) and special frequencies (cm^{-1}) of the infrared absorption bands for atmospheric gases measured using NDIR systems.

Gas	Wavelength (μm)	Frequency (cm^{-1})
CO	4.6, 4.67, 4.8	2174, 2141, 2083
CO_2	4.26, 2.7	2347, 3703
SO_2	7.35, 19.25	1360, 519
N_2O	4.3, 4.4	2326, 2278

These wavelength ranges also represent the wavelength band passes of the filters used in these systems.

cell and passes through an IR bandpass filter in front of the detector. The bandpass filter isolates the wavelengths where the CO absorption occurs and eliminates all other wavelengths. The IR beam passes through a rotating fan blade, called a chopper, before entering the sample absorption cell. The rotating blade of the chopper pulses the IR beam, generating an alternating current (AC) signal at the detector. The AC signal from the detector is compared to the chopping frequency, removing any signal components that are not chopped, including instrument noise or source fluctuations.

NDIR systems are used in most large buildings as CO detectors for safety purposes, instead of the simple, inexpensive electrochemical CO detectors used in homes. This is because the electrochemical methods have numerous interferences and low sensitivities when compared to the NDIR systems. So, they are likely to give false positive signals. NDIR systems are also used for measurements of other atmospheric gases, including CO_2, SO_2, and N_2O, which have characteristic IR absorption bands that can easily be used in NDIR systems. In some systems the IR bandpass filter is replaced with a filter wheel, which can hold several different bandpass filters. The filter wheel turns to place different filters in the beam path, allowing for the measurement of more than one gas sequentially. Table 7.1 lists the wavelengths of the absorption bands for these gases. The wavelengths also represent the wavelength bandpass of the filters used in the NDIR systems for the measurement of the different gases.

The FRM for SO_2 is a colorimetric method based on the reaction with pararosaniline ($[(H_2NC_6H_4)_3C]Cl$), known as the West–Gaecke method, which is commonly used in autoanalyzers. The samples are collected with an impinger train using a potassium tetrachloromercurate (K_2HgCl_4) reagent solution. Ethylenediaminetetraacetic acid (EDTA) is added to the impinger reagent solution to remove interferences from heavy metals by complexation. The SO_2 reacts with $KHgCl_4$ to form a stable dichlorosulfitomercurate complex:

$$K_2HgCl_4 + SO_2 + H_2O \rightarrow K_2[HgCl_2(SO_3)] + HCl$$

After sample collection, pararosaniline, formaldehyde, and an acetate buffer are added to the solution. The $K_2[HgCl_2(SO_3)]$ complex reacts with pararosaniline to form the highly colored pararosaniline methyl sulfonic acid dye:

$$[(H_2NC_6H_4)_3C]^+Cl^- + K_2[HgCl_2(SO_3)] \rightarrow [(H_2NC_6H_4)_3C]^+CH_2SO_3^-$$

The absorption of the pararosaniline methyl sulfonic acid dye, measured at 530 nm, is directly proportional to the atmospheric SO_2 concentration.

Sulfur dioxide has characteristic absorption bands in the UV wavelength range between 300 and 320 nm, as shown in Figure 7.13, which have also been used to measure this gas by an

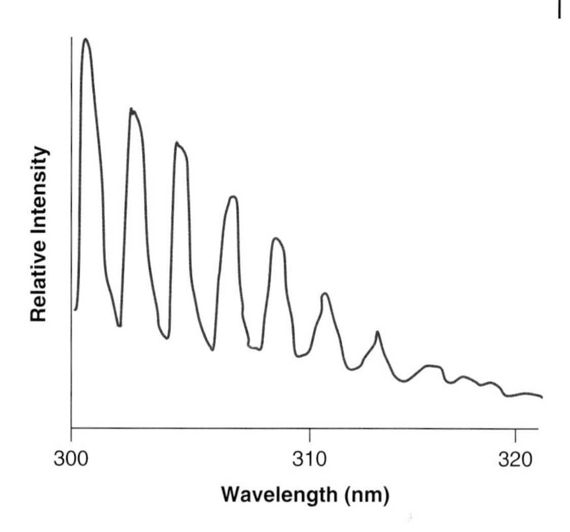

Figure 7.13 The UV absorption spectrum of SO_2.

open-path absorption system called a correlation spectrometer. The correlation spectrometer operates much like the Dobson meter. High-resolution spectra are obtained by using a grating spectrometer with the Sun as the light source. Since the SO_2 absorption bands are narrow, a spectral mask is used at the exit of the monochromator to isolate the wavelength ranges of the bands. The light intensity measured with the mask in place is compared with the light intensity with the mask moved off the SO_2 absorption lines. These two measurements represent the light intensity before and after sample absorption (I/I_o). The SO_2 concentrations are then determined from Beer's Law in the same manner as with the Dobson meter. The correlation spectrometer has also been used for the measurement of NO_2 concentrations, which has the same type of characteristic absorption bands in the visible wavelength region.

7.2.3 The Organics: Volatile Organic Hydrocarbons, Aldehydes, Ketones, and Halogenated Hydrocarbons

The initial measurements of atmospheric VOCs were made with a simple flame ionization detector (FID). In the FID, the sample is pulled into a H_2 flame where all organic compounds in the sample are combusted at high temperature. At the temperature of the hydrogen flame (2254 °C) the organic compounds form positive ions, which are collected onto a negatively charged plate. The ions induce a current in an external circuit that is a measure of the number of carbon atoms hitting the plate. The signal is proportional to the concentrations of the hydrocarbons in the gas stream. Different hydrocarbons have different response factors, which are directly proportional to the number of carbon atoms in the molecule. Oxygenated hydrocarbons and species that contain other heteroatoms have lower response factors. The measurement from the FID is reported as total hydrocarbons (THC), in units of ppm carbon. The major VOC species in any air mass is typically CH_4, at a concentration of 2 ppm or higher. The measurement of hydrocarbons other than CH_4 is accomplished by pulling the air through a short nonpolar column, which separates methane from the rest of the hydrocarbons. The signal is then proportional to the concentration of non-methane hydrocarbons (NMHC). The system can be operated in THC mode or in NMHC mode. In NMHC mode the sample analysis is delayed for about 20 s to allow the CH_4 to exit the pre-column. The measurement after the short time delay is reported as total NMHC in ppm carbon. This method gives a measurement of the total amount of organics present but does not give any information about the reactivity of the compounds in the air sample or the composition of the NMHC.

The EPA-approved method for the measurement of speciated VOCs in air uses high-resolution capillary GC to determine the concentrations of the specific hydrocarbon compounds. The GC detector is typically a positive ion mass spectrometer (GC–MS) or an FID (GC–FID) for the determination of nonpolar compounds or partially oxidized organics such as aldehydes or ketones. The ECD, described in Section 7.2.1, or a negative ion mass spectrometer is used for compounds containing halogens, which are highly electronegative and resistant to combustion in the FID. Samples are collected in glass-lined stainless steel containers, which are heated to remove any organic contamination prior to sample collection. The inside of the containers is lined with a passivated glass surface that has been treated to minimize sample adsorption. Air samples are pressurized inside the containers at 1–2 atm in order to obtain sufficient sample for the trace organic species. Another approach is to pull the air through a thermally stable adsorbent such as Tenax (a polyphenyl ether resin) to concentrate the hydrocarbons onto a small cartridge. Once collected, the sample is then heated rapidly to about 300 °C to release the adsorbed compounds into the GC column for analysis. Because of the very large number of hydrocarbon species present in the atmosphere, the hydrocarbon speciation is typically limited to compounds containing less than 15 carbon atoms.

The GC–MS, GC–FID, or GC–ECD give an integrated VOC measurement over the time period of the sample collection. However, proton transfer mass spectrometry (PTMS) allows for real-time analysis of the lighter hydrocarbons (C_3 to C_{12}) along with a number of aldehydes and ketones. PTMS is based on the reaction of H_3O^+ with hydrocarbons to form an $(M+1)^+$ ion, where M is the parent molecule. Water vapor is injected into an ion source at low pressure (about 2 mbar), where electron bombardment produces the H_3O^+ ions. The H_3O^+ ions are then mixed with the sample air in a drift tube where the proton transfer reaction occurs. The gas-phase VOC molecules are ionized by the transfer of a proton from H_3O^+ to the VOC molecules:

$$VOC + H_3O^+ \rightarrow VOC - H^+ + H_2O$$

The chemical ionization reaction does not produce much, if any, molecular fragmentation, and so the mass spectrum is primarily made up of the parent ions with a proton added: $(M+1)^+$. The $(M+1)^+$ ion is then analyzed using either time of flight mass spectrometry or quadrupole mass spectrometry to obtain compound-specific signals. This method has been very successful in rapidly measuring a number of VOCs, including formaldehyde, acetaldehyde, and small ketones in air at low ppt levels (Blake et al., 2009).

Other methods for the measurement of specific organic species include high-resolution Fourier transform infrared (FTIR) with long-path gas cells. An example of this application is the determination of ethene by FTIR. Ethene was initially identified as an important air pollutant because of its role in crop damage. It is a potent plant growth hormone and at low levels is used to ripen fruit, which is transported green to prevent bruising. It is most commonly used as a mixture with CO_2 to ripen bananas, and this practice gives the gas mixture the common name "banana gas." Ethene is produced from incomplete combustion of fuels such as biomass and ethanol. PTMS is not useful for the determination of ethene, because its mass (28.05 g mol^{-1}) is similar to that of CO (28.01 g mol^{-1}) and N_2 (28.02 g mol^{-1}), which are both in much higher concentrations in air. High-resolution FTIR with long-path-length gas cells can determine ethene concentrations from the IR absorption bands, as shown in Figure 7.14, with little or no interference. This same method can be applied to other important atmospheric organics, including aromatics, aldehydes, and ketones. The sensitivity of the method depends on the path length of the gas cell and the presence of interferences. The main interference is the IR absorption from water vapor. However, the air sample can first be dried to remove the interfering absorption lines from water.

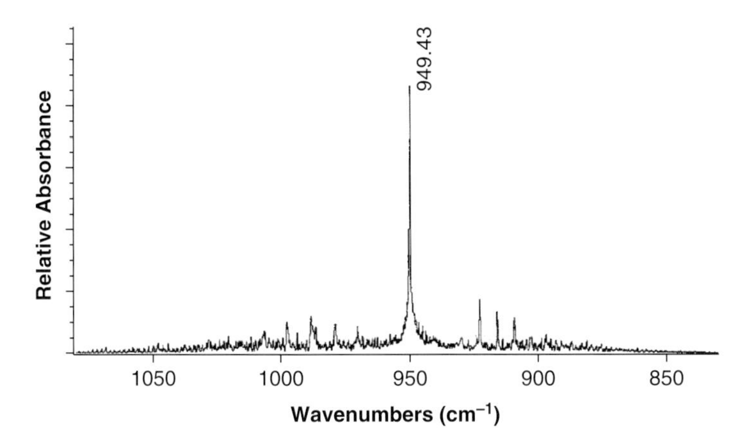

Figure 7.14 Long-path FTIR spectrum of 100 ppb ethene taken at a resolution of 0.5 cm^{-1} using a 20 m multi-pass gas cell.

The EPA-approved method for the measurement of aldehydes and ketones is based on their determination by chemical derivatization using 2,4-dinitrophenylhydrazine (DNPH). The samples are collected using an impinger sampling train containing the derivatization reagent solution (acidified DNPH) and an equal amount of isooctane. The reaction between an aldehyde or ketone with DNPH proceeds by nucleophilic addition of the $-NH_2$ group of DNPH to the carbonyl group, followed by elimination of water to form the hydrazone. The reaction of DNPH with a ketone proceeds as

$$RR'{=}O + (NO_2)_2C_6H_3NH{-}NH_2 \rightarrow (NO_2)_2C_6H_3NH{-}N{=}C{-}RR' + H_2O$$

where R′ is H for the reaction with an aldehyde. Acid is required to protonate the carbonyl so that the DNPH nucleophile can attack the electron-deficient carbon atom of the carbonyl, which allows for the removal of water. The hydrazones formed by the reaction are slightly soluble in the acidified water but are very soluble in the isooctane. They are recovered by extracting the aqueous layer with a mixture of hexane and methylene chloride. The extraction solution and the isooctane layer are combined and evaporated to dryness. The dry product is then dissolved in methanol and measured using reversed-phase HPLC with a UV absorption detector at 370 nm.

The DNPH can also be coated onto a solid sorbent and packed into a cartridge. The air is then drawn through the cartridge where the hydrazones produced by the reaction with aldehydes or ketones are immobilized on the solid DNPH. The hydrazones are extracted from the cartridge with acetonitrile and analyzed by HPLC with UV detection. The DNPH method has shown difficulties in the measurement of unsaturated ketones such as acrolein. This is likely due to the formation of isomeric compounds that vary in yields depending on the reaction conditions. The DNPH methods also give a measurement that is integrated over the time of sample collection. They are being replaced with high-resolution PTMS, OP–FTIR, or DOAS for real-time in-situ measurements. The DOAS measurement method uses the characteristic UVA absorption bands of the carbonyl compounds.

The ozone chemiluminescence reaction has been used successfully for the measurement of alkenes. As discussed in Section 5.4.4, the addition of ozone to an alkene forms a molozonide, which is unstable. The molozonide rapidly decomposes to form an excited aldehyde or ketone and a Criegee biradical:

$$> C{=}C < +O_3 \rightarrow > C{=}O^* + > C^+COO^-$$

The substituents on the carbonyl products depend on the substituents on the original alkene. The major product from the terminal alkenes, including isoprene, is excited formaldehyde (HCHO*), which emits light from 300 to 550 nm. In rural areas, where the major alkene is isoprene, this chemiluminescent reaction has been used as an isoprene monitor. The isoprene system operates in a similar manner as the NO monitor, shown in Figure 7.10, without the thermal converter. It also uses a blue–green cut-off transmission filter to pass the chemiluminescent signal produced by the excited formaldehyde and other excited carbonyl compounds, while blocking the chemiluminescent emission from any NO present in the sample. As with the NO monitor, the chemiluminescence signal can be maximized by maintaining the cell at low pressure, which minimizes collisional quenching of the HCHO*. The resulting light emission from the chemiluminescence reaction is proportional to the ozone–alkene reaction rate, and so can be used as a measure of the total alkene reactivity in the air sample. When this system is used as a GC detector, the specific alkenes can be measured with detection limits similar to that of the FID detector (Marley and Gaffney, 1998).

The CFCs and other halogenated hydrocarbons have low FID signals, since they form negative ions more readily than they form positive ions. The compounds were initially characterized by GC–TCD, but the sensitivities were not adequate to measure ambient concentration levels. In addition, the saturated halocarbons do not have large absorption band strengths in the UVB or UVA wavelength regions, and so cannot be measured by DOAS. Since the CFCs have strong IR absorption bands, long-path IR methods have been used to measure them in industrial atmospheres where they are in the high concentrations (ppb to ppm range). Today, the halogenated organics – including chlorinated and brominated pesticides and polychlorinated biphenyls (PCBs) – are measured using GC–ECD or GC with negative ion mass spectrometry detection. The less volatile higher molecular weight compounds, such as the halogenated pesticides, have been measured using HPLC with negative ion mass spectrometry detection.

When using negative ion mass spectroscopy of a chlorinated or brominated hydrocarbon, the peak pattern for the parent ion is very distinctive and reflects the relative abundances of the natural stable isotopes. The natural abundances of the chlorine and bromine stable isotopes are ^{35}Cl (75.8%), ^{37}Cl (24.2%), ^{79}Br (51%), and ^{81}Br (49%). So the ratio of the abundances of the natural isotopes of the chlorine atom is 3/1 (^{35}Cl/^{37}Cl), while the ratio of the abundances of the bromine isotopes is 1/1 (^{79}Br/^{81}Br). These natural abundances influence both the number and relative intensities of the peaks from the parent ion of an organic compound containing chlorine or bromine. For example, the mass spectrum of an organic compound containing one Cl atom will have two peaks from the parent ion with an intensity ratio of 3 : 1. Similarly, an organic compound with one Br atom will have two peaks from the parent ion with an intensity ratio of 2 : 1. The more halogens in the compound, the more complicated the peak patterns become. However, they can be used to determine the number of chlorine and bromine atoms in the molecule.

The dominant perfluorocarbons in the atmosphere are CF_4 and C_2F_6, which are produced from high-temperature electrochemical production of aluminum. These compounds are very inert, with lifetimes estimated to be >50 000 years. Because of the presence of the strongly electronegative fluorine atoms, they can be measured using GC–ECD or GC–negative ion MS with very high sensitivity. Since they both have strong IR absorption bands, they can also be measured by long-path FTIR. Although the higher analog perfluorocarbons have not been used in industry and have negligible or no backgrounds in the atmosphere, they have been used as atmospheric tracers in both indoor and outdoor pollution studies. Because of their high sensitivities for GC–ECD detection, they can have very low detection limits with this method. This allows them to be introduced into the atmosphere in very low amounts for the purpose of tracking air parcels. For example, the release of a perfluorocarbon into a stack exhaust allows the air mass from that stack to be tracked for thousands of kilometers as the perfluorocarbon travels with the air mass. Also, the release of a small amount indoors allows for the determination of air exchange rates in homes and buildings. The mixing of air in complex urban street environments can be determined by using a series of simple perfluorocarbon gases released at different areas in an urban environment.

7.3 Aerosols

7.3.1 Sample Collection

As discussed in Section 5.3.1, the EPA standards for aerosols are based on the aerosol mass in size ranges of <10 μm (PM_{10}) and <2.5 μm ($PM_{2.5}$). There is also a recommended standard for the size range of <10 to >2.5 μm called particulate matter coarse (PM_c), but the FRM has not yet been established for the direct measurement of particles in this size range. The cut-off of the size range of an aerosol sample is determined by the flow rate and inlet systems of the high-volume aerosol samplers. So these parameters must be adjusted in order to collect the sample in the mandated size range. The samples are collected on a nonreactive, non-disintegrating glass fiber, quartz, or polymer filter material that does not have an organic binder. The filter must also have a collection efficiency of at least 99.95%, or <0.05% penetration as determined by 0.3 μm dioctyl phthalate particles. Water vapor varies in the atmosphere and can be adsorbed onto the samples during collection, which affects the mass measurements. In order to correct for this interference, the filters are weighed in a constant humidity chamber. After the sample is collected it is allowed to equilibrate again in the weighing chamber, along with the blank filter, before weighing. The difference in weight is recorded for the various size ranges to obtain the masses of the PM_{10} and $PM_{2.5}$ samples. Currently, PM_c is being determined as the difference between the PM_{10} and $PM_{2.5}$ measurements, until a direct method of measurement has been approved by the EPA.

When large particles move through the air in a sampling line, their moment of inertia causes them to move forward in a straight line so that a turn in a sampling line will cause them to impact on the surface of the tube at the turn and not be collected by the sampler. At the same time, the smaller particles will be able to make the turn and be collected by the aerosol sampler. The size of the aerosols that are stopped by impaction and those that continue with the air flow are dependent on how fast the particles are traveling and the sharpness of the turn. This basic principle of aerosol impaction, causing a change in the direction of the flow, depends on the flow rate and the radius of curvature of the change in the directional flow. Cascade impactors with slotted filters, shown in Figure 7.15, can be used to collect samples of aerosols in various size ranges as a function of effective aerodynamic diameter before the final collection of the ultrafine particles that have passed through all the stages of the cascade impactor (>0.1 μm). These cascade aerosol sampling systems usually have up to six or seven size cuts before the final size collection. This is accomplished by using multiple plates each with increasingly narrow slots that are set off-axis from the preceding plate. The slots in the plates are closer together as the aerosols move through the stages, and at each stage the largest aerosols are impacted on the filter collection material while smaller particles pass through the slots to the next collection stage. This same principle can be used with an aerosol sampler that uses holes instead of slots. Like the slotted sampler, the holes are set off-axis from each other to obtain a cascade effect for the sizing of the aerosols. Standard-sized particles, which are usually composed of polystyrene beads, are used to characterize the accuracy and cut-off diameters for the aerosol samplers. The size of the atmospheric particles collected can also be verified using optical spectroscopy.

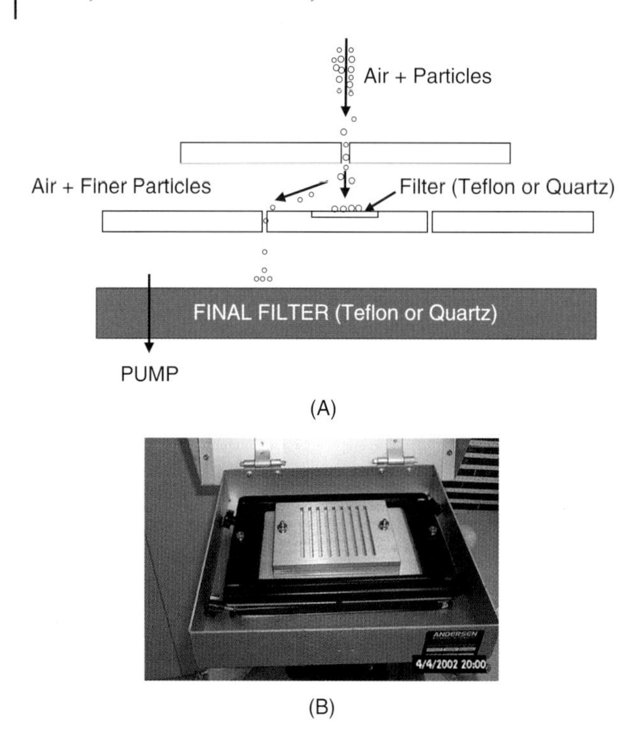

Figure 7.15 (A) The principle of operation of a slotted impactor aerosol collection system. (B) A photograph of a slotted impactor system used to collect aerosol samples as a function of aerodynamic diameter.

As discussed in Section 7.1, filter materials should be chosen carefully depending on the analysis to be performed. Teflon filters are used for the collection of samples for inorganic analysis, since Teflon is made of polytetrafluoroethylene (PTFE) and is free of inorganic material. In some cases, a thin coating of mineral oil, composed of alkanes and cycloalkanes, is used on the surface of the Teflon filters when sampling dry inorganic dusts to minimize bounce-off from the filter material. Quartz-fiber filters with no binding agents are used for organic analyses, since they have very low carbon content. To further minimize the organic background, the quartz filters can be heated before use to remove any adsorbed gases.

7.3.2 Aerosol Composition

Sulfate and nitrate ions are measured using ion chromatography after the samples are extracted with acidified high-purity water. Elemental analysis can be performed directly on the filter samples using X-ray fluorescence, neutron activation, or proton-induced X-ray emission (PIXE) methods. Inductively coupled plasma (ICP), with or without mass spectrometry detection (ICP–MS), has also been used for the detection of element concentrations in the aerosol samples with high sensitivity and low detection limits. Indeed, the EPA-approved method for the measurement of lead in atmospheric PM is ICP–MS. The elemental concentrations determined on the filter samples are then combined with the total air flow measurement to yield an atmospheric concentration in nanograms or micrograms per cubic meter.

Thermal evolution methods have been used as a means of determining the total amounts of the more volatile organic carbon species from the more recalcitrant organic carbon material. The more recalcitrant carbon fraction of the aerosols was initially called elemental carbon (EC). This fraction is operationally defined as the carbon that is volatilized only at very high temperatures. It should not be assumed that the EC is an elemental form of carbon such as graphite, since the samples do not behave thermally like graphite or have the spectral characteristics of graphite (Gaffney et al., 2015). In fact, it has been shown that the EC fraction contains primarily black carbon (BC) aerosols. The fraction that is volatilized at low temperatures was called organic carbon (OC). These two terms are operationally defined and are used to differentiate between the two carbon fractions that are separated by the thermal evolution methods.

The separation and analysis of OC and EC in the thermal evolution methods involves two heating cycles. During the first cycle, the quartz filter sample is heated to a defined low temperature, usually 400 °C, while passing an inert gas over the sample. The volatile organics are then carried by the inert carrier gas through a copper oxide tube at 400 °C, which acts as an oxidative catalyst to convert the organics to CO_2 and H_2O. The CO_2 can then be determined directly using NDIR or determined by FID after conversion to CH_4. The amount of carbon measured in this first cycle is the amount of OC in the aerosol sample in ppm carbon. Once the first cycle is complete, O_2 is added to the inert gas stream and the temperature is raised to a defined high temperature, usually 700 °C. The amount of CO_2 formed in the second cycle is the amount of EC in the aerosol sample as ppm carbon, but it has been found that the heating rate in thermal analysis can affect the accuracy of the measurements. If the heating rate is not rapid enough during OC determination, charring of the sample can occur leading to a lower OC measurement and a higher EC measurement (Tanner et al., 1982). This charring artifact is minimized if rapid heating rates are used. Sample charring can be monitored by measuring the light transmission through the filter at 632.8 nm generated by a HeNe laser. The decrease in transmission is directly proportional to the formation of BC from sample charring and can be used to correct for charring effects on the OC and EC measurements.

For the determination of total carbon, as well as the determination of $^{13}C/^{12}C$ ratios and ^{14}C content, the sample is collected on low-carbon quartz filters and combusted to CO_2. After total combustion of the sample to CO_2, the CO_2 can be cryogenically trapped for measurements of the carbon isotopes by mass spectrometry. The stable carbon isotope ratios ($^{13}C/^{12}C$) are measured on the CO_2 directly using an isotope ratio mass spectrometer. For measurements of radiocarbon (^{14}C) using accelerator mass spectrometry (AMS), the CO_2 must first be reduced to graphite using the Bosch reaction:

$$CO_2 + 2H_2 \rightarrow C + 2H_2O$$

The reaction is carried out at 500–600 °C with a metal catalyst, usually iron or cobalt. The ^{14}C in the graphite is then measured by AMS, which employs high-energy physics methods to obtain low detection limits.

The measurement of specific organic carbon species in the aerosol samples requires extraction of the filter samples with high-purity organic solvents. The extraction solutions are analyzed using methods specifically designed for the analyte. For example, polycyclic aromatic hydrocarbon (PAH) compounds can be measured by HPLC with fluorescence detection or with HPLC–MS, both giving low detection limits and high sensitivity. Polar organic species can also be measured using HPLC–MS. In some cases, the organic species of interest can be converted to more volatile compounds for measurement by GC–MS when HPLC–MS is not available. An example of this would be the derivatization of organic acids to form the more volatile organic esters for GC–MS analysis.

With the advent of improved mass spectrometry methods, the real-time analysis of atmospheric aerosols as small amounts of particles or even single particles has become possible. Two types of instruments based on mass spectrometry have been designed for this purpose. The first design uses a quadrupole mass analyzer while the second design uses a time of flight mass

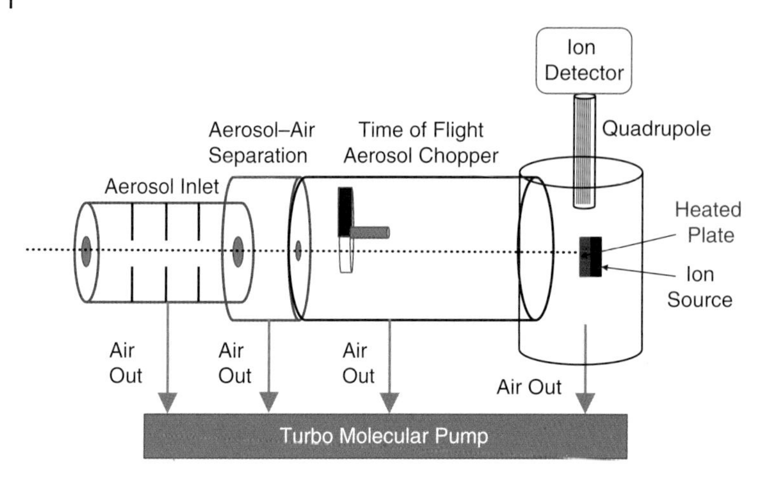

Figure 7.16 A schematic diagram of an aerosol mass spectrometer developed by Aerodyne Research, Inc.

analyzer. The quadrupole system, shown in Figure 7.16, was pioneered by Aerodyne Research, Inc. It uses an aerosol inlet coupled to differential pumping to separate the aerosol from the air. The aerosol stream is then pulled into a time of flight tube which contains a mechanical aerosol chopper that pulses the aerosol stream. The aerosol pulses hit a heated plate at the end of the time of flight tube, where they are vaporized and ionized. The ions are then pulled into the quadrupole mass analyzer, which consists of four parallel cylindrical rods that generate an oscillating electric field. The quadrupole filters the ions based on their mass to charge ratio (m/z). Since both positive and negative ions can be produced from the particles, the instrument can operate in either positive or negative ion mode by changing the polarities of the ion source and the quadrupole mass analyzer. This instrument is fairly compact and has been used successfully in a mobile van to characterize vehicle aerosol emissions by following vehicles on the road. It has also been used successfully onboard aircraft for ambient aerosol characterizations aloft.

The aerosol time of flight mass spectrometer (ATOFMS) system, shown in Figure 7.17, was pioneered by the Prather research group at the University of California, Riverside and has continued to develop into airborne and mobile versions by this group at the University of California, San Diego. The ATOFMS also uses an aerosol inlet with differential pumping to separate the aerosol from the air. The aerosol inlet can either use a converging nozzle to sample aerosols in the size range of 0.2–3.0 μm or an aerodynamic focusing lens to select aerosols in the size range of 0.050–1.0 μm. The movement of the aerosols is tracked by two diode-pumped Nd:YAG lasers at 532 nm, which are coupled to two PMTs (PMT 1 and PMT 2). The two PMTs monitor the laser light scattered by the particles, indicating their position in the air stream. When the particle falls into position, a pulsed UV laser both vaporizes and ionizes the aerosol particle. Ion extraction plates are placed at negative and positive potentials on either side of the ionized particle to draw both the negative ions and positive ions into the time of flight tube in opposite directions. The potential of the extraction plate determines the kinetic energy of the ions as they enter the flight tube. Since the kinetic energy is related to the mass and the velocity of the ion, the time that it takes for the ions to reach the detector at a known distance (time of flight) can be used to determine the mass of the ion. The smaller ions move fastest and are detected first, followed by the heavier ions.

The resolution of the ATOFMS is determined by the path length of the ions in the tube. So ion reflector plates, called reflectrons, are placed at the end of the tube to return the ions to detectors placed in front of the reflector plates. This results in a doubling of the path length

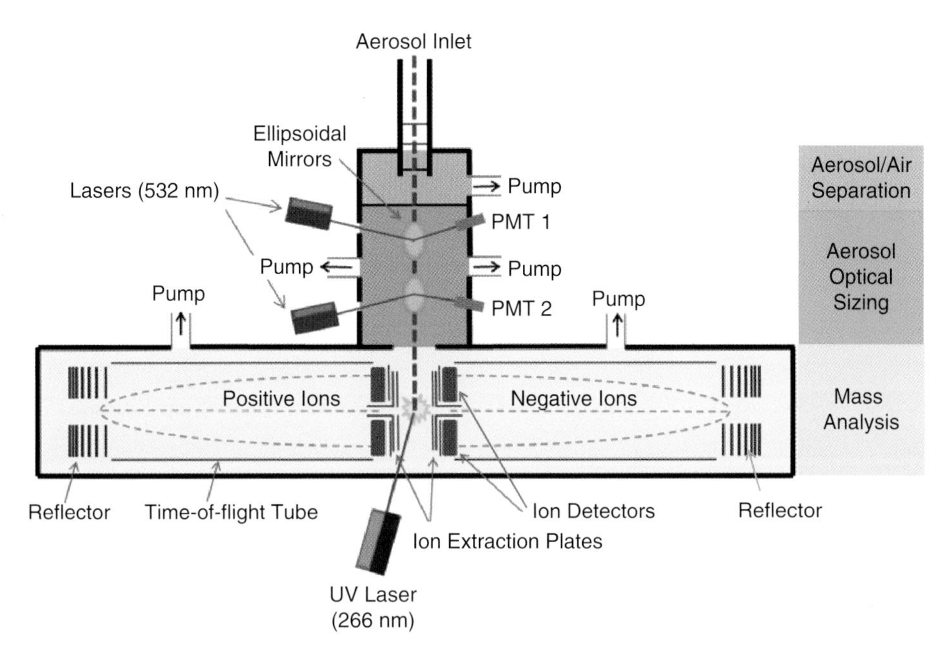

Figure 7.17 A schematic diagram of an aerosol time of flight mass spectrometer (ATOFMS). Source: Adapted from Benjamin Haywood, Wikimedia Commons.

that the ions travel, increasing the resolution. Also, since the negative and positive ions are directed into the time of flight tube toward ion detectors in opposite directions, both negative ion and positive ion mass spectra can be obtained simultaneously on the aerosols in real time. The AOTFMS has been used to determine mass spectral aerosol "fingerprints" from specific aerosol sources. It has also been used in the characterization of bioaerosols such as bacteria and viruses in the medical and environmental health fields. In this application, the analysis of bioaerosols results in a characteristic "fingerprint" of the cells and viral material which can be used in the identification of health hazards.

7.4 Aerosol Optical Properties

As discussed in Section 6.6, aerosols can both scatter and absorb light. The scattering of solar light can cool the atmosphere while the absorption of solar and IR radiation can warm the atmosphere. So it is important to determine the aerosol light-scattering and absorption coefficients in order to determine their effects on the radiative balance of the Earth. Light scattering by atmospheric aerosols depends on the size of the particles relative to the wavelength of light. Light-scattering measurement systems, called nephelometers, are used to determine the size of aerosols, the number of particles, and the amount of light scattered by the aerosols in an air sample. The air sample is drawn into a tube where it is illuminated by a light beam, which can be a white light, a HeNe laser at 623.8 nm, or a set of up to three light-emitting diodes in the visible wavelength range. The scattered light is measured by a photodetector placed at an angle of 90° to the incident light. The intensity of the scattered light is then proportional to the aerosol number concentration. Three wavelength nephelometers can also be used to determine aerosol size ranges by measuring the intensity of the scattered light as a function of wavelength. While nephelometers are used for the real-time measurement of aerosol particle number concentrations and aerosol size ranges, they are also used to obtain the fine aerosol scattering strengths for radiative transfer calculations.

Most inorganic salts that contribute to atmospheric aerosols are either present as a solution in a water droplet or as a white solid. Although these aerosols are major contributors to light scattering, they do not contribute to aerosol light absorption since they do not have strong UV or visible absorption strengths. The majority of aerosol light absorption is due to carbonaceous aerosols. BC aerosols have significant absorption coefficients in the UVA, visible, and IR wavelength regions. Organic carbon aerosol species such as PAHs, large carbonyl-containing compounds, and nitrated phenols also have strong absorptions in the visible and UVA wavelength regions. Aerosol HULIS, sometimes called brown carbon, has especially strong absorption strengths in these regions.

Real-time aerosol absorption measurements have been carried out by an absorption system called an aethalometer (Hansen et al., 1982). The name aethalometer derives from the Greek word *aethalos* meaning smoky flame or thick smoke. In this system the air sample is pulled through a strip of white filter paper tape, creating a spot on the tape where the aerosol accumulates with time. Light of specific wavelengths is transmitted sequentially through the filter to a photodetector on the other side. As the sample accumulates on the filter it absorbs the light and the light transmission decreases. The decrease in light intensity is measured as a function of time and is used to determine the concentration of the absorbing aerosol species by using an assumed wavelength-dependent absorption coefficient ($m^2\ g^{-1}$) of aerosol. Once the sample has accumulated to the point where the light transmission decreases past a minimum allowed point, the filter tape is advanced to a new position and sampling is reinitiated. The aethalometer uses either two light sources at wavelengths of 370 and 880 nm, or seven light sources at wavelengths of 370, 450, 571, 615, 880, and 950 nm. Although the instrument software gives the final values in the concentration of absorbing aerosol in $ng\ m^{-3}$ of air, the total aerosol absorption can also be obtained from the aethalometer measurements. This can be used to obtain the absorption coefficients by measuring the mass of the aerosol deposited on the filter.

An aerosol absorption system that operates in a manner similar to that of the aethalometer is the particle soot absorption spectrometer (PSAP). However, the PSAP operates in a double-beam mode. The sample is collected on a small circular filter and a second filter is used as a reference to correct for light scattering from the filter surface. Light, generated by a light-emitting diode at 565 nm, is transmitted through both filters and the aerosol absorption is determined by the difference between the light transmission through the sample filter and the reference filter. Although light scattering is an interference in absorption measurements, the scattering from the filter surface will decrease as the sample is collected. This may result in the reference measurement overcorrecting for samples primarily consisting of BC. In addition, the system requires that the sample filter be changed manually when the sample absorption becomes too high. So, unlike the aethalometer, the system cannot be operated unattended for long periods of time, especially in areas of high aerosol loadings.

Both the aethalometer and the PSAP obtain measurements that are integrated over the sample collection times. Aerosol absorbance can be measured in real time by using photoacoustic methods. The photoacoustic effect is the production of sound waves in a sample by the absorption of pulsing light. Alexander Graham Bell discovered the photoacoustic effect when he observed a sound produced by sunlight shining through a rapidly rotating thin slotted disc. Light absorption photoacoustic systems are based on this photoacoustic effect. Heat is generated when light is absorbed by an absorbing material. If the light source is pulsed, the heat that is generated will also be pulsed. This pulsing heat wave generates a pulsing pressure wave in the sample, which can be detected with sensitive microphones. In a photoacoustic system, the sample air is pulled into a photoacoustic cell where modulated laser light in the visible wavelength

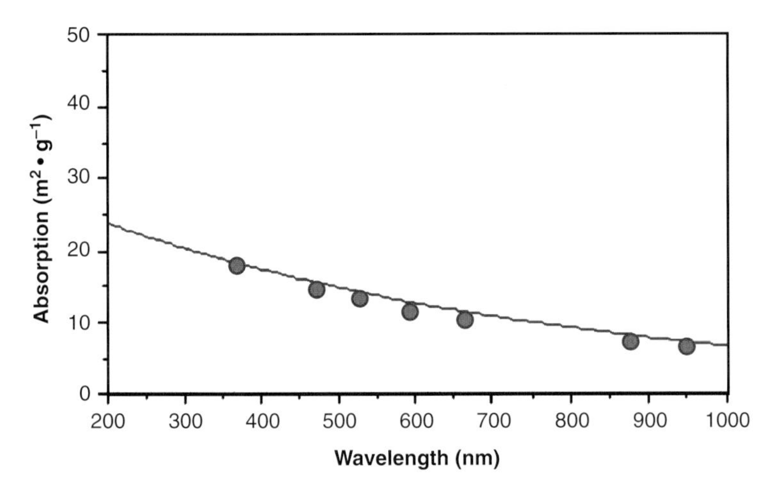

Figure 7.18 Aerosol absorption coefficient of a sample of NIST standard diesel soot (SRM 1650) obtained as a function of wavelength using an integrating sphere coupled to a scanning optical spectrometer (red) compared to those determined by a seven-wavelength aethalometer (blue circles).

range passes through the sample. The laser can be either power modulated or optically modulated using a chopper. The resulting photoacoustic signal is measured by sensitive microphones placed off-axis to the laser light. This photoacoustic signal can be analyzed to give the amount of light absorbed by the aerosols in the sample. The aerosol absorption coefficient can also be determined after the aerosol number concentration is measured by nephelometry.

The measurement of aerosol absorption as a function of wavelength can be accomplished by diffuse reflectance using an integrating sphere. The integrating sphere consists of a hollow spherical cavity with an interior that is coated with a diffusely reflective material. This coating produces a uniform diffuse reflection of light in the interior of the sphere. Light that hits any point on the coated inner surface is distributed equally to all other points through multiple reflections. The sample is mounted at a small sample port on one side of the sphere. Light which is dispersed by a scanning monochromator is directed onto the sample surface through another small hole opposite the sample port. A photodetector placed on the sphere off-axis detects the intensity of light inside the integrating sphere. This light intensity is measured with and without the sample in place to determine the amount of light that is absorbed by the sample. The aerosol absorption coefficients are obtained from Beer's Law after the measurement of the sample mass per unit area on the filter surface. The absorption coefficients as a function of wavelength for a diesel soot standard (NIST SRM 1650) are shown as the red line in Figure 7.18. These are compared to the absorption coefficients of a BC aerosol sample obtained by the seven-wavelength aethalometer (blue circles).

7.5 Method Selection

When determining the best method to be used for the measurement of atmospheric gases or aerosols, it is important to consider how the data is going to be used and who the users will be. For example, environmental measurements can be made for research purposes or they can be part of a legislatively mandated monitoring program. Because the measurement of criteria pollutants may eventually be used in a legal context, the sampling, measurement, data analysis, and reporting must adhere to certain guidelines. Measurements made for pollution monitoring or enforcement purposes require that specific protocols be followed, which determine

Table 7.2 The Federal Reference Methods (FRM) and number of approved equivalent methods for the measurements of the criteria pollutants.

Species	FRM	Equivalent methods	Comments
PM_{10}	HI-vol filter collection, mass measurement	3	
$PM_{2.5}$	HI-vol filter collection, mass measurement	5	
PM_C	$PM_{10} - PM_{2.5}$ difference	3	
SO_2	Tetrachloromercurate impinger collection, pararosaniline colorimetric detection	1	EDTA and NH_2SO_3H added to remove interference from metals and NO_x
O_3	Chemiluminescent reaction with ethene	4	not commercially available
CO	NDIR	1	
NO_2	Chemiluminescent reaction with O_3 using heated NO_2 converter to give NO	5	red cut-off filter to remove alkene interference
Pb	HI-vol filter collection, acid extraction ICP/MS detection	2	

the instrumentation and measurement methods to be used. In contrast, measurements made for research purposes are not intended to be used to assess the adherence to environmental regulations but are intended to develop a better fundamental understanding of the processes occurring in the atmosphere. So the choice of methods depends more on the detection limits, sensitivity, and response time required for the particular application.

The primary methods accepted for monitoring the levels of criteria pollutants are the FRMs. Although there are accepted equivalent methods for many of the FRMs, they must be standardized using the approved FRM. The methods required for monitoring the EPA National Ambient Air Quality Standards for gases and PM, along with the number of accepted equivalent methods, are listed in Table 7.2 (Gilliam and Hall, 2016). Commercial instruments intended for use as a FRM must first undergo EPA review and official approval before they can be used to measure the criteria pollutants. In all cases, except for ozone, commercial instruments are available for use in the FRMs. In the case of O_3, no commercial ethene chemiluminescence instruments are available due to fire hazards involved with handling the pure ethene required for the reaction. In this case, the EPA has recommended that approved equivalent methods be used for ozone determinations.

The EPA has developed and approved a number of methods for the measurement of air toxics. The currently approved air toxic methods are listed in Table 7.3. These methods are given identification numbers specific to the method, preceded by the letter designation "TO" (for toxic) followed by a number. A number of older methods have undergone review since they were originally approved, and in some cases have been revised. In this case the amended method is given the same ID number followed by the letter A. Details of the methods can be found on the EPA Air Toxics website (EPA, 2018a). The method lists the sampling protocols as well as analytical method requirements, and calibration procedures for each approved method. Many of the approved methods are for total VOCs, which include the air toxics.

When performing research in atmospheric chemistry, the measurement methods for the species of interest are not mandated and the environmental scientist has more flexibility in determining the methods to be used. The choices are often based on the specifications of the

Table 7.3 The EPA-approved measurement methods for the air toxics.

Species	Approved method	Method ID
Nonpolar VOCs (bp 80 to 200 °C)	Tenax collection, GC–MS	TO-1
Nonpolar VOCs (bp −15 to 120 °C)	molecular sieve collection, GC–MS	TO-2
VOCs (bp −10 to 200 °C)	cryogenic collection, GC–FID, GC–ECD	TO-3
Pesticides and PCBs	PUF[a] collection, GC–FID, GC–ECD, GC–MS, HPLC–UV	TO-4A
Aldehydes and ketones	impinger collection with DNPH, HPLC–UV	TO-5
Phosgene ($Cl_2C{=}O$)	impinger collection, aniline reaction, HPLC–UV	TO-6
N-nitrosodimethylamine	thermosorb collection, GC–MS	TO-7
Phenol and methyl phenols (Cresols)	impinger collection, OH^- reaction, HPLC–UV fluorescence, electrochemical detection	TO-8
Br–, Cl– dibenzo-*p*-dioxins, dibenzofurans	HI-vol collection, ^{13}C internal standards, high-resolution GC–MS	TO-9A
Pesticides and PCBs	PUF[a] collection, GC–FID, GC–ECD, GC–MS	TO-10A
Formaldehyde	DNPH[b] cartridge collection, HPLC–UV	TO-11A
Non-methane organic compounds	cryogenic collection, GC–FID	TO-12
PAHs	cartridge sampling, GC–MS	TO-13A
VOCs	canister collection, GC–FID, GC–ECD, GC–MS	TO-14A
VOCs	canister collection, GC–MS	TO-15
Atmospheric gases	OP–FTIR	TO-16
VOCs	cartridge collection, GC–MS	TO-17

a) Polyurethane foam.
b) 2,4,-Dinitrophenylhydrazine.

method, which are accuracy, precision, detection limits, sensitivity, resolution, and response time. Other concerns are interferences, size and weight limitations, difficulty of measurement and calibration procedures, and safety concerns. Probably the most important of the method specifications for measurements of atmospheric gases are detection limits and sensitivity. This is because most atmospheric gas species are in ppb to ppt concentration ranges. The detection limits of the measurement method must be sufficient to measure the species of interest at the expected concentrations, and the sensitivity must be sufficient to differentiate between concentrations at low levels. The accuracy and precision of a measurement method is controlled by calibration and replication of the measurements. Problems with accuracy and/or precision can often be corrected by using these procedures. Resolution is most important in spectroscopic and chromatographic methods and can be optimized by adjusting the experimental conditions. Response time is most important for mobile measurements, such as in aircraft or moving vehicles. It is especially important for aircraft operations due to the large ground speeds. In these cases, slow response times will result in spatially integrated measurements over a large region. Size and weight limitations are also important for field measurements when shipping equipment or when space is limited. Weight and power requirements are especially important in aircraft measurements. The method interferences need to be evaluated depending on the site.

For example, PAN measurements by GC–ECD will have a faster response time and higher sensitivity at an arid site than a humid site, because of the lower interference from water vapor. This would not be the best choice for PAN measurements in the Amazon rainforest, where the high humidity will cause the measurement to take much longer. Also, instruments that require thermal or vibrational stability are not suitable for use outside a climate-controlled building.

In many cases, the individual environmental chemist has limited equipment, availability, or training, as well as a lack of the funding required to make all the measurements needed to evaluate the complex atmospheric chemistry. For this reason, most atmospheric field research studies are conducted by a team of scientists. These teams allow for highly trained personnel and equipment to be pooled together to obtain a more complete data set, which is shared among the participating scientists for analysis. These data sets are inherently large in size and require a data management team to place the data into well-documented online files for easy access. Most of these field studies have been funded by federal research agencies and data sets are made available after the study to outside researchers, particularly to the environmental modeling community, for analysis and atmospheric chemical model verification.

7.6 The Importance of Baseline Measurements

Baseline measurements of gases and aerosols are defined as an accurate measurement of species concentrations over time before any change occurs. Baseline measurements can be used as a standard for measuring future concentrations to help identify any problems as they occur. They also allow for the determination of the effectiveness of any solution that is applied to an identified problem. Baseline measurements of a gas or aerosol species allow for any temporal variance to be evaluated. These measurements are very important to determine if the atmospheric concentrations of pollutants or greenhouse species are increasing or decreasing over time. Without baseline measurements, there is no way to recognize if the gas or aerosol species is above natural or mean background levels for a specific area.

Originally, baseline measurements implied background levels and in many cases researchers, as well as state and federal agencies, focused on the determination of background concentrations of atmospheric species by measuring them at remote locations. It was assumed that these remote locations were not impacted by local or regional sources. For example, the National Oceanic and Atmospheric Administration (NOAA) has operated and maintained sampling stations at Mauna Loa, HI, American Samoa, Barrow, AK, and the South Pole for decades. These sites have measured greenhouse gases and the stratospheric depleting CFCs continuously, demonstrating their global increase. These data have given us a better understanding of the changes in the concentrations of the greenhouse gases on a global level. In some cases, they have discovered and confirmed that gases, such as PAN, and aerosol species, such as BC, were being carried over global distances.

Continuous measurements of air pollution levels at urban sites have allowed for local air chemistry to be evaluated and control strategies to be assessed. Without the continuous measurements of pollutant species, it would not be possible to determine if the air quality was improving or worsening as different control strategies were implemented to help clean the air. These control strategies included post-combustion exhaust gas and aerosol controls on both mobile and stationary sources. They also included adding oxygenated compounds such as methyl-*t*-butyl ether (MTBE) and ethanol to motor vehicle fuels to decrease the reactivity of the emissions and lower the CO levels. In the case of stationary sources, such as coal-fired power plants, control strategies included the use of low-sulfur-content fuel or changing to petroleum or natural gas to lower SO_2 emissions.

Regional sites have been established by university researchers at Harvard Forest, MA, Niwot Ridge, CO, and Mt. Bachelor, OR. A number of sampling towers have been operated in North Carolina, WI and in Texas, along with other sites in polluted regions. NOAA is implementing a long-term sampling station at Trinidad Head, CA on the west coast of the United States to allow for monitoring of the long-range transport of air pollutants from Asia. Plans are to obtain measurements of aerosols and trace gases to establish baselines for the better understanding of these contributions to local and regional pollution levels. The DOE Atmospheric Radiation Measurement (ARM) research program, which has focused on the measurement of the optical properties of clouds and aerosols, has operated a site in the Great Plains for decades. This site has been collecting long-term measurements of aerosol and cloud optical data along with some greenhouse gas measurements, such as CO_2. The site was set up to give baseline information on an area the size of a general circulation model grid cell (1000 km × 1000 km) and covers approximately 9000 mi^2 in north central Oklahoma and southern Kansas. Although this site was set up to collect data important to climate modeling, it has also provided information about the transport of air pollutants in the region, along with important information on the impact of biomass-burning aerosols on cloud formation and radiative forcing.

With the addition of spectroscopic methods onboard satellites, which can measure specific gases as well as characterize optical depths in the atmosphere, the ground station baseline determinations have been invaluable to be able to "ground truth" the satellite data. An excellent example of the validation of satellite observations are the comparisons of ground-based measurements of ozone column density in Antarctica with results from the total ozone mapping satellite (TOMS). Although the initial cost of satellites is high, they have finite lifetimes, and they require reinvestment. Satellite remote-sensing abilities from space platforms yield large amounts of data obtained over regional and global scales. They are extremely useful for total air column measurement of gases, clouds, and aerosols. But there are still problems with remote measurements from space platforms, including their inability to differentiate the concentrations of atmospheric species in the lower troposphere, particularly in the boundary layer. Since the exposure of humans and ecosystems to air pollutants occurs in the boundary layer, continual operation of ground stations is needed to determine the present and future changes in air quality.

The EPA has operated criteria pollutant monitoring sites since the passing of the Clean Air Act. However, these sites were only intended to measure concentrations of the criteria pollutants. The currently active EPA air monitoring stations that report data for CO, lead, O_3, PM_{10}, and $PM_{2.5}$ are shown in Figure 7.19 (EPA, 2018b). In many cases, the measurement systems are not co-located, nor are all of the criteria pollutants being monitored at every site. Although there were a large number of stations originally, many are no longer active. In most cases, the monitoring site was discontinued due to the cost of maintenance. In some cases, the decision was made to stop acquiring data on a specific pollutant because values were well below the standards and did not appear to merit continuation. This is especially the case for the criteria pollutants CO and lead, since control strategies for these pollutants have successfully reduced their atmospheric levels. The addition of oxidative catalysts to motor vehicle exhaust systems substantially lowered CO emissions. The addition of tetra-alkyl lead compounds to gasoline was halted nationwide and unleaded gasoline was used in its place, which substantially lowered lead emissions from automobiles. Since the leaded gasoline additive was the largest contributor to airborne lead levels, the atmospheric lead concentration dropped with the use of unleaded

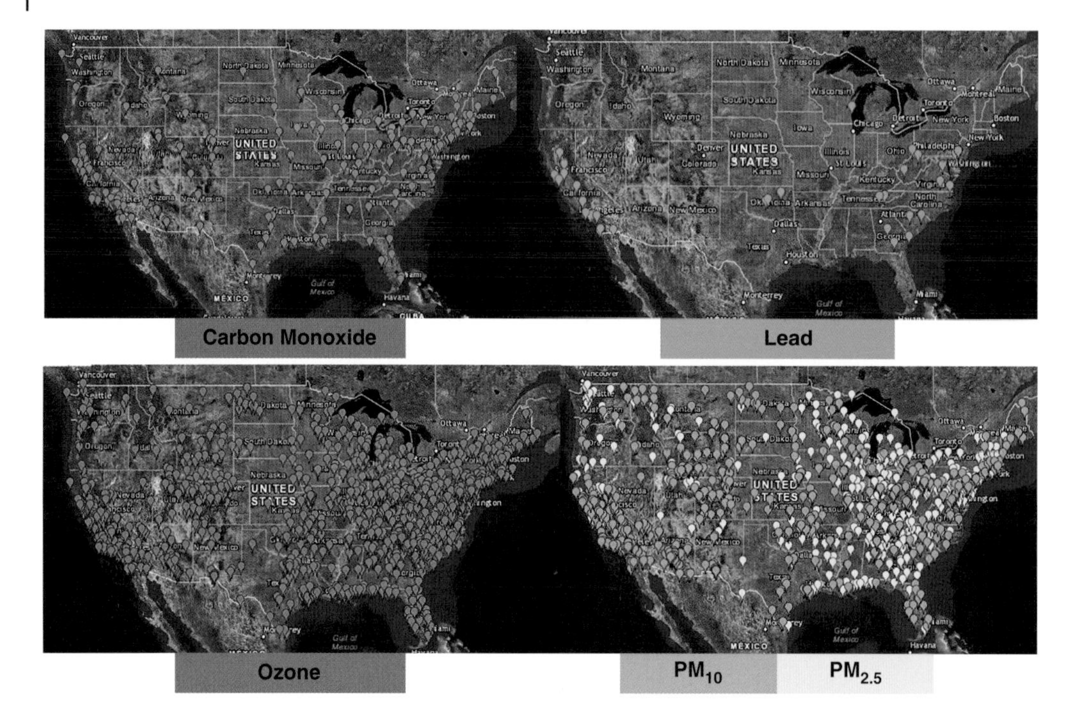

Figure 7.19 The currently active monitoring sites for the EPA criteria pollutants; carbon monoxide (blue), lead (green), ozone (pink), PM_{10} (orange), and $PM_{2.5}$ (yellow).

gasoline. The reduction in monitoring of these two pollutants is clearly indicated in Figure 7.19. There are currently many more active sampling stations for ozone and PM than there are for CO and lead, since the ozone and PM levels continue to be above the EPA standard levels in many areas.

Over the years, some urban sampling stations were moved, or the measurement methods were updated. These changes can cause an apparent increase or decrease in the observed pollutant levels and need to be taken into account when evaluating the EPA data sets. In addition, complete criteria pollutant data sets are more likely to be available in areas where there are high population densities and/or high industrial activity. California and the eastern United States have numerous sampling stations, while the mid-western, northern, and southern regions have fewer stations. While there is an abundance of data in the urban regions, there is a sparsity of sampling stations in rural areas. In some cases, the entire state may have only a few monitoring stations or as little as one. With the continuing control of pollutant emissions in urban areas, the rural agricultural areas are increasing in their importance as pollutant sources. Agricultural activity contributes significantly to atmospheric pesticide levels, and fertilizer use is becoming a more important source of airborne NO_x. The agricultural burning of crop debris can produce large amounts of CO, VOC, NO_x, and carbonaceous aerosols. These emissions will continue to become more important and will likely require that these source regions be better monitored in the future.

The acquisition of baseline measurements will be important for determining the impacts of long-range transport of pollutants. One such strategy would be to consider the location of OP–FTIR systems using the EPA TO-16 methodology, along with PM measurements at state universities throughout the country. State universities are located in both urban and rural areas and would give baseline measurements in a wide variety of locations throughout all of the 50 states. These systems would be able to obtain data on the criteria pollutant gases and air toxics, as well as on greenhouse gases. This type of regional data will be needed in the future in order to evaluate mitigation strategies for climate change and to determine their effectiveness when implemented.

References

Blake, R.S., Monks, P.S., and Ellis, A.M. (2009). Proton transfer mass spectrometry. *Chem. Rev.* 109: 861–896.

EPA 2018a *Air Toxics Monitoring Methods*. Available online at https://www3.epa.gov/ttnamti1/airtox.html (accessed February 20, 2018).

EPA 2018b *Interactive Map of Air Quality Monitors*. Available online at https://www.epa.gov/outdoor-air-quality-data/interactive-map-air-quality-monitors (accessed February 20, 2018).

Gaffney, J.S., Fajer, R., and Senum, G.I. (1984). An improved procedure for high purity gaseous peroxyacetyl nitrate production: Use of heavy lipid solvents. *Atmos. Environ.* 18: 215–218.

Gaffney, J.S., Marley, N.A., Steele, H.D. et al. (1999). Aircraft measurements of nitrogen dioxide and peroxyacyl nitrates using luminol chemiluminescence with fast capillary gas chromatography. *Environ. Sci. Technol.* 33: 3285–3289.

Gaffney, J.S., Marley, N.A., and Smith, K.J. (2015). Characterization of fine mode aerosols by Raman microscopy and diffuse reflectance FTIR. *J. Phys. Chem.* 119: 4524–4532.

Gilliam, JH & Hall, ES 2016 *Reference and Equivalent Methods Used to Measure National Ambient Air Quality Standards (NAAQS) Criteria Air Pollutants, Volume I*. Available online at https://cfpub.epa.gov/si/si_public_record_report.cfm?dirEntryId=321491 (accessed June 3, 2018).

Hansen, A.D., Rosen, H., and Novakov, T. (1982). Real-time measurement of the absorption coefficient of aerosol particles. *Appl. Opt.* 21: 3060–3062.

Hargrove, J., Wang, L., Muyskens, K. et al. (2006). Cavity ring-down spectroscopy of ambient NO_2 with quantification and elimination of interferences. *Environ. Sci. Technol.* 40: 7868–7873.

Jagerska, J., Jouy, P., Tuzson, B. et al. (2015). Simultaneous measurement of NO and NO_2 by dual wavelength quantum cascade lasers. *Opt. Express* 23: 1512–1522.

Kok, G.L., Holler, T.P., Lopez, M.B. et al. (1978). Chemiluminescent method for determination of hydrogen peroxide in the ambient atmosphere. *Environ. Sci. Technol.* 12: 1072–1076.

Lazrus, A.L., Kok, G.L., Gittin, N., and Lind, J.A. (1985). Automated fluorometric method for hydrogen peroxide in atmospheric precipitation. *Anal. Chem.* 57: 917–922.

Marley, N.A. and Gaffney, J.S. (1998). A comparison between flame ionization and ozone chemiluminescence for the detection of atmospheric hydrocarbons. *Atmos. Environ.* 32: 1435–1444.

Marley, N.A., Gaffney, J.S., White, R.V. et al. (2004). Fast gas chromatography with luminol chemiluminescence detection for the simultaneous determination of nitrogen dioxide and peroxyacetyl nitrate in the atmosphere. *Rev. Sci. Instrum.* 75: 4595–4605.

Nielsen, T., Hansen, A.M., and Thomsen, E.L. (1982). A convenient method for preparation of pure standards of peroxyacetyl nitrate for atmospheric analysis. *Atmos. Environ.* 16: 2447–2450.

Stephens, E.R., Burleson, F.R., and Holtzclaw, K.M. (1969). A damaging explosion of peroxyacetyl nitrate. *J. Air Pollut. Control Assoc.* 19: 261–264.

Tanner, R.L., Gaffney, J.S., and Phillips, M.F. (1982). Determination of organic and elemental carbon in atmospheric aerosol samples by thermal evolution. *Anal. Chem.* 54: 1627–1630.

Further Reading

Finlayson-Pitts, B.J. and Pitts, J.N. Jr. (2000). *Chemistry of the Upper and Lower Atmosphere: Theory, Experiments, and Applications*. San Diego, CA: Academic Press, chapter 11.

Study Problems

7.1 (a) What is measurement precision?
(b) What is it often called?
(c) What kind of errors cause low precision?

7.2 (a) What is measurement accuracy?
(b) What kind of errors cause low accuracy?

7.3 (a) What is measurement sensitivity?
(b) What is the detection limit of a measurement?

7.4 (a) What is an FRM?
(b) What is it used for?

7.5 (a) What is an impinger?
(b) What is its function?

7.6 (a) What types of atmospheric gases are most susceptible to loss on the surfaces of the sampling system?
(b) What material is commonly used to reduce losses of oxidants in sampling lines?
(c) What is the chemical nature of this material?

7.7 (a) As organic species become concentrated by sample collection they become more susceptible to what kind of chemical reactions?
(b) What kinds of errors does this cause in the final measurement?

7.8 What filter material is used to remove nitric acid from an air stream by surface reaction?

7.9 How do passive and active air sampling differ?

7.10 (a) What was the first chemical reaction method used for measuring ozone concentrations in air?
(b) How is this method used today?

7.11 (a) What chemical reaction is the basis for the FRM for ozone?
(b) What chemical reaction is the basis for an approved FEM method?
(c) What is the most commonly used FEM method?

7.12 What is the major difference between the function of a DOAS that uses the Sun as a source and the function of a Dobson meter?

7.13 (a) What is commonly used as the light source for the long-path-length UV measurement of ozone?
(b) What are the common interferences with this method?
(c) How are they removed?

7.14 What is commonly used as an ozone scrubber?

7.15 (a) What does an ozonesonde measure?
(b) What system is responsible for transmitting the signal from an ozonesonde?

7.16 (a) How are OH radical concentrations usually determined?
(b) OH radicals can be measured directly by what two methods?

7.17 (a) What chemical reaction is the basis for a frequently used method of measurement of H_2O_2?
(b) How is the product detected?
(c) What other atmospheric species are also detected by this method?
(d) How is H_2O_2 separated from these other species?

7.18 (a) What is the FRM for CO determination in air?
(b) What is used to select the absorption wavelength in this method?
(c) How is the IR beam pulsed?
(d) Why is it pulsed?

7.19 What four gases are measured using NDIR?

7.20 (a) What is the FRM for SO_2?
(b) What two species can interfere with this method?
(c) How are they removed?

7.21 (a) What two species are commonly determined by correlation spectroscopy?
(b) How does the correlation spectrometer isolate the absorption wavelengths?
(c) What is the light source?

7.22 What was the original FRM for NO and NO_2?

7.23 (a) What is the FRM for NO in air?
(b) What is commonly used to convert NO_2 to produce NO in a NO_x Box?

7.24 What is the FRM for NO_2 in air?

7.25 (a) What is NO_y?
(b) Give some examples of NO_y.

7.26 What does the NO_x Box use to make sure there is no interference from ozone chemiluminescent reactions with alkenes?

7.27 (a) What method has been used successfully for decades to measure PAN in the atmosphere with detection limits of about 10 ppt?
(b) What are some of the problems in using this method for aircraft measurements?
(c) What two atmospheric species cause problems with this method?

7.28 What chemiluminescent reaction coupled with GC can give PAN detection limits of 10 ppt with an analysis time under 1 minute?

7.29 The synthesis of PAN in solution is by the strong acid dehydration of what two acids?

7.30 How can HNO_3 be measured using the NO_x Box?

7.31 What are the two most common methods for determining light halocarbon compounds that include CFCs and HCFCs?

7.32 (a) What two perfluorocarbons are readily determined using GC–ECD?
(b) What are their sources?

7.33 (a) What was the original detector used for the measurement of total VOCs?
(b) What is used to separate methane from the non-methane hydrocarbons using this detector?

7.34 (a) What is the current EPA-approved method for the measurement of VOCs in air?
(b) What detector is used for nonpolar organics? What detector is used for halogenated compounds?

7.35 (a) The EPA-approved method for the measurement of aldehydes and ketones is based on what reaction?
(b) How is the product detected?

7.36 (a) What reaction is used for the measurement of alkenes?
(b) A commercial instrument uses this reaction to measure what natural hydrocarbon?

7.37 What method uses H_3O^+ ions as a reagent to measure light organics and small aldehydes and ketones in real time?

7.38 What can be determined from the pattern of the peaks from the parent ion in the mass spectrum of an organic compound containing chlorine or bromine?

7.39 (a) What two mass spectrometry methodologies are used for the real-time analysis of aerosols?

7.40 (a) What mass spectrometry method is used to measure ^{14}C content in carbonaceous aerosols?
(b) How is the sample pretreated for analysis by this method?

7.41 Aerosol PAHs can be measured by what two methods after filter solvent extraction?

7.42 What two carbon fractions are measured in aerosol samples by thermal evolution?

7.43 (a) Aerosol light scattering is measured by what measurement systems?
(b) Three-wavelength light-scattering systems measure what three quantities?

7.44 What instrument determines the absorption of an aerosol sample after sampling the air through a movable filter tape?

7.45 What instrument measures the absorption of an aerosol on a filter sample by using a double-beam method to correct for scattering?

7.46 What is the photoacoustic effect?

7.47 What is the EPA-approved method for the measurement of lead in aerosol samples?

7.48 (a) What type of filter medium is used for the measurement of inorganic species in aerosols?
(b) For carbon measurements in aerosols?

7.49 What is zero air?

7.50 What are baseline measurements?

7.51 What are two major advantages to maintaining baseline measurements?

8

Chemistry of Surface and Ground Waters

There are two major chemical differences between the Earth and its neighboring planets. The first is the presence of oxygen in the atmosphere, which plays a major role in both the chemistry and the physics of the environment. The second is the very large amount of surface water that is present on the planet. The Earth is commonly called the "water planet," since nearly 71% of the surface is water. The water in the oceans gives the Earth a deep blue color that is seen from space contrasted against the red deserts and green forested regions, as shown in Figure 8.1.

There are two plausible explanations for the presence of such large amounts of water on the planet. The first is that it was brought to Earth by comets and other water-containing extraterrestrial bodies. The second is that it was brought to the surface by internal outgassing of volcanic activity. Both of these mechanisms are reasonable and they both probably contributed to the majority of the water found on Earth. It is known that most of the liquid water was present on the Earth very close to its beginning. This is determined from the ^{238}U dating of zircon minerals ($ZrSiO_4$), which require large amounts of liquid water to form. The procedure for dating minerals using ^{238}U will be discussed later in Chapter 12. The results of this dating method found that the zircon minerals were about 4.4 billion years old. So, the major oceans are estimated to be around 4 billion years old. The best estimate of the Earth's age is 4.55 billion years, so the oceans must have developed over a period of about 550 million years. This means that liquid water and the oceans have been an integral part of the Earth's environment from the beginning of its history.

Biological sources, including early photosynthetic purple sulfur bacteria, also likely contributed to the amount of water now found on the Earth. Purple bacteria have been shown to produce water as a product of the photosynthetic reaction;

$$CO_2 + 2H_2S + h\nu \rightarrow CH_2O + H_2O + 2S$$

where the H_2S was likely present in the atmosphere from early volcanic activity. Purple bacteria use sulfide (S^{2-}) or thiosulfate ($S_2O_3^{2-}$) as the electron donor in their photosynthetic path way. The sulfur is oxidized in the reaction to produce granules of elemental sulfur, which can be further oxidized to form H_2SO_4. Chemotrophs, early bacteria that produce energy from electron capture instead of photon capture (phototrophs), also produce water from H_2S by the chemosynthetic reaction;

$$12H_2S + 6CO_2 \rightarrow C_6H_{12}O_6 + 6H_2O + 12S$$

Chemotrophs are commonly found on the dark ocean floors near hydrothermal vents, since they are not dependent on sunlight for energy. They are also found in hot springs, volcanic fumaroles, and geysers. Nearby volcanic activity provides heat for warmth as well as the necessary H_2S for chemosynthesis. The early presence of both purple bacteria and chemotrophic bacteria would slowly add to the total water content of the planet.

Chemistry of Environmental Systems: Fundamental Principles and Analytical Methods, First Edition.
Jeffrey S. Gaffney and Nancy A. Marley.

Figure 8.1 Images of Earth taken by the MODIS imaging spectrometer on board the Terra satellite at a distance of about 700 km. Source: Reto Stöckli and Robert Simmon, NASA.

8.1 The Unique Properties of Water

Pure water is tasteless, colorless, and does not conduct electricity. Water is a polar molecule with the ability to form an unusually high number of hydrogen bonds with itself. Each liquid water molecule can form hydrogen bonds with four other water molecules, two through the two lone electron pairs on the oxygen atom and two through the two hydrogen atoms. This large number of hydrogen bonds between water molecules is responsible for increasing the boiling point to 100 °C at atmospheric pressure. It also leads to the unusually high freezing point and the large drop in density as the water freezes. Water expands when it freezes as it transitions from the large number of tightly clustered hydrogen bonds in the liquid form (Figure 8.2, right) to an open cage-like tetrahedral crystal structure in the solid form (Figure 8.2, left). Because of this highly ordered structure in ice, there are less H_2O molecules in a given volume than in liquid water. Thus, the density of ice is about 8.3% lower than that of water at the same temperature.

The difference in density of liquid water and ice results in ice floating on the surface of liquid water. It is also responsible for the natural weathering of rocks. Liquid water can seep into cracks in the rocks, widening the cracks as the ice expands when the temperature drops to freezing point at night. This natural diurnal rise and fall in temperatures results in an expansion–contraction cycle which widens cracks and increases weathering and erosion.

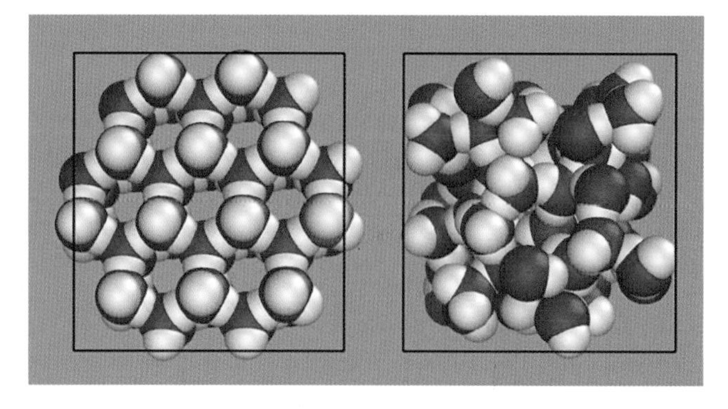

Figure 8.2 The structures of ice (left) and liquid water (right). Source: Adapted from P99am, Wikimedia Commons.

The density of pure liquid water above 4 °C is inversely proportional to the temperature. The density of liquid water at 4 °C is 1000.0 kg m^{-3} while the density at 25 °C is 997.0 kg m^{-3}. This decrease in density with increasing temperature is due to an increase in the kinetic energy of the water molecules, which decreases the strength and number of the hydrogen bonds between water molecules. Since colder water is denser, it can sink below warmer water, creating mixing in aqueous systems. This change in the density of water with temperature is responsible for mixing in lakes and ocean water. However, the density of water is at a maximum at 4 °C. If the water temperature drops below 4 °C, the density decreases as the water molecules begin to reorient themselves in preparation for freezing. So, the colder water below 4 °C will rise above the 4 °C water, creating a thermal inversion. This assures that the bottom of cold freshwater lakes will not freeze and will remain at about 4 °C. As the water warms up or cools down through the 4 °C temperature point (maximum density), considerable mixing occurs in aqueous systems, leads to vertical transport of nutrient, and dissolved oxygen (DO) concentrations.

Water is an excellent solvent and is often called "the universal solvent," since it has the ability to dissolve more types of substances than any other solvent. It is for this reason that pure water is never found naturally on Earth. Water always occurs with dissolved species including dissolved salts, which allow it to conduct electricity and also give it its taste. Water and the dissolved natural organics are also important in the transport of inorganic species in the natural and polluted environments. Water is needed to sustain both plant and animal life on Earth. Since water is a liquid at room temperature, it both provides a living environment for aquatic life and provides the liquid environment inside all cells. Inside cells, the water acts as both a solvent for biochemical reactions and a transport medium for dissolved compounds into and out of the cell. Liquid water has a very high specific heat capacity, which helps the aqueous environment inside organisms to resist damaging temperature changes. Rapid temperature changes could inhibit enzyme activity, since the enzyme reactions occur only in a narrow range of temperatures. Water also has a high heat of vaporization and a high heat of fusion, as a result of the extensive hydrogen bonding. The high heat of fusion helps prevent the liquid environment inside cells from freezing under normal conditions, which would tear the cells apart as the frozen water expands.

As with all natural water, the water inside cells is not pure. It contains ions and large organic molecules that act to reduce the normal freezing point of liquid water. The solutes (ions and molecules) strongly associate with the water molecules as they become solvated. This association interferes with the normal crystal formation of the water as it begins to freeze. Thus, freezing occurs at lower temperatures than for pure water. The depression of the freezing point is directly proportional to the concentration of the solutes. This freezing point depression is the reason that, while fresh water freezes at 0 °C, sea water freezes at about −2.0 °C. However, when sea water freezes only the fresh water forms ice, while most of the salt is left behind. Since the ice contains very little salt, it can be melted down to use for drinking water. Some organisms that live in cold environments are capable of producing high concentrations of chemical compounds in their cells, such as sorbitol and glycerol, when exposed to extreme cold. These solutes decrease the freezing point of the water inside the cells, preventing the organism from freezing as the water around them freezes. Some examples of these organisms are the Arctic rainbow smelt and the spring peeper frog.

Terrestrial organisms are especially dependent on fresh water. The amount of water contained in biological systems is equal to about half of all the water contained in the rivers of the world. Humans contain about 60% water by mass, and all biological organisms are made up of 60–90% water. All organisms must maintain a fluid balance that maintains the concentration of the electrolytes (salts) in the body fluids within a range needed for healthy biochemistry. This means that the amount of water lost by respiration, perspiration, urination, etc. must equal the amount of fresh water taken in. Samuel Taylor Coleridge expressed the need for fresh water on a planet mostly made up of sea water in *The Rime of the Ancient Sea Mariner* as

> Water, water, everywhere, And all the boards did shrink;
> Water, water, everywhere, Nor any drop to drink.

8.2 The Hydrological Cycle

Water is present in many different types of reservoirs, which are listed in Table 8.1 (Shiklomanov, 1993). Saltwater oceans account for almost 97% of the water found on the Earth, because of water's ability to solubilize materials from the rocks in the terrestrial environment. About 69% of the total fresh water is present as ice in the form of ice caps, glaciers, and permanent snow located at the poles and on very high mountain ranges. Ground ice and permafrost represent 0.9% of the fresh water, which can reform liquid water when warmed. Lakes and ponds account for only 0.3% of the surface fresh water. The Great Lakes, located on the border of the United States and Canada, which are 22 671 km^3 in volume, account for about 25% of the global fresh water contained in lakes.

Most of the readily available fresh water (30%) is located in groundwater reservoirs, even though about half (55%) of the water contained in groundwater reservoirs is saline. The worldwide groundwater reservoirs are commonly used for crop irrigation, mining operations, industrial processes, and also as a source of clean drinking water. Most of the available fresh water in groundwater reservoirs is being depleted faster than it can naturally recharge. Freshwater sources in general are a very important but limited resource, which is not evenly distributed across the globe. Some regions of the Earth are fresh water rich, such as areas near the Great Lakes, while others are extremely fresh water poor, such as the deserts in Northern Africa and the Middle East. With the growing world population, the demand for fresh water is increasing

Table 8.1 Sources of water on the Earth.

Water source	Volume (km^3)	% Freshwater	% Total
Oceans, seas, and bays	1 338 000 000	—	96.54
Ice caps, glaciers, and permanent snow	24 060 000	68.6	1.74
Groundwater, Total	23 400 000	—	1.69
Groundwater, Fresh	10 530 000	30.1	0.76
Groundwater, Saline	12 870 000	—	0.93
Soil moisture	16 500	0.05	0.001
Ground ice and permafrost	300 000	0.86	0.022
Lakes and ponds, Total	176 400	—	0.013
Lakes and ponds, Fresh	91 000	0.26	0.007
Lakes and ponds, Saline	85 400	—	0.007
Atmosphere	12 900	0.04	0.001
Swamp water	11 470	0.03	0.0008
Rivers	2120	0.006	0.0002
Water in biological systems	1120	0.003	0.0001

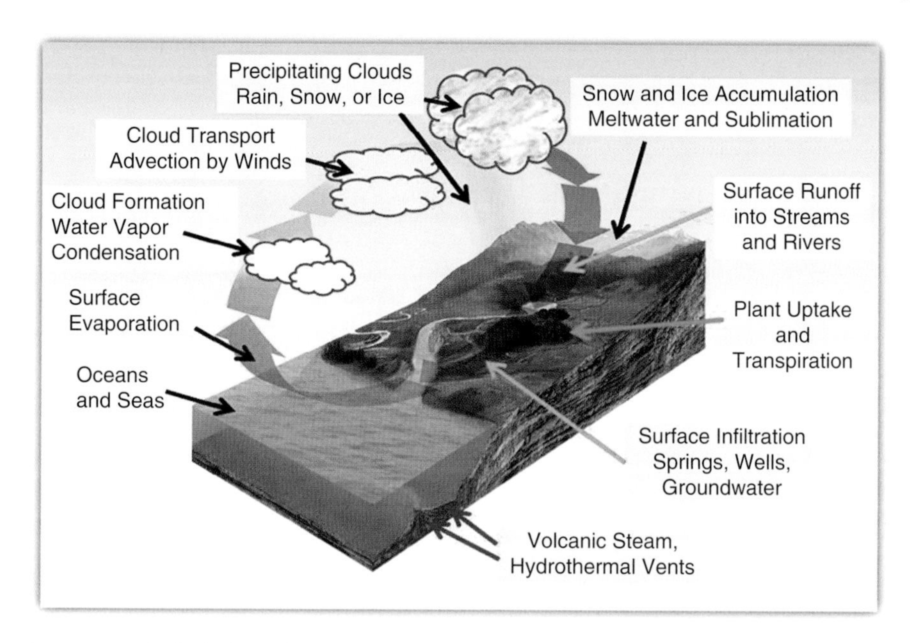

Figure 8.3 The hydrological cycle. Source: Adapted from Ehud Tul, Wikimedia Commons.

while the supply is decreasing. This need for fresh water worldwide is rapidly becoming a crisis (Gleick, 1993).

All of the water reservoirs are connected together by the hydrological cycle. The hydrological cycle, shown in Figure 8.3, is the sequence of processes that control the circulation of water throughout the hydrosphere, the combined sources of water in all its forms found on, under, and above the surface of the Earth. Water enters the atmosphere as water vapor through evaporation from the various surface water reservoirs, from wet ground surfaces, and from plants. Plants release water vapor through transpiration, which is the process of water movement through a plant, including the evaporation from plant surfaces such as leaves, stems, and flowers. The evaporation process is strongly dependent on air temperature, humidity, and pressure, which is dependent on altitude. After evaporation, the air containing the water vapor is less dense than drier air and so it rises up into the cooler regions of the atmosphere. As the surrounding air becomes cooler, the water vapor condenses on aerosols to form clouds, as discussed in Section 6.2. Both the water vapor and the cloud water can be transported long distances. During transport the clouds can undergo re-evaporation and condensation until the cloud droplet density and size become large enough for gravitational settling of the water droplets to occur as precipitation. Evaporation, like distillation, purifies the water so the precipitation replenishes the surface of the Earth with fresh water. The precipitation can be in the form of rain, snow, or ice, depending on the altitude of the cloud and the air and surface temperature.

In many cases the precipitation simply falls back into the oceans. Under warm temperature conditions, the water that falls on land is taken up into the soils and biota. In some areas, this water can flow deeply into porous soils and rocks to replenish groundwater resources in subsurface aquifers. If the rainfall is more than the rate of soil uptake can handle, the excess rain becomes surface runoff from the top of the saturated soils. This excess water can flow to lower elevations under the force of gravity, where it enters small streams that connect to larger streams and rivers, and eventually flows into the oceans. In the case of very heavy rainfalls, flash flooding can occur as the rivers and streams suddenly overflow their banks. This flash flooding can lead to rapid erosion of soils. During colder months the precipitation is in the form of snow or ice. In regions where rising spring temperatures allow the ice and snow to melt, the gradual thawing produces a slow runoff from the mountains into the rivers. If a rapid rise in temperature occurs, the snow or ice thaws more rapidly, producing faster runoff accompanied again by flooding and erosion.

Rivers and streams can carry large amounts of dissolved and suspended solids with them. If the river flow is sufficiently slow, the suspended particulates can be deposited in the river bottom resulting in a reduction of depth and an increase in width of the river. This increases the probability of flash flooding during periods of heavy rain. In larger, more rapidly flowing rivers the suspended particulates are usually deposited at the mouth of the river where it flows into the ocean. The deposition of particulates at the river mouth creates a large flat area known as a delta. The river deltas are typically rich in nutrient materials that were deposited in the sediments with the particulates. Thus, the soil in the delta regions is usually quite fertile. Examples of this are the Nile Delta in Egypt and the Ganges Delta located in India and Bangladesh, which are some of the most fertile regions in the world. Over time, the sediment deposition can result in a change in the height and course of the river system, causing it to meander slowly back and forth. Flooding can periodically occur along the banks of the river in the delta areas as the course of the river changes. In the United States, levees are commonly built to prevent any natural change in a river's position in order to prevent flooding along its banks. The largest of these levee systems is that along the Mississippi River, which consists of 3500 miles of levees averaging approximately 25 ft in height. The presence of the levees along the Mississippi prevents the river from depositing the sediment naturally in the delta region, forcing it to carry the suspended particulates and nutrients into the Gulf of Mexico. This causes both subsidence of the delta area and deposition of sediments and nutrients in an area off-shore where the river water mixes with the Gulf. This deposition of sediment and nutrients in the Gulf has led to the reduction in dissolved oxygen (DO) levels near the bottom, as will be discussed in Section 10.5.2.

The soil, biota, and underlying bed rock control the water flow through a system. If the soil is porous, it can absorb a large quantity of the water, which will continue to seep down through the subsurface until it reaches an area where the bedrock is no longer porous. The collection of water in these areas forms groundwater reservoirs, called aquifers, which can store fresh water for long periods of time. The water in these aquifers can often move underground, creating a subterranean river or stream, which can eventually reappear on the surface as a spring. Water can flow underground for long distances, even in areas where there is little rainfall. As it flows through soil and rock, it will dissolve any soluble minerals along the way. When the mineral concentrations of the water become high enough, it can be transformed from a freshwater system with a low salt content into a saltwater or brine system with a high salt content. The dissolved salts increase the mass of the water, resulting in an increase in density. So, the water with high mineral content will usually sink below the less dense fresh water, forming a deeper layer in the aquifer, while the fresh water forms a layer on top of the more dense brine water. When large amounts of fresh water are withdrawn for use in agricultural, industrial, or mining operations, the groundwater level drops and the remaining water in the aquifer becomes more brackish and eventually only the higher-density brine remains.

Like petroleum reservoirs, the freshwater aquifers have taken many hundreds or thousands of years to form by the natural hydrological cycle. The average residence times for water in the various large reservoirs are listed in Table 8.2 (Pidwirny and Jones, 2006). The residence time, which is the average time that a water molecule will spend in the reservoir before returning to the hydrological cycle, can also be considered to be the measure of the average age of the water in the reservoir at the present time. Shallow aquifers are estimated to be 100–200 years old and deeper aquifers can be about 10 000 years old. Once depleted, these freshwater aquifers are very

Table 8.2 The average residence times for water in the important reservoirs.

Water source	Residence time	Unit
Oceans, seas, and bays	3200	years
Ice caps, glaciers, and permanent snow	20–100	years
Groundwater, shallow	100–200	years
Groundwater, deep	10 000	years
Soil moisture	1–2	months
Ground ice and permafrost	20 000	years
Lakes	50–100	years
Atmosphere	9	days
Rivers	2–6	months

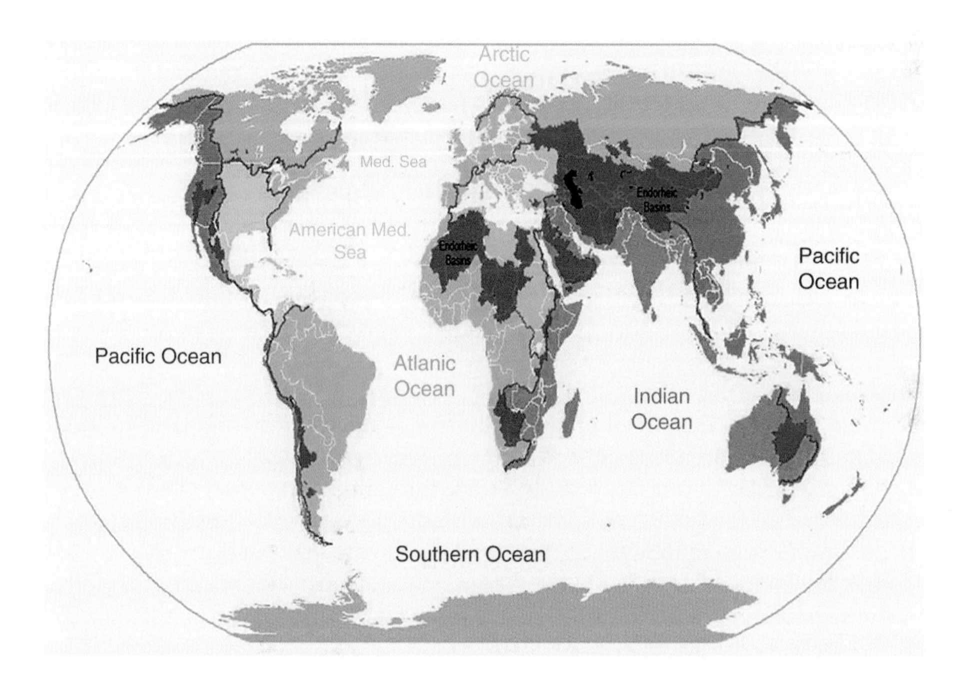

Figure 8.4 The major drainage basins of the world, including endorheic basins (dark gray), Arctic Ocean (blue green), Atlantic Ocean (green), Pacific Ocean (purple), Indian Ocean (red), the Southern Ocean (brown), the Mediterranean Sea (blue), and the American Mediterranean Sea (olive). Source: Adapted from Citynoise, Wikimedia Commons.

slow to recover. Considering this, many important large-scale groundwater systems should be considered unrenewable resources.

Unlike the air contained in the atmosphere, which knows no real boundaries, liquid water in the hydrosphere is confined to regions known as watersheds or drainage basins. The size and position of a drainage basin depends on the land topography, which determines the direction that the water will flow as it moves from higher to lower elevations. The major drainage basins of the world, shown in Figure 8.4, are divided into two types: 1) those that drain into an ocean or sea and 2) those that are landlocked. Landlocked basins, called endorheic basins, are indicated in Figure 8.4 in dark gray. In these closed drainage basins the water converges to a single point inside the basin, which may be a permanent lake, a dry lake, or the transition of surface water to a subterranean river. The drainage basins that drain into an ocean or sea are color coded in Figure 8.4. The major ocean drainage basins include the Arctic Ocean, Atlantic Ocean, Indian Ocean, Pacific Ocean, Southern Ocean, Mediterranean Sea, and the American Mediterranean Sea, which includes the Gulf of Mexico and the Caribbean Sea. The precipitation falling in each of these drainage basins will eventually flow, through rivers and streams, into the ocean or sea for which the basin is named. For example, all the water located in the land area color coded olive green in Figure 8.4 will flow into the American Mediterranean Sea (Gulf of Mexico and the Caribbean Sea).

The hydrological cycle has always been controlled by climate. During cold climatic periods, more ice and glaciers are formed, which decreases the amount of water in other parts of the hydrological cycle. The reverse is true during warm climatic periods. During the last ice age, which began about 2.6 million years ago and lasted until about 11 700 years ago, glaciers covered almost one-third of the land mass. This resulted in the oceans being 400 ft lower than they are today. During the last global warm period, which was about 125 000 years ago, the oceans were about 18 ft higher than they are today. Projections are for the hydrological cycle to intensify throughout the twenty-first century due to anthropogenic climate change and melting of the polar ice caps (Vahid et al., 2009). This intensification of the hydrological cycle doesn't mean that precipitation will increase in all areas. Precipitation is projected to decrease in subtropical land areas, which are already relatively dry, with drought strongest near the Mediterranean Basin, South Africa, southern Australia, and the southwestern United States. Precipitation is projected to increase most strongly in the near equatorial regions and at high latitudes. According to the 4th Assessment of the Intergovernmental Panel on Climate Change (IPCC), the increased variability of the hydrological cycle has now and will continue to have a profound effect on water availability and water demand at the global, regional, and local levels.

Although the hydrological cycle is itself a biogeochemical cycle, the flow of water over and beneath the surface of the Earth is a key component of other biogeochemical cycles. Rainwater runoff is responsible for the transport of eroded sediment and nutrients from land to water reservoirs. The salinity of the oceans is derived from erosion and transport of dissolved salts from the land. Runoff also plays a role in the carbon cycle through the transport of eroded rock and soil. It is also responsible for the transport of nutrients necessary for plants. Water is a major factor in the slow erosion of mountain regions, where the rocky surfaces are slowly worn away by the force of flowing water and the broken fragments and particles are carried to lower regions as soil and sedimentary material. This process is the first step in the formation of soils.

8.3 Ocean Currents and Circulation

The salinity and temperature of ocean water, along with the positions of terrestrial land masses, which act as flow restrictions, determine how and where the ocean currents circulate. Surface currents are strongly dependent on surface winds, tidal forces, and the Coriolis force, discussed in Section 2.3, while deep-water currents are primarily determined by differences in water density. The large-scale ocean circulation pattern is driven by global density gradients, which are caused by changes in water temperature, dissolved salts, and freshwater fluxes. Since most of the ocean salt is chloride, this large ocean circulation is commonly called the thermohaline circulation, also known as the "global oceanic conveyor belt." The thermohaline circulation was examined briefly in Section 2.5 as it related to the surface wind patterns. In relation to the hydrological cycle, the thermohaline circulation, shown in Figure 8.5, links the major surface and deep-water currents in the Atlantic, Indian, and Pacific Oceans.

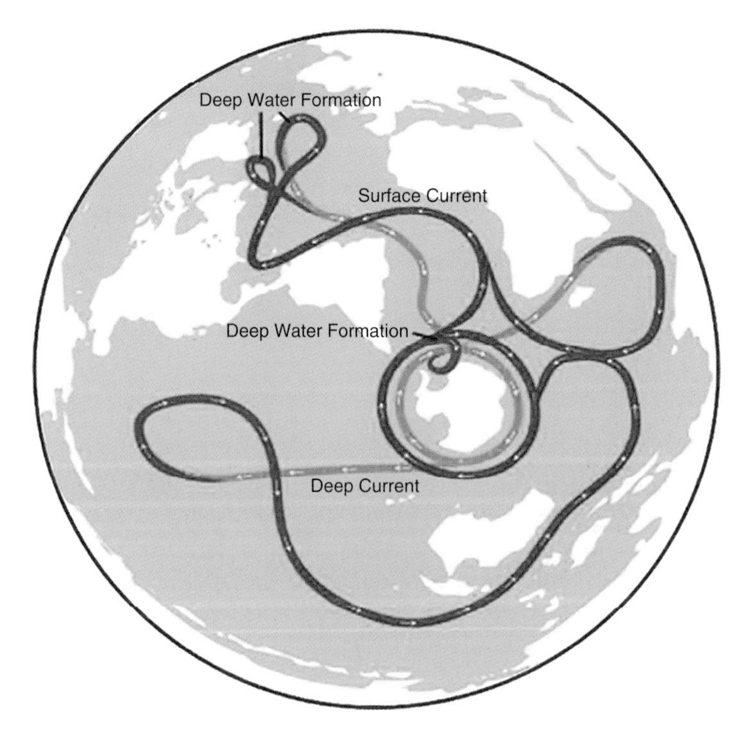

Figure 8.5 The thermohaline circulation, known as the global oceanic conveyor belt, including surface currents (red), deep currents (blue), downwelling or deep-water formation currents (fuchsia), and upwelling currents (fuchsia). Source: Avsa, Wikimedia Commons.

Colder waters with a higher salt content are denser than less salty warm waters. This leads to the formation of layers in the ocean that are similar to those in the atmosphere, with the less dense water in the layer above the denser water. Under these stable density conditions, the water movement is primarily in the horizontal direction. Downwelling of the surface water occurs when the water density increases due to a decrease in temperature and/or an increase in salinity. The higher-density water then sinks below the lower-density water flowing at deeper levels. These deep, density-driven currents move along submarine valleys toward the deepest parts of the ocean until they mix with warmer waters and begin to rise and work their way back to the surface. This global circulation pattern mixes the waters of the world's oceans, turning the ocean reservoirs into one single, large, interconnected system.

Near Greenland in the North Atlantic and around Antarctica in the Southern Ocean, the density of the ocean surface water increases due to a drop in temperature caused by the cold winds. The salinity also increases due to evaporation of water by the winds and the removal of fresh water in the freezing polar ice. The decrease in temperature and increase in salinity results in increased density, which causes the water to sink toward the ocean floor, as shown in Figure 8.5 (fuchsia). The dense water from the North Atlantic moves southward along the ocean floor and joins the sinking waters of the Southern Ocean in the far South Atlantic. Formations in the ocean floor cause the circulation to turn east where it splits, with part flowing along the east coast of Africa into the Indian Ocean and part flowing into the Pacific Ocean. The two branches of the circulation begin to mix with the less dense, warmer waters in the tropical Pacific, causing an upwelling in the current as shown in Figure 8.5 (fuchsia). After upwelling, the near-surface currents in the thermohaline circulation flow toward the North Atlantic and the Southern Ocean near the areas of the downwelling, dense water replacing the sinking waters and closing the current flow.

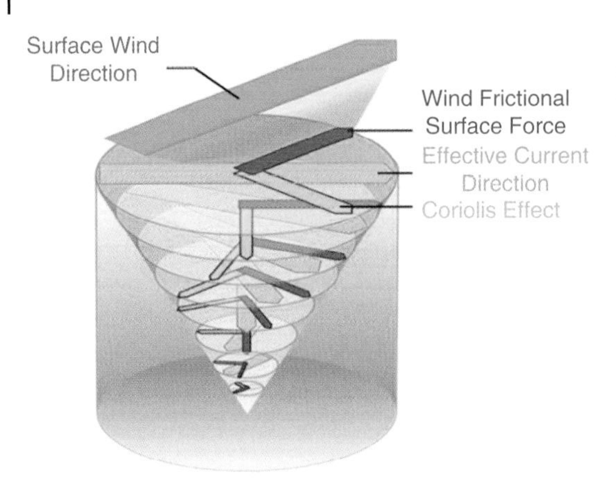

Figure 8.6 An Ekman spiral caused by the effects of surface winds interacting with the Coriolis force to create a spiral upwelling event. Source: Adapted from Timer, Wikimedia Commons.

The upwelling of the deep-water currents in the thermohaline circulation is aided by local winds interacting with the Coriolis force, called Ekman transport after the Swedish oceanographer Vagn Ekman who proposed the interaction in 1902. The effect of this interaction between the Coriolis force and the surface wind friction alters the direction of the surface water flow. In the Northern Hemisphere, the current flow is moved to the right of the wind direction and in the Southern Hemisphere, the current flow is moved to the left of the wind direction due to the Coriolis force (see Figure 2.7). Ekman transport causes the surface layer of the ocean water to move at about a 45° angle from the direction of the wind. The friction between the surface layer and the layer beneath it causes successive layers to move in the same direction. This process continues downward leading to a Coriolis-driven spiral motion in the water column, called an Ekman spiral, as shown in Figure 8.6. The Ekman spiral motion causes the mixing of the surface water layers with deeper, more dense layers, reducing their density and causing the upwelling events. As the deeper water moves upward it is replaced by water below it, resulting in a continuous upward movement of the deep-water currents. The rate of upward movement depends on the wind speed and length of time that the wind continues to flow. Typical upward movement of the water column is at a rate of about 5–10 m per day.

The best-known Ekman spiral-type upwelling occurs in coastal areas where it brings cold nutrient-rich waters up from the deeper ocean regions to the surface waters. These nutrient-rich waters increase the biota in the coastal areas, making them important to the fishing industry. The deep ocean waters contain detritus, which is dead particulate organic material that includes the bodies or fragments of dead organisms. These organisms consist mostly of phytoplankton and zooplankton. The decay of the detritus produces phosphate, nitrate, and silica in the form of silicic acid ($Si(OH)_4$). The phosphate and nitrate are important nutrients needed by phytoplankton. Thus, the upwelling of this deep ocean water leads to higher surface phytoplankton concentrations in the area of the upwelling. The silicic acid is a nutrient needed for zooplankton, which feed upon the phytoplankton, so the increased phytoplankton concentrations are usually associated with high zooplankton populations. The ocean food chain begins with the phytoplankton, which are in turn fed upon by zooplankton and filter feeders such as krill. The filter feeders are fed upon by fish, which are fed upon by larger fish, marine mammals, and birds. Since the upwelling ocean waters bring with them high concentrations of nutrients, phytoplankton, and zooplankton, the rest of the marine food chain follows. The coastal upwelling areas, shown in Figure 8.7, account for about half of the ocean biological productivity and contain some of the most productive fisheries in the world.

Large-scale upwelling also occurs due to the constant wind flow patterns just north or south of the equator. These trade winds of the two hemispheres converge near the equator. The

Figure 8.7 The major coastal upwelling areas of the world (red). Source: NOAA, Wikimedia Commons.

interaction of the wind patterns with the Coriolis force pulls the water in different directions in the Northern and Southern Hemispheres. The water north of the equator will be directed to the right while the water south of the equator will be directed to the left. This Ekman transport of the equatorial surface waters away from the equator gives rise to a divergence of the surface currents, which acts to pull the nutrient-rich deep ocean water to the surface. The equatorial upwelling is responsible for bringing up the deep-water currents of the thermohaline circulation in the Pacific Ocean. The upwelling of the African branch of the thermohaline circulation is mostly driven by coastal upwelling between the coast of Africa and the island of Madagascar.

The phytoplankton concentrations in the upwelling areas can be represented by the measurement of chlorophyll-a, which is present in phytoplankton that is brought up by the upwelling. The global measurements of chlorophyll-a obtained by the AQUA satellite using the MODIS (Moderate Resolution Imaging Spectroradiometer) imaging system in July of 2017 are shown in Figure 8.8. The highest concentrations of chlorophyll-a are seen in the coastal upwelling regions

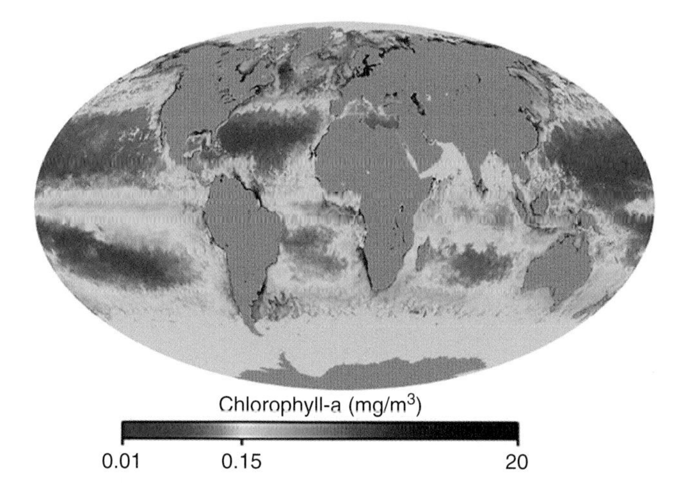

Figure 8.8 Measurements of chlorophyll-a obtained by the AQUA satellite using the MODIS imaging system in July 2017. Source: Adapted from NASA.

and in the cold waters near the Arctic. The higher productivity in the cold-water areas is due to the increased solubility of CO_2, which acts as a nutrient for phytoplankton. High concentrations of chlorophyll-a are also clearly seen in the Pacific and Atlantic Oceans near the equator due to equatorial upwelling. The chlorophyll-a concentrations due to equatorial upwelling in the Indian Ocean are not seen as clearly as in the Pacific and Atlantic Oceans in Figure 8.8, because of the variations in the trade winds in July.

8.4 The Structure of Natural Aquatic Systems

8.4.1 The Oceans

The ocean waters are divided into three zones according to the available sunlight penetration, as shown in Figure 8.9. These three zones are the euphotic zone, the dysphotic zone, and the aphotic zone. The euphotic zone is the area where photosynthesis occurs. The average depth of the euphotic zone is about 200 m, depending on the clarity of the water. The temperature ranges from 40 to −2.8 °C, dependent on location. The pressure ranges from 1 atm at the surface to about 20 atm at the bottom of the zone. The dysphotic zone extends from the bottom of the euphotic zone to about 1000 m. In this zone the sunlight intensity drops off significantly, creating a twilight region. There is enough light to see in the dysphotic zone during the day but not enough light for photosynthesis to take place, so no plants can exist in this zone. The water in the dysphotic zone averages from 5 to 3.9 °C, decreasing with depth. The pressure is high (20–100 atm) and increases with depth. The animals that live in the dysphotic zone are adapted to life in near darkness, cold temperatures, and high pressures. Many have large eyes to enable them to see in the nearly dark waters. The aphotic zone is a region where no sunlight can penetrate. It extends from about 1000 m, the bottom of the dysphotic zone, to the ocean floor. It is completely dark except for the occasional occurrence of bioluminescence. Temperatures in the aphotic zone are near freezing (0–3.9 °C) and decreases with depth. Pressure is extremely high, from 100 atm at the bottom of the dysphotic zone to greater than 1100 atm near the ocean floor. Organisms that live in the dysphotic zone must be able to live in complete darkness and in close to freezing water. Unusual and unique creatures have been discovered living in this large region of darkness, including the giant squid, anglerfish, gulper eel, and vampire squid.

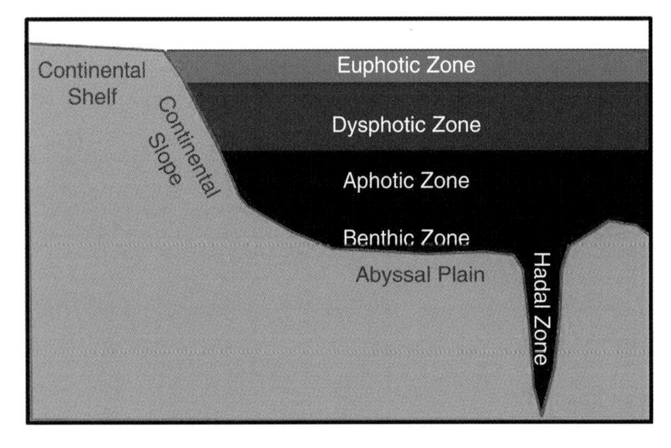

Figure 8.9 The ocean topographical zones (continental shelf, continental slope, abyssal plain, and hadal zone) and water zones classified according to the amount of light penetration (euphotic zone, dysphotic zone, aphotic zone, and benthic zone).

The bottom of any body of water, fresh or salt water, is called the benthic zone, or benthos. The benthic region of the ocean begins at the shore line and extends along the surface of the continental shelf out to the open ocean. The continental shelf is a gently sloping benthic region that extends away from the land mass. The edge of the continental shelf, known as the continental slope, begins a steeper descent that drops down to the deep ocean floor, which is called the abyssal plain. The abyssal plain is typically about 4000 m deep. Both submarine ridges and deep trenches, known as the hadal zone, interrupt the flat abyssal plain. The benthic substrate can include materials such as sand, coral, rock, or mud.

The surface of the bottom substrate in the benthic zone is called the benthic boundary layer. The composition of the boundary layer influences the biological activity that takes place there. The organisms that live in the benthic zone are known as benthic organisms, and they are quite different from those that live in the water column. The energy source for the benthic organisms is typically dead or decaying organic matter (detritus) that has drifted down from higher up in the water column. The number density of aggregated detritus particles (agg) sinking to the ocean floor can average 307 000 agg/(m^2 day) (Shanks and Trent, 1980). However, the benthos in a shallow region will have more available food than the benthos in the deep ocean. Most benthic organisms are scavengers or detritivores, but some microorganisms use chemosynthesis to produce their biomass, as discussed in Section 8.1. The benthic organisms generally live in close relationship with the bottom substrate, with many permanently attached to the boundary layer. Benthic organisms are divided into two categories based on where they live in the benthic zone. Those organisms that live on the surface of the benthic boundary layer are called epifauna, while those that live an inch or two below the boundary layer within the benthic substrate are called infauna.

8.4.2 Freshwater Systems

Freshwater systems are defined as those having a low salt concentration, usually <1%. Fresh surface waters are separated into three types: ponds and lakes, streams and rivers, and wetlands. Lakes and ponds are inland bodies of still or slowly moving water that are localized in a basin surrounded by land. Although they cover only 2% of the land area, they contain most of the available surface fresh water. Lakes and ponds are divided into three zones determined by depth and distance from the shoreline. There are two surface zones, the littoral zone, which is near the shore line, and the limnetic zone, which is further from the shore. The littoral zone is the warmest surface zone since it is shallow and can absorb more heat from the Sun. It sustains a fairly diverse community, which can include algae, rooted and floating plants, snails, clams, insects, crustaceans, fish, and amphibians. The limnetic zone is the main photosynthetic body of the lake. It is occupied by a variety of phytoplankton, including algae and cyanobacteria as well as zooplankton and small crustaceans, which support the lake's consumers. The lake's deep water below the limnetic zone is the profundal zone. Since little light can penetrate completely through the limnetic zone to reach the profundal zone, plants cannot grow in this zone. Thus, oxygen content in the profundal zone is dependent only on the extent of mixing in the lake. Also, since it is the deepest zone, it is the coldest. The types of organisms that live in the profundal zone must be adapted to low oxygen concentrations and low temperatures, for example leeches, annelid worms, insect larvae, and some species of crabs and mollusks.

The temperature of lakes and ponds varies by season, as shown in Figure 8.10. The temperature in deeper lakes can range from 4 to 22 °C in summer and from 0 to 4 °C in winter, depending on the depth. In the spring, any ice that covered the lake during the winter melts and the colder surface water above sinks and mixes with the water below, up to a temperature of about 4 °C. During the summer months the water absorbs solar energy, heating the surface to about 20 °C.

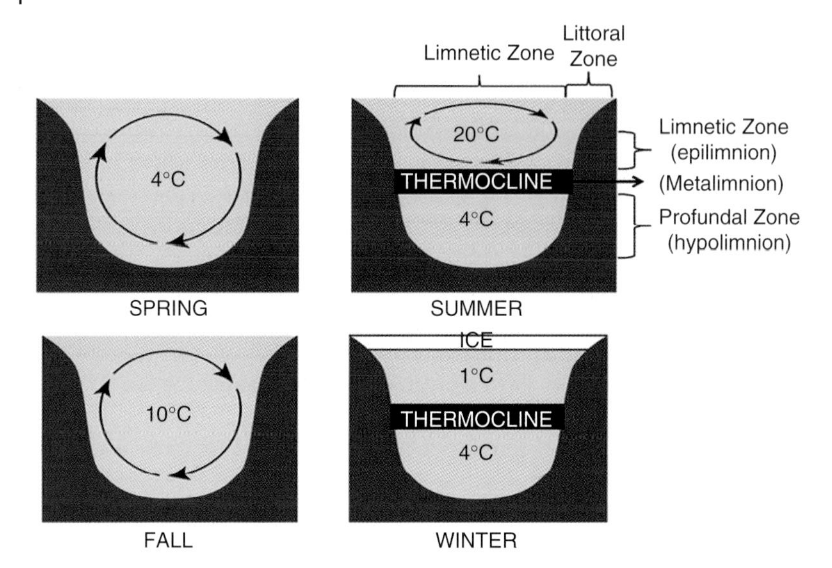

Figure 8.10 The seasonal changes in the structure of deeper lakes due to water temperature and density changes. Source: Adapted from Hydrated, Wikimedia Commons.

Since the warmer water is less dense than the colder water below, it remains at the surface leading to stratification. A thermocline develops between the limnetic and profundal zones, where the temperature of the water changes rapidly. This stratification in deep lakes can cause waters at the bottom of the lake in the summer to become anoxic due to the loss of DO from bacterial decomposition of bottom detritus. In the fall, winds blow across the lake creating surface friction, which induces vertical mixing. This leads to cold water upwelling and mixing, with the warmer water at the surface breaking the thermal stratification. In the winter, the water cools and ice can form at the top of the lake, cooling the surface water to near freezing. The 1 °C water is slightly less dense than the 4 °C water, as discussed in Section 8.1. So the colder water remains at the surface, reforming the thermocline and creating a thermal inversion. The winter thermocline is not as strong as the summer thermocline due to the smaller temperature difference between the surface and deep water.

All large lakes can stratify in the summer and winter, followed by upwelling and mixing events driven by strong winds in the spring and fall. Smaller lakes and ponds do not undergo stratification in summer and winter because they do not have large temperature differences with depth. Since the thermal stratification is an important physical characteristic, which controls the fauna, flora, sedimentation, and chemistry of the lake, the structure of a stratified lake is classified according to the stratification layers: hypolimnion, metalimnion, and epilimnion. The colder, denser water that forms the layer near the bottom is the hypolimnion while the warmer, less dense water at the surface is the epilimnion. The area in between the two layers is the metalimnion, otherwise known as the thermocline.

Rivers and streams are inland bodies of flowing water moving in one direction. Rivers begin at the source, or headwaters, which can include springs, snowmelt, or lakes. Many river beds can appear to be dry but actually have a substantial amount of water flowing underneath the surface. They follow a path, called a course, and end at a mouth, which can be another flowing body of water or, most commonly, an ocean or sea. The water in a river is usually confined to a channel consisting of a river bed between two banks. Larger rivers also have a flood plain that is wider than the river bed and is defined by flood waters that have at some time overflowed the banks of the river.

The structural characteristics of a river or stream change from the headwaters to the mouth. Rivers can be divided into three primary zones that make up the course. These are the crenon, the rhithron, and the potamon. The crenon is the zone at the source of the river. It is characterized by cooler temperatures, reduced oxygen content, low suspended materials, and slow-moving water. The rhithron is the upstream portion of the river that follows the crenon. It is characterized by cool temperatures, high oxygen levels, and fast, turbulent flow. The potamon is the downstream portion of the river that follows the rhithron, which is characterized by warmer temperatures, higher suspended particulates, lower oxygen levels, slow flow, and sandier bottoms. Near the mouth of the river or stream the water becomes cloudy due to increased suspended particulates that have become mobilized from sediments along the course. The suspended particulate burden increases with the flow rate of the river or stream. Because light cannot easily penetrate the murky waters, both DO and species diversity decrease near the mouth.

Rivers can generally be classified as alluvial, bedrock, or a mixture of the two. Alluvial rivers have channels and floodplains that are composed of mobilized sediment and/or soil. Their channels are shaped by the erosion of their banks during floods and the deposition of the sediment on their floodplain. Alluvial rivers can be classified by their channel pattern as meandering (sinuous curves), braided (multiple streams), anastomose (flowing around vegetative islands), or straight. These channel patterns are created by the supply of available sediment, substrate composition, topography, and vegetation. Bedrock rivers form when the river cuts through the bottom sediments into the underlying bedrock. This typically occurs in areas that have undergone uplift, which increases the river gradients, causing very rapid flow across the downward gradient. Most bedrock rivers are not pure forms. They typically consist of a combination of a bedrock channel and an alluvial channel.

Wetlands, including marshes, swamps, bayous, bogs, and fens, are distinct ecosystems that are either permanently or seasonally flooded with water. Wetlands can be fresh water, salt water, or brine, but the primary factor that distinguishes freshwater wetlands is the wide variety of aquatic plants, called hydrophytes, that are adapted to the very humid conditions and hydric soil. These include pond lilies, cattails, sedges, tamarack, cypress, black spruce, and gum. Wetland hydrology is associated with the spatial and temporal dispersion and flow of surface and ground water in its reservoirs. The sources of water flowing into a freshwater wetland are precipitation, surface water, and ground water. The water can leave the wetland by evaporation, transpiration, surface runoff, and subsurface flow. Wetlands are often considered to be the transition between dry land and another body of water. Based on this view, freshwater wetlands can be classified as riverine (associated with a river), lacustrine (associated with a lake), and palustrine (isolated). Saltwater wetlands are classified as marine (associated with an ocean) or estuarine (associated with an estuary).

Wetlands have the highest species diversity of all aquatic ecosystems, serving as home to a wide range of plant and animal life. Many species of amphibians, reptiles, birds, and small mammals can be found in the wetland areas. Wetlands serve many essential functions in an ecosystem, including water filtration and storage, flood and erosion control, and providing food and habitat for fish and wildlife. Wetlands can absorb and slow flood waters, which helps to alleviate property damage and save lives. They also absorb excess nutrients, sediments, and other pollutants before they reach rivers, lakes, and oceans. The largest wetlands in the world include the Amazon River Basin, the West Siberian Plain, the Pantanal in South America, and the Sundarbans in the Ganges Delta.

8.5 The Composition of Natural Aquatic Systems

The oceans contain significant amounts of dissolved inorganic species, especially NaCl, which is responsible for most of the salinity. The term saline is defined as the concentration of dissolved salts in water, usually expressed in parts per thousand by weight. The total salinity of sea water is about 3.5%, or 35 parts per thousand. However, the average ocean concentration of dissolved NaCl in ocean water dominates the dissolved salt concentrations, at about 0.5 M or 29 parts per thousand. This is why a solution of sodium chloride salt dissolved in water is commonly called a saline solution. There are other important inorganic ions that are present in significant concentrations in ocean waters, which are listed in Table 8.3 (Dickson and Goyet, 1994). The Mg^{2+} and SO_4^{2-} ion concentrations are a factor of 10 less than the Na^+ and Cl^- concentrations. These are followed by Ca^{2+}, K^+, and HCO_3^-, which have concentrations that are about a factor of 100 less than those of Na^+ and Cl^-.

Fresh water is chemically defined as containing a concentration of dissolved salts less than two parts per thousand. The concentrations of the individual ions are usually measured in parts per billion (ppb). Some average values for dissolved ions in fresh waters are Ca^{2+} at 15 ppb, Mg^{2+} at 4 ppb, Na^+ at 6 ppb, SO_4^{2-} at 11 ppb, and Cl^- at 6 ppb. Although sea water contains much more dissolved ions than any freshwater system, the relative concentrations of the different ionic species differ widely. For example, the concentration of HCO_3^- in ocean water is 0.14% of the total ion concentration, while it is 48% of the total ion concentration in river water. This difference in the relative ion concentrations is a consequence of the large concentration of Na^+ and Cl^- in sea water.

The water chemistry of both fresh and saltwater systems is strongly connected to the dissolved atmospheric gases: O_2, CO_2, and N_2. The DO concentration is important to most all forms of aquatic life. The concentration of dissolved CO_2 in water will become important in discussions of the carbon cycle, as well as pH-dependent aqueous chemistry. The concentration of dissolved N_2 will be important in discussions of the nitrogen cycle. Both of these important cycles control

Table 8.3 Concentrations, in parts per thousand and moles per liter, of dissolved inorganic species in ocean water at a salinity of 35.

Name	Chemical formula	ppt	mol l^{-1}
Chloride ion	Cl^-	19.4	5.46×10^{-1}
Sodium ion	Na^+	10.8	4.69×10^{-1}
Magnesium ion	Mg^{2+}	1.3	5.28×10^{-2}
Sulfate ion	SO_4^{2-}	2.7	2.82×10^{-2}
Calcium ion	Ca^{2+}	0.4	1.03×10^{-2}
Potassium ion	K^+	0.4	1.02×10^{-2}
Bicarbonate ion	HCO_3^-	0.1	1.77×10^{-3}
Bromide ion	Br^-	7.0×10^{-2}	8.44×10^{-4}
Boric acid	$B(OH)_3$ or H_3BO_3	2.0×10^{-2}	3.2×10^{-4}
Carbonate ion	CO_3^{2-}	1.6×10^{-2}	2.6×10^{-4}
Tetrahydroxy borate	$B(OH)_4^-$ or $H_4BO_4^-$	7.0×10^{-3}	1.0×10^{-4}
Strontium ion	Sr^{2+}	8.0×10^{-3}	9.1×10^{-5}
Fluoride ion	F^-	1.0×10^{-3}	6.8×10^{-5}
Hydroxide ion	OH^-	2.0×10^{-4}	1.0×10^{-5}

the amount of nutrients in both soil and aqueous systems. The concentrations of these gases in water are dependent on their aqueous solubilities, as described by Henry's Law discussed in Section 6.2.3. Recall that Henry's Law can be expressed in two forms: $H_x = [X(aq)]/P_x$ and $H_x = P_x/[X(aq)]$, where $[X(aq)]$ is the concentration of gas in water and P_x is the partial pressure of X above the water. Each of these forms of Henry's Law has a different value and different units for the Henry's Law constant. The first form (in units of M atm^{-1}), represents the uptake of the gas-phase species into an aqueous environment, while the second form (in units of atm M^{-1}), represents the aqueous-phase species leaving the aqueous system into the atmosphere. The Henry's Law constants for the three important gases in water are given in Table 6.2 as

$$O_2 = 1.3 \times 10^{-3}\ \text{M atm}^{-1} \text{ and } 7.7 \times 10^{2}\ \text{atm M}^{-1}$$
$$CO_2 = 3.4 \times 10^{-2}\ \text{M atm}^{-1} \text{ and } 2.9 \times 10^{1}\ \text{atm M}^{-1}$$
$$N_2 = 6.1 \times 10^{-4}\ \text{M atm}^{-1} \text{ and } 1.6 \times 10^{3}\ \text{atm M}^{-1}$$

8.5.1 Dissolved Oxygen

None of the important dissolved gases has very high solubility in water. The equilibrium concentration of O_2 in the surface water at an atmospheric partial pressure of 0.21 atm at 298 K according to Henry's Law is:

$$[O_2(aq)] = (H_{O2})(P_{O2}) = (1.3 \times 10^{-3}\ \text{M atm}^{-1})(0.21\ \text{atm}) = 2.7 \times 10^{-3}\ \text{M}$$

The O_2 concentration increases with decreasing temperature, as shown in Figure 8.11 (Engineering Toolbox, 2005). Although the dissolved gases are nonpolar, they still form weak intermolecular attractive forces with the polar water molecules through induced dipoles in the gas molecules, called London dispersion forces. The increasing gas solubility with decreasing temperature is due to a decrease in kinetic energy, which increases the strength of the London dispersion forces between the gas molecules and the water molecules, allowing more of the gas to become dissolved. Conversely, an increase in temperature increases the kinetic energy, allowing the gas molecules to overcome the London dispersion forces and escape the solution.

The O_2 concentration also decreases with increasing salinity. The salinity of sea water is due to dissolved salts in the water, including NaCl, $NaHCO_3$, KNO_3, $MgSO_4$, and NaBr. When ionic

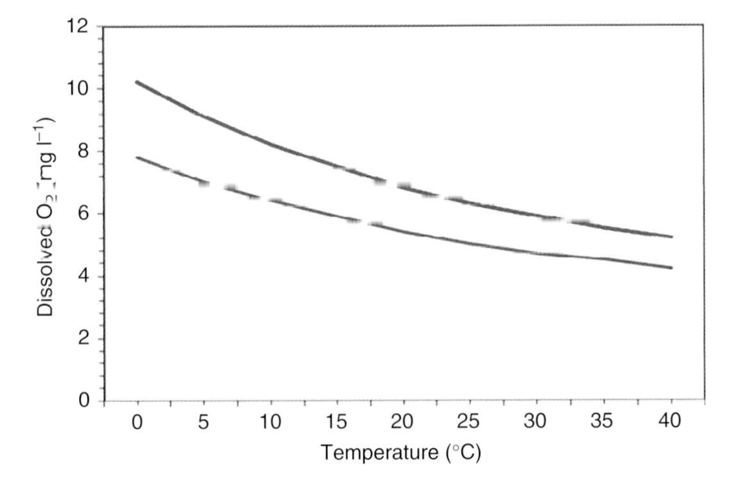

Figure 8.11 The solubility of O_2 (mg l^{-1}) in fresh water (red) and salt water (blue) as a function of temperature (°C).

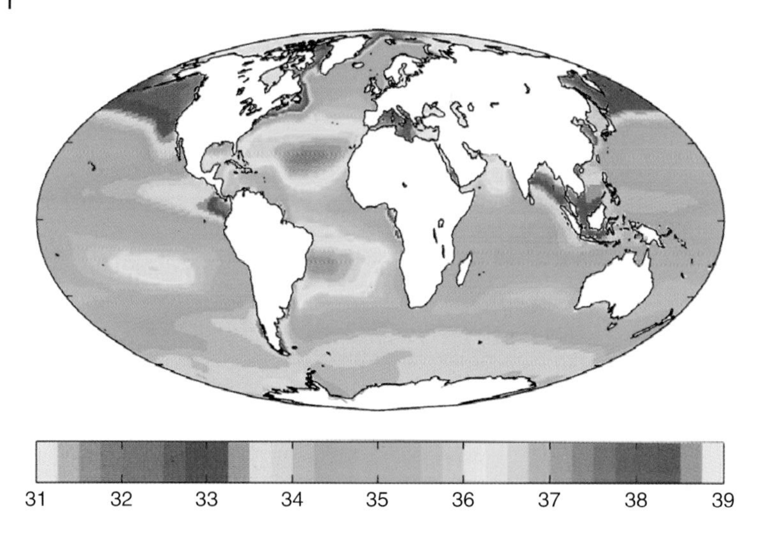

Figure 8.12 Annual mean ocean surface salinity (g kg^{-1}) in 2009. Source: Adapted from Plumbago, Wikimedia Commons.

salts dissolve in water, the polar water molecules form a solvation shell around the charged ions. This disrupts a portion of the extended hydrogen bonding between the water molecules. It also disrupts the London dispersion forces between the nonpolar gas molecules and the water molecules, decreasing their solubility. The surface salinity in the oceans varies from about 33 to 39 g kg^{-1}, as shown in Figure 8.12. The surface water salinity can be increased due to increased amounts of salt input from atmospheric deposition, from land runoff, and from increased surface water evaporation. The surface salinity can also be decreased due to a variation in surface freshwater inflow. Global water salinities vary from 31–33 g kg^{-1} in the cold waters of the Arctic to 36.5–37.5 g kg^{-1} in the subtropics and a maximum of 38.5 g kg^{-1} in the Mediterranean Sea. The low salinities in the Arctic are partially due to the decrease in solubility of the salts at cold temperatures. They are also due to the more recent inflow of fresh water from melting of Arctic ice. The very high salinities in the Mediterranean Sea are due to high evaporation rates, coupled with very little mixing of the sea water with open ocean waters in the nearly landlocked Mediterranean Sea.

The DO levels of both freshwater and saltwater systems are usually found to be lower than the Henry's Law equilibrium value in the water layers below the surface. The only sources of O_2 in water systems are from the equilibrium with the atmosphere at the surface and from photosynthetic plants. The surface DO concentrations can be transported to lower depths by vertical mixing. However, the loss of DO with depth also occurs due to the uptake of O_2 by biochemical respiration and by chemical oxidation reactions in the water column. The photosynthetic production of O_2 from aqueous plants has depth limits because of the loss of sunlight with depth. So, the photosynthetic production of O_2 in the ocean occurs only in the photic zone. In freshwater lakes it occurs in the limnetic and littoral zones. The DO below these levels in both ocean and freshwater lakes depends on the extent of vertical mixing, as well as biochemical and chemical oxidation processes.

8.5.2 Nitrogen and Phosphorus

Two important elements in both aqueous and soil systems are nitrogen and phosphorus. Nitrogen is found in water as NH_3, NH_4^+, NO_3^-, NO_2^-, and organic nitrogen. The biogeochemical

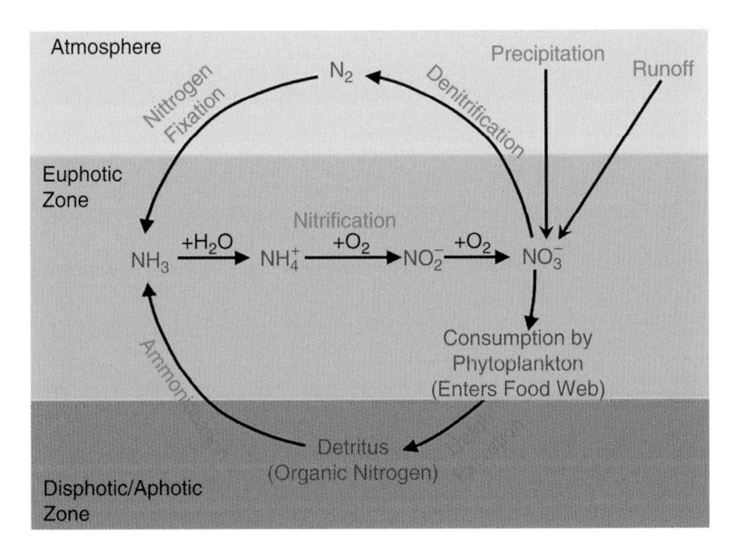

Figure 8.13 The nitrogen cycle (N-cycle) in aqueous systems.

cycle that describes the conversion of nitrogen into multiple chemical forms as it travels through the atmosphere, hydrosphere, and biosphere is known as the nitrogen cycle (N-cycle), shown in Figure 8.13. The paths for conversion of nitrogen into the various species can be carried out through both biological and physical processes. Nitrogen enters the water through precipitation (as NO_3^- in acid rain), runoff (as NO_3^- from fertilizer), or directly from the atmosphere (as N_2). The equilibrium concentration of N_2 in the surface water at an atmospheric partial pressure of 0.78 atm at 298 K according to Henry's Law is:

$$[N_2(aq)] = (H_{N2})(P_{N2}) = (6.1 \times 10^{-4}\ M\ atm^{-1})(0.78\ atm) = 4.8 \times 10^{-4}\ M$$

The N_2 concentration also increases with decreasing temperature and decreases with increasing salinity, as does O_2. NH_3 can be produced by direct conversion of atmospheric N_2 by cyanobacteria in a process called nitrogen fixation. The ammonia can then form NH_4^+ by an equilibrium reaction with water:

$$N_2 + 3H_2 \rightarrow 2NH_3$$
$$NH_3(aq) + H_2O(l) \rightarrow NH_4^+(aq) + OH^-(aq)$$

NH_3 is toxic to freshwater organisms at concentrations ranging from 0.53 to 22.8 mg l^{-1}, depending on the pH and temperature of the water. The toxicity increases as both pH and temperature increases. Aquatic plants are more tolerant to ammonia toxicity than animals, and invertebrates are more tolerant than fish.

The $NH_3(aq)$ and $NH_4^+(aq)$ formed will be oxidized further by the aerobic bacteria during the nitrification process. Nitrosomonas bacteria convert $NH_3(aq)$ to NO_2^- and then nitrogen-fixing bacteria in the genus *Nitrobacter* convert the NO_2^- to NO_3^- by the biochemical reactions:

$$2\ NH_3(aq) + 3\ O_2(aq) \rightarrow 2\ NO_2^-(aq) + 2\ H^+(aq) + 2\ H_2O(l)$$
$$2\ NO_2^-(aq) + O_2(aq) \rightarrow 2\ NO_3^-(aq)$$

Aquatic plants, including algae and phytoplankton, consume the NH_4^+ and NO_3^-, which are then converted into nitrogen-containing organic molecules, such as amino acids and DNA. After death and decomposition of the aquatic organisms, specialized decomposing bacteria can convert the organic nitrogen species in the detritus back into NH_3 and NH_4^+ in a process called ammonification. They are then converted back into N_2 by anaerobic bacteria, which reduce the oxidized forms of nitrogen in a process called denitrification. Denitrification generally proceeds through a series of reduction reactions, resulting in the following net reaction:

$$2NO_3^- + 10e^- + 12H^+ \rightarrow N_2 + 6H_2O$$

The N_2 is released back into the atmosphere, completing the N-cycle.

Phosphorus is found in aqueous systems as organic phosphates or inorganic phosphate (PO_4^{3-}). The P-cycle describes the movement of phosphorus through the geosphere, hydrosphere, and biosphere. The phosphorus cycle (P-cycle) is the only biogeochemical cycle that does not have an atmospheric component, since phosphorus does not form gaseous products. Although phosphates move quickly through plants and animals, they move much more slowly through the soil and aquatic systems. So, the P-cycle is one of the slowest biogeochemical cycles. Phosphorus is an essential element in both animal and plant biochemistry, since it is a key component of molecules that store energy, such as adenosine triphosphate (ATP). In addition, it is required for the formation of the DNA double helix, which is linked by a phosphate ester bond. Calcium phosphate is also the primary component of mammalian bones and teeth, insect exoskeletons, and phospholipid membranes of cells.

The largest reservoir for phosphorus is that contained in sedimentary rocks in the form of apatite ($Ca_5(PO_4)_3OH$). Organic acids formed in the upper soils during the decomposition of organic detritus can react with the apatite to release the soluble phosphate ions. When it rains the phosphates are transported through the soil and into aquatic systems. So phosphorus occurs most often in aquatic systems as solvated PO_4^{3-} ions, commonly known as orthophosphate. It is the orthophosphate that is required for phytoplankton and algae, which are at the bottom of the aquatic food chain. Phosphates can also enter the water in runoff from fertilizers in the form of ammonium phosphates or as organic phosphates in sewage or industrial treatment plant effluents. The organic phosphates tend to settle on ocean floors and lake bottoms, and are recycled by bacteria, which convert organic phosphate into the more soluble orthophosphate in a process known as mineralization. Phosphorus is a limiting nutrient for aquatic organisms. This is partly due to the fact that the large amount of phosphorus contained in subterranean sedimentary rock is not readily available and partly due to the slow chemistry of phosphorus in the environment.

8.5.3 Sulfur

Another important nutrient in aqueous systems is sulfur, which is found in aqueous systems as SO_4^{2-}, dissolved H_2S, organic sulfur, and mineral S^{2-}. The sulfur cycle (S-cycle), shown in Figure 8.14, is important in biochemistry because sulfur is a constituent in many proteins and enzyme cofactors. It is a component of the amino acid cysteine and so is involved in the formation of disulfide bonds in proteins, which determines their folding patterns and thus their functions. Sulfur occurs in the atmosphere principally as SO_2 from both anthropogenic and natural sources. The SO_2 is rapidly oxidized to H_2SO_4, which is incorporated into cloud water, as discussed in Section 6.2.2. The H_2SO_4 is carried into the oceans in acid rain. The main reservoirs for sulfur are in the oceans as $SO_4^{2-}(aq)$, in sedimentary rocks as $SO_4^{2-}(s)$ or $S^{2-}(s)$, and in ocean sediments as $S^{2-}(s)$. The amount of SO_4^{2-} in the oceans is controlled by input from land runoff and atmospheric precipitation, sulfate reduction to organosulfur compounds or H_2S, sulfide reoxidation on continental shelves and slopes to SO_4^{2-}, and burial of anhydrite and pyrite in the oceanic crust as S^{2-}.

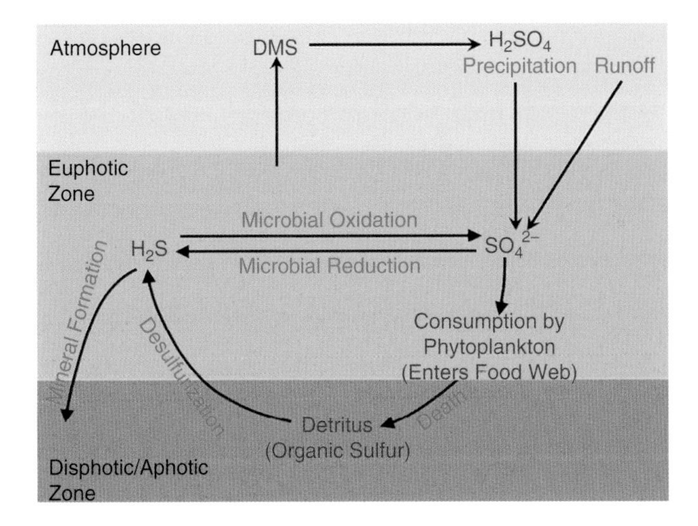

Figure 8.14 The sulfur cycle (S-cycle) in aqueous systems.

The SO_4^{2-} is assimilated by aquatic plants, algae, and phytoplankton, where it is reduced and incorporated into organosulfur compounds. Upon the death and decomposition of the aquatic organisms the organosulfur compounds are converted to inorganic sulfur in the form of H_2S, a process called desulfurization. The H_2S can be reoxidized to SO_4^{2-} by sulfur-oxidizing bacteria or it can react with iron to form the mineral pyrite (FeS_2), which settles to the ocean floor. Marine algae produce a compound called dimethylsulfonium propionate (DMSP), which is enzymatically cleaved to yield the volatile compound dimethylsulfide (DMS). DMS is oxidized in the marine atmosphere to various sulfur-containing compounds, including H_2SO_4, SO_2, dimethyl sulfoxide or DMSO ($(CH_3)_2SO$), and methanesulfonic acid (CH_3SO_3H). Also, carbonylsulfide or COS (O=C=S) is produced photochemically from dissolved organic matter in sea water. The natural release of these volatile organic sulfur compounds from the ocean acts to transport sulfur from the ocean to the atmosphere, geosphere, and biosphere. However, the oceans remain the primary reservoir for sulfur.

8.5.4 Carbon

The carbon cycle is the biogeochemical cycle that describes the exchange of carbon among the atmosphere, hydrosphere, biosphere, and geosphere. The major carbon reservoirs include the atmosphere (CO_2, CH_4, aerosols), terrestrial biosphere (plants, soils), oceans (dissolved inorganic carbon, dissolved organic carbon, living organisms, detritus), sediments (soils, fossil fuels, minerals), and the Earth's interior (mantle and crust). The carbon moves between the reservoirs by chemical, physical, geological, and biological processes. The oceans contain the largest reservoir of carbon on the surface of the Earth. This includes the dissolved inorganic carbon species (DIC): dissolved CO_2, H_2CO_3, HCO_3^-, and CO_3^{2-}. It also includes dissolved organic carbon species (DOC), particulate organic carbon (POC), and organic carbon in living and dead organisms, as well as inorganic carbon in sediments and particulate matter. About 95% of the total active carbon pool on the surface of the Earth is stored in the ocean, primarily as DIC.

Photosynthesis by aqueous organisms requires CO_2. Although CO_2 is more soluble in water than O_2, the atmospheric concentration of CO_2 is much lower than that of O_2. For an atmospheric concentration of 400 ppm at 298 K, the equilibrium concentration of CO_2 in the surface water determined by Henry's Law is

$$[CO_2(aq)] = (H_{CO2})(P_{CO2}) = (3.4 \times 10^{-2}\ M\ atm^{-1})(4.0 \times 10^{-4}\ atm) = 1.36 \times 10^{-5}\ M$$

which is about 20 times lower than the equilibrium concentration of O_2. However, unlike O_2, CO_2 can react reversibly with water to form carbonic acid ($H_2CO_3(aq)$):

$$CO_2(g) + H_2O(l) \rightleftarrows H_2CO_3(aq) \qquad K_{eq} = \frac{[H_2CO_3]}{[CO_2]} = 1.2 \times 10^{-3}$$

Carbonic acid is a diprotic acid which rapidly ionizes in water to lose up to two protons. It forms first the bicarbonate ion (HCO_3^-) and then the carbonate ion (CO_3^{2-}) by the following reactions:

$$H_2CO_3(aq) + H_2Ol \rightleftarrows H_3O^+(aq) + HCO_3^-(aq) \qquad K_{eq} = \frac{[HCO_3^-][H_3O^+]}{[H_2CO_3]} = 9 \times 10^{-7}$$

$$HCO_3^-(aq) + H_2O(l) \rightleftarrows H_3O^+(aq) + CO_3^{2-}(aq) \qquad K_{eq} = \frac{[CO_3^{2-}][H_3O^+]}{[HCO_3^-][CO_3^{2-}]} = 6.5 \times 10^{-10}$$

In the oceans, both inorganic and biological $Ca^{2+}(aq)$ can react with $CO_3^{2-}(aq)$ to form solid calcium carbonate, $CaCO_3(s)$, usually in the form of calcite:

$$CaCO_3(s) \rightleftarrows Ca^{2+}(aq) + CO_3^{2-}(aq) \qquad K_{eq} = [Ca^{2+}][CO_3^{2-}] = 6.4 \times 10^{-7}$$

The formation of the species HCO_3^- and CO_3^{2-} in water is dependent on the H_3O^+ concentration. The concentration of dissolved CO_2 in water, in the form of H_2CO_3, is dependent on its equilibrium with HCO_3^-. Thus, the concentrations of H_2CO_3, HCO_3^-, and CO_3^{2-} are all a function of pH, as shown in Figure 8.15.

The pH of sea water ranges from 7.5 to 8.4. Most of the ocean surface water is buffered at a pH of about 8.2, because of large deposits of $CaCO_3$ in limestone sediments, shellfish, and corals. The pH of the oceans varies due to differences in temperature, salinity, and mineral carbonate abundance. As the water temperature decreases, the dissolved CO_2 concentration decreases and so the formation of carbonic acid decreases and the pH increases. The combustion of fossil fuels, which increase CO_2 emissions into the atmosphere, has increased the ocean surface acidity, lowering the pH from 8.2 to 8.1. The pH of 8.2 is equivalent to 6.31 nanomolar H_3O^+, while 8.1 is 7.94 nanomolar H_3O^+. Since the pH unit is a log function, this change of 0.1 pH unit is equivalent to an increase in surface water acidity of about 26%. Natural, unpolluted rain water will have a pH of about 5.6 due to the equilibrium with $CO_2(g)$ from the air and the equilibrium reactions with the acidic carbonate species. This is about the same pH expected for surface lake

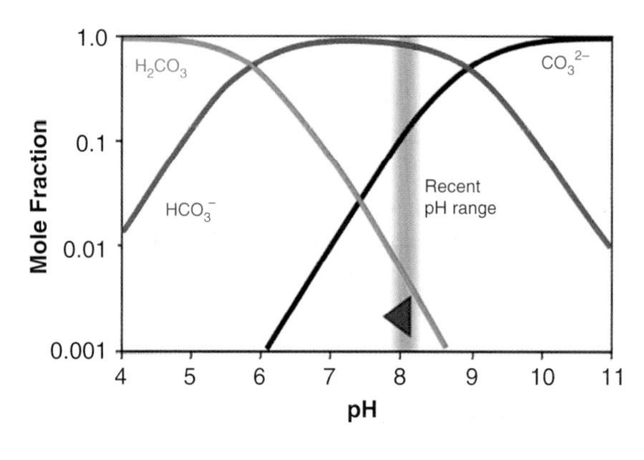

Figure 8.15 The mole fraction of each of the species H_2CO_3 (green), HCO_3^- (red), and CO_3^{2-} (black) in water at 25 °C as a function of water pH. The pH range highlighted in blue represents the change in ocean pH due to the increased levels of CO_2 in the atmosphere from fossil fuel combustion since 1780. Source: Adapted from Karbonatsystem_Meerwasser_de.svg, Wikimedia Commons.

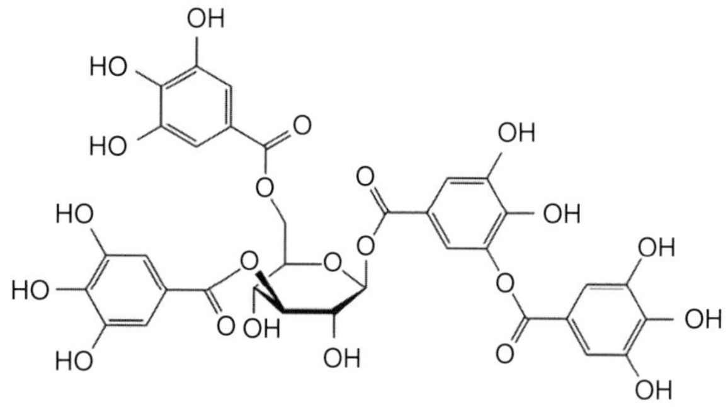

Figure 8.16 The unit structure of tannic acid found in both ocean and fresh waters. Source: Adapted from Mikhal Sobkowski, Wikimedia Commons.

waters that do not have dissolved minerals to add alkalinity. Streams and lakes that have sediment or bedrock containing carbonate, such as limestone ($CaCO_3$) or dolomite ($Ca(Mg)CO_3$), will have a higher pH due to the dissolution of the carbonate minerals. The normal range for pH in surface fresh waters is 6.5–8.5 and that for ground water is 6.0–8.5, depending on the mineral carbon content of the water system.

The DIC and DOC concentrations in the ocean are usually about the same (about 2 mM). The majority of the DOC is colored and is commonly referred to as colored dissolved organic matter (CDOM). The CDOM concentrations are higher near the mouths of rivers than they are in the open ocean, which indicates that a significant amount of the CDOM is of terrestrial origin. The CDOM is derived from decaying organic material and leaching of tannins from terrestrial plants. Tannins are complex polyphenolic compounds that are widely distributed in most species of plants. The unit structure of one of the most common types of tannin, tannic acid, is shown in Figure 8.16. The tannic acid molecules are composed of from 2 to 12 structural units and the tannins derived from them range in molecular weight from about 3000 to 20 000 Daltons (Da) depending on their source.

Tannins are resistant to bacterial degradation due to their aromatic structure and phenolic character. They are highly colored due to their extended conjugation, ranging from light orange to dark brown. The large number of phenol groups makes them good complexing agents for metals and for esterification reactions with other organic compounds. Tannins are highly water soluble and can leach out of plants into soil water or nearby streams. From there they can be transported into ground water, lakes, and oceans. Tannins are found in the highest concentrations in bogs, swamps, small streams, and slow-moving rivers. These waters with the highest tannin concentrations tend to be very dark brown in color, as shown in Figure 8.17, and are commonly called blackwater rivers.

The other major contributor to the DOC in fresh and ocean waters comes from decaying organic matter in soils. Water percolating through the soils leaches out complex soluble organic compounds as well as organic colloidal materials. This generic organic material is called humic substances. Humic substances come from the chemical and bacterial oxidation of plant and animal detritus in soils and sediments. They have a wide range of molecular weights, composition, and chemical structure, depending on the origin and age of the materials. They have a strong absorption below 400 nm, giving them an intense yellow to orange color similar to that of tannic acid. Historically, the soil organics that could not be classified in any other group of organic compounds, such as polysaccharides, proteins, lipids, or carbohydrates, were classified as humic substances.

Early studies of soil organics separated humic substances into three classes depending on their solubility in aqueous solutions, with different pHs. The organics were extracted from soil using a NaOH solution (pH = 12), which converted the organic acids to their soluble sodium salts. Some soil organics did not dissolve in this highly alkaline solution because they did not

Figure 8.17 The Upper Tahquamenon River Falls in Michigan, which is colored by tannins leaching into the cedar swamps that drain into the river. Source: Bhasker Garudadri, Wikimedia Commons.

contain a large amount of acid groups. These insoluble soil organics were called humin. The colored alkaline solution was then acidified to pH 2, which converted the acid salts back to the acid form, causing a portion of the organics to precipitate. The organics that formed a precipitate at pH 2 were called humic acids and the colored organics that remained in solution at pH 2 were called fulvic acids. The differences in solubilities of the humic and fulvic acids at pH 2 are likely due to differences in the molecular weight, polarity, and aromatic content of the two groups. Humic acids are generally much larger than fulvic acids, with a higher aromatic content, as shown in Figure 8.18. Both humic and fulvic acids were found to contain multiple functional groups, including carboxylic acids, carbonyls, phenols, amines, as well as ether and ester linkages.

Humic acids are typically colloidal in fresh and saltwater systems, while most fulvic acids are completely solvated. This is due to the fulvic acids being a smaller, less aromatic species. Both humic and fulvic acids are large polyelectrolytes, polymers with structural units that contain electrolyte groups. The electrolyte groups in humic and fulvic acids are the carboxylic acid groups, which can dissociate in aqueous systems. The multiple carboxylate and phenolic groups allow them to complex and transport inorganic species in aqueous systems, including toxic metals and radionuclides. Humic and fulvic acids also act as surfactants, since they have both polar and nonpolar character. As colloidal surfactants, humic acids have been shown to associate with nonpolar pesticides, assisting in their aqueous transport. In the absence of the humic acids the nonpolar pesticides would not be highly water soluble, and would not be transported long distances in aqueous systems.

Humic acids isolated using hollow-fiber ultrafiltration methods, discussed in Section 9.3, are found to range in size from about 10 000 to 300 000 Da, while fulvic acids are found to range from 250 to 3000 Da. The smaller fulvic acids are not as highly colored as the larger humic acids due to their lower aromatic content and less extensive conjugation (Gaffney et al., 1996). For example, a sample of water from Volo Bog, located in northern Illinois, was concentrated into molecular size fractions of 0.1 μm–100 000 Da, 100 000–30 000 Da, 30 000–3000 Da, and 3000–500 Da by hollow-fiber ultrafiltration. The colors of the first two molecular size fractions (0.1 μm–30 000 Da), shown in Figure 8.19, are deep yellow to orange, with the largest size range (0.1 μm–100 000) being the darkest. These molecular size ranges are classified as humic acids. The smallest size fraction (3000–500 Da) contains the fulvic acid fraction, which is very light

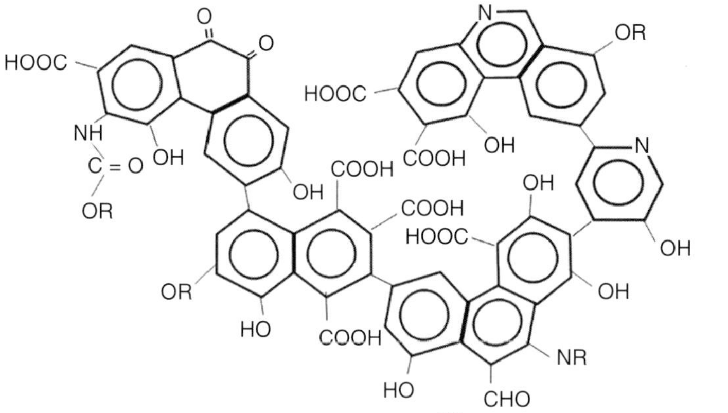

(A)

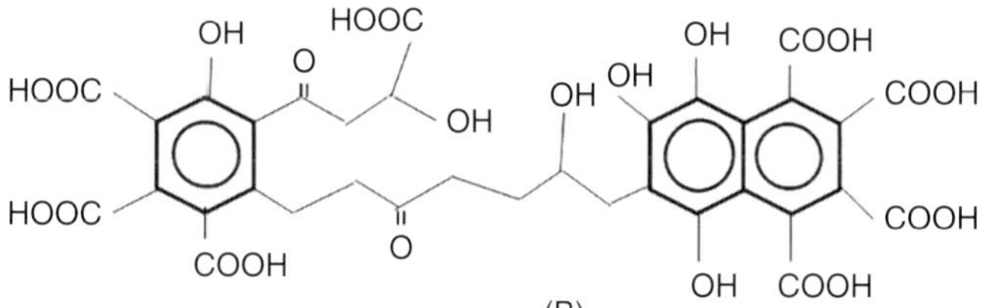

(B)

Figure 8.18 Generalized structures for (A) humic acid and (B) fulvic acid.

Figure 8.19 Size-fractionated concentrates of water obtained from Volo Bog in northern Illinois: (A) 0.1 μm to 100 000 Da; (B) 100 000 to 30 000 Da; (C) 30 000 to 3000 Da; (D) 3000 to 500 Da. Source: Photo by N. Marley and J. Gaffney.

colored. The third size fraction (30 000–3000 Da), which is the darkest in color, contains predominantly the highly colored tannins. Volo Bog has no surface inlet or outlet, with the only recharge coming from precipitation. It is thus not surprising that 60% of the DOC in the bog water was due to these highly colored tannins in the 30 000–3000 Da size range. Most bog water typically has a large DOC input from dissolved peat tannins. The remaining DOC in the Volo Bog water was 25% humic acids and 15% fulvic acids.

Because the aqueous humic substances are complex mixtures that include humic acids, fulvic acids, and tannins, their concentrations are typically determined by combustion methods, discussed in Section 9.2. They are reported as total DOC in units of ppm C. The concentrations in freshwater systems range from 0.1 ppm C in clear freshwater lakes to 50 ppm C in blackwater rivers, swamps, and bogs. Groundwater systems range from 0.1 to 10 ppm C. The DOC concentrations in ground water depend on the type of overlying soil and sediments. High organic soils above the aquifer will result in higher DOC percolating into the ground water. Humic substances in open ocean waters range from 0.5 to 1.2 ppm C.

Since the humic and fulvic acids are fully oxidized compounds produced from the bacterial degradation of the humus materials, they do not contain any significant nutrient content. Because of this, they are fairly resistant to further bacterial processing. The only chemical reactions that can produce further oxidation of humic and fulvic acids are OH radical abstraction or addition reactions. These reactions can occur in surface waters from photochemical processes that produce OH radical. The OH radical reactions are fairly slow, due to the significant aromatic character and high oxygen content of the humic and fulvic acids. However, their bacterial degradation is even slower than the OH radical reactions. Thus, the humic and fulvic acids have long lifetimes in aqueous systems.

Other carbon species that occur naturally in aqueous systems are PIC and POC, which are interconnected with DIC and DOC in a complex carbon cycle involving inorganic and biochemical processes. DIC can be converted to PIC by precipitation as $CaCO_3$. The $CaCO_3$ precipitates can form spontaneously during changes in the water temperature, known as "whiting events." $CaCO_3$ can also be precipitated by certain types of marine bacteria. Increased marine fish can also excrete $CaCO_3$ particulates in order to regulate osmotic pressure, a process called osmoregulation. POC is composed of living or dead organisms, their fecal matter, and detritus. DIC can be converted to POC through photosynthesis and it can be converted back to DIC through ingestion followed by respiration in aquatic organisms. DOC can be converted into POC through ingestion followed by excretion. It can also be converted into POC by the death and decomposition of a secondary or tertiary consumer. POC can be converted to DOC through disaggregation or dissolution of the molecules within the particulates. However, POC is generally converted to DIC through ingestion by secondary consumers, followed by the production of DIC through excretion and respiration.

8.6 Water Pollution

Water is one of the most effective solvents for many natural and pollutant chemical species. It is especially good at dissolving salts, such as nitrates, sulfates, and phosphates, which are used in agricultural fertilizers and are produced during wastewater treatment. Water is also especially good at the dissolution of polar organic compounds, such as organic pesticides, allowing them to be incorporated into the food chain where they can be bioaccumulated in aquatic consumers. However, nonpolar organics are not highly soluble in water. Many of the nonpolar organics, such as those in found in petroleum, are less dense than water and so tend to float on the water surface, spreading out to create a thin barrier between the water and the atmosphere. This presence of nonpolar organics on the water surface affects the equilibrium of the dissolved gases with the atmosphere, interfering with the water chemistry.

Pollution that originates from a single source, such as industrial or domestic wastewater discharge, chemical and oil spills, large livestock farms, and coal fly ash storage facilities, is called "point source pollution." Pollution that originates from diffuse sources, such as agricultural runoff, stormwater runoff, or atmospheric deposition and windblown debris, is called "nonpoint source pollution." Nonpoint source pollution most often originates from more than one location, as with stormwater runoff. It is difficult to regulate because, due to its diffuse nature, the exact location of the major sources of the pollution is not easy to identify. Because of this, nonpoint source pollution remains the leading cause of water pollution in the United States (NRDC, 2018). Pollution that originates in one country or state and is transported to another country or state is called transboundary pollution. This type of pollution is most difficult to control, due to problems that occur when dealing with different governmental entities.

8.6.1 Point Sources

Both domestic (household) waste and industrial waste is pumped to a waste treatment plant where it is treated to remove the pollutant species. The goal of wastewater treatment is to produce a water discharge, called an effluent, which will be as free of harmful pollutants as possible. This includes the removal of soluble and particulate organic substances, nitrates, phosphates, sulfates, and heavy metals. Complete treatment of the waste water involves three stages: primary, secondary, and tertiary. Primary treatment focuses on the removal of heavy solids by sedimentation and the removal of floating material by skimming. Secondary treatment removes dissolved and suspended organic materials by natural bacterial decomposition. Tertiary treatment can involve several different processes aimed at the final removal of the remaining nitrogen and phosphorus, along with heavy metals and bacterial-resistant organic compounds. Because of the cost, tertiary treatment is commonly used only in cases where the effluent is to be discharged into more fragile ecosystems. This means that most wastewater effluents still contain elevated levels of the nutrients NO_3^- and PO_4^{3-}, as well as some resistant organic compounds such as pharmaceuticals.

The final stage in wastewater treatment is disinfection to kill pathogens prior to discharge into surface waters. Methods of disinfection include ozonation, chlorination, or photolysis (UV). Chlorination is the most common method of wastewater disinfection in North America, due to its low cost and long-term history of use. Chlorination of waste water can be achieved by the direct addition of Cl_2 or by addition of sodium hypochlorite ($NaClO_2$) to the effluent prior to discharge. The addition of chloramines is restricted to use in drinking water only. One disadvantage of chlorination is that chlorine can react with organic compounds in the waste water to produce chlorinated organics, which can be harmful to aquatic ecosystems and are suspected carcinogens.

The most common chlorination biproducts (CBPs) are chloroform ($CHCl_3$), trichloroacetic acid (CCl_3COOH), and dichloroacetic acid ($CHCl_2COOH$). N-nitrosodimethyamine or NDMA ($C_2H_6N_2O$) can also be formed in waste water with a high nitrate content. Most waste water also contains bromide from domestic waste, road salt, and vehicle emissions. In the presence of this bromide, the chlorination of waste water also produces bromoform ($CHBr_3$), dihalomethanes, dihalogenated acetic acids, and the trihalomethanes (THMs); $CHCl_3$, $CHBr_3$, $CHBrCl_2$, $CHClBr_2$, and $CHCl_2Br$. The addition of chlorine can also transform the side chains on natural humic and fulvic acids into small chlorinated hydrocarbons, such as chloromethane (CH_3Cl), methylene dichloride (CH_2Cl_2), and vinyl chloride ($CH_2{=}CHCl$) along with the THMs. The U.S. Environmental Protection Agency (EPA) has set drinking water standards for the THMs at $80\,\mu g\,l^{-1}$ and the total haloacetic acids at $60\,\mu g\,l^{-1}$. Both classes of halogenated organics are suspected carcinogens at higher levels.

Many of the pollutants in the waste become concentrated in the solids, which are removed in the primary treatment process. The wet solid waste, called sludge, is treated separately from the water. The goal of primary sludge treatment is the reduction of solid material. Primary treatment processes include aerobic digestion, anaerobic digestion, and composting. Secondary sludge treatment is aimed at water removal and can involve centrifugation, filtration, evaporation, or drying. After treatment, sludge disposal methods include land application, landfill disposal, and incineration. Incineration is usually used only in special cases because of the cost of the fuel and the required treatment of the combustion gases to prevent air pollution. Both land application and landfill disposal offer opportunities for the concentrated pollutants to enter aqueous systems through runoff or soil percolation. Although elevated levels of NO_3^- and PO_4^{3-} in surface waters are commonly linked to domestic waste effluent and fertilizer runoff, they can also come from runoff from sludge land disposal.

Many heavy metals and other toxic elements find their way into freshwater systems from mining or industrial waste sludge. Some of these include Cd, Cr, Hg, Pb, Sb, As, Sr, U, and Th. These elements are used in a number of industrial applications, including metal plating, paints, batteries, light bulbs, electric switches, and electronics. During and after the production processes, a portion of these elements end up in the industrial waste water. Large industrial production facilities may have their own wastewater treatment plants, which can utilize special tertiary treatment to remove heavy metals and other pollutants specific to the industrial processes. However, it is not uncommon for industrial waste to be routed to the nearest domestic waste treatment plant. In these facilities, the heavy metals typically become concentrated in the sludge. The sludge treatment processes, which include dewatering and drying, produce humic materials from the organics. These humic materials can complex the heavy metals when exposed to rain water, causing them to be transported into surface and groundwater systems. Some of the more important elements of environmental concern found in industrial and domestic waste sludge are listed in Table 8.4, along with the maximum concentrations allowed in drinking water by the EPA. Many of these metals play important roles in biological systems at trace levels and so are considered to be essential elements. However, at higher levels they can be toxic to plants and animals.

Table 8.4 Elements commonly found in industrial and domestic sludge with their allowed concentration limits in drinking water.

Elements	Sources	Concentration limit
Sb^{3+}, Sb^{5+}	petroleum refinery, electronics, fire retardants	6 ppb
As^{3-}, As^{3+}, As^{5-}	coal fly ash, smelters, pesticides, electronics, glass	10 ppb
Ba^{2+}	drilling, fracking fluids, metal refineries	2 ppm
Be^{2+}, Be^{0}	coal fly ash, smelters, aerospace, defense	4 ppb
Cd^{1+}, Cd^{2+}	galvanized pipe, battery, paint	5 ppb
Cr^{3+}, Cr^{6+}	steel mills, paper mills	100 ppb
Cu^{1+}, Cu^{2+}	copper pipes used in plumbing	1.3 ppm
Pb^{2+}, Pb^{4+}	batteries	15 ppb
Hg^{0}, Hg^{1+}, Hg^{2+}	electronics, petroleum refinery, gold mining	2 ppb
Se^{2-}, Se^{2+}, Se^{4+}, Se^{6+}	mining, petroleum refinery	50 ppb
Tl^{1+}, Tl^{3+}	mining, glass, pharmaceuticals, electronics	2 ppb

Some organic pollutants that are not removed by conventional wastewater treatment processes include surfactants, prescription and nonprescription drugs, ingredients in personal care products, organophosphate flame retardants, and plasticizers. These compounds act as endocrine disruptors (EDCs). They can alter the normal functions of hormones, changing natural hormone levels and leading to reproductive effects in aquatic organisms. Although the concentrations of these substances in wastewater effluents are quite low (ppb to ppt), long-term exposure can cause harm to aquatic organisms. Common surfactants are long-chain fatty acids or sulfonates that are used as soaps and high-strength industrial cleaners. Of particular interest are the nonylphenol ethoxylates (NEPs) and quaternary ammonium compounds (Quat). As cleaners, these surfactants are designed to have adverse effects on bacteria, so the long-chain carbon backbones in surfactants are very difficult to decompose during secondary treatment and are commonly released intact in the wastewater effluent. Many of the pharmaceuticals and ingredients in personnel care products (PPCPs) can also inhibit bacterial decomposition and so are resistant to removal by conventional wastewater treatment processes.

Some industrial production facilities, including oil refineries, pulp and paper mills, fertilizer manufacturers, and chemical, electronics, and automobile manufacturers, can often discharge one or more pollutants in their effluents directly into a body of water with little or no treatment. The impact on the water system depends on the type of pollutant and its concentration levels. The CN^- ion is contained in wastewater effluents from steel mills, fertilizer and plastic factories, metal plating plants, and mining activities. The concentration limit for CN^- in drinking water in the United States is limited to 200 ppb. Although fluoride is commonly added to drinking water to improve dental health, fluoride levels well above the allowed 4 ppm can be released into surface waters in the industrial waste waters from aluminum and fertilizer production plants. The chronic ingestion of high levels of fluoride has been linked to the development of dental fluorosis in humans and animals, a disorder characterized by hypomineralization of tooth enamel. In humans, it occurs most commonly in rural areas where contaminated drinking water is obtained from shallow wells.

Large farms that raise livestock, such as cows, pigs, and chickens, are also sources of point source pollution. These farms, known as concentrated animal feeding operations (CAFOs), can house anywhere from hundreds to millions of animals, most often dairy cows, hogs, or chickens. A small CAFO can typically produce urine and feces equivalent to that produced by 16 000 humans (Sierra Club, 2018). They typically do not treat the animal waste, allowing it to run off into nearby streams or rivers as raw sewage. In addition to the contamination of surface waters with excessive nutrients, microbial pathogens, and pharmaceuticals present in the waste, the CAFOs emit 168 gases to the atmosphere, including NH_3, H_2S, and CH_4. In addition, airborne particulates can carry disease, causing bacteria, fungus, or other pathogens.

Other common sources of point source pollution are chemical and oil spills. Oil spills, such as the Deepwater Horizon and the Exxon Valdez accidents and their impacts on water quality are discussed in detail in Section 10.5.1. Oil tankers are just one of the many sources of oil spills. According to the U.S. Coast Guard, only 13.4% of the oil spilled in the United States between 1995 and 2004 was from oil tankers, 26.2% was from tanker barges, 21.5% from other kinds of vessels, and 24.6% from petroleum factories and land containers (USDOT, n.d.). Oil spills at sea are generally much more ecologically damaging than those on land, since they can spread for hundreds of miles in a thin surface oil slick. In addition, land-based oil spills are more readily containable by building a temporary dam around the spill site.

Another energy-related point source of water pollution is the accidental release of coal fly ash slurry from storage ponds, as well as the routine release of pond effluent. Since coal contains trace levels of elements such as As, Be, B, Cd, Cr (including Cr^{+6}), Co, Pb, Mn, Hg, Mo, Se, Sr, Th, and V, fly ash obtained after combustion of the coal contains enhanced concentrations of these

Figure 8.20 A 25-ft wall of fly ash slurry located approximately 1 mile from the original impoundment at the Kingston Fossil Plant. Source: Brian Stansberry, Wikimedia Commons.

elements. In addition, fly ash contains low concentrations of dioxins and polycyclic aromatic hydrocarbons (PAHs), along with around 10–30 ppm uranium. Coal fly ash is disposed in over 700 large ponds, called surface impoundments, located on-site at coal-fired power plants. These ponds can discharge millions of gallons of contaminated waste water into rivers and streams, which can cause huge environmental damage. The waste sites are known to have contaminated ground water, wetlands, creeks, or rivers. The largest discharge of fly ash slurry in U.S. history occurred at the Kingston Fossil Plant, a 1.7 GW coal-fired power plant near Kingston, TN. On Monday December 22, 2008, an ash impoundment dike ruptured, releasing 1.1 billion gallons of coal fly ash slurry. The volume of fly ash slurry released was about 101 times larger than the volume of oil released in the Exxon Valdez oil spill. The fly ash slurry, shown in Figure 8.20, traveled across the Emory River to the opposite bank, covering up to 300 acres of land, damaging homes, and flowing into tributaries of the Tennessee River. River water near the spill showed elevated levels of Pb and Th, with detectable levels of Hg and As. In January 2009, about 1 year later, significantly elevated levels of toxic metals, including As, Cu, Ba, Cd, Cr, Pb, Hg, Ni, and Th, were found in water samples of the Tennessee River. There are 194 known cases of fly ash water contamination and 27 known cases of fly ash spills in the United States as of 2014.

Most pollution from fracking operations, as discussed in Section 10.5.1, is considered to be a nonpoint source due to the number of wells typically present in a fracking field and the transport of the fluids within a groundwater aquifer. However, the discharge of the flowback fluids and produced waters into rivers, lakes, and streams from a discharge facility is classified as a point source. It is estimated that an average fracking field produces about 20 million gallons of waste water per day. The discharge of treated or untreated waters from fracking operations can cause lasting environmental damage to surface water systems. These waste waters contain very high levels of salt and organic chemicals. They also contain toxic metals such as Hg and radioactive species such as Ra, U, and Th, which are released and concentrated from the subsurface source rock during the fracking process. These species have been found in surface waters and sediments tens of kilometers away from the legal discharge point of treated fracking waste water, and they persist in the sediments 5–10 years after discharge. The storage, treatment, and disposal of fracking waste water remains one of the more significant environmental concerns of the natural gas fracking process.

8.6.2 Nonpoint Sources

The link between the atmosphere and water systems is important to the introduction of pollutant species into the hydrosphere as a nonpoint source. Precipitation is an important method of pollutant transport from the atmosphere into surface water systems, particularly for highly soluble species such as HNO_3 and H_2SO_4. These are formed from the oxidation of NO_x and SO_2 in air and cloud water, as discussed in Section 6.2. The presence of these strong acids in rain water results in acid rain, which lowers the pH of the rain water below the natural CO_2 equilibrium value of 5.2. Acid rain affects aquatic systems more than most other ecological systems, by lowering the pH of rivers, lakes, and wetlands. Lakes and streams become more acidic when their soils and sediments do not contain enough calcite ($CaCO_3$) or dolomite ($CaMg(CO_3)$) minerals to neutralize the acid deposition. For example, the surface waters in the Adirondacks and Catskills in New York State have soils with low carbonate mineral content and are thus more susceptible to the ecological effects of acid rain.

Lower pH levels in surface waters can increase the solubility of heavy metals. As the level of hydrogen ions increases, metal cations such as Al, Pb, Cu, and Cd are released into the water instead of being absorbed into the sediment. For example, the percolation of acid rain through aluminum-containing soils causes the dissolution of Al^{3+}. Under normal pH conditions, the aluminum hydroxide ($Al(OH)_3$) present in soils is insoluble. However, at lower pH values the $Al(OH)_3$ dissolves by the following reaction:

$$Al(OH)_3(s) + 3H^+(aq) \rightarrow Al^{3+}(aq) + 3H_2O(l)$$

The result is increased levels of Al^{3+} ions in ground water and nearby surface waters. As the concentrations of heavy metals in the aqueous systems increase, their toxicity also increases. The Al^{3+} ion is toxic to plants and at high concentrations dissolved Al^{3+} can limit growth and reproduction, as well as increasing mortality rates of aquatic organisms. It is also toxic to crayfish, clams, fish, and other aquatic animals.

Most nonpoint source pollution occurs as a result of runoff and the largest source of runoff pollution is stormwater runoff. It is the number one cause of surface water pollution in urban areas. Stormwater runoff is caused by rainfall that flows over impermeable surfaces that do not allow the rain water to soak into the ground. These surfaces include roads, driveways, parking lots, rooftops, and other paved or concrete areas. All the rain falling on these surfaces eventually flows into storm drains, carrying with it any contaminants present on the surfaces, such as organic sediments, bacteria, oil and grease, trash, pesticides and fertilizers, and soluble metals. The goal of urban storm drain systems is to ensure that the storm water is efficiently collected through pipe networks to prevent flooding. The collected water is discharged into nearby streams, rivers, or other bodies of water. The runoff of large stormwater flows can erode the banks of waterways and cause flooding in areas downstream of the discharge point. In some areas, polluted runoff from roads and highways is the largest source of water pollution. For example, about 75% of the toxic chemicals in Puget Sound are carried in stormwater runoff from paved roads, parking lots, and driveways (Dicks, 2010). This high concentration of runoff pollution in Puget Sound is killing the local Coho Salmon before they can spawn. The death rates are so high that some populations of wild Coho are at risk of local extinction.

Agriculture is the second largest source of runoff pollution. It is the biggest consumer of freshwater resources worldwide, and is a leading cause of water pollution due to runoff of contaminated irrigation water. In the United States it is the main source of pollution of rivers and streams, the second biggest source of pollution in wetlands, and a major source of groundwater pollution (NRDC, 2018). The overuse of inorganic fertilizers in agriculture leads to runoff of NO_3^-, SO_4^{2-}, and PO_4^{3-} into rivers and streams. The main crop nutrients supplied in fertilizers

Figure 8.21 Eutrophication of the Potomac River caused by elevated nutrients, which created a dense bloom of cyanobacteria. Source: Alexandr Trubetskoy, Wikimedia Commons.

are NO_3^-, PO_4^{3-}, and K^+. Nitrate is supplied as NH_4NO_3, $NaNO_3$, KNO_3, or $Ca(NO_3)_2$. Phosphate is usually supplied in the form of $Ca(H_2PO_4)_2$, commonly known as superphosphate. Potassium is supplied as potash, a mixture of potassium minerals made up of KCl, K_2SO_4, K_2CO_3, or KNO_3. Although SO_4^{2-} is not a necessary nutrient, it is added as $(NH_4)_2SO_4$ to adjust the pH of alkaline soils. It is also found as a byproduct in the manufacture of the fertilizer additive $Ca(H_2PO_4)_2$, as $CaSO_4$.

While the increases in concentrations of PO_4^{3-} and NO_3^- are not toxic to aquatic organisms, they can lead to the overgrowth of algae, which is called an algal bloom. Algal blooms can cause eutrophication of lakes and slow-moving streams, as shown in Figure 8.21. Eutrophication is a state of excessive richness of nutrients in a lake or other body of water, which causes a dense growth of plant life. The dense algae growth in an algal bloom prevents the sunlight from penetrating into the water, so the plants beneath the algal bloom die due to lack of sunlight for photosynthesis. Eventually, the algae also die and sink to the bottom of the water column. Bacteria then begin to decompose the dead biomass, using up oxygen in the process. This causes a lowering of the DO levels and the creation of a state of hypoxia. As discussed in Section 8.5.2, PO_4^{3-} is the limiting nutrient species in aqueous systems. It is also the limiting factor for eutrophication, so any excess availability of PO_4^{3-} promotes excessive plant growth and decay, while favoring the growth of simple algae and phytoplankton over more complicated plants.

Other pollutants from agricultural runoff are herbicides and pesticides. The potential environmental and ecological impacts of pesticides became evident during the widespread use of dichlorodiphenyltrichloroethane (DDT), beginning in 1945. DDT was originally synthesized in 1874 and soon after the Swiss chemist Paul Hermann Müller discovered that it was a potent insecticide. It was used widely during WWII to control mosquitos, which carried malaria and typhus, and Müller was awarded the Nobel Prize in Medicine in 1948 for this discovery. Like many of the insecticides, DDT was easily bioaccumulated throughout the food chain after entering the water system, causing unintentional impacts on ecosystems. The growing evidence that DDT was a potential carcinogen, as well as the cause of some declining bird populations, was chronicled in 1962 in the book entitled *Silent Spring*, authored by Rachel Carson. The

bioaccumulation of DDT in aquatic birds caused their egg shells to be underdeveloped, which threatened the loss of many avian species, especially raptors. Further studies led to the banning of DDT internationally, except in exceptional cases where insect disease carriers could not be controlled by any other means. As new herbicides and pesticides are developed, they now require more detailed research into their potential for environmental impact on aquatic and terrestrial ecosystems before they are approved for use.

Pesticides and herbicides commonly used in agriculture and detected in agricultural runoff are listed in Table 8.5, along with their allowable concentration limits for drinking water in the United States (USEPA, 2018). The chemical structures of these compounds can be found in Appendix IV. Although the insecticide Toxaphene was banned for use by the EPA in 1990, and the herbicide 2,4,5-TP was banned in 1985, they are listed in Table 8.5 because they still persist in surface and ground waters. Many pesticides and herbicides used commonly in agriculture are resistant to bacterial degradation. Their chemical structures commonly contain multiple aromatic rings, as well as organic chlorine or bromine bonds, which are difficult to break. This leads to their stability in aqueous systems. Pesticides have been detected in surface waters in all regions of the United States. However, herbicides have been detected more frequently than pesticides, consistent with their greater use. The most frequently detected herbicides in surface waters include atrazine, alachlor, and 2,4-D. These compounds are among the highest used in agriculture currently. Because of their resistance to biological and abiotic degradation,

Table 8.5 Pesticides and herbicides commonly used in agriculture and present in agricultural runoff along with their allowed concentration limits in drinking water in units of ppb.

Compound	Chemical name	Use	Limit
Alachlor	2-chloro-*N*-(2,6-diethylphenyl)-*N*-(methoxymethyl)acetamide	herbicide	2
Atrazine	1-chloro-3-ethylamino-5-isopropylamino-2,4,6-triazine	herbicide	3
Carbofuran	2,2-dimethyl-2,3-dihydro-1-benzofuran-7-yl methylcarbamate	pesticide (rice, alfalfa)	40
2,4-D	2,4-dichlorophenoxy acetic acid	herbicide	70
Dinoseb	2-(*sec*-butyl)-4,6-dinitrophenol	herbicide	7
Diquat	1,1′-ethylene-2,2′-bipyridyldiylium dibromide	herbicide	20
Endothall	3,6-endoxohexahydrophthalic acid	herbicide	100
Glyphosate	*N*-(phosphonomethyl)glycine	herbicide	700
Lindane	hexachlorocyclohexane	insecticide	0.2
Methoxychlor	p,p′-dimethoxydiphenyltrichloroethane	insecticide	40
Oxamyl	*N*,*N*-dimethyl-2-methyl-carbamoyloximino-2-(dimethylthio) acetamide	insecticide (orchards)	200
Pichloram	4-amino-3,5,6-trichloro-2-pyridinecarboxylic acid	herbicide	4
Simazine	6-chloro-*N*,*N*′-diethyl-1,3,5-triazine-2,4-diamine	herbicide	4
Toxaphene	1,4,5,6,7,7-hexachloro-2,2-bis(chloromethyl)-3-methylidenebicyclo[2.2.1]heptane	insecticide, banned 1990	3
2,4,5-TP (Fenoprop)	2-(2,4,5-trichlorophenoxy)propanoic acid	herbicide, banned 1985	50

pesticides and herbicides have fairly long lifetimes in aqueous systems. Pesticides have been found to persist for decades in ground water. The long lifetimes of these compounds lead to long-term exposures, which amplifies their effects on aquatic organisms.

8.7 Contaminant Transformation

The chemical transformation of water pollutant species can occur by photolysis, hydrolysis, ion exchange, oxidation, reduction, complexation, and microbial transformation. Many of these changes are important in determining the aqueous solubility and volatility of the pollutant species, and can thus control their transport in environmental systems. Inorganic species that are released into the aqueous environment by atmospheric deposition or wastewater discharge enter the system in the oxidizing environment of the water surface, and so the primary route of transformation is oxidation. Water-soluble oxidants such as H_2O_2, O_3, and OH found in cloud and surface waters can oxidize both organic and inorganic pollutant species, leading to increased water solubility. This increase in solubility also makes them more susceptible to aqueous microbial transformation. Deeper in the water column, natural bacteria can act to either oxidize or reduce inorganic species. In some cases, these microbial processes can produce toxic compounds, which can be bioaccumulated and biomagnified in the food chain. For example, the oxidation of the volatile Hg^0 by H_2O_2 in cloud water forms the highly water-soluble Hg^{2+}, which is then transformed into the highly toxic methyl mercury ($HgCH_3^+$) by microorganisms in oceans and surface waters.

Mercury is released into the atmosphere naturally as Hg^0 from volcanos. It is also released from mining operations, especially in artisanal gold mining, where mercury is used to extract gold from the ore by forming an amalgam. The Hg–Au amalgam is then heated to remove the Hg^0, which is vented to the atmosphere. This type of gold mining accounts for 37% of the Hg^0 released into the atmosphere globally (Gaffney and Marley, 2014). Mercury is also released to the atmosphere during coal combustion. It forms a fairly stable sulfide mineral known as cinnabar (HgS), which is present in coal. When the coal is combusted, the mercury is released primarily as Hg^0 or Hg^{2+} in the gas phase. Coal combustion accounts for 33% of the mercury released to the atmosphere. The Hg^0 is not highly water soluble and can be transported in the atmosphere for up to a year. However, it can react with H_2O_2 in cloud water to form the oxidized form Hg^{2+}. The Hg^{2+} is much more water soluble, with an atmospheric residence time of less than two weeks. Therefore, the Hg^{2+} can be rapidly taken up in rain water or snow and deposited to surface waters by wet deposition.

Once in surface waters, the Hg^{2+} enters a complex cycle shown in Figure 8.22. The Hg^{2+} can be converted to $HgCH_3^+$ by methanogenic and sulfate-dependent bacteria under low oxygen conditions. It can also be reduced to Hg^0 and revolatilized back into the atmosphere. The $HgCH_3^+$ is ingested by aquatic plants and animals and is biomagnified through the aquatic food chain, with the highest concentrations found in the top predators. The $HgCH_3^+$ is highly lipid soluble, with about 95% absorbed in the gastrointestinal tract of the organism. It reacts with sulfhydryl groups, interfering with cellular structure, enzyme function, and protein synthesis. It is slowly broken down to Hg^{2+} by demethylation in the intestines, which results in the elimination of mercury as Hg^{2+}. The half-life of $HgCH_3^+$ in the body is reported to be 70–80 days. Over time, the Hg^{2+} can recombine with sulfur and become deposited in the sediments as HgS, which can also be released slowly by resuspension. The DOC and pH levels have a strong effect on the ultimate fate of mercury in an ecosystem. The Hg^{2+} is known to bind strongly to natural humic materials and lower pH and higher DOC levels can facilitate the complexation, enhancing the mobility of Hg^{2+} in aqueous systems.

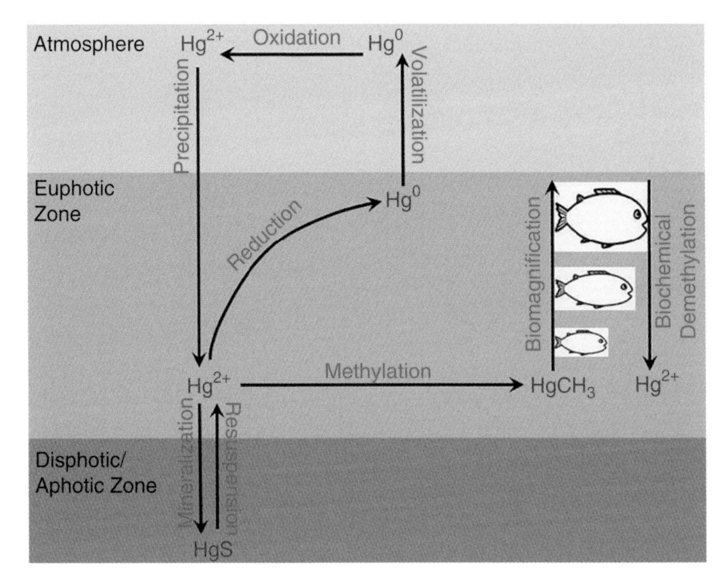

Figure 8.22 The mercury cycle in aqueous systems.

The direct photolysis of aqueous pollutant species requires that they absorb light strongly in the UVA or UVB, where there is sufficient energy to break chemical bonds. Aqueous photolytic transformation processes occur in the atmosphere in cloud water and aerosols, on exposed surfaces, and in shallow surface waters where the intensity of the UV light is highest. The importance of photolysis for the transformation of aqueous contaminants is limited for most of the inorganic pollutants, because of their lack of strong absorption bands in the UVA to UVB spectral range. The exception to this is the nitrate ion. NO_3^- can absorb radiation in the UVA and UVB to produce the following reactions in surface waters:

$$NO_3^-(aq) + h\nu \rightarrow NO_2^-(aq) + O(aq)$$
$$NO_3^-(aq) + h\nu \rightarrow NO_2(g) + O^-(aq)$$

The second reaction produces insoluble NO_2, which is released into the atmosphere.

Aqueous photolysis is more important for organic contaminants than for inorganic contaminants. An example of the direct photolysis of a more resistant organic pollutant is the photolytic transformation of the highly chlorinated PCBs, shown in Figure 8.23. The chemical formula of PCBs is $C_{12}H_{10-n}Cl_n$, where n ranges from 1 to 10. Only the highly chlorinated PCBs ($n \geq 4$) can absorb radiation in the UVB, which has sufficient energy to break the Ar–Cl bond. This photolysis leads to the formation of a biphenyl with one less chlorine atom in the molecule. It can also result in phenolic derivatives, which have higher water solubilities. The transformation of the highly chlorinated PCBs to those with lower chlorine content allows for their microbial degradation. The PCBs with lower chlorine content are biodegraded faster than those with high chlorine content. However, the rate of degradation depends on the positions of the chlorine atoms in the aromatic ring. This is because microbial transformation occurs by either chlorine elimination or aromatic oxidations, and the rates of both these mechanisms depend on the stability of the substituted aromatic ring. The degradation is accelerated if the chlorine atoms are in meta or para positions due to the resonance destabilization over the ortho-substituted species.

Photolytic reactions in cloud or surface waters can lead to the production of radicals, which are more effective in the decomposition of organic pollutants than is direct photolysis. Many of these radical decomposition reactions involve the OH radical, which is as important to the

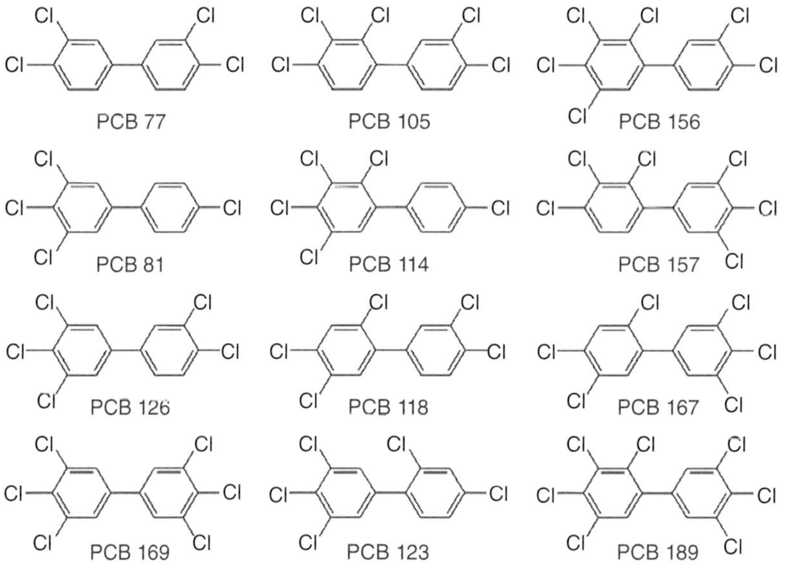

Figure 8.23 Chemical structures of highly chlorinated polychlorinated biphenyls (PCBs) found in surface waters. Source: Leyo, Wikimedia Commons.

oxidation of pollutant species in aqueous systems as it is in the atmosphere. However, in many cases, abstraction or addition reactions with OH radical also results in long-lived products. This is because the lower oxygen content in water systems compared to that of the atmosphere limits further oxidative degradation reactions. The PCBs were designed to be resistant to combustion for safety reasons. PCBs are used as insulating fluids in capacitors where very low flammability is required. However, they are also resistant to other types of oxidation processes due to the combination of stable aromatic rings with electronegative chlorine functional groups. Thus, PCBs have no removal mechanisms (with the exception of photolysis of the higher chlorine-containing species followed by microbial degradation).

The PCBs are listed as persistent organic pollutants (POPs), which are a group of hazardous organic chemical compounds that are resistant to biodegradation and thus remain in the environment for a long time. Since the POPs have long lifetimes, they can bioaccumulate, with potential adverse effects on human health and the environment. The long lifetimes also lead to long-term exposures, which amplifies their effects on aquatic organisms. The POPs were identified at a United Nations-sponsored meeting held in Stockholm. The Stockholm Convention Treaty was adopted in 2001 and entered into force in 2004. The Stockholm Convention originally identified 12 compounds and, since that time, 8 more POPs have been added to the list. In addition to the PCBs, the POPs include 9 insecticides (Aldrin, Chlordane, Dieldrin, Endrin, Heptachlor, Mirex, Toxaphene, DDT, and Endosulfan), 3 pesticides (Kepone, Lindane, and pentachlorobenzene), 1 fungicide (hexachlorobenzene), 4 industrial chemicals (tetrabromodiphenyl ether, pentabromodiphenyl ether, perfluorooctanesulfonic acid, hexabromocyclodecane), and 2 combustion products (dioxins and polychlorinated dibenzofurans). The POP compounds and their environmental effects are listed in Table 8.6. Their chemical formulas and structures can be found in Appendix D.

Most all of the POPs contain chlorine or bromine in cyclic alkane or aromatic structures. These structures are difficult to degrade chemically, biochemically, or photolytically, which gives them their high chemical and biochemical resistance and thus higher toxicities. Many of the compounds are nonpolar and so have low water solubilities and high fat solubilities. This

Table 8.6 The POPs, including the first identified by the United Nations Stockholm Convention on Persistent Organic Pollutants (first 12 entries).

Compound	Use	Environmental effects
Aldrin	insecticide	toxic to avian and aquatic animals, and humans
Chlordane	insecticide	toxic to avian and aquatic animals, carcinogen
Dieldrin	insecticide	toxic to avian and aquatic animals, carcinogen
Endrin	insecticide	toxic to aquatic animals and humans
Heptachlor	insecticide	bird reproduction, toxic to humans, carcinogen
Hexachlorobenzene	fungicide	toxic to humans, especially infants
Mirex	insecticide	toxic to plants and animals, carcinogen
Toxaphene	insecticide	highly toxic to fish, human carcinogen
PCBs	high T oil	transformers, toxic to fish and humans
DDT	insecticide	bird reproduction, toxic to animals, carcinogen
Dioxins	combustion	highly toxic, carcinogen, teratogen
Polychlorinated dibenzofurans	combustion	highly toxic, carcinogen, teratogen
Kepone	pesticide	toxic to aquatic organisms, carcinogen
Lindane & isomers	pesticide	toxic, carcinogenic
Pentachlorobenzene	pesticide	highly toxic to aquatics, toxic to humans
Tetrabromodiphenyl ether	industrial chemical	bioaccumulated
Pentabromodiphenyl ether	industrial chemical	bioaccumulated
Perfluorooctanesulfonic acid and salts	polymer production	bioaccumulated.
Endosulfan	insecticide	toxic to water and land animals and humans
Hexabromocyclodecane	flame retardant	bioaccumulated

results in their bioaccumulation in plant and animal fats, and biomagnification in the aquatic food chain. In many cases, the POPs are either identified as human carcinogens or are suspect carcinogens. Although many of the POPs included in the Stockholm Convention are no longer produced in the United States, there is still an exposure risk due to their long lifetimes and environmental persistence. New environmental contamination can also occur from POP waste sites, existing stockpiles, and air or water transport from other countries. Although most developed nations have taken swift action to control the production and use of POPs, a large number of developing nations have only recently begun to restrict their production, use, and release.

Moderately soluble organic compounds that have chlorine attached to an alkyl group can undergo elimination reactions in water or on wet surfaces. The elimination reaction follows an E1 mechanism and requires a tertiary hydrogen to stabilize the carbocation intermediate. For example, the formation of 1,1-dichloro-2,2-bis(*p*-chlorophenyl)ethane (DDE) from the insecticide DDT follows an E1 reaction mechanism shown in Figure 8.24. The rate-determining step is the loss of the tertiary hydrogen from DDT to form the intermediate carbocation, followed by loss of a chlorine to form the double bond of DDE. Other reactions that form stable DDT degradation products include the direct reductive dechlorination, which removes

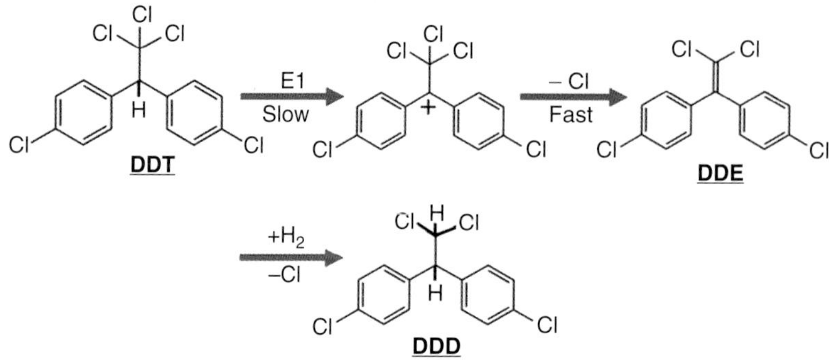

Figure 8.24 DDT transformation to DDE by an E1 elimination and transformation to DDD by reductive dechlorination. Source: Adapted from Leyo, Wikimedia Commons.

a Cl atom from the trichloromethyl group on DDT and replaces it with a hydrogen atom to form 1,1-dichloro-2,2-bis(*p*-chlorophenyl)ethane (DDD), shown in Figure 8.24. This reductive dechlorination can proceed by a chemical or biochemical mechanism. Both DDE and DDD are lipophylic and so can bioaccumulate in aquatic organisms. Biodegradation of DDE and DDD in water is expected to be very slow and so, as with DDT, they are expected to have long lifetimes. The DDT and DDE biodegradation lifetimes are estimated to be from 10 years in tropical regions to 60 years in temperate zones.

DDT derivatives have been produced that do not contain the para-chlorophenyl functional groups, such as Methoxychlor (1,1,1-trichloro-2,2-bis(*p*-methoxyphenyl)ethane):

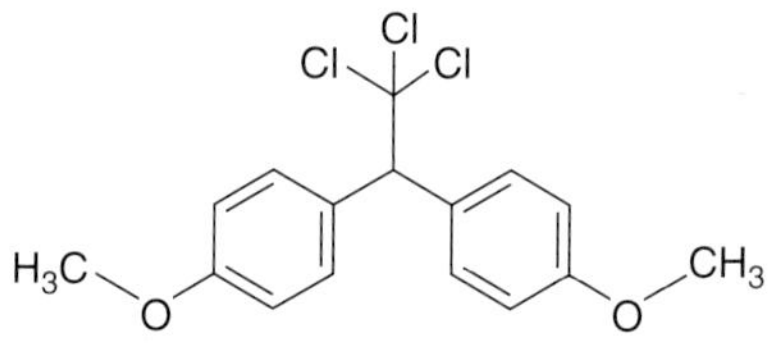

This minor structural change allows for more rapid biodegradation in aqueous systems. The methoxy groups (CH_3O—) are readily converted to phenol groups (OH—) by cytochrome P-450-type enzymes. The presence of these enzymes in insects increases the biodegradability but decreases the toxicity, so that about 20 times the amount of Methoxychlor is required to obtain the same effect as DDT. Since Methoxyclor is a di-ether, it is more water soluble than DDT and has a 40 ppb limit set by the EPA in drinking water.

Smaller aromatic hydrocarbons, such as benzene, toluene, and the xylenes, which are released from gasoline and petroleum storage tanks, can react with OH radicals on water surfaces in the presence of oxygen to form alcohols, ketones, aldehydes, and organic acids. These transformation products are much more water soluble than the reactants, and can be transported longer distances in aqueous systems. These photolytically driven reactions with OH radicals are much slower for the larger organic compounds found in crude oil and petroleum. The lack of surface degradation processes is one reason for the existence of long-lived oil slicks such as those produced by the Exxon Valdez or Deepwater Horizon accidents discussed in Section 10.5.1.

Petroleum-based plastics are also resistant to chemical oxidation and so have significantly long lifetimes in the environment. Plastic containers, packing materials, and coatings that are not recycled or incinerated end up in landfills and aquatic environments. Plastic bottles, cups,

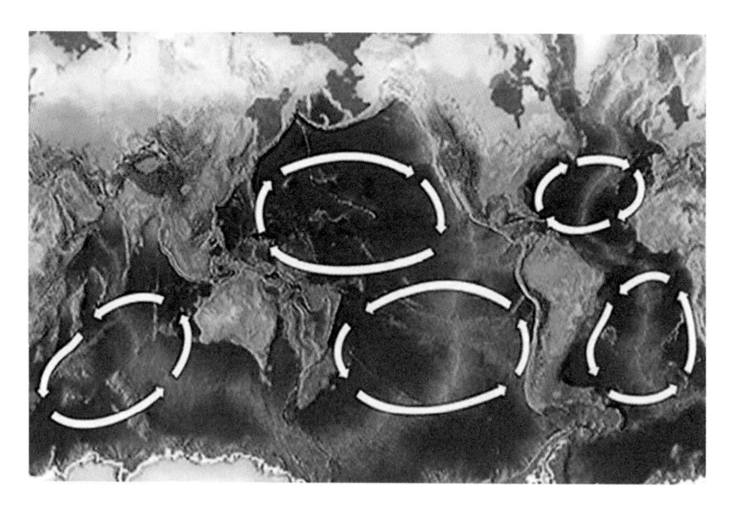

Figure 8.25 Areas of increased microplastics in the Atlantic, Indian, and Pacific Oceans. Source: NOAA, Wikimedia Commons.

and other products discarded into the marine environment can exist intact for 50–500 years or more. Plastic nets discarded from fishing boats are a major problem in the oceans, causing the loss of marine mammals that become entangled in the nets. These plastic materials are low in density and so float on the water surface and over time begin to slowly undergo photooxidation. The photooxidation breaks up the plastics into small particles called microplastics (<5 mm), which are resistant to further degradation. Another type of microplastics, known as microbeads, are very tiny pieces of polyethylene added to health and beauty products, which can easily pass through water treatment systems and end up in the ocean, increasing the problem. These microplastics are transported across the ocean surface by winds and currents, concentrating them in certain areas of the oceans, as shown in Figure 8.25. These areas of microplastic concentration are called "garbage patches" by the U.S. National Oceanic and Atmospheric Administration (NOAA).

Areas in the North and South Pacific, the North and South Atlantic, and the Indian Oceans have very high concentrations of microplastics floating on the surface. These areas are very large in size and are continuing to grow due to the discharge of plastics in trash from commercial and military vessels. Currently, the concentrations are at about four particles per cubic meter. Since the microplastics have very long lifetimes in the aquatic environment, they can be ingested and accumulated in the bodies and tissues of marine organisms and excreted in fecal matter. The fecal matter then sinks to the bottom, contaminating sediments with the microplastics. The microplastics can also adsorb synthetic organic compounds in the environment, such as POPs, which can be transferred into an organism's tissues after ingestion, increasing the level of contamination. In addition, additives added to the plastics during manufacture to increase their stability may leach out upon ingestion, potentially causing serious harm to an organism.

Most of the plastics derived from petroleum are composed of long alkyl chains, such as polyethylene, or are polyaromatics, such as polystyrene. These structures are very difficult to degrade biologically and so their lifetimes in landfills are extremely long. For this reason, there have been efforts to develop biodegradable plastics, which are easier to oxidize and have much shorter lifetimes. Currently, there are two types of biodegradable plastics: (1) those that are made from biological sources and so are renewable and (2) those that are made from petroleum and have had their structures altered to allow them to be more readily oxidized. The type 1 biodegradable plastics can be designed for either aerobic or anaerobic degradation. So, the rate of biodegradation depends on the type of disposal the plastic undergoes. For example, an anaerobic degradable plastic, which is released from a seagoing vessel, will have a long lifetime. Conversely, an aerobic degradable plastic may have a long lifetime if it is buried quickly in landfill in an anaerobic environment. The type 2 biodegradable plastic typically breaks down into microplastics upon degradation. These plastics ultimately behave in a manner similar to a petroleum-based polymer and still have very long lifetimes in the environment.

8.8 Contaminant Transport

The rate of transport of a pollutant species in a water system is dependent on both the rate of physical transport and the chemical and biological reactivity. The rate of physical transport depends on the type of water body and the speed of the water flow. Contaminants can exist in aqueous systems as dissolved, including complexed species, or adsorbed onto particles and colloids. The dissolved substances move at essentially the same rate as the bulk water. The movement of an adsorbed pollutant depends on the size of the particle or colloid. The particles move in the water flow in much the same way as an aerosol particle moves with the flow of air, as discussed in Section 6.2.4. Small particles and colloids can move with the flow of the water or can aggregate to form larger particles. The larger particles are typically removed by gravitational settling, becoming part of the sediment. Resuspension of the particle-bound contaminants from the sediment back to the water system can also occur. The deposition and resuspension of both mineral and organic contaminants in bottom sediments occur continuously in any water system and are an important mechanism for controlling the movement of the adsorbed species.

The residence time of a dissolved species in a water system, defined as the time between the arrival and departure of a contaminant in the system, depends not only on the rate of transport through the system but also on various chemical, physical, and biological processes that occur. If a contaminant is nonvolatile, chemically and biologically stable, and highly water soluble, it will have a residence time similar to that of the water itself. However, the residence time of soluble species may be decreased by chemical reactions that form insoluble or volatile products. For example, the oceanic residence time for soluble inorganic species, such as Na^+, has been estimated to be about 10^8 years. Meanwhile the residence time of iron in the ocean is estimated to be about 50–100 years due to the formation of insoluble particulate and colloidal forms. The residence time can also be decreased by incorporation into sediments after biological uptake, followed by excretion or death of the organism. The residence time of particle-bound pollutants is typically short, due to the tendency of the particles to become incorporated into the sediment.

The residence times of organic contaminants are determined by their aqueous solubility and their Henry's Law constants, which control the partitioning between the hydrosphere and the atmosphere. Since any pollutant in the gas phase will travel faster than the same species in the liquid phase, the transport of surface water pollutant species around the globe depends on the time spent in the hydrosphere compared to the time spent in the atmosphere. The more volatile species will travel further than the less volatile species. For example, a number of POPs with low water solubilities can easily be revolatilized into the atmosphere, especially during summer months. Once in the atmosphere, they can condense onto aerosol particles which can form clouds, as shown in Figure 8.26. They can also undergo atmospheric transport followed by wet deposition back to the surface. As this process repeats itself, these volatile compounds become widely spread throughout the globe. This slow but continual transport process has allowed POPs, such as the PCBs and Lindane, to be found in the Arctic and Antarctic thousands of miles from where they were released into the environment. This repeating cycle, which acts to carry persistent pollutants from the warmer regions to the colder regions, is called "the global

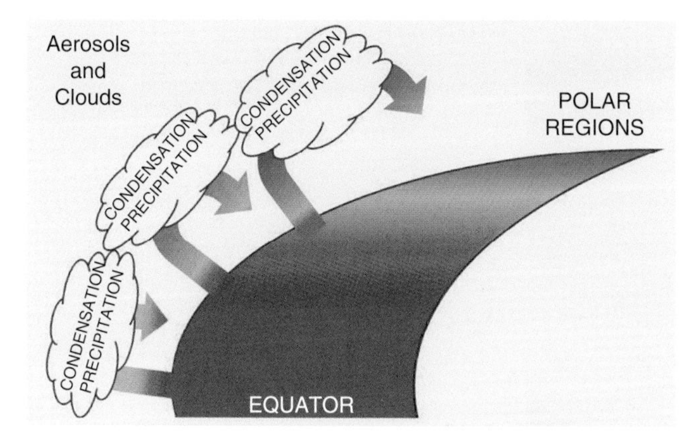

Figure 8.26 The global distillation mechanism for the long-range transport of persistent organic pollutants (POPs) from warmer regions (red) to colder regions (blue).

distillation effect." This process allows for long-range transport of fairly insoluble pollutants from aqueous systems at the equator and temperate zones to the poles.

In surface waters the denser organic compounds and insoluble inorganic compounds tend to settle to the bottom of the water column and become part of the sediment. The water sediment is a porous layer of solid material that forms at the bottom of a water system by the deposition of mineral and organic particles. Surface water sediments have at least two distinct layers. The top layer is characterized by a high degree of chemical and biological activity. The deeper layer has less potential for chemical reactions since it is isolated from the water column and typically has a very low DO. However, it can have the potential for anaerobic biological activity. The deeper the water system, the more likely that contaminants will remain in the sediments due to the lower DO and colder water temperatures reducing biological and chemical activity, as well as water flow at the lower depths. The resuspension of pollutant species bound to the upper layer of sediment is dependent on the force exerted by the water flow and the size, volume, density, and shape of the sediment grains. Stronger flows will increase the lift on the particle, causing smaller particles to rise and travel with the water flow. Under the same flow, larger or denser particles typically will not be resuspended but will remain with the sediment.

Sedimentary material is classified by both its physical and chemical properties. It can range in size from very small colloidal particles to clay, silt, sand, and gravel. Gravel is a loose aggregation of rock fragments ranging from 2 to 64 mm. It has a very low ability to retain moisture and mineral nutrients. However, due to its large size, water can percolate through thin layers of gravel by gravitational flow. Sand is a naturally occurring granular material composed of finely divided rock and mineral particles. It is typically defined by its grain size (125 μm to 2 mm) and is commonly composed of silica (SiO_2) with little organic matter. Silts are typically made up of quartz (SiO_2) or feldspar ($KAlSi_3O_8$, $NaAlSi_3O_8$, $CaAl_2Si_2O_8$) and can contain some organic matter. They are classified as between sand and clays, with a size range of 3.9–62.5 μm. Clay consists of the smallest particles (<3.9 μm) that make up sediments. Clay is a finely grained natural mineral or soil material that combines hydrated phyllosilicate (sheet-like silicates) minerals mixed with quartz, metal oxides, and organic matter. Colloidal particles in sediments are considered to be less than 1 μm in size and are the most mobile size fraction of the sediments.

Pollutant transport in groundwater systems depends on the properties of the overlying soil and the sedimentary materials that surround the aquifer. Soil is composed of sand, silt, and clay in different proportions. The percentage of these three components regulates the soil porosity, permeability, infiltration rate, and water-retaining capacity. Soils that have equal amounts of

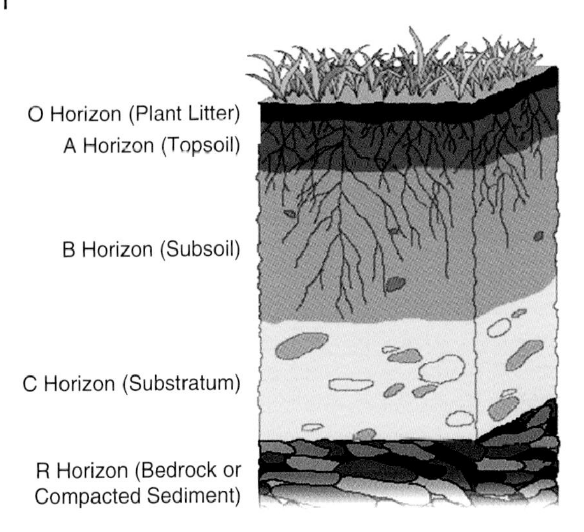

Figure 8.27 A typical soil profile consisting of five layers, called horizons, with different compositions and grain sizes. Source: Wilsonbiggs, Wikimedia Commons.

sand, silt, and clay are called loam soils. A typical soil profile, shown in Figure 8.27, consists of five layers called horizons. The top layer, or O horizon, consists of a layer of plant litter in relative undecomposed form. The surface soil (topsoil) is called the A horizon. It contains the most organic material and microorganisms and so has the most clay. It also contains the most water-soluble constituents, such as iron, aluminum, clay, and organic compounds. The subsoil is the B horizon. It is composed of a variable mixture of small particles of sand, silt, and clay. However, it has a much lower percentage of organic matter and humus than the A horizon. The C horizon is a layer of weathered and broken bedrock, called the substratum. The substratum is also known as the "parent rock," because it has a large influence on the nature of the resulting soil. For example, a soil composed mostly of clay is derived from the weathering of shale, while a sandy soil is derived from the weathering of sandstones. The R horizon is typically bedrock, a layer of partially weathered rock at the base of the soil profile. However, in some areas it can also be composed of a well-compacted sediment layer. The bedrock is a continuous mass of hard rock, preventing the further transport of water downward, but compacted sediment can allow for the slow movement of water vertically to create groundwater aquifers.

As water moves through the soil and sediments, it collects soluble salts from the minerals, which are then transported into groundwater systems. The dissolution and transport of these inorganic salts is dependent on their aqueous solubility. However, while many ionic compounds are considered to be insoluble, they can actually dissolve in very small amounts. Water flowing over and through mineral surfaces continually removes the material with time, even for those with very low solubilities. The solubilities determine how fast this process can occur and varies widely under normal environmental conditions from very slow transport, such as in the case of cinnabar (HgS), to fairly rapid transport, such as in the case of $PbCl_2$. The solubilities of some sparingly soluble salts important to aqueous systems and contaminant transport are listed in Table 8.7. The K_{sp} values listed in the table are for freshwater systems at 25 °C. In sea water or brine systems with high chloride concentrations, the solubility of the metal chlorides ($PbCl_2$ and AgCl) will be decreased due to the common ion effect.

In general, the salts of Na^+, K^+, and NH_4^+ are all very water soluble. All Cl^-, Br^-, and I^- salts are very water soluble with the exception of Ag, Pb, and Hg, which have very low solubilities. All NO_3^- salts are water soluble and SO_4^{2-} salts are soluble with the exception of $BaSO_4$ and $PbSO_4$. Also, CO_3^{2-}, CrO_4^{2-}, PO_4^{3-}, and SiO_4^{2-} have very low solubilities with the exception of the Na^+, K^+, and NH_4^+ salts and $MgCrO_4$, which are all very soluble. All sulfides have very low water solubilities with the exception of $(NH_4)_2S$, Na_2S, K_2S, BaS, CaS, and MgS. Most all

Table 8.7 The solubility product constant (K_{sp}) for some inorganic salts in water at 25 °C and 1 atm pressure.

Compound	Chemical formula	K_{sp}
Aluminum hydroxide	$Al(OH)_3$	1.9×10^{-33}
Copper (I) bromide	CuBr	6.3×10^{-9}
Iron carbonate	$FeCO_3$	2.1×10^{-11}
Lead chloride	$PbCl_2$	1.2×10^{-5}
Mercury (II) sulfide (cinnabar)	HgS	1.5×10^{-54}
Calcium carbonate (calcite)	$CaCO_3$	6.4×10^{-7}
Silver chloride	AgCl	1.8×10^{-10}
Barium sulfate	$BaSO_4$	1.0×10^{-10}
Lead carbonate	$PbCO_3$	7.4×10^{-14}
Aluminum phosphate	$AlPO_4$	9.8×10^{-21}
Calcium hydroxide	$Ca(OH)_2$	4.7×10^{-6}

of the hydroxide salts are insoluble with the exception of NH_4OH, NaOH, KOH, CsOH, and RbOH. Also, $Ba(OH)_2$, $Ca(OH)_2$, and $Sr(OH)_2$ are a little more soluble than most of the other inorganic hydroxide salts.

The aqueous solubility of ionic solids is determined experimentally and the solubility product constant (K_{sp})is calculated from the equilibrium equation as

$$A_nB_m(s) \rightleftarrows nA^{x+}(aq) + mB^{y-}(aq)$$
$$K_{sp} = [A^{x+}]^n[B^{y-}]^m$$

However, if one or both of the product ions is strongly complexed by an organic ligand, such as the natural humic acids ($RCOO^-$) or fulvic acids, the equilibrium will become

$$A_nB_m(s) + RCOO^- \rightleftarrows nA^{x+}(RCOO^-)(aq) + mB^{y-}(aq)$$
$$K_{sp} = \frac{[nA^{x+}[RCOO^-]]^n[B^{y-}]^m}{RCOO^-}$$

Thus, a dissolution equilibrium that involves the formation of a complex with one of the product ions usually overwhelmingly favors the formation of products because the majority of the free A^{x+} ions are essentially removed from the equilibrium. Then, according to Le Chatelier's Principle, the position of the equilibrium shifts to the right, resulting in an increase in the solubility of $A_nB_m(s)$. The same effect will occur if one or more of the ions form an insoluble precipitate.

The surface area to volume ratio of the soil particles, as well as the ionic charge and organic content, determine the cation exchange capacity (CEC) of the soil. The CEC of soil is a measure of how many cations can be retained on the soil particles. It is defined as the amount of positively charged ions that can be exchanged per mass of soil. Positively charged ions in the water can bind to negatively charged species on the surfaces of the soil particles, such as ionized organic acid or hydroxyl groups. Once bound to the soil, they can exchange with other positively charged particles in the water. The CEC affects many aspects of soil chemistry, and is used as a measure of soil fertility since it determines the soil's capacity to retain positively charged plant nutrients (K^+, NH_4^+, Ca^{2+}). It also indicates the capacity of the soil to retain positively charged pollutants, such as heavy metals.

Metals in the percolation water can bind to the humic materials containing —COOH groups or to clay minerals containing ≡SiOH and ≡$Al(OH)_2$ groups. If the pH of the percolating water decreases, these sites can exchange the H^+ for cations in the water. The aluminum and silicon minerals in clays can also acquire negative charge if the aluminum or silicon atoms are replaced by elements with lower charge, such as Mg^{2+}. This acquired charge does not involve deprotonation and so is independent of water pH. After binding to the negatively charged sites on the surface of soil particles, the metals can later be displaced by H^+ in more acidic percolating waters or they can be removed by binding to soluble humic and fulvic acids in the percolating water, which have stronger binding constants than the soil humic matter. Once in the water they can be transported into the groundwater system. Uncharged organic pollutants can also be adsorbed onto the surface of soil particles. They can become reversibly associated with the natural organics in clays, followed by association with soluble humic and fulvic acids in percolation water, which transports them into the groundwater system. These transport mechanisms result in the concentration of both ionic and organic contaminants on the soil surfaces, followed by transport of the concentrated species into groundwater aquifers.

Precipitation falls onto the soils and percolates into the porous soil and sediments to form the groundwater systems as shown in Figure 8.28. The resulting groundwater aquifers can be confined or unconfined. Unconfined aquifers are those that are recharged from water percolation through the ground surface directly above the aquifer. The water surface of an unconfined aquifer, called the water table, is at atmospheric pressure. Unconfined aquifers are recharged by rain or stream water infiltrating directly through the overlying soil. The discharge of water from an unconfined aquifer can be due to pumping of water out of the aquifer, as well as loss from natural spring formation, which is usually associated with a surface stream. The ground water is sometimes discharged in large quantities using large pumping stations for agricultural irrigation of crops in regions that do not have sufficient rainfall during the growing season. This type of groundwater discharge by pumping for drinking and irrigation often exceeds the recharge rates for the aquifers, lowering the water table.

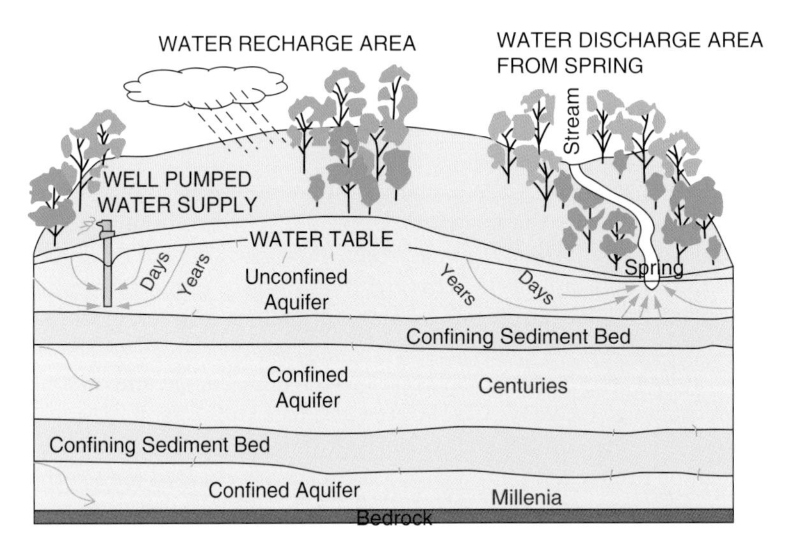

Figure 8.28 The structure of confined and unconfined aquifers along with the transport and residence times (red) of water through the aquifers. Source: Adapted from T. C. Winter, J. W. Harvey, O. L. Franke, and W. M. Alley, Wikimedia Commons.

A confined aquifer is covered by an impermeable layer, such as rock or compacted sediment, which prevents water from recharging the aquifer from the soil surface located directly above the aquifer. The water in a confined aquifer is under pressure and will rise up spontaneously inside a well borehole drilled into the aquifer. Confined aquifers can be recharged by rain or stream water percolating through the soil at some distance away from the aquifer where the impermeable layer no longer exists. Shallow aquifers tend to be unconfined while deeper aquifers are typically confined by bedrock. The residence times for water in these underground aquifers range from days in the shallow, unconfined aquifers to millennia in deeper, confined aquifers.

Transport within the groundwater systems depends on the rate of water flow, interactions of the contaminants with soil and sediments, complexation or adsorption of contaminants with soluble organics, and interactions with microorganisms or other contaminant species. Inorganic contaminants are commonly carried into ground water as organometallic complexes in humic and fulvic acid colloids. These natural organics can complex elements which are normally considered to exist in an insoluble form. This is of particular importance for some toxic metals and radionuclides, such as Hg, U, Th, and Pu. For example, the oxides of U, Th, and Pu were thought to be strongly bound to clay soils containing charged natural organics ($R—COO^-$). However, field measurements have shown that these normally insoluble radionuclides can be mobilized by ground waters containing very low concentrations of humic and fulvic acids. The complexation with the natural organic colloidal materials allows them to be transported in flowing subsurface groundwater systems much further than predicted by their water solubilities alone (Marley et al., 1993).

Organic contaminants can also be transported in groundwater systems adsorbed onto natural organic colloidal materials. The carboxylic and phenolic hydroxyl group content, as well as the aromaticity of the humic and fulvic acids, are found to be most important in determining the strength of the adsorption of pesticides. The binding between the organic contaminants and the humic colloids is thought to be multiphase and can include ion–dipole, dipole–dipole, and hydrogen bonding. Since the association of the organics with the humic substances is most likely not a covalent bond or an electron transfer complex, the association is most likely reversible. In this case it is represented by an equilibrium with the free fraction available to adsorb onto the soil or sediment, slowing transport. This is most important in unconfined aquifers, where the water is in contact with the soil subsurface.

References

Dicks, D 2010 *Puget Sound's slow oil spill*. Available online at https://www.seattletimes.com/opinion/puget-sounds-slow-oil-spill (accessed September 1, 2018).

Dickson, AG & Goyet, C 1994 *Handbook of Methods for the Analysis of the Various Parameters of the Carbon Dioxide System in Sea Water*. Available online at https://www.nodc.noaa.gov/ocads/oceans/DOE_94.pdf (accessed June 16, 2018).

Engineering Toolbox 2005 *Oxygen – Solubility in Fresh Water and Sea Water*. Available online at https://www.engineeringtoolbox.com/oxygen-solubility-water-d_841.html (accessed June 14, 2018).

Gaffney, J.S. and Marley, N.A. (2014). In-depth review of atmospheric mercury: Sources, transformations, and potential sinks. *Energy Emission Control Technol.* 2: 1–21.

Gaffney, J.S., Marley, N.A., and Orlandini, K.A. (1996). The use of hollow-fiber ultrafilters for the isolation of natural humic and fulvic acids. In: *Humic and Fulvic Acids: Isolation, Structure, and*

Environmental Role (ed. J.S. Gaffney, N.A. Marley and S.B. Clark), 26–40. Washington, D.C.: American Chemical Society.

Gleick, P.H. (1993). An introduction to global fresh water issues. In: *Water in Crisis: A Guide to the World's Fresh Water Resources* (ed. P.H. Gleick), 3–12. New York: Oxford University Press.

Marley, N.A., Gaffney, J.S., Orlandini, K.A., and Cunningham, M.M. (1993). Evidence for radionuclide transport and mobilization in a shallow, sandy aquifer. *Environ. Sci. Technol.* 27: 2456–2461.

NRDC 2018 *Water Pollution: Everything You Need to Know*. Available online at https://www.nrdc.org/stories/water-pollution-everything-you-need-know (accessed August 26, 2018).

Pidwirny, M & Jones, S 2006 *The Hydrologic Cycle*. Available online at http://www.physicalgeography.net/fundamentals/8b.html (accessed August 6, 2018).

Shanks, A. and Trent, J. (1980). Marine snow: Sinking rates and potential role in vertical flux. *Deep Sea Res.* 27A: 137–143.

Shiklomanov, I. (1993). World fresh water resources. In: *Water in Crisis: A Guide to the World's Fresh Water Resource* (ed. P.H. Gleick), 13–24. New York: Oxford University Press.

Sierra Club 2018 *Why are CAFOs Bad?* Available online at https://www.sierraclub.org/michigan/why-are-cafos-bad (accessed August 29, 2018).

USDOT n.d. *Petroleum Oil Spills Impacting Navigable U.S. Waters*. Available online at https://www.bts.gov/content/petroleum-oil-spills-impacting-navigable-us-waters (accessed August 29, 2018).

USEPA 2018 *National Primary Drinking Water Regulations*. Available online at https://www.epa.gov/ground-water-and-drinking-water/national-primary-drinking-water-regulations#Organic (accessed June 6, 2018).

Vahid, A, Qaddumi, HM, Dickson, E, Diez, SM, Danilenko, AV, Hirji, RF, Puz, G, Pizarro, C & Jacobsen, M 2009 *Water and climate change: Understanding the risks and making climate smart investment decisions*. Available online at http://documents.worldbank.org/curated/en/362051468328208633/Water-and-climate-change-understanding-the-risks-and-making-climate-smart-investment-decisions (accessed June 6, 2018).

Further Reading

Gaffney, J.S., Marley, N.A., and Orlandini, K.A. (1992). Evidence for thorium isotopic disequilibria in natural waters due to organic complexation: Geochemical implications. *Environ. Sci. Technol.* 26: 1248–1250.

Gleick, P.H. (ed.) (1993). *Water in Crisis: A Guide to the World's Fresh Water Resources*. New York: Oxford University Press.

USGS 2018 *National Water Quality Assessment (NAWQA) Project*. Available online at https://water.usgs.gov/nawqa (accessed July 4, 2018).

Study Problems

8.1 Water covers approximately 71% of the Earth's surface. Under ambient conditions on the Earth, in what states is water present?

8.2 Water has a MW of 18.0 and though lightweight and polar, its boiling point is 100 °C. What is the reason for this elevated boiling point?

8.3 Why is absolutely pure water not present naturally on the Earth?

8.4 Of the total water reservoirs, salt water makes up 97% of the water on the Earth. Why?

8.5 Where does most of the fresh water on the planet originate?

8.6 In what state does most of the fresh water on the planet reside?

8.7 Is there more fresh water in lakes or groundwater reservoirs?

8.8 Is there more fresh water in lakes or rivers?

8.9 There are two types of drainage basins where the precipitation striking the land flows via rivers and streams. Drainage basins usually have the water flowing into an ocean or sea. However, there are landlocked systems where the water cannot drain to a sea or ocean. What is the term used to describe these types of drainage basins?

8.10 The thermohaline circulation is the name used to describe the ocean currents which are driven primarily by changes in water density. What are the two major factors in determining the ocean water density?

8.11 Fresh water flowing in rivers and streams into the oceans carries both solid and dissolved material into the seas and oceans. Salinity is the concentration of dissolved ions.
(a) What are the two ions found in sea water with the highest concentrations?
(b) What is a solution of these ions in water called?

8.12 Along with many other dissolved salts the oceans contain dissolved organic matter that is in solution and in colloidal form. The dissolved organic matter is highly colored and is given an abbreviation.
(a) What is the abbreviation?
(b) Are higher concentrations of this organic matter found in the oceans or at the mouth of rivers?
(c) What does this infer about the primary source of this organic matter?

8.13 What naturally occurring organic compounds are responsible for most of the deep brown color in natural waters?

8.14 What two general names are given to the naturally occurring complex organic acids that are either in colloidal or solid form in natural waters and are derived from the breakdown of natural humus found in soils?

8.15 What is another name for still water lake systems?

8.16 (a) What is often found to occur where oceanic upwelling occurs?
(b) Why?

8.17 The Ekman spiral that causes upwelling near coastal regions is due to surface winds interacting with another effect. Name that effect.

8.18 Is dissolved oxygen more likely to be at higher concentrations at the surface or the bottom of a water body?

8.19 What is the term used to describe very low levels of dissolved oxygen or lack of oxygen at the bottom of a freshwater lake or saltwater body?

8.20 Why is the surface of freshwater lakes usually at a pH of about 5.3, while ocean surface waters are at a pH of about 8.2?

8.21 What are the main types of nitrogen-fixing organisms in the ocean environments?

8.22 What two ionic chemical forms of nitrogen are used as fertilizers?

8.23 In the phosphate cycle is there a gaseous form of phosphorus that can enter the atmosphere?

8.24 What nitrogen-containing compound is used as a fertilizer and also adjusts soil pH?

8.25 What effect has the increasing concentration of CO_2 in the atmosphere due to fossil fuel combustion had on the surface ocean pH?

8.26 The chemical transformations of polychlorinated biphenyls or PCBs is very slow in the environment. PCBs are considered to be POP-type compounds. For what words does the acronym POP stand?

8.27 The pesticide DDT undergoes an E1 elimination to form a compound that has lost HCl. What is the abbreviation for this chlorinated hydrocarbon derivative?

8.28 Many metallic cations can be complexed by polyelectrolytes in water. What are the two classes of naturally occurring organic compounds that are excellent metal chelating agents?

8.29 Surfactants contain an ionic group along with a long chain alkyl group that allows them to carry nonpolar materials in water. What are the two types of natural organic materials that have alkyl and aromatic backbones as well as many —COOH and —OH groups that can also act as surfactants and transport nonpolar organic pollutants?

8.30 What are the major sources of arsenic in drinking water supplies?

8.31 The use of chlorination to sterilize drinking water produces what bi-products from reactions with dissolved organic compounds?

8.32 Many organic pollutants – especially plastics – are slow to undergo photochemical or other chemical reactions in the environment. Biological degradation can occur. What two types of organisms are the most important in biodegradation?

8.33 When plastics do break down they form very small particles of organic debris that are very small and called micro-debris. The material tends to be carried into the surface oceans and is being found concentrated in regions of the major oceans in surface waters. What term has the NOAA used to describe this ocean pollution problem?

8.34 To make DDT more biodegradable, the para-substituted chlorinated phenyls on the molecule were replaced with methoxy groups —OCH_3. What is the industrial name for this insecticide?

8.35 Soils can hold a significant amount of water and also act to complex and absorb organic pollutants. They are commonly classified by the size and composition of the material.
(a) What are the smallest-sized particles in soil commonly called (<3.9 μm)?
(b) What are the particles in the size range from 3.9–62.5 μm called?
(c) What are soil components called that are above these size ranges?

8.36 POP compounds that have very low water solubility tend to remain on surfaces. Their slow evaporation and condensation onto aerosol particles, followed by rainout back to the ocean and terrestrial surfaces, leads to long-range transport. This leads to these pollutants that are released into the warmer equatorial and temperate regions moving toward the polar regions. What is the term used to describe this transport phenomenon?

8.37 Soils that are considered to be in frigid regimes are at what temperature range?

8.38 What is the soil moisture regime called where there is sufficient water for plants all year round and the regions have high humidity?

8.39 Groundwater aquifers can be unconfined or confined. What are the water residence times for (a) unconfined aquifers and (b) confined aquifers?

8.40 The transport in water systems for many inorganic salts and compounds depends on their water solubility. What is the term used to describe the equilibrium between the solid and the dissolved inorganic species in water?

9

Analytical Methods for Water Analysis

The determination of pollutant concentrations, as well as the properties of surface and ground water, is accomplished by both chemical and physical methods. The methods used depend on the type and use of the water system. Analytical methods have been developed for determining inorganic and organic pollutants in freshwater systems, including precipitation, rivers and streams, lakes and ground water, as well as in brine and marine systems that have high salinities. These methods are also applied to a wide range of concentrations, from high levels in industrial waste water to trace levels of pollutants in drinking water. They are used to monitor water quality as well as to follow the transport and transformation of the water pollutants in the aqueous environment. Also, because water is such a good solvent for all polar and polarizable species, including gases, it is important to recognize and avoid potential sources of contamination while sampling and analyzing water samples.

Most of the analytical methods reviewed here have been approved by the U.S. Environmental Protection Agency (EPA) as part of the Clean Water Act, and the approved procedures for these methods have been detailed for each specific chemical species in *Clean Water Act Analytical Methods: Approved CWA Chemical Test Methods* (USEPA, 2017a). Water quality measurements include biological, chemical, and physical characterization of the water system. Here the focus is on the chemical and physical characterization, focusing on sampling methodology, analytical measurement, and contaminant issues. The approved methods for biological characterization of a water system are detailed by the EPA Environmental Protection Agency in *Clean Water Act Analytical Methods: Approved CWA Microbiological Methods* (USEPA, 2017b).

9.1 Sampling Methods

The sampling methods for various types of water systems require that the water sample be collected without contamination and represent the body of water being sampled. There are many types of water systems, as discussed in Section 8.4. For still or slow-moving shallow water systems, a single sample taken just below the water surface at a specific time or over as short a period as is feasible, called a grab sample, will adequately represent the water system. Grab samples are also adequate for sampling most waste water effluents. However, in some cases, more than one sample is required to represent the variability in the water system due to spatial inhomogeneities and temporal changes in the water chemistry and current flow. For large bodies of water, such as a lake or river, it is required to analyze samples obtained at different locations separately. But samples obtained at one sample point over a specific time frame can be mixed together to produce a composite sample, which would represent an average of the water conditions over the period of time sampled. This compositing procedure is also used for sampling waste water effluents that are expected to change over a period of time.

Chemistry of Environmental Systems: Fundamental Principles and Analytical Methods, First Edition.
Jeffrey S. Gaffney and Nancy A. Marley.

In most cases, the water system to be sampled is reasonably large, such as a river, lake, wetland, or bay. In these cases, sampling of the water system takes place well below the surface. Unless a depth profile is desired, one sample is typically obtained at mid-depth. Samplers commonly used for obtaining water samples at depths of tens to about one hundred feet are the Kemmerer sampler (Figure 9.1(A)) and the Van Dorn sampler (Figure 9.1(B)). Either sampler can be constructed of stainless steel, polyvinyl chloride, or acrylic pipe, depending on the type of contaminant being measured. For example, the sampling of trace organic compounds would typically require a stainless steel sample tube, while sampling for trace metals would use a polyvinyl chloride or acrylic sample tube. After lowering the sampler to the desired depth, a messenger weight is released to activate the trip line, which seals the sample tube. The messenger weight is usually a narrow brass or stainless steel cylinder weighing 0.5 kg. After activating the trip line, the stoppers at either end of the sampler move into place to seal the sample tube, capturing the water sample inside. Typical sample volumes are 2, 3, or 5 l. The Van Dorn sampler can be used in either the horizontal or vertical sampling modes. Horizontal bottles are often used for sampling at the thermocline or at other stratification levels, where the water chemistry may change over a narrow vertical distance.

A number of parameters are to be measured in the field at the time the sample is taken. These include temperature, pH, dissolved oxygen (DO), turbidity, sample depth, and total depth. The pH and DO are measured on the sample before it is transferred to a sample collection bottle, while turbidity, sample depth, and total depth are measured in situ. Sample depth and total depth can be obtained by length measurements of the sampler line as it is dropped to obtain the sample. Temperature is taken by thermocouple, while pH and DO are measured with electrodes, as described in Section 9.2. Samples obtained for the analysis of trace metals are placed into a "trace clean" glass bottle using the drain valve on the sampler. The samples are acidified after transfer to prevent precipitation or adsorption of the metal ions on the walls of the sample container. Samples for organic and nutrient analyses are drained into a "trace clean"

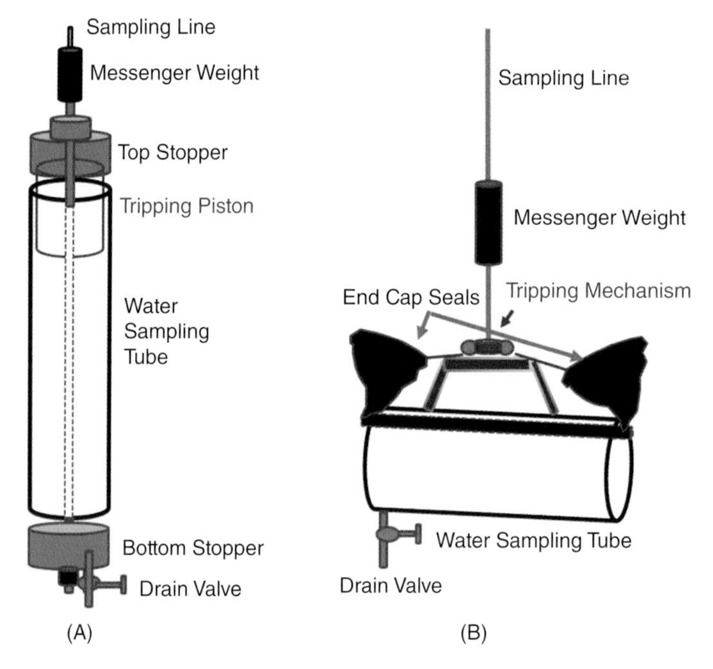

Figure 9.1 Schematic diagrams of (A) the Kemmerer water sampler and (B) the Van Dorn water sampler used for obtaining samples at moderate depths.

polyethylene bottle and placed on ice for transport to the laboratory. The samples must be kept at a temperature below 20 °C and in the dark until analysis, in order to stop bacterial decomposition or transformation processes.

The simplest means of determining water turbidity in the field is with a Secchi disk. The Secchi disk was originally developed by Angelo Secchi in 1865 as an easy way to determine water transparency. There are two types of Secchi disks in use today. The disk used in marine waters is a plain white 30 cm diameter disk, similar to that originally designed by Angelo Secchi. The disk used for fresh water has alternating quarters of black and white on a 20 cm diameter disk. The disk is lowered into the water slowly on a pole or line until it can no longer be seen. The depth in meters at this point is called the Secchi depth. Since the Secchi depth measurement can be affected by the light reflected off the surface of the water, it is recommended that measurements be taken between 9 a.m. and 3 p.m. from the shady side of the boat or dock. The measurement is usually taken in two steps. The first determination is measured during the Secchi disk descent at the point where it disappears from sight. The second measurement is obtained by lowering the disk past the Secchi depth and recording where it can first be seen when bringing it back to the surface. The two measurements are averaged and recorded as the Secchi depth (m). The inverse Secchi depth ($1/S$) (in units of m^{-1}), is a measure of the light extinction coefficient of the water. The light extinction coefficient for the atmosphere was discussed in detail in Section 6.6. The light extinction coefficient is the sum of light scattering by suspended particles and light absorption by dissolved species, while the water turbidity is due only to light scattering by particulates. So, the Secchi depth is related to the turbidity as:

$$T = \frac{1}{b_{abs} + S}$$

where b_{abs} is the absorption coefficient of the water, which is primarily due to the presence of colored dissolved organic matter (CDOM).

Turbidity can be measured directly using a turbidity meter or turbidity sensor for use in situ. These instruments measure the amount of light scattered by the particles in the water and operate on the same principle as the nephelometers used to measure aerosols in air, described in Section 7.4. Light from a tungsten source is directed through the water. A photodetector placed at 90° to the source light measures the intensity of the light scattered by the particles in the water. The intensity of the scattered light also depends on the wavelength of light and the concentration of light-absorbing species in the water. However, due to the short path lengths in turbidity meters, the interference due to absorption is only significant for highly colored waters. Turbidity meters and turbidity sensors that use light-emitting diode (LED) sources in the near-infrared region (860 nm) are less affected by light absorption than those that use a tungsten light source. The turbidity meters and field sensors are calibrated with standards of known turbidity made with Formazin or a styrene (vinylbenzene) polymer, and the results are reported in nephelometry turbidity units (NTU).

The collection of composite samples may be required for the monitoring of storm water, industrial discharge, waste water effluent, and surface waters where the chemistry is likely to change over a period of time. Composite sampling consists of a collection of individual discrete samples taken at regular intervals over a period of time, usually 24 hours. The water being sampled is collected automatically in a common container over the sampling period. In some cases, such as the monitoring of bottom waters in the Gulf of Mexico, where anoxia is known to occur seasonally, long-term continuous sampling may be required. Continuous sampling is usually accomplished by placing a sampling hose or pipe near the sample site and pumping water automatically into sample bottles. The long-term continuous sampling of large bodies of water requires specialized samplers that are more expensive to build and operate. One type

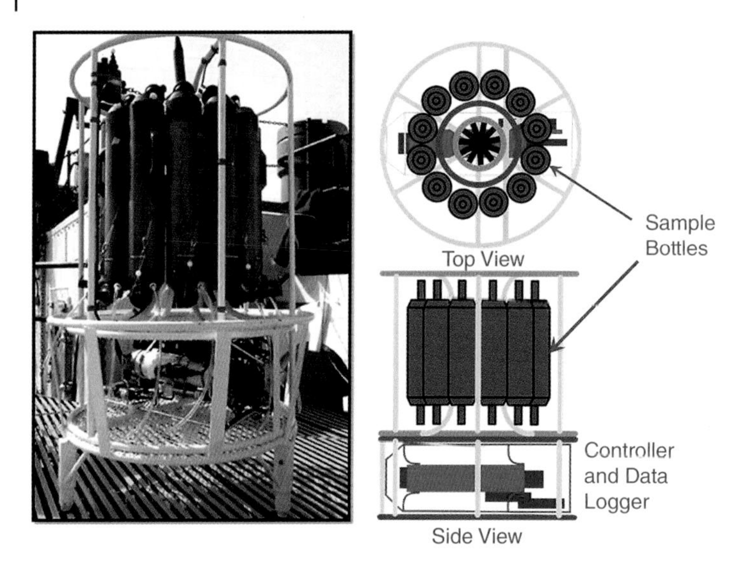

Figure 9.2 The rosette water sampler with 12–36 sample bottles (1.2–30 l), digital controller, and data logging system. Source: Photograph from USEPA, Wikimedia Commons. Diagram from BrnBld, Wikimedia Commons.

of continuous sampler that is typically used for this purpose is the rosette sampler, shown in Figure 9.2. The rosette sampler is very heavy and requires a reasonably large vessel equipped with an electric wench with steel wire cable to place and recover the sampler. The sampler can be placed on the sediment floor for bottom sampling or it can be moored to a buoy using a strong wire cable. The rosette sampler holds from 12 to 36 sample bottles, each with a volume of from 1.2 to 30 l, depending on the required sample volume. The sampler includes a waterproof electronic controller and data logger for operation and data storage. The data logger can also include simple instruments for the continuous measurement of water parameters, such as temperature, pH, DO, conductivity, and turbidity. The water samples can be collected at regular intervals over a period of time, or they can be collected when changes in the measured water parameters occur.

9.2 Dissolved Species

Natural water systems contain dissolved species, including inorganic ions, organic compounds, and gases, as well as suspended particulates. The major sources of dissolved species in freshwater systems are agricultural runoff, stormwater runoff, leaching of minerals from soils and sediments, and waste water discharge. Macromolecular and colloidal natural organics such as tannins, fulvic acids, and small humic acids (<2.0 μm) are also considered to be dissolved species. The amount of dissolved solid species can be determined as total dissolved solids (TDS). The TDS measurement is obtained by first filtering the sample through a 2.0 μm glass-fiber filter to remove all suspended particulates. All chemical species that pass through this filter are considered to be dissolved. After the filtration, 100 ml of the filtrate is placed in a pre-weighed ceramic crucible and dried at 103 °C for approximately 24 hours. The temperature is then increased to 180 °C for 8 hours to remove any occluded water. The sample is cooled in a desiccator and reweighed to obtain the mass of the TDS. In most cases the TDS is due to dissolved salts in the water. The most common species in freshwater TDS are Ca^{2+}, PO_4^{3-}, NO_3^-, Na^+, K^+, and Cl^-. However, the TDS measurement cannot reveal any information about the type of dissolved solids, their chemistry, or their environmental effects. Although the TDS measurement can serve as an indication of the water quality of freshwater systems, further analysis is required to determine the type and environmental impact of the dissolved species in the water.

9.2.1 Electrochemical Methods

Many dissolved species can be measured using a specific ion electrode (SIE) or ion selective electrode (ISE). An ISE is an electrode constructed with a permeable membrane that passes the ion of interest. The types of permeable membranes are glass (H^+, Na^+, Ag^+), crystal (F^-, Cl^-, Br^-, I^-, Cu^{2+}, Pb^{2+}, CN^-), liquid (NH_4^+, Ca^{2+}, NO_3^-, K^+), or polymer (NH_3 and DO). Glass membranes are made from a type of glass that acts as an ion exchanger. Glass-membrane ISEs have good selectivity for singly charged cations. The most common type of glass-membrane electrode is the pH electrode. Crystal membranes are made from crystals of a single substance. The electrodes respond only to ions that can introduce themselves into the crystal lattice. So they respond only to the cation and anion of the ionic compound that forms the crystal. They also do not require an internal electrolyte. The most widely used crystal-membrane electrode in water analysis is the F^- electrode. Liquid membranes consist of an ion exchanger or ion carrier dissolved in a viscous organic liquid membrane. The membrane is held in place by a porous disk. The most commonly used liquid membrane electrode is the Ca^{2+} electrode, which uses a calcium dialkyl phosphate membrane. Polymer membranes are composed of a polymer, such as Teflon or polyvinylchloride (PVC), with the ion carrier or exchanger embedded into it. The most commonly used polymer membrane electrode in water analysis is the DO electrode.

The ISE measures the concentration of one ion in the presence of other ions in a water solution by converting the ion concentration into an electrical potential. The electrical potential (E) of the galvanic cell formed by the ISE and a reference electrode is theoretically dependent on the logarithm of the ion concentration in an oxidation–reduction (redox) equilibrium according to the Nernst equation:

$$E = E_o + \frac{0.059\ \text{V}}{n^{e-}} \log \frac{[\text{ox}]}{[\text{red}]}$$

where E_o is the standard potential of the galvanic cell, [ox] is the concentration of the species being oxidized, [red] is the concentration of the species being reduced, and n^{e-} is the number of electrons exchanged in the redox equilibrium reaction. The electrode is calibrated by comparing the response to that of known standards. Since the Nernst equation is typically not linear for potentials less than 200 mV, the measurement of low concentration measurements requires a nonlinear multipoint calibration. Also, the membranes can become fouled by organics in the water and must be cleaned or replaced periodically.

The measurement of pH is routinely accomplished in the field using a glass electrode and a standard reference electrode attached to a portable pH meter. The glass electrode is a type of ISE, which is selective for H^+ ions. The H^+ sensor electrode is constructed of a silver wire with a partial coating of solid AgCl, which is immersed in a 1×10^{-7} M HCl solution. It functions as a redox electrode with the half reactions:

$$Ag^+(aq) + e^- \rightleftarrows Ag(s)$$

$$AgCl(s) + e^- \rightleftarrows Ag(s) + Cl^-(aq)$$

The potential of this Ag—AgCl electrode is dependent on the concentration of the $Cl^-(aq)$ ions:

$$E = E_o - 0.059\ \log[Cl^-(aq)]$$

The end of the glass electrode is a very thin glass membrane coated with a hydrated gel on either side, which can exchange ions with the solution. The difference in H^+ ion concentrations between the internal solution (1×10^{-7} M) and the external solution causes H^+ ions to migrate in and out of the internal gel, changing the H^+ concentration in the internal solution. This change in concentration of the H^+ ions causes a change in the concentration of $Cl^-(aq)$ ions, creating a change in potential of the sensor electrode that is read in millivolts (mV) versus the stable potential of the reference electrode. The Nernst equation for the glass pH electrode is expressed as:

$$E = E_o - 0.059 \log[H^+(aq)]$$

In most cases, the glass and reference electrodes are combined into one combination electrode as shown in Figure 9.3. In the combination electrode, the H^+ sensor electrode is housed in an inner glass tube containing the 1×10^{-7} M HCl solution (pH 7) while the reference electrode, which is also an Ag—AgCl electrode, is housed in an outer glass tube containing a 0.1 M KCl solution. The sensor electrode has contact with the sample through the glass membrane, while the reference electrode has contact with the sample through a porous ceramic junction on the side of the outer tube. The potential between the two electrodes is converted to pH units by comparison with the potential generated by two standard solutions of known pH. The standards are typically pH = 4 and pH = 10 for a sample with $pH > 4$ and pH = 2 and pH = 4 for a sample with $pH < 4$. The electrode should be rinsed with distilled water before and after each measurement to prevent cross-contamination. In addition, the glass membrane must remain wet to maintain hydration of the conductor gel. So, when not in use the pH electrode is stored in a solution of saturated KCl.

The most common crystal membrane ISE used for water analysis is the F^- electrode. The crystal membrane in the F^- electrode is made from a EuF_2-doped LaF_3 crystal. Doping the crystal with the larger Eu^{2+} ion creates lattice vacancies in the crystal. This allows the F^- ions in the solution to move into the vacancies and migrate through the crystal. The F^- ISE operates in much the same manner as the glass pH electrode. The sensor electrode is a Ag—AgCl electrode immersed in a solution of 0.3 M KCl + 0.001 M KF. The internal solution has a F^- concentration ($19\ mg\ l^{-1}$) that is much higher than that expected from any water sample. This difference in F^- concentration between the internal solution and the water sample causes F^- ions to migrate

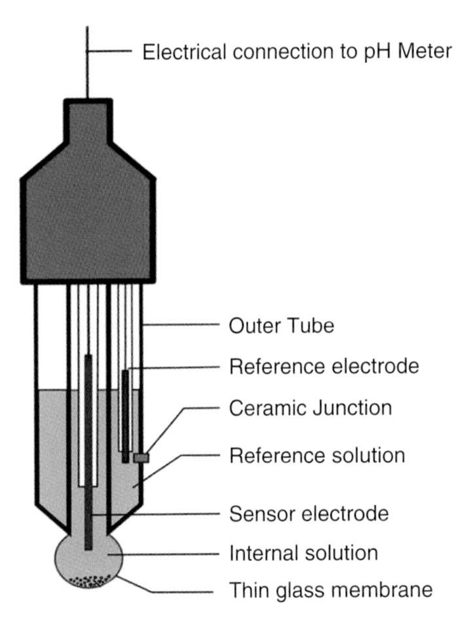

Figure 9.3 The combination pH electrode. Source: Adapted from Kaverin, Wikimedia Commons.

through the crystal from the higher concentration to the lower concentration, decreasing the F^- concentration in the internal solution. This decrease in F^- concentration is accompanied by an increase in Cl^- concentration, which creates the change in potential of the sensor electrode. The change in potential is read in millivolts versus the stable potential of a Ag—AgCl reference electrode, and is related to the concentration of F^- in the water sample as:

$$E = E_o - 0.059 \log[F^-(aq)]$$

Electrodes are calibrated using a series of KF standards that span the expected concentration range of the water samples. The electrode calibration curve generated from the standards is a plot of E as a function of log $[F^-]$ with a slope of −0.059 and an intercept of E_o. The F^- concentrations of the water samples are then calculated from the calibration curve.

The most important interfering species for the F^- electrode is the OH^- ion. The OH^- ion is the same size as the F^- ion and can also migrate through the doped LaF_3 crystal. The response of the F^- ISE will therefore be affected by changes in the sample pH. In addition, if the pH of the water is below pH 5, the H^+ will react with F^- to form HF and HF_2^-:

$$H^+ + F^- \rightleftarrows HF$$

$$HF + F^- \rightleftarrows HF_2^-$$

which are both too large to enter the membrane vacancies. Most manufacturers offer a total ionic strength adjustment buffer (TISAB), which contains acetic acid, NaCl, NaOH, and cyclohexanediaminetetraacetic acid (CDTA). The TISAB maintains the pH at 5.5 to eliminate both H^+ complexation and OH^- interference. It is added to both standards and samples to adjust the ionic strength, so that the background ionic strength is high and constant relative to the variable concentrations of F^-. In addition, it adds a complexing agent to disrupt any complexation of the F^- by Al^{3+}, Si^{4+}, and Fe^{3+}, which would prevent the F^- ion from migrating into the crystal.

The most important membrane ISE used for both field and laboratory water analysis is the DO electrode. The most common type of DO electrode, also known as a DO probe, is the Clark-type polarographic electrode shown in Figure 9.4. It uses a platinum or gold sensor electrode and a Ag—AgCl reference electrode, which are both immersed in a KCl electrolyte

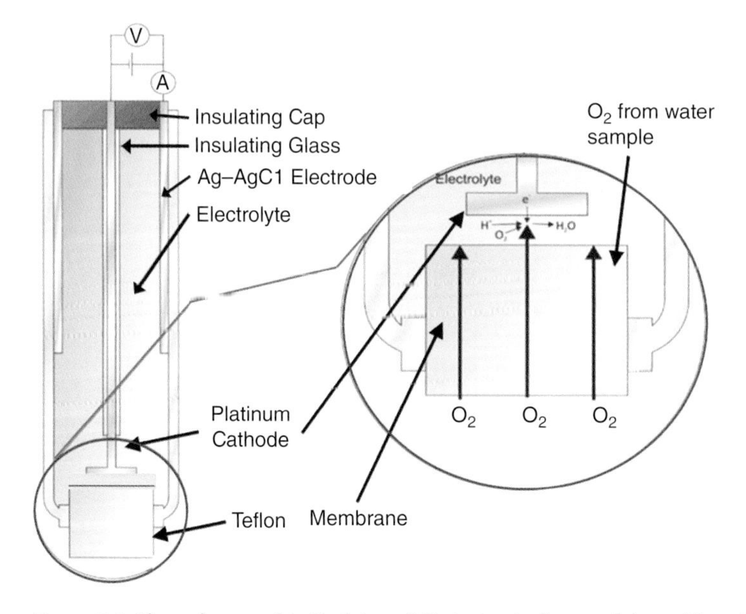

Figure 9.4 The polarographic Clark-type DO electrode. Source: Adapted from Larry O'Connell, Wikimedia Commons.

solution. The electrodes are isolated from the water sample by a thin Teflon membrane. Dissolved O_2 molecules in the water sample diffuse through the membrane to the sensor electrode. A constant polarizing voltage is applied across the electrodes, which causes the O_2 molecules to be reduced at the platinum electrode, producing an electrical signal that travels from the sensor electrode to the reference electrode with the following redox half reactions:

$$4Ag + 4Cl^- \rightarrow 4AgCl + 4e^-$$

$$O_2 + 4e^- + 2H_2O \rightarrow 2OH^-$$

Since the O_2 is rapidly reduced once it enters the electrolyte solution, the partial pressure inside the membrane is essentially zero. So the amount of oxygen diffusing through the membrane is proportional to the partial pressure of oxygen outside the membrane, and the current is directly proportional to the DO concentration in the sample in mg l^{-1} (ppm).

The DO in the sample is consumed during the measurement, so the concentration of O_2 near the electrode membrane is constantly being depleted. Because of this, the sample must be continuously stirred at the sensor tip to obtain a reliable and steady measurement. Also, the sensor requires a 10-min warm-up period to polarize the electrodes before use. From time to time the Teflon membrane requires replacement as it can adsorb organics in the water, which interferes with the gas diffusion through the membrane. The electrode is calibrated using a sample that is O_2 saturated (8.4 ppm) and a sample that is O_2 depleted (0 ppm). The O_2 saturated sample is obtained by bubbling air through a sample of tap water for 1 hour. The O_2 depleted sample is obtained as a freshly made solution of 1 g l^{-1} of sodium sulfite, which removes the O_2 in the water by the reaction:

$$2Na_2SO_3 + O_2 \rightarrow 2Na_2SO_4$$

Another type of electrochemical DO sensor operates as a galvanic cell instead of a polarographic cell. The sensor electrode in the galvanic-type O_2 ISE is constructed of a silver-sensing electrode and a zinc reference electrode. The difference in electrical potential between the two electrodes is large enough for the redox reaction to run spontaneously with the half reactions:

$$2Zn \rightarrow 2Zn^{2+} + 4e^-$$

$$O_2 + 2H_2O + 4e^- \rightarrow 4OH^-$$

The galvanic-type DO sensor does not require the constant polarizing voltage or the 10-min warm-up period required by the polarographic-type sensor. The DO concentration is determined by the Nernst equation as

$$E = E_o + 0.059 \log[O_2]$$

The DO electrodes are also used for the determination of biochemical oxygen demand (BOD). BOD is a measure of the amount of oxygen consumed by bacteria during the decomposition of organic material present in a water sample. It is determined by the measurement of the decrease in DO concentration over a given period of time in water samples stored at 20 °C. The determination of BOD involves the measurement of the initial DO in the water sample after the addition of a small amount of seed bacteria. The Winkler sample bottles are incubated at constant temperature in the dark, to ensure that no photosynthetic oxygen is produced. After a 5-day incubation period, the DO concentration is measured again. The 5-day BOD (BOD_5) is calculated as the difference between the initial DO (DO_0) and the final DO (DO_5) and is reported in milligrams per liter.

For water samples with a high organic content, the sample is diluted for the BOD_5 measurement. Measurements are also made on the diluted seed after the addition of 198 ± 30.5 mg l^{-1}

of glucose glutamic acid (GGA) as nutrient. This measurement evaluates the effectiveness of the seed as well as providing a value for the BOD_5 added to the sample. In this case, the final BOD_5 result is corrected for the dilutions and addition of seed as:

$$BOD_5 = \frac{(DO_0 - S_0) - (DO_5 - S_5)}{P}$$

where S_0 is the initial DO of the diluted seed, S_5 is the 5-day DO of the diluted seed, and P is the decimal volumetric fraction of the water sample used. A BOD_5 is also determined on the dilution water to make sure that it does not contain any significant available organic material. Samples with high aerobic bacteria content, such as domestic waste influent, do not require bacterial seeding.

The total BOD of wastewater is composed of two components; a carbonaceous biochemical oxygen demand (CBOD) and a nitrogenous biochemical oxygen demand (NBOD). The CBOD component represents the conversion of organic carbon to CO_2. The nitrogenous component represents the conversion of organic nitrogen, NH_3, and NO_2^- to NO_3^-. The nitrogenous oxygen demand typically begins in most water systems after about 6 days. However, for some waste water effluents nitrification can occur in less than 5 days when ammonia, nitrite, and nitrifying bacteria are present. In order to determine the CBOD independent of the NBOD on these water samples, 0.1 g of 2-chloro-6-(trichloromethyl) pyridine is added to the sample bottle before the incubation period. This inhibits the nitrifying bacteria in the sample. The result is then reported as CBOD.

The measurement of water conductivity is accomplished by measuring the resistance of a water sample by using a simple electrical cell with two electrodes. Conductivity is the measure of the ability of a material to pass an electrical current. It is the reciprocal of the resistivity, which is the ability of a substance to resist the flow of an electrical current. Both conductivity and resistivity are intrinsic properties, which are only dependent on the material and are independent of the amount of material acting as a conductor or resistor. However, conductance and resistance are extrinsic properties, which depend on the amount of material.

Since the charge-carrying species in a water sample are the dissolved ions, the conductivity is directly related to the concentration of dissolved ions in the water. The more dissolved ions present in the water, the higher the conductivity of the water. In many cases the conductivity can be directly correlated to the TDS. The standardized way of reporting conductivity is as specific conductance, which is the conductivity measured or corrected to 25 °C. A specific conductance temperature coefficient is used to convert conductivity measurements made at temperatures other than 25 °C to specific conductance. The temperature coefficient can vary from 0.0191 to 0.02, depending on the measured temperature and ionic composition of the water. Many conductivity meters measure both conductivity and temperature, and make the temperature corrections automatically.

The conductivity of freshwater samples is routinely determined from measurements of current and potential. A conductivity probe is connected to a conductivity meter, which generates an electrical current and measures the amount of current flowing through the water sample and the electrical potential generated by that current. The conductivity probe consists of two electrodes constructed of a conductive material, such as graphite, stainless steel, or platinum, which are connected to the conductivity meter as shown in Figure 9.5. The surface area and distance between the electrodes define the volume of water that is acting as the conductor. The ratio of the distance (d) between the electrodes to the area (a) of the electrodes gives a cell constant ($K = d/a$). The current applied between the electrodes is an alternating current (AC) in order to prevent electrolysis reactions from occurring on the electrode surfaces during the measurement.

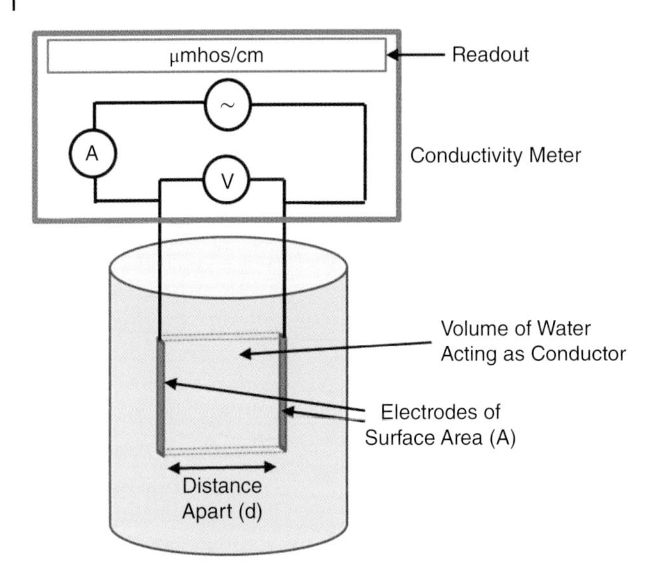

Figure 9.5 A conductivity meter with a two-electrode conductivity probe. The conductance is the amount of electrical current passed by the volume of water between the two electrodes and the conductivity is calculated from the cell constant ($K = d/A$).

The potential (V) and current (I) are both measured in order to determine the resistance of the sample ($R = I/V$) and calculate the conductance ($G = 1/R$). The conductivity meter then uses the conductance (G) and the cell constant (K) to display the conductivity (C) in units of microsiemens per meter ($\mu S\ cm^{-1}$) or micromhos per centimeter ($\mu mhos\ cm^{-1}$):

$$C\,(\mu S\ cm^{-1}) = K\,(cm^{-1}) \times G\,(\mu S)$$

A solution of known resistance is used to calibrate the conductivity cell and determine the cell constant, which the conductivity meter then uses to calculate the sample conductivity. A common calibration standard is 0.1 M KCl with a conductivity of $1410\ \mu S\ cm^{-1}$ at 25 °C.

9.2.2 Spectroscopic Methods

Most of the spectroscopic methods for the measurement of dissolved species in water are absorbance methods based on Beer's Law, which was discussed in Section 3.2. Beer's Law relates the absorbance (A) of a narrow wavelength of light to the concentration of the absorbing species (c) as:

$$A = \ln\left(\frac{I_0}{I}\right) = abc$$

where I_0 is the light intensity before passing through the sample, I is the light intensity after passing through the sample, b is the path length of the light through the sample in cm, c is the concentration of the absorbing species in mg l^{-1}, and a is the absorptivity of the absorbing species at the wavelength used for the absorption measurements in units of l $mg^{-1}\ cm^{-1}$. The concentrations of the analyte in the water samples are determined from a plot of absorbance versus concentration for a series of analyte standard solutions. This plot is the calibration curve for the measurement method, which is a straight line with a slope of $a \times b$ and a theoretical intercept of zero. The unknown sample concentrations are then determined as:

$$c = \frac{A}{ab} \text{ or } \frac{A}{\text{slope}}$$

Any absorbance by the reagents, as determined by the reagent blank, should be subtracted from the absorbances of the standards before the preparation of the calibration curve. It is also subtracted from the absorbances of the water samples before determining the analyte concentrations. Nonzero intercepts, other than those from reagent absorption, will sometimes occur due to instrumental factors such as stray light. In these cases, the concentrations of the water samples at low concentrations should be determined directly from the calibration curve instead of using the straight-line calculation. In addition, deviations from linearity often occur at high analyte concentrations due to interactions between the absorbing species or alterations in the refractive index of the solution. In these cases, the samples must be diluted to bring the concentrations into the linear portion of the calibration curve.

The nutrients, nitrogen (NH_3, NH_4^+, NO_3^-, NO_2^-, and organic nitrate), phosphate (PO_4^{3-} and organic phosphate), as well as SO_4^{2-} are routinely monitored by spectroscopic methods. The original measurements of NH_3 and NH_4^+ were accomplished spectroscopically by addition of Nessler's reagent to the water sample. Nessler's reagent is a 0.09 M solution of potassium tetraiodomercurate(II) ($K_2(HgI_4)$) in an alkaline solution of 2.5 M KOH. It reacts with NH_4^+ in the water to produce an enhanced yellow color by the reaction:

$$NH_4^+ + 2HgI_4^{2-} + 4OH^- \rightarrow HgO \cdot Hg(NH_2)I + 7I^- + 3H_2O$$

The concentration of the product $HgO{\cdot}Hg(NH_2)I$ is directly proportional to the concentration of NH_4^+ and is determined by the measurement of the absorbance at 425 nm according to Beer's Law. At high NH_4^+ concentrations a brown precipitate may form due to secondary reactions that produce $3HgO{\cdot}Hg(NH_3)_2I_2$ and $NH_2{\cdot}Hg_2I_3$. When this occurs, the sample must be diluted to obtain accurate results.

Another spectroscopic method for the measurement of NH_3 and NH_4^+ has more recently been adopted due to the high toxicity of the mercury complexes in Nessler's reagent. This method involves the reaction of Bertholet's reagent, which is an alkaline solution of phenol (C_6H_5OH) and the hypochlorite ion (ClO^-), with NH_3 to form the highly colored dye indophenol blue. The water sample is first buffered to pH 9.5 with borate (BO_3^{3-}), which converts all the NH_4^+ to NH_3 and reduces the interference from organic nitrogen compounds. The NH_3 is then distilled into a H_3BO_3 solution, followed by the addition of Bartholet's reagent along with sodium nitroferricyanide, which acts as a catalyst to ensure the reaction goes to completion. The overall reaction, shown in Figure 9.6, takes place in the following three steps:

$$NH_3 + OCl^- \rightarrow NH_2Cl + OH^-$$

$$C_6H_5OH + NH_3Cl \rightarrow p\text{-}OC_6H_4NCl$$

$$\text{p-}C_6H_4ONCl + C_6H_5OH \rightarrow OC_6H_4NC_6H_4OH$$

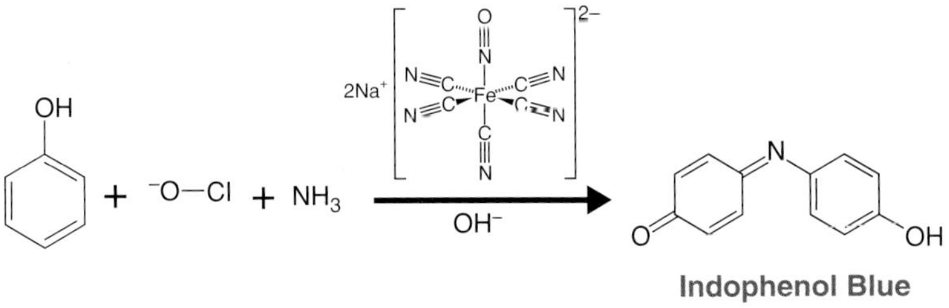

Figure 9.6 The reaction of Bertholet's reagent, which includes hypochlorite and phenol, with ammonia to form the intensely blue-colored dye indophenol blue. The reaction takes place in basic solution in the presence of a sodium nitroferricyanide catalyst.

The first step involves the reaction of NH_3 with ClO^- to form monochloramine (NH_2Cl). The NH_2Cl then reacts with C_6H_5OH to form a para-substituted benzoquinonechlorimine (p-OC_6H_4NCl). The p-OC_6H_4NCl reacts with another C_6H_5OH to form the cross-linked product indophenol blue. The light absorption of the indophenol blue at 630 nm is directly proportional to the total concentration of $NH_3 + NH_4^+$ in the water samples.

Bertholot's reagent is also used in the determination of total organic nitrogen, called total Kjeldahl nitrogen (TKN) after the Danish chemist Johan Kjeldahl who originally developed the procedure. The determination of TKN involves the digestion of the sample in H_2SO_4 with the addition of K_2SO_4, which increases the boiling point of the solution from the original 330 °C up to 400 °C. The sample is digested for 60–90 minutes to oxidize the organic compounds and convert the nitrogen to $(NH_4)_2SO_4$. The NH_4^+ is then converted to NH_3 by addition of NaOH and the NH_3 is distilled into an HCl solution. The amount of NH_4^+ in the distillate is determined by the addition of Bertholet's reagent followed by the measurement of absorption at 630 nm. The results are reported in milligrams per liter TKN, which is the sum of NH_3 nitrogen and organic nitrogen. It does not include other forms of inorganic nitrogen such as NO_3^- or NO_2^-. During the procedure it is important to prevent NH_3 contamination of the samples from laboratory and human sources, since any NH_3 present in the surrounding air can easily be absorbed into the solutions during the procedure.

The concentration of NO_3^- in water is determined by reaction with brucine sulfate ($(C_{23}H_{26}N_2O_4)_2 \cdot H_2SO_4$) and sulfanilic acid (4-aminobenzene-1-sulfonic acid; $H_2NC_6H_4SO_3H$) in 13 M H_2SO_4 to form the yellow-colored product cacotheline ($C_{21}H_{21}N_{13}O_7$), as shown in Figure 9.7. The NO_3^- ion is converted to HNO_3 under strong acidic conditions. Brucine then reacts with the HNO_3 to form cacotheline. The concentration of NO_3^- is determined by measuring the absorption of cacotheline at 410 nm. Any NO_2^- present in the water reacts with the sulfanilic acid to give a diazo product that absorbs at a different wavelength than cacotheline. The intensity of the color development is dependent on the temperature and the time of development, so these two factors must be carefully controlled during the procedure. The reaction of brucine with HNO_3 is temperature controlled by placing the solution tubes in a boiling water bath for 25 minutes. After the reaction is complete, the sample tubes are quickly cooled to room temperature using a cold-water bath and the absorption measurements are obtained immediately.

The natural color of some surface waters containing humic substances can also give an absorbance at 410 nm. In addition, some organic compounds can form colored products in the hot concentrated acid solution. The contribution from these interferences is removed by subtracting the absorbance of a sample blank that does not contain the brucine sulfate reagent. The method is calibrated by using a series of NO_3^- standards that span the concentration range of the samples. The color development follows Beer's Law only at NO_3^- concentrations of 0.1–1 mg l^{-1}. Deviations at higher concentrations are due to interactions between the

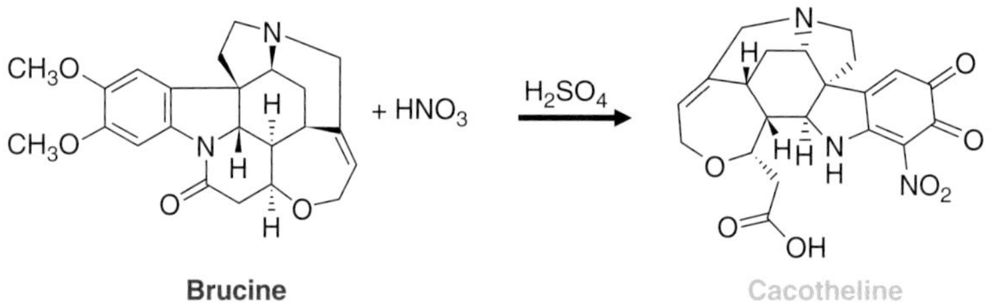

Figure 9.7 The reaction of brucine with nitric acid which is formed from nitrate in strong sulfuric acid solution, to produce the colored product cacotheline.

cacotheline molecules, which cause the absorbance to be lower than predicted by Beer's Law. Samples with NO_3^- concentrations >1 mg l^{-1} should be diluted to bring the concentration within the linear portion of the calibration curve.

The NO_3^- in water can also be determined by reduction to NO_2^- followed by reaction with *N*-(1-naphthyl)-ethylenediamine to form a red-colored diazo dye. Reduction of the NO_3^- is accomplished by passing the filtered water sample through a column containing granulated copper-coated cadmium metal. The NO_2^- then reacts with sulfanilamide (4-aminobenzesulfonamide; $C_8H_8N_2O_2S$) and *N*-(1-naphthyl)-ethylenediamine (NED) in two steps to form the diazo dye as shown in Figure 9.8. The sum of the concentrations of $NO_3^- + NO_2^-$ is determined by the absorbance of the colored product measured at 540 nm. The NO_2^- concentration is determined by reaction with the sulfanilamide and NED without cadmium reduction. The NO_3^- concentration is then obtained by subtracting the NO_2^- concentration from the total $NO_3^- + NO_2^-$ concentration:

$$[NO_3^-] = [NO_3^{2-}(red)] - [NO_2^-]$$

where $[NO_3^{2-}(red)]$ is the $NO_3^- + NO_2^-$ concentration, which is obtained from the reduced sample. The method is calibrated using a series of NO_3^- standards that span the concentration range of the samples. Two NO_2^- standards of different concentrations should also be compared to the results obtained from NO_3^- standards of the same concentration to determine the efficiency of the cadmium reduction.

As discussed in Section 8.5.2, phosphorus is found in aqueous systems as organic phosphates and inorganic phosphate (PO_4^{3-}). It occurs most often in aquatic systems as solvated PO_4^{3-} ions, commonly known as orthophosphate. Orthophosphate is determined by reaction with ammonium molybdate ($(NH_4)_2MoO_4$) and antimony potassium tartrate ($K_2Sb_2(C_4H_2O_6)_2$) in acidic solution to form the polyatomic phosphomolybdate complex 12-molybdophosphoric acid (MDA) as:

$$PO_4^{3-} + 12MoO_4^{2-} + 27H^+ \rightarrow H_3PO_4(MoO_3)_{12} + 12H_2O$$

Under acidic conditions, an excess of MoO_4^{2-} ions forms larger molybdate anionic structures, which are in equilibrium with the MoO_4^{2-} ions:

$$7MoO_4^{2-} + 8H^+ \rightleftarrows Mo_7O_{24}^{6-} + 4H_2O$$

In this manner the phosphomolybdate complex self-assembles into a Keggin structure of composition $P^{n+}Mo_{12}O_{40}^{(8-n)-}$. The Keggin structure is a heteropoly anion with the transition

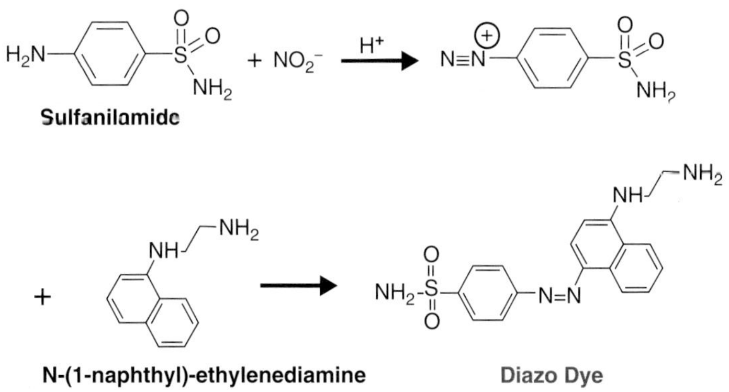

Figure 9.8 The reaction of NO_2^- with sulfanilamide to form the *p*-diazonium sulfanilamide intermediate, which undergoes a para-addition of *N*-(1-naphthyl)-ethylenediamine (NED) to form the colored diazo dye.

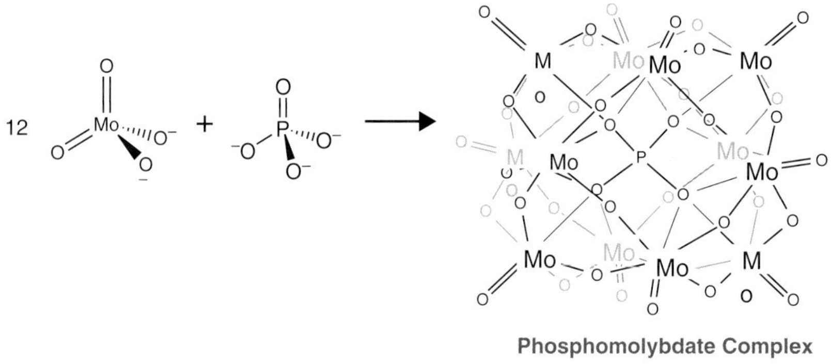

Phosphomolybdate Complex

Figure 9.9 The reaction of molybdate anions with orthophosphate to form the phosphomolybdate complex, which self-assembles into a Keggin structure. Source: Adapted from Myasein, Wikimedia Commons.

metal oxyanions linked together by shared oxygen atoms to form a cage-like three-dimensional framework as shown in Figure 9.9. At the center of the structure is the phosphorus surrounded by four tetrahedral oxygen atoms. This tetrahedral phosphate is surrounded by 12 octahedral MoO_6 units linked to one another by neighboring oxygen atoms. There are a total of 24 bridging oxygen atoms that link the 12 molybdenum atoms. The 12 octahedral MoO_6 units are arranged on a sphere in four Mo_3O_{13} units, which gives the entire structure an overall tetrahedral symmetry.

The phosphomolybdate complex is reduced by ascorbic acid ($C_6H_8O_6$) to an intensely blue-colored complex called "molybdenum blue." The near colorless phosphomolybdate complex can accept more electrons when in the Keggin structure to form a mixed-valence molybdate complex as expressed in the half reaction:

$$PO_4Mo_{12}{}^{(VI)}O_{36}{}^{3-} + 4e^- \rightleftarrows PMo_4{}^{(V)}Mo_8{}^{(VI)}O_{40}{}^{7-}$$

The absorbance of the molybdenum blue complex at 880 nm is directly proportional to the concentration of the orthophosphate in the water sample. Total phosphorus (TP) is determined by digesting the sample in H_2SO_4 and $K_2S_2O_8$ at 100 °C for 60 minutes, which decomposes all of the organic phosphates to H_3PO_4. Total dissolved phosphorus (TDP) is determined by filtering the sample before digestion. Insoluble phosphorus (IP) is determined by subtraction (IP = TP − TDP). Both iron and arsenate ($AsO_4{}^{3-}$) can interfere with the measurement of orthophosphate by the molybdenum blue method. Arsenate in the oxidation state As(V) can also form a molybdenum complex which absorbs at 880 nm, causing positive interference. This is prevented by first reducing the As(V) to a lower oxidation state by the addition of sodium bisulfite ($NaHSO_4$). Although iron does not form a molybdenum complex, it can be reduced by ascorbic acid. So, high concentrations of iron in the samples results in a reduction of the ascorbic acid concentration, causing negative interference. However, the iron interference can also be removed by the addition of $NaHSO_4$.

Since the $SO_4{}^{2-}$ ion does not form many colored complexes, the determination of $SO_4{}^{2-}$ in water is based on the spectroscopic determination of excess reactant after the formation of an insoluble $SO_4{}^{2-}$ product. The $SO_4{}^{2-}$ reacts with an excess amount of $BaCl_2$ to form the insoluble $BaSO_4$ precipitate:

$$xSO_4{}^{2-}(aq) + yBa^{2+}(aq) \rightarrow xBaSO_4(s) + (y - x)Ba^{2+}(aq)$$

The amount of unreacted Ba^{2+} is then determined by reaction with methylthymol blue (MTB) to form a blue complex that absorbs at 460 nm:

$$(y - x)\,Ba^{2+}(aq) + yMTB^{5-}(aq) \rightarrow (y - x)BaMTB^{3-}(aq) + xMTB$$

However, the MTB ($C_{37}H_{44}N_2O_{13}S$) can complex other divalent metal ions, including Ba^{2+}, to form products that also absorb at 460 nm. So metals must be removed from the sample prior to analysis by using an ion exchange resin. In addition, to prevent the MTB from complexing the Ba^{2+} before the precipitate is formed, $BaCl_2$ and MTB are added to the water samples at a pH of 2.5–3.0. At this low pH the MTB is completely protonated, as shown in Figure 9.10(A). In this completely protonated form the MTB cannot complex the Ba^{2+}. After precipitation the pH is raised to 12.5–13.0. At this pH the MTB becomes completely ionized (MTB^{5-}), as shown in Figure 9.10(B). In this ionized form the MTB will complex the unreacted Ba^{2+} to form the blue-colored product $BaMTB^{3-}$, which absorbs at 460 nm.

The original concentration of SO_4^{2-} in the water sample is then equal to the amount of $BaSO_4$ precipitate, which is calculated as the difference in the initial concentration of Ba^{2+} and the amount of unreacted Ba^{2+} as $BaMTB^{3-}$:

$$[SO_4^{2-}] = [Ba^{2+}] - [BaMTB^{3-}]$$

Cyanide (CN^-) is used in the mining of gold and silver, in electroplating, and in the steel and chemical industries. It can be discharged into surface waters in industrial waste water effluents from these activities. Cyanide can exist in water as the free CN^- ion, as hydrocyanic acid (HCN), or as metal-bound CN^-. It can be determined as free CN^- ($CN^- + HCN$) or as total CN^-. For the measurement of total CN^-, the metal-bound CN^- is released as HCN by reflux under strong acidic conditions. The measurement of free CN^- is then accomplished by distilling an acidified sample and trapping the HCN gas in NaOH solution. The CN^- ion is converted to a colored product by a series of reactions, shown in Figure 9.11. The CN^- is converted to CNCl by reaction with the strong oxidizing agent chloramine-T. The CNCl then reacts with pyridine by first adding the C≡N group to the pyridine nitrogen, which then reacts with Cl^- ion causing the cyclic pyridine to undergo ring opening to form the acyclic glutaconaldehyde. The glutaconaldehyde then undergoes a condensation reaction with two molecules of barbituric acid to form an unsaturated organic compound that has strong light absorption at 570 nm. The CN^- concentration in water is directly proportional to the absorbance of the product of

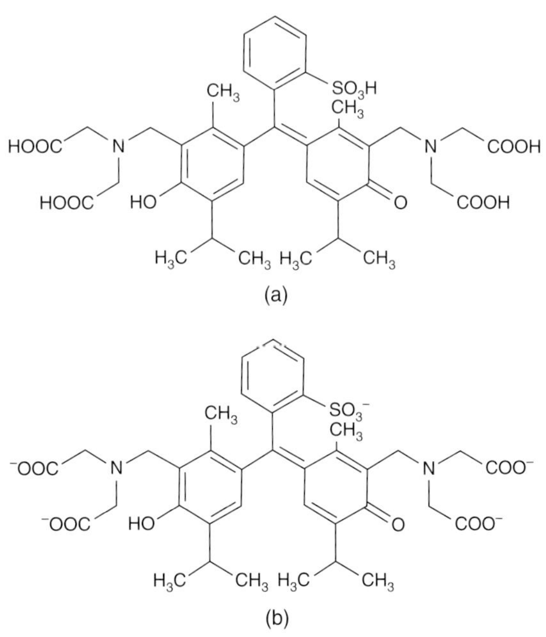

Figure 9.10 The chemical structure of methylthymol blue at (A) pH = 2.5–3.0 and (B) pH = 12.5–13.0.

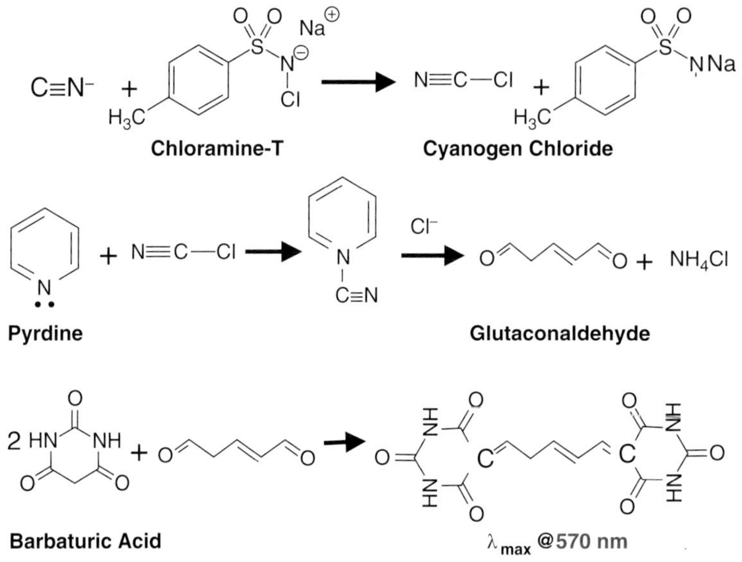

Figure 9.11 The reaction of CN^- with chloramine-T, followed by pyridine and barbituric acid to form a compound that absorbs at 570 nm for the spectroscopic determination of CN^- in water.

the barbituric acid reaction, which is determined by absorption measurements according to Beer's Law.

A spectroscopic measurement used most commonly for ground water and drinking water is the determination of water hardness. The hardness of water is defined as the amount of Ca^{2+} and Mg^{2+} salts in the water. Hardness is usually reported as mg l $CaCO_3$, because most methods of analysis cannot readily distinguish between Ca^{2+} and Mg^{2+} and most hardness is caused by $CO_3{}^{2-}$ minerals. Water hardness is an esthetic issue due to poor lathering of soap and spots left on dishes after cleaning. It is also a mechanical issue causing mineral build-up in pipes, plumbing fixtures, and hot water heaters. However, it is not a health or environmental issue. The measurement of water hardness is based on the complexation of Ca^{2+} with ethylenediaminetetraacetic acid (EDTA), which is stronger than the Mg^{2+} complex. A solution of an Mg^{2+}–EDTA complex is added to the water sample and allowed to equilibrate with the Ca^{2+} in the water. Since the Ca^{2+} forms a stronger bond to EDTA than Mg^{2+}, it will exchange with the Mg^{2+} bound to the EDTA, releasing it into the solution as shown in Figure 9.12(A). The free Mg^{2+} is then measured by the addition of the azo organic dye calmagite ($C_{17}H_{14}N_2O_5S$). The calmagite forms a red-colored complex with the free Mg^{2+} in an alkaline solution as shown in Figure 9.12(B). The concentration of Ca^{2+} in the water is directly proportional to the absorbance of the Mg^{2+}–calmagite complex measured at 520 nm.

The determination of dissolved organic carbon (DOC) and total organic carbon (TOC) are bulk measurements of the amount of organically bound carbon in a water sample. The measurement of DOC requires that the sample is first filtered through a 0.45 μm filter to remove any organic particulate matter. The measurement of TOC in samples with high particulate levels requires that the unfiltered water sample must first be homogenized. Any interference from dissolved inorganic carbon species (DIC) can be eliminated by acidifying samples to pH 2 or less to convert inorganic carbon species to CO_2. The CO_2 from the DIC is then removed from the sample by purging with an inert gas or by vacuum degassing. However, this procedure also removes any volatile organic carbon in the sample and so the results will represent non-purgeable organic carbon.

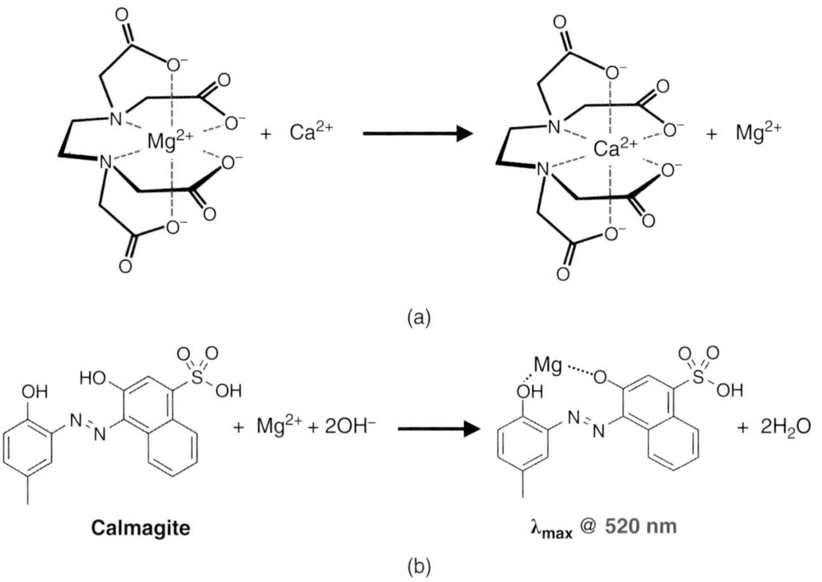

Figure 9.12 The displacement reaction of Mg^{2+} bound to ethylenediaminetetraacetic acid (EDTA) by free Ca^{2+} ion followed by the complexation of the free Mg^{2+} by calmagite to form a red-colored product.

The measurement methods for DOC and TOC are based on the spectroscopic determination of CO_2, which is generated from the organic carbon by either high-temperature combustion or low-temperature chemical oxidation. In the high-temperature combustion method, a small measured portion of the sample is injected into a heated reaction chamber at 680 °C, which is packed with an oxidative catalyst such as copper oxide, cobalt oxide, barium chromate, or platinum. The water is first vaporized then the organic carbon is oxidized to CO_2 and H_2O. In the low-temperature oxidation method, the organic carbon in the sample is oxidized to CO_2 by persulfate ($K_2S_2O_8$) in the presence of UV light. The UV lamp is submerged in a continuously gas-purged reactor filled with a constant feed of persulfate solution. In either case, the samples can be introduced serially into the reactor by an autosampler or injected manually. After oxidation of the organic carbon, the CO_2 is purged from the sample, dried, and transferred into a nondispersive infrared spectrometer (NDIR), which has been described in Section 7.2.2. The organic carbon concentrations in the water sample are directly proportional to the absorbance of CO_2 in the carrier gas stream measured at 4.26 μm.

The instruments are calibrated using a sample blank and a series of standards of potassium hydrogen phthalate (KHC_8O_4 or KHP), which is easily oxidized. Standards of benzoquinone ($C_6H_4O_2$) can also be used for organics that are more difficult to oxidize. Sample concentrations are typically calculated automatically from the calibration standards. Alternately, the instrument response of the standards can be converted to a calibration curve for manual calculation of sample concentrations in mg C l^{-1}. Neither DOC nor TOC provides any information about the chemical structures or composition of the complex mixture of organic compounds in the water sample. However, they provide a measure of the amount of organic species in the water sample. Measurements of DOC and TOC are used in many areas as regulatory requirements for water quality of waste water effluents in a similar manner as BOD measurements. They are also useful in providing an estimate of the volume of water needed for analysis of specific organic compounds.

DO can also be determined by an optical luminescent method, which operates on the principle that dissolved O_2 quenches both the lifetime and intensity of the fluorescence from some chemical dyes. The lifetime and intensity of the fluorescence emission are at their maximum when there is no O_2 present and are decreased as the concentration of O_2 increases. So, the lifetime and intensity of the fluorescence are inversely proportional to the DO concentration and can be determined by the Stern–Volmer equation:

$$\frac{I_0}{I} = 1 + k_q t_0 [O_2]$$

where I_0 is the maximum fluorescence intensity, I is the fluorescence intensity in the presence of O_2, k_q is the quenching rate coefficient, and t_0 is the fluorescence lifetime of the dye.

As shown in Figure 9.13, a short-wavelength (blue) light from an LED is directed onto a polymer sensor that has a fluorescent dye, such as rhodium or platinum porphyrin, embedded into it. The dye emits a long-wavelength (red) fluorescence light, which is monitored by a photodiode detector. The O_2 in a water sample passes through an opaque permeable membrane to reach the polymer dye sensor. Unlike with the DO electrode sensors, the O_2 is not depleted on the internal side of the membrane. Instead, it rapidly reaches equilibrium with the DO in the water. Either the intensity or the lifetime of the fluorescence from the dye can be measured to give the DO concentration in mg l^{-1} using the Stern–Volmer equation. However, the lifetime method is more commonly used because it is less sensitive to degradation of the dye than the intensity measurements. The excitation light is modulated, creating a modulated fluorescence emission signal, and the fluorescence lifetime is determined from the phase shift between the fluorescence signal and the excitation light. The optical luminescence DO sensor is calibrated using the same two-point calibration that is used for the electrochemical DO sensor described in Section 9.2.1. The optical luminescence DO sensor is typically used for low DO levels. This is because the fluorescence of the dye is rapidly quenched by O_2, leading to a very small signal at high O_2 concentrations. The small signals result in poor measurement precision and accuracy at high DO levels.

The three main spectroscopic methods used for the analysis of metals and some metalloids in water are based on atomic absorption spectroscopy (AAS), atomic emission spectroscopy (AES), and mass spectroscopy (MS). Table 9.1 lists the elements that are measured in natural waters by each of these methods, along with their detection limits. Methods based on AAS measure the light absorption of the free analyte atoms created in a graphite furnace. The graphite furnace is a pyrolytically coated graphite tube open at both ends, which produces a volatilized

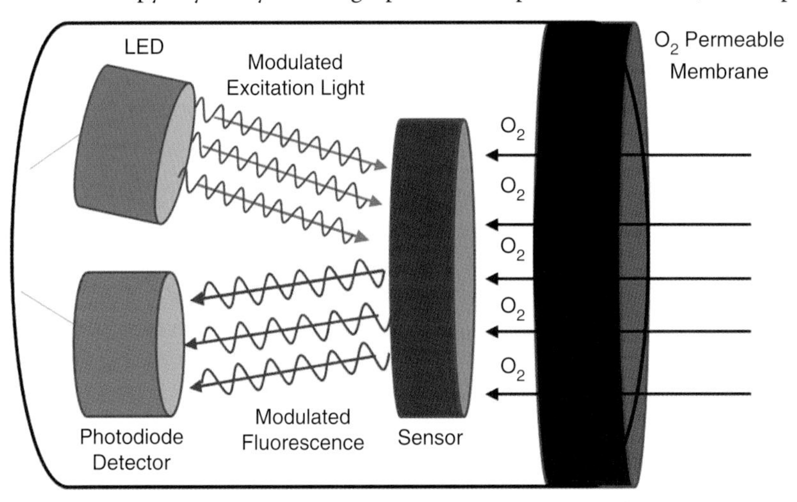

Figure 9.13 The optical luminescent DO sensor.

Table 9.1 The elements measured by graphite furnace AAS, ICP–AES, and ICP–MS along with the typical detection limits in water for each method.

Element	Graphite furnace AAS (μg l^{-1})	ICP–AES (μg l^{-1})	ICP–MS (ng l^{-1})
Aluminum	—	2.2	1.0
Antimony	0.05	0.9	0.5
Arsenic	0.05	1.4	1.0
Barium	0.35	0.05	0.5
Beryllium	0.008	0.02	0.5
Boron	—	0.3	50
Cadmium	0.002	0.1	0.5
Calcium	—	0.02	5.0
Chromium	0.004	0.2	0.5
Cobalt	—	0.2	0.5
Copper	0.014	0.3	0.5
Iron	0.1	3.3	2.0
Lead	0.05	1.1	0.5
Lithium	—	0.3	5.0
Magnesium	—	0.02	1.0
Manganese	0.005	0.06	0.5
Molybdenum	0.03	0.5	0.5
Nickel	0.07	0.6	2.0
Phosphorus	—	0.5	1000
Potassium	—	1.0	50
Selenium	0.05	1.3	2.0
Silicon	—	10.0	90
Silver	0.005	0.5	0.5
Sodium	—	0.4	10
Strontium	—	0.01	1.0
Thallium	—	0.2	1.0
Tin	—	0.5	10
Titanium	—	0.1	1.0
Vanadium	0.1	0.2	0.5
Zinc	0.02	0.4	1.0

Detection limits for AAS and ICP–AES (μg l^{-1}); for ICP–MS (ng l^{-1}).

population of free analyte atoms. This is accomplished in three steps: 1) a drying step which removes the solvent; 2) a pyrolysis step which removes the organic and inorganic matrix materials; and 3) an atomization step which creates the free atoms in a confined zone within the furnace. The absorbance of the free atoms is then measured by a line spectrum emission lamp, either a hollow cathode lamp (HCL) or an electrodeless discharge lamp (EDL), coupled to a photodetector as shown in Figure 9.14. The absorbance is directly proportional to the concentration of the metal in the water sample according to Beer's Law, and the concentrations are determined from a calibration curve.

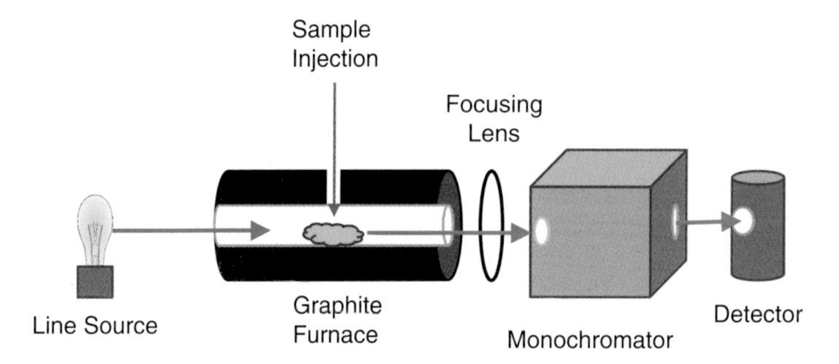

Figure 9.14 A graphite furnace atomic absorption spectrometer for determining the concentrations of metals and some metalloids in water samples.

A small amount (20–100 μl) of the acidified water sample is placed into the graphite furnace through an injection port on top of the graphite tube. The graphite tube is heated slowly to a temperature of 100–120 °C to evaporate the water, which is carried out of the furnace by a gas flow of 95% Ar and 5% H_2 at 250 ml min^{-1}, leaving the dissolved solids behind. The temperature is then increased to 420 °C to decompose and volatilize the inorganic and organic matrix components from the sample, leaving behind the sample element in a less complex matrix. After pyrolysis of the sample matrix, the sample it heated to a temperature of 1400 °C with no gas flow through the graphite tube. This volatilizes the analyte element and concentrates it in the path of the light beam for absorption measurements. A monochromator is placed after the graphite furnace to isolate the appropriate emission line from the source and prevent scattered light from reaching the detector.

The sample should be filtered prior to measurement of dissolved species. However, measurements of total species concentrations require prior oxidation by conventional acid digestion (HNO_3 + HCl) before analysis. Samples containing large amounts of organic materials may also require acid digestion to minimize interferences from broad band absorption if the instrument used does not have adequate background correction. Background correction is commonly applied using a continuum deuterium (D_2) source or by using a Zeeman magnetic field. In D_2 background correction, the source lamp and the D_2 lamp are sequentially pulsed. The D_2 lamp records any broad band absorption, which is then subtracted from the absorption at the source line. Zeeman background correction uses a magnetic field to split the absorption line into three components. Since the different components have different polarization, the absorption line can be removed when passed through a light polarizer. This gives a background measurement at exactly the same wavelength as the absorption line. It is particularly useful when analyzing for As in the presence of Al, and when analyzing for Se in the presence of Fe, as these elements have absorption lines close enough to interfere with each other. The AAS instruments that use Zeeman background correction can more effectively overcome interfering spectral emission lines while D_2 background correction is specifically for broadband absorption interferences.

Methods based on AES use an inductively coupled plasma (ICP) torch to create a population of analyte atoms and ions while at the same time promoting them into an excited state. The wavelength of the light emitted by the excited analyte atoms as they return to the ground state is characteristic of the particular element. The intensity of the emission (I_e) is directly proportional to the number of atoms in the excited state (N^*):

$$I_e = k_e N^*$$

where k_e is the efficiency of the electronic transition to the excited state and N^* is directly proportional to the concentration of atoms in the sample ($N^* = kc$). Thus, the concentration of an element in a sample can be determined in the same way as with absorption methods, by obtaining a calibration curve of I_e as a function of c. The concentration of the element in a water sample is then determined from I_e and the slope of the calibration curve. Calibration curves for AES can also suffer from interferences at high analyte concentrations, resulting in nonlinearity. This is caused by the transfer of energy from one analyte atom to another, reducing the light emission intensity. Water samples with high concentrations of the elements of interest should be diluted to bring them into the linear portion of the calibration curve. The water samples are filtered prior to analysis for determination of dissolved species and the determination of total analyte concentrations requires sample digestion with HNO_3 and HCl to solubilize suspended solid material. The sample should be diluted so that the TDS is $<2\%$ in order to reduce potential interference from high concentrations of dissolved salts.

The line emissions from the multiple analytes in the sample can be separated in two ways: a scanning monochromator can be used to sequentially focus each line onto a single exit slit, or a polychromator can be used to focus a series of emission lines onto several exit slits at the same time, as shown in Figure 9.15. In either case, the intensity of each emission line is measured by the appropriate number of photodetectors. The polychromator system can measure a number of analytes simultaneously, dramatically reducing analysis time. The ICP/AES can measure many elements in the sample, in a reduced amount of time over the graphite furnace AAS system, which requires changing lamps for each element to be measured. The ICP/AES also has the ability to handle both simple and complex sample matrices with a minimum of interference due to the very high temperature of the ICP torch.

The ICP plasma consists of argon gas, which is ionized by an intense electromagnetic field created by an oscillating radiofrequency (RF) generator. The RF field causes the gaseous ions to oscillate with the field, resulting in extreme heat. A stable high-temperature plasma of 7000–10 000 K is created as a result of the high-energy collisions between the ionized argon atoms. The torch consists of three concentric quartz tubes. A cooling gas flows through the outer tube to cool the outer portion of the torch and isolate the plasma, which forms in the middle tube. The sample is aspirated in an argon carrier gas into a small inner tube of about 1–2 mm diameter. As the sample exits the injection tube it enters the plasma, where the atoms collide

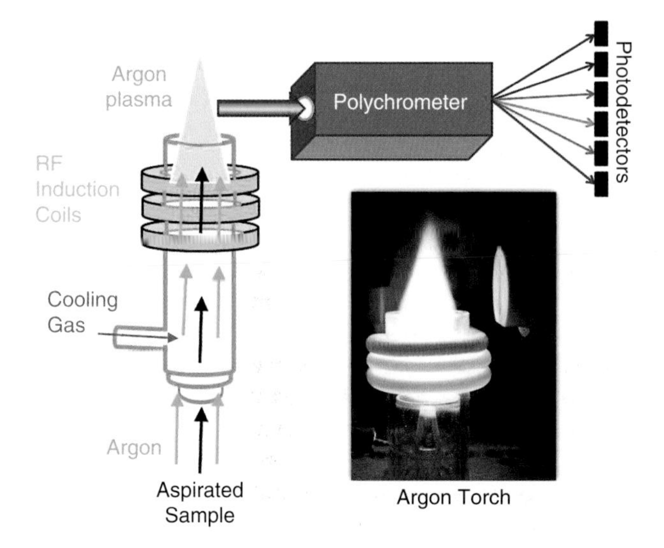

Figure 9.15 An inductively coupled plasma atomic emission spectrometer (ICP/AES) with an argon ion torch and a polychromator to isolate the emission lines for detection by a series of photodetectors. A photograph of an argon plasma ICP in operation at lower right. Source: Wblanchard, Wikimedia Commons.

with the argon ions and free electrons. These high-energy collisions cause the sample molecules to break up into atoms, which then lose electrons and recombine repeatedly in the plasma.

An ICP torch can also be coupled to a mass spectrometer for separation and detection of the analyte ions (ICP–MS). The analyte ions formed in the plasma at atmospheric pressure can be pulled into the inlet of a mass spectrometer at $<1 \times 10^{-5}$ Torr by transporting them through an interface region between two metal interface cones as shown in Figure 9.16. The interface cones, consisting of a sampler cone and a skimmer cone, sample the center portion of the ion beam coming from the ICP torch through a small 1 mm hole in the center of each cone. A light stop placed behind the skimmer cone blocks the emission light from the ICP. The ions from the ICP are focused into the entrance slit of the mass spectrometer by electrostatic lenses. After the ions enter the mass spectrometer they are separated by their mass/charge (m/e) ratio by a mass filter. The most commonly used type of mass filter is a quadrupole mass filter, which uses four metal rods placed symmetrically in a tube shape, allowing the ions to travel down a path between the four rods as shown in Figure 9.17. Alternating AC and DC voltages are rapidly switched between two pairs of the four rods, so that the rods opposite each other have the same charge. The positively charged ions are attracted toward the negatively charged rods and are repelled when the charge changes to positive. This causes the ions to spiral as they move down the length of the quadrupole. Stable ion paths, which allow the ions to reach the ion detector, depend on the m/e ratio, the charge on the rods, and the rate of charge oscillation. Unstable paths cause the ions to collide with the quadrupole rods. Only ions with a specific m/e ratio can make it through the quadrupole to the ion detector. A positively charged plate placed at the end of the rods acts to deflect the ions into a high-voltage electron multiplier that detects the ion signal. The signal is then sent to a computer and stored with time to give a mass spectrum of the ion pulse intensity versus ion mass, assuming the analyte is a singly charged ion.

The ICP/MS systems have much lower detection limits than the graphite furnace AAS or the ICP–AES, as shown in Table 9.1. Detection limits for most metals range from 0.002 to 0.35 $\mu g\, l^{-1}$ for graphite furnace AAS and 0.01–3.0 $\mu g\, l^{-1}$ for ICP–AES. The detection limits for ICP–MS range from 0.5 to 2 $ng\, l^{-1}$. Since the element detection is based on the m/e ratio, ICP–MS can also provide information on the different isotopes of an analyte species. In addition, the ICP–MS can determine multiple analytes simultaneously, resulting in a much higher sample throughput than the graphite furnace AES.

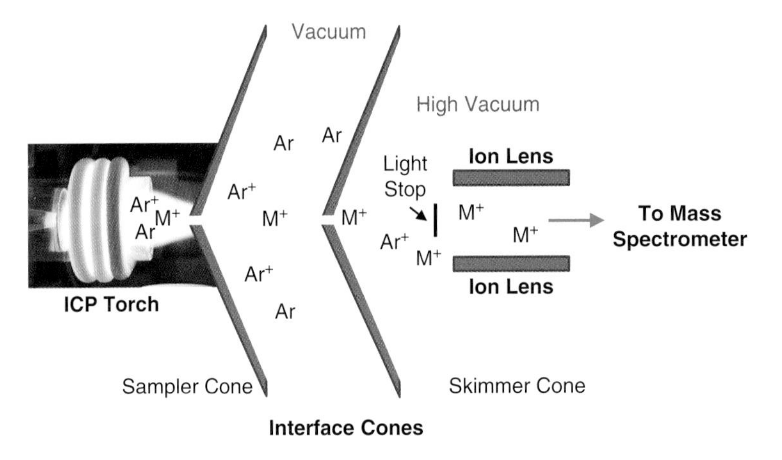

Figure 9.16 An ICP–MS interface that allows sampling the atmospheric pressure plasma and directs it into a high-vacuum mass spectrometer using two cones, the sampling cone and the skimmer cone.

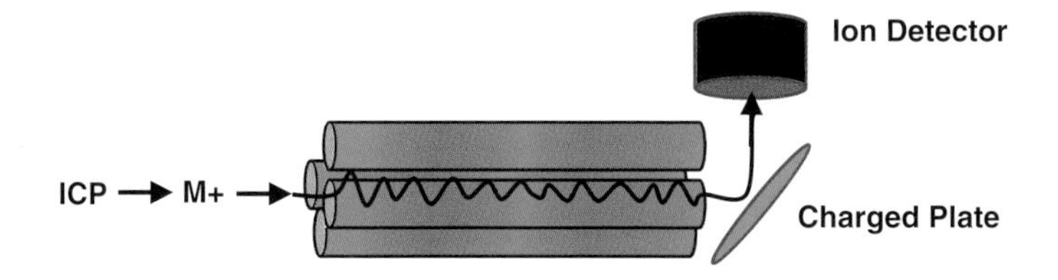

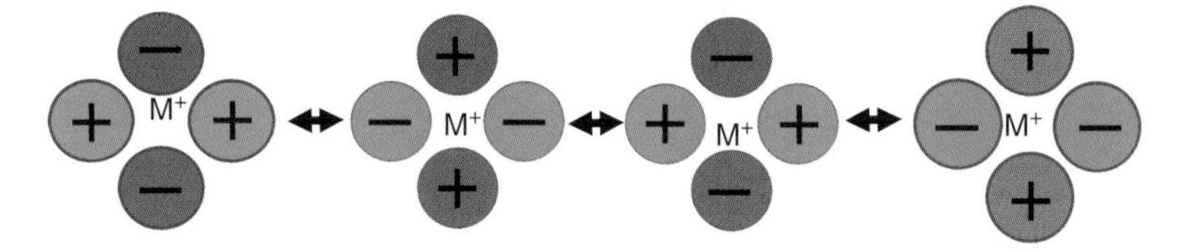

Figure 9.17 A quadrupole mass analyzer (top) creates a spiral ion path by using an alternating charge between two opposite pairs of rods (bottom).

Most interferences in ICP–MS measurements are caused by atomic or molecular ions that have the same *m*/*e* ratio as the analyte. The interferences caused by isotopes of different elements with the same mass number but different atomic numbers, such as ^{40}Ca and ^{40}Ar or ^{87}Sr and ^{87}Rb, are easily predicted and can easily be corrected by the instrumental software. However, interferences from polyatomic ions created from atoms in the sample matrix, reagents used for sample preparation, plasma gases, or entrained atmospheric gases are numerous and more difficult to predict. Examples include $^{40}Ar^{35}Cl$ interference for ^{75}As, $^{37}Cl^{16}O$ interference for ^{52}Cr, and $^{40}Ar^{16}O$ interference for ^{56}Fe. Once identified, corrections can be made for this type of interference by using the natural abundances of the different isotopes to calculate the amount of the expected interferences. However, these calculations can become complicated very quickly if multiple interferences occur.

The use of reagent blanks and standards, along with standard additions, can be used to help identify the types of interferences. In a standard addition, a standard is added directly to a sample and the result of the sample is compared to the result of the same sample with the standard added. This method is used in situations where the sample matrix can contribute to the signal, known as a matrix effect. Reagent interferences, such as those attributed to polyatomic ions containing chlorine, can be eliminated by changing the sample preparation procedure. Comparing the results obtained from multiple isotopes of the analyte can not only identify the presence of interferences, but sometimes identify an isotope with little or no interference, which is best for analyte determination.

Mercury is one element that is not routinely measured by either graphite furnace AAS, ICP–AES, or ICP–MS. This is because mercury is a volatile element that is typically lost during the work-up and sample handling procedures required for these measurement methods. Mercury is commonly determined as a cold elemental vapor using AAS or atomic fluorescence. The AAS instrumentation operates in the same way as a graphite furnace AAS, shown in Figure 9.14, except that a flow-through quartz absorption cell replaces the furnace and the light source is a low-pressure mercury lamp that emits UV light at 253.7 nm. The atomic fluorescence is measured in much the same way as molecular fluorescence, discussed in Section 3.4. The Hg^0 atoms in a flow-through quartz fluorescence cell absorb light at 253.7 nm and emit fluorescent light at 253.7 nm. The detector is placed off-axis, usually at 90° from the sample, as shown in Figure 9.18 (purple outline). This prevents the light from the excitation lamp from reaching the

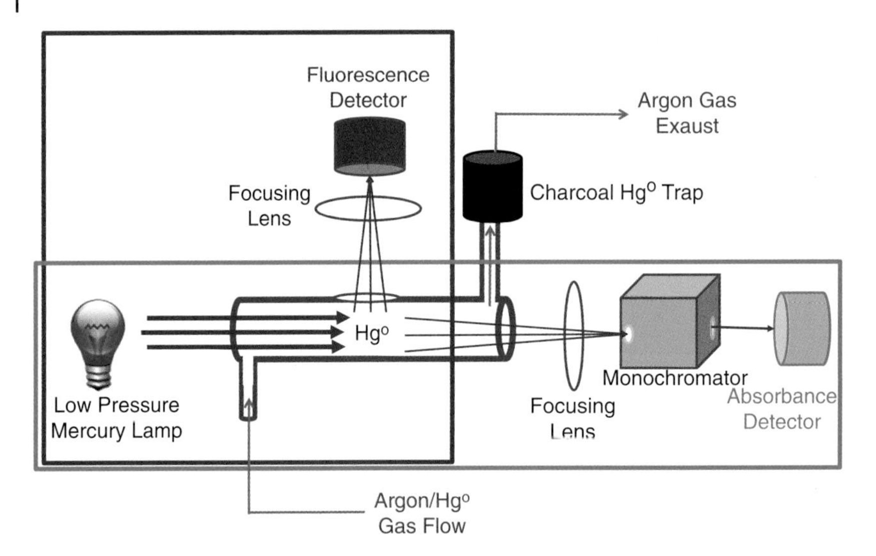

Figure 9.18 An atomic absorption cold vapor mercury spectrometer (green outline) and an atomic fluorescence cold vapor mercury spectrometer (purple outline) for the determination of mercury in water.

detector. Both AAS and atomic fluorescence are commonly used, but the atomic fluorescence method has a wider operating range and is more sensitive than AAS.

Mercury can be present in water as the dissolved Hg^0 and Hg^{2+}, organically complexed Hg^{2+}, and covalently bound organomercury compounds, such as CH_3HgCl, $(CH_3)_2Hg$, and $C_6H_5HgOOCCH_3$. It can also be in particulate form. For the determination of dissolved Hg^0 and Hg^{2+}, the sample is first filtered through a 0.45 μm capsule filter to remove particulates. Samples are then placed in a reaction vessel with the addition of 1–2 ml of 20% $SnCl_2$ in concentrated HCl. The $SnCl_2$ reducing agent is added in excess in order to ensure that all of the Hg^{2+} is converted to Hg^0 as:

$$Sn^{2+}(aq) + Hg^{2+}(aq) \rightarrow Sn^{4+}(aq) + Hg^0$$

After an equilibration time of 1.5–2 minutes, the Hg^0 is purged into the vapor phase by bubbling high-purity argon gas through the solution. The argon vapor containing Hg^0 is then dried before passing it through the flow-through quartz tube.

All organic and particulate forms of mercury require digestion for 2 hours at 95 °C prior to analysis. The digestion solution consists of concentrated $H_2SO_4 + HNO_3$ with the addition of $KMnO_4$ until the purple color persists, followed by the addition of potassium persulfate ($K_2S_2O_8$). Since the absorbance or fluorescence intensity is measured on the sample as it flows through the quartz cell, either peak height or peak area can be used for calibration and determination of mercury concentration in the water samples. Since the cold vapor methods have baselines with very low noise levels, scale expansion can be used to determine the height or area of very low concentrations. Results are typically reported as ng l^{-1} Hg.

9.2.3 Chromatographic Methods

Ion chromatography (IC), also known as ion exchange chromatography, is another commonly used method for determining ion concentrations in water samples. IC can determine the anions F^-, Cl^-, Br^-, NO_2^-, NO_3^-, SO_4^{2-}, PO_4^{3-}, CN^-, and AsO_4^{3-} and the cations Li^+, Na^+, NH_4^+, Ca^{2+}, and Mg^{2+} in the ppb concentration range. The ions are separated based on their interactions with a stationary phase and a mobile phase, called the eluent. The ion chromatograph,

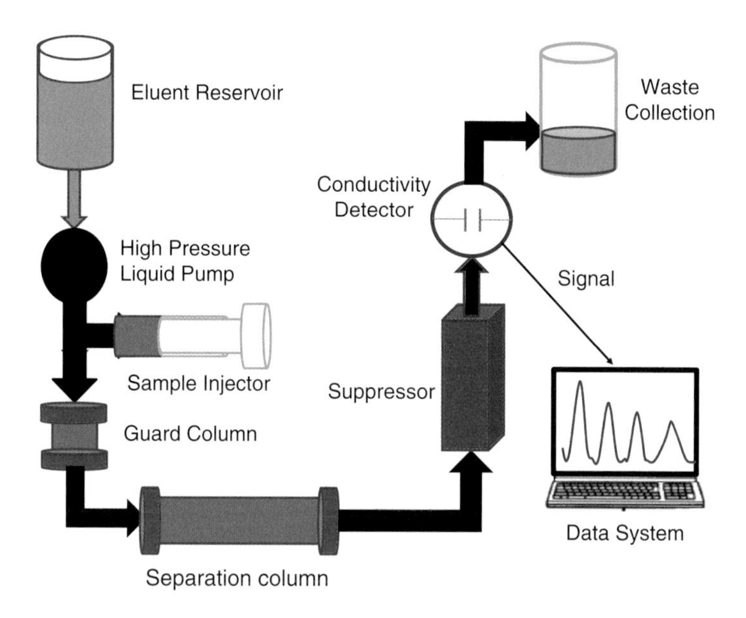

Figure 9.19 An ion chromatography system used for determining either anions or cations in water.

shown in Figure 9.19, consists of a separation column containing the stationary phase and an eluent that carries the ions through the column. The stationary phase is usually an organic resin that has a surface functional group with either a negative or positive charge, depending on the charge of the ions to be separated. The eluent used to move the ions through the column is usually high-purity water or an aqueous buffer solution.

After injection, the sample is first passed through a guard column, which removes any particles that could clog and damage the separation column. After the guard column, the sample flows onto the separation column where the charged analytes become attached to the oppositely charged surface of the resin. The ions will move through the separation column at different speeds, depending on their affinity for the resin. As the ions pass through the separation column, those with a weaker affinity for the resin will move through the column faster and will exit the column first, while those with a stronger affinity for the resin will move through the column more slowly and will exit the column last. As the ions exit the column they pass through a suppressor, which removes any background ions in the eluent and replaces them with nonionic species before passing through the detector. Since the detector is a conductance cell, similar to that discussed in Section 9.2.1, the suppressor reduces the background signal and lowers the detection limits for all ions. As the ions pass through the conductance detector, a chromatogram is produced as the conductance (μS) versus elution time (min). The peaks are then integrated by a data system, giving the total number of counts for each ion peak. A calibration curve is generated for each ion as a plot of total counts as a function of ion concentration in $\mu g\ l^{-1}$ (Figure 9.20).

The conductance detector is the most commonly used IC detector because it can respond to all ions with reasonable sensitivity and detection limits. However, other detectors have become available that offer lower detection limits for a limited number of ions. A UV–vis absorption detector can be used for a variety of ions, either directly or after post-column reaction. Species that can be determined directly by a UV–vis detector are the colored ion CrO_4^{2-} (365 nm) or the ions that have UV absorption at 200–227 nm, including I^-, IO_3^-, Br^-, BrO_3^-, NO_2^-, NO_3^-, ClO_3^-, HS^-, $S_2O_3^{2-}$, SCN^-, metal chloride complexes, and metal cyano complexes. Other ions require conversion to a colored complex by post-column reaction. Although the detection limits with a UV–vis detector are significantly lower than with conductance detection, the number

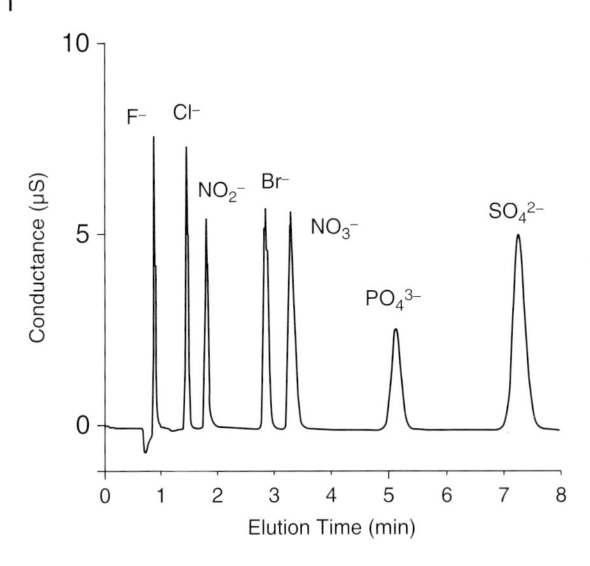

Figure 9.20 A typical anion chromatogram obtained from an ion chromatograph showing the different elution times for the various anions. Source: Adapted from EPA.

of ions detected in a single sample is limited to those that absorb in the wavelength range of the detector or those that respond to the post-column reaction, which are typically limited. One important application to IC–UV–vis with post-column reaction is the determination of lanthanide ions by complexation with Arsenazo III.

An amperometric detector can be used for IC with ions that can be oxidized or reduced. The detector uses a three-electrode electrochemical cell consisting of a working electrode, a reference electrode (Ag—AgCl), and a counter-electrode. The electrochemical reaction, which can be either an oxidation or a reduction reaction, occurs at the working electrode and the resulting flow of electrons to the reference electrode is detected as a current that is directly proportional to the analyte concentration. The purpose of the counter-electrode, which is usually made of glassy carbon, is to maintain the potential. The IC chromatogram is displayed as the electrochemical current as a function of time. The detection limits for an amperometric detector can reach the nanograms per liter range. A choice of working electrodes (Ag, Pt, carbon paste, and glassy carbon) can expand the range of its applicability, however, it is still limited to ions that can be oxidized or reduced in the electrochemical system. Ions measured by electrochemical detectors include CN^-, HS^-, I^-, Br^-, SCN^-, and $S_2O_3^{2-}$ with the Ag working electrode; SO_3^-, OCl^-, AsO_2^-, and N_2H_4 with the Pt working electrode; NO_2^-, ClO_2^-, and $S_2O_3^{2-}$ with the carbon-paste working electrode; and phenols with the glassy carbon electrode.

The identification and measurement of purgeable volatile organic compounds in surface water, ground water, and drinking water is accomplished by gas chromatography with mass spectroscopy detection (GC–MS). The method is applicable to a wide range of organic compounds, which have sufficiently high volatility and low water solubility to be removed from water samples with purge and trap procedures. The detection limits are compound, instrument, and matrix dependent, and typically vary from 1.6 to 0.02 $\mu g\,l^{-1}$. The concentration range is about 200–0.02 $\mu g\,l^{-1}$ with a thick film column and 20–0.02 $\mu g\,l^{-1}$ with a thin film column. The volatile organic compounds with low water solubilities are purged from the water sample by bubbling ultrahigh purity nitrogen (UHP–N_2) through the sample. The volatile compounds are carried away by the N_2 and are trapped in a tube containing a sorbent, such as Tenax or Sulpelco K-trap. The sorbent tube is heated and backflushed with UHP–N_2 to desorb the volatile compounds onto a temperature-programmed capillary GC column. The volatile compounds eluting from the GC column enter the MS by way of a jet separator, as shown in Figure 9.21. The end of the capillary GC column is placed close to the capillary

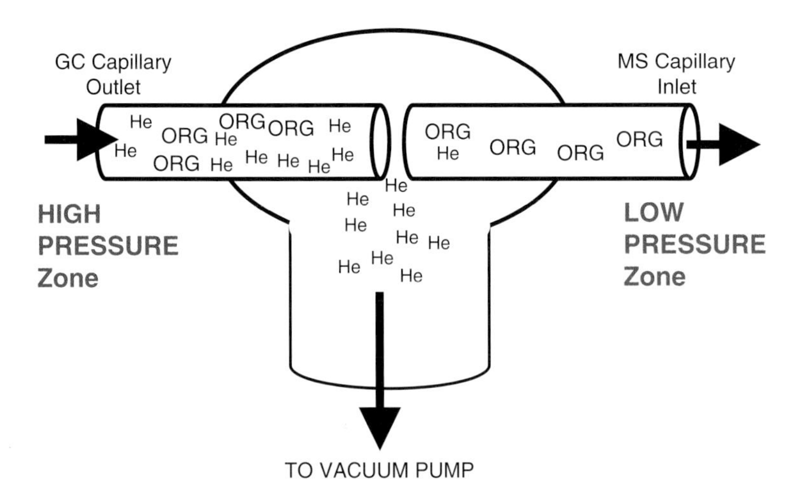

Figure 9.21 A jet separator used as an interface between the high-pressure capillary gas chromatograph and the low-pressure mass spectrometer (GC–MS).

inlet of the MS. Both tubes are contained in a closed system that is maintained at decreased pressure by pumping through a downstream orifice. The GC effluent expands into the reduced pressure region where the low-mass carrier gas molecules (typically He) are pumped away from the higher-mass analyte molecules, leaving the sample stream enriched in the analyte.

The analyte molecules are identified and measured by the MS. The organic compounds are identified by comparing their retention times to the retention times of reference spectra and calibration standards. The concentration of each volatile compound is determined from a multi-point calibration curve of the mass signal as a function of standard concentration. Multiple calibration curves can be obtained simultaneously by using multiple dilutions of a commercial standard that contains all of the analytes of interest.

Nonvolatile organic compounds can be determined in water by a liquid–solid extraction (LSE) followed by GC–MS detection. The organic compounds are extracted from the water sample by passing a known volume of the water sample through a cartridge containing a solid phase with a chemically bonded C_{18} organic phase, which extracts the organics from the water and concentrates them onto the cartridge. The organic compounds are eluted from the LSE cartridge with small quantities of ethyl acetate followed by methylene chloride. This extract is concentrated further by evaporation of a portion of the ethyl acetate–methylene chloride solvent. The sample components are separated, identified, and measured by injecting an aliquot of the resulting solution into the GC–MS, equipped with a high-resolution fused silica capillary column. The organic compounds are identified by comparing their retention times to the retention times of reference spectra and calibration standards. The concentration of each identified compound is then determined from a calibration curve of the mass signal as a function of standard concentration.

Other methods used for the determination of specific organic compounds in water samples after extraction and concentration using LSE include GC–ECD for organohalogens, pesticides, and insecticides. As discussed in Section 7.2, the detection limits of the ECD are typically in the low ppt range for compounds that contain highly electronegative atoms, such as oxygen and the halogens. High-performance liquid chromatography with UV detection (HPLC–UV) is used for routine analysis of the herbicides diaquat ($C_{12}H_{12}N_2Br_2$) and paraquat ($[(C_6H_7N_2)_2]Cl_2$). Although the retention times of the two compounds are not significantly different (diaquat = 2.03 minutes and paraquat = 2.25 minutes), the separate concentrations of the two compounds can be determined due to their different absorptions at 308 nm (diaquat) and 257 nm (paraquat). This is accomplished by analyzing the sample twice with the UV detector set at the two different wavelengths. At 308 nm only diaquat is measured and at 257 only paraquat is measured.

Carbonyl compounds can be determined in ground water, surface water, or waste water by reversed-phase HPLC–UV after dinitrophenylhydrazine (DNPH) derivatization, as discussed in Section 7.2.3. Reversed-phase HPLC uses a hydrophobic stationary phase on the HPLC column (C_{18} or C_8) instead of the normal hydrophilic stationary phase, such as silica or alumina. Instead of the normal hydrophobic-solvent mobile phase, reversed-phase HPLC uses a polar mobile phase, typically 70% methanol/30% water. Since the stationary phase has a stronger affinity for hydrophobic compounds, the hydrophilic molecules in the mobile phase will exit the column before any hydrophobic compounds. Reversed-phase HPLC–MS is used for the determination of benzadines (1,1′-biphenyl-4,4′-diamine and its derivatives) and nitrogen-containing pesticides. The HPLC is coupled to the MS by an electrospray ionization (ESI) interface developed by John Fenn in 1984, for which he was awarded the Nobel Prize in Chemistry in 2002. The ESI interface uses a Taylor cone, named after Sir Geoffrey Taylor, who discovered that a fluid flowing from a tube will form a cone when placed in a high-voltage electric field. When a threshold voltage is reached, a liquid jet known as a cone jet is emitted from the tip of the cone, as shown in Figure 9.22. As the voltage is increased past the threshold, the liquid jet breaks up into droplets. In the ESI interface, the droplets formed from the Taylor cone are pulled into a vacuum chamber where the solvent begins to evaporate, concentrating the analyte charge in the droplet. The charge within the droplet continues to increase as it evaporates, causing it to explode. The ions formed by the exploding droplet are pulled into the MS by a charged inlet system. The organic compounds are identified by comparing the retention times of the ions formed in the interface to the retention times of reference spectra and calibration standards. The concentration of each identified compound is then determined from a calibration curve of the mass signal of the ions as a function of standard concentration.

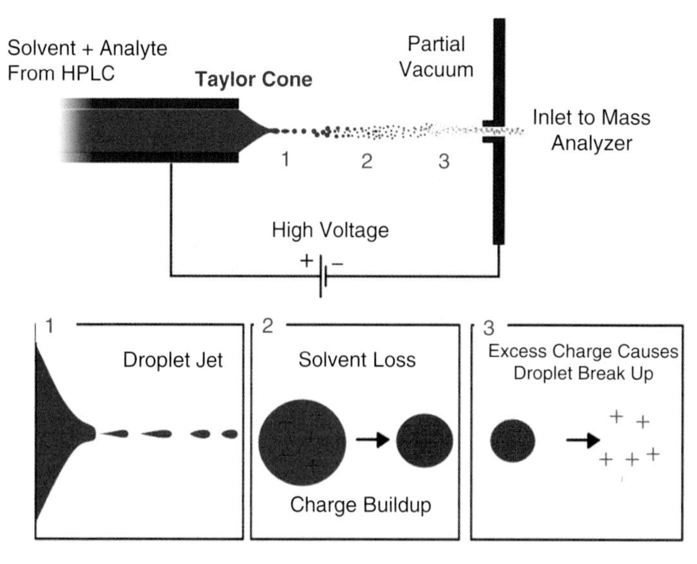

Figure 9.22 The three steps in the electrospray ionization process that lead to the formation of ions which are identified and measured by a mass spectrometer. Source: Adapted from Evan MAson, Wikimedia Commons.

9.2.4 Titration Methods

The standard technique for the measurement of DO in fresh and salt water is a titrimetric determination called the Winkler method. The DO concentration is fixed by adding a solution of $MnSO_4$ followed by 15% KI in 70% KOH. The Mn^{2+} reacts with OH^- to form a pinkish-brown precipitation of $MnO(OH)_2$:

$$2Mn^{2+}(aq) + 4OH^-(aq) + O_2 \rightarrow 2MnO(OH)_2(s)$$

After mixing and allowing the precipitate to settle, the solution is acidified to dissolve the precipitate and the Mn^{4+} ion then reacts with the I^- to form I_2:

$$MnO(OH)_2(s) + 4H^+(aq) \rightarrow Mn^{4+}(aq) + 3H_2O(l)$$

$$Mn^{4+} + 2I^-(aq) \rightarrow Mn^{2+}(aq) + I_2(aq)$$

The I_2 concentration is determined by titration with a 0.025 M standard solution of sodium thiosulfate ($Na_2S_2O_3$) using a starch indicator:

$$2S_2O_3^{\ 2-} + I_2 \rightarrow S_4O_6^{\ 2-} + 2I^-$$

At the point that the dark blue starch solution turns clear, all of the I_2 has been converted back to I^-. For a 200 ml sample, the volume of titrant (ml) used to reach the end point of the titration is equal to the DO concentration in mg l^{-1}.

The main difficulty with using the Winkler method for the determination of DO concentration is the potential for O_2 contamination from the air or reagent solutions used in any of the various steps of the titration process. During all stages of the procedure, O_2 cannot be introduced or lost from the sample. This is accomplished by taking the original sample in a Winkler BOD bottle, which is specifically designed to prevent air contact with the sample. Winkler bottles have conical tops with a close-fitting ground-glass stopper to exclude air bubbles once the top is sealed. Other common interfering species include oxidizing and reducing agents, NO_2^-, Fe^{2+}, and suspended solids containing organic matter. Various modifications of the Winkler method have been developed to eliminate most of these interferences. The Alsterberg modification (addition of NaN_3) is commonly used to eliminate the NO_2^- interference, the Rideal–Stewart modification (addition of $KMnO_4$ followed by KHC_2O_4) eliminates the Fe^{2+} interference, and the Theriault modification (alum flocculation) is used to remove interferences from suspended solids.

Another common titration method used routinely in the analysis of water samples is chemical oxygen demand (COD). The COD is a measure of the amount of oxygen required to oxidize both soluble and particulate organic matter in water. It is determined by the measurement of the amount of oxidant required to convert all of the oxidizable organic species to CO_2, NH_4^+, and H_2O. This is accomplished by the addition of an excess amount of the strong oxidant potassium dichromate ($K_2Cr_2O_7$) under strongly acidic conditions (concentrated H_2SO_4). The amount of $Cr_2O_7^{\ 2-}$ required to oxidize the generic organic compound ($C_nH_aO_bN_c$) is given by:

$$C_nH_aO_bN_c + dCr_2O_7^{\ 2-} + (8d + c)H^+ \rightarrow nCO_2 + \frac{a + 8d - 3c}{2} H_2O + cNH_4^{\ +} + 2dCr^{3+}$$

where $d = \frac{2n}{3} + \frac{a}{6} - \frac{b}{3} - \frac{c}{2}$.

Several inorganic species can interfere with the $Cr_2O_7^{\ 2-}$ oxidation, including Fe^{2+}, Cl^-, and NO_2^-. The most common interference is by Cl^-, which can be oxidized by $Cr_2O_7^{\ 2-}$ as:

$$6Cl^- + Cr_2O_7^{\ 2-} + 14H^+ \rightarrow 3Cl_2 + 2Cr^{3+} + 7H_2O$$

The interference from Cl^- is eliminated by the addition of mercuric sulfate ($HgSO_4$), which forms the oxidation-resistant mercuric chloride complex:

$$Hg^{2+} + 2Cl^- \rightarrow HgCl_2$$

The NO_2^- interference is removed by the addition of sulfamic acid (H_3NSO_3), which reacts to form N_2 and H_2SO_4. Potassium hydrogen phthalate (KHP) is typically used as an organic standard for $Cr_2O_7^{2-}$ oxidation and can also be used as a standard addition to water samples in order to determine the amount of interference in a water sample.

After the oxidation is complete, the amount of unreacted $Cr_2O_7^{2-}$ is determined by titration with ferrous ammonium sulfate ($Fe(NH_4)_2(SO_4)_2$) as:

$$6Fe^{2+} + CrO_7^{2-} + 14H^+ \rightarrow 6Fe^{3+} + 2Cr^{3+} + 7H_2O$$

The end point of the titration is signaled by a color change of the redox indicator ferroin from blue green to a reddish brown. Ferroin ($Fe(o\text{-phen})_3^{3+}$) is a coordination compound of Fe and 1,10-phenanthroline (*o*-phen). It is oxidized to $Fe(o\text{-phen})_3^{2+}$ by $Cr_2O_7^{2-}$ after all the organic compounds in the sample are oxidized, resulting in the color change. The difference between the amount of Fe^{2+} in ml used to titrate a sample blank and the amount of Fe^{2+} used to titrate the sample after $Cr_2O_7^{2-}$ oxidation gives the COD in mg $l^{-1}$by the formula:

$$\text{COD (mg l}^{-1}) = \frac{8000\,(B - S)[Fe^{2+}]}{V}$$

where V is the volume (ml) of sample oxidized, B is the volume (ml) of titrant required for the sample blank, S is the volume (ml) of titrant required for the sample, and $[Fe^{2+}]$ is the concentration (M) of iron in the titrant. The value 8000 in the calculation is used to convert the amount of $Cr_2O_7^{2-}$ used to the amount of O_2 that would be required for the oxidation. This is because COD is to be reported as the mass of O_2 that would be consumed per volume of water (mg l^{-1} O_2).

The amount of $Cr_2O_7^{2-}$ remaining after oxidation of the sample can also be determined by spectroscopy. Since $Cr_2O_7^{2-}$ and Cr^{3+} are both colored species with absorptions at different wavelengths, their concentrations before and after sample oxidation can be determined by Beer's Law. The amount of $Cr_2O_7^{2-}$ remaining after sample oxidation can be determined by absorption at 420 nm and the amount of Cr^{3+} formed during the reaction can be determined by absorption at 600 nm using calibration curves, as discussed in Section 9.2.2.

9.2.5 Radiochemical Methods

The three major types of radiation emitted from nuclear reactions are gamma rays (γ), alpha particles (α), and beta particles (β). These are described in detail in Chapter 12. Briefly, an α particle is a helium nucleus (two protons and two neutrons) with a 2+ charge. A beta particle has the mass of an electron and can have either a negative charge (electrons) or a positive charge (positrons). Gamma rays are energetic electromagnetic waves similar to X-rays. The measurement of total α, β^-, and γ emissions from water samples is typically accomplished by evaporating a large volume of water (1 l or larger) onto a thin stainless steel disk called a planchet. The planchet is placed inside a lead shielded counter and the α, β^-, and γ radiation emitted from the sample are separated using charged plates, as shown in Figure 9.23. The doubly charged α particles are attracted to a negatively charged plate, while the negatively charged β^- particles are attracted to a positively charged plate. Separate detection systems count the energetic α and β^- particles, while the undeflected γ emissions are detected by a γ-sensitive crystal. This γ-sensitive crystal produces electrons when it is struck by the high-energy γ rays and the electrons are measured as a voltage pulse. The instrument is calibrated using radioactive emission standards.

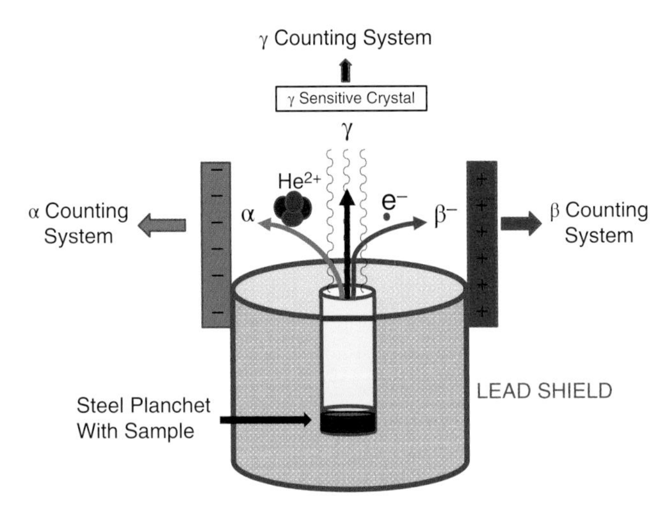

Figure 9.23 A lead-lined radioactive emission counter, which separates α and β particles with charged plates and detects γ rays with a γ-sensitive crystal detector.

The major contributor to radioactivity in most water systems is ^{226}Ra in the form of $^{226}Ra^{2+}(aq)$, which decays by α emission with a 1.3×10^3-year half-life to form the ^{222}Rn gas with a half-life of 3.8 days. Radium-226 is determined in water after concentration and separation by coprecipitation with the mixed $BaSO_4$—$PbSO_4(s)$. This is accomplished by the addition of $(NH_4)_2SO_4$, $BaCl_2$, and $Pb(NO_3)_2$, which forms $BaSO_4(s)$, $PbSO_4(s)$, and $RaSO_4(s)$. The precipitate is redissolved and the Pb^{2+} is removed from the solution by complexing with EDTA. This procedure also removes the α emitters $^{210}Pb^{2+}$ and $^{212}Pb^{2+}$, which can present interferences. The precipitate is then reformed as $BaSO_4$ by addition of $(NH_4)_2SO_4$ and transferred to a planchet.

The concentration of ^{226}Ra is then determined by monitoring the production of the radiochemical decay product $^{222}Rn(g)$ from the sulfate precipitate. The planchet with the precipitated sample is placed in a sealed bubbler and stored to allow the ingrowth of $^{222}Rn(g)$. When the short-lived ^{222}Rn daughters are in equilibrium with the parent (about 4 hours), the ^{222}Rn gas is purged into a liquid scintillation cell in a liquid scintillation counter. The liquid scintillation cell contains organic compounds that emit light when struck by the energetic α particles from the ^{222}Rn decay. The intensity of light is directly proportional to the amount of ^{222}Rn decay in the sample. This can be used to determine the initial concentration of ^{226}Rn (D) in the sample in pCi l^{-1} as

$$D = \frac{Ce^{0.693\,(\Delta t/91.8)}}{2.22EV}$$

where C is the measured α decay rate in counts per minute (cpm), E is the scintillation cell counting efficiency for ^{222}Rn, V is the volume of sample used in liters, and Δt is the delay time in hours between the sample collection and the end of activity counting.

9.3 Particulates and Colloids

Water clarity, which is affected by suspended particulates as well as light-absorbing species, is evaluated in the field using the Secchi disk, as discussed in Section 9.1. Water turbidity, which is primarily due to light scattering by particles, is also measured in the field by nephelometry.

The measurement of both water clarity and turbidity is used to evaluate the amount of suspended particles and large colloidal materials present in the water at the time of sampling. The suspended particulates in water, called total suspended solids (TSS), are determined in the laboratory by gravimetric analysis. The TSS are separated from the water sample by passing a known volume of water through a 2.0 μm glass-fiber filter. The filter is dried and weighed before and after sample filtration and the total weight of the solids collected on the filter is determined relative to the volume of water filtered. The results are reported as TSS in mg l^{-1}.

Another measurement of the particulate concentration in water, which is important for industrial or domestic waste water, is settleable solids. The settleable solids are any solid particulates that do not remain suspended or dissolved in water not subject to motion. The settleable solids can include both larger particulates and insoluble molecules. Settleable solids are measured as the volume of solids in 1 l of sample that will settle to the bottom of an Imhoff cone during a specific amount of time. The Imhoff cone is a clear cone-shaped flask, usually made of Pyrex, that is marked with gradations from the bottom of the cone. The graduated marks are calibrated to measure the total volume of the solids that settle to the bottom of the cone. The results give an indication of the volume of solids that can be removed in sedimentation tanks, clarifiers, or settling ponds during waste treatment. Performing this measurement before and after primary and secondary waste treatment gives an indication of whether the waste treatment processes are functioning properly.

Settleable solids and TSS are measurements of the amount of solid material that is suspended in water, while TDS is the measurement of the amount of chemical species that form a solution in water. An aqueous solution is defined as a homogeneous mixture of one or more solutes dissolved in water. In addition, the solute interacts chemically with the water, a process called solvation. In the solvated state, the solute is surrounded or complexed by the water molecules, which leads to stabilization of the solute species in the water. An aqueous suspension is defined as a heterogeneous mixture in which some of the particles settle out of the water upon standing. An aqueous colloid is a heterogeneous mixture in which the particles are intermediate in size between those of a solution and a suspension and do not settle out upon standing. They also do not interact chemically with the water molecules. Aqueous colloidal materials, such as humic and fulvic acids, are too small to be included in the TSS determination. Although they are included in the TDS determination, there is no simple procedure that can separate them from the truly dissolved species, such as the dissolved salts. For this reason, stepwise ultrafiltration procedures have been developed that allow for the separation of colloidal materials from both suspended solids and dissolved species in water. These ultrafiltration methods were first developed in the pharmaceutical industry for the separation and purification of macromolecular pharmaceuticals. Although they are not routinely used as part of water quality assessments, they are often used for research purposes to evaluate the role of colloidal materials in the transport of both inorganic and organic pollutants in aqueous systems. They are also increasingly being used in tertiary waste water and drinking water treatment processes.

After the TSS is separated from the water using a 2 μm filter, the filtered water samples are passed through a 0.45 μm filter to remove any very small suspended material, including small phytoplankton and bacteria. What remains after the filtration through the 0.45 μm filter are the colloidal materials in the size range of 1000 Da to 0.45 μm, along with the dissolved species <1000 Da. The samples are further separated using flat ultrafiltration membranes in stirred cells or with hollow-fiber ultrafiltration membranes. A comparison between the flat membrane and hollow-fiber ultrafiltration systems is shown in Figure 9.24. The sample is pulled vertically through the flat membrane (Figure 9.24(A)). Dissolved species and colloidal materials, which are smaller than the membrane pore size, flow through the membrane while colloidal materials larger than the pore size are concentrated above the membrane. The concentrate above the

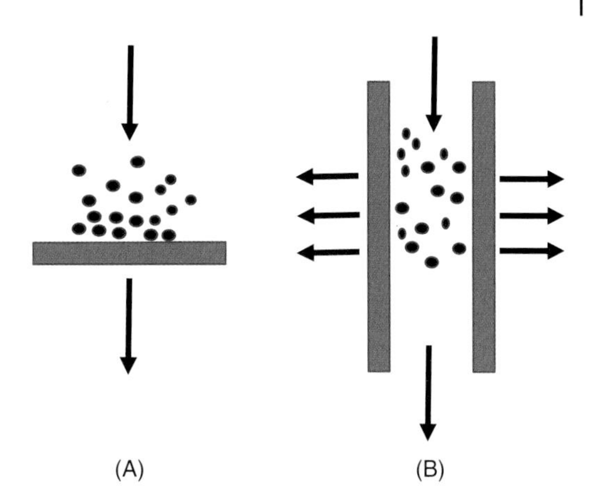

Figure 9.24 (A) A flat ultrafiltration membrane operated in a stirred cell and (B) a hollow-fiber ultrafiltration membrane using cross-flow pumping.

membrane is continuously stirred to minimize build-up of the large colloids on the membrane surface. This build-up, called polarization, can alter the effective pore size of the membrane and limit the rate of filtration by blocking the membrane pores. The hollow-fiber ultrafilters are operated by cross-flow pumping (Figure 9.24(B)), which pulls colloids that are smaller than the ultrafilter pore size horizontally through the membrane. Colloids larger than the pore size will flow vertically through the interior of the hollow fiber.

Concentrated colloidal materials can be obtained in specific size ranges by passing the sample through a series of ultrafilters with decreasing pore sizes. Typical colloidal size ranges are 0.45–0.1 μm, 0.1 μm–100 000 Da, 100 000–30 000 Da, 30 000–3000 Da, and 3000–1000 Da. Table 9.2 shows some of the characteristic pore sizes available for the hollow-fiber and flat-membrane ultrafilters (Gaffney et al., 1996). The hollow-fiber ultrafilters are best for the larger colloidal size ranges, because the cross-flow filtration helps to prevent polarization. For size fractionation below 3000 Da, the samples can be processed by using flat-membrane ultrafilters in a stirred cell, which are available in 1000–500 Da pore sizes. The water processing rates through the hollow-fiber ultrafilters are limited by the pore sizes. The sample throughput decreases from 1.0 l min^{-1} for a pore size of 100 nm (>1×10^6 Da) to 0.2 l min^{-1} for a pore size of 1 nm (3×10^3 Da). Since the stirred cells are pressurized, the sample flow rate through the flat-membrane ultrafilters depends on the pressure used in the cell. After the sampling and separation is complete, the hollow-fiber ultrafilters are cleaned by backflushing with clean water or with a weak acid solution, depending on the composition of the sample. Some large humic materials can adsorb to the membrane surfaces and require acid washing to

Table 9.2 Characteristic pore sizes of hollow fiber ultrafilters and water processing rates.

Ultrafilter type	Pore size			Molecular weight (Da)	Flow (l min^{-1})
	(μm)	(nm)	(Å)		
Hollow fiber	0.10	100.0	10 000	>1×10^6	1.0
Hollow fiber	0.005	5.0	50	100 000	1.0
Hollow fiber	0.0025	2.5	25	30 000	0.8
Hollow fiber	0.002	2.0	20	10 000	0.3
Hollow fiber	0.001	1.0	10	3000	0.2
Flat membrane	0.0003	0.3	3	1000	—
Flat membrane	0.0001	0.1	1	500	—

remove. Because of this, they must be acid washed before reuse when processing high humic waters.

The use of ultrafiltration membranes for the isolation of the colloidal humic and fulvic acids by molecular size has advantages over some more commonly used methods of separation. One method achieves separation by passing them through an ion exchange resin. Since the humic and fulvic acids are polyionic in water, the ion exchange resin separates them based on the amount of charge on the organic colloidal materials, assuming that larger colloids will have more sites of ionization. However, the available charge on the humic and fulvic acids will be affected by the amounts of inorganic cations that are complexed to them. The classical separation of humic and fulvic acids, discussed in Section 8.5.4, involves extraction with a strong base followed by acidification to a pH of 2. Some procedures also include washing the extract with HCl and HF to remove the complexed metals. These harsh chemical separation procedures are not recommended for the isolation of humic and fulvic materials, since the base and acid treatments are likely to result in decarboxylation, deamination, and dehydration reactions, which drastically alter their chemical structures. The structural alterations will likely result in chemical behavior that is not representative of the natural organic colloidal materials. For these reasons, the physical ultrafiltration methods are preferred to any chemical separation techniques before further chemical and elemental characterizations are to be performed.

Once the colloidal materials are separated according to their molecular size, they can be further characterized using a variety of methods, such as AAS, ICP–MS, or α, β, and γ counting to determine the inorganic metals and radionuclide content as a function of colloid size. For example, Figure 9.25 shows the amount of three metals (Fe, Mg, and Mn) and select radionuclides (Am, Pu, Th, and U) in colloids from Volo Bog, located in northern Illinois (Gaffney et al., 1996). The colloids have been separated into five size ranges: 0.45–0.1 μm, 0.1 μm–100 000 Da, 100 000–30 000 Da, 30 000–3000 Da, and <3000 Da. The carbon content (DOC) is measured as an indication of the colloid concentrations in each size range. The correlations between the inorganic species and carbon content in the different size ranges can be useful in evaluating the ability of each size to complex the inorganic species. Some elements are preferentially concentrated in the larger colloidal size ranges, while others are concentrated in the smaller sizes. Most of the radionuclides are found predominantly in the two smallest

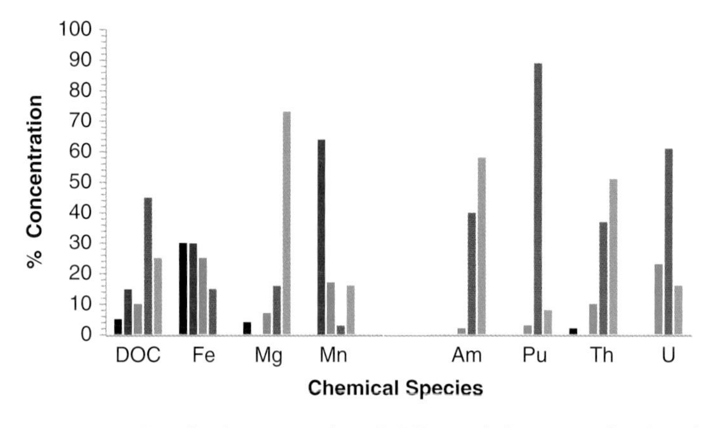

Figure 9.25 Dissolved organic carbon (DOC), metals (Fe, Mg, and Mn), and radionuclides (Am, Pu, Th, and U) in colloidal size ranges: 0.45–0.1 μm (black), 0.1 μm–100 000 Da (red), 100 000–30 000 Da (green), 30 000–3000 Da (purple), <3000 Da (blue), in water from Volo Bog located in northern Illinois.

size ranges (30000–3000 Da), while the metals are more evenly distributed between the five size ranges. This implies that the radionuclides are likely complexed with fulvic acids, which are most abundant in the small colloidal sizes.

Organic structural information can also be obtained to determine the predominant functional groups in each colloidal size range. Typical analysis methods commonly used for this purpose include pyrolysis GC, Fourier transform infrared spectroscopy (FTIR), and nuclear magnetic resonance spectroscopy (NMR) (Gaffney et al., 1996; Marley et al., 1996). These techniques have shown that the small colloids contain significant amounts of carboxyl groups, which is consistent with the observation of strong metal and radionuclide complexation in the smaller size ranges. The determination of nonpolar organic pollutants, such as pesticides and herbicides, in the colloidal size fractions can also determine the potential for their adsorption onto organic colloidal materials, which could facilitate their transport in aqueous systems. Although not routinely performed, this type of size-fractionated colloidal analysis can aid researchers in understanding the importance of colloidal materials in transporting pollutant species in the hydrosphere.

9.4 Contaminant Issues

The detection limit for any method of measurement is defined as the lowest amount of a chemical species that can be detected by the method. It is equal to a measurement signal that is three times the standard deviation of the signal obtained on a sample blank. So the detection limits for measurement methods, which determine the lowest concentration of analyte that can be measured, are directly related to the signal of the sample blank. The signal of the sample blank is directly affected by contamination, which can occur with the use of inadequately purified water or insufficiently cleaned laboratory glassware used to prepare the reagents and blanks. In addition, the precision and accuracy of the measurement are also affected by the same contamination. The use of impure water to prepare reagents can introduce systematic errors in the measurements of both samples and standards, which affects the accuracy of the method. The use of unclean glassware can cause random errors, which affect the correlation value of the calibration curve and the precision of the measurement.

Since water is one of the best solvents, and many gases present in the laboratory air are soluble in water, obtaining pure water for analyses is not a simple task. Distillation has been used as a classical method for producing purified water. To reduce the energy required for distillation, the pressure over the water is typically reduced so that the water will reach boiling point at a lower temperature. Distillation effectively removes dissolved salts and suspended particulates. It will not remove organic contaminants that have low boiling points. However, most organic compounds with boiling points lower than 150 °C can effectively be distilled without reduced pressure. To increase water purity, the distilled water can be redistilled a second or third time to produce doubly or triply distilled water. After purification, the water must be stored so that contaminants do not re-enter the water.

Deionized water, commonly called DI water, is typically used for the analysis of inorganic ions in water. The production of DI water is less energy intensive and much faster than distillation. A DI water system, shown in Figure 9.26, includes a series of anion and cation exchange resin columns, which exchange H^+ and OH^- ions for dissolved ions in the water. The cation resin removes the positively charged ions such as Ca^{2+}, Mg^{2+}, Na^+, Fe^{2+}, Fe^{3+}, and K^+, while the anion filter removes negatively charged ions such as Cl^-, F^-, NO_3^-, SO_4^{2-}, and PO_4^{3-}. A mixed cation–anion resin is used to capture any residual ions not removed by the separate cation and anion resins. The efficiency of the ion exchange resins is determined by the conductivity

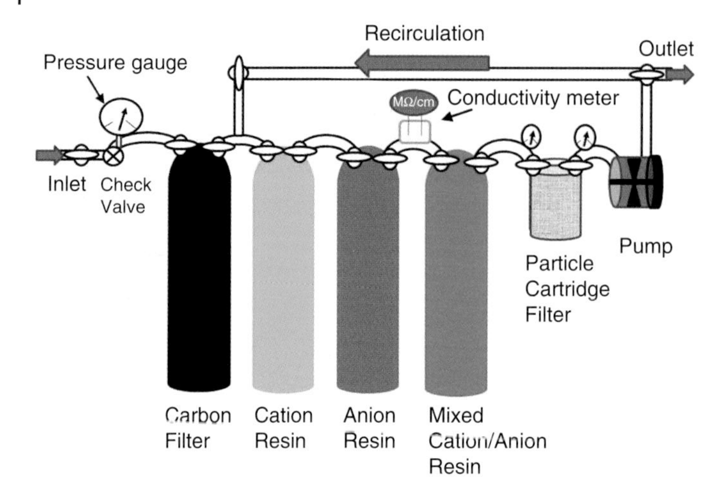

Figure 9.26 A deionized water system used to produce high-purity water for water analysis.

of the water in $M\Omega\,cm^{-1}$. Ultrapure water (UPW) has a conductivity of 18.18 $M\Omega\,cm^{-1}$ at 298 K. One hundred parts per trillion of Cl^- in the water will lower this conductivity to a value of 18.11 $M\Omega\,cm^{-1}$. When conductivity of the water begins to fall, signaling ion saturation of the resins, the ion exchange resins can be regenerated by flushing with H^+ to remove trapped cations, and OH^- to remove trapped anions. The water pressure is also measured to determine if the filters have become clogged, causing a reduction in pure water output.

Since the ion exchange resins do not remove uncharged organic molecules, an activated carbon prefilter lowers the concentrations of the organic material by adsorption on activated surface sites. The activated carbon filter must be replaced on a regular basis due to saturation of the adsorption sites, which reduces the effectiveness of organic removal. A particle cartridge filter is used before the outlet of the DI system to catch any particulate matter that may be released from the ion exchange resins and the carbon filter. When not in use, the water is usually continuously recirculated through the system instead of storing the purified water in a tank. Tank storage can result in recontamination of the water from atmospheric absorption of gases and leaching of tank materials. However, allowing the water to constantly recirculate in the system provides the water with multiple passes through ion exchange resins and results in improved water quality. Enhanced water purity can also be achieved by the addition of prefiltration with 0.45 or 0.2 μm filters to remove viruses and bacteria, which can decompose to produce dissolved organic compounds over time.

In cases where the DI water does not have adequately low levels of dissolved gases or organics, UPW can be produced by the ultrafiltration methods described in Section 9.3, coupled with inert gas purging to remove the dissolved gases. The UPW must be stored in Pyrex glass bottles to minimize leaching of contaminants from the containers. If stored in plastic bottles, the UPW will leach plasticizers and other organics from the plastic into the water, introducing contaminants that would cause errors in the measurement of dissolved organics. For gas-specific measurements such as pH, DO, or NH_3, the UPW must also be stored such that it is not exposed to the atmosphere, since soluble gases can easily be absorbed from laboratory air. Dust can also contaminate distilled water if left open to the atmosphere for any length of time. This can add trace metals and solids to blanks, standards, and samples.

In addition to the use of adequately pure water, the analytical methods must be conducted in a clean environment to ensure that reagents, blanks, and water samples are not exposed to contamination. The analyte or analytes to be measured, as well as the required detection limits, determine the level of precautions required to prevent contamination. The analysis of analytes commonly found in a laboratory environment requires steps to assure that environmental contamination does not occur. For example, if the water pH is to be measured in an open container, increased levels of atmospheric CO_2 from combustion sources or human respiration will be absorbed into the water, altering the pH. This can easily occur from the analyst breathing over an open beaker while taking the measurement. For very accurate pH measurements, glove boxes are used which maintain an inert atmosphere over the samples to minimize contamination by soluble gases.

Dust in the air and on surfaces in many laboratories can be a source of contamination for the trace analysis of metals or organic pollutants. For the measurement of very low concentrations of some analytes such as metals or radionuclides, a clean room environment may be necessary. In this case the analyst is required to wear clean, dust-free, disposable outer clothing when entering the clean room. Filtered air flow is used to maintain a positive pressure to minimize any contaminant transport into the area when the laboratory is entered. Trace metal and organic contamination can come from many sources, so care is taken to assure that all of the surfaces of the equipment and glassware used to perform the measurements are kept clean and are made of materials that are not likely to produce analyte contamination.

Sample replicates, reagent blanks, and both commercial and freshly prepared standards are an important part of determining the presence and sources of contamination. To ensure that inaccuracies have not been introduced into the method of analysis from instrumental, environmental, or chemical errors, special calibration standards are used for all analytes. These are usually obtained in the United States from the National Institute for Standards and Testing (NIST). Approved standards should be used to recalibrate instruments and to check chemical measurements to ensure that the results obtained fall within acceptable uncertainties. The occurrence of reagent blank or standard results that fall outside the acceptable limits can be due to instrumental drift or reagent contamination. In either case, the source needs to be determined before further analyses are conducted. Instrument calibration drift typically results in a systematic error, which would be observed as consistently high or low standard results. Instrument drift that appears random is a signal that the problem is more serious than a simple calibration can correct.

Contamination could result in either systematic or random errors. Common sources of contamination are improperly cleaned glassware, DI water systems requiring maintenance, contaminated stock reagents, or an environmental source of the analyte. Improperly cleaned glassware would result in randomly high blank, standard, and replicate values. Both impure water and contaminated reagents would result in systematically high standard, blank, and replicate values. A problem with the DI water system can easily be identified by checking the water conductivity. Contaminated reagents would result in a high reagent blank. However, for most measurement methods, it is difficult to determine which reagent or reagents are the source of the contamination. In this case, all reagents should be replaced.

References

Gaffney, J.S., Marley, N.A., and Orlandini, K.A. (1996). Hollow-fiber ultrafilters for isolation of natural humic and fulvic acids. In: *Humic and Fulvic Acids: Isolation, Structure and Environmental Role* (ed. J.S. Gaffney, N.A. Marley and S.B. Clark), 26–40. Washington, D.C.: American Chemical Society.

Marley, N.A., Gaffney, J.S., and Orlandini, K.A. (1996). Characterization of aquatic humic and fulvic materials by cylindrical internal reflectance spectroscopy. In: *Humic and Fulvic Acids:*

Isolation, Structure and Environmental Role (ed. J.S. Gaffney, N.A. Marley and S.B. Clark), 96–107. Washington, D.C.: American Chemical Society.

USEPA 2017a *Clean Water Act Analytical Methods: Approved CWA Chemical Test Methods.* Available online at https://www.epa.gov/cwa-methods/approved-cwa-chemical-test-methods (accessed August 23, 2018).

USEPA 2017b *Clean Water Act Analytical Methods: Approved CWA Microbiological Methods.* Available online at https://www.epa.gov/cwa-methods/approved-cwa-microbiological-test-methods (accessed August 23, 2018).

Study Problems

9.1 (a) What is a grab sample?
(b) What are the names of the two sampling bottles that are commonly used to obtain water samples at depths of tens to about one hundred feet?

9.2 (a) What six measurements are taken in the field at the time the sample is collected?
(b) Which of these are measured in situ?

9.3 What do the following abbreviations stand for in water chemistry:
(a) DO?
(b) BOD?
(c) COD?

9.4 (a) What water parameter is measured in the field with a Secchi disk?
(b) The results from the Secchi disk are reported in what units?

9.5 (a) What is composite sampling?
(b) What large composite sampler can collect from 12 to 36 water samples over a period of time in deep waters such as near the bottoms of a lake or in ocean waters?
(c) What field parameters can this sampler measure continuously?

9.6 (a) Water turbidity can be measured directly by what measurement sensor?
(b) What does this sensor actually measure?
(c) It operates on the same principle as what instrument used to measure aerosols in air?

9.7 (a) What is the most commonly used glass electrode?
(b) What is the most common ion that is detected using a crystal membrane electrode?

9.8 (a) What are the two pH standard solutions used to calibrate a glass electrode pH meter for samples with pH <7?
(b) For samples with pH >7?

9.9 (a) What two instrumental methods are used for the determination of DO concentrations in water samples?
(b) What is the standard method for DO determination?

9.10 (a) The determination of BOD requires the measurement of what species before and after a 5-day incubation period?
(b) The BOD result is reported as what?

9.11 (a) The total BOD of waste water is composed of what two components?
(b) What compound is added to the sample to inhibit nitrifying bacteria for determination of CBOD?

9.12 (a) What is the conductivity of a material?
(b) How are the conductivity and resistivity related?

9.13 (a) The original standard method for the measurement of NH_3 and NH_4^+ was based on what measurement method?
(b) What was the reagent used?
(c) What is its chemical composition?

9.14 (a) The addition of an alkaline solution of phenol (C_6H_5OH) and the hypochlorite ion (ClO^-) in the presence of ammonia produces what colored dye?
(b) What is the common name for the hypochlorite and phenol reagent used in this method for ammonia determination?

9.15 What is the name of the method of measurement for total organic nitrogen that involves the conversion of organic nitrogen to ammonia by digestion in H_2SO_4?

9.16 (a) The concentration of NO_3^- in water is determined by reaction with what reagent?
(b) How is the concentration determined?
(c) What two factors must be carefully controlled during the reaction?

9.17 (a) NO_3^- in water can also be determined by reduction to what species?
(b) How is the reduction accomplished?
(c) How is the NO_3^- concentration determined?

9.18 NO_3^- in water is measured by reacting it with what two reagents to form a colored diazo dye?

9.19 (a) What two reagents are used for the determination of PO_4^{3-} in water?
(b) What is the complex formed?
(c) What is the function of ascorbic acid in this reaction?

9.20 (a) What precipitation reaction is the basis for the analysis of SO_4^{2-} in water?
(b) How is the SO_4^{2-} concentration determined?

9.21 (a) What is the first step for the chemical determination of cyanide in water?
(b) What is the final reagent in the CN^- reaction that forms a colored product?

9.22 (a) In the measurement of water hardness, what ions are exchanged in the EDTA complex?
(b) How is the free ion measured?
(c) How is water hardness reported?

9.23 (a) What measurement is the determination of TOC and DOC based on?
(b) How does the TOC measurement differ from the DOC measurement?
(c) How is the data reported?

9.24 (a) What are the three main spectroscopic methods for the analysis of metals in water?
(b) Which has the lowest detection limits?

9.25 (a) What are the two types of atomic line source lamps used in atomic absorption spectroscopy?
(b) What is used to create free analyte atoms from a water sample in atomic absorption spectroscopy?
(c) What are the two types of background correction used in atomic absorption spectroscopy?

9.26 (a) How are the free analyte atoms created in atomic emission spectroscopy?
(b) What two methods can be used to separate emission lines?

9.27 (a) What two methods are used for mercury analysis in water samples?
(b) What is used as the reducing agent to convert Hg^{2+} to Hg^0?

9.28 (a) What is the most common detector used in ion chromatography?
(b) What is used to reduce the background signal and lower the detection limits in ion chromatography?
(c) How does it function?

9.29 (a) What method is used for the measurement of purgeable volatile organic compounds in water?
(b) What method is used for the measurement of nonvolatile organic compounds in water?
(c) What method is used for the measurement of organohalogens, pesticides, and insecticides in water?
(d) What method is used for the measurement of carbonyl compounds in water?

9.30 What is the difference between normal HPLC and reversed-phase HPLC?

9.31 (a) How is a GC coupled to a mass spectrometer?
(b) How is an HPLC coupled to a mass spectrometer?

9.32 (a) What strong oxidant is used in excess in the determination of COD in a water sample?
(b) What is the main interference for the determination of COD in water?
(c) What is added to remove this interference?

9.33 (a) What is the major contributor to radioactivity in most water systems?
(b) How is it most commonly measured?
(c) What method is used?

9.34 What is the major difference between hollow-fiber ultrafiltration and flat-membrane ultrafiltration?

9.35 (a) What radioactive particle is a helium atom nucleus with a +2 charge?
(b) What radioactive particle is an electron or positron?

9.36 Why are ultrafiltration methods preferred over chemical separation methods for the separation of colloidal materials into size fractions?

9.37 (a) What two methods are commonly used to provide purified water for laboratory analysis?
(b) Which method is faster and less energy intensive?
(c) What method can be used to provide ultrapure water?

9.38 What are the two common gases that can contaminate and change water chemistry when samples are exposed to air?

9.39 What is the resistivity of ultrapure water at 25 °C?

9.40 (a) What is present in laboratory air that can add trace metals and solids to blanks, standards, and samples if left open to the atmosphere for any length of time?
(b) What laboratory facility is designed to prevent this type of contamination?

9.41 (a) How would instrumental calibration drift affect standard measurements?
(b) How would improperly cleaned glassware affect blank values?
(c) How would contaminated reagents affect blank values?

10

Fossil and Biomass Fuels

10.1 Combustion Chemistry

Organic compounds are abundant and play important roles in the chemistry of the environment. Organic carbonaceous materials are present naturally in the air, surface waters, soils, and ground water. They come from both natural sources, such as the biological activity of plants and animals, and anthropogenic sources, such as the combustion of fossil and biomass fuels for energy production, agricultural practices, and industrial activities. Both plants and bacteria produce volatile organic carbon compound (VOC) emissions and solid organic materials, which are the major sources of natural organic compounds found on Earth. Most of these naturally occurring organic compounds in the environment play important roles in maintaining the ecosystems by acting as fertilizers for plants and food for animals, bacteria, and fungi, which make up the biological ecosystems.

The term organic was derived to describe all of the materials that are formed from living organisms. In fact, it is defined as "relating to or derived from living matter." Organic materials include a wide number of different types of organic compounds that are present in the environment and are the basis for the most studied area of chemistry, organic chemistry. Organic molecules are formed primarily from the covalent bonding of the elements: carbon, hydrogen, oxygen, and nitrogen. Other elements that also play important roles in biological systems – such as calcium, iron, phosphorus, and silicon – are considered to be inorganic.

The electronic structure of carbon is $[He]2s^22p^2$. So it needs to share four electrons with other atoms in order to obtain the stable electronic configuration of neon ($1s^22s^22p^6$). Also, since the *s* and *p* orbitals of carbon are close in energy they can interact, giving carbon the ability to hybridize these valence orbitals into *sp*, sp^2, and sp^3 combination orbitals. This means that the carbon atom has the ability to form four covalent bonds, which will always contain from one to four of the hybrid orbitals. Organic compounds usually have multiple carbon atoms linked together with these types of bonds. In most cases the carbon atoms are also bonded to hydrogen atoms but they can also bond to other atoms, most commonly nitrogen, oxygen, and the halogens, to form an astronomical number of compounds. Some of these many possible organic compounds formed from carbon bonding are isomers, compounds with the same molecular formula but different chemical structures.

Organic chemistry uses a method of classifying compounds with similar structures and composition that allows them to be able to generalize and predict the chemical behavior and reactivity of specific organic molecules. The hydrocarbons, compounds that contain only hydrogen and carbon, are divided into compound classes as listed in Table 10.1. The compounds in each of these hydrocarbon classes share the same chemical and physical properties of the compound class, due to their similar molecular structures. Organic compounds that contain atoms other than carbon and hydrogen are classified according to functional groups, which are listed in

Chemistry of Environmental Systems: Fundamental Principles and Analytical Methods, First Edition.
Jeffrey S. Gaffney and Nancy A. Marley.

Table 10.1 Common classes of hydrocarbons important to environmental chemistry.

Class	Chemical formula	Structural traits
Alkanes	C_nH_{2n+2}	straight chain
Branched alkanes	C_nH_{2n+2}	single or multiple branches
Cyclic alkanes	C_nH_{2n}	isomeric with alkenes
Alkenes	C_nH_{2n}	carbon double bond, straight chain or branched
Cis-alkenes	C_nH_{2n}	groups on same side of double bond (isomeric with trans-)
Trans-alkenes	C_nH_{2n}	groups on opposite side of double bond (isomeric with cis-)
Dienes	C_nH_{2n-2}	two double bonds (isomeric with alkynes)
Alkynes	C_nH_{2n-2}	carbon triple bond
Aromatics	C_nH_n	one or more ring structures (benzene, PAHs)

Table 10.2 Common organic functional groups important to environmental chemistry.

Functional group	Formula	Structural or chemical traits
Alcohols	R—OH	primary, secondary, or tertiary
Phenols	Ar—OH	weak acids
Ethers	R—O—R	R can be the same or different
Alkyl halides	R—X	X = F, Cl, Br, or I
Aromatic halide	Ar—X	X = F, Cl, Br, or I
Aldehydes	R—CH=O, Ar—CH=O	important in photochemical reactions
Ketones	R—CO—R, Ar—CO—R	important in photochemical reactions
Organic acids	R—CO—OH, Ar—CO—OH	weak acids unless R contains F or Cl atoms
Esters	R—CO—OR	R can be the same or different
Amides	R—CO—NH_2	organic acid derivative
Secondary amides	R—CO—NHR	R can be the same or different
Tertiary amides	R—CO—NR_2	R can be the same or different
Amines	R—NH_2	
Secondary amines	R—NRH	R can be the same or different
Tertiary amines	R—NR_2	R can be the same or different
Mercaptans	R—SH	odorant added to natural gas
Sulfides	R—S—R	R can be the same or different
Disulfides	R—S_2—R	R can be the same or different

R— = alkyl group; Ar— = aromatic group.

Table 10.2. The notation "R—" used in the table represents an alkyl group that is bonded to the functional group. In compounds that contain more than one R— group, the R— groups can be the same or different alkyl groups. The notation "Ar—" represents a cyclic aromatic group, the simplest being the phenyl group (C_6H_5—). Organic compounds that contain the same functional group have similar structures and chemical properties, and the functional group controls the chemistry of the compounds within the group. Similarly, the chemistry of the functional groups can be used to predict the possible chemical reactions, physical properties, and impacts on the environment of the organic compounds.

Fossil fuels are typically made up of a complex mixture of hydrocarbon molecules. Fossil petroleum is typically composed of alkanes, branched alkanes, and cyclic alkanes, with some aromatic hydrocarbons, depending on the nature of the petroleum. These organic hydrocarbons are combusted to release their energy in the form of the heat of combustion, also known in thermodynamics as the enthalpy of combustion (ΔH_c). The standard enthalpy of combustion of a compound is the enthalpy change that occurs when 1 mol of the compound is burned completely in oxygen, with all reactions and products in their standard states. The standard state of a chemical substance is often called "room conditions," because it is at 25.0 °C and 1 atm pressure, with each element in the physical state that occurs normally under these conditions. The main products of a complete combustion reaction are CO_2 and H_2O. For example, the complete combustion of a generic hydrocarbon with the formula C_aH_b proceeds as:

$$C_aH_b + O_2 \rightarrow aCO_2 + b/2\, H_2O$$

Under conditions of high temperature and pressure, combustion can also produce NO from the reaction of N_2 and O_2 in the combustion air. Molecular oxygen and N_2 do not react at ambient temperatures. But at high temperatures (>1600 °C), the N_2 and O_2 in the combustion air disassociate into their atomic states and participate in a series of endothermic reactions, called the Zel'dovich mechanism, which forms NO:

$$N_2 + O \rightarrow NO + N$$

$$N + O_2 \rightarrow NO + O$$

$$N + OH \rightarrow NO + H$$

Also, if sulfur is present in the fuel, SO_2 is formed by the reaction:

$$S + O_2 \rightarrow SO_2$$

The complete combustion of fuel requires a rich supply of air, so that oxygen is in excess and the elements in the fuel react completely with the oxygen, producing only the products CO_2 and H_2O. Incomplete combustion of the fuel occurs when the supply of oxygen is insufficient for the complete reaction to occur. In this case, water is still produced, but CO and carbon are produced instead of carbon dioxide. Carbon produced by incomplete combustion takes the form of char. The charring process that occurs during incomplete combustion includes cracking processes that remove the hydrogen during the formation of the char. This results in most of the char being composed primarily of carbon and released as carbon soot particulates.

Biomass fuels are usually partially oxidized organic compounds, most commonly alcohols or esters. The complete combustion of biofuels also produces CO_2 and H_2O. For example, the complete combustion of a generic alcohol proceeds as:

$$C_aH_bOH + O_2 \rightarrow a\, CO_2 + (b + 1)/2\, H_2O$$

The incomplete combustion of a biofuel also produces H_2O, CO, and carbon soot. However, under low-temperature conditions cracking of the fuel can also occur, producing partially oxidized products along with small hydrocarbons. For example, ethanol (C_2H_5OH) can form acetaldehyde (CH_3CHO), methane (CH_4), ethene (C_2H_4), and acetic acid (CH_3COOH) under these conditions.

The combustion of biofuels generates a different amount of energy than the combustion of hydrocarbon fossil fuels, because of the differences in the chemical bonds that are being broken during the combustion reaction. Although the enthalpy of combustion is measured experimentally using a bomb calorimeter, it can be estimated using the enthalpy of formation of the reactants and products. The enthalpy of formation is the change in enthalpy when 1 mol of

Table 10.3 The heat of combustion (MJ kg^{-1}) for some common fuels.

Fuel	Heat of combustion
Methane	55
Ethane	52
Propane	50
Butane	49
Ethanol	30
Natural gas	37–39
Gasoline	44–46
Coal (anthracene)	31–32
Coal (bituminous)	25–33
Coal (sub-bituminous)	21
Coal (lignite)	16
Wood	15

a substance in the standard state is formed from its pure elements also in their standard states. It is a measure of the total sum of the bond energies that make up the substance. For example, the combustion of ethane:

$$C_2H_6 + 3\tfrac{1}{2}O_2 \rightarrow 2CO_2 + 3H_2O$$

can be estimated from the heats of formation for C_2H_6 (−84.7 kJ mol^{-1}), CO_2 (−393.5 kJ mol^{-1}), and H_2O (−285.8 kJ mol^{-1}). The enthalpy of formation of O_2 is 0 kJ mol^{-1}, since it is already in its standard state and so there is no change when it is formed. Any element in its standard state has an enthalpy of formation of 0 kJ mol^{-1} by definition. The enthalpy of combustion for ethane is calculated as

$$\begin{aligned}\Delta H_c &= \Sigma\Delta H_f(\text{products})\text{–}\Sigma\Delta H_f(\text{reactants})\\ &= 2(-393.5\text{ kJ mol}^{-1}) + 3(-285.8\text{ kJ mol}^{-1})\text{–}(-84.7\text{ kJ mol}^{-1})\text{–}2(0\text{ kJ mol}^{-1})\\ &= -1559.7\text{ kJ mol}^{-1}\text{ or}-51.9\text{ MJ kg}^{-1}\end{aligned}$$

The heats of combustion for some common fuels are listed in Table 10.3. An example of a direct comparison between the heat of combustion of a hydrocarbon fuel with that of an oxygenated biofuel can be seen as the comparison of the heat of combustion of ethane (52 MJ kg^{-1}) with that of ethanol (30 MJ kg^{-1}). The oxygenated compound has 42% less energy content than its parent alkane.

10.2 Formation and Recovery of Fossil Fuels

Fossil fuels, which include coal, oil, and natural gas, are currently the largest source of energy in the world. They are formed from prehistoric organic materials by geological processes that occurred over millions of years. Fossil fuels have been at the center of the world's economic development since the introduction of coal during the Industrial Revolution. However, they are a finite resource, which has become more difficult to recover as their supply has decreased. As the use of fossil fuels has continually increased and their supply has steadily decreased, the methods required for recovery have become more extreme. Their impacts on the environment have also increased along with their increased use and the changing methods of recovery.

10.2.1 The Formation of Fossil Fuels

Fossil fuels were formed by complex geochemical processes over tens of millions to hundreds of millions of years. These processes began with the release of organic matter from dead organisms into sediments. In order for the organic matter to accumulate and for the geochemical processes to take over, it must be deposited into a low-oxygen environment where the natural decay processes can be slowed or halted. Since the decaying organic matter can be deposited into either terrestrial or marine environments, it is believed that the best environment for anoxic conditions is aquatic. Once the organic matter reaches the bottom of a body of water, the dissolved oxygen (DO) drops to very low levels, which slows the oxidation processes. This could occur in marine environments such as inland seas or on shallow shelf regions of the oceans. It could also occur in terrestrial environments where large lakes are fed by rivers or stream channels that carry the organic matter to the bottom of a deep lake. This type of organic material terrestrial deposit is called a lacustrine deposit.

After the organic material is deposited in the low-oxygen environment, it is slowly buried over time by sediment accumulating on top of it. During this burial process, the organic material is removed further from the oxygen in the atmosphere and exists in a totally anoxic chemical system. In this anoxic environment, anaerobic bacteria begin to chemically alter the organic material. The bacteria tend to consume most of the nitrogen-containing compounds, leaving the least digestible materials, such as fatty acids, behind. This complex mixture of recalcitrant waxy material is commonly called humin. Humin is operationally defined as an inhomogeneous dark-brown class of organic compounds that are insoluble in water at all pHs. As the humin is further buried, the weight of the earth above the organic deposit causes a geothermal gradient, which is the rate of increasing temperature with respect to increasing depth. The change in temperature of the humin as a function of burial depth is shown in Figure 10.1 for three different geothermal heating rates. The heating rates vary depending on the type and density of the overburden material. The variation in density of the sediment results in a variation in weight of the overburden, which results in a variation in the thermal gradient. A 40 °C km^{-1} gradient would result from an overburden made up of very dense material, while a 10 °C km^{-1} would result from a less dense material.

The gradual heating due to the geothermal gradient causes the humin to undergo a thermal alteration process. The humin then forms an organic carbonaceous material called kerogen. Kerogen is operationally defined as organic material in rocks that cannot be taken up into solution by the use of normal organic solvents because of its high molecular weight. Kerogen is a mixture of organic materials whose chemical composition can vary widely from sample to sample. It contains varying amounts of carbon and hydrogen, depending on the source of the original organic material. The kerogen then undergoes catagenesis, a cracking process which results in the conversion of organic kerogens into hydrocarbons. Catagenesis requires heating in the range of 50–150 °C, which would require a depth of 2–6 km for the 25 °C km^{-1} thermal gradient. These high temperatures cause chemical bonds to break, generating liquid hydrocarbons. Secondary cracking of the liquid hydrocarbons occurs at the highest temperatures, generating hydrocarbon gases. There is a significant range of temperatures that the kerogen material will experience as it is gradually buried deeper with time. For example, 100 °C is reached at depths of 8, 3.2, and 2 km, for each of the 10, 25, and 40 °C km^{-1} thermal gradients shown in Figure 10.1. These varied heating rates, combined with the differences in kerogen H/C ratios, leads to the formation of the different fossil fuel products that are now found in the subsurface.

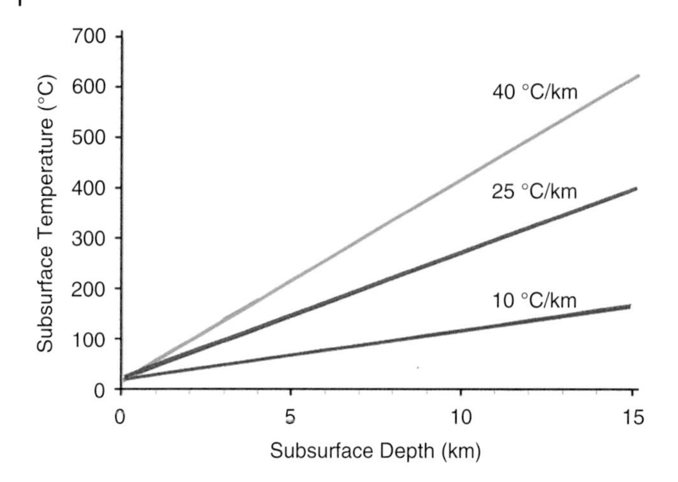

Figure 10.1 Subsurface temperatures as a function of depth caused by the weight of an overburden sediment.

Table 10.4 The four types of kerogen, organic ecosystem sources, and the fossil fuels produced from them.

Type	Source	Organic material	Fossil fuel	Hydrogen content
I	Lacustrine	algae	oil	high H/C ratio
II	Marine	phytoplankton, zooplankton	oil and gas	medium H/C ratio
III	Terrestrial	plants	gas	low H/C ratio
IV	Terrestrial	woody plants	coal	inert, little, or no H

There are four recognized types of kerogen, which are listed in Table 10.4. These types of kerogen are classified according to the type of organic material that created them and thus by their H/C ratios. The kerogen H/C ratios vary significantly if the organic materials originated from lacustrine, marine, or terrestrial environments. In lacustrine environments the primary organic material is believed to be from algae, which have a high H/C ratio. The kerogens that form in this lacustrine environment are classified as type I. The fossil fuel formed from type I kerogen is primarily oil, because of the high H/C ratio. The primary organic material in marine environments is phytoplankton and zooplankton, which is deposited in the form of marine detritus. This material has a lower H/C ratio than type I kerogen formed in lacustrine environments. Marine kerogens are classified as type II and they can form oil and gas depending on the thermal gradients they are exposed to. The two other types of kerogens (type III and type IV) are derived from terrestrial material. Type III kerogen has a very low H/C ratio and produces natural gas. Type IV kerogen is called vitrinite, a coal-like material formed from the original deposition of woody plants. Vitrinite is also terrestrially derived but it has little or no hydrogen and contains inert carbon only.

The oil and gas produced from the first three types of kerogen tend to migrate out of the source rocks since they are less dense than the rocks. The oil and gas then migrates through porous rock formations, such as sandstone, until they are trapped by nonporous rock formations above them, such as clays or slates. The entire process of the formation of oil and gas is termed diagenesis, which is the physical and chemical changes that occur during the conversion of organic sediment to sedimentary rock. The subsurface depths where the oil and gas

Figure 10.2 The La Brea tar pits in Los Angeles, CA showing subsurface oil and gas reaching the surface to form oily tars and gas bubbles. Source: Photo by Nancy Marley, August 2017.

deposits are found depend on the processes involved in the diagenesis, including the original burial location, the temperature history, the type of kerogen, and the possible migration into a subsurface reservoir. These differences in the diagenesis processes lead to a wide variety of fossil fuel deposits. These include the formation of tar sands, which are not as far along in the process, as well as typical petroleum and natural gas deposits. If the oil and gas are produced in a reservoir that has no nonporous capping material, they can rise to the surface. An example of this can be seen at the La Brea tar pits in Los Angeles, CA – shown in Figure 10.2. The surface tar pits are well known for their trapping and preserving the remains of many extinct mammals, such as the woolly mammoth.

Deposits of organic material in marine environments have occurred over a longer period of time than deposits in terrestrial environments. This is because terrestrial plants evolved after marine organisms. During the Carboniferous era, which spanned about 60 million years from approximately 300–360 million years ago, terrestrial plants thrived and the organic deposits that resulted from them led to the formation of large deposits of coal and gas (type III and IV kerogen). Different types of coal are formed from the type IV kerogen, depending on its depositional history. Since coals are formed from the geothermal heating of vitrinite, they are chemically and physically altered from a wet soft material to a hard dry material. This process is called coalification, which converts the vitrinite into coal of increasingly higher rank with anthracite as the final product. The density of vitrinite begins to decrease slowly with increasing coalification. This is because the reduction in the oxygen content is greater than the increase in the carbon content, and the atomic weight of oxygen is larger than the atomic weight of carbon. But after the majority of oxygen is removed, the density of coal increases sharply with further increase in coalification, because of the increasing amount of more compact aromatic structures.

Coals are classified into four types, listed in Table 10.5, depending on their carbon content (fixed and volatile), water content, and hardness. The values given in Table 10.5 are typical of each coal type, however, they can vary widely depending on the source of the coal. Anthracite, the highest grade of coal, is very hard (4 on the Mohs scale) and brittle. It has a very low water

Table 10.5 Average values for the properties and composition of the four general types of coal.

Property	Unit	Anthracite	Bituminous	Sub-bituminous	Lignite
Hardness	Mohs	4	3.8–3	~3.5	2.5–2
Density	$g\,cm^{-3}$	0.8–0.9	0.7–0.9	0.7–0.8	0.6–0.7
Vitrinite reflectance	%	2.5	1.1	0.5	0.4
Fixed carbon content	%	85	75	45	30
Water content	%	<3	5	15	40
VOC content	%	5	20	35	35
Sulfur content	%	<1	1–4	2	<1
Ash content	%	9	7	6	4
Energy content	$MJ\,kg^{-1}$	31–32	25–33	21	16

(<3%) and VOC (5%) content. Anthracite also has a very high fixed carbon content (85%), which accounts for its black luster and high energy content. The lowest grade of coal is lignite, sometimes called brown coal, which is formed from peat at shallow depths and low temperatures (<100 °C). Although the Mohs hardness is difficult to measure due to its tendency to crumble, it is assigned a hardness of 2.5–2. This is compared to talc, which is the lowest on the Mohs scale with an assigned value of 1. Lignite has a very low fixed carbon content of 30%, a VOC content of about 35%, and a water content of about 40%. The heat of combustion of lignite (16 $MJ\,kg^{-1}$) is similar to that of wood (15 $MJ\,kg^{-1}$).

The United States classifies coal between the two extremes of anthracite and lignite into two categories: bituminous and sub-bituminous. The properties and composition of these two types of coal are intermediate between anthracite and lignite. However, some countries further divide these two classes into a total of six grades, and the engineering field has as many as 22 classes of coal after anthracite. These more detailed systems of coal classification represent a progressive transition from anthracite to lignite, and are an indication of the difficulty of identifying distinct species between the two extremes. Bituminous coal is assigned a Mohs hardness of 3.8–3. It is classified as coals with a fixed carbon content of about 75%, a VOC content of about 20%, and a water content of about 5%. Sub-bituminous coal has a Mohs hardness of about 3.5, a fixed carbon content of about 45%, a VOC content of about 35%, and a water content of about 15%.

The three major types of coal are shown in Figure 10.3, along with a hypothetical structure for the highly aromatic polymeric material in coal. The vitrinite reflectance, listed in Table 10.5, is the proportion of incident light that is reflected from a polished surface of coal. It is a standard method for calculating the relative amount of coalification that a coal has undergone. So the amount of vitrinite reflectance of the coal increases progressively from lignite to anthracite. Anthracite is a shiny black material with a metallic-like luster, but duller and less reflective than a metal. Bituminous coal is shiny and black in color, with some bituminous coals being dark gray. Sub-bituminous coal is considered to be a black coal even though its appearance varies from bright black to dull dark brown. It is an intermediate between bituminous and lignite, so its consistency ranges from hard and strong to soft and brittle. It is sometimes called "black lignite." Lignite is brownish black to brown in color, with a rough texture and essentially no measurable vitrinite reflectance.

The proposed chemical structure for coal – shown in Figure 10.3 – consists primarily of aromatic ring structures, many of them polycyclic. These aromatic structures give coals their low hydrogen content. As coalification proceeds and the coal undergoes geothermal heating,

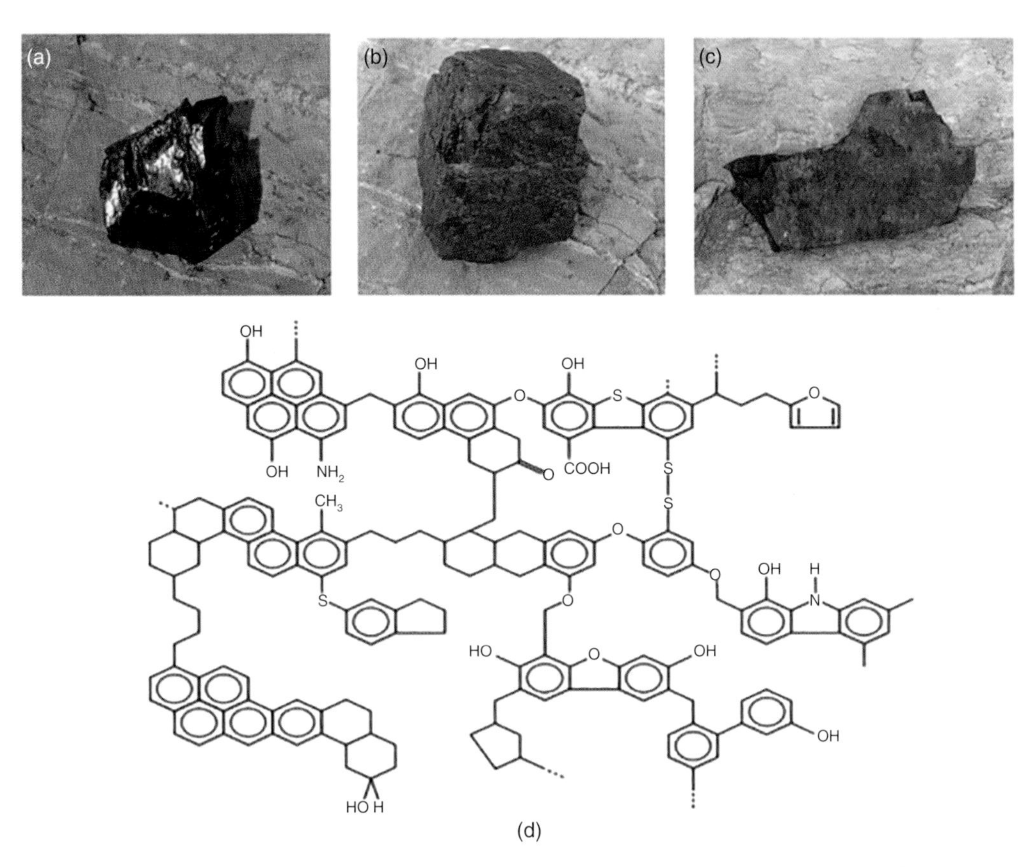

Figure 10.3 The three major types of coal: (A) anthracene; (B) bituminous; (C) lignite; including (D) an idealized chemical structure of coal. Source: (A–C) Donna Pizzarelli, USGS; (D) Karol Glab, Wikimedia Commons.

it matures from lignite to the higher-ranked coals. During this natural heating process, the aromaticity of the material increases as the amount of phenoxy (—OH), methoxy (—OCH_3), and carboxylic acid (—COOH) groups decrease, thus increasing the density of the coal. These groups are lost during the geothermal heating process in the form of water or gas (CH_4 and CO_2). During the thermal aging process, the carbonyl groups and any organic nitrogen groups tend to form ring structures, such as pyridines, quinolones, carbazoles, and pyrroles, increasing the polycyclic aromatic rings in the coal structure as hydrogen is lost with time.

10.2.2 Coal Mining

Coal is found in both near-surface and below-surface deposits, so surface mining and deep underground mining are the most common methods of coal mining. The choice of mining method depends on the depth of the coal seam, thickness and density of the overburden, and thickness of the coal seam. Coal seams that are relatively close to the surface (<180 ft) are usually surface mined, while those from 180 to 300 ft are usually deep mined. Many of the near-surface deposits were once at lower depths but have undergone erosional loss of the sediments above the coal bed. There are three types of surface mining: strip mining, contour mining, and mountain-top mining. All three types of surface mining require the removal of top soil and overburden using drag lines and power shovels to expose the coal seam. The method used depends on the type of terrain and the location of the coal seam. Strip mining is most commonly

Figure 10.4 A coal contour mining operation in Wyoming. Source: Wikimedia Commons.

used in flat areas where the depth of the coal does not vary. The overburden is removed in long successive strips and is deposited on top of the previous strip after mining is complete. The overburden from the first strip is deposited in an area outside the planned mining area. Contour mining, also called bench mining, is commonly used in areas of rolling or steep terrain. The overburden is removed in a pattern following the contour of a ridge or hillside. This is done in stages, leading to a series of shelves or benches in the side of the mining pit. An example of a contour coal mining operation in the state of Wyoming is shown in Figure 10.4. The overburden was once deposited on the downslope side of the latest bench. This practice consumed additional land and created severe landslide and erosion problems. Now the overburden is routinely used to fill the previously mined area. Mountain-top mining methods involve the removal of the top of a mountain to expose coal seams. The removed overburden is deposited in nearby valleys. This type of mining leaves the mountains or hilltops looking like flattened plateaus.

Deep underground mining is the most common type of coal mining, because most coal is too deep to reach using surface mining methods. About 60% of the coal production worldwide comes from underground mining operations. There are five types of underground mining methods: longwall mining, continuous mining, room and pillar mining, blast mining, and retreat mining. Longwall mining and continuous mining both use sophisticated mining equipment called the longwall shearer and the continuous miner machine. These are cutting machines that automatically remove the coal from the underground seam using a cutting head. However, mining with this automatic cutting equipment is not entirely continuous, due to the time needed to install roof supports and remove coal that has been cut free. Still, they account for 95% of underground coal production. Room and pillar methods mine coal by cutting a network of rooms into the coal seam. Pillars of coal are left behind in order to support the roof of the room, so coal removal is not as complete as with the other methods available. In order to more completely remove all the coal from an underground seam, retreat mining is used. This is a form of room and pillar mining where the coal pillars are removed as the miners work back toward the mine entrance. This practice allows the mine roof to collapse and so is one of the most dangerous forms of underground mining, since the time when the roof will collapse is unpredictable. Blast mining is an older mining technique that uses explosives to loosen the coal from the seam.

Other safety hazards involved in deep underground mining are the occurrence of pockets of damp gas (CO, CO_2, N_2), which can lead to suffocation, fires and/or even explosions if the damp gas contains methane. Powdered coal dust is also highly flammable and can cause fires and explosions. Once a fire is ignited in a coal seam it is nearly impossible to put it out. In Centralia, PA an underground mine has been on fire since 1962 and continues to burn. Thousands of coal seam fires occur throughout the world. The most famous of these is Burning Mountain, Australia where the coal seam fire has continued to burn for an estimated 6000 years. It is not known if the fire was started naturally by a lightning strike or forest fire, or by aboriginal activity. The more longwall and continuous mining methods routinely monitor gas and dust levels to minimize fires and explosions. They also make use of stronger pillars made of steel to support the mine roof and prevent collapse. Underground mining often leads to subsidence or sinking of the land above or near the mining operation, which can cause significant damage to buildings and roads in the mining area.

10.2.3 Oil and Gas Recovery

Most of the productive oil and gas fields are in areas that were once covered by large lakes, inland seas, or ocean shelves, where significant organic sediments were once deposited. Past oil and gas exploration focused on the most accessible deposits. Exploratory gas and oil wells averaged a depth of 1.1 km in 1949 and have gradually increased in depth with time, up to depths of 1.8 km by 2007, as shown in Figure 10.5 (USEIA, 2018a). The general trend for both oil (red) and gas (blue) wells is an increase in well depth during this time. It became necessary to drill deeper as the more shallow deposits became depleted. Also, the gas wells are generally deeper than the oil wells, since the secondary cracking of hydrocarbons to form gas requires higher temperatures, which occurs with a deeper overburden. In the United States, more gas and oil wells are now being drilled in marine shelf regions, particularly in the Gulf of Mexico off the coast of the southern states. This drilling requires special floating drilling rigs. The most famous of these was the Deepwater Horizon, which successfully drilled one of the deepest wells (10.6 km) in September of 2009, before having an uncontrollable explosive blowout at another drilling site in late April 2010. That accident led to the oil drilling rig sinking, accompanied by the largest crude oil release into U.S. waters to date.

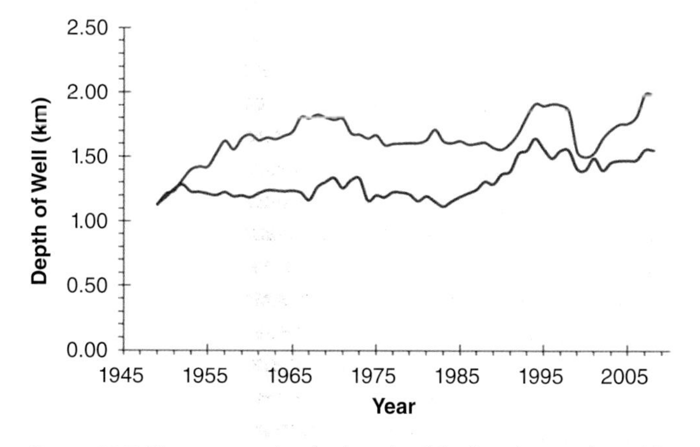

Figure 10.5 The average depth of crude oil (red) and natural gas (blue) wells as a function of the year drilled.

The methods of recovery of crude oil from oil fields include primary, secondary, and enhanced recovery methods. During primary oil recovery the oil is extracted using the natural rise of hydrocarbons to the surface. Since the oil is pressurized in the source rock, the driving force to the surface is the pressure difference between the source rock formation and the well bore, which is open to the atmosphere. Pumps can also be used to further reduce the pressure in the well bore, to increase the pressure differential and increase oil recovery. There are several factors that limit the amount of oil that can be recovered by this method, including rock permeability, subsurface water saturation, and the nature of the subsurface formation. Only about 10% of the total available crude oil in any deposit can usually be recovered using the primary methods.

All of the secondary and enhanced recovery methods require fluids to be injected into the oil field through several injection wells in order to enhance the mobility of oil that remains in the reservoir after primary methods have been exhausted. The oil is then recovered through nearby production wells. Secondary recovery techniques involve the injection of external fluids into the reservoir to increase reservoir pressure and to displace the oil toward the production wellbore. One common method of secondary oil recovery is "water flooding." In this method the oil reservoir is pressurized by the injection of brackish water which then pushes the oil to the production wellbore. The water replaces the loss of reservoir pressure caused by the removal of oil using primary recovery processes. The oil recovered in a water flood usually contains both water and oil, with the amount of water increasing over time. The oil is separated from the water at the surface and the oil is sent to pipelines or holding tanks while the water is recovered for reinjection. This secondary recovery method typically produces an additional 10–20% over the oil produced by the primary methods.

Enhanced oil recovery (EOR) methods, or tertiary methods, are designed to alter the properties of the oil itself in order to increase oil recovery. There are three main methods of EOR: thermal injection, gas injection, and chemical injection. Thermal injection methods use different techniques to heat the crude oil in the reservoir. This decreases the viscosity and reduces the surface tension of the oil, which results in an increase in the permeability through the source rock. One thermal injection method is steam flooding, shown in Figure 10.6. During steam flooding, steam is injected into the reservoir in a similar manner as is done during water flooding. The steam condenses to hot water, which causes the oil to expand, decreasing the viscosity

Figure 10.6 The steam flooding method of enhanced oil recovery. Source: U.S. DOE, Wikimedia Commons.

and increasing the permeability of the oil. Since the hot water eventually cools, the steam flooding process must be cyclical to continue to be effective. This is done by repeating the steam injection when the flow of oil at the production well decreases, indicating that the water is no longer hot enough. Another method of thermal injection is fire flooding, which involves igniting a fire at the interface of the oil formation and the injection wellbore. A continuous injection of air maintains the flame front, which moves through the reservoir toward the production wells. The heat produced by the fire reduces the oil viscosity and vaporizes any water in the reservoir. The steam that is produced, along with the combustion gas, drives the oil in front of the fire toward production wells.

Gas injection is the most commonly used method of EOR. A gas, usually CO_2, is injected into the reservoir to maintain the reservoir pressure and reduce the interfacial tension between the oil and water in the reservoir. CO_2 is commonly used over other gases, because of an anomalous phase behavior of the CO_2–oil mixture, which results in a reduction of the oil viscosity. However, the viscosity reduction is also strongly dependent on the reservoir temperature, pressure, and composition of the crude oil. Chemical injection is currently the least commonly used EOR method. It involves the injection of dilute solutions of chemicals targeted to reduce the surface tension or capillary pressure, which prevents oil from moving through the reservoir. The general application of the chemical injection methods is limited by the cost of the chemicals and their ability to adsorb onto the source rocks, thus restricting movement in the reservoir.

Most of the chemical injection methods use dilute aqueous solutions of surfactants, which are surface-active compounds that contain at least two functional groups: one that is water soluble and one that is oil soluble. They are well known to reduce the surface tension between two immiscible liquids, typically oil and water. The most commonly known surfactants are soaps and detergents, because they are able to transport oily materials into the aqueous phase. The injection of an alkaline solution ($NaOH(aq)$) into an oil reservoir which contains hydrocarbons with organic acid functional groups, called naphthenic acids (NAs), results in the in-situ production of surfactants. The NAs are a mixture of cyclopentyl and cyclohexyl carboxylic acids, with chemical formula $C_nH_{2n-z}COOH$, where z depends on the number and type of cycloalkyl groups in the hydrocarbon chain. The alkaline solution reacts with the acid functional groups to produce the sodium salts of the acids ($RCOO^-$ Na^+). While the acid salt on the molecule is water soluble, the long chain hydrocarbon on the same molecule is oil soluble. So the product of the reaction is a surfactant. The organic content of crude oil varies widely and the success of the in-situ formation of surfactants depends on the NA content of the reservoir oil. For oils that do not contain sufficient NAs, a surfactant solution can be injected directly into the reservoir.

Raw natural gas is also recovered by drilling. It can be recovered from several types of reservoirs, as shown in Figure 10.7. The raw natural gas recovered from oil wells, called associated gas, can exist as a gas cap above the oil. In this case, the gas can flow naturally from the well before the oil is recovered. The natural gas can also exist in the reservoir dissolved in the crude oil. This associated gas is usually viewed as an unwanted byproduct of oil production and is either vented to the atmosphere or flared. Natural gas found in coal seams is usually in the form of coal-bed methane, which can be captured when coal is mined. Coal-bed methane can be added to natural gas pipelines without any special treatment.

As with crude oil recovery, primary gas recovery is based on the natural flow of gas due to pressure differentials between the source rock and the wellbore. This can be increased by decreasing the pressure in the wellbore or by increasing the pressure in the reservoir. Reservoir pressure can be increased by injection of a fluid if the source rock is permeable enough. The reservoirs that consist of shallow deposits of gas only, oil and gas, and coal and gas are highly permeable to moderately permeable reservoirs. The natural gas can be recovered from these reservoirs by vertical wells using primary recovery methods. However, these highly permeable and moderately permeable gas reservoirs are rapidly being depleted. The remaining reservoirs

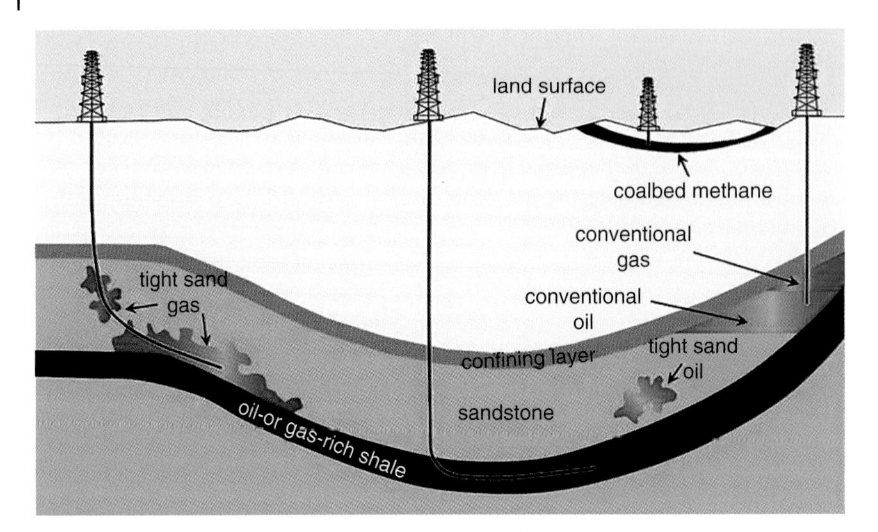

Figure 10.7 A schematic diagram of the types of natural gas sources and their geological environments. Source: Wikimedia Commons.

now being tapped are low permeability reservoirs in sandstone source rock, called tight sand, and ultralow permeability reservoirs in shale source rock.

Natural gas in both sandstone and shale reservoirs is held as small bubbles in the source rock. The release of the natural gas from these low to ultralow permeability reservoirs requires a combination of methods, including directionally drilling multiple wells and using stimulation techniques to induce release of the gas from the source rock. The most well known stimulation technique is called fracking. During the fracking process, a directional well is drilled through the gas reservoir to depths of 6000–10 000 ft. Large volumes of fresh water (about 6 MG), fracking fluid, and sand are then pumped into the well under high pressure and the fluids are forced out through small holes in the well casing. The force of the fluids in the reservoir fractures the source rock and releases the gas bubbles in the rock. In some formations, acidizing is used as a stimulation technique instead of, or along with, fracking. In this method, from 9600 to 15 600 gallons of HCl are used instead of the fracking fluid. The HCl dissolves any limestone, dolomite, and calcite that exists in the source rock to initiate crack formation. In some reservoirs, HF can also be used to dissolve quartz, sand, and clay. Acid inhibitor additives are also introduced into the well to prevent the strong acids from corroding the steel casing in the well. This acidizing can be done under low pressure to induce crack formation in the source rock or it can be done under high pressure, which fractures the source rock. After release from the source rock, the raw natural gas can then flow up the well to the surface, along with "flowback fluid," which consists of injected fluids, salt-saturated water, drilling muds, and/or brine solutions. They are characterized as having high salinity and total dissolved solids (TDS). The flowback fluids are pumped into retention ponds or storage tanks and either recycled or disposed of.

Raw natural gas consists primarily of methane but commonly contains C_2 to C_6 alkanes in varying amounts, along with some CO_2, N_2, helium, H_2S, methyl mercaptan, ethyl mercaptan, mercury, and radon. The smaller alkanes (C_2 and C_3) are typically in higher abundance than the C_4 to C_6 alkanes. Natural gas wells that have a significant amount of H_2S and CO_2 are called "sour gas" wells. The raw natural gas must be processed to remove water vapor, N_2, CO_2, H_2S, mercaptans, mercury, and radon, before it can be used as a commercial product. The acidic H_2S and CO_2 are removed by bubbling the gas through an aqueous solution of a basic amine, where the acidic gases are trapped while the nonpolar CH_4 and other alkanes pass through unaffected. Cryogenic separation or low-temperature distillation are used to separate the N_2 and helium. Helium is usually present at levels of 1–2% and is recovered as a valuable side product. Cryogenic or adsorption methods are used to separate the higher alkanes (>C_2), which are sold commercially as liquefied petroleum gas (LPG).

10.3 Fossil Fuel Use

The industrial use of fossil fuels began with coal. The first reported use of coal was about 1000 BCE in China, where it was used to smelt copper. By the Middle Ages, coal was actively mined throughout Europe for use in metal forges, lime burners, and breweries. Coal did not become a domestic source of heating until after the invention of firebricks in 1822 by William Weston Young in Wales. Fire bricks, composed of a refractory ceramic material, were built primarily to withstand the high temperatures that coal can produce. However, they also have a low thermal conductivity, providing high energy efficiency. The Chinese were also credited with the first industrial use of natural gas in 200 BCE, when it was used to heat brine to produce salt. This process was so successful that they began drilling for natural gas and transporting it by using bamboo pipes in order to allow for the production of salt on a large sale. By 100 CE, the Persians were using natural gas for domestic heating.

The widespread use of coal began in the Industrial Revolution because of design improvements made to the steam engine at that time. This new availability of a dependable source of power made possible mass production in factories, mills, and mines. Thus began the major worldwide consumption of fossil fuel which has been growing continually since about 1800. By the twentieth century, the use of fossil fuels became more varied with coal declining from 96% in 1900 to less than 30% in 2000. Today, crude oil has the highest consumption rate at about 39%, followed by coal at 33%, and natural gas at 28%. The top 10 consumers of fossil fuels by country in the year 2015 are shown in Figure 10.8 (USEIA, 2017a). China is the largest coal

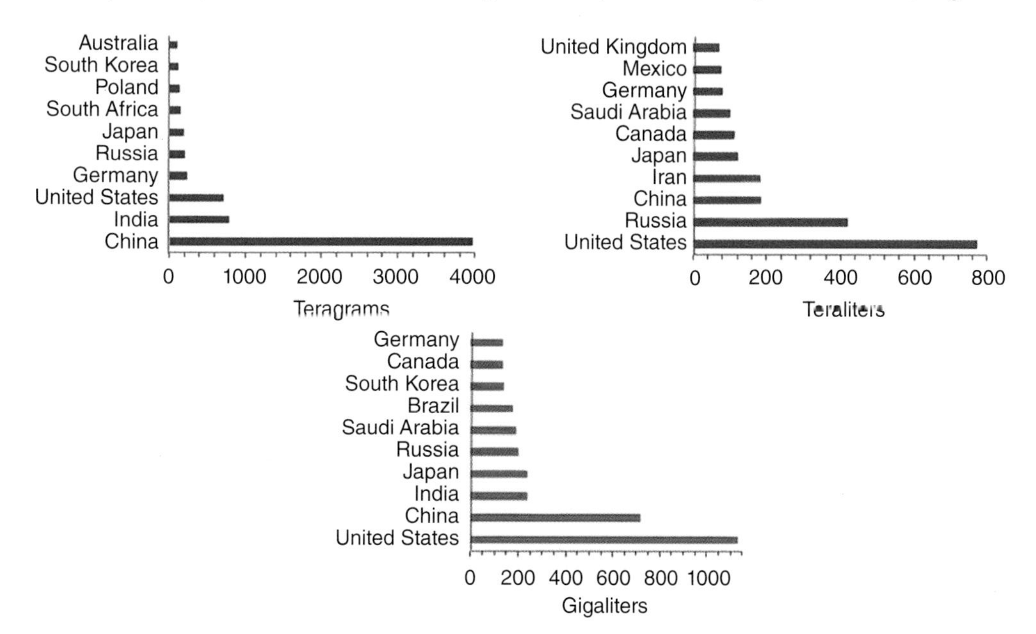

Figure 10.8 The top 10 consumers of coal (blue) in teragrams, natural gas (red) in tetraliters, and crude oil (orange) in gigaliters in 2015.

consumer, making up about half of the world's total coal consumption, followed by India and the United States. Coal consumption in India has more than doubled from 1990 and has now passed that of the United States. This increasing consumption of coal in India is driven by their continuing growth in population and economic development. The single largest oil consumer is the United States, followed by China, India, and Japan. The United States is also the largest natural gas consumer, followed by Russia, China, and Iran. Natural gas consumption has seen significant growth across all regions over the last few decades. This is because both high and low-income nations are steadily shifting from coal to natural gas as a transition toward a less polluting, low cost, and more sustainable energy source.

The combustion of fossil fuels is the major source of energy for vehicles and for stationary power plants used to generate electricity. Hydrocarbons from fossil fuels are also used as solvents and lubricants, and as reactants in the industrial production of plastics, chemicals, and pharmaceuticals. Today, the primary use of coal is in the generation of electricity. It is first milled to a fine powder in order to increase the surface area and allow it to burn more quickly. The powdered coal is then blown into the combustion chamber of a boiler where it is combusted at high temperatures. The heat from the coal combustion converts the water in the boiler to steam, which is used to spin turbines. The mechanical energy from the turbine is then converted to electricity by a generator. The efficiency of this process is as low as 25% in some older coal-fired power plants. However, the efficiencies of newer supercritical and ultra-supercritical steam cycle turbines, which operate at temperatures >600 °C and pressures >266 atm, are >45% when using anthracite coal.

Other important uses of coal are in cement manufacturing, the production of coke for iron smelting, and for the synthetic production of liquid hydrocarbon fuels. Coal is used in the cement industry to provide the energy to heat high-temperature kilns to a melt temperature of 1450 °C. Coke, used for the production of pig iron for the steel industry, is formed by baking coal in an oven without oxygen at temperatures as high as 2000 °C. This high-temperature baking causes the release of volatile components while the fixed carbon and residual ash are fused together. Coke produced from coal is gray, hard, and porous, with a heat of combustion of 30 MJ kg^{-1}. It is used as a reducing agent for smelting iron in a blast furnace. The iron produced by this process is called pig iron, which is primarily used in steel production. But since it is rich in carbon (3.8–4.7%), it must be pretreated to reduce the carbon content before it can be used in the production of steel.

Another proposed use of coal is in the production of synthetic fuels, called synfuels. The most common method of producing synfuels involves the production of a mixture of CO and H_2, called syngas, from coal. The process of producing syngas from coal is called coal gasification. During gasification, the coal is heated in the presence of O_2 and steam. The concentration of O_2 must be kept low enough to prevent complete combustion of the coal but high enough to convert the majority of the carbon in the coal to CO:

$$3C(s) + O_2(g) + H_2O(g) \rightarrow 3CO(g) + H_2(g)$$

The syngas can then be converted to liquid hydrocarbon fuels by the addition of a catalyst at temperatures of 150–300 °C and pressures of one to several tens of atmospheres. This reaction was first demonstrated by the German chemists Franz Fischer and Hans Tropsch, and so is called the Fischer–Tropsch reaction. Any sulfur in the coal gases must first be removed, because sulfur containing impurities will poison the catalyst. Chemical reactions are then used to adjust the H_2/CO ratio for optimum production of the synfuel. The most commonly used reaction is the water gas shift reaction, which gets its name from the mixture of H_2 and CO_2, known as "water gas." The water gas shift reaction increases the H_2 concentration while decreasing the CO concentration, depending on the progress of the following reaction:

$$H_2O(g) + CO(g) \rightleftarrows H_2(g) + CO_2(g)$$

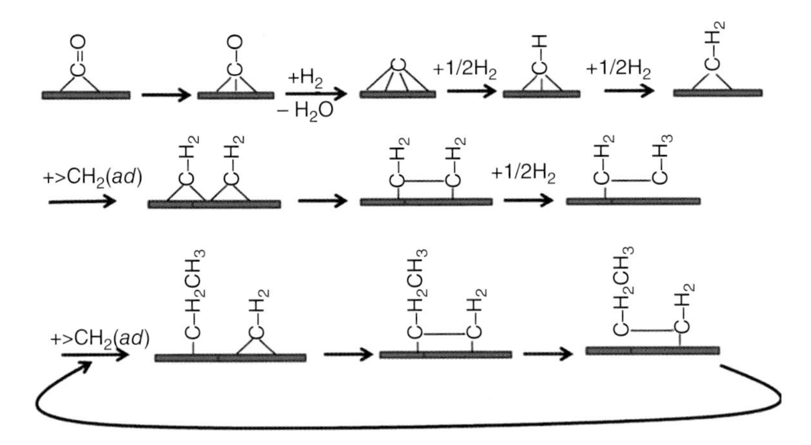

Figure 10.9 The mechanism of the Fischer–Tropsch reaction, a surface polymerization reaction that involves a series of reaction steps with reactants CO and H_2 on the surface of a catalyst (blue), producing a variety of straight chain alkanes. The notation "*ad*" indicates the species is adsorbed to the surface of the catalyst.

Since the equilibrium of this reaction has a significant temperature dependence, the resulting H_2/CO ratio can be controlled by adjusting the reaction temperature.

The Fischer–Tropsch reaction is a surface polymerization reaction involving a series of steps that produce a variety of straight chain alkanes (C_nH_{2n+2}). The general chemical equation for the production of alkanes by the Fischer–Tropsch reaction is:

$$(2n+1)\,H_2 + nCO \rightarrow C_nH_{2n+2} + nH_2O$$

where n is typically 10–20. The majority of the mechanism takes place on the surface of the metal catalyst, as shown in Figure 10.9. It involves the adsorption of the CO on the surface of the metal catalyst, forming a metal carbonyl (M=C=O). The oxygen is removed from the CO(*ad*) by reaction with H_2 forming water (*ad* = adsorbed). At the same time, the carbon atom forms four bonds to the metal catalyst. The carbon atom then adds two hydrogen atoms, resulting in >CH_2(*ad*) bonded to the catalyst. The >CH_2(*ad*) groups are able to migrate on the surface of the catalyst, forming C—C bonds and adding to the length of the carbon chain with each step. The length of the hydrocarbon chain depends on the concentrations of CO and H_2 supplied for the reaction, as well as the surface area of the catalyst. While no coal synfuel plants are currently under operation in the United States, South Africa has been operating synfuel plants for over 30 years.

Crude oil is a complex mixture of hundreds of different hydrocarbons from gases to very heavy waxes. The composition varies significantly, depending on the nature of the formation processes. The hydrocarbons that make up this crude oil mixture have wide ranges of different sizes, molecular weights, and boiling points. It is impractical to attempt to separate the crude oil into all of its pure components, because of the complexity. However, the first step in refining is typically to separate the crude oil into more useful fractions by fractional distillation, as shown in Figure 10.11(A). In fractional distillation, the crude oil is heated in a boiler to a high temperature of about 400 °C. The hot liquid and vapor are then pumped from the boiler into a large industrial fractional distillation tower, shown in Figure 10.11(B). These industrial-sized distillation columns can range from about 1–10 m in diameter and 10–60 m or more in height. The hot gases rise up the tower until they reach their condensation points. As the gases cool, the different components condense back into several distinct liquid fractions. Each fraction is a mixture of limited components with a restricted boiling point range, and similar physical properties.

The refinery products are typically grouped into three classes in order of increasing boiling point: light distillates (gases, naphtha, and gasoline), middle distillates (kerosene and diesel),

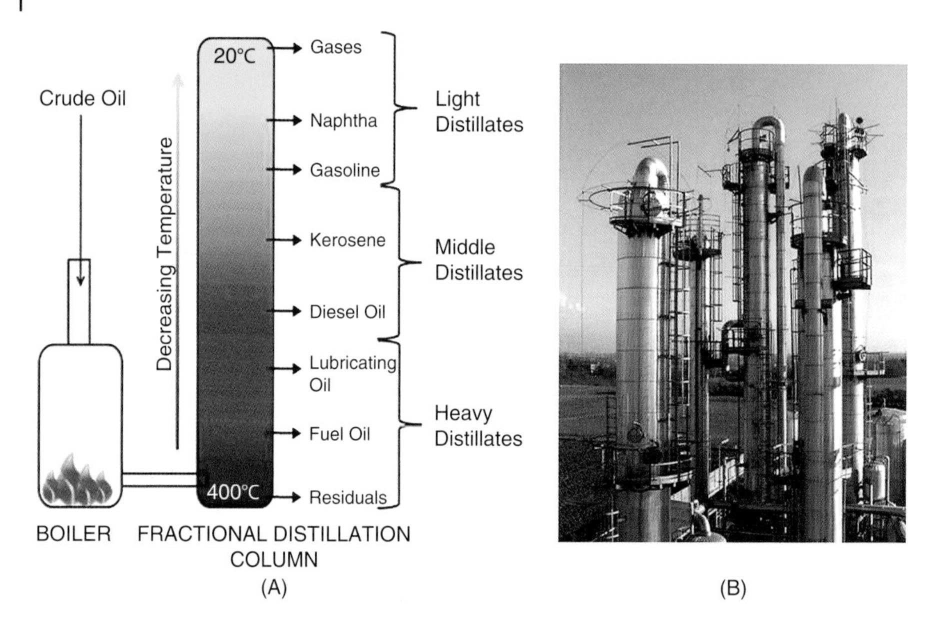

Figure 10.10 (A) The fractional distillation of crude oil into petroleum distillate fractions. (B) Industrial fractional distillation columns (towers) used in petroleum refining. Source: Luigi Chiesa, Wikimedia Commons.

and heavy distillates (lubricating oil, fuel oil, and the solid residuals). Any gases dissolved in the crude oil, including methane, ethane, propane, and butane, are collected at the top of the tower. All of these except methane are often liquefied under pressure to form LPG, which is used as a residential heating and cooking fuel. It is also used as a propellant, refrigerant, vehicle fuel, and as petrochemical feedstock. The first liquid distillate, naphtha, is most commonly used as an additive to boost the octane rating of high-octane gasoline. The next liquid fraction after naphtha is gasoline, followed by the middle distillates: kerosene and diesel oil. The heaviest and highest boiling liquid fractions are lubricating oil and fuel oil. The residuals, which are left in the bottom of the tower, include coke, asphalt, tar, and waxes. They are commonly used as adhesives, roofing compounds, and asphalt. Many of the refined fractions are further chemically processed to make other more profitable products. For example, only 40% of the crude oil is recovered directly as gasoline. So other fractions are further chemically processed into gasoline in order to increase the yield (Figure 10.10).

Processed raw natural gas is now one of the primary fuels used in stationary power plants for electricity generation in the United States. It currently accounts for about 35% of the electrical power consumed, along with heating, cooking, hot water heaters, and clothes dryers. Since the processed natural gas is odorless, odorants are added to natural gas lines so that leaks can readily be detected. The odorant is usually a mixture of *t*-butyl mercaptan or TBM (2-methyl-2-propanethiol, $C_4H_{10}S$) and dimethyl sulfide or DMS ((methylthio)methane, $(CH_3)_2S$). The odor of DMS is described as cabbage-like and is highly disagreeable, even at very low concentrations. The odor threshold of DMS is from 0.02 to 0.1 ppm. TBM has an odor threshold of 0.1 ppb and can cause nausea at levels of 2–3 ppm. These odorants are typically added to natural gas at concentrations of low ppm and are intended to rapidly alert anyone to the existence of a potential gas leak.

Both natural gas and LPG have been used as alternative fuels for gasoline and diesel fuel-powered motor vehicles in regions where the use of low-sulfur fuels is required to meet environmental air quality standards. Natural gas is also used in the petrochemical industry, particularly for the production of ethene. Ethene is used as the starting material for numerous synthetic organic compounds that are important in both industrial and pharmaceutical applications. It is also used to make polyethylene plastics $(C_2H_4)_n$, which polymerize upon contact with a metal oxide or metal chloride catalyst:

$$nH_2C = CH_2 + \text{catalyst} \rightarrow -[CH_2{-}CH_2]-_n$$

Natural gas has also been used in the production of low-sulfur content liquid synfuels. The natural gas is converted directly into CO and H_2 in a steam reformer, a processing device that promotes the reaction of steam with the natural gas. At high temperatures (700–1100 °C) and in the presence of a metal catalyst, such as nickel, the steam reacts with methane to give CO and H_2:

$$CH_4(g) + H_2O(g) \rightarrow CO(g) + 3H_2(g)$$

The products are then used as reactants for the Fischer–Tropsch formation of synfuels.

10.4 Biomass Fuels

Biomass fuels are produced through contemporary biological processes instead of by geologically processed prehistoric organic matter. Fossil fuels are considered to be nonrenewable resources, since their natural production requires hundreds of millions of years. Biomass fuels are considered to be renewable, since they are produced from natural resources that are replenished on a human time scale. Prior to the Industrial Revolution, biomass fuels provided nearly all of the world's energy requirements. The burning of wood or dried animal dung was the principal energy source for heating and cooking before the 1860s. The United States used biomass, in the form of wood, for nearly 91% of all energy consumption prior to the 1800s. Since the Industrial Revolution, the worldwide use of fossil fuel resources continues to increase and, more recently, the demand for low-sulfur fuels with a high energy content and lower environmental impact is also increasing. Because of this, the potential use of renewable biomass fuels, such as biogas, bioalcohol, and biodiesel, has been investigated by a number of countries as an alternative energy source.

Biomass fuels are derived from plants, animal byproducts, and municipal waste. The plant biomass can be grown as a crop, such as corn and sugar cane, to produce ethanol. It can also be produced from naturally available biological materials, such as algae or forest-derived wood products. Although biomass fuels are considered to be renewable, they are limited by the available amounts of water, nutrients, and suitable land for growing. Forest products such as wood are also limited by the time it takes to grow the trees. Also, biomass fuels have low energy densities compared to fossil fuels. This means that they have a lower heat of combustion and a significantly larger volume of biomass fuel is required to generate the same energy as a smaller volume of fossil fuel. This low energy density means that the cost of fuel production and transportation can rapidly exceed the value of the fuel.

Each year it is estimated that the total carbon biomass production in the world, which exists as approximately 50% in the oceans and 50% on land, is about 105 Pg yr^{-1} (petagrams per year) as carbon (Field et al., 1998). This can be compared to the worldwide use of fossil fuels, which is about 8 Pg yr^{-1} as carbon. So the total replacement of fossil fuels with biomass fuels would require about 8% of the total carbon biomass produced on the planet each year, including all carbon forms. Also not considered is that a significant amount of the 105 Pg yr^{-1} of carbon is cycled from the plant kingdom into the animal kingdom as food.

10.4.1 Biomass Fuel Production

There are two types of biomass fuels, those that are produced from food crops and those that are produced from non-food crops and other non-food processes. The biomass fuels that are produced from food-based crops are first-generation biomass fuels. These include ethanol, which is produced from corn or sugar cane, and biodiesel, which is produced from oil-producing seeds or nuts, such as rapeseed, soybean, sunflower, safflower, or palm. The production of first-generation biofuels has important limitations due to the fact that there is a production threshold set by the need to also produce food from the same crops. The first-generation biomass fuels are also not cost competitive with existing fossil fuels and, in most cases, require subsidies to make production cost effective.

The biomass fuels that are produced from non-food biomass, such as non-food crops, wood, plant waste, or industrial and municipal waste, are known as second-generation biomass fuels. The production of second-generation biofuels focuses on the use of non-food feedstocks due to the concern that using food crops for the production of first-generation biofuels could result in a depletion of the world's food supply. This makes the first-generation biomass fuels less viable as a sustainable energy source as the population continues to increase and the food demand continues to rise. One goal of second-generation biomass fuels is to extend the amount of bio-fuel that can be produced from a food crop by using the biomass from the non-food portion of the crops. This would include the stems, leaves, and husks left behind after the food portion is harvested. It also includes the use of other non-food crops, such as switchgrass, grass, and cereals that bear little grain. However, this approach does not address the land use issue. The usable land available for crop growing is limited. Crops which are grown solely for the production of fuel reduce the amount of land that can grow food.

The first-generation ethanol fuels are produced by microbial fermentation of sugars or starches in the biomass. The sugars can be used directly by extraction with water. However, the starches require first hydrolyzing them with the enzyme diastase, which converts the starch to the disaccharide sugar maltose:

$$2(C_6H_{10}O_5) + nH_2O \rightarrow n(C_{12}H_{22}O_{11})$$

When yeast is added to the maltose, it produces the enzyme maltase, which converts the maltose to glucose:

$$C_{12}H_{22}O_{11} + H_2O \rightarrow 2C_6H_{12}O_6$$

In a similar manner the yeast can also convert the plant sugars – disaccharides (sucrose, trehalose), fructans, and raffinose oligosaccharides – to glucose by the enzyme invertase. The glucose is then converted to ethanol and CO_2:

$$C_6H_{12}O_6 \rightarrow 2C_2H_5OH + 2CO_2$$

This conversion proceeds by a two-step biochemical process known as glycolysis, which converts the glucose into pyruvate (CH_3COCOO^-). The pyruvate is then decomposed into ethanol and CO_2 by the enzyme pyruvate decarboxylase.

The fermentation process takes place in an aqueous solution with a final ethanol content of about 15%. It also produces side products such as acetic acid and glycols, which are removed along with the majority of the water by distillation. The final purity of the ethanol is limited to 95–96% due to the formation of a low boiling water–ethanol azeotrope, a constant boiling point mixture composed of 96.5% ethanol to 3.5% water. This final mixture of ethanol and 5–4% water is called hydrous ethanol. If needed, the remaining water can be removed in further treatment to produce anhydrous ethanol. This is accomplished by adding benzene or cyclohexane to the hydrous ethanol, which forms a higher boiling point azeotropic mixture. Distillation

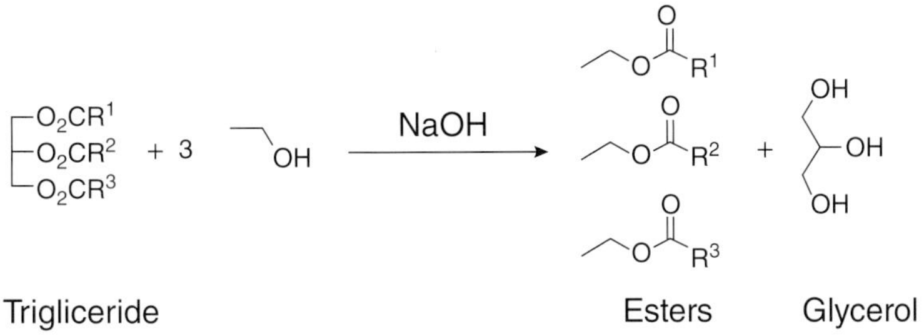

Figure 10.11 The transesterification reaction of a triglyceride with ethanol to produce a mixture of ethyl esters, which is the basic composition of biodiesel. Source: Wikimedia Commons.

then removes the water–benzene or water–cyclohexane in the vapor phase, leaving the anhydrous ethanol in the liquid phase. However, since the ethanol is normally blended with gasoline, complete dehydration of the ethanol prior to blending is not always needed. The remaining water can be removed after blending using mixer–settler tanks. The water separates from the gasoline–ethanol blend in the settling tanks due to the liquid–liquid phase equilibria.

Biodiesel is produced from vegetable oils or animal fats by transesterification, which is a chemical reaction that exchanges an alkyl group of an ester (RCOOR) with an alkyl group of an alcohol (ROH). Animal fats and vegetable oils are composed of triglycerides, which are esters of three fatty acids, and glycerol (OHC_2OHCOH). The triglycerides react with methanol or ethanol in the presence of a base (NaOH) to give a mixture of the methyl or ethyl esters and glycerol in the transesterification reaction shown in Figure 10.11 for the reaction with ethanol. The R^1, R^2, and R^3 groups are hydrocarbon chains with typically 16, 18, and 20 carbon atoms. The glycerol product is easily separated from the mixture of esters by decanting due to their different densities and chemical properties. The NaOH and excess alcohol will also be removed with the glycerol.

Biodiesel is composed of a mixture of mono-alkyl esters of long chain fatty acids produced from this transesterification reaction. The most common alcohol used in the reaction is methanol, which produces a mixture of methyl esters, commonly called fatty acid methyl esters (FAME). Ethanol is also used to produce a mixture of ethyl esters, commonly called fatty acid ethyl esters (FAEE).

Second-generation biomass includes wood and wood byproducts, which is the largest source of biomass energy worldwide. It is used for heating, cooking, and power generation. The energy content of solid biomass fuels is dependent on the moisture content of the fuels. Most fresh wood products have about 50% moisture content by weight. In order to improve the energy content, the wood products can be dried and compressed into pellets. But this process requires energy, since the water is typically bound in the wood fibers at the cellular level. Other second-generation biomass fuel sources include the combustion of municipal waste, landfill-generated gas, and non-wood waste products. Second-generation liquid biomass fuels, also known as advanced biofuels, are mainly produced from wood waste and plant byproducts. There are several second-generation technologies under study to produce biofuels from plant materials, such as gasification followed by the Fischer–Tropsch reaction, pyrolysis to bio-oil, and freeing the sugar molecules from cellulose for fermentation.

Pyrolysis is the decomposition of organic material at elevated temperatures in the absence of oxygen. Biomass can be used as a feedstock for this procedure to produce liquid oily products. However, this bio-oil typically requires significant additional treatment for it to be used in refineries as a replacement for crude oil. Also, these fuels do not meet the standards of either diesel or biodiesel. The most successful second-generation technology in use today is the separation of the sugar molecules from cellulose using enzymes. All plants are composed of two carbohydrate polymers – cellulose and hemicellulose – along with the aromatic polymer lignin. Hemicellulose is composed mainly of xylose ($C_5H_{10}O_5$) and cellulose is composed of glucose ($C_6H_{12}O_6$). The removal of these sugars is very difficult, because the cellulose and hemicellulose fibers are the structural building blocks of the plant, which are tightly bound to the aromatic lignin. The techniques of releasing the sugars from the cellulose and hemicellulose under consideration are acid hydrolysis, enzymatic hydrolysis, organic dissolution, or supercritical hydrolysis. Once released from the plant fibers, the sugars can then be fermented to produce ethanol in the same way as the first-generation bioethanol production. The fuel produced from this process is known as cellulosic ethanol.

10.4.2 Biomass Fuel Use

The first-generation biomass fuels in use today are ethanol (from corn and sugar cane) and biodiesel (from rapeseed, soybeans, and palm). The most common use of ethanol as a fuel is in ethanol–gasoline blends, called gasohol. The most common blends are E10 (10% ethanol, 90% gasoline), E15 (10.5–15% ethanol), and E85 (51–83% ethanol, 40–17% gasoline). More than 98% of the gasoline sold in the United States is E10, which is used to boost octane, meet air quality standards, and satisfy the Renewable Fuel Standard (RFS). The U.S. EPA is responsible for developing and implementing the RFS in order to ensure that transportation fuel sold in the United States contains a minimum volume of renewable fuel. All gasoline vehicles can use E10. Only light-duty vehicles with a model year 2001 or newer can use E15. E85 is a gasoline–ethanol blend containing a varying composition of 51–83% ethanol. The exact percentage of ethanol depends on the geographic location and the season when it is sold. The ethanol content of E85 is adjusted during colder months to improve vehicle performance during cold start and warm-up periods. The volatility of E85 is also adjusted seasonally and geographically by increasing the proportion of light hydrocarbons during colder months. Only flexible-fuel vehicles (FFVs) can use E85. At E15 and above, the automotive engine and fuel delivery system have to be modified in order to use the higher ethanol content, because of the fuel's ability as a solvent to interact with polar materials, such as the polymers used in fuel lines. Gasoline tank filling caps of unmodified vehicles are often posted with warnings against using ethanol blends higher than E10, as shown in Figure 10.12. There are some states where E85 is sold, however it is not consumed on a large scale due to the potential for vehicle damage.

While biodiesel is produced mainly from soybeans in the United States, it is produced from rapeseed in Europe, and palm oil in countries with warmer climates such as Malaysia. Biodiesel fuel has chemical characteristics similar to petroleum-based diesel. So it can be used as a direct substitute for diesel fuel. However, as with ethanol, biodiesel is commonly sold as biodiesel–petroleum diesel blends. The most commonly available blends are B20 (20% biodiesel, 80% petroleum diesel), B5 (5% biodiesel, 95% petroleum diesel), and B2 (2% biodiesel, 98% petroleum diesel). Any diesel engine can use B2 or B5. These blends are popular fuels with the trucking industry because they can increase engine performance due to their lubricating properties. Public fueling stations that sell biodiesel blends to the public are available in nearly every state. Also, some federal and state government fleets, such as school and transit buses, snowplows, garbage trucks, mail trucks, and military vehicles, use biodiesel.

Properly manufactured and blended B2 and B5 biodiesel is a superior fuel to diesel produced from petroleum, petrodiesel. However, problems arise when the consumer tries to manufacture their own biodiesel from used animal fat or oils. If the consumer mixes or filters the fuel incorrectly, then damage may occur to the engine. Both the alcohol and the catalyst used to convert the triglycerides into biodiesel are highly corrosive. If too much alcohol is used or if

Figure 10.12 The gas cap of a 2012 Toyota Camry showing a warning against using gasoline–ethanol blends higher than E10 in the vehicle. Source: Mario Roberto Durán Ortiz, Wikimedia Commons.

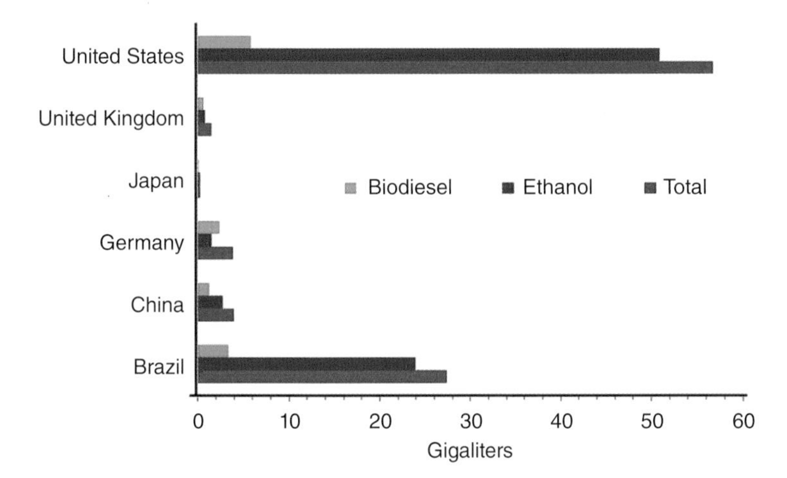

Figure 10.13 The major consumer countries of first-generation biomass fuels in 2014, including ethanol (red), biodiesel (green), and total first-generation biomass fuels (blue) in gigaliters of fuel.

the catalyst (NaOH) is not fully recovered, the resulting biodiesel will be corrosive, resulting in degradation, corrosion, and rust of the engine and fuel system. Also, the glycerol is very hygroscopic. If not totally removed from the fuel it will absorb and condense water from the air. This leads to bacterial and fungal growth, which live in the top of the water layer and feed on the fuel, appearing as a black slime in the fuel tank and on the fuel filter. Because of this, there are no vehicle manufacturers that recommend using a biodiesel blend greater than B5. If the consumer chooses to use higher biodiesel blends or pure biodiesel, the engine warranty will be void.

Figure 10.13 shows a comparison of the major consumers of first-generation biomass fuels in 2014 (USEIA, 2018b). The United States and Brazil are by far the largest users of first-generation biofuels, with most of this being ethanol. Brazil was the first country to use ethanol as a fuel additive. They use E18 to E27.5 on a mandated basis to lower reliance on foreign fossil petroleum fuels. Currently in the United States, E10 is used as an unleaded gasoline nationwide and is mandated in seven states. This accounts for most of the U.S. ethanol biofuel consumed. Other countries throughout the world are typically using E5 to E10 as a means of extending their petroleum reserves or lowering the importation of petroleum from other countries. In the United States, about 5% of total energy use in 2017 was provided by biomass fuels. About 47% of this was first-generation biofuels, with most of that coming from ethanol used in gasoline blends. Second-generation biomass fuels accounts for about 53% of the total biomass fuels, with 43% from wood and wood-derived biomass and 10% from the combustion of municipal waste materials (USEIA, 2017b).

Second-generation biomass fuels currently in use consist mostly of wood waste and wood byproducts. Wood burning is the largest source of biomass energy worldwide. In 2017 about 2% of total U.S. energy consumption was from wood and wood waste, including bark, sawdust, wood chips, wood scrap, and paper mill waste. Wood is used for residential heating as cordwood in fireplaces and wood burning stoves and as pellets in pellet stoves. It is used in industry, electric power plants, and commercial businesses. The wood and paper industries use the wood waste to produce steam and electricity. Some coal-burning power plants include wood chips with the coal to reduce the total SO_2 emissions.

Municipal solid waste can also be used to produce energy at waste-to-energy plants constructed at landfills. These waste-to-energy plants use the heat from controlled combustion of the waste to produce steam for generating electricity. In 2015 there were 71 of these waste-to-energy power plants. In addition to directly burning the waste at landfills to produce energy, the gas produced at the landfill, called biogas, can also be collected as a second-generation biofuel. Anaerobic bacteria in landfills decompose the organic waste to produce biogas, a mixture of different gases produced by the breakdown of organic matter in the absence of oxygen. The biogas is primarily CH_4 (50–75%) and CO_2 (25–50%), with small amounts of H_2 (0–1%) and H_2S (0.1–0.5%). Although the typical CH_4 content of biogas is about 50%, advanced waste treatment technologies can produce biogas with 55–75% CH_4 and this can be increased to 80–90% using gas purification techniques. Although some landfills still dispose of the CH_4 emissions by flaring in a similar manner as done with the associated gas in oil recovery, many landfills do collect and purify the biogas as a substitute for natural gas. The biogas is then either sold for fuel or used to generate electricity in a waste-to-energy plant at the landfill.

Table 10.6 lists the heat of combustion and the oil equivalent volumes (OEVs) for some of the biomass fuels (EBIA, 2017). The OEV is a measure of the energy equivalent volumes for different kinds of fuels. It is equal to the volume of fuel required to obtain the same energy as 1 m^3 of fuel

Table 10.6 The heat of combustion (ΔH_c) in MJ kg^{-1}, oil equivalent volumes (OEV) in m^3, and mass density in kg m^{-3} for oil, coal, and biomass fuels.

Fuel	Moisture (%)	ΔH_c (MJ kg^{-1})	OEV (m^3)	Mass density (kg m^{-3})
Fuel oil	0	41.9	1.0	—
Coal, sub-bituminous	0	25	1.6	—
Wood, pelleted	8	17.5	3.5	650
Cordwood	50	9.5	7.0	460
Softwood, chips	50	9.3	13.1	420
Softwood, chips	20	15.2	12.5	370
Softwood, chips	30	13.2	12	327
Hardwood, chips	30	13.4	9.3	383
Straw, chopped	15	15.0	45.9	70
Straw, baled	15	14.3	19.7	130

oil. Fuel oil has the most energy content, with a heat of combustion of about 42 MJ kg^{-1}, since it is essentially composed only of carbon and hydrogen. Coals vary in energy content depending on the carbon concentration relative to the inorganic content of the coal. Anthracite, the coal with the highest carbon content, has a heat of combustion of 31–32 MJ kg^{-1}. Lignite, with the lowest carbon content, has a heat of combustion of 16 MJ kg^{-1}, which is similar to that of wood pellets (17.5 MJ kg^{-1}). The OEVs for the coals will also vary according to the differences in energy content, as determined by the heat of combustion.

Wood varies in energy content according to its density and moisture content. The heat of combustion of wood and straw is dependent on the moisture content, since they are all composed of cellulose with the same type of chemical bonds. Dry cellulosic materials all contain about 50% carbon, 6% hydrogen, 40% oxygen, and 4% nitrogen, phosphorus, and bound metals. An increase in the moisture content of the biomass decreases the energy content, as measured by the heat of combustion. Pelleted wood, with the lowest moisture content (8%), also has the highest heat of combustion of the biomass products (17.5 MJ kg^{-1}). It also has the highest density of the biomass products (650 kg m^{-3}), because it is composed of compressed wood. Both cordwood and softwood chips have the lowest heat of combustion (9.3–9.5 MJ kg^{-1}) and the highest moisture content (50%). The energy density increases with the mass density of the fuel and the OEV generally increases as the density of the fuel decreases. Since pelleted wood has the highest mass density of the biomass fuels, it also has the lowest OEV (3.5 m^{3}). Conversely, chopped straw has the highest OEV (45.9 m^{3}) and the lowest density (70 kg m^{-3}) of the biomass products.

Since the heats of combustion for the biomass products are substantially lower than oil or sub-bituminous coal, woody biomass fuels are often combined with coals to both improve energy efficiency of the biomass fuel and lower the amount of nonrenewable fuel used. The first-generation biomass fuels – ethanol and liquefied biogas – also have a lower energy content compared to gasoline or diesel fuel. A gallon of ethanol (E100), with a heat of combustion of about 30 MJ kg^{-1}, has only 66% of the energy content of a gallon of gasoline, with a heat of combustion of about 45 MJ kg^{-1} (Gaffney and Marley, 2009). The measure of energy equivalent volumes for liquid transportation fuels is the gasoline gallon equivalent (GGE), which is the amount of non-gasoline fuel that is required to equal the energy content of 1 gallon of gasoline. Ethanol (E100) has a GGE of 1.5 and E85 has a GGE of 1.4. This means that 1.4 gallons of E85 must be combusted in order to obtain the same vehicular mileage as 1 gallon of gasoline. Either liquefied biogas or liquefied natural gas requires 1.54 gallons to obtain the same mileage as 1 gallon of gasoline.

Both first and second-generation biomass fuels require the growth of the biomass, which includes the need for fresh water, fertilizer, pesticides, and available land with suitable soils for growing the biomass. There is an energy cost associated with growing, harvesting, and transporting the biomass. There is also energy associated with the production of the first-generation fuels, ethanol and biodiesel, and with some second-generation fuels such as pelleted wood and biogas. So there must be a source of the energy required to produce the fuels, which is usually the combustion of either a second-generation biomass fuel or a fossil fuel. This energy requirement for the production of biomass fuels lowers the energy efficiency and the renewable properties of the fuels. The consideration of the energy used to grow and produce the fuel compared to the energy obtained in the combustion of the fuel is known as the fuel lifecycle.

The time required to grow the biomass is also a factor in replacing fossil fuels with biomass fuels. There has been considerable work done over the last few decades to identify fast-growing species that yield higher energy biomass products in a shorter time frame with less energy input. For example, the controlled growth of sweetgum and switchgrass has been proposed as second-generation biomass fuel sources due to their fast rate of growth and lower water requirements than other second-generation biomass alternatives.

10.5 Impacts on Water Quality

The recovery and processing of fossil fuels has a great impact on the environment and its natural resources, especially water. Fossil fuel extraction from the various reservoirs can have impacts on both ground and surface waters. There are many ways that the water can become contaminated during energy production, including contamination during recovery processes with a variety of pollutants from mine sediment to synthetic chemicals. Ground and surface water contamination can also occur during fuel manufacture and storage. There is a growing recognition of the close connections between water and energy production. Water is used in every step of fossil fuel recovery and processing. It is also used in every step of biomass production and both first and second-generation biomass fuel manufacture. Because of this close connection between water and energy, our increased energy consumption will put stress on the world's finite water supplies.

10.5.1 Fossil Fuels

In many areas where gas and oil wells are in operation, there has been contamination of the ground water from well seepage. In some cases, sour gas wells, which contain significant amounts of H_2S, can contaminate groundwater supplies, leading to both health and odor problems in nearby drinking water wells. Leaching and erosion from coal strip-mining operations can contaminate both ground and surface waters with metals such as arsenic, selenium, magnesium, iron and mercury, as well as uranium and its radiochemical daughters. Erosion can carry mine debris for significant distances, where it can be deposited in surface waters resulting in changes in both stream chemistry and water flow patterns. The fly and bottom ash left over from coal combustion can cause similar contamination of water supplies from the erosion of storage mounds.

One of the most significant sources of both ground and surface water contamination from energy production is during the extraction of raw natural gas from shale reservoirs, as shown in Figure 10.14. During this process, fracking fluids are injected into the shale to expand and fracture the source rock, releasing the natural gas that is trapped in the rock, as discussed in Section 10.2.3. This fracking process uses very large volumes of surface water, which are mixed with fine drilling sand and proprietary drilling chemicals. Drilling and fracking of large shale deposits can require 3–6 million gallons of water mixed with 15–60 000 gallons of chemicals. After high-pressure injection of the fluids, the pressure in the well is released, reversing the fluid flow. The water and fracturing fluids flow back through the wellbore to the surface as flowback fluids, which then consist of injected fluids, salt-saturated water, and drilling muds. Some of the returned water is actually water that occurs naturally in the shale reservoir, called produced water. The flowback fluids and produced water are pumped into retention ponds or storage tanks and later either recycled or treated for waste disposal.

The major environmental issues with the fracking process have to do with the chemicals that are injected into the reservoir with the water and sand. A detailed list of the fracking chemicals and their purposes is given in Appendix E (FracFocus, 2018). In many cases the major chemical is an organic alcohol, such as methanol or isopropanol, which is used to reduce freezing of the fluids during winter-time operation and act as a corrosion inhibitor of the piping and well casing. Hydrochloric acid and HF are also used, as discussed in Section 10.2.3, to help initiate

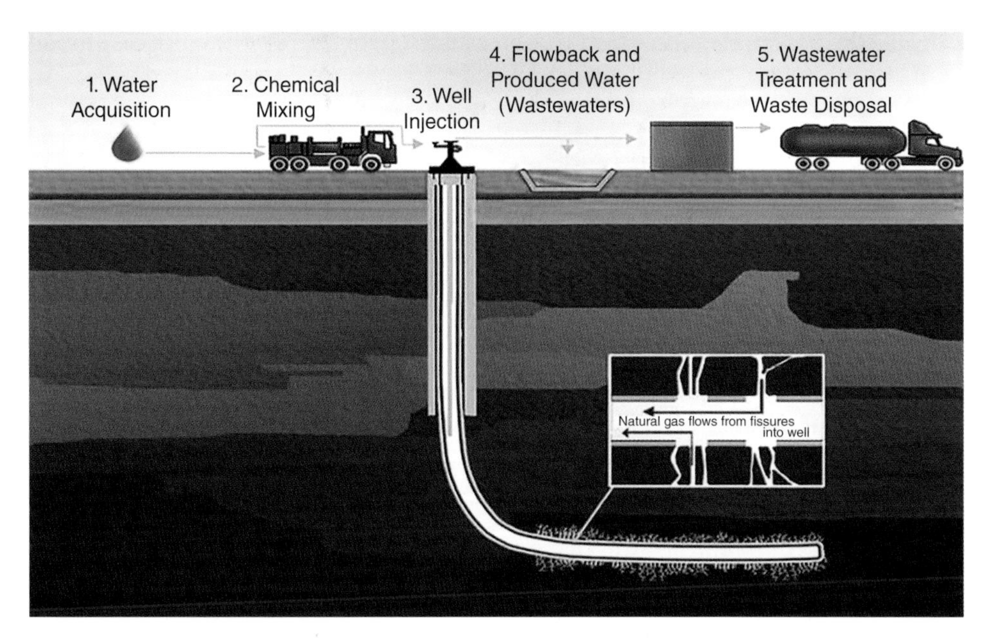

Figure 10.14 A schematic diagram of a fracking process for natural gas recovery from a shale reservoir. Source: U.S. Environmental Protection Agency, Wikimedia Commons.

crack formation in the shale by mineral dissolution. Petroleum products and gelling materials are injected to carry the sand into the initial cracks formed by the acid injection. Bactericides such as glutaraldehyde and quaternary ammonium salts are added to prevent bacterial growth that can cause corrosion in the well piping.

As the well is operated, the chemical and physical properties of the flowback fluids are compared to those of the injection fluids to determine if changes in the injection fluids need to be made. Adjustments are made to the injection fluids by the addition of buffering agents, bactericides, friction reducers, gel stabilizers, and viscosity agents, in order to maintain the proper subsurface conditions for the fracking process to operate successfully. These additives are generally carried by methanol, isopropanol, or ethylene glycol and are all potential contaminating pollutants for groundwater aquifers that lie above the shale formations. Leakage into the groundwater aquifer can occur directly from the shale formation if it should lose integrity during fracking or from cracks in the piping or well casing, which occur from corrosion and frictional wear. The contamination of ground water by methanol and ethylene glycol is of particular concern due to their toxicity and high water solubility. Also, the contaminated flowback fluids and produced waters are typically stored in large retention ponds, which are allowed to evaporate to concentrate the chemicals before the waste is either buried or processed. These contaminated ponds can create a serious hazard for migrating water birds, wildlife, or domestic animals looking for water. Secondary and EOR methods, discussed in Section 10.2.3, also contribute to water pollution in the same way as the fracking of shale reservoirs, since the methods used are similar.

As oil and gas resources are depleted, these more aggressive processes are being used more often to recover the remaining oil and gas resources. In addition, gas and oil exploration has increased in ocean shelf areas such as the Gulf of Mexico. In these cases, the seepage of oil and drilling fluids from pipes and well casings leads to contamination of the marine environment. When a well fails, there can be a catastrophic release of very large volumes of petroleum. The largest of these petroleum releases was caused by a drilling accident on the Deepwater Horizon operated by the British Petroleum company (Deepwater Horizon, 2018). The oil release began in late April of 2010 and continued until mid-September when the well was declared sealed. An estimated 5 million barrels of crude oil were released into the Gulf of Mexico and transported throughout the region during that time.

Once the oil was released, the less dense components rose to the surface, creating a film of oil over a large area of the surface water called an oil slick. These oil slicks can prevent oxygen transport into the upper water layer, causing damage or death to organisms living in this layer. The surface oil was carried by currents and tides onto the shoreline and wetlands, causing damage to biota. Attempts to contain the Deepwater Horizon oil spill included the use of physical barriers and skimmer ships to collect the surface oil. Controlled burns were also attempted to control the oil slick, which contributed to air pollution. Over 1.8 million gallons of dispersing chemicals were used to break up the crude oil into small droplets. These dispersing chemicals, mixtures of surfactants and solvents, were released both at the surface to break up the oil slick and at the point of release in an attempt to induce the oil to sink to the bottom. A satellite image of the large oil slick is shown in Figure 10.15 about a month after the blowout occurred. The oil slick can be seen in the satellite image as a white area due to the reflection of sunlight from the surface oil. Notice the white areas in the coastal Louisiana marshes, swamps, and barrier islands – indicating the infiltration of oil into these regions.

The heavier crude oil components tend to disperse on or near the marine shelf floor. This low-level material moves slowly with currents and tidal motions and was transported onto shore long after the well was sealed. The shoreline in Louisiana was impacted by about 1000 tons of oil in 2012 and over 2000 tons in 2013, which was removed from 55 miles of beachfront. Similar impacts were seen in the surrounding states of Alabama, Mississippi, and Florida. Increased levels of polycyclic aromatic hydrocarbons (PAHs) were observed in the waters and also in both fish and shellfish. The dispersing chemicals caused the oil to become colloidal. These oil colloids could more easily enter the food chain to be ingested by marine organisms and migratory birds, causing mutagenic effects. Methane and the heavy oils released during the event resulted in anoxic regions in the Gulf of Mexico, which are believed to have led to fish and shellfish kills. This region is still undergoing study for long-term effects of this oil release.

Figure 10.15 The release of crude oil from the Deepwater Horizon well blowout (white reflection) as seen from the MODIS satellite on May 24, 2010. Source: NASA/GSFC, MODIS Rapid Response, Wikimedia Commons.

The transport of petroleum from the oil wells to the refineries can also result in accidental release and contamination of surface waters. Transport across land is normally accomplished by pipelines, while transport across oceans is accomplished by large tanker ships. Pipeline ruptures and oil tanker accidents lead to the release of large amounts of oil into both terrestrial and marine environments, which can cause serious impacts to ecosystems. The best known of the tanker accidents was the grounding of the Exxon Valdez on a reef in Prince William Sound, Alaska. The tanker released almost 10 million gallons of crude oil into the Prince William Sound, and caused significant short and long-term damage to marine and shore life in the region (Exxon Valdez, 2018).

While these large oil spills receive the most attention due to the size and scope of the contamination, surface waters and ground waters are also continually impacted by leakage from underground and above-ground petroleum storage tanks. Of particular concern is the leakage from gasoline and diesel fuel storage tanks placed under commercial fueling stations, at airports, and in residential areas. The main health concern from gasoline storage tanks is from the aromatic hydrocarbons, such as benzene and toluene, used in gasoline as an octane enhancer. Besides groundwater contamination, the petroleum products released from underground storage tanks (USTs) can also contaminate soils. There are a few million underground petroleum fuel storage tanks in the United States, which store a number of fuels including gasoline, heating oil, jet fuel, diesel fuel, and kerosene. While there are regulations in the United States regarding the placement and use of USTs, the older tanks were typically made from steel which could corrode over time, resulting in fuel leakage. The corrosion of steel is an oxidation–reduction process where the iron metal in the steel is oxidized to Fe^{2+} and O_2 from the air is reduced to OH^- in the presence of water:

$$2Fe^0(s) \rightarrow 2Fe^{2+}(aq) + 4e^- \text{ (oxidation)}$$

$$O_2(g) + 2H_2O(l) + 4e^- \rightarrow 4OH^-(aq) \text{ (reduction)}$$

The Fe^{2+} and OH^- ions then combine to form $Fe(OH)_2(s)$, which reacts further with O_2 and water to form rust (Fe_2O_3). The EPA has confirmed the existence of 540 000 corroded and leaking USTs. Their cleanup has been underway for over a decade, with the remediation of over 473 000 contaminated sites and about 67 000 sites remaining to be addressed (EPA, 2018).

As part of the site remediation plan, the leaking USTs are replaced with improved storage tanks as shown in Figure 10.16. To prevent corrosion of the steel, the new tanks have an external coating of a polymeric material to seal off the metal from the O_2 and H_2O required for the corrosion reactions. As an additional precaution, should the polymeric coating become damaged, sacrificial metal bars are placed on the outside of the tank using a joint that conducts electrons to the steel. The metal bar is chosen to be more readily oxidized than the iron in the steel. The metals commonly used for this purpose are those that have a standard reduction potential (SRP) with a fairly large negative value, such as magnesium (SRP = −2.37 V) or zinc (SRP = −0.76 V). The iron in the steel (SRP = +0.77) then becomes the cathode and the magnesium or zinc becomes the anode in a galvanic cell. This process, called cathodic protection, prevents oxidation of the protected metal (steel) as long as the electrons can flow through the joint between the two metals. However, as the reaction progresses, the sacrificial metal oxidizes to the ionic form and loses mass until it eventually ceases to function. Should the corrosion continue until the sacrificial metal is exhausted, the tanks also have interior linings of fiber glass or other inert materials to prevent corrosion from reaching the inside of the tank.

The best estimates of the relative amounts of oil contamination in the marine environment worldwide from 1990 to 1999 are listed in Table 10.7 (Committee on Oil in the Sea, 2003). Natural seeps, which account for 47.3% of the oil input into the marine environment, are from reservoirs that do not have a natural cap to contain the oil or from reservoirs whose cap has been

Figure 10.16 A polymer-coated steel petroleum underground storage tank (UST) showing white metal bars attached to the steel, which act as sacrificial anodes for cathodic protection of the tank. Source: U.S. Environmental Protection Agency, Wikimedia Commons.

Table 10.7 The relative amounts of oil contamination in the marine environment worldwide from 1990 to 1999.

Source type	Specific source	Contributions	
		(kT)	(%)
Natural extraction	natural seeps	600	47.3
	platforms	0.9	0.1
	atmospheric deposition	1.3	0.1
	produced waters	36	2.8
Transportation	pipeline spills	12	0.9
	tanker spills	100	7.9
	cargo washing	36	2.8
	coastal facility spills	5	0.4
	atmospheric deposition	0.4	0.05
Consumption	land-based runoff	140	11.0
	spills	7.1	0.6
	large vessel discharge	270	21.3
	atmospheric deposition	52	4.1
	jettisoned aircraft fuel	7.5	0.6

compromised by earthquake activity. The next largest source of oil to the marine environment is by discharges from large oceangoing vessels, at 21.3%. This is primarily due to the release of oil-contaminated bilge water and ballast water. The bilge well is the most important residual collection tank of the entire engine room. Oil collects in the bilge well, mainly from the fuel oil purifiers, leakage in fuel lines and spills. Since both fresh and sea water also collect in the bilge, it must be periodically emptied to prevent overfill. Although the bilge water is first passed through a separator to remove oily particles, much of the oil still finds its way into the marine environment. Another source of fresh and marine water contamination is from the release of motor oil and lubricant products into the environment after their use. About half of the 11.0% attributed to land-based runoff into the marine environment is due to the release of lubricating oils used in vehicles.

10.5.2 Biomass Fuels

In biofuel production, the major water quality issues are associated with the production of the first-generation fuels and the agricultural practices used in growing the crops. All the crops used for the production of biomass fuels require a reliable freshwater source, which is usually supplied by irrigation from wells located near the irrigation systems. Corn and soybeans in particular are water-intensive crops, which require larger amounts of water than some second-generation non-food crops, such as switchgrass. Fertilizers and pesticides are applied to the crops to stimulate growth and protect the crop from insect damage and loss. The excess fertilizers and pesticides contaminate the irrigation water runoff, leading to nitrogen, phosphorus, and pesticide pollution of rivers and streams, and ultimately the marine environment.

As the demand for biomass fuels – ethanol and biodiesel – increases, the areas of corn and soybean crops in the upper Mississippi River Valley increase. These vast areas discharge increasingly larger amounts of fertilizer into the Mississippi River and finally into the Gulf of Mexico. The largest source of nutrients to the Mississippi River comes from farms in Illinois, Iowa, Ohio, and southwest Minnesota. In these areas, the wet soils are drained by plastic pipes that crisscross the fields underground in order to make the soil arable and dry enough to plant the corn and soybean crops in the spring. However, these pipes also continuously flush the fertilizer and pesticides into tributaries that flow into rivers and eventually into the Gulf of Mexico. The nitrogen and phosphorus from farm runoff acts as a natural fertilizer for marine phytoplankton, such as cyanobacteria, green algae, dinoflagellates, coccolithophores, and diatoms. The sudden increase in nutrients during the summer months, when the crops in the midwest are in their strongest growing season, causes the phytoplankton to greatly increase in numbers. This large unnatural growth of phytoplankton is commonly called an algae bloom. It can easily be seen from satellites as a change in the color of the water, as demonstrated in Figure 10.17.

The phytoplankton that are not consumed by zooplankton or fish accumulate in the water, die, and sink to the bottom. As bacteria degrade this biomass of phytoplankton, primarily

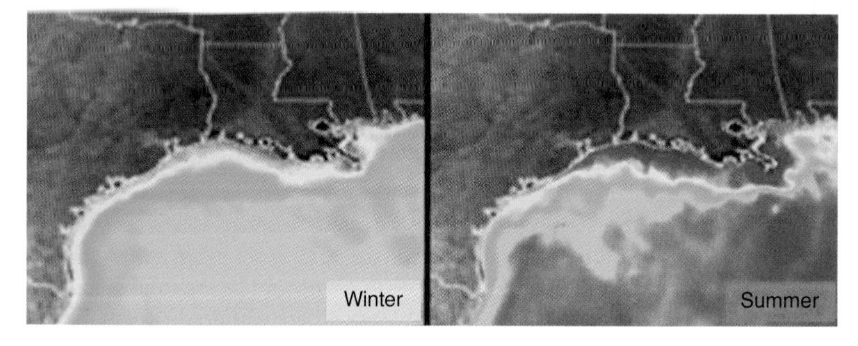

Figure 10.17 Observations of ocean color from the MODIS/Aqua satellite showing large blooms of phytoplankton extending from the mouth of the Mississippi River all the way to the Texas coast. The red and orange colors used on the photographs represent the areas that have been identified as having high concentrations of phytoplankton. Source: NASA, 2007.

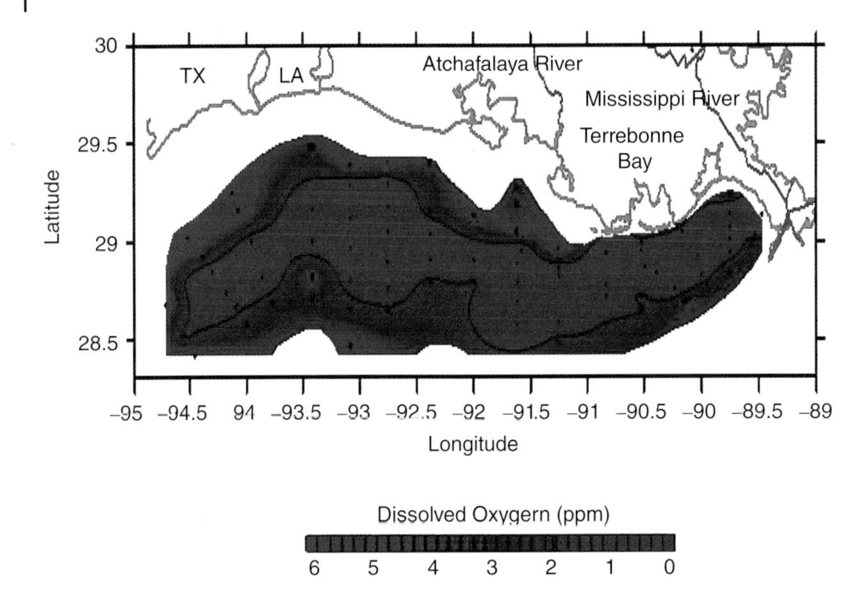

Figure 10.18 Bottom-water dissolved oxygen (DO) concentrations in ppm in the dead zone of the Gulf of Mexico from June to July 2008. Source: NOAA Photo Library, Wikimedia Commons.

cyanobacteria, they consume the oxygen in the water, creating a state of hypoxia near the bottom. Surges of fresh water from the Mississippi River amplify the problem by creating a layering effect in the water. The less dense fresh water forms a layer above the denser salt water, acting as a cap preventing oxygen from the surface water reaching the deeper anoxic water. Under these conditions, the oxygen concentrations in the bottom layer of the Gulf drop so low that sea life must either leave the area or die from suffocation. This area of hypoxia, which is severe enough to be unable to support most marine life in the bottom and near-bottom water layers, is called the "dead zone." The area of hypoxic bottom water, which occurs during the summer months off the coast of Louisiana in the Gulf of Mexico, is the largest recurring hypoxic zone in the United States. The DO values for the summer of 2008 in the dead zone are shown in Figure 10.18. Hypoxia in the northern Gulf of Mexico is traditionally defined as a concentration of dissolved oxygen less than 2 ppm.

In 2017 the fertilizer discharge to the Gulf of Mexico was estimated to be 1.7×10^6 tons, with the dead zone reaching a size of 8776 square miles, an area about the size of New Jersey. This is the largest dead zone ever measured in the Gulf of Mexico since 1985, when measurements of the dead zone began. The large area covered by the dead zone causes a loss of fish habitat, which will force them to move to other areas to survive. It can also cause a decrease in reproductive capabilities and a reduction in the average size of shrimp caught in the Gulf. The adaptation of better agricultural practices designed to keep fertilizers out of fresh water may help to reduce the water quality problems associated with the cultivation of corn and soybeans and the production of corn ethanol and biodiesel. However, in order to protect our water resources while reducing dependence on fossil fuels, ethanol and biodiesel production cannot continue to rise. This means a transition to second-generation crops and waste materials that are more diverse and more environmentally friendly.

Water is also required for the operation of ethanol production facilities for the grinding, liquefaction, fermentation, separation, and drying procedures. These processes require high-purity water, which is obtained mainly from confined groundwater aquifers. In the distillation process, which purifies the ethanol from the fermentation liquid, the ethanol production plants currently use about 4 gallons of water to produce 1 gallon of fuel. This is more than twice the rate of water used in a gasoline refinery. Several other types of water are required for the production of ethanol, including boiler feed water and cooling tower make-up water. The rate of water use at the ethanol production facilities will have a major impact on the groundwater aquifers that supply them. For example, the Jordan aquifer in Iowa has already been pumped down by 150–200 ft by the large number of ethanol production facilities in that area (NRC, 2008). In addition, the ethanol production plants are built close to the corn and soybean-producing crop lands, competing with them for groundwater resources.

Ethanol production in the United States reached a level of 1.4×10^{10} gallons in 2011, which produced about 10 gallons of waste water for every gallon of ethanol produced. After fermentation, the biomass is removed by flocculation and settling. The remaining water in the biomass is typically removed by evaporation and the biomass is incinerated. This can result in air quality problems. After distillation, the remaining water contains very high levels of organic carbon and dissolved solids. The organic materials in the waste water include unconverted polysaccharides, reduced sugars, lignin, proteins, glycerol, and acetic acid. Inorganic materials include nitrate, potassium, phosphates, magnesium, calcium, sodium, potassium, chloride, silica, and sulfate. Since the ethanol production facilities are located in agricultural areas, most are not connected to municipal wastewater treatment plants, and so each plant must have their own water treatment processes. The approved wastewater treatment process is anaerobic digestion followed by membrane filtration to remove colloidal materials. Together, these processes are expensive and difficult to operate with large volumes of waste water. After treatment, the wastewater can still contain pollutants such as pesticides, aromatic hydrocarbons, glycols, and salts, including nitrate, that can contaminate drinking water and endanger fish and other aquatic life.

After the biodiesel is produced it is often "washed" with water to remove the water-soluble contaminants and reaction byproducts. These include soap, glycerol, methanol, and catalyst (NaOH). This washing process can result in as much as 1 gallon of waste water per gallon of biodiesel produced. In its pure form, glycerol is a valuable industrial chemical that is used in many different products. But the glycerol produced in the transesterification process is typically 50% in water, mixed with other contaminants. Many small biodiesel producers do not have the facilities to refine the glycerol to the 80% that is required by glycerol refineries. Because the production waste contains methanol in addition to glycerol, it cannot be released into the environment. This presents a problem for proper disposal of the process waste. The only approved method of waste treatment of the process biodiesel waste is anaerobic digestion. The only other option is transporting the process waste to a larger biodiesel plant that is equipped to refine the glycerol.

The storage of ethanol in USTs is also a concern. Ethanol contains up to 3 ppm acetic acid as a byproduct of fermentation. Also, since it can absorb water it can foster the growth of the bacteria *Acetobacter aceti*, which converts ethanol to acetic acid. This enhanced concentration of acetic acid can accelerate corrosion of steel tanks and cracking of pipelines. Also, since ethanol is a very good organic solvent, it can have corrosive effects on polymers, rubber, plastics, and elastomers. It can also dissolve glues and sealants, including the resin that binds fiberglass. This could accelerate corrosion in older storage tanks and present problems for new storage tanks once the cathodic protection ceases to function. The result is a potential for releases into the environment that could contaminate ground water and surface water. As discussed in Section 10.4.2, ethanol is generally mixed with petroleum fuels to lower the amount of fossil fuel used. Although gasoline is not readily soluble in water, the addition of ethanol to gasoline changes its solvation properties and can act to increase the transport of the fuel during a leakage event. This effect is called co-solvency: the increase in aqueous solubility of hydrocarbons due to the presence of other compounds in water that serve as a co-solvent. Since ethanol is infinitely water soluble it would act as the co-solvent, increasing the water solubility of the gasoline hydrocarbons. The effect of co-solvency is concentration dependent and so would probably not be important for E10 fuels. However, it is expected to become important for E25 or higher ethanol–gasoline blended fuels.

The storage of biodiesel–petroleum diesel blends in USTs can also cause corrosion problems similar to those encountered with ethanol blends. Biodiesel can contain glycerol, a byproduct of the transesterification process. Although most of the glycerol is removed during the purification process, a small concentration remains after purification, as allowed by the fuel standard. Glycerol can also foster the growth of bacteria species closely related to *Acetobacter*. These bacteria produce small molecular weight organic acids such as propionic, lactic, or glyceric acids, which are highly corrosive. By producing organic acids, the bacterial metabolism can result in a lowering of the pH in the UST fluids. This is thought to contribute to the rapid and severe episodes of corrosion that have been observed in steel tanks storing ultra-low sulfur diesel (ULSD)–petroleum diesel blends.

10.6 Impacts on Air Quality

The primary impacts of both fossil and biomass fuels on air quality are due to the combustion of the fuels. The combustion of any fuel can produce CO, VOCs, and NO_x. It can also release SO_2, particulates, and other materials depending on the chemical composition of the fuel. In addition, the combustion of any fossil or biomass fuels will produce CO_2, which is of concern as a greenhouse gas. The nonrenewable fuels – coal, petroleum, and natural gas – produce emissions from both stationary sources, such as electrical power plants, and mobile sources, such as motor vehicles and seagoing vessels. The combustion temperature and pressure conditions, along with the fuel characteristics, determine the type of emissions and their relative concentrations.

10.6.1 Fossil Fuels

The emissions from the combustion of coal are strongly dependent on the combustion temperature. At flaming conditions (>673 K) the alkyl and ether linkages, shown in the idealized coal structure in Figure 10.3(D), are the first bonds to be broken. This is quickly followed by functional group loss that leads to the release of CO, CO_2, SO_2, NO, NO_2, and CH_4. However, if the temperature is less than 673 K, the coal smolders and a number of organic compounds can be released, such as light PAHs. Water vapor is also released depending on the moisture content of the coal. The approximate moisture contents of the different types of coal are listed in Table 10.5: anthracite <3%, bituminous 5%, sub-bituminous 15%, and lignite 40%. The material that is not combustible is left behind as a mineral ash at the bottom of the combustion chamber.

Coal-fired power plants release significant amounts of NO_x and SO_2 into the troposphere. The combustion of fossil fuels can produce NO_x either directly from the nitrogen contained in the fuel or indirectly from the high temperature reactions between N_2 and O_2 in the air supplied for combustion. Although the combustion of coal occurs at high temperatures, it occurs at lower pressures compared to the combustion of gasoline in a motor vehicle engine. So most of the NO_x produced from the combustion of coal is formed directly from the nitrogen content of the fuel, accounting for about 80% of the total NO_x emitted. The NO_x produced from coal combustion accounts for almost 20% of the anthropogenic NO_x emissions in the United States, second only to motor vehicle emissions.

The SO_2 is formed from the sulfur contained in the coal, which varies widely depending on the type of coal. The approximate sulfur contents in the different types of coal are listed in Table 10.5: anthracite <1%, bituminous 1–4%, sub-bituminous 2%, and lignite <1%. Although bituminous coal has the highest sulfur content, it has the second highest fixed carbon content and is the most readily available coal, making up 45% of U.S. coal production. So it is the most used coal in electrical power plants. Coal-fired power plants account for 60% of U.S. anthropogenic SO_2 emissions. The SO_2 can be oxidized to H_2SO_4 during the transport of the power plant plume. Conversion rates are approximately 1–6% per hour, which can result in approximately 40% conversion of the SO_2 to H_2SO_4 during plume transport. In the same power plant plumes, the conversion rate of NO_2 to HNO_3 is about 3–6% per hour. In both cases, these secondary products can either add to the $PM_{2.5}$ levels or be taken up into clouds, leading to the formation of acid rain.

The coal is normally pulverized before combustion in high-temperature coal-fired power plants. Under these conditions, about 80% of the ash leaves the combustion chamber as fly ash in the exhaust gas. Electrostatic precipitators are used to collect the fly ash from the exhaust gas before it is released to the atmosphere. These electrostatic precipitators apply a high-voltage electrostatic charge to the particles, which are then collected on charged plates. Although electrostatic precipitators have efficiency removal rates approaching 99.9%, this does not mean that no fly ash is released to the atmosphere. For example, a 1 GW coal-fired power plant will combust approximately 12 kT of sub-bituminous coal per day and the average sub-bituminous coal will produce approximately 2.4 kT of fly ash per day. Assuming that the electrostatic precipitators are working to full capacity, approximately 1 kT of fly ash is still released into the atmosphere per year (Gaffney and Marley, 2009). Most of this fly ash is released as $PM_{2.5}$, which is likely to undergo long-range transport, impacting both local and regional air quality. The fly ash that is removed by the electrostatic precipitators is typically stored in ponds to prevent winds from carrying the fly ash back into the atmosphere. In some cases, the collected fly ash has been used in the manufacture of concrete for road surfaces. This application reduces the amount of fly ash that needs to be stored.

The composition of the fly ash varies depending on the type of coal burned, since the elemental composition of the coal depends on the type of sediment, the regional geochemistry, and the coalification history. Figure 10.19 shows the average percent composition of the major inorganic species present in fly ash from bituminous (blue), sub-bituminous (red), and lignite

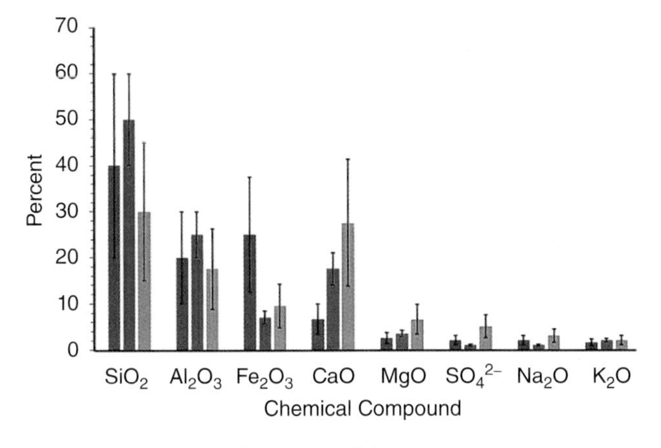

Figure 10.19 The weight percent of the major inorganic components in bituminous (blue), subbituminous (red), and lignite (gray) coal with error bars that indicate the natural variability within the coal types.

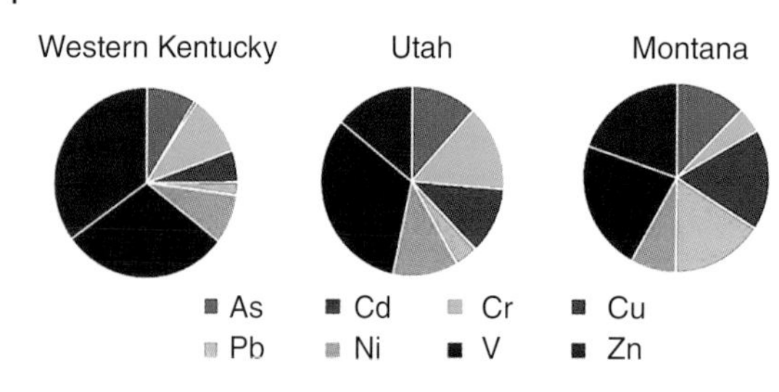

Figure 10.20 The trace metal content (μg g^{-1}) in sub-bituminous coal fly ash from western Kentucky, Utah, and Montana.

(gray) coals (Gaffney and Marley, 2009). The error bars shown in the figure indicate the variability of the inorganic species between the coal sources. The SO_4^{2-} in the fly ash is produced from SO_2 oxidation to SO_3, which then reacts with the MgO and CaO in the hot gases to form SO_4^{2-}. In addition to the compounds shown in Figure 10.19, the coals contain trace toxic metals, including arsenic, cadmium, chromium, copper, lead, nickel, vanadium, and zinc. These trace metal species also vary depending on the source of the coal. The trace metals in fly ash from the combustion of subbituminous coal from western Kentucky, Montana, and Utah are shown in Figure 10.20. The coal from western Kentucky has the highest concentration of all the toxic metals, except for copper and lead. Although the concentrations of copper did not vary significantly between the three areas, lead was highest in the coal from Montana. The coal fly ash also contains varying amounts of the radioactive elements, which are usually measured as total radioactivity content. The three sub-bituminous fly ash samples from Kentucky, Montana, and Utah contain 243, 218, and 293 mBq g^{-1} of total gamma radioactivity, respectively. This radioactivity is primarily due to ^{238}U, ^{226}Ra, ^{210}Pb, and ^{228}Ra, which are concentrated in the fly ash during the combustion process relative to their initial concentrations in the coal.

Mercury is another toxic metal that is emitted from coal-fired power plants. The mercury emissions from coal-fired power plants in the United States have been estimated to be 50 tons per year, which is 40% of the total atmospheric burden. Mercury is emitted from power plants in four forms: mercury vapor, oxidized mercury, organically complexed mercury, and mercury adsorbed onto $PM_{2.5}$. Each of these forms has different chemistries and therefore has different atmospheric fates. Elemental mercury vapor can be transported over long distances with atmospheric lifetimes of a few months to up to a year. Oxidized mercury, complexed mercury, and mercury adsorbed onto particles have shorter lifetimes. They are removed from the atmosphere in a few days or weeks by wet or dry deposition, which results in the contamination of surface waters and soils. After deposition, the mercury compounds each follow chemical pathways that can recycle them back into the atmosphere. This process of deposition and re-emission leads to long-range transport of mercury in the various forms.

Coal has the highest carbon content of all the fossil fuels and, when combusted, yields the highest CO_2 emission rate. The combustion of coal in electrical power plants generates 1 kg of CO_2 per kilowatt hour of electricity. In 2017, the emission of CO_2 from the combustion of coal in electric power plants was 1.74×10^9 metric tons, which was about 34% of the total U.S. energy-related CO_2 emissions. Along with CO_2, coal combustion also produces emissions of CH_4 and N_2O, all of which are greenhouse gases. About 40% of the worldwide electricity generation is produced from coal combustion. So the use of coal has been a major focal point in discussions of control strategies for greenhouse gas emissions. This is one reason why there has been a move toward the use of natural gas instead of coal as the fuel for combustion-driven electric power plants.

When crude oil is processed in a petroleum refinery it is divided into heavy, middle, and light distillate fractions, as discussed in Section 10.3. The heavy distillates, which are burned as fuels in some power plants, only account for about 3% of U.S. electricity generation. The heavy

Figure 10.21 Number 6 heavy distillate fuel oil known as Bunker C. Source: Glasubruch 2007, Wikimedia Commons.

distillates are given numbers from 1 to 6 to distinguish the grade of the fuel. The fuels with lower numbers are the lubricating oils with lower boiling points, while those with higher numbers are the heavier fuel oils with higher boiling points. The volatility, viscosity, nitrogen, sulfur, and ash content of the oil increase with the increasing boiling point of the oil fraction. Low boiling point distillates, including kerosene and diesel fuels, have little or no nitrogen content, and usually have less than 0.3% sulfur content. The high boiling point distillates contain a high ash, nitrogen, and sulfur content, which when combusted give high NO_x, SO_2, and PM emissions.

The heavy distillate number 6, also known as Bunker C, is shown in Figure 10.21. It gets its name from the large bunkers on ships used to transport it. Bunker C is generally used in large cargo and container ships, oil tankers, and large cruise ships. As a high boiling distillate, Bunker C is high in sulfur content and the ships that burn it are a significant worldwide source of SO_2, accounting for about 15% of global SO_2 emissions. These SO_2 emissions can account for a high proportion of the sulfate aerosol found in the $PM_{2.5}$. For example, shipping is estimated to account for more than 50% of the sulfate aerosol found over the Mediterranean Sea. Large ships are also high emitters of NO_x, typically producing approximately 70 g of NO_x per kilogram of fuel combusted. They are also high emitters of carbonaceous soot from incomplete fuel combustion. Over 130 Gg yr^{-1} of carbonaceous aerosols are estimated to be emitted from the combustion of high boiling distillate fuel oil by ships, which is about 2% of global emissions. Because of the high emissions, many nations now require that large ships burning Bunker C fuel use cleaner fuels, such as diesel, when they are in or near ports. They are also encouraged to use land power lines when docked, instead of continuing to run their engines for onboard power. This practice is often called "cold ironing," since it allows the engines to cool to ambient temperature when docked.

Of particular concern are the carbonaceous soot emissions from large tanker and container ships operating in the Arctic and Antarctic. An increase in shipping traffic is expected in the Arctic Ocean because of the reduction of sea ice due to climate change. As new shipping lanes open, large ships will be able to decrease travel distances by taking advantage of a northwestern passage from the northern Atlantic Ocean through the Arctic Ocean into the Pacific Ocean. Increased ship traffic in this region will result in increased releases of carbonaceous soot aerosols. This is anticipated to add to atmospheric warming in the Arctic region, because of the strong absorption of incoming solar radiation by black carbon (Gaffney and Marley, 2009). Also, the dry deposition of black carbonaceous aerosols onto sea and land ice will change its albedo, decreasing reflection and increasing surface absorption of solar radiation. This will likely accelerate the melting of Arctic ice.

Diesel fuel is a petroleum distillate collected between about 220 and 350 °C that contains some sulfur and nitrogen. It is used in diesel engines, which do not thoroughly mix the air with the fuel before ignition. The design of the engine inherently leads to combustion regions that are fuel rich, resulting in incomplete combustion and the formation of diesel soot. Diesel engines produce $PM_{2.5}$, mostly in the form of black carbon, along with CO, NO_x, and SO_2, which are emitted into the atmosphere with the engine exhaust as shown in Figure 10.22. Large diesel engines produce about five times the amount of NO_x as gasoline engines for the same mass of fuel combusted. For example, although California has estimated that heavy duty diesel trucks make up only 1% of all on-road vehicles in the state, they account for over 20% of NO_x emissions (Gaffney and Marley, 2009). These heavy duty diesel vehicles also produce about 75% of particulate matter emissions from mobile sources, mostly in the form of diesel soot. Diesel soot is fine-mode aerosol that contributes to the $PM_{2.5}$ atmospheric loading. It is composed primarily of nonpolar components, including PAHs and unburned hydrocarbons. This hydrophobic nature, along with the fact that the particulates are in the 0.1–0.3 μm size range, allows it to be transported long distances compared to the more polar particulates such as sulfate.

Diesel engine exhaust systems use an oxidative catalyst to remove VOCs and reduce CO emissions. However, these catalysts cannot oxidize the diesel soot. Diesel soot control can be accomplished by using diesel particulate filter (DPF) systems, also called soot traps. Some DPFs are single-use filters, intended for replacement after the filter becomes clogged, while others are

Figure 10.22 The emission of a large amount of diesel soot from a heavy duty diesel truck during engine start-up. Source: USEPA, Wikimedia Commons.

designed for reuse after the accumulated particulate matter is removed. The removal of particulate matter from a reusable filter can be accomplished using an oxidizing catalyst heated by the exhaust gases. It can also be accomplished by heating the filter to soot combustion temperatures using an external burner.

Kerosene distillates are produced in the temperature region of 150–220 °C. The organic compounds that are collected in this temperature range typically have a very low sulfur and nitrogen content and are used as jet fuels. So, private, military, and commercial jets are negligible contributors to the anthropogenic SO_2 emissions because of the low sulfur content of the fuel. However, jet engines operate at high temperatures and during takeoff the combustion of the jet fuel can generate 30–45 g of NO_x per kilogram of jet fuel burned, which is 10–20 times that of a typical gasoline-powered automobile. This can create air pollution problems in the vicinity of airports. However, once the jets reach cruising speeds, the NO_x emission rates fall to about 10 times lower, reducing their impacts on regional tropospheric chemistry. Overall, the NO_x contributions from jet aircraft are considered to be less than 1% of the total NO_x emissions into the troposphere over North America. However, the release of the NO_x from jet planes flying in the stratosphere has and continues to be a concern for stratospheric ozone depletion. An examination of aircraft cruising altitudes with tropopause heights indicates that 20–40% of all jet NO_x emissions are released into the stratosphere (Karol et al., 1999).

Gasoline is a mixture of lighter-weight distillates produced in the temperature region of 90–150 °C. It is one of the most important fossil fuel products produced in the oil refineries. Gasoline is used to fuel most all motor vehicles as well as smaller ships, lawn mowers, and small gasoline engine-powered equipment including backup generators. The engine performance when using gasoline depends on the type of hydrocarbons in the mixture. It is typically compared to that of octane and reported as an octane rating. The octane rating is a standard measure of the performance of an engine when using the fuel. It is a measure of the tendency of the fuel to burn in a controlled manner, rather than exploding in an uncontrolled manner. Some fuels will self-ignite under pressure in the engine cylinder before the ignition flame reaches them, causing engine knocking. The higher the octane number, the more compression the fuel can undergo before self-igniting. The hydrocarbon octane is used as a standard for controlled ignition and has an octane rating of 100.

Since the early gasoline products had highly variable octane ratings, alkyl lead compounds (tetramethyl lead and tetraethyl lead) were added to the gasoline to improve the fuel combustion characteristics and reduce engine knocking, as discussed in Sections 1.5 and 5.3.1. A small amount of alkyl chlorides and bromides were also added, so that the lead was emitted principally as particulate lead halides along with some organic lead compounds. Lead was identified as a criteria air pollutant in the early 1970s, and the addition of lead to gasoline was banned in the United States in 1996 because of the extreme toxicity of organic and inorganic lead compounds. Leaded fuel is still allowed in the United States for use in off-road vehicles, including farm, marine, racing, and aircraft engines. With the ban on lead additives in on-road vehicles, the amount of lead in atmospheric aerosols in the United States has been dramatically reduced. However, leaded gasoline is still used in some countries, including Algeria, Iraq, Yemen, Myanmar, North Korea, and Afghanistan. Compounds added to gasoline to replace lead as an anti-knocking agent include aromatics, branched alkanes, and oxygenates (primarily ethanol).

The major emissions from the combustion of gasoline are NO_x, VOCs, and CO, which have impacts on local and regional air quality. The internal combustion engine, which uses either a spark or electronic ignition, is designed to ignite a well-mixed gasoline–air mixture after it is injected into the engine cylinder. The high temperatures and pressures produced in the cylinder form significant amounts of NO_x from reactions of N_2 and O_2 in the combustion air. The NO_x,

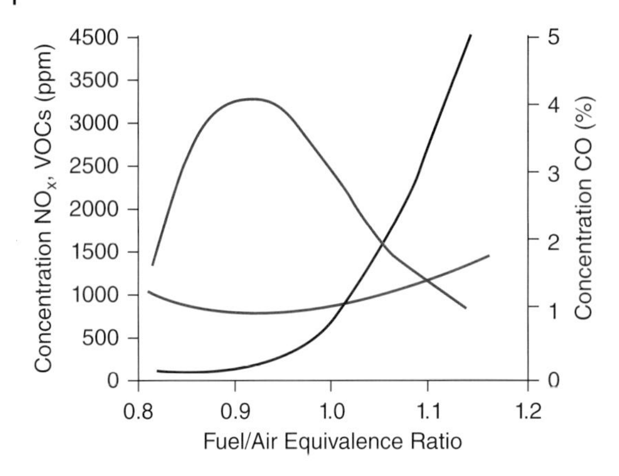

Figure 10.23 The concentrations of NO_x (blue) and VOCs (red) in ppm, and CO (black) in percent in the exhaust gas of an internal combustion engine as a function of the fuel-to-air equivalence ratio.

CO, and VOC emissions in an internal combustion engine vary as a function of the fuel-to-air equivalence ratio, as shown in Figure 10.21. A fuel-to-air equivalence ratio less than one is fuel lean, while a fuel-to-air equivalence ratio greater than one is fuel rich. The optimal performance of most gasoline vehicles occurs when the fuel-to-air ratio is near the stoichiometric value of one. However, the engines are usually operated under slightly fuel-lean conditions in an attempt to lower the CO emissions. Both the CO and VOC emissions are lower at fuel equivalence ratios of 0.9–0.95, and they are further reduced by the use of an oxidative catalytic converter. However, while the CO and VOC emissions are minimized under these fuel-lean combustion conditions, the NO_x emissions (principally NO) are at their highest. This is the primary reason that motor vehicles using gasoline-driven internal combustion engines tend to produce significant NO_x emissions, being responsible for over 50% of the anthropogenic NO_x produced in the United States. The reduction of NO_x emissions is accomplished by reducing the engine operating temperature. This temperature reduction is achieved by recycling a small amount of the exhaust gas into the combustion air, a system called exhaust gas recycling (EGR) (Figure 10.23).

The majority of the emissions from gasoline-powered mobile vehicles are from older vehicles. As vehicles age the EGR systems become less effective and must be repaired to maintain reduced NO_x emissions. Also, the oxidative catalyst may become poisoned by substances such as sulfur, manganese, phosphorus, and zinc, which begin to leak from the engine oil in older vehicles. When this happens the catalyst is no longer effective in oxidizing the CO and VOCs to CO_2 and water, and the catalytic converter must be replaced to maintain reduced emissions. Internal combustion engines also emit substantially higher levels of CO and VOCs after a cold start. This is because the oxidative catalyst must reach a temperature of 430 °C in order to be effective. It takes approximately 15 minutes after a cold start for the catalyst in the exhaust line to reach the optimum operating temperatures required for the complete oxidation of the CO and VOC emissions.

VOCs can be emitted from gasoline directly into the troposphere from storage and refueling facilities. With millions of vehicles being fueled each day, this source of VOCs can be significant. In order to reduce these emissions, the EPA mandated the use of specialized equipment to capture the gas vapors and redirect them back into the storage tank. This procedure is used both when filling the underground tank from a gasoline tank truck and when filling the gas tank of a vehicle from the underground tank. During the filling of underground tanks, the gas vapors coming from the underground storage tanks are captured and directed back into the tank truck. In addition to preventing the vapors from escaping into the atmosphere, this procedure also effectively "vapor balances" the two tanks. When refueling vehicles from a gas pump, gas vapors are pushed out of the vehicle's gas tank by the incoming fuel. These vapors are captured

Figure 10.24 A gasoline dispensing nozzle with a rubber gasket used to capture gasoline vapors and return them to the underground storage tank. Source: Mroach, Wikimedia Commons.

by special fuel-dispensing nozzles at the pump, which have a rubber gasket that forms an airtight seal against the tank opening, as shown in Figure 10.24. The vapors are then captured by the special nozzle and directed into the underground storage tanks, where they are stored until a delivery is made.

10.6.2 Biomass Fuels

The addition of ethanol to gasoline can raise the vapor pressure of the fuel substantially, leading to enhanced evaporative VOC emissions. The evaporative emissions from a vehicle using E10 fuel will increase by about 30% under non-operating conditions compared to gasoline alone. The evaporative emissions will be even higher under operating conditions when the engine and fuel lines are hot. The change in vapor pressure of the fuel, which results in changes in evaporative emissions, is dependent on the ethanol content of the fuel. Pure ethanol has a lower vapor pressure than gasoline due to hydrogen bonding between the ethanol molecules. However, when ethanol is blended into gasoline at relatively low concentrations (5–30%) the gasoline molecules disrupt the ethanol hydrogen bonds, allowing both the ethanol and the gasoline to easily evaporate, raising the vapor pressure of the fuel blend. The E10 fuel has a vapor pressure that is about 1 psi higher than the vapor pressure of pure gasoline (NREL, 2012). However, the increase in vapor pressure is dependent on the ethanol concentration. The vapor pressure increases are less for ethanol blends above about 20%, because the higher ethanol content allows the ethanol molecules to partially re-establish the hydrogen bonds. Fuel blends above about 50% have vapor pressures that are actually less than that of the gasoline.

The average VOC and CO emissions are decreased in vehicles using E10 fuel. This is because the higher oxygen content of the fuel causes the engine to run fuel lean, increasing complete fuel combustion. But the higher oxygen content also causes the engine to run at higher temperatures, producing higher NO_x emissions. So the lean fuel-to-air ratio obtained with E10 leads to the observed decrease in CO and VOC emissions with an increase in NO_x emissions, as shown in Figure 10.23. Incomplete ethanol combustion, which is particularly high during cold-start conditions, results in the emission of methane, ethene, acetaldehyde, and acetic acid. Field studies in Brazil, Albuquerque, NM, and Denver, CO have reported significant atmospheric levels of acetaldehyde when ethanol–gasoline fuel blends are used (Gaffney and Marley, 2009). In Norway, E95 buses were found to emit very high levels of acetaldehyde (>150 ppm), along with the oxidized product acetic acid (20–30 ppm). These high-level emissions caused both eye irritation and odor problems in the areas where the E95 was being used (Lopez-Aparicio and Hak, 2013). The primary emission of acetaldehyde leads to the photochemical production of peroxyacetyl nitrate (PAN), acetic acid, and peracetic acid. The increase in PAN formation can result in impacts on regional ozone production, since PAN can act to transport NO_2 over long distances. In general, the higher the ethanol content of the fuel, the higher the level of ethene and acetaldehyde emissions, which – along with the increased NO_x emissions – will lead to increased levels of tropospheric ozone (Gaffney and Marley, 2009).

The addition of ethanol to gasoline increases the fuel octane rating, while reducing the overall mileage performance. Ethanol is an octane booster with an octane rating of 113 compared to regular unleaded gasoline with an octane rating of 87. Since ethanol has an octane rating higher than 100, it resists self-ignition slightly better than octane. Aromatics such as benzene, toluene, and xylene have been used to increase the octane rating of gasoline fuels. Because of the increased octane rating of ethanol–gasoline blends, lower amounts of aromatics are needed to boost the octane rating of the fuel. This leads to some reductions in the emissions of the air toxics benzene and toluene. The reduction in mileage of ethanol-blended fuels is because ethanol has a lower carbon content and a lower energy content than gasoline. Ethanol has a heat of combustion of about 30 MJ kg^{-1}, while gasoline has a heat of combustion of about 45 MJ kg^{-1}. So the heat of combustion of ethanol is 33% lower than that of gasoline and the heat of combustion of E10 (41 MJ kg^{-1}) would be about 9% lower than that of gasoline. This is why the ethanol–gasoline blended fuels yield fewer miles per gallon (MPG) than gasoline alone, with the MPG being inversely proportional to the amount of ethanol in the fuel. The ethanol-blended fuels will consequently have higher emissions of CO and VOCs per mile than that of gasoline, even though their emissions per gallon of fuel combusted are less.

Biodiesel fuels cause the diesel engine to run a little hotter than petroleum diesel due to the added oxygen content in the esters that make up the biodiesel. The combustion of biodiesel cracks the methyl ester in the fuel to formaldehyde, which is released into the atmosphere. Formaldehyde is a carcinogen, an eye irritant, as well as a photochemically active species. It is also highly water soluble and can be adsorbed onto wet aerosols as the hydrated species methylene glycol:

$$CH_2O + H_2O \rightleftarrows CH_2(OH)_2$$

with an equilibrium constant of 10^3. The combustion of biodiesel fuels produces about the same amount of NO_x, CO, and VOC emissions as the combustion of petroleum diesel, even though they usually contain less nitrogen. However, SO_2 emissions are reduced substantially when biodiesel or biodiesel blends are used, due to the lack of sulfur compounds in the fuel. However, the biodiesel has been found to produce significantly different particulate emissions than petroleum diesel. The particulate emissions from the combustion of biodiesel are composed of a larger number of smaller particles with higher benzene and soluble organic content than that of petroleum diesel (Agarwal et al., 2011). The number of particulates in all size ranges is higher for non-blended biodiesel (B100). However, the peak particle concentrations were shifted toward smaller size particles. The particulate matter emitted from soybean biodiesel was also found to contain an increased amount of octanedioic acid, which is a dicarboxylic acid produced from the combustion of the methyl esters (Dutcher et al., 2011). Biodiesel particulates contain a lower heavy metal content than petroleum diesel. This is because the heavy metals enter the exhaust from engine wear and the lubricating properties of the biodiesel reduce engine wear compared to petroleum diesel. Also, since biodiesel contains no sulfur compounds, the particulate sulfate levels in biodiesel blends are also lower than with petroleum diesel.

Similar to the case of ethanol-blended fuels, biodiesel fuels have less carbon content and so have a lower energy content than petroleum diesel. This means that more fuel must be burned to travel the same distance. The heat of combustion for the transesterified biodiesel fuel is about 34 MJ kg^{-1} compared to the heat of combustion of petroleum diesel of 43 MJ kg^{-1}. So the heat of combustion and thus the energy content of the biofuel is approximately 21% lower than that of petroleum diesel. The decrease in energy content of a biodiesel–petroleum diesel-blended fuel is directly proportional to the amount of the biodiesel content of the fuel. The higher the biodiesel content, the more fuel must be combusted to travel the same distance as with petroleum diesel. This larger volume of fuel required per mile traveled also results in larger emissions per mile of NO_x, VOCs, CO, and CO_2.

10.7 Gasoline Additives: Lessons Learned

As previously discussed, the octane ratings of gasoline can vary substantially. This was the reason that tetraethyl lead was added to gasoline as a means of increasing and maintaining a stable octane equivalent ratio. It was later accepted that adding a toxic metal to a combustion fuel was not a good idea and the lead additives were finally banned in the United States for use in on-road vehicles because of the toxicity of their emissions. In January of 1995, Congress mandated the use of oxygenated fuel additives in nine major cities in the United States in order to reduce the ambient levels of CO and the ozone-forming compounds NO_x and VOCs. The oxygenated fuel additives methanol, ethanol, methyl-*t*-butyl ether (MTBE), ethyl tertiary-butyl ether (ETBE), or tertiary-amyl methyl ether (TAME) were to be blended with gasoline in order to result in an oxygen content of 2% by weight. The resulting fuels were called reformulated gasoline (RFG). At the same time, the use of benzene and other aromatics as octane boosters was restricted. So the branched chain additives MTBE, ETBE, and TAME became more popular in RFG because of their octane-enhancing properties. The RFG fuels lower CO emissions because the higher oxygen content in the fuel produces fuel-lean combustion conditions. However, they also increase NO_x and aldehyde emissions, which increase urban ozone levels.

The standard 8-hour average ambient CO levels measured in urban areas gradually fell after the use of oxygenated fuel additives was implemented in 1995, as shown in Figure 10.25 (blue) (USEPA, 2017a). However, the use of oxidative catalytic converters, which effectively oxidized

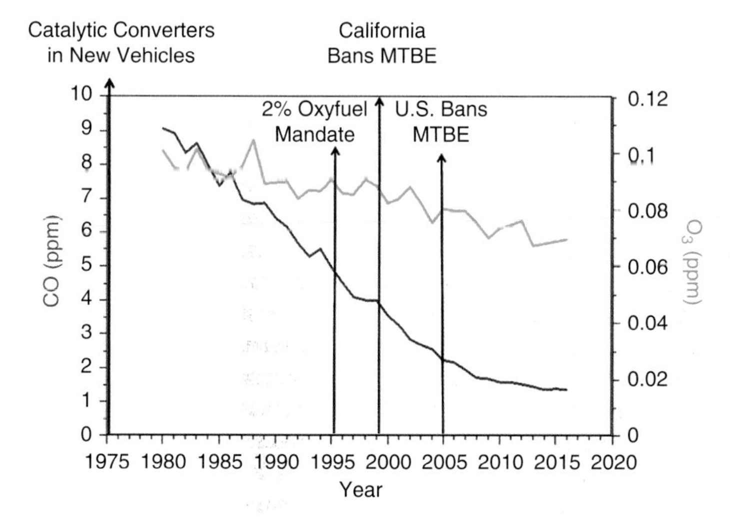

Figure 10.25 Carbon monoxide 8-hour annual means (blue) taken from 62 urban sites and ozone 8-hour annual means (green) taken from 206 urban sites in the United States.

CO to CO_2 in the exhaust stream, was also mandated in new vehicles in 1975. A careful examination of the records showed that the fall in CO levels actually began earlier than 1975 and correlated well with the number of on-road vehicles with catalytic converters. It then became apparent that the observed reduction in CO levels was almost entirely due to the slow replacement of older vehicles with newer models equipped with catalytic converters. At the same time, O_3 levels were not falling as significantly as expected after the use of RFG fuels, as shown in Figure 10.25 (green) (USEPA, 2017b). It was predicted that the widespread use of fuel additives with a low OH radical reactivity would lower the ozone-forming potential of the emissions. Although NO_x and VOC emissions were not much different in the RFG than with gasoline alone, the emissions of aldehydes increased with RFG, which increased the ozone-forming potential of the fuel emissions.

The most used oxygenated fuel additive in RFG was MTBE. What was not considered when the mandate was discussed was that the branched chain ethers – MTBE, ETBE, and TAME – are all manufactured from an addition reaction of an alkene and an alcohol. The MTBE is produced from the addition reaction of methanol with isobutene:

$$CH_3OH + (CH_3)_2CCH_2 \rightarrow CH_3OC(CH_3)_3$$

When MTBE undergoes incomplete combustion, it can crack to produce equal amounts of the starting reactants isobutene and methanol. The methanol is then oxidized to formaldehyde by oxygen in the combustion air to give the emission products isobutene and formaldehyde:

$$CH_3OC(CH_3)_3 + \tfrac{1}{2}O_2 \rightarrow (CH_3)_2C = CH_2 + CH_2O + H_2O$$

both of which are highly OH-reactive compounds that can enhance ozone production.

Although gasoline is not very water soluble, MTBE is a fairly water-soluble ether. When the MTBE–RFG was placed into older leaking USTs used for gasoline, the MTBE quickly contaminated groundwater supplies and nearby wells used for drinking water. Complaints about drinking water with bad odor and taste were rapidly identified as being due to MTBE contamination. Health studies followed, which indicated that consumption of MTBE at high concentrations was a potential carcinogenic hazard and could cause liver damage. Based on the number of complaints and the potential health risk involving the use of MTBE, California banned the use as a gasoline additive in March 1999, a little over 4 years after the use was mandated in Los Angeles and San Diego. California completely eliminated the use of MTBE as a fuel additive by 2002. Other cities where the MTBE additive was mandated included Houston, TX, Philadelphia, PA, Milwaukee, WI, Baltimore, MD, Hartford, CT, and Chicago, IL. All of these areas and surrounding suburbs used MTBE–RFG with similar reports of contaminated groundwater supplies. By May 2006 the use of MTBE as a gasoline additive was abandoned in the United States and replaced by ethanol. This led to the widespread use of E10 in most parts of the United States.

The MTBE case is similar that of the tetraethyl lead additives: not enough was known about the health and environmental effects before they were used in large scale-applications. The water solubility and toxicity of all the proposed oxygenated additives should have been evaluated before their widespread use to prevent or minimize environmental and health impacts. These types of "quick fix" mistakes in the past have led to the development of the safety data sheets (SDSs) that were initially required in the United States for hazardous materials in 1986. Now SDSs, also known as material safety data sheets (MSDSs), are required for all chemicals and materials. These information sheets contain what is known currently about each chemical species and are updated as our knowledge increases. Environmental scientists and policy makers should make use of this valuable resource when considering the environmental and health impacts of any chemical species.

The mandated use of MTBE as a fuel additive is a case where an intended solution to an air quality problem led to a water quality problem. This occurred because the possibility of leakage from gasoline USTs was not adequately considered before the mandate was implemented. Problems such as this continue to arise because environmental scientists view the environment as compartmentalized. Atmospheric chemists often do not adequately consider impacts on water, soil, or biota when working to solve an atmospheric pollution problem. The same can be said for scientists working in other environmental areas such as water quality or ecology. This situation with MTBE could have been prevented by carefully evaluating the impacts on both water and air at the same time. Since the environment is made up of complicated and interrelated systems, this can best be accomplished by involving environmental chemists with specialties in the different environmental areas in the evaluation of proposed solutions to environmental problems.

References

Agarwal, A.K., Gupta, T., and Kothari, A. (2011). Particulate emissions from biodiesel vs diesel fuelled (*sic*) compression ignition engine. *Renewable Sustainable Energy Rev.* 15: 3278–3330.

Committee on Oil in the Sea (2003). *Oil in the Sea III: Inputs, Fates and Effects.* Washington, D.C: National Academies Press.

Deepwater Horizon 2018 *Wiki article.* Available online at https://en.wikipedia.org/wiki/Deepwater_Horizon_oil_spill (accessed March 22, 2018).

Dutcher, D.D., Pagels, J., Bika, A. et al. (2011). Emissions from soy biodiesel blends: A single particle perspective. *Atmos. Environ.* 45: 3406–3413.

European Biomass Industry Association 2017 *Biomass characteristics.* Available online at http://www.eubia.org/cms/wiki-biomass/biomass-characteristics (accessed November 12, 2018).

Exxon Valdez 2018 *Wiki article.* Available online at https://en.wikipedia.org/wiki/Exxon_Valdez_oil_spill (accessed November 12, 2018).

Field, C.B., Behrenfeld, M.J., Randerson, J.T., and Falkowski, P. (1998). Primary production of the biosphere: Integrating terrestrial and oceanic components. *Science* 281: 237–240.

FracFocus 2018 *What Chemicals are Used: Groundwater Protection Council & Interstate Oil and Gas Compact Commission.* Available online at https://fracfocus.org/chemical-use/what-chemicals-are-used (accessed November 12, 2018).

Gaffney, J.S. and Marley, N.A. (2009). The impacts of combustion emissions on air quality and climate – from coal to biofuels and beyond. *Atmos. Environ.* 43: 23–36.

Karol, I.L., Kelder, H., Kirchhoff, V.W.J.H. et al. (1999). *Aviation and the Global Atmosphere.* Cambridge: Cambridge University Press.

Lopez-Aparicio, S. and Hak, C. (2013). Evaluation of bioethanol fueled buses on ambient air pollution screening and on-road measurements. *Sci. Total Environ.* 452&453: 40–49.

National Renewable Energy Laboratory 2012 *Effect of Ethanol Blending on Gasoline RVP.* Available online at http://www.ethanolrfa.org/wp-content/uploads/2015/09/RVP-Effects-Memo_03_26_12_Final.pdf (accessed November 12, 2018).

National Research Council (2008). *Water Implications of Biofuels Production in the United States.* Washington, D.C: National Academies Press.

USEIA 2017a *International Energy Outlook* 2017. Available online at https://www.eia.gov/outlooks/ieo (accessed November 12, 2018).

USEIA 2017b *Biomass Explained.* Available online at https://www.eia.gov/energyexplained/?page=biomass_home (accessed November 12, 2018).

USEIA 2018a *Petroleum and Other Liquids.* Available online at https://www.eia.gov/dnav/pet/PET_CRD_WELLDEP_S1_A.htm (accessed November 12, 2018).
USEIA 2018b *Renewables and Alternative Fuels: Biofuels Overview.* Available online at https://www.eia.gov/renewable/data.php#biomass (accessed November 12, 2018).
USEPA 2017a *Carbon Monoxide Trends.* Available online at https://www.epa.gov/air-trends/carbon-monoxide-trends#coreg (accessed November 12, 2018).
USEPA 2017b *Ozone Trends.* Available online at https://www.epa.gov/air-trends/ozone-trends#oznat (accessed November 12, 2018).
USEPA 2018 *Underground Storage Tanks (USTs).* Available online at https://www.epa.gov/ust/cleaning-underground-storage-tank-ust-releases (accessed November 12, 2018).

Study Problems

10.1 Organic chemistry is defined as what?

10.2 What two types of organic compounds contain the R—CO—R functional group, where R can be an alkyl group, a hydrogen, or an aromatic group?

10.3 The complete combustion of fuel requires what reactant to be in excess?

10.4 What conditions are needed for the accumulation of organic material in a marine or lacustrine environment?

10.5 After initial bacterial digestion of organics, what is the complex mixture of recalcitrant waxy material called that is left in the sediments?

10.6 (a) Upon burial, the variation in the weight of the sediment overburden results in the variation of what?
(b) What is this variation called?

10.7 (a) The waxy organic substances that are buried in the sediments gradually form an insoluble organic material called what?
(b) This substance undergoes what process that converts it into hydrocarbons?

10.8 What are the types of organic materials that are deposited in the following environments:
(a) marine?
(b) terrestrial?
(c) lacustrine?

10.9 What type of fossil fuel is formed from kerogen with the following H/C ratios:
(a) low H/C?
(b) high H/C?
(c) very low H/C?

10.10 (a) The entire process of the formation of oil and gas is called what?
(b) The processes that form coal are called what?

10.11 What are the four ranks of coal from the lowest to the highest?

10.12 During the thermal aging of coals:
(a) what organic functional groups decrease?
(b) what organic functionality increases?

10.13 (a) What are the three types of coal surface mining techniques?
(b) What do they all have in common?

10.14 (a) What are the three methods of crude oil recovery?
(b) What are the differences between them?

10.15 The release of natural gas from ultralow permeability reservoirs by a stimulation technique is called what?

10.16 Natural gas wells that have a significant amount of H_2S and CO_2 are called what?

10.17 (a) What country is the leading consumer of both petroleum and natural gas?
(b) What country is the leading consumer of coal?

10.18 (a) What method is used to separate helium and nitrogen from natural gas?
(b) What method is used to remove H_2S and CO_2?

10.19 (a) What is the most common reaction used to produce synfuels?
(b) What is the process for producing the reactants for this reaction?
(c) What are the products of this reaction?

10.20 (a) The first stage of crude oil refining is done by what process?
(b) The products are grouped into what three classes?

10.21 Processed natural gas has what two odorant compounds added to it for safety?

10.22 (a) What are the two types of biomass fuels?
(b) How are they different?
(c) Which type is soybean biodiesel fuel?
(d) Which type is landfill gas?

10.23 (a) By what biochemical reaction is ethanol made?
(b) By what chemical reaction is biodiesel made?

10.24 What two countries are the major consumers of the first-generation fuel ethanol?

10.25 The second-generation biomass fuels currently in use are composed of what?

10.26 (a) What fuel has the highest energy content?
(b) What wood-based fuel has the highest energy content?
(c) Wood products vary in energy content according to what properties?

10.27 What are the shorthand designations for the following:
(a) a 25% by volume ethanol–gasoline blended fuel?
(b) pure ethanol?
(c) a 5% by volume biodiesel–petroleum diesel blended fuel?

10.28 What are the main components in fracking fluids?

10.29 (a) What was the name of the mobile drilling rig that caused the largest oil spill in the Gulf of Mexico?
(b) What was the name of the petroleum tanker that leaked a huge amount of oil off the coast of Alaska?

10.30 (a) What two species must be present to promote the corrosion of iron in steel?
(b) Why are steel underground fuel storage tanks required to have zinc or magnesium bars attached to the outside?

10.31 First-generation biomass production causes water pollution from what two main sources?

10.32 (a) What is the largest source of oil contamination in the marine environment?
(b) What is the largest source of oil contamination during extraction?
(c) During transportation?
(d) During consumption?

10.33 (a) What defines the dead zone in the Gulf of Mexico?
(b) What is its cause?

10.34 (a) What type of coal has the highest sulfur content?
(b) Why is it most widely used for the generation of electricity?

10.35 (a) The combustion of what fossil fuel is responsible for most of the anthropogenically emitted SO_2?
(b) SO_2 can be oxidized to what compound during transport of a plume?
(c) What is the fate of this oxidation product?

10.36 (a) What are the two sources of nitrogen that lead to NO_x formation during the combustion of fossil fuels?
(b) Where does most of the NO_x come from during coal combustion?
(c) Where does most of the NO_x come from in gasoline-powered spark ignition engines?
(d) What is the major NO_x product from the combustion of gasoline in a spark ignition engine?

10.37 What three greenhouse gases are emitted during the combustion of coal?

10.38 How is particulate matter removed from diesel exhaust?

10.39 What types of organic compounds are now typically used to increase the octane ratings of gasoline?

10.40 (a) Which has the lower vapor pressure, ethanol or gasoline?
(b) Why?
(c) Which has the higher vapor pressure, E10 or gasoline?
(d) Why?

10.41 The incomplete combustion of ethanol results in the emission of what two ozone-forming compounds?

10.42 How do the particulate emissions of biodiesel differ from those of petroleum diesel?

10.43 (a) The heavy distillate used by large cargo and container ships is commonly known by what name?
(b) This fuel has very high emissions of what three important species?
(c) Which of these species is of most concern to the Arctic environment?
(d) Why?

10.44 What compounds are produced from incomplete combustion of the gasoline additive MTBE?

11

Climate Change

Although climate and weather are related, they differ in the time frames that they cover. Climate is defined as the average condition of the weather at a given region over a long period of time, as exhibited by temperature, wind velocity, and precipitation. It is typically considered to be the average weather expected for a region based on 20, 100, and 500-year time frames. However, weather is the state of the atmosphere at a particular place and time with respect to meteorological conditions, such as wind, temperature, cloudiness, moisture, pressure, etc. Weather varies over the course of a year depending on the season, and can be very different from region to region at the same time. In contrast, climate describes what we expect the weather to be at a particular time in the future.

There are five different climate groups on the Earth, as first described by Russian climatologist Wladmir Köppen in 1900. These groups are: tropical, dry, temperate, continental, and polar. Each of these groups is subdivided according to seasonal precipitation types and level of heat, resulting in a total of 29 possible climate types. Each climate type is designated by a two or three-letter designation according to the climate group, precipitation type, and heat level, as listed in Table 11.1. The first letter identifies the climate group, the second letter indicates the seasonal precipitation type, and the third letter indicates the level of heat. For example, the designation BWh is a hot desert climate while BWk is a cold desert climate.

The climate classifications and their designations for regions across the globe are shown in Figure 11.1 (Peel et al., 2007). Tropical climates (A) are located at about 15°–25° latitude north and south of the equator. In these climates all months have an average temperature of >18 °C and an annual precipitation >1500 mm. Dry climates (B) extend from 20° to 35° latitude and in large continental regions of the mid-latitudes often surrounded by mountains. They are defined by small amounts of precipitation where evaporation and transpiration dominate. Temperate climates (C) extend from 30° to 50° latitude, mainly on the eastern and western borders of most continents. The defining weather features are the mid-latitude cyclone during the winter and convective thunderstorms during the summer. Continental climates (D) extend from the temperate climates poleward to the polar climates (E), predominantly in the interior regions of large land masses. They are defined by severe winters with temperatures of less than −30 °C and strong snow storms. Polar climates are found on the northern coastal areas of North America, Europe, and Asia, as well as most of Greenland and Antarctica. They have cold temperatures year round, with the warmest month less than 10 °C. A highland climate group (H) is sometimes added to Köppen's climate classification system in order to account for climate differences above 1500 m (about 4900 ft).

The climate types are classified according to seasonal precipitation and temperature patterns. Climate change is concerned with the potential for long-term changes in these patterns in each of the climate regions. The regional climate can become warmer or colder, wetter or drier, as it undergoes change. In some regions, such as the tropics, the climate may change

Chemistry of Environmental Systems: Fundamental Principles and Analytical Methods, First Edition.
Jeffrey S. Gaffney and Nancy A. Marley.

Table 11.1 Climate groups, precipitation type, and heat levels for each possible climate type, designated by a combination of the letters in parentheses.

Group	Precipitation types	Precipitation patterns	Heat levels	Heat patterns
Tropical (A)	Rainforest (f)	No dry season		
	Monsoon (m)	Short dry season		
	Savannah (w)	Winter dry season		
Dry (B)	Desert (W)	Dry arid	Hot (h)	Coldest 0 °C
	Steppe (S)	Dry semi-arid grassland	Cold (k)	Coldest <0 °C
Temperate (C)	Mediterranean (s)	Summer dry season	Hot summer (a)	Warmest >22 °C
	Subtropical (w)	Winter dry season	Warm summer (b)	Warmest <22 °C
	Marine (f)	No dry season	Cool summer (c)	1–3 mo <22 °C
Continental (D)	Mediterranean (s)	Summer dry season	Hot summer (a)	Warmest >22 °C
	Subtropical (w)	Winter dry season	Warm summer (b)	Warmest <22 °C
	Marine (f)	No dry season	Cool summer (c)	1–3 mo <22 °C
			Very cold winter (d)	Winter <−38 °C
Polar (E)	Tundra (T)	Permafrost		
	Ice cap (E)	Permanent snow cover		

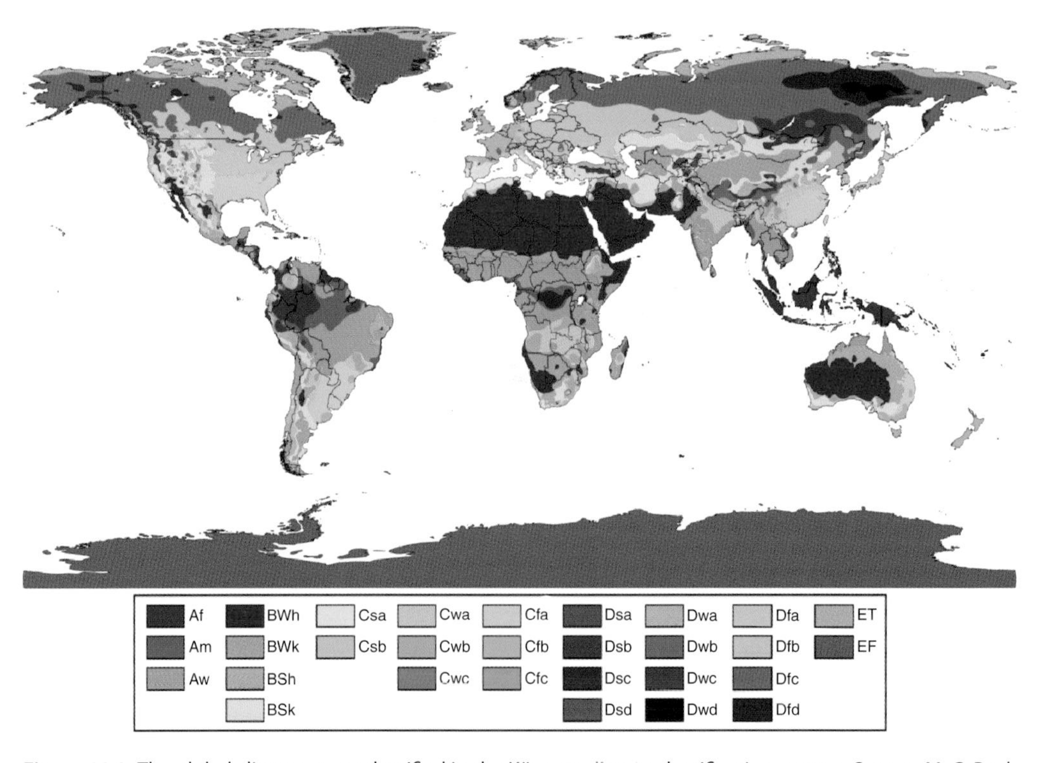

Figure 11.1 The global climate types classified in the Köppen climate classification system. Source: M. C. Peel, B. L. Finlayson, and T. A. McMahon, University of Melbourne, Wikimedia Commons.

very little while in other regions, such as the polar areas, the climate can undergo dramatic changes due to climate-forcing parameters. The total climate of the Earth is typically expressed as a mean global temperature. However, this approach can be misleading, since there is significant variation in the mean temperature of the Earth's surface across the globe. The average global monthly surface daytime temperatures from the Moderate Resolution Imaging Spectroradiometer (MODIS) on the NASA Terra satellite, shown in Figure 11.2, vary from −25 to 45 °C, with a range of 70 °C. The mean annual surface temperature for the Earth in 2010 was determined as 14.86 °C, which was 0.96 °C above the mean temperature for the twentieth century (NOAA, 2011). However, there is considerable variation in the surface temperature as a function of season and latitude. The regional temperature rise is much higher in many of the climate types compared to the approximately 1 °C rise in the annual mean temperature reported for the Earth in 2010.

Climate change has occurred naturally over the lifetime of the Earth, but it usually occurs gradually over time scales of thousands of years. The anthropogenic combustion of fossil fuels since the Industrial Revolution has led to the release of CO_2 into the troposphere, which is dramatically increasing the level of this potent greenhouse gas over a fairly short time period compared to the natural climate change drivers. The increased concentrations of CO_2, along with other greenhouse gases and aerosols, have led to serious concerns that global climate change is now occurring on a much shorter time scale than natural. This rapid climate change is affecting regional climates, resulting in a number of serious impacts including sea-level rise from the melting of polar land ice. Climate change is increasing severe weather events, wildfires, floods, and droughts, which are affecting ecosystems as well as mankind. Global climate change has already had observable effects on the established climate regions. Many of the indicators used by Köppen to define the climate types are changing, including shrinking glaciers, earlier melting of river and lake ice, changing precipitation patterns, shifting of plant and animal ranges, and earlier flowering of trees. There is high confidence that global temperatures will continue to rise for decades to come, while the effects on the individual climate regions will vary over time.

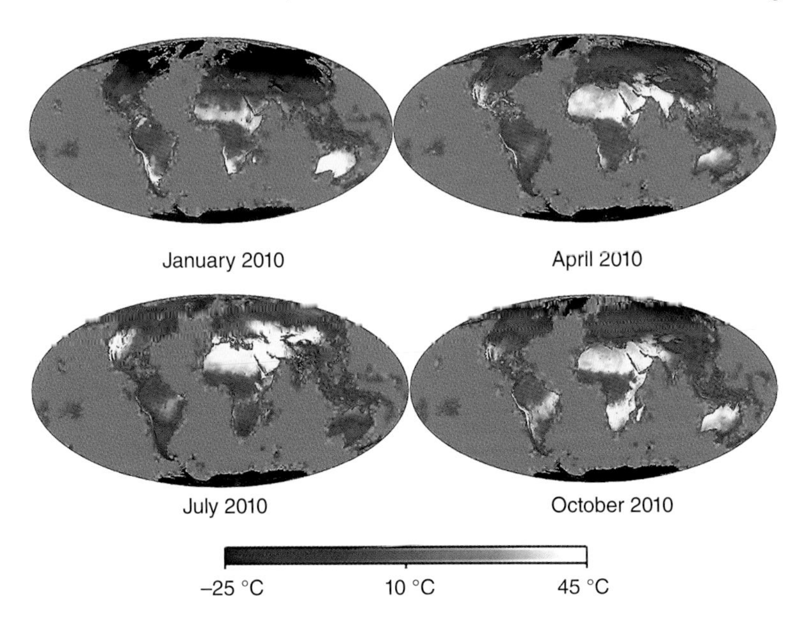

Figure 11.2 The monthly global average daytime surface temperatures from the MODIS on the NASA Terra satellite for the months of January, April, July, and October 2010. Source: Adapted from NASA Earth Observatory.

11.1 Prehistoric Climates

The Earth's climate has changed many times throughout its history. There have been cold periods where most of the planet was covered in ice and warm periods when the ice levels receded. The study of how the Earth's climate has changed over its history is called paleoclimatology. Paleoclimatology makes use of a number of methods to determine the temperature of the Earth over geological time scales. Most of these methods use the $^{18}O/^{16}O$ and $^{2}H/^{1}H$ stable isotopic ratios to determine the variation in mean ocean temperatures over time. Water molecules containing the heavier isotopes (^{2}H and ^{18}O) evaporate at a higher temperature than water molecules containing the lighter isotopes (^{1}H and ^{16}O). The increase in evaporation of the heavier elements at warmer temperatures results in an enrichment of the heavier elements in frozen precipitation which forms the ice sheets, as demonstrated in Figure 11.3. The difference in evaporation rates due to the isotopic composition of water is called temperature-dependent fractionation, which represents a kinetic isotope effect (KIE) during the evaporation process.

Both the $^{2}H/^{1}H$ and $^{18}O/^{16}O$ isotopic ratios are determined using stable isotope ratio mass spectrometry on thin slices of ice core samples. The ice cores are obtained in very cold regions, such as Greenland in the Northern Hemisphere and Antarctica in the Southern Hemisphere, where the ice has never melted and has continued to accumulate over thousands of years. The isotopic ratios are then used to estimate the change in temperature of the ocean water. Due to the KIE, there are lower values of $^{18}O/^{16}O$ and $^{2}H/^{1}H$ in the ice when the ocean water is colder and higher values of $^{18}O/^{16}O$ and $^{2}H/^{1}H$ in the ice when the ocean water is warmer. Thus, the ocean temperature is determined to have become warmer when the $^{18}O/^{16}O$ and $^{2}H/^{1}H$ has increased and colder when the $^{18}O/^{16}O$ and $^{2}H/^{1}H$ has decreased. The same ice sample can contain trapped air bubbles, which can yield the greenhouse gas concentrations in the atmosphere at the time the temperature is measured. Other materials in the ice are used to give approximate dates of the stable isotope measurements. These include the annual layering of pollen and the presence of volcanic ash, which can mark eruption events. The ice core record obtained in this manner is useful from the present back in time to about 800 000 years.

The geological sediment record can be used to obtain data on temperature changes further back in time than 800 000 years. The sediment record has been used to obtain data back in time

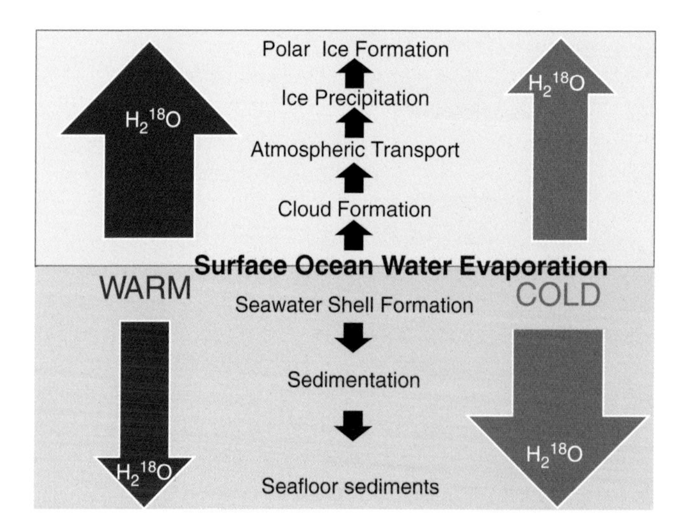

Figure 11.3 A diagram of the effects of temperature-dependent fractionation of ^{18}O on its concentration in ice and carbonate shells in seafloor sediments. The size of the red and blue arrows is a reflection of the relative concentrations of the ^{18}O isotope under warm (red) and cold (blue) climate conditions.

to about 200 000 million years from measurements of the $^{18}O/^{16}O$ in macroscopic and microscopic shells found in the sediments. Since the shells are formed from minerals in the ocean water, the isotopic composition of the shells is directly proportional to the isotopic composition of the ocean water. Thus, there are higher values of $^{18}O/^{16}O$ and $^{2}H/^{1}H$ in the shells when the ocean water is colder and lower values of $^{18}O/^{16}O$ and $^{2}H/^{1}H$ in the sediments when the ocean water is warmer. As demonstrated in Figure 11.3, the evaporation of seawater under warmer climate conditions leads to an increase in the $^{18}O/^{16}O$ ratio in ice and a decrease in the $^{18}O/^{16}O$ ratio in the shells contained in ocean sediments, while in colder climates the ice has a lower $^{18}O/^{16}O$ ratio and the shells have a higher $^{18}O/^{16}O$ ratio. The relative errors associated with the temperature estimates increase in the sediment data as the age of the sediments increases. The 200 000 million year limit for the sediment temperature estimates is tied to crustal plate tectonics. Continental drift causes subduction and crustal folding, which leads to heating and chemical alteration of the sediments over these very long time periods. This limits the usefulness for temperature estimates to about 200 000 million years.

The change in ocean temperature (ΔT) for the last 5 million years, as determined from $^{18}O/^{16}O$ ratios in sediment carbonates, is shown in Figure 11.4. This ΔT represents the temperature difference in average recorded values from 1960 to 1970. The results for the last 1000 years are not included, because temperatures determined from isotope ratios of sediment carbonates are less reliable in the near past due to the slow rate of sediment accumulation. The results show that the ocean temperature has varied considerably over the last 5 million years, with an overall cooling trend. This can be compared to the change in temperature over the last 400 000 years determined from $^{18}O/^{16}O$ and $^{2}H/^{1}H$ ratios in ice cores from Vostok, Antarctica, shown in Figure 11.5(a). The change in temperature is compared to CO_2 concentrations in Figure 11.5(b) and particulate concentrations in Figure 11.5(c), which were found in the ice cores over the same time frame. Most of the dust was likely from volcanic activity and large wildfires. The warmer periods are strongly correlated with higher CO_2 levels and the cooler periods are associated with higher dust concentrations. Although the temperatures are strongly correlated with CO_2 levels, the CO_2 never exceeded 300 ppm during this period.

More recently, atmospheric measurements of CO_2, which were initiated by David Keeling in the 1950s, have been used to supplement the measurements obtained from ice cores. The CO_2 rose above 300 ppm for the first time in 400 000 years in 1950, when it was measured at 310 ppm.

Changes in the Earth's past climate are attributed to a number of causes, including changes in the Earth's orbit, changes in the solar radiation reaching the Earth, volcanic eruptions, and the beginning of plant photosynthesis. However, these past climate-forcing events all occurred

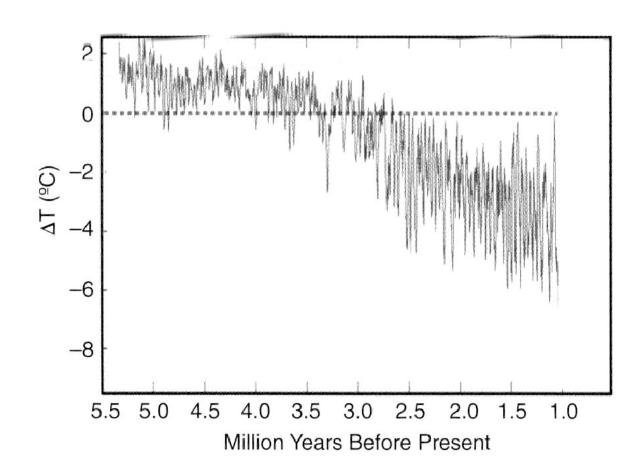

Figure 11.4 The changes in the ocean temperature (ΔT) determined from measurements of $^{18}O/^{16}O$ in deep-sea sediment cores. Source: Adapted from Dragons flight, Wikimedia Commons.

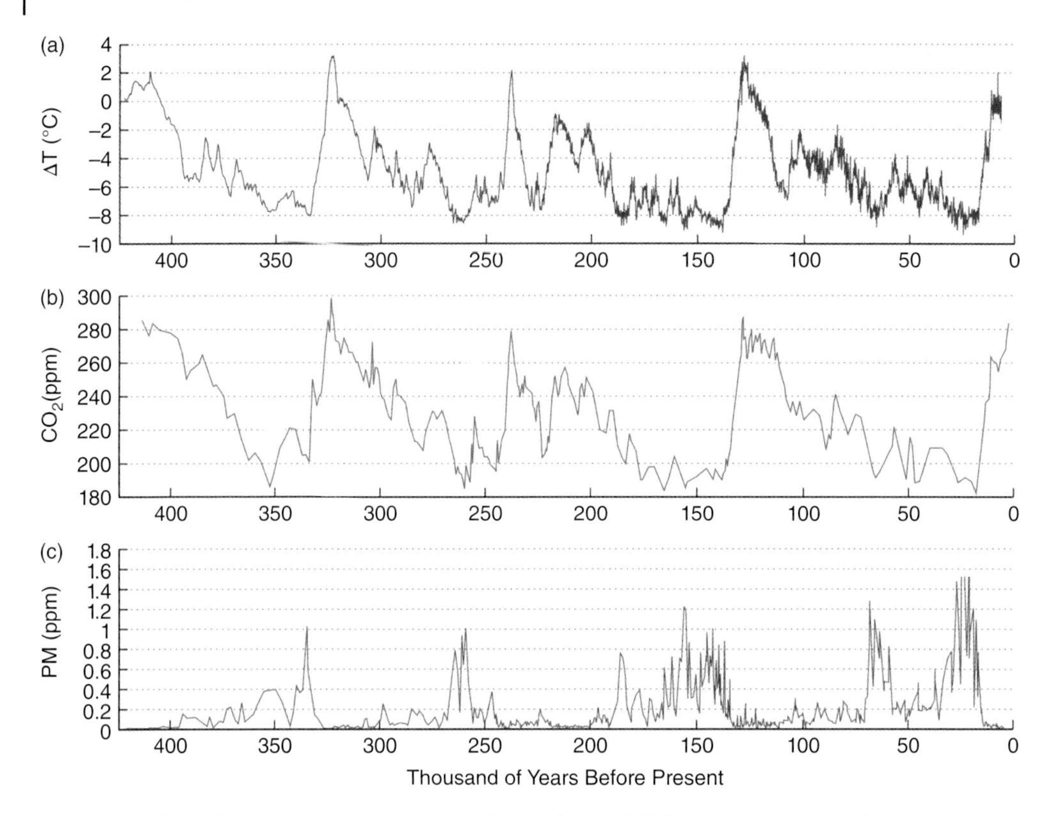

Figure 11.5 The variation in ocean temperature (a) over the last 400 000 years determined from measurements of $^{18}O/^{16}O$ and $^{2}H/^{1}H$ ratios in ice cores from Vostok, Antarctica compared to (b) CO_2 and (c) particulate concentrations. Source: Adapted from NOAA, Wikimedia Commons.

at fairly slow rates over long periods of time. Currently, the global climate is changing at a rate not seen in the past. This recent rapid climate change is due to anthropogenic activities, which was not a factor in past climate, and is occurring on a much shorter time scale than any of the naturally driven climate changes. These activities are primarily related to the large-scale combustion of fossil fuels, as well as the increased conversion of land areas for agriculture.

11.2 Causes of Climate Change

Recall from Section 2.4 that warming of the atmosphere is dependent on the concentration of atmospheric greenhouse gases that absorb IR radiation emitted from the surface of the Earth, which has been heated by incoming solar radiation. Thus, heat is trapped in the atmosphere, warming the planet. Any change in the incoming solar radiation or the concentration of greenhouse gases in the atmosphere can affect climate. For example, the period of cooling that occurred during 1650–1850, known as the Little Ice Age, is believed to have been caused by decreases in solar radiation. During this time period, which covered two centuries, glaciers were observed to increase in the Swiss Alpine regions and in other areas of the world. The available evidence indicates that the solar radiation has been constant or increased slightly since that time. If the solar irradiance has been increasing significantly in recent years, the temperature throughout the atmosphere should be increasing. However, this is not what has been observed. The atmospheric temperature increases since the Little Ice Age have generally been restricted to the troposphere. In addition, atmospheric modeling, which includes current solar radiation levels, cannot explain the observed increases in mean global temperature without considering the increasing levels of greenhouse gases. The conclusion is that the main cause for the current rapid rise in global surface temperature is due to rapidly increasing greenhouse gas concentrations, which trap heat in the troposphere.

The greenhouse gases in the troposphere are H_2O, CO_2, CH_4, N_2O, tropospheric O_3, CFCs, and HCFCs. While water vapor is the strongest greenhouse gas, it is also the only greenhouse gas that can be removed naturally from the atmosphere by condensation to form clouds followed by precipitation. Table 11.2 lists the increases of CO_2, CH_4, and N_2O in the troposphere since 1790 based on their maximum concentrations in ppm as determined from ice core measurements. Atmospheric concentrations of CO_2 have increased by 46%, while CH_4 concentrations have increased by 166% and concentrations of N_2O have increased by 22%. Figure 11.6 shows the atmospheric concentration profiles in the troposphere from 1979 to 2017

Table 11.2 The current atmospheric concentrations (ppm) as of June 2018 of the three major greenhouse gases – CO_2, CH_4, and N_2O – compared to concentrations before 1750.

Gas	Concentration before 1750	Concentration in June 2018	Increase (ppm)	Increase (%)
CO_2	280	410	130	46
CH_4	0.70	1.86	1.16	166
N_2O	0.27	0.33	0.06	22

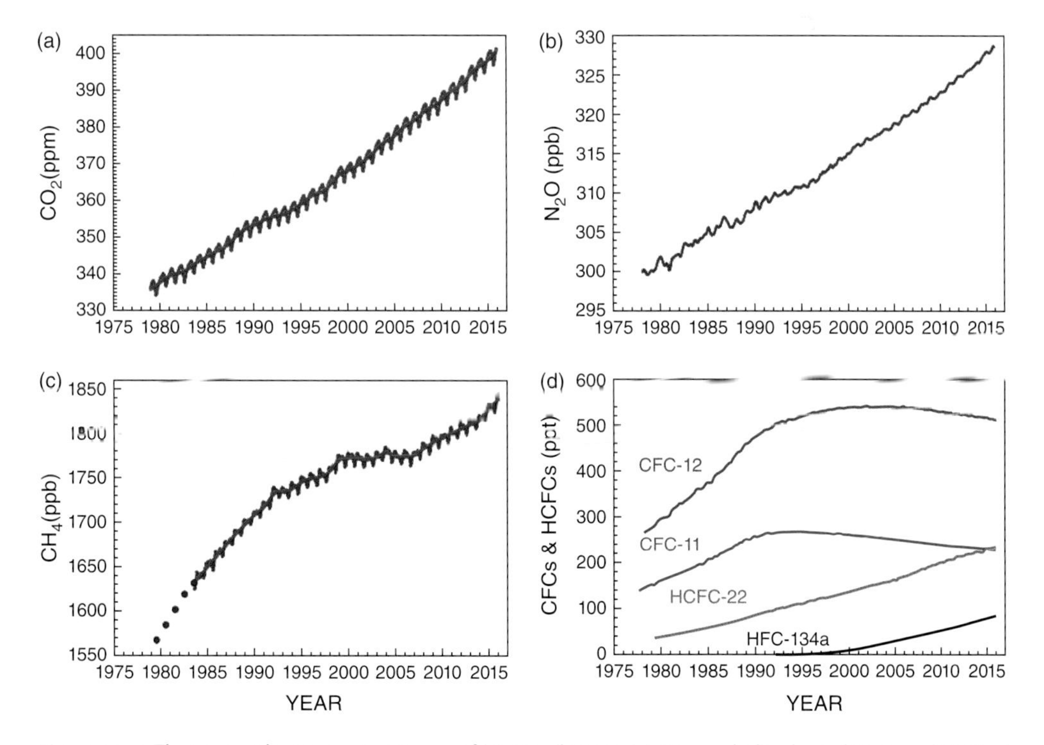

Figure 11.6 The tropospheric concentrations of (a) CO_2, (b) N_2O, (c) CH_4, and (d) selected CFCs and HCFCs from 1979 to 2017. Source: Adapted from NOAA, Earth System Research Laboratory, Global Monitoring Division.

for CO_2, CH_4, N_2O, two of the CFCs (CFC-11 and CFC-12) and two of the HCFCs (HCFC-22 and HCFC-134a). Both CO_2 and N_2O have shown a steady increase from 1979 to 2017. Methane also showed a steady increase in the atmosphere from 1979 until 1993, when the increase began to slow. This is thought to be due to changes in rice cultivation practices. Prior to 1993, organic fertilizers were most commonly used in rice farming. These organic materials undergo bacterial decomposition to produce CH_4 in the wet rice fields. After about 1993, inorganic nitrate fertilizers, which do not undergo bacterial decomposition, became more popular for rice production. This decrease in CH_4 emissions from rice fields resulted in a stabilization of atmospheric CH_4 concentrations. Methane emissions again began to increase steadily after about 2007 because of the increased use of natural gas, as discussed in Section 10.3. The increased demand for natural gas resulted in increased emissions from natural gas wells and transportation facilities.

The CFCs and HCFCs are important greenhouse species since they are present in the low to a few hundred parts per trillion (ppt), they have long atmospheric lifetimes, and they have large IR absorption cross-sections in the atmospheric spectral window regions. Both the CFCs and the HCFCs have no natural sources, while the major greenhouse gases – CO_2, CH_4, and N_2O – have both natural sources and anthropogenic sources. The atmospheric concentrations of the CFCs increased rapidly up until about 1990. This increase in the atmospheric concentrations of CFCs has ceased since the 1990s due to the banning of their use by the Montreal Protocol. Conversely, the atmospheric concentrations of HCFCs, which are used as replacements for CFCs, are steadily increasing.

11.2.1 Global Warming Potentials

The global warming potential (GWP) of a greenhouse gas is the measure of how much IR energy will be absorbed by 1 ton of the gas in the atmosphere over a given period of time (20, 100, or 500 years) relative to 1 ton of CO_2, where the GWP of CO_2 is defined as 1. The larger the GWP, the more the greenhouse gas will warm the atmosphere compared to CO_2 over the same time period. The magnitude of the GWP depends on the strength of the IR absorption bands, their spectral location, and the atmospheric lifetime of the greenhouse gas. A greenhouse gas that has a large IR absorption and a long atmospheric lifetime will have a high GWP. The effect of the wavelength of the IR absorption on the GWP takes into consideration the relative impact of an IR absorption band that is located in an atmospheric window region compared to an absorption band that is located in a wavelength region where other major gases also absorb.

The GWPs are used to determine the relative impacts of an increase in the concentration of a greenhouse gas on climate. The GWPs for some of the major greenhouse gases are listed in Table 11.3 (Forster et al., 2007). The longest lived of the three major greenhouse gases is N_2O with an atmospheric lifetime of 114 years. The greenhouse gas sulfur hexafluoride (SF_6) is used as an insulating gas in high-voltage facilities such as high-energy physics particle accelerators. The current atmospheric concentration of SF_6 is about 9 ppt with an atmospheric lifetime of 3200 years. The two perfluorocarbon compounds listed in Table 11.3 (PFC-14 and PFC-116) are very long-lived gases with atmospheric lifetimes of 50 000 and 10 000 years. They are produced during the electrochemical production of aluminum from Al_2O_3. The PFC-114 is currently at a concentration of about 35 ppt, while the PFC-116 is at about 0.1 ppt. The atmospheric concentrations of the CFCs and HCFCs are also in the ppt concentration range. Although the main climate-forcing species are CO_2 at 410 ppm, CH_4 at 1.9 ppm, and N_2O at 330 ppb, the combined impact of the CFC, HCFC, and HFC compounds also makes an important contribution to the total GWP of the atmosphere.

Table 11.3 The global warming potentials for some major greenhouse gases for the climate intervals of 20, 100, and 500 years along with their atmospheric lifetimes in years.

Greenhouse gas	Formula	Lifetime	20 year	100 year	500 year
Carbon dioxide	CO_2	30–95	1	1	1
Methane	CH_4	12	72	25	7.6
Nitrous oxide	N_2O	114	289	298	153
Sulfur hexafluoride	SF_6	3 200	16 300	22 800	32 600
CFC-12	CCl_2F_2	100	11 000	10 900	5 200
HCFC-22	$CHClF_2$	12	5 160	1 810	549
HFC-23	CHF_3	270	12 000	14 800	12 200
PFC-14	CF_4	50 000	5 210	7 390	11 200
PFC-116	C_2F_6	10 000	8 630	12 200	18 200

11.2.2 Greenhouse Gas Sources and Sinks

Carbon dioxide is the best studied of the greenhouse gases. It is the second strongest greenhouse gas after water vapor. The annual carbon cycle, which includes the atmospheric CO_2 cycle, is shown in Figure 11.7. The CO_2 reservoir and flux values shown in the figure are average values for the period 2000–2006. The natural carbon emissions include plant respiration at 60 $GtC\,yr^{-1}$, microbial respiration and decomposition at 60 $GtC\,yr^{-1}$, and air–sea gas exchange at 90 $GtC\,yr^{-1}$. These emissions are balanced by photosynthetic uptake, which totals 120 $GtC\,yr^{-1}$, and ocean uptake, which is equal to 90 $GtC\,yr^{-1}$. Anthropogenic carbon emissions include the combustion of fossil fuels, cement production, and land use changes, such as deforestation, which account for about 9 $GtC\,yr^{-1}$.

The increases in CO_2 emissions from these anthropogenic sources have led to an increased atmospheric level of CO_2, which acts as a growth enhancer to photosynthetic plants and phytoplankton. The increased plant growth has resulted in an increase in photosynthetic uptake of CO_2 to about 3 $GtC\,yr^{-1}$ with a net annual atmospheric CO_2 increase of 6 $GtC\,yr^{-1}$ and an atmospheric CO_2 reservoir of 800 GtC. This is equivalent to a 0.5% increase in atmospheric CO_2 concentration per year. In addition, the combustion of fossil fuel has increased from about 6.6 $GtC\,yr^{-1}$ in 2000 to about 9 $GtC\,yr^{-1}$ in 2017 due to the increasing demand for energy worldwide. This has led to a continuing rise in the atmospheric concentration of CO_2, which closely follows the fossil fuel combustion rates. Thus, the primary source of the increase in atmospheric CO_2 is fossil fuel combustion, with some additional CO_2 coming from deforestation and the manufacture of concrete as global urbanization continues to increase.

Methane is the next strongest greenhouse gas after CO_2. Methane has a number of significant natural sources from the biosphere, as shown in Figure 11.8. About 26% of the methane in the troposphere is produced naturally from anaerobic bacteria. These bacteria live in wetlands where the DO levels are low. They are also present in the gastrointestinal tracts of animals and some species of termites. The anaerobic bacteria consume organic material and produce CH_4 by enteric fermentation. Some species of termites make use of these anaerobic bacteria to aid in the digestion of wood. When formed in a cold climate, CH_4 can be trapped in ice as gas bubbles, forming a clathrate. Methane clathrates ($CH_4{\cdot}5.75H_2O$ or $4CH_4{\cdot}23H_2O$), also called methane hydrates, are a solid similar to ice in which a large amount of CH_4 is trapped inside the crystal structure of water. Most of these methane clathrates are present in very cold areas such as the tundra regions. Upon warming, the methane trapped in the clathrate structure is

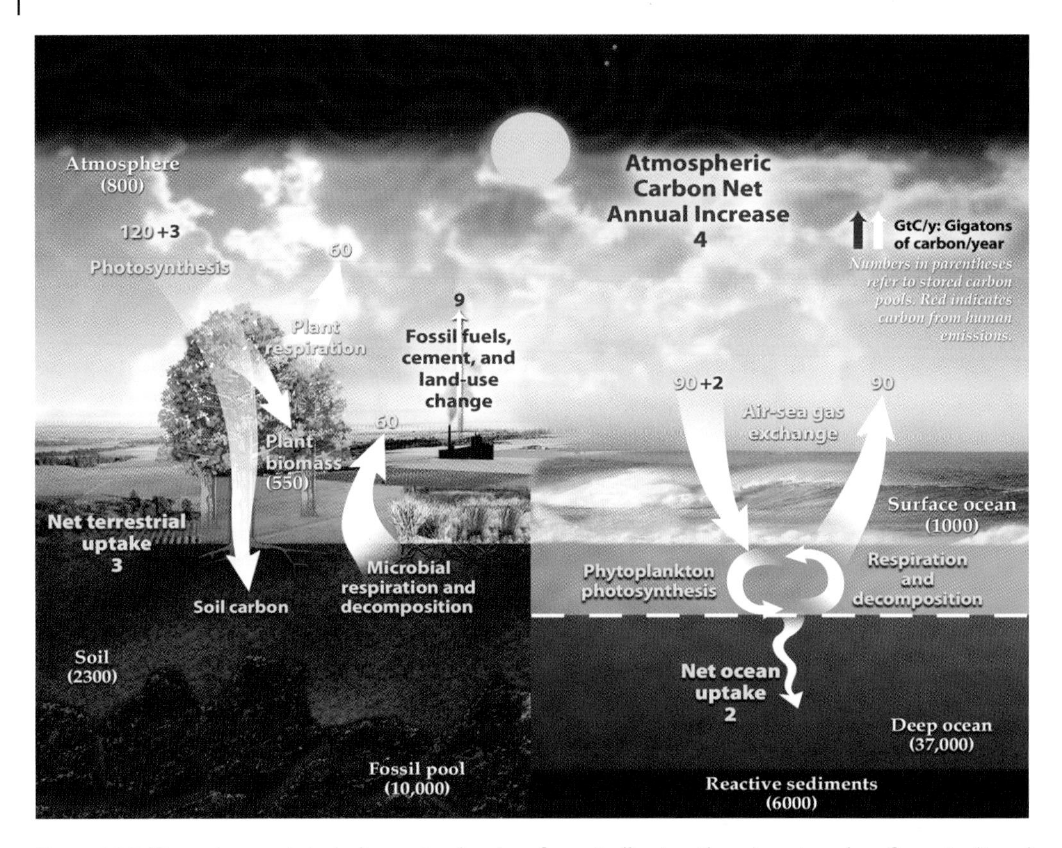

Figure 11.7 The carbon cycle including natural carbon fluxes (yellow), anthropogenic carbon fluxes (red), and stored carbon (white) in gigatons of carbon per year (GtC year^{-1}). Source: U.S. DOE, Biological and Environmental Research Information System, Wikimedia Commons.

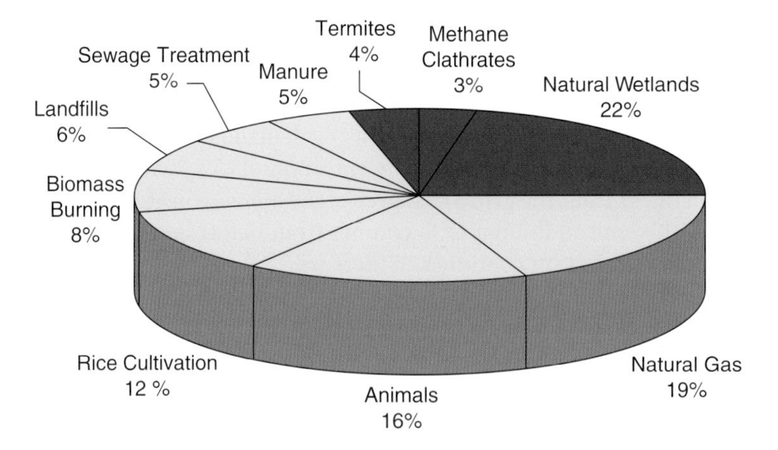

Figure 11.8 Natural (blue) and anthropogenic (yellow) sources of methane in the troposphere. Source: Adapted from Harvey Augenbraun, Elaine Matthews, and David Sarma, Wikimedia Commons.

released into the troposphere. Methane clathrates can also be formed at high pressures in the deep ocean. These ocean clathrates are slowly moved to the surface by ocean currents. The decrease in pressure, as the clathrates are transported to the upper water column, causes the CH_4 to be released from the clathrates into the atmosphere.

Methane also has a significant number of anthropogenic sources that are a result of agricultural activities. These include agricultural burning, which contributes to biomass burning, and animal husbandry, which leads to manure production. Ruminant animals, including beef cattle, dairy cows, and sheep, also produce methane in their gastrointestinal systems. Enteric anaerobic bacteria help to break down the plant matter that these animals feed on and convert acetate in the animals' stomachs to methane. Rice cultivation is also a major source of methane, especially when organic fertilizers are used. This is due to the fact that rice fields are flooded as part of the rice production process, artificially turning the fields into wetlands where anaerobic bacteria can produce methane. The methane production is increased when organic fertilizers are used due to the increased bacterial nutrient levels. Other sources of anthropogenic methane include natural gas, which is lost to the atmosphere during extraction, purification, and transport.

There are four removal processes for tropospheric CH_4, which are shown in Figure 11.9. The major removal process is the reaction with OH, accounting for almost 90% of the CH_4 loss in the troposphere. The reaction of CH_4 with OH produces H_2O and the CH_3 radical, as discussed in Section 5.4.3:

$$CH_4 + OH \rightarrow H_2O + CH_3$$

The second largest loss mechanism for CH_4 involves slow mixing into the stratosphere, where CH_4 is rapidly oxidized to form CO_2 and H_2O. Other loss mechanisms include oxidation by methanotrophic bacteria in soils. The tropospheric air that contains methane infiltrates the top soils where these organisms use the methane as a source of energy. This depletes a small amount of methane from the troposphere near the soil surface. About 1% of the tropospheric CH_4 is also lost by reaction with Cl radical. There is a small amount of chlorine gas that is released to the troposphere, which rapidly photolyzes to form Cl radicals. These Cl radicals react with CH_4 to form HCl and the CH_3 radical. Since the reaction with OH is by far the most important loss mechanism, the atmospheric lifetime of methane is determined by the OH concentration. Thus, any changes in the OH concentration of the troposphere will result in a change in methane loss rate. A decrease in atmospheric OH concentration will lead to longer methane lifetimes and higher concentrations of CH_4 in the troposphere.

Figure 11.9 Removal processes for tropospheric methane.

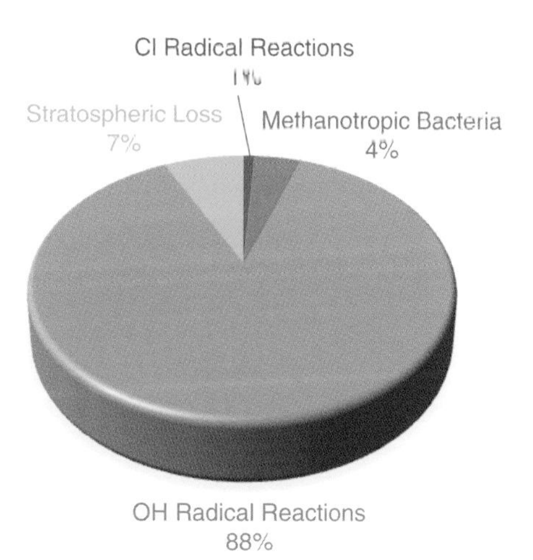

The third major greenhouse gas is N_2O. The majority of atmospheric N_2O is due to anaerobic bacterial decomposition of organic materials in wetlands, similar to the production of methane. This natural source is fairly constant and is responsible for about 60% of N_2O emissions. The remaining 40% is due to anthropogenic sources. Most of the anthropogenic sources (31%) are due to agricultural practices, including the use of nitrate fertilizers and agricultural burning associated with crop waste management. The steady increase in tropospheric levels of N_2O is attributed to the increased use of nitrate fertilizers, particularly in Asia, where they are replacing organic fertilizers for rice cultivation. About 4% of the N_2O is produced from combustion in stationary power plants and mobile sources. Another 2% is associated with the chemical industrial production of HNO_3 and nylon, which produces N_2O as a side product. Manure management in animal farms and in residential areas accounts for another 2%. An additional 1% of the tropospheric N_2O comes from other sources, such as propellants in spray cans for some food products (Figure 11.10).

Nitrous oxide is slow to react with OH radical in the troposphere. It can be absorbed into soils where it is decomposed by bacteria. However, this is a very slow process and only affects the small concentration near the soil surface. Thus, the only major loss process is mixing into the stratosphere. Once in the stratosphere, N_2O can undergo reaction with $O(^1D)$ to form NO, which initiates the NO_x ozone depletion cycle discussed in Section 4.3.3. Thus, the atmospheric lifetime of N_2O is primarily determined by the stratospheric mixing rate, which is estimated to be 114 years.

The final set of greenhouse gases, which are typically considered as a group, are the CFCs, HCFCs, and PFCs. The CFCs slowly mix into the stratosphere where they react to give Cl radicals causing stratospheric ozone depletion, as discussed in Section 4.4.1. The atmospheric concentrations of CFCs are on the decline due to the Montreal Protocol, which restricted their production and use. However, the loss rate of existing tropospheric CFCs is not fast due to the slow mixing rate from the troposphere into the stratosphere. The concentrations of the HCFCs, which are used as replacements for the CFCs, are continuing to rise slowly in the troposphere. Unlike the CFCs, the HCFCs can react with OH in the troposphere by abstraction of their hydrogen, resulting in shorter lifetimes. However, they still have lifetimes on the order of years to decades. The PFCs have very long lifetimes of tens of thousands of years due to the very stable C—F bonds. Their gradual increase in the troposphere is primarily from one source. The two major PFCs (CF_4 and C_2F_6) are produced as a byproduct of the production of aluminum by electrochemical reduction of Al_2O_3. This reduction reaction involves the use of carbon electrodes and takes place in molten cryolite (Na_3AlF_6), which acts to dissolve the Al_2O_3. The process

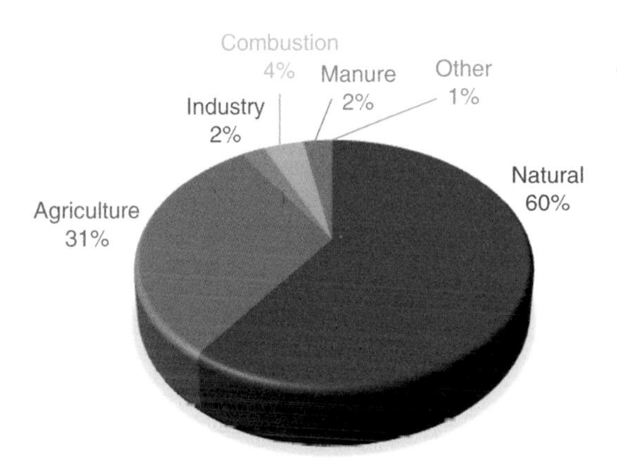

Figure 11.10 Natural and anthropogenic sources of nitrous oxide in the troposphere.

produces CF_4 and a small amount of C_2F_6 as side products. Sulfur hexafluoride has no known natural sources. It is produced industrially as an insulating gas in high-voltage operations.

The major greenhouse gases – CO_2, CH_4, and N_2O – all have major sources and sinks that are affected by land use. Deforestation can add to CO_2 levels by removing plants that normally would take up and store CO_2. The tilling of agricultural fields leads to loss of soil carbon, resulting in further CO_2 production. Rice production is a major source of both CH_4 and N_2O. Agricultural burning of field stubble and sugar cane bagasse is an important source of all three major greenhouse gases, as well as carbonaceous aerosols. Livestock farms, which are often located on deforested lands, are important sources of CH_4 and N_2O from increased manure production as well as enteric digestive fermentation. In the final analysis, the emissions of all of the anthropogenic greenhouse gases are related to increased population growth. An increase in population means increasing demand for water, food, and energy, all of which lead to increasing greenhouse gas emissions.

11.2.3 Radiative Forcing

The energy balance of the Earth is the equilibrium between the amount of energy from sunlight absorbed by the Earth and its atmosphere and the amount of energy that is lost back to space. Any change in this energy balance is radiative forcing, or climate forcing. The atmospheric warming from the increases in greenhouse gases is a positive radiative forcing. This warming due to increased greenhouse gases leads to faster evaporation of water from lakes and oceans. The increased water evaporation adds to the level of atmospheric water vapor, which is the strongest greenhouse gas. Water vapor interacts with aerosols to form clouds, which can result in both positive and negative radiative forcing. If the clouds are thick, the incoming radiation can be backscattered, resulting in atmospheric cooling, which is a negative radiative forcing. If the clouds are optically thin, they can act to increase warming due to their strong IR absorption, resulting in a positive radiative forcing. Thus, the hydrological cycle, which includes evaporation as well as the process of evapotranspiration from plants, has multiple effects on the Earth's energy balance.

Atmospheric aerosols can also have multiple effects on the energy balance. As described in Section 6.6, atmospheric aerosols can scatter incoming solar radiation, resulting in negative radiative forcing, and absorb both incoming solar radiation and outgoing IR radiation, resulting in positive radiative forcing. The magnitude of these two opposing climate effects depends on the aerosol composition. The aerosols that absorb incoming light the strongest are BC and OC. Primary inorganic aerosols such as wind-blown dust and fly ash, and secondary inorganic aerosols such as nitrate and sulfate, act primarily to scatter incoming radiation. These hygroscopic aerosols also act as cloud condensation nuclei (CCN), which aid in the formation of clouds. Biomass burning is an important source of BC and OC aerosols, as well as inorganic nitrate. A major feedback between climate change and biomass burning occurs in climatic regions with wet springs and dry summers. These regions are predicted to have wetter springs, longer growing seasons, and drier hotter summers, as climate change increases. These climate effects will likely lead to increasing numbers of large-scale wildfires, due to increased growth in the spring followed by lack of rain and soil moisture in the summer.

The Earth's energy balance is also affected by changes in the amount of incoming solar radiation and changes in the planetary albedo, which can increase or decrease the solar radiation at the Earth's surface. The changes in incoming solar radiation due to the various solar cycles, as discussed in Section 2.4.1, can cause both positive and negative radiative forcing, depending on the stage of the cycle. The variation in solar radiation due to the solar cycles occurs on fairly long time scales. In addition, current measurements indicate that only a very slight warming of the

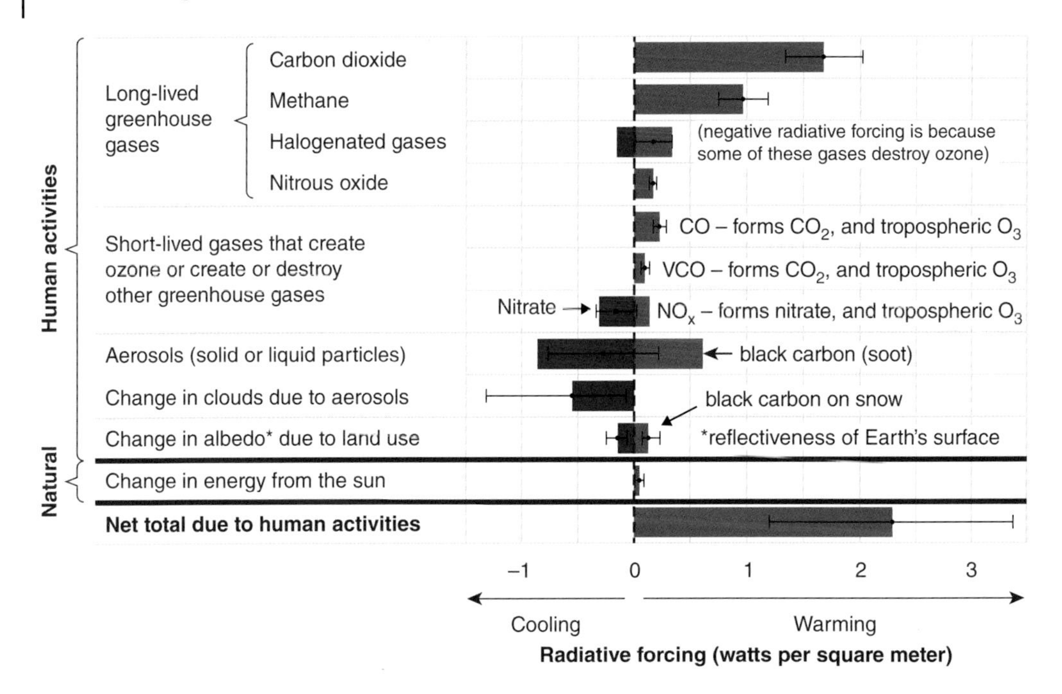

Figure 11.11 The estimated radiative forcing in watts per square meter between 1750 and 2011 along with the uncertainties in the measurements (thin black bars) for the greenhouse gases, aerosols, clouds, albedo, and solar energy input. Source: Adapted from EPA.

planet can be attributed to changes in incoming solar radiation due to the solar cycles. Thus, the incoming solar radiation plays a very small role in the currently observed rapid climate change. Predictions indicate that a reduction of ice on the planet would reduce the amount of sunlight reflected back to space, resulting in a positive radiative forcing. As BC is deposited onto snow and ice surfaces, the BC will absorb solar radiation, reducing the amount of solar light reflected by the ice. Since the BC absorbs solar radiation, it will also warm the surface, causing the ice and snow to melt, exposing the surface below. This reduces the surface albedo even further.

All of the known influences on the Earth's energy balance have been examined thoroughly over the years by the Intergovernmental Panel on Climate Change (IPCC), with the findings published in detailed reports on a regular basis since 1990. The latest of these reports was released in 2016 (IPCC, 2016) and includes the estimated impacts of each of the radiative forcing influences as a relative value in watts m^{-2} as shown in Figure 11.11. Most of the positive radiative forcing is due to CO_2, CH_4, and BC, while the negative radiative forcing is due to light-scattering aerosols and clouds. The overall uncertainty of the total radiative forcing in the system is primarily due to a lack of understanding of the short-term aerosol and cloud-forcing values. This is due in part to a lack of understanding of the atmospheric lifetimes of the aerosols, as well as how the different aerosols affect cloud formation and evaporation.

11.3 Climate Models

The physics of the atmosphere is controlled by the absorption of solar radiation and the evaporation and transport of water vapor in the air, which are responsible for the general circulation patterns of wind and precipitation. This physics is modeled by general circulation models (GCMs), which were the first models used for the prediction of radiative forcing. These models are mathematical representations of the atmospheric gas and oceanic liquid movements using the Navier–Stokes equations. The fluid movements of the air and water are placed on a rotating sphere and the models add equations to consider the heat flow that arises from the incoming solar radiation and the latent heat of the fluids. Because the time scales for motion in the atmosphere and oceans are quite different, both atmospheric general circulation models (AGCMs) and oceanic general circulation models (OGCMs) were developed to treat each system separately. The first AGCM was developed in 1956 by Norman Phillips, which depicted the monthly and seasonal variations of weather in the troposphere. Further model development led to the coupling of the AGCMs and OGCMs in the late 1960s at the NOAA Geophysical Fluid Dynamics Laboratory located at Princeton University in New Jersey. A community atmosphere model was developed during the early 1980s at the National Center for Atmospheric Research, which included vegetation and soil types. Other major climate models have been developed at the Hadley Centre for Climate Prediction and Research located at Exeter in the United Kingdom. As these models have developed, they have been coupled to other models including carbon cycles, emissions of pollutants, and tropospheric and stratospheric chemistry, in order to better predict weather and climate. Although these combined ocean–atmosphere global climate models are much more advanced than the original general circulation models, they are still commonly referred to as GCMs.

Initially, atmospheric modeling was quite simple due to the lack of large high-speed computers available for complex mathematical computing. These first atmospheric models were based on simple box models used to place greenhouse gases in reservoirs, which were assumed to be uniformly mixed. The reservoirs were linked together by fluxes that represented the sources and sinks from the various reservoirs. With the advent of supercomputers, which allowed high-speed computing and large data storage, the simple modeling was replaced by the three-dimensional modeling used today in advanced GCMs. The AGCMs use a set of boundary conditions for the air–ocean interface, while the coupled GCMs combine separate calculations from the air and the oceans. These models have to restrict the differential equations to limits based on the grid cells described in Section 2.5. The grid cells are interconnected to give a full representation of the air and ocean temperatures and circulations. They have a horizontal resolution of 300 km × 300 km or larger. The GCMs usually have 10–20 vertical layers in the atmosphere and as many as 30 vertical layers in the oceans.

The GCMs do a fairly good job at predicting global climate change, however, they do not perform well in predicting local climate change, because of the large grid scales. For this reason, smaller-scale regional climate models (RCMs) have been developed. The RCMs have the advantage of being able to incorporate terrain changes such as elevation, soil type, albedo, vegetation, and emission sources, on a grid size of 10 km × 10 km. They allow for the assessment of some of the complex interrelated issues important to climate change, such as sea-level changes, air–ice interactions, and atmosphere–biosphere interactions, as shown in Figure 11.12. The RCMs are more detailed in their calculations and contain modules that allow for incorporation of ecosystem types, agricultural and industrial activities and emissions, atmospheric chemistry, the hydrological cycle, and the carbon cycle. The shorter time scales and smaller physical grid sizes also allow for the assessment of the regional impacts of short-lived atmospheric species, leading to the local production tropospheric ozone and secondary aerosol formation. The results from the RCMs are used to develop parameterizations that can be included in the GCMs to improve their accuracy on global scales. The performance of both the GCMs and the RCMs continues to improve due to advances in our knowledge and improvements in supercomputing speed and data storage capacity. These continued improvements will allow for more complex computations that will be needed for more accurate and timely climate change predictions.

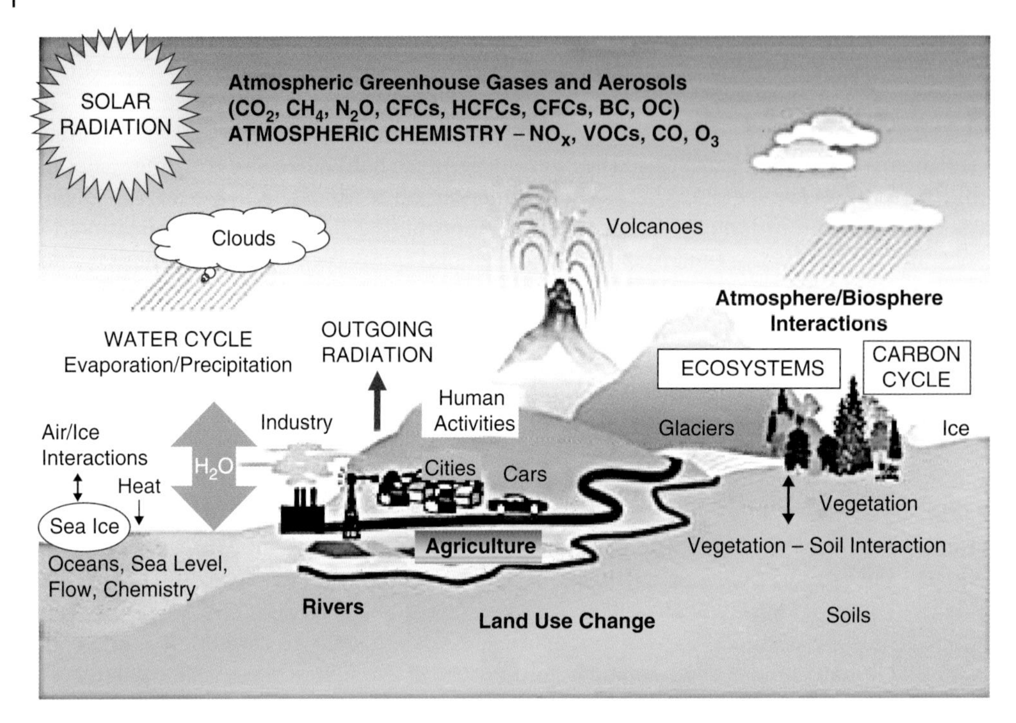

Figure 11.12 The complex climate-forcing interactions that can be incorporated into regional climate models (RCMs) with smaller grid sizes. Source: Adapted from NOAA, Wikimedia Commons.

The RCMs use boundary conditions calculated by GCMs for long-term climate forecasting. To achieve acceptable accuracy, the RCMs require that boundary conditions be calculated over time periods on the order of 6 hours or less, depending on the issues being addressed. For example, shorter time periods are required for tropospheric chemical modeling, while longer time periods are suitable for transport descriptions. This means that the GCM must store outputs on these short time periods required by the RCMs, which was not typically done due to the large data storage requirements. As more models are coupled together to obtain regional scale accuracy, care must be taken to make sure that the computer codes are synchronized on the required time scales. This has led to the use of stacked GCMs, where high-speed calculations are carried out on short time scales and averaged, then connected to longer-time-scale processes for global climate predictions. The continuing development of the GCMs and RCMs requires increasingly faster computing with increasingly larger storage capacity in order to be able to obtain predictions on annual as well as decadal time scales.

11.4 Predictions of Future Climate Change

The United Nations Environment Programme (UNEP) and the World Meteorological Organization (WMO) established the IPCC in 1988, which was later endorsed by the United Nations General Assembly. The IPCC is responsible for providing reports to the United Nations Framework Convention on Climate Change (UNFCC). The objective of the UNFCC is to stabilize greenhouse gas levels that are on the rise due to anthropogenic activities, making use of the best available scientific data. The IPCC and UNFCC have essentially worldwide participation and support in their efforts. The IPCC has made use of GCMs and RCMs, along with the continuous measurements of greenhouse gas and aerosol concentrations in order to predict the probabilities of global and regional climate change and to release these predictions to the public in periodic assessment reports (ARs). These ARs involve worldwide participation from scientific specialists to assess the current status of the Earth's climate, predict the extent of change, and identify areas of risk. They also address the mechanisms and causes of the predicted climate change. In addition, the ARs allow for the consideration of mitigation strategies by all of the countries involved. The first AR was published in 1990 with a supplement in 1992. Since then there have been four more reports, with the most recent being AR5, completed in 2014. AR6 is scheduled for release in 2022.

The AR5 report is organized into three working groups and a task force. The working groups are: Working Group I, which deals with the physical science behind climate change; Working Group II, which examines climate change impacts, adaptation, and vulnerability; and Working Group III, which deals with possible mitigation strategies of climate change. The Task Group develops methodologies for assembling and reporting national greenhouse gas inventories.

Working Group I examined the various causes and potential impacts of radiative forcing and identified the forcing agents, both natural and anthropogenic. They set probability levels for their conclusions regarding the causes and predicted impacts of climate change. These probability levels are described in the report by the terms listed in Table 11.4. The general findings from the Working Group I report are:

1. The atmosphere and ocean systems are unequivocally warming. Associated impacts, such as sea-level rise occurring since 1950, have happened at historically unprecedented rates.
2. Humans are clearly influencing climate.
3. It is extremely likely that anthropogenic activities are the dominant cause. The longer the wait for greenhouse gas emission reductions, the higher the cost of the impacts.

AR5 scientists reported that it was likely that the period from 1983 to 2013 was the warmest 30 years in the past 1400 years. It is virtually certain that the upper ocean levels have warmed over the period of 1971–2010, and it is with high confidence that energy accumulation accounts for the observed warming. It is also with high confidence that the Greenland and Antarctic ice sheets have been losing mass over the last 20 years, and the Northern Hemisphere spring snow

Table 11.4 Probability levels used to assess the reported impacts of climate change used in the IPCC AR5 Working Group I report.

Descriptive term	Probability level
Virtually certain	≥99%
Extremely likely	≥95%
Very likely	≥90%
Likely	≥66%
More likely than not	≥50%
About as likely as not	33–66%
Unlikely	≤33%
Very unlikely	≤10%
Extremely unlikely	≤5%
Exceptionally unlikely	≤1%

cover has continued to decrease. Working Group I reported with high confidence level that sea-level rise since 1950 has been greater than the mean sea-level rise over the past 2000 years. The major greenhouse gases – CO_2, CH_4, and N_2O – have all risen to unprecedented concentrations not seen for over 800 000 years. They found that the total radiative forcing relative to 1750 is positive and that the major responsible forcing agent is CO_2.

The AR5 re-evaluated the GCMs and RCMs and found that the models had improved their abilities to reproduce available data for both land and sea-surface temperatures. This improved ability included being able to reproduce the rapid warming observed since 1950, as well as the short cooling periods, which are known to follow large volcanic eruptions. This improved ability to reproduce past temperature data from the measurements of atmospheric greenhouse gas and aerosol concentrations obtained at the time gives higher confidence that the model predictions for future climate change and associated impacts are becoming more and more reliable. The GCM predictions are given for four different future greenhouse gas concentration profiles. These four future profiles, referred to as representative concentration pathways (RCPs), are shown in Figure 11.13. The four RCPs – RCP2.6, RCP4.5, RCP6, and RCP8.5 – are named for the possible changes in radiative forcing in W m^{-2} by the year 2100. For example, RCP2.6 is the RCP that will lead to a 2.6 W m^{-2} change in radiative forcing in the year 2100.

By using the four RCPs, the best available GCM models have been used to predict the mean global temperature and sea-level increases for the years 2045–2065 and 2081–2100, as shown in Table 11.5 (IPCC, 2016). The ranges given in the table represent the variability in the results of the GCMs. A major outcome from the AR5 report is that the mean global temperature is projected to increase in the late twenty-first century by 0.3–4.8 °C and the mean sea level is projected to rise by 0.17–0.82 m under all RCPs. The time frames for temperature increase are: from 1.0 to 2.0 °C between 2045 and 2065 and 1.0 to 3.7 °C between 2081 and 2100. Thus, global warming will continue and is strongly tied to greenhouse gas emissions.

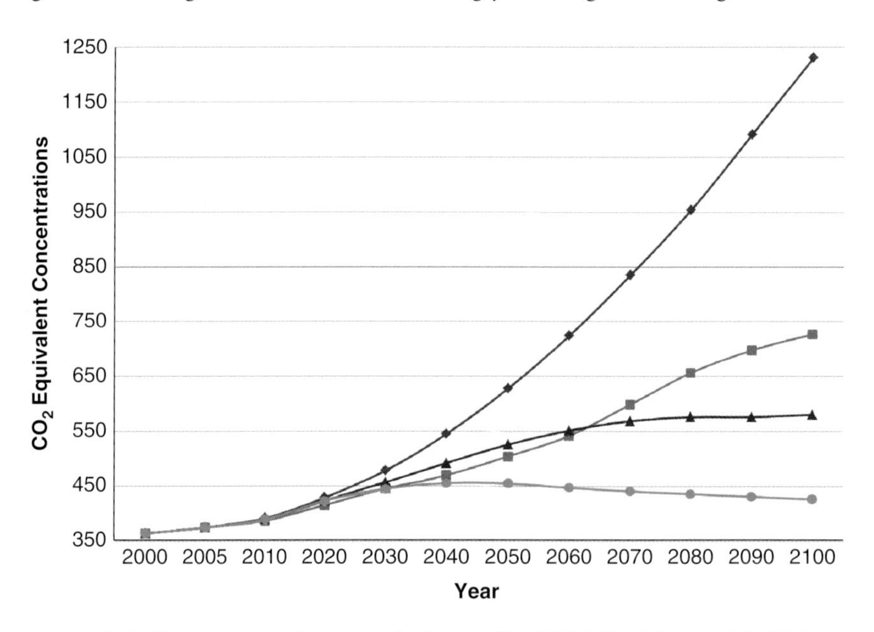

Figure 11.13 The representative concentration profiles (RCPs): RCP2.6 (green), RCP4.5 (purple), RCP6.0 (orange), and RCP 8.5 (red), for equivalent CO_2 greenhouse gas concentrations that lead to positive radiative forcing of 2.6 W m^{-2}, 4.5 W m^{-2}, 6.0 W m^{-2}, and to 8.5 W m^{-2} by 2100. Source: Adapted from Illini, Wikimedia Commons.

Table 11.5 The mean global temperature change (ΔT) in °C, the likely temperature range in °C, the sea-level rise (ΔSL) in m, and the likely sea-level rise range in m, predicted using GCMs for the periods 2045–2065 and 2081–2100 for the four RCPs.

	2045–2065		2081–2100		2045–2065		2081–2100	
	ΔT (°C)	Range (°C)	ΔT (°C)	Range (°C)	ΔSL (m)	Range (m)	ΔSL (m)	Range (m)
RCP2.6	1.0	0.4–1.6	1.0	0.3–1.7	0.24	0.17–0.32	0.40	0.26–0.55
RCP4.5	1.4	0.9–2.0	1.8	1.1–2.6	0.26	0.19–0.33	0.47	0.32–0.63
RCP6.0	1.3	0.8–1.8	2.2	1.4–3.1	0.25	0.18–0.32	0.48	0.33–0.63
RCP8.5	2.0	1.4–2.6	3.7	2.6–4.8	0.30	0.22–0.38	0.63	0.45–0.82

11.5 Impacts from the Predicted Temperature Rise

The projected temperature increases will have strong impacts on the global water cycle. The predictions are that arid desert regions will increase in size, while flooding in wet as well as wet–dry regions will occur more frequently. The warming of the surface ocean waters will gradually spread to the lower colder water layers, leading to changes in ocean circulation patterns. Decreases in the Arctic sea ice, glacial ice volumes, and spring snow cover in the Northern Hemisphere are projected to continue. The melting of land ice will lead to increased rates of sea-level rise, which are very likely to exceed rates observed during the last 40 years. The uptake of CO_2 by land and oceans will increase due to the higher atmospheric concentrations. The uptake by the ocean will cause increasing ocean acidification, which is especially a concern for reef ecosystems. The AR5 states that even if CO_2 emissions were stopped today, many of the surface temperature changes and impacts on the terrestrial systems will continue due to the increased overburden of CO_2 relative to pre-1850 levels.

Some potential effects of the predicted temperature rise are positive climate-forcing feedback, which contributes directly to further global warming and extreme weather events. Chemical reaction rates, such as oxidation reactions in soils and surface waters, are directly proportional to temperature. Rising temperatures will act to increase the chemical oxidation rates of organic carbon species stored in surface reservoirs, leading to further increases in CO_2 levels. At the same time the CO_2 uptake by plants will increase in spring and summer, leading to faster growth, provided there are sufficient water and nutrients available. The increasingly large deciduous tree populations will result in an increased leaf litter deposition in the fall, with decomposition rates increasing during warmer winter periods with less snow cover. The increase in plant growth during wetter springs leads to increased amounts of dry biomass available for large-scale wildfires. The western United States has experienced an increase in the number and size of wildfires since the 1980s, which is very likely a result of these climate change effects. The large-scale wildfires, such as that shown in Figure 11.14, which burned over 1000 km^2 of forest near the Yosemite National Park in 2013, have feedback beyond the release of CO_2 stored in the biomass. Wildfires also release NO_x, and CO, along with aerosols and VOCs. Among the VOC emissions are large amounts of aldehydes, ethene, and methane. The aldehydes and ethene are very reactive toward OH and have large ozone-forming potential. The wildfire aerosols are in the form of BC, which is usually coated with hydrophobic organic tars. These emissions contribute to both local and regional air pollution problems, since the submicron wildfire aerosols can be transported long distances due to their hydrophobicity.

The aftermath of wildfires can also lead to soil erosion, increasing CO_2 emissions from the increased oxidation of carbon in the exposed soil. After wildfires, heavy rainfall events cause

Figure 11.14 A large-scale wildfire that burned 1000 km^2 of forest near the Yosemite National Park in California in 2013. Source: U.S. Department of Agriculture, Wikimedia Commons.

mudslides and landslides due to soil erosion. Flooding also occurs more readily, as the forest ecosystem is no longer able to help uptake water efficiently. Water runoff lowers the amount of water that percolates into soil and groundwater aquifers, reducing groundwater recharge. Ultimately, the number of precipitation events is lowered over the affected region due to a loss of humidity from the deforestation. Thus, the wildfires impact both the carbon cycle and the terrestrial hydrological cycle. In addition, increased erosion changes the landscape by the movement of silt leading to the production of steep ravines, which are susceptible to landslides.

Although warming temperatures lead to sea-level rise, they also lead to changes in ocean surface temperature and salinity. Warm ocean waters are less dense than cold waters. In addition, the melting ice adds fresh water, further changing the density of the surface ocean waters. This causes a slowdown of the ocean currents in the thermohaline circulation, discussed in Section 8.3. This large-scale ocean circulation pattern is driven by changes in water temperature, dissolved salts, and freshwater fluxes. The deep-water currents are created by downwelling of the surface water as the water density increases. An increase in water temperature and a decrease in salinity in these areas could result in a slowing or halting of the downwelling currents, which would also result in a slowing of the upwelling currents.

The surface currents of the thermohaline circulation in the northern Atlantic Ocean are known as the Atlantic Meridional Overturning Circulation (AMOC), shown in Figure 11.15. The North Atlantic Current (Figure 11.15, red) and the Norwegian Atlantic Current (Figure 11.15, orange) are responsible for terrestrial regions such as Northern Europe and the British Isles having milder climates than anticipated from their latitudinal positions. The warmer currents located near the coast of Great Britain have a significant influence on the regional energy balance of the area, leading to a much milder climate than that of Nova Scotia, which is near the colder, southerly flowing current. Recent observations of higher than normal sea-level rise in the regions of Northern Europe and the British Isles indicate the beginning of a slowdown of the ocean currents in these regions. If a complete shutdown of the thermohaline circulation should occur, there will be a cooling in the areas of Northern Europe and the British Isles. This would have dramatic effects on the British Isles, Europe, and the Nordic regions, where the climate change would lead to significant cooling instead of warming. The North Atlantic Current also has an important effect on North America, where the North Atlantic Drift causes some warming. Other general impacts of changes in the ocean current systems include

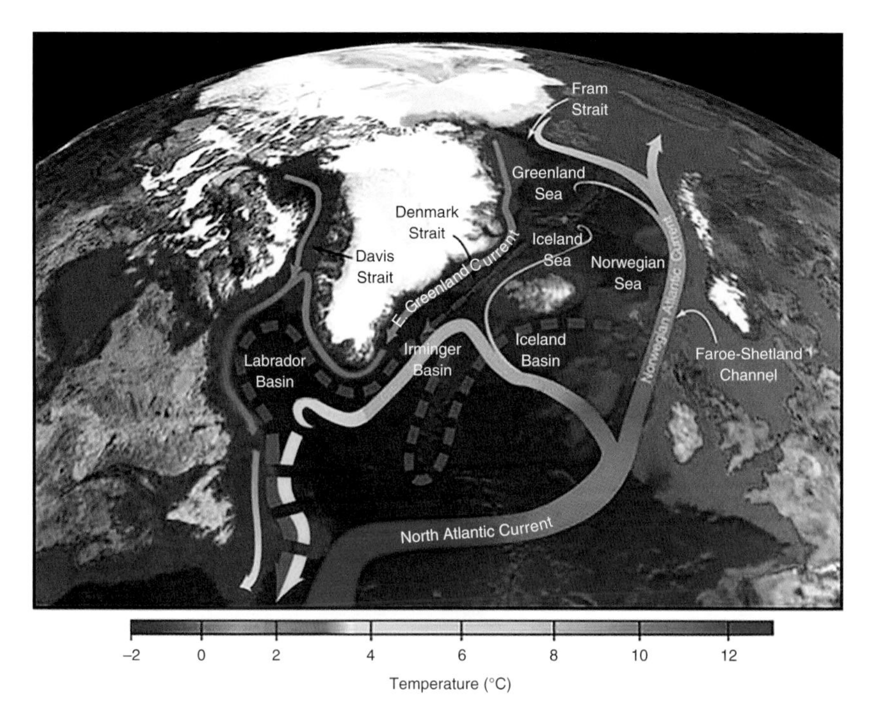

Figure 11.15 The Atlantic Meridional Overturning Circulation (AMOC) showing the warmer surface ocean currents in solid colors and the colder subsurface ocean currents in dashed colors. Source: R. Curry, Woods Hole Oceanographic Institution, Wikimedia Commons.

increased precipitation and flooding, more frequent and severe hurricanes, and changes in phytoplankton numbers and location. Phytoplankton are the basis of the aquatic food web, providing an essential ecological function for all aquatic life. Thus, the oceanic phytoplankton are tied to fisheries, and decreases in fish populations could cause food shortages.

Warming of the surface oceans may very likely shorten the Pacific warm pool cycling time, which drives the El Niño–Southern Oscillation (ENSO) phenomenon. The ENSO is a periodic variation in wind and sea temperatures over the eastern tropical Pacific Ocean. The periodic warming of the surface ocean waters over the equatorial Pacific Ocean is referred to as an El Niño event, while a periodic cooling of the waters is known as a La Niña event. The terms originally came from names used for temperature changes in the surface waters observed by Peruvian fishermen off their coastal shores in late December, around Christmas time. The original name for warming of the water was "El Niño de Navidad" or "the little boy of Christmas." The observation of cooling currents was then given the name of a little girl (La Niña).

The El Niño is associated with a band of warm ocean water that develops in the central and east-central equatorial Pacific, as shown in Figure 11.16. Under normal conditions, a warm pool in the western Pacific drives atmospheric convection winds. Hot air rises in the western Pacific, then travels eastward until it cools and descends over South America. During an El Niño event, the sea surface is warm in the central and eastern Pacific. Hot air rises in the central Pacific, where it travels east and west before descending. The warm ocean waters in the equatorial Pacific create a high-pressure area over the western Pacific and a low-pressure area over the eastern Pacific, bringing warm and very wet weather along the coasts of northern Peru and Ecuador during the months of April through October. During a La Niña event, warm water is further west than usual, causing the equatorial water to be cooler than under normal conditions. Heavy rains occur over Malaysia, the Philippines, and Indonesia.

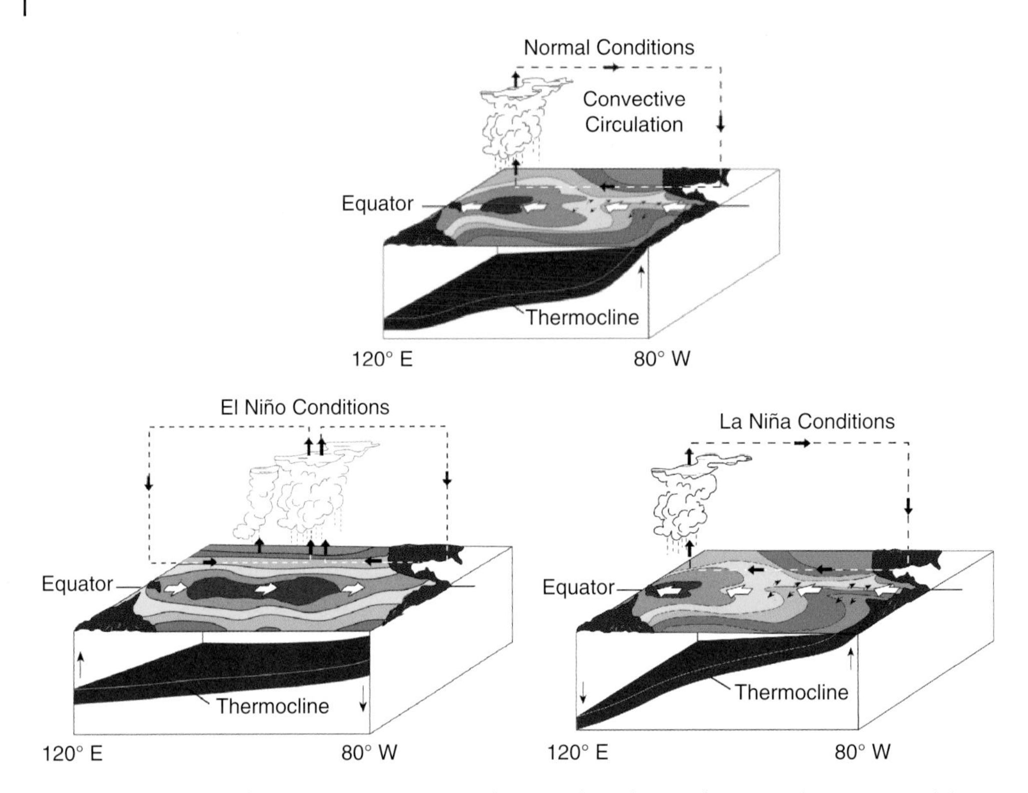

Figure 11.16 The Pacific Ocean temperatures under normal conditions, during an El Niño year and during a La Niña year. Source: Adapted from Fred the Oyster, Wikimedia Commons, NOAA.

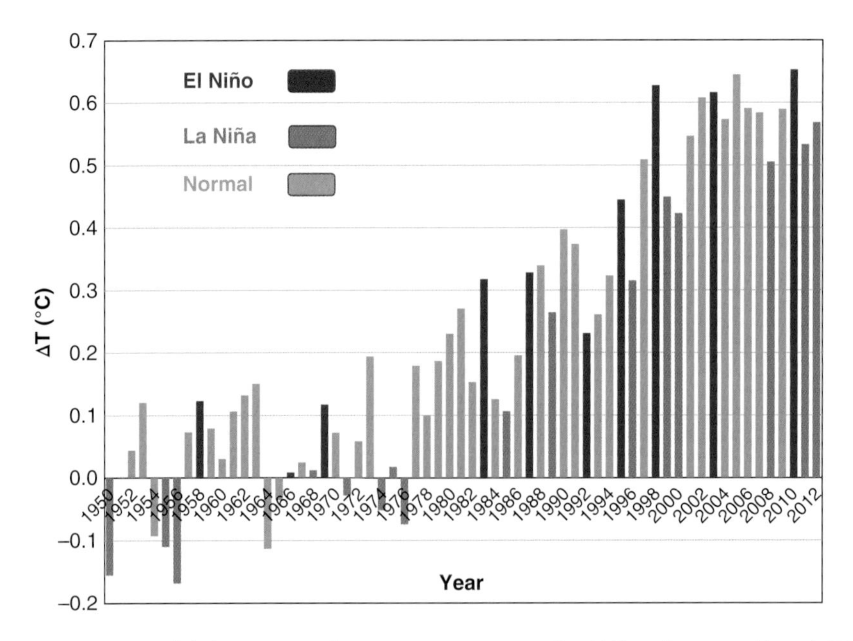

Figure 11.17 Global average surface temperature anomalies (ΔT) under normal (gray), El Niño (red), and La Niña conditions from 1950 to 2012. Source: Adapted from NOAA, Wikimedia Commons.

In the ENSO cycle, both the El Niño and the La Niña cause changes in air temperature and rainfall pattern. Thus, the frequency and intensity of the El Niño and La Niña events have impacts on global weather patterns and climate. The frequencies of both the El Niño and La Niña patterns have increased since 1980, as shown in Figure 11.17. However, the global temperature anomalies during the El Niño years are not significantly above those under normal conditions, because the temperature of the Pacific Ocean is elevated even in normal years since 1980. Also, the ocean temperatures during La Niña events are no longer significantly cooler than normal, as they were prior to 1980. This trend could further weaken the La Niña events, causing land areas in the western Pacific to become drier.

11.6 Climate Effects on Air Quality and Health

Although air quality and health are usually treated as separate and distinct issues from climate change, they are closely coupled. The impacts of climate and air quality on health have been analyzed in detail by the United States Global Change Research Program (USGCRP) and reported in national climate assessment reports (Melillo et al., 2014). One of the major air quality and health concerns connected to climate change by the USGCRP is the increased production of ground-level ozone. Tropospheric ozone is a greenhouse gas whose concentration is driven by the NO_x–VOC chemistry discussed in Sections 5.4 and 5.5. Climate change projections for temperate regions predict warmer periods in summer months, leading to more stable high-pressure meteorological conditions. Under these stable high-pressure conditions, continuous emissions from combustion sources concentrate in the stagnant air masses. This leads to enhanced formation of O_3 over the higher emission regions, such as urban centers.

With increasing temperatures during the summer in the temperate zones, increased energy use is projected due to increased use of air-conditioning systems in urban and residential areas. This increases the emissions from coal and natural gas-fired power plants. Automobiles operating air-conditioning systems get fewer miles per gallon and will increase their emissions during the hotter summer periods. Burning more fossil fuels in stationary and mobile sources increases emissions of both greenhouse gases as well as the ozone precursors. Thus, climate change is predicted to increase summer temperatures in the temperate zones, leading to increased air-pollutant emissions and poorer air quality. The increased combustion of fossil fuels will also produce higher emissions of CO_2 during these periods, causing another connection between air quality and climate change.

Increasing levels of CO_2 in the troposphere, along with earlier and warmer temperate-zone springs, will lead to longer growing seasons with increased risk of wildfires. Wildfires increase health impacts due to enhanced ozone and aerosol pollution. In addition, the later fall frosts and earlier springs, which result in longer growing seasons, lead to an increase in pollen levels. Climate change in the temperate zones is currently causing higher pollen counts that last longer. The increases in allergens, such as ragweed, are well documented and strongly correlated with climate change in temperate regions (Ziska et al., 2011). Since 1995, the length of the ragweed pollen season has increased in areas above 44°N latitude by as much as 13–27 days. Pollen allergies can compound the impacts of exposure to air pollutants, such as ozone and $PM_{2.5}$, and pose a high risk for asthma sufferers.

Higher temperatures, in combination with shorter, less severe winters in the temperate zones, are changing the potential risk of contracting diseases that were restricted to the tropics in the past. Most of these diseases, such as West Nile virus, malaria, dengue fever, and Lyme disease, are carried by mosquitos. Mosquitos, along with ticks and fleas, infect the host by transferring viral material directly into the blood during feeding. All of these insect carriers have low cold tolerance and cannot survive at low temperatures, which typically occur in late fall and winter in the temperate and polar latitudes. The increasing temperatures in the higher latitudes, which result in shorter winters and later falls, allow the insects to survive and multiply over longer periods, increasing their potential health risks. Waterborne diseases such as cholera, which are caused by bacteria, are also expected to increase with climate change in areas where clean fresh water is not readily available. The warmer temperatures allow for rapid growth of the infectious bacteria. Increases in waterborne diseases are also likely due to increases in flooding events, which typically result in contamination of drinking water sources with waste. In poorer areas of the world, more severe droughts will act to concentrate these disease carriers in water supplies that are not treated, adding to existing water quality problems in these regions.

Climate change has been linked to increasing severe weather events, such as hurricanes, typhoons, and tornados, associated with severe thunderstorms. These events cause power outages, safe drinking water shortages, flooding, wind damage, and other risks to populations. Summer heat waves, where daily temperatures are well above average, are considered to be severe weather events due to the increased risk of mortality. During the summer of 2003, the hottest summer on record in Europe since 1540, a severe heat wave occurred in Europe during the months of July and August. The differences in average temperature recorded in Europe during July 20 to August 20, 2003 are shown in Figure 11.18. The heat wave led to a health crisis

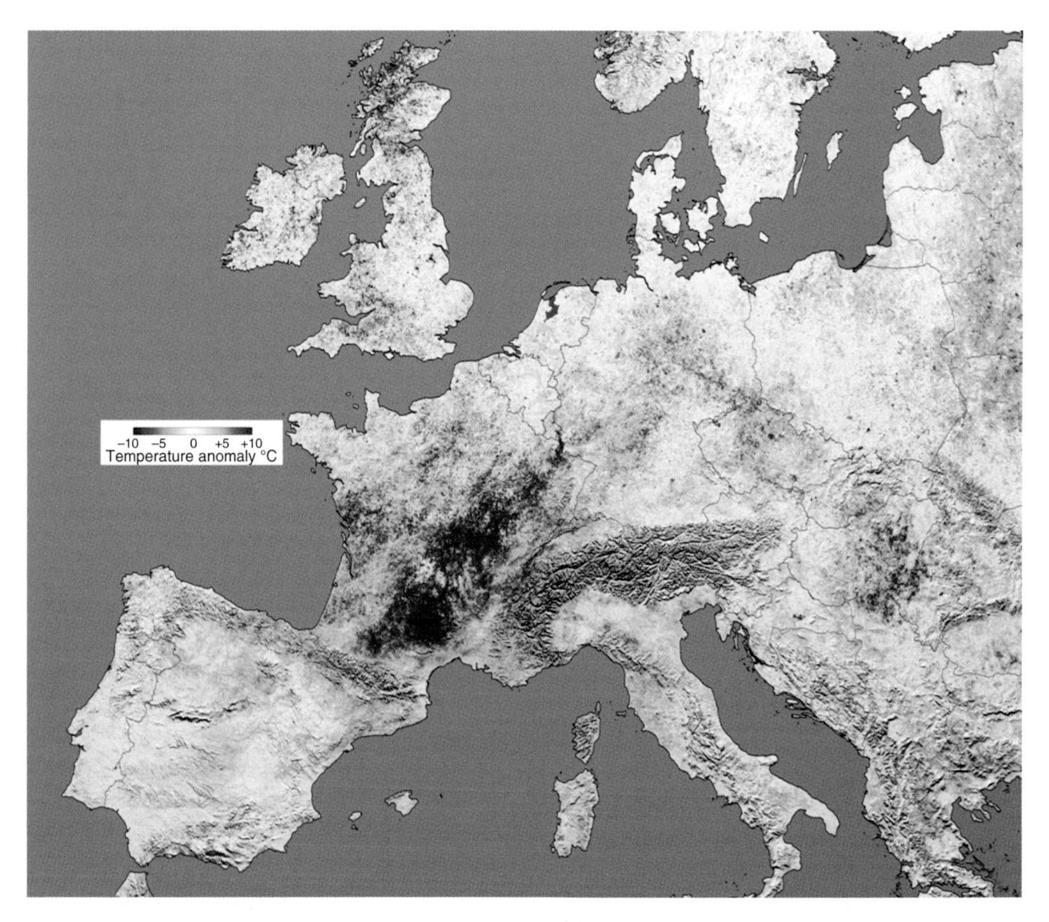

Figure 11.18 The difference in average temperature (determined from the years 2000–2004) observed in July 20 to August 20, 2003 over Europe during late July to early August in 2003 during the severe heat wave. Source: Reto Stockli and Robert Simmon, based on data provided by the MODIS Land Science Team, Wikimedia Commons.

in several countries and, combined with drought, created a crop shortfall in parts of Southern Europe. The European death toll was recorded at more than 70 000. The heat wave was partly a result of the influence of the warm Atlantic Current in combination with hot continental air and strong southerly winds.

Heat waves and increasing summer temperatures lead to the need for air conditioning of indoor air. The efficient use of air conditioning in buildings for minimum energy use is mainly accomplished by reducing the exchange rate of the cooled indoor air with the hot outdoor air. Similarly, lowering energy use for heating buildings during the winter is also accomplished by recycling the heated air and reducing the intrusion of colder outside air. The reduction of air-exchange rates in buildings depends on reducing air leaks through windows and doors and using more efficient insulation. This leads to an increase in the concentration of indoor air pollutants, because of the reduction in outside air exchange. These indoor air pollutants can be a mixture of anthropogenic pollutants as well as naturally occurring pollutants such as radon gas, discussed in Chapter 12. Thus, indoor air pollution could increase if energy conservation through lowering air-exchange rates in buildings is used as a mitigation strategy for reducing the emissions of greenhouse gases.

11.7 Mitigation and Adaption Strategies

It is now clear that climate change is occurring and human activities are a major contributing factor, if not the major factor. Climate change mitigation is the planned efforts to reduce or prevent the emission of greenhouse gases, which are the major climate-forcing agents. Mitigation strategies consider possible options for reducing the rate of climate change, which can include using new technologies and renewable energies, making older equipment more energy efficient, changing management practices, and changing consumer behavior. Adaption strategies are actions taken to help communities and ecosystems cope with the impacts of changing climate. These can include the design and construction of wind-resistant buildings and the use of dikes and pumping to protect against sea-level rise. As pointed out in AR5, there are many global constraining factors that limit our ability to establish effective mitigation and adaptation strategies to slow climate change and prepare for its consequences. Some of the most important constraints include the increasing global population and urbanization causing more demand for energy and resources, shortages of knowledge and education about climate change risks, difficulties in obtaining agreements between nations with divergent views and cultures, lack of funding to address the capital investments needed nationally and internationally, and inadequate technologies in many countries.

The shortage of resources, including food, water, and energy, is a major problem related to climate change and increasing population densities. Some countries will be more vulnerable than others to the effects of climate change, and differences in economic wealth will affect how well different countries can adapt. Although greenhouse gas emissions affect the entire planet, the amount of greenhouse gases emitted by different countries varies significantly. Climate change is a problem that has higher risks in longer time frames, which will increase the impacts on future generations. In a global economy, which is largely concerned with the present, addressing global change will require leadership that is focused on establishing long-term sustainability through stricter resource management. Thus, climate change is a global problem that will affect the quality of living of future generations and so requires long-term solutions. Dealing with these issues will require overcoming the global societal and political issues that stand in the way of implementing mitigation and adaptation strategies worldwide.

Carbon dioxide is the major anthropogenic greenhouse gas, which is primarily increasing due to the combustion of fossil fuels. Fossil fuels are not renewable and are not sustainable, so one obvious mitigation strategy is the reduction in the combustion of fossil fuels. The conservation of energy use at all levels will require changes in energy consumption patterns. For example, the implementation of car pooling and the development and use of mass transit for transportation has been proposed to conserve energy. However, making sure that these energy conservation methods are used is not an easy task. In many cases, single-driver gasoline vehicles continue to remain the dominant method of transportation even in areas where mass transit is available. The implementation of fast lanes that are reserved for cars with one or more passengers has encouraged car pooling in some states. One alternative to gasoline-powered vehicles is electric cars, which do not emit greenhouse gases and reduce the amount of fossil fuels burned. However, the electricity which powers the vehicles is usually supplied by fossil-fueled power plants. Thus, the electric cars do not have zero CO_2 emissions.

Alternative energy sources to fossil fuel combustion include wind, solar, hydro, woody biomass combustion, geothermal, landfill gas, biofuels, and nuclear power. Each of these sources of power has its drawbacks. Wind, solar, hydro, and geothermal energy sources require suitable conditions for operation. That is, they are dependent on the environment in the region where they are to be used. Wind and solar are more likely to supply intermittent power since they are dependent on changing weather conditions. Since wind and solar power sources do not produce a steady supply of electricity, they require reliable electric storage devices with a large capacity and improved efficiencies to store power when it is produced. Both solar and wind power also require the dedication of large land areas in order to supply large amounts of power. The widespread use of wind and hydroelectric power generation can result in significant impacts on ecosystems, particularly on avian and fish populations. The biomass fuel alternatives still rely on the combustion of carbon fuels and so produce CO_2. They also produce air pollutants, which have their own negative impacts, as discussed in Section 10.4. The use of multiple types of sustainable energy sources can help to reduce the use of fossil fuel and slow climate change. However, no single alternative fuel can produce enough power to replace the amount currently generated by fossil fuels. In addition, while the use of natural gas in power plants as an alternative to coal reduces the CO_2 emissions, it is not sustainable and also leads to increased CH_4 emissions. Thus, it also is not a long-term solution.

When comparing the various energy alternatives, it is important to examine the complete lifecycle of the energy-producing system. This includes the fuels consumed in the production of the materials used in the construction of the power plant, routine and long-term maintenance, handling of any waste materials, and decommissioning the plant. In 1999, the Vattenfall Utility Company in Sweden provided a lifecycle comparison for a number of electric power-generating alternatives, including coal, gas, gas combined cycle, solar photovoltaics, wind, hydro, and nuclear (Vattenfall, 1999). The gas combined cycle uses a gas-powered turbine and waste heat to run a steam-powered turbine as a means of increasing efficiency. A comparison of the grams of CO_2 produced per kilowatt hour (g CO_2/kWh) for the various electric power alternatives used in Sweden, Japan, and France is shown in Figure 11.19. This study found that nuclear was the lowest CO_2 emitter over its lifecycle, at 5 g CO_2/kWh compared to 400 g CO_2/kWh for natural gas and 700 g CO_2/kWh for coal. In a similar study, the World Nuclear Association (WNA, 2011) reported the average lifecycle emissions for nuclear energy at 29 g CO_2/kWh (1–130 g CO_2/kWh), which was 7% of the emission intensity of natural gas and only 3% of the emission intensity of coal-fired power plants. The variation in the lifecycle emissions for nuclear was primarily dependent on the purity of the fuel used. The average lifecycle emissions of nuclear power were found to be consistent with the renewable energy sources: hydroelectric (26 g CO_2/kWh) and wind (26 g CO_2/kWh). Lifecycle emissions from biomass

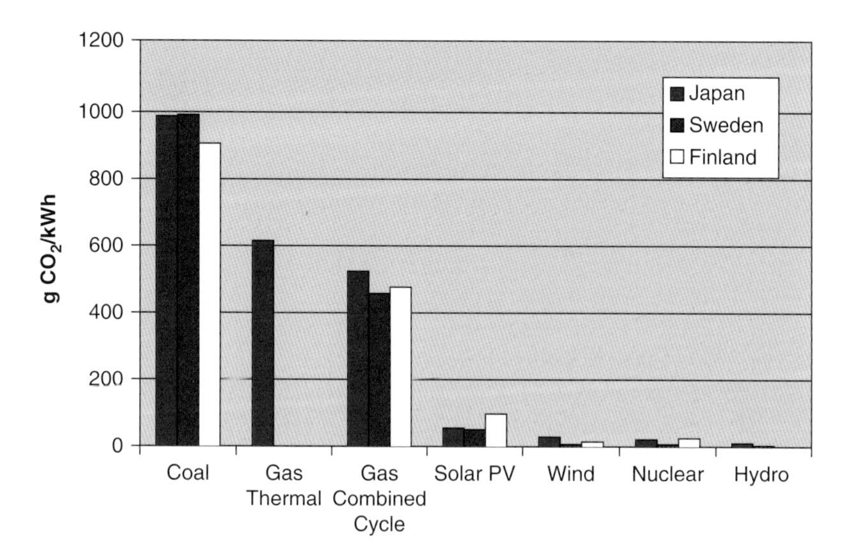

Figure 11.19 The lifecycle CO_2 emissions in g kWh^{-1} of the various electrical power-generating technologies in Japan, Sweden, and Finland. Source: Adapted from Wikimedia Commons.

were a little higher at 45 g CO_2/kWh with a range of 10–101 g CO_2/kWh depending on the biomass used.

Thus, nuclear power is an option for electrical energy production that would dramatically reduce greenhouse gas emissions while resulting in a continuous power output, unlike solar, wind, or hydroelectric. Also, power densities are much higher than for fossil fuels. Nuclear power could be used as an alternative power source for electric cars as well as mass transit, residential, and industrial power. The major opposition to nuclear power is the fact that plutonium, which is produced as a side product in nuclear fission, can be used in nuclear weapons. However, plutonium is fissionable and can also be used as a fuel in nuclear plants, extending uranium fuel resources. The separation of plutonium from spent uranium fuel rods for use as a nuclear fuel is the basic idea behind the "breeder reactor." The breeder reactor uses uranium as a nuclear fuel. The uranium generates plutonium as it is consumed. The plutonium side product is then separated from the spent uranium and used in the reactor to generate energy. The fission of plutonium in the reactor alters it into non-fissionable materials, eliminating the need for the safe, secure storage of the original plutonium waste. Currently, France is routinely using breeder reactor technology. The United States does not currently use breeder reactor technology to recycle the spent fuel rods and so has to deal with the long-term storage of the plutonium side product. Nuclear power designs have been improved to the point where they are safer than other power plant designs, which will be discussed in Chapter 12. However, these new developments in nuclear plant design need to be effectively communicated to the public before nuclear power is generally accepted as a main alternative to fossil fuel combustion. Significant capital investments will also be required if breeder reactors are to be used on a large scale.

The large-scale use of biomass energy sources will need land that otherwise could be used for food production. With a growing population, the need for food and water will become an increasing problem and biomass production will have to compete with that need. While biomass fuels can be used to extend the current fossil fuel reserves, their production and combustion leads to continued air pollution problems tied to NO_x, VOC, and CO emissions that lead to tropospheric ozone formation and air quality problems. The very large amounts of energy that will be needed for future use cannot be supplied by biomass fuels alone, especially considering the growing need to use land for food production. The widespread production of biomass fuels will also lead to increased water pollution from increased fertilizer and pesticide use, as well as the production of other greenhouse gases such as N_2O and CH_4 from the increased use of fertilizers.

One method of CO_2 mitigation is carbon sequestration, which involves the capture of CO_2 from power plant flue gases for the purpose of placing it into long-term storage. One option is to convert the CO_2 to carbonate by reaction with a metallic oxide base. The inorganic carbonate can then be stored in a geological formation. Another option is to inject the CO_2 exhaust gases directly into geological formations as a supercritical fluid. A supercritical fluid exists at a temperature and pressure above its critical point, where distinct liquid and gas phases do not exist. Because of this, the fluid can effuse through solids like a gas and dissolve materials like a liquid. Thus, supercritical CO_2 can act as a gas and travel through porous rock formations, or it can act as a liquid to dissolve sedimentary materials creating space for storage. Deep ocean storage of CO_2 has also been proposed. In this method, CO_2 would be injected directly from power plants into deep ocean systems by using pressurized pipelines. At the bottom of the ocean, the cold temperatures and high pressures would cause CO_2 clathrates to form. This methodology has not been pursued because the clathrates would eventually be transported from the deep ocean by currents where they would result in the release of the CO_2 back into the atmosphere on a time frame that could negatively impact future generations. Some nations have considered transporting and dumping the CO_2 into surface ocean waters, which are basic. This storage option has been banned due to impacts on ocean acidity and marine life. Storage of CO_2 in deep lakes has also been considered. This has been abandoned since the natural release of CO_2 from Lake Nyos in the Cameroons due to volcanic activity released about a cubic kilometer of CO_2. This catastrophic release of CO_2 caused the asphyxiation deaths of 1700 people, demonstrating the serious potential hazard associated with this storage mechanism near populated areas.

The inorganic capture of CO_2 as carbonate rock is the only storage proposal that could transform the CO_2 into a form that would remain stable over long time periods. This requires the use of metal oxides for the reaction with CO_2. The metal oxides that have been considered are listed in Table 11.6, along with their heats of formation and water solubilities. Calcium and magnesium carbonates are stable and have low water solubilities, so the stored CO_2 would not be easily released back into the atmosphere from the geosphere. The alkali metal carbonates (K_2CO_3 and Na_2CO_3) are soluble in water at 25 °C. Thus, after formation, they would need to be precipitated with either Fe^{2+}, Ca^{2+}, or Mg^{2+} ions in an aqueous solution to form the least-soluble carbonates ($FeCO_3$, $CaCO_3$, and $MgCO_3$). The carbonate capture methods require the collection of the CO_2 followed by reaction with the metal oxide and

Table 11.6 Metal oxides that could be used for conversion of CO_2 to a carbonate for sequestration and geological storage along with their heats of formation (ΔH) in kcal mol^{-1} and water solubilities (K_{sp}) at 25 °C.

Metal oxide	% in crust	Carbonate	ΔH	K_{sp}
Fe_2O_3	2.6	$FeCO_3$	27	3.13×10^{-11}
K_2O	2.8	K_2CO_3	−94	soluble
FeO	3.5	$FeCO_3$	−20	3.13×10^{-11}
Na_2O	3.6	Na_2CO_3	−77	soluble
MgO	4.4	$MgCO_3$	−28	6.82×10^{-6}
CaO	4.9	$CaCO_3$	−43	3.36×10^{-9}

the removal and storage of the carbonate. This requires energy and it is estimated that use of carbonate capture would likely double the energy cost compared to an energy plant without carbonate capture. The required energy would come from the power plant, reducing its power-generating capacity by as much as 25%. This would require burning more fossil fuel to obtain the same amount of energy as a power plant without carbonate capture.

The sequestration of the CO_2 as a supercritical fluid would require the collection and separation of the CO_2 from water vapor to produce CO_2 in a supercritical fluid state. The supercritical CO_2 is then injected into geological formations for long-term storage. One concern with this option is the potential for a build-up of pressure within the geological formation, causing seismic activity with potential release of the CO_2 back into the atmosphere. This type of sequestration would typically require transport of the supercritical CO_2 from the power plant to the storage areas through pressurized pipelines. In order for the CO_2 to remain in the supercritical state during transport, it must be maintained at pressures above 1070 psi (>83.8 bar) and temperatures above 304 K. This means that the pipelines must be able to withstand high pressures and be kept warm. The potential for leaking due to aging or the occurrence of seismic events is a significant risk and could pose a hazard to populations close to the pipelines.

As the proposed impacts of climate change have become more likely, more drastic geoengineering methodologies have been proposed to reduce radiative forcing and the impacts of climate change. Geoengineering, or climate engineering, is the deliberate and large-scale intervention in the Earth's climate system with the aim of mitigating the adverse effects of climate change. The proposed geoengineering approaches fall into two categories: the removal of greenhouse gases from the atmosphere or the reduction of incoming solar radiation. Direct removal of CO_2 from the air has been proposed by using a strong base to absorb the CO_2. However, this approach has not been proven to be successful. Other proposals involve the reduction of incoming solar radiation to offset the warming effects of the increasing greenhouse gases. A number of proposals have been made that involve the forced injection of sulfate aerosol precursors or alumina aerosols (Al_2O_3) into the stratosphere to cool the troposphere. Proposed methods of injecting the aerosols into the stratosphere include the use of high-altitude aircraft, very tall towers, continuous artillery shells, or tethered high-altitude balloons. The presence of the sulfate or alumina aerosols in the stratosphere would lead to a cooling of the troposphere by scattering the incoming solar radiation. The effect would simulate the known impacts of large-scale volcanic events, which lead to cooling of the Earth.

Since the stratospheric aerosols have about a 3–4-year lifetime, continuous injection of sulfate or sulfate precursors into the stratosphere would be required to maintain the effect. This idea has many potential large-scale impacts, including the increase in stratospheric ozone depletion, which is known to be catalyzed by the presence of aerosols. In addition, the light scattering by stratospheric aerosols would interfere with ground-based astronomy. With the reduction of incoming solar radiation, solar-powered electrical generation capacity would be reduced. In addition, any large volcanic eruptions would inject additional aerosols into the stratosphere, causing more rapid cooling than anticipated, which would have sudden adverse effects on weather and climate. However, if aerosol injection is suddenly stopped, rapid warming would follow. The costs of this geoengineering project are estimated in the tens of billions of dollars annually. And lastly, there would be major ethical and political issues regarding what country or countries would have the responsibility of operating this artificial planetary thermostat.

Another suggested approach to lowering the incoming solar radiation would be to add light-scattering sulfate aerosols over the open oceans in the lower troposphere. This would be accomplished by requiring tankers to burn high-sulfur fuel to increase sulfate precursor emissions. This would presumably lead to light scattering and cloud formation in the troposphere, with impacts that would be more localized than stratospheric aerosol injection. This approach assumes that the aerosols would not be transported long distances and that they would not have significant impacts on cloud formation, which would affect precipitation and weather. Another major drawback of this approach is that the acidic sulfate aerosols would lead to increased rates of ocean acidification, which would result in a lowering of the oceans' natural ability to take up CO_2, as well as causing marine ecosystem damage. Also, sulfate aerosols have been shown to strongly absorb IR radiation (Marley et al., 1993). Thus, they may act as greenhouse species in the troposphere, offsetting the expected cooling effect.

Other very large-scale extraterrestrial geoengineering approaches to reducing the incoming solar radiation have been proposed. These include using mirror-covered satellites that would orbit the Earth and reflect radiation back into space, mining moon dust to create a dust cloud around the planet, and placing a huge diffraction grating in space between the Earth and the Sun to disperse sunlight. All of these extraterrestrial geoengineering approaches would be extremely expensive, and it is not clear if they would even be feasible considering the difficulties that the global community currently has simply maintaining the existing small orbiting international space station. These very large geoengineering projects would face numerous problems beyond the huge expense and large resources needed to build and maintain them. Problems include meteorite and radiation damage of the materials over long periods of time, which would require maintenance in space. Other planetary approaches involve changing the albedo of the Earth's surface to be more reflective, thus lowering the absorbed incoming radiation. Some of these approaches include adding reflective sheets to large areas of land to decrease light absorption by the surface, and painting roofs and buildings white to increase the reflection of solar light.

All the geoengineering approaches to reducing the impacts of climate change are likely to cause other problems in their attempts to control climate change. For example, using white reflective materials over urban centers would act to double the rate of OH radical photochemistry over the cities by doubling the light path. This would increase the production of NO_x, VOCs, CO, and tropospheric ozone, decreasing air quality in urban centers. Also, there would have to be agreement by all nations if any of these larger-scale geoengineering approaches were to be implemented. The whole Earth system needs to be well understood before attempts are made to alter the Earth's energy budget on a global scale. Most medical treatments have side-effects that are typically unforeseen before they are tested. The development of drugs and medical procedures requires testing of large numbers of humans to evaluate their effectiveness. However, there is only one Earth and only one chance to succeed. Any drastic attempts to alter the whole Earth system need to be thoroughly and carefully examined before they are attempted in order to prevent another worldwide problem such as that which occurred with CFC stratospheric ozone depletion.

Since the impacts of climate change are expected to continue for many decades even if we immediately stop greenhouse gas emissions, adaptation to climate change will quickly become a necessity. Adaptation, like mitigation, will require capital investment, especially in those areas that are most susceptible to climate change impacts – such as coastal areas. In many areas the coastal cities are already below sea level, requiring dikes and continual pumping to hold back the ocean waters. The Netherlands is one of the best examples of adaptation to sea-level rise. Approximately 17% of this very low and flat nation is located under sea level and is kept dry by the use of a system of dikes and pumps. The current land area of the Netherlands is shown in Figure 11.20 (right). This is compared to the land area of the Netherlands that would be below ocean water if the land reclamation efforts were not in effect (Figure 11.20, left). This would include many major cities – including Amsterdam, the largest city and capital of the Netherlands. Only about 50% of the Netherlands is higher than 1 m above sea level, making it very susceptible to flooding with continued sea-level rise. The Netherlands is currently working

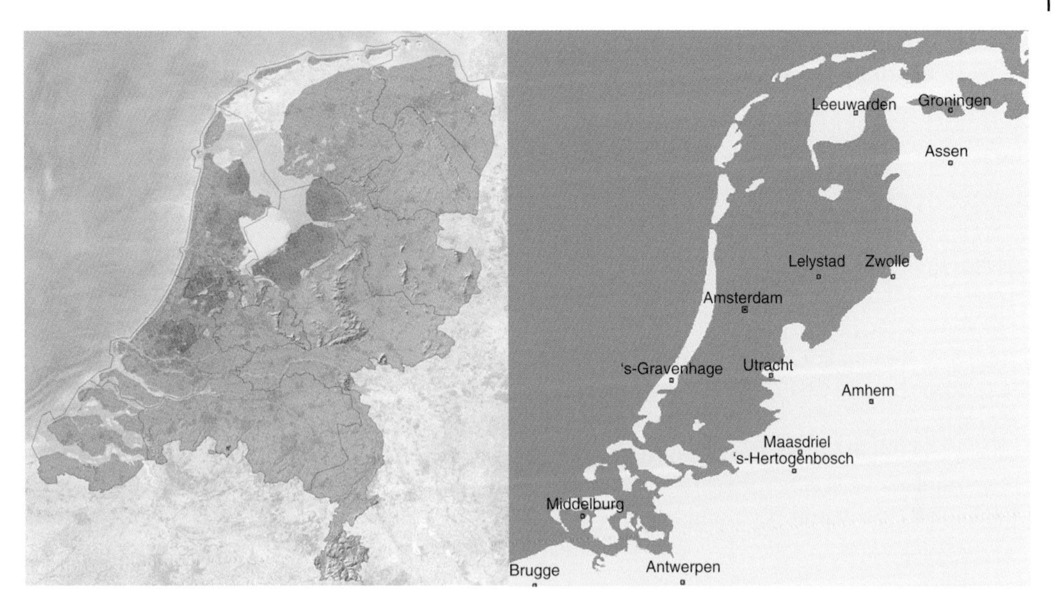

Figure 11.20 A topographical map of the Netherlands (left) showing the current land area compared to the amount of land that would be covered by ocean water (right) without the use of dikes and pumping to reclaim the land. Source: Adapted from Jan Arkesteijn (right) and Janwillemvanaalst (left), Wikimedia Commons.

on strengthening its dikes and flood control systems to adapt to climate change sea-level rise, requiring major investments.

With sea-level rise, many groundwater aquifers in coastal areas will become more brackish, leading to freshwater shortages. Freshwater shortages are also predicted to increase due to longer and more frequent droughts in many regions. Warmer temperatures will lead to faster evaporation of surface freshwater reservoirs commonly used for drinking water, industry and agriculture. Groundwater recharge may be affected, depending on the region and frequency of droughts. A hotter and drier climate will require agricultural areas to depend on heat and drought-resistant crops. Irrigation and flood control systems will have to be established or increased for continued and reliable food production. In some cases, where fresh water is scarce, desalination plants could be used to produce fresh water from ocean or brackish groundwater sources. Most desalination plants produce salt or brine, which is usually released into the ocean. Large amounts of salt can cause local damage to marine ecosystems. One issue with current desalination processes is that the water produced contains no iodine. This has led to health problems due to iodine deficiency in the diet in Israel, where desalination is used to produce about 50% of the water supply.

New and improving adaptation technologies will continue to be developed. However, the implementation of these technologies may be slowed by societal and political resistance, driven by high costs and lack of understanding. Climate change will increase demand on key resources, including energy, food, and water. Mitigation and adaptation together will be required to establish a sustainable future. Since resources and energy use are not uniformly distributed across the globe, the development of sustainable mitigation and adaptation policies will be location dependent. Some areas will be impacted more than others, with the temperate, polar, and low-lying coastal regions at the highest risk. Another area of major concern is India, where the Ganges River is recharged by glacier melt. Since these glaciers are receding, freshwater shortages are projected to occur in this high-population-density area. These types of impacts are likely to intensify decades from now if mitigation and adaptation are not implemented in a timely manner. This will require global cooperation and long-term commitment if conflicts between nations over diminishing resources are to be avoided. The quality of life of future generations depends on our ability to reach a sustainable environmental system.

References

Forster, P., Ramaswamy, V., Artaxo, P. et al. (2007). Changes in atmospheric constituents and in radiative forcing. In: *Climate Change 2007: The Physical Science Basis. Contribution of Working Group I to the Fourth Assessment Report of the Intergovernmental Panel on Climate Change* (ed. S. Solomon, D. Qin, M. Manning, et al.), 129–234. Cambridge: Cambridge University Press.

IPCC 2016 Climate Change 2013: The Physical Science Basis. Available online at http://www.climatechange2013.org (accessed April 26, 2018).

Marley, N.A., Gaffney, J.S., and Cunningham, M.M. (1993). Aqueous greenhouse species in clouds, fogs, and aerosols. *Environmental Science & Technology* 27: 2864–2869.

Melillo, J.M., Richmond, T.C., and Yohe, G.W. (eds.) (2014). *Climate Change Impacts in the United States: The Third National Climate Assessment*. Washington, D.C: U.S. Global Change Research Program.

NOAA 2011 *National Centers for Environmental Information, State of the Climate: Global Climate Report for Annual 2010*. Available online at https://www.ncdc.noaa.gov/sotc/global/201013 (accessed April 16, 2018).

Peel, M.C., Finlayson, B.L., and McMahon, T.A. (2007). Updated world map of the Köppen-Geiger climate classification. *Hydrology and Earth System Sciences* 11: 1633–1644.

Vattenfall 1999 *Life Cycle Assessment: Vattenfall's electricity generation in the Nordic countries*. Available online at https://corporate.vattenfall.com/globalassets/corporate/sustainability/reports/life_cycle_assessment.pdf (accessed May 3, 2018).

WNA (2011). *Comparison of Lifecycle Greenhouse Gas Emissions of Various Electricity Generation Sources*. London: World Nuclear Association.

Ziska, L., Knowlton, K., Rogers, C. et al. (2011). Recent warming by latitude associated with increased length of ragweed pollen season in central North America. *Proceedings of the National Academy of Science of the United States of America* 108: 4248–4251.

Study Problems

11.1 (a) What is climate?
(b) What are the three different time frames commonly used to evaluate climate?

11.2 How do weather and climate differ?

11.3 (a) What are the five different climate groups on Earth?
(b) On what basis are the five groups subdivided?

11.4 (a) What is the average global monthly surface daytime temperature range on the Earth?
(b) What was the mean annual surface temperature of the Earth in 2010? How does this differ from the mean temperature of the twentieth century?

11.5 How is current climate change different from past climate change without man's influence with regard to time scales?

11.6 (a) What two stable isotopic ratios were used to determine the variation in mean ocean temperatures over time as determined from their measurements in ice cores?
(b) What is the physical basis for this determination?

11.7 (a) How do the stable isotopic ratios change if the ocean temperature decreases?
(b) How do they change if the ocean temperature increases?

11.8 (a) What stable isotope ratio is used to determine the past ocean temperature from ocean sediment cores?
(b) Specifically, what part of the sediment core is used to measure the isotope ratios?

11.9 (a) How is the stable isotopic ratio in sediment cores related to ocean temperature changes?
(b) Why does this differ from the relationship in ice cores?

11.10 (a) How far back before the present can ice core data be used to determine temperature changes?
(b) How far back before the present can ocean sediment data be used?

11.11 (a) What was the maximum CO_2 concentration during the past 400 000 years?
(b) In what year was it determined that the atmospheric level of CO_2 was 310 ppm?
(c) What was the level of CO_2 in 2018?

11.12 (a) What was the Earth's climate called in the period between 1650 and 1850?
(b) What was responsible for this climate?

11.13 (a) Why are the atmospheric temperature increases since 1750 not attributed to changes in solar radiation?
(b) What is the major cause of the temperature changes?

11.14 (a) What are the greenhouse gases in the atmosphere?
(b) What is the strongest greenhouse gas?
(c) How is it removed from the troposphere?

11.15 What are the next three strongest gases important in climate change?

11.16 (a) How much have CH_4 levels increased in the atmosphere since 1750?
(b) CO_2 levels?
(c) N_2O levels?

11.17 (a) What greenhouse gases have no natural source?
(b) Why are the atmospheric levels of the CFCs beginning to decrease?
(c) Why are the levels of the HCFCs slowly increasing?

11.18 (a) What is a global warming potential (GWP)?
(b) What determines the magnitude of a GWP?

11.19 (a) What are the natural CO_2 sources?
(b) What are the natural sinks for CO_2?
(c) What are the major anthropogenic sources for CO_2?

11.20 (a) What is the major source of CH_4?
(b) What are the major anthropogenic sources of tropospheric CH_4?
(c) What is the most important sink for CH_4?

11.21 (a) What is the major source of N_2O?
(b) What are the major anthropogenic sources of N_2O?

11.22 What it the major sink for N_2O?

11.23 (a) What is the major source of the PFCs in the troposphere?
(b) What is the source of SF_6?

11.24 (a) What is climate forcing?
(b) What species cause positive climate forcing?
(c) What species cause negative climate forcing?

11.25 (a) What components of the atmospheric aerosols are the strongest absorbers of incoming solar radiation?
(b) This results in what kind of climate forcing?
(c) What atmospheric aerosols scatter the incoming solar radiation?
(d) This results in what kind of climate forcing?

11.26 What are the two factors contributing most to the uncertainty in determining climate forcing?

11.27 (a) What is the abbreviation for the name of the global-scale computer models used for predicting atmospheric temperature changes?
(b) What is the abbreviation for the name of the global-scale models used for predicting oceanic temperature changes?
(c) What is the simple abbreviation used for the combined atmosphere/ocean global-scale climate models?

11.28 (a) How do GCMs differ from RCMs?
(b) What type of information is needed from GCMs to run the RCMs?

11.29 (a) What are the four representative concentration pathways (RCPs) used in GCM models to predict the climate change by 2100?
(b) What are they named for?

11.30 (a) What is the IPCC?
(b) What is the latest assessment report from the IPCC called?

11.31 According to the latest IPCC assessment, what would be the outcome if anthropogenic CO_2 emissions were stopped today?

11.32 What is the range of temperatures and mean sea-level rise predicted by the late twenty-first century in all RCPs?

11.33 (a) The increase in plant growth during wetter springs, along with increased amounts of dry biomass available during dry summers, is projected to result in what?
(b) How will this add to current air pollution problems?

11.34 The surface currents of the thermohaline circulation in the northern Atlantic Ocean are known as what?

11.35 (a) What is climate change mitigation?
(b) What is climate change adaptation?

11.36 What are the important alternative energy sources to fossil fuel?

11.37 (a) When comparing different energy power sources, what is important to consider?
(b) What does this include?

11.38 (a) What three energy sources have the lowest CO_2 emissions during their lifecycle?
(b) Which of these is the most reliable?

11.39 What is a breeder reactor?

11.40 (a) What is carbon sequestration?
(b) What is the most effective proposed method for long-term storage of CO_2?

11.41 (a) What is geoengineering?
(b) What are the two categories of geoengineering projects?

11.42 The reduction of incoming solar radiation by geoengineering projects would have a direct impact on what type of power generation?

12

Nuclear Energy

Nuclear energy, which is based on nuclear fission, is a reliable source of energy, with low CO_2 emissions. Initially, nuclear fission was researched during World War II for the purpose of developing powerful nuclear weapons. After the war, above-ground nuclear weapons testing labeled the environment with radioisotopes, which resulted in the recognition of the potential health issues related to nuclear fallout. The availability of nuclear fission eventually led to the development of nuclear power plants based on the fission reactions of uranium. About 10^8 times more energy is released during a single fission event than is released in a combustion reaction. Thus, nuclear fission-powered reactors require much less fuel to obtain the same amount of energy as coal or natural gas-fired power plants. This makes them an attractive alternative to fossil fuel combustion in reducing the amount of greenhouse gas emissions, air pollution, and waste materials.

12.1 Radioactivity

Recall that the different isotopes of an element have the same number of protons and therefore the same number of electrons. However, the isotopes differ in the number of neutrons and thus their mass. For example, the simplest of the isotopes would be a hydrogen atom with one proton and one electron. Hydrogen has two additional isotopes; deuterium and tritium. In isotopic notation the mass of the isotope, which is the number of protons plus the number of neutrons, is written as a superscript to the left of the element symbol (^{A}X). The symbol for the element (X) designates the number of protons and electrons of the element. Therefore, the isotopic notation for the stable and most abundant hydrogen isotope is ^{1}H, which is an atom of hydrogen with one proton. Deuterium has the isotopic notation ^{2}H, which is an atom of hydrogen with one proton and one neutron. Deuterium is sometimes given the symbol "D." It is found in a very low abundance of only 0.015% of the total hydrogen concentration. Tritium has the isotopic notation ^{3}H, which is an atom of hydrogen with one proton and two neutrons. It is sometimes given the symbol "T."

The ^{3}H isotope of hydrogen is radioactive. This means that its atomic nucleus is unstable and can undergo a nuclear decay process, resulting in the release of energy. The energy released from the nucleus is usually in the form of a high-energy particle along with a high-energy photon and is known as radioactivity. Radioactivity is defined as the spontaneous emission of high-energy particles, electromagnetic radiation, or both, caused by the disintegration of atomic nuclei. In the case of ^{3}H one of the neutrons decays to form a proton, releasing a high-energy electron called a beta particle and a small particle called an antineutrino. After the decay, the ^{3}H then becomes a stable ^{3}He atom with one neutron and two protons. This is one type of radioactive decay process called beta decay. There are other types of radioactive decay processes, which will

Chemistry of Environmental Systems: Fundamental Principles and Analytical Methods, First Edition.
Jeffrey S. Gaffney and Nancy A. Marley.

be discussed in Section 12.2. An isotope is also called a nuclide. When the isotope is unstable and undergoes a radioactive nuclear decay process, the prefix "radio" is used before either name, for example radioisotope or radionuclide. Both of these terms are used to describe unstable atomic isotopes. However, while the high-energy physics community uses the term radionuclide, environmental chemists generally use the term radioisotope. In this chapter the term radioisotope will be used to describe unstable nuclei that are radioactive.

Radioactivity was first discovered by Antoine Henri Becquerel in the late nineteenth century, during his study of phosphorescent materials. His work followed studies of X-rays by Wilhelm Conrad Röntgen in 1895. Röntgen used cathode-ray tubes to accelerate energetic electrons onto the surface of the glass tubes. Röntgen realized that some invisible rays were coming from the tubes, which could pass through the cardboard surrounding the tubes. He also found that these rays could pass through books and papers on his desk. The actual source of this radiation was not known at the time. Hence, he gave the unknown ray the name "X-ray." Röntgen used photographic plates to yield the first medical X-ray of his wife's hand with a large ring on it, shown in Figure 12.1.

Becquerel thought that the X-rays produced by the cathode-ray tubes were a form of phosphorescence coming from phosphorescent mineral coatings on the inside of the glass tubes. He tried to duplicate the results of Röntgen's X-ray experiments by placing a number of known phosphorescent materials on top of photographic plates covered with two sheets of a thick opaque black paper. The black paper was used to prevent exposure of the plates when they were placed in sunlight to induce phosphorescence. His experiment was designed to determine if light from the phosphorescing materials could penetrate through the paper and cause the photographic plate to be exposed. Becquerel obtained negative results from these experiments, until he tried some uranium salts. He performed the same experiment with phosphorescent uranium salts and uranium salts that were not phosphorescent. What Becquerel discovered was that both phosphorescent and non-phosphorescent uranium would cause the photographic

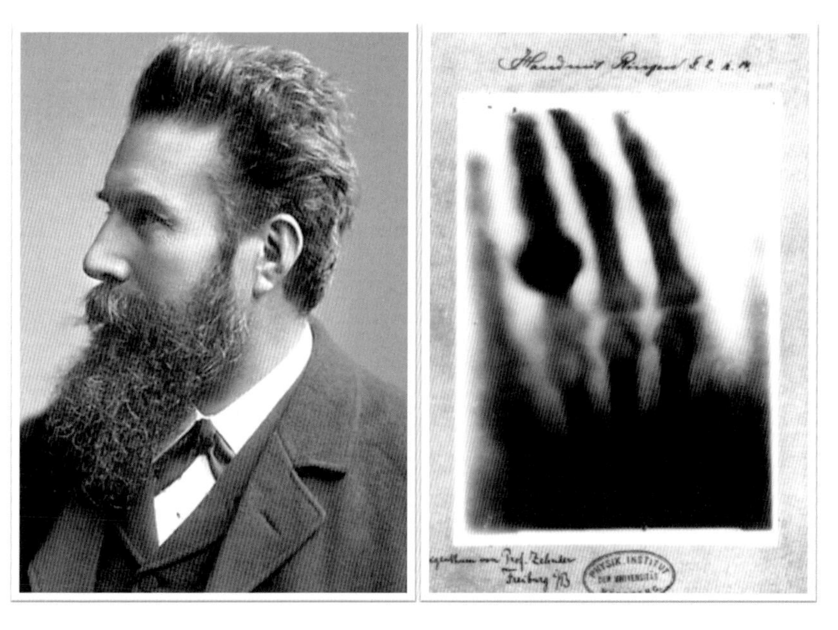

Figure 12.1 Wilhelm Conrad Röntgen (left) and the first medical X-ray taken of his wife, Anna Bertha Ludwig's hand showing her finger bones and a large ring taken in 1895. Source: left, Nobel Foundation; right, Wilhelm Röntgen, Wikimedia Commons.

plates to be exposed. In addition, no external light source was needed to produce the exposure of the photographic plates in Becquerel's experiments. As discussed in Section 3.4, phosphorescence requires an external light source to first produce an excited electronic state in the phosphorescent material. Thus, the invisible rays that exposed the photographic plates were not due to phosphorescence.

Becquerel conducted a number of experiments to determine the properties of the invisible rays that were emitted from the uranium to expose the photographic plates. These experiments included placing the uranium salt $K_2UO_2(SO_4)_2$ on top of a photographic plate in the dark. Beneath the uranium salt he placed a Maltese cross made of metal. The resulting images on the photographic plates showed the image of the uranium salt as well as an image of the metal cross under the salt, as shown in Figure 12.2 (right). Becquerel concluded that the invisible radiation, which was emitted from the uranium, could pass through the paper and expose the plate but could not pass through the metal cross.

This same methodology was used by others to find new materials that would produce the same penetrating rays that Becquerel discovered. The German chemist Gerhard Carl Schmidt and Marie Curie discovered thorium in 1898. Further study by Pierre and Marie Curie isolated polonium and radium, two new elements associated with uranium ores that also emitted the penetrating rays. After the discovery of radium and polonium, they coined the term "radioactivity" to describe the penetrating rays and the materials that produced the radioactivity were said to be radioactive. Their research on radioactivity of uranium and the discovery of radium launched an era of using radium for the treatment of cancer. Their exploration of radium can be seen as the first peaceful use of nuclear energy and the start of modern nuclear medicine.

Ernest Rutherford, working with Robert Bowie Owens, found that thorium and uranium emitted radiation at different rates. They also discovered radon, a short-lived radioactive gas.

Figure 12.2 Antoine Henri Becquerel (left), the experimental design of his exposure of a photographic plate by a uranium salt (right top), and the resulting image he obtained (right bottom). Source: left, Paul Nadar; lower right, Henri Becquerel, Wikimedia Commons.

Figure 12.3 Pierre and Marie Curie (left) and Ernest Rutherford (right). Source: left, Smithsonian Institute; right, George Grantham Bain Collection, Library of Congress, Wikimedia Commons.

Continued work by Becquerel and Rutherford demonstrated that there were two types of energetic particles emitted from the known radioactive elements. Rutherford adopted the names "alpha rays" and "beta rays" to describe the radiation, based on the ability of the two types of rays to penetrate through different thicknesses of aluminum foil. The alpha rays were the least penetrating and the beta rays were more penetrating. Later, Rutherford adopted the name "gamma rays" for the very penetrating, high-energy light emissions discovered by Paul Villard. Further work using electric and magnetic fields by Rutherford and coworkers showed that alpha rays were particles containing two protons and two neutrons with a +2 charge. He also demonstrated that beta rays were negatively charged energetic electrons.

In 1903, the Nobel Prize in Physics was awarded to Antoine Henri Becquerel and to Pierre and Marie Curie, shown in Figure 12.3, for the discovery of radioactivity. Marie Curie was the first woman to receive a Nobel Prize. Marie Curie was also awarded the Nobel Prize in Chemistry in 1911 for her discoveries of polonium and radium, another first for women chemists. In 1908, Ernest Rutherford was awarded the Nobel Prize in Chemistry.

12.2 Radioactive Emissions and Decay Kinetics

The nuclei of radioactive isotopes spontaneously disintegrate, accompanied by the emission of high-energy particles and/or gamma rays. The change in the nuclei after decay depends on what type of particle is emitted. There are many types of radioactive decay process. However, there are five major radioactive decay pathways. These five processes include alpha decay, β^- decay, β^+ decay (positron emission), gamma-ray emission, and electron capture (EC). There are other radioactive decay processes that are not as common, which include the emission of a proton (p), the emission of a neutron (n), and self-fission (SF).

Alpha decay involves the loss of an alpha particle from the nucleus of the radioisotope. An alpha particle (α) is an energetic helium nucleus with two protons and two neutrons. Since it has no electrons, it has a positive charge of 2+. Thus, the decaying isotope loses four mass units and produces an isotope of a different element that has two protons less than the parent radioisotope. Figure 12.4 shows the α decay of a ^{238}U atom, which produces an α particle with an energy of 4.2 MeV and the product isotope ^{234}Th:

$$^{238}U \rightarrow {}^{234}Th + \alpha$$

A radioisotope that is produced from a radioactive decay process is called the daughter, while the original radioisotope before decay is called the parent. In Figure 12.4, the parent ^{238}U undergoes α decay to produce the daughter ^{234}Th. Since the parent has lost two protons, the daughter element formed from α decay is two positions to the left of the parent on the periodic table.

Beta (β) decay can occur by two processes: β^- decay and β^+ decay, which are shown in Figure 12.5. The most common β decay process is β^- decay, which involves the conversion of a neutron in the nucleus of the radioisotope to a proton by the emission of an energetic electron (e^-). The least common β decay process is β^+ decay, which involves the conversion of a proton in the nucleus of the radioisotope to a neutron by the emission of a positively charged

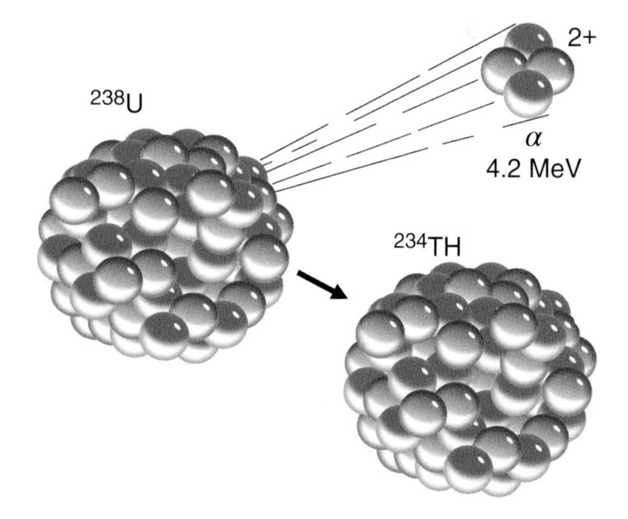

Figure 12.4 Alpha decay of the parent ^{238}U radioisotope with the emission of an α particle to form the daughter ^{234}Th. Source: Adapted from Inductiveload, Wikimedia Commons.

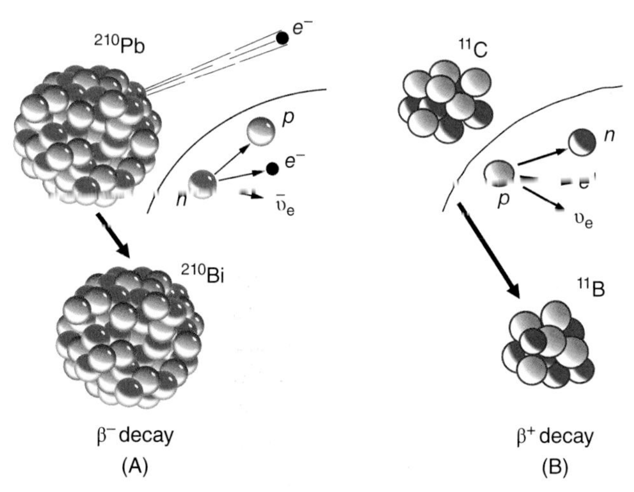

Figure 12.5 (A) The β^- decay of ^{210}Pb to form ^{210}Bi by the conversion of a neutron to a proton with the emission of an energetic electron (e^-) and an antineutrino ($\bar{\upsilon}_e$). (B) The β^+ decay of ^{11}C to form ^{11}B by the conversion of a proton to a neutron with the emission of a positron (e^+) and a neutrino (υ_e). Source: Adapted from Inductiveload, Wikimedia Commons.

electron (e^+), known as a positron. The β^- decay process results in the loss of a neutron and the addition of a proton. Thus, the daughter isotope does not change mass, but it does change the number of protons and therefore becomes a different element. By the addition of a proton, the daughter becomes the element that is one step to the right of the parent in the periodic table. The β^+ decay process results in the loss of a proton and the addition of a neutron. There is also no mass change from the parent to the daughter, but the daughter loses a proton and becomes the element that is one step to the left of the parent on the periodic table.

The β^- decay process is shown in Figure 12.5(A) as the decay of the parent radioisotope ^{210}Pb with the emission of an energetic electron (e^-) and an antineutrino ($\overline{\upsilon}_e$). The reaction is written as:

$$^{210}\text{Pb} \rightarrow {}^{210}\text{Bi} + \beta^- + \overline{\upsilon}_e$$

This results in the conversion of a neutron into a proton in the nucleus of the ^{210}Pb and produces the daughter radioisotope ^{210}Bi. The Bi daughter, with one more proton and one less neutron than the parent Pb, is one step to the right of Pb on the periodic table. The β^+ decay process is shown in Figure 12.5(B) as the decay of the parent radioisotope ^{11}C with the emission of a positron (e^+) and a neutrino (υ_e), which are released during the conversion of a proton to a neutron in the ^{11}C nucleus. This produces the daughter ^{11}B, which has one more neutron and one less proton than the parent ^{11}C. This reaction is written as:

$$^{11}\text{C} \rightarrow {}^{11}\text{B} + \beta^+ + \upsilon_e$$

The B daughter is thus one step to the left of C on the periodic table. The υ_e and $\overline{\upsilon}_e$ are extremely small subatomic particles that do not interact with most materials.

Gamma rays (γ-rays) are energetic photons. The γ-ray decay process does not involve a change in the mass or particle composition of the parent radioisotope. It involves the emission of γ-rays from a nucleus that is in an excited state. The excited nucleus emits the high-energy γ-rays in order to return to the nuclear ground state. This is similar to the emission of light from an electronically excited molecule to return to the electronic ground state as discussed in Section 3.4. The γ-rays emitted in the nuclear deactivation are very energetic, ranging from tens of kiloelectron volts to about 10 MeV. An example of a γ decay process, shown in Figure 12.6, is the decay of the metastable technetium radionuclide ^{99m}Tc, which is in a nuclear excited state. The ^{99m}Tc undergoes a nuclear isomerization, releasing a γ-ray with an energy of 145.5 keV to form the ^{99}Tc in the nuclear ground state:

$$^{99m}\text{Tc} \rightarrow {}^{99}\text{Tc} + \gamma$$

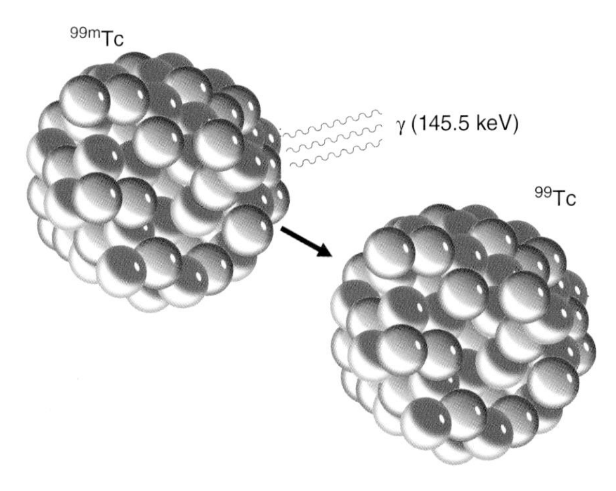

Figure 12.6 Gamma ray emission from the isomerization of the excited ^{99m}Tc to form the daughter ^{99}Tc in the nuclear ground state. Source: Adapted from Inductiveload, Wikimedia Commons.

The EC decay process involves the capture of an electron by a proton in the nucleus of a radioisotope, which converts the proton to a neutron with the emission of a γ-ray. This results in loss of the proton and gain of a neutron in the nucleus of the radioisotope. The EC decay process effectively changes the nucleus of the radioisotope in the same way as a β^+ decay process, since a proton is converted to a neutron in both decay processes. Both EC and β^+ nuclear processes result in a daughter that has one less proton than the parent and is located one step to the left on the periodic table from the parent. The lightest element to undergo an EC process is ^{7}Be. After EC decay, the ^{7}Be produces ^{7}Li accompanied by the emission of a γ-ray with an energy of 477.5 keV. The reaction is written as:

$$^{7}\text{Be} \rightarrow {}^{7}\text{Li} + \gamma$$

The driving force behind the nuclear decay process is the number of neutrons relative to the number of protons in the nucleus of an isotope. A plot of the number of neutrons (N) versus the number of protons, or the atomic number (Z), for the known radioisotopes is shown in Figure 12.7 (National Nuclear Data Center, n.d.). The stable isotopes are shown in black, known as the area of stability. All other isotopes undergo nuclear decay. The isotopes with an equal number of protons and neutrons ($Z = N$) are those that fall on the red line. Above $Z = 82$ and $N = 128$ all of the known isotopes are radioactive. Also, all isotopes above $Z = N = 20$ require more neutrons than protons to be stable. The relationship of a radioisotope to the area of stability (black) determines its decay mechanism. If the radioisotope has more protons than neutrons and is above the area of stability, it will tend to undergo EC or β^+ decay so that it will lose a proton and gain a neutron, converting it into a stable isotope. The opposite occurs when the radioisotope is below the area of stability. Since the radioisotopes below the area of stability have more neutrons than protons, they are more likely to undergo β^- decay, which converts a neutron into a proton, to achieve stability. Also, α decay is more common for the heavier isotopes ($N > 128$). This moves the heavy radioisotope back toward the area of stability by losing two protons and two neutrons. When a heavy radioisotope such as ^{238}U decays, it

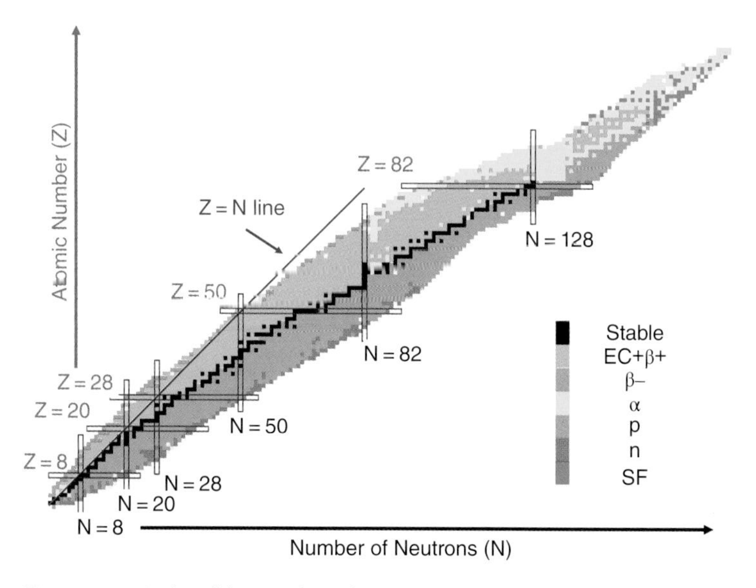

Figure 12.7 A plot of the number of protons (Z) versus the number of neutrons (N) for the known radioisotopes, including stable isotopes (black) and those undergoing α decay (yellow), β^- decay (pink), β^+ or EC decay (blue), p decay (orange), n decay (purple), and SF decay (green). Source: Adapted from National Nuclear Data Center.

usually does so through a series of decay steps, producing a series of daughter isotopes, before it finally reaches a stable isotope such as ^{206}Pb. The series of decay processes that converts a radioactive element into a series of different elements until it produces a stable isotope is called a decay series.

The three types of radiation – α particles, β^- particles, and γ-rays – are emitted during these decay processes. Alpha particles are produced during α decay. Beta particles are either energetic electrons or positrons and are produced from β^- and β^+ decay, respectively. Gamma rays are produced when the other decay processes (α, β^-, β^+, p, or n) leave the nucleus in an excited state. They are also produced when positrons interact with electrons and undergo annihilation. The annihilation of the positron and the electron results in the production of energy in the form of two γ-rays emitted at 180° from each other with equal energies of 0.511 MeV. Alpha particles have a charge of +2 with the mass of a helium nucleus. Beta particles can have a charge of −1 or +1, both with the mass of an electron. Gamma rays are a form of high-energy electromagnetic radiation and are not charged. Alpha particles are the largest and, because of their size, they can be stopped by thin materials. Beta particles are smaller than α particles, so they are stopped by slightly thicker materials. Gamma rays are not readily stopped by normal materials. Gamma radiation requires lead shielding or thick metal shielding to absorb the high-energy electromagnetic radiation.

The different forms of radiation produced by a radioisotope during the decay process can be identified by placing the radioactive material in a magnetic or electric field. The most common forms of radiation – α particles, β^- particles, and γ-rays – can be detected by separating them according to charge, as discussed in Section 9.2.5. The charged particles are deflected by using charged plates while the γ-rays are not deflected. Radiation detectors placed at the points of deflection determine the relative amounts of each form of radiation released during the decay process, as shown in Figure 12.8(A). These most common forms of radiation can also be identified by using different materials placed in front of a detector. Each material is selected to test the penetrating power of the radiation, as shown in Figure 12.8(B).

Radioactive decay processes observe simple first-order kinetics. For example, the decay rate of ^{238}U to produce the ^{234}Th daughter was experimentally determined to be first order in the concentration of ^{238}U:

$$\text{decay rate} = -\frac{d[^{238}U]}{dt} = \lambda[^{238}U]$$

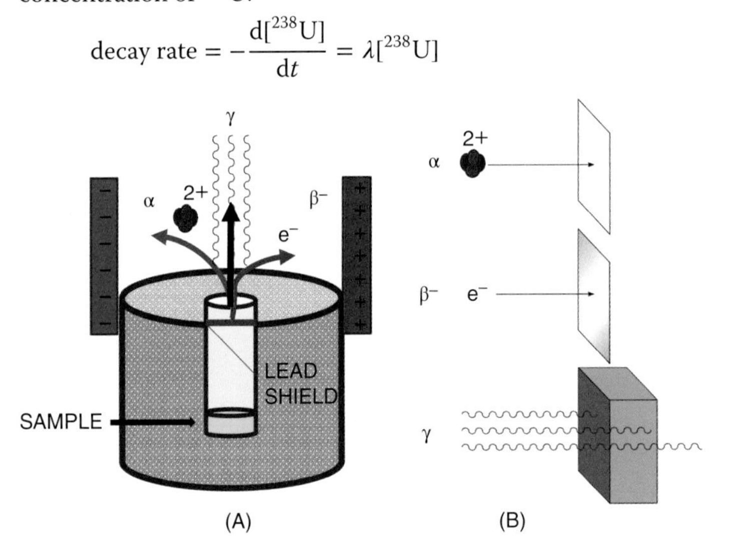

Figure 12.8 The identification of α, β^-, and γ using (A) an electric field or (B) different thicknesses of materials. Source: Adapted from Stannered, Wikimedia Commons.

where [^{238}U] is the concentration of ^{238}U and λ is the rate constant for the nuclear decay process. The decay rate, which is also called the activity of the isotope, is directly proportional to the concentration of the radioisotope, which is directly proportional to the number of radioactive particles in the sample (N):

$$-\frac{dN}{dt} = \lambda N$$

Thus, the activity of a radioisotope ($-dN/dt$) is equal to the amount of the radioisotope (N) times the decay rate (λ). This expression describes the nuclear decay rate as a function of the change in amount of the radioisotope over time (dN/dt). The integrated form of the decay rate is:

$$N = N_0 e^{-\lambda t}$$

where N is the amount of the radioisotope at time t and N_0 is the initial amount of the radioisotope. This expression describes the decay rate as a function of the initial amount of the radioisotope and the amount of the radioisotope after an amount of time has passed. The average lifetime (τ) of the radioisotope is equal to $1/\lambda$. Since τ is usually determined in seconds, λ has units of s^{-1}. The half-life ($t_{1/2}$) of a radioisotope is the amount of time that it takes for half of the parent radioisotope to decay ($N_0/2$). The half-life is related to the decay rate by:

$$t_{1/2} = \frac{\ln 2}{\lambda} = \frac{0.693}{\lambda}$$

Thus, $t_{1/2}$ for a radioisotope decay process is a constant.

The activity of a radioisotope depends only on the number of decays per second, not on the type of decay. The most commonly used unit for radioisotope activity is the becquerel (Bq), named after Antoine Henri Becquerel. A becquerel is defined as the activity of a quantity of a radioisotope which results in the decay of one nucleus per second. The becquerel is therefore equivalent to the number of disintegrations of the radioisotope per second (s^{-1}). An older activity unit is the curie, named after Pierre and Marie Curie. One curie is approximately the activity of 1 g of radium and equals 3.7×10^{10} becquerel.

12.3 Sources of Radioisotopes

There are three sources of radioisotopes in the environment. These are primordial, cosmogenic, and anthropogenic. Both primordial and cosmogenic radioisotopes are formed naturally, while anthropogenic radioisotopes are artificially produced. Anthropogenic radioisotopes are those that are not normally found on Earth. Most of them are produced by nuclear reactions in nuclear reactors or high-energy nuclear accelerators. The nuclear reactions include nuclear fission, neutron capture, and nuclear fusion. Primordial radioisotopes are those that have existed in their current form since before the Earth was formed. They have very long half-lives and thus have not decayed away since the formation of the Earth. One example of a primordial radioisotope is ^{238}U, with a half-life of 4.5×10^9 years. The Earth is believed to be about 4.5 billion years old, so about half of the initial amount of primordial ^{238}U is still present on the planet today. The energy released from the decay of ^{238}U is responsible for about 40% of the geothermal heat produced within the Earth. The parent ^{238}U decays through a series of 14 steps, which are listed in Table 12.1, to reach the stable isotope ^{206}Pb. The ^{238}U decay series produces ^{222}Rn, known commonly known as Radon, in step 6. Radon-222 gas decays through a series of four short-lived daughter species to produce ^{210}Pb, which can attach to fine-mode aerosol particles in the troposphere. This behavior allows the ^{210}Pb to act as an aerosol source tracer, as discussed

Table 12.1 The ^{238}U decay series including the daughter produced in each successive decay process, the type of decay process (α or β^-), and the half-life of the parent radioisotope.

Parent	Decay process	Daughter	Half-life
^{238}U	α	^{234}Th	4.5×10^9 yr
^{234}Th	β^-	^{234}Pa	24.2 d
^{234}Pa	β^-	^{234}U	1.16 min
^{234}U	α	^{230}Th	2.4×10^5 yr
^{230}Th	α	^{226}Ra	7.5×10^4 yr
^{226}Ra	α	^{222}Rn	1.6×10^3 yr
^{222}Rn	α	^{218}Po	3.8 d
^{218}Po	α	^{214}Po	3.1 min
^{214}Po	β^-	^{214}Bi	26.8 min
^{214}Bi	β^-	^{214}Po	20 min
^{214}Po	α	^{210}Pb	1.6×10^{-4} s
^{210}Pb	β^-	^{210}Bi	22.2 yr
^{210}Bi	β^-	^{210}Po	5 d
^{210}Po	α	^{206}Pb (stable)	138 d

in Section 6.3. Radon is also considered to be a naturally occurring radiation hazard when in high concentrations, since ^{222}Rn and its daughters emit α and β^- particles.

Another primordial radioisotope is ^{40}K ($t_{1/2} = 1.25 \times 10^9$ years), which decays by two alternate pathways. The decay of ^{40}K proceeds by β^- decay about 89% of the time, leading to the formation of stable ^{40}Ca. The second pathway proceeds by an EC about 11% of the time to yield the stable isotope ^{40}Ar. This decay of ^{40}K by EC explains the 1% ^{40}Ar that occurs in the atmosphere. The measurements of the isotopic ratios of both $^{40}K/^{40}Ar$ and $^{238}U/^{206}Pb$ have been used as geochronometers for dating the oldest rocks on Earth.

Another important primordial radioisotope is ^{232}Th, which decays through a series of 11 steps to reach the stable isotope ^{208}Pb, as listed in Table 12.2. While ^{232}Th is a very long-lived radioisotope, with a half-life of 1.4×10^{10} years, its decay series produces fairly short-lived daughters with half-lives of 0.15 s (^{216}Po) to 5.7 years (^{228}Ra). Thorium-232 also produces a radioisotope of radon (^{220}Rn) in decay step 6. The ^{220}Rn is commonly called Thoron to distinguish it from the longer-lived ^{222}Rn known as Radon. The radioisotope ^{220}Rn only has a half-life of 55.6 s compared to the half-life of 3.8 days for ^{222}Rn. Thus, when ^{220}Rn is formed in soil or water there is not much time for it to diffuse into the atmosphere before it decays to the ^{216}Po ($t_{1/2} = 55.6$ s) and ^{212}Pb ($t_{1/2} = 0.15$ s) daughters. It therefore has a lower environmental contribution to the natural atmospheric radioactivity than does the longer-lived ^{222}Rn.

Cosmogenic radioisotopes are created by high-energy cosmic rays interacting with the nucleus of a naturally occurring stable atmospheric isotope, causing protons and neutrons to be expelled from the atom, a process known as spallation. Once expelled, these particles can interact with other atomic nuclei, producing a shower of particles in the upper atmosphere. One of the more abundant of the cosmogenic radioisotopes is ^7Be, with a half-life of 53.1 days. This radioisotope is continually produced by collisions of protons and neutrons with N_2 in the lower stratosphere and upper troposphere. Once produced, it attaches to fine aerosols in the atmosphere and has been used to trace aerosol transport and deposition, as discussed

Table 12.2 The ^{232}Th decay series including the daughter produced in each successive decay process, the type of decay process (α or β^-), and the half-life of the parent radioisotope.

Parent	Decay process	Daughter	Half-life
^{232}Th	α	^{228}Ra	1.4×10^{10} yr
^{228}Ra	β^-	^{228}Ac	5.7 yr
^{228}Ac	β^-	^{228}Th	6.13 h
^{228}Th	α	^{224}Ra	1.9 yr
^{224}Ra	α	^{220}Rn	3.64 d
^{220}Rn	α	^{216}Po	55.6 s
^{216}Po	α	^{212}Pb	0.15 s
^{212}Pb	β^-	^{212}Bi	10.6 h
^{212}Bi	α	^{208}Tl	1.1 h
^{212}Bi	β^-	^{212}Po	1.1 h
^{212}Po	α	^{208}Pb (stable)	20 min

in Section 6.3. Another example of a cosmogenic radioisotope important to environmental chemistry is ^{14}C, commonly called radiocarbon. Carbon-14 is produced by a neutron colliding with ^{14}N, which produces ^{14}C and a proton. It has a half-life of 5.7×10^3 years and undergoes β^- decay to reform ^{14}N.

12.4 Nuclear Fission

During the scientific exploration of radioactivity, researchers began to make use of specific radioisotopes as sources of α, β^-, p, and n particles to study nuclear reactions. Rutherford discovered in 1917 that α particles could cause a nuclear reaction that converted ^{14}N atoms into ^{17}O atoms:

$$^{14}N + \alpha \rightarrow {}^{17}O + p$$

This was the first known nuclear reaction. Following the discovery of neutrons in 1932 by James Chadwick, physicists began using them to bombard uranium in order to produce heavier elements from neutron capture reactions. These experiments were successful and resulted in a new set of elements, called the transuranic elements. In 1934, German chemist Ida Noddack proposed the possibility of producing smaller fragments from neutron bombardment of heavier atoms such as uranium. However, it was not until 1938 that nuclear fission was demonstrated. German chemists Otto Hahn and Fritz Strassman bombarded ^{235}U and ^{238}U with a neutron source and observed the formation of ^{141}Ba, which could only have been produced from nuclear fission of the uranium isotopes. Further study showed that approximately three neutrons were released for every fission event. In 1933, Leo Szilard proposed that a nuclear chain reaction could occur if each of the neutrons produced from a fission event could produce another fission event. Szilard was not sure what material could produce this chain reaction, although he knew it had to be neutron rich. If the chain reaction was controlled, the large release of energy from the nuclear fission could be used for power generation. If uncontrolled, it could be used to produce a military weapon known as a nuclear bomb.

Careful work later showed that the ^{238}U was not fissionable with neutron bombardment. The isotope that produced nuclear fission was the lighter and less naturally abundant ^{235}U radioisotope. Once the fission research on ^{235}U was published, Szilard, along with Edward Teller and Eugene Wigner, discussed the potential for a nuclear chain reaction with Albert Einstein and convinced him to write a letter to President Franklin Roosevelt in October of 1939 warning him of the potential to produce a nuclear fission weapon. In 1939, Roosevelt started a modest project to investigate the possibility of producing such a nuclear fission weapon. After the attack on Pearl Harbor in December 1941, Roosevelt ordered the beginning of a project, initially called "Development of Substitute Materials," to develop a nuclear fission weapon. The name of the project was changed to the "Manhattan Project" after the location of the headquarters of the U.S. Army Corps of Engineers.

At that point there were considerable obstacles to making a nuclear fission weapon. The only known fissile material at that time was ^{235}U. However, ^{235}U would have to be enriched from the natural ores by a substantial fraction to sustain the chain reaction. Natural uranium ore contains a number of isotopes, which are listed in Table 12.3. The most abundant of these natural uranium isotopes is ^{238}U at 99.274%, due to its long half-life. The next most abundant isotope is ^{235}U at 0.73%. Three other uranium isotopes – ^{236}U, ^{234}U, and ^{233}U – are present in amounts of 0.005% or less. In order to obtain enough ^{235}U for the production of a successful nuclear chain reaction, it needed to be separated from the ^{238}U. This separation was not an easy task, since both ^{235}U and ^{238}U have the same chemistry and fairly similar masses. Enrichment of ^{235}U was attempted by four approaches: centrifugation, electromagnetic separation, gaseous diffusion, and thermal diffusion. Centrifugation was abandoned due to the difficulties of large-scale operations. The other three methods were all developed successfully, and all contributed to the separation of the required amount of ^{235}U needed to produce a nuclear weapon. Uranium hexafluoride (UF_6), the only known gaseous compound of uranium, was produced from the uranium oxide ore, enriched by diffusion, and processed into the metallic form. This was used to produce one of the first nuclear weapons, called "Little Boy."

As the first neutron strikes a ^{235}U atom it will cause nuclear fission, which generates three neutrons along with the non-fissionable products ^{92}Kr and ^{141}Ba and a large amount of energy, as shown in Figure 12.9(A). The neutrons released in the first fission can then either hit a ^{238}U atom forming the fissile ^{239}Pu, escape from the material unreacted (N.R.), or strike another ^{235}U releasing three more neutrons, as shown in Figure 12.9(B). If the ^{235}U concentration is high enough, the chance of a neutron striking another ^{235}U is increased and the chain reaction will continue. If a neutron moderator is placed in the system, the neutron concentration will be lowered sufficiently that the nuclear reaction will occur at a steady pace. This controlled nuclear chain reaction is the basis for nuclear power generation. If the chain reaction is not controlled and the fissile material concentration is high enough, energy is released explosively

Table 12.3 Naturally occurring uranium isotopes.

Isotope	Abundance (%)	Half-life (yr)	Decay process	Daughter
^{238}U	99.274	4.5×10^9	α	^{234}Th
^{236}U	trace	2.3×10^7	α	^{232}Th
^{235}U	0.73	7.0×10^8	α	^{231}Th
^{234}U[a]	0.005	2.5×10^5	α	^{230}Th
^{233}U	trace	1.6×10^5	α	^{229}Th

a) Produced in the decay series of ^{238}U.

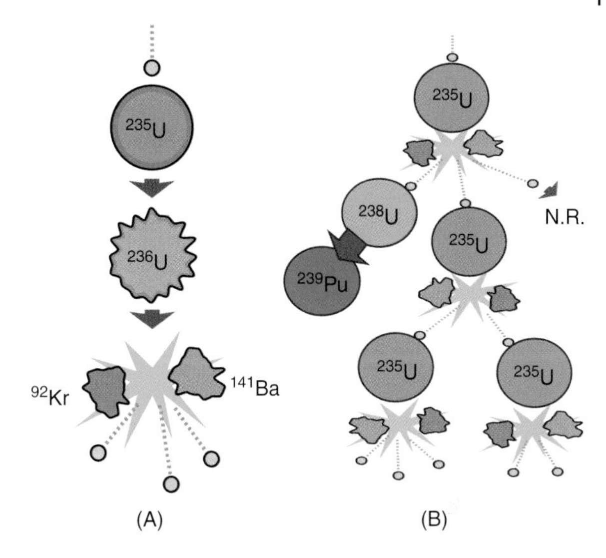

Figure 12.9 Nuclear fission of ^{235}U by neutron bombardment. (A) A single fission event producing two non-fissionable fragments, three neutrons, and a burst of energy. (B) Three possible interactions of the three neutrons, including no reaction (N.R.), capture by ^{238}U to form the fissile material ^{239}Pu, or striking another ^{235}U releasing three more neutrons. Source: Adapted from fastfission, Wikimedia Commons.

and a nuclear explosion occurs. This uncontrolled nuclear chain reaction is the basis of a nuclear bomb.

Since the separation of ^{235}U in sufficient quantities to produce more than one nuclear device was not practical in the days after Pearl Harbor, it was proposed to use the ^{239}Pu, which was produced during the fission of ^{235}U, for a second device. The separation of ^{239}Pu from ^{238}U was fairly straightforward, since they are different elements. However, the production of sufficient ^{239}Pu required a controlled nuclear chain reaction of the ^{235}U. Enrico Fermi led a group at the University of Chicago, as part of the Manhattan Project, to construct a viable nuclear reactor capable of sustaining the nuclear fission chain reaction. Very pure carbon, in the form of graphite bricks, was used as the neutron moderator. The purpose of the neutron moderator was to reduce the speed of fast neutrons, reducing their kinetic energy and converting them into thermal neutrons, which can be absorbed more easily by an atomic nucleus. The reactor was constructed into a pile of graphite and fuel, by adding one layer after another with a neutron source at the bottom of the pile. It was called the Chicago Pile number one (CP-1).

Fermi's group successfully demonstrated a controlled nuclear fission reaction in December 1942. The work was accomplished at the Metallurgical Laboratory, which was moved from the University of Chicago to a Chicago suburban site and later named Argonne National Laboratory (ANL). The graphite reactor design was later adapted and plutonium-producing facilities were built in the state of Washington at the Hanford site. The Hanford graphite reactors produced the ^{239}Pu that was used for the second nuclear weapon, called "Fat Man." This site later became Pacific Northwest National Laboratory (PNNL). Both ANL and PNNL are now Department of Energy National Laboratories.

12.5 Nuclear Weapons Testing and Fallout

A device similar to Fat Man was used for the first above-ground test of a nuclear weapon. The test, called "Trinity," was carried out at a remote site on the U.S. Army Air Force Alamogordo Bombing and Gunnery Range near Socorro, NM in the early morning of July 16, 1945. A mushroom-shaped cloud, shown in Figure 12.10, formed 10 s after the explosion was initiated. This first test began the atomic age. Both Fat Man and Little Boy were dropped on the cities of Hiroshima and Nagasaki, Japan on August 6 and 9, 1945, respectively. Shortly after,

Figure 12.10 The Trinity nuclear fission bomb test carried out at dawn on July 16, 1945 marking the beginning of the atomic age. Source: Federal government of the United States, Wikimedia Commons.

Japan surrendered, ending World War II. This was the only time that nuclear bombs have been used in warfare. The explosive energy of a nuclear weapon is measured by comparison to that generated by an equivalent number of tons of the standard chemical explosive trinitrotoluene (TNT). Each of these first three nuclear explosions was estimated to have the equivalent energy of about 20 kT of TNT.

All these initial devices were nuclear fission weapons, which were known as "atomic bombs." However, during the Manhattan Project, it was proposed that a fission explosion could provide enough energy to initiate a nuclear fusion reaction, which would produce significantly more energy than a fission reaction. Nuclear fusion is a nuclear reaction in which atomic nuclei of low atomic number are fused together to form a heavier nucleus with the release of energy. A new nuclear device was developed based on nuclear fusion, which involved combining two isotopes of hydrogen (deuterium and tritium) to form helium, with the release of about 18 MeV of energy and a neutron, as shown in Figure 12.11. The new fusion devices were called "hydrogen bombs" or "thermonuclear weapons," a reference to the high temperatures required to initiate the reaction.

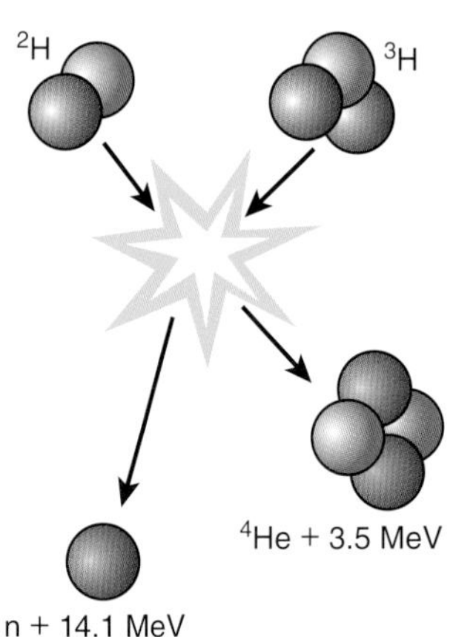

Figure 12.11 Nuclear fusion of deuterium (^{2}H) and tritium (^{3}H) to form a helium atom with the release of a neutron (n) and 17.6 MeV of energy. Source: Wykis, Wikimedia Commons.

The fusion weapons were first demonstrated in an above-ground test in November 1952 over the Enewetak Atoll in the Marshall Islands in the Pacific Ocean by the United States. This fusion bomb released energy equivalent to 10.4 MT of TNT, which was much more than the energy of 20 kT of TNT that was generated by the first nuclear fission devices. Other countries, including the former Soviet Union, the United Kingdom, and France, also developed thermonuclear capabilities and began above-ground testing. The above-ground nuclear testing continued worldwide from 1952 until 1963, when three of the four countries agreed to stop above-ground testing due to concerns about the health effects of the radioactivity released by the explosions. However, testing was continued below ground, with France and China conducting limited above-ground tests. In 1996, the United Nations General Assembly adopted the Comprehensive Nuclear Test Ban Treaty (CTBT), a multilateral treaty that bans all nuclear explosions for both civilian and military purposes in all environments. However, India, Pakistan, and North Korea still conduct limited testing of nuclear weapons since the CTBT was opened for signature in 1996.

The above-ground testing released a significant amount of anthropogenically produced radioactive isotopes into the atmosphere. Nuclear devices larger than 10 kT lofted radioactive material into the stratosphere and into the local area around the blast sites. During the above-ground testing, meteorologists obtained high-altitude samples and set up ground stations, which monitored both wet and dry deposition of materials from the nuclear bomb tests. These measurements showed that radioactive isotopes, as well as fission fragments from the nuclear explosion, were being carried into the upper atmosphere, transported aloft, and deposited globally through both wet and dry deposition. This process was described by the term "nuclear fallout," because it "falls out" of the sky after the explosion and the shock wave have passed. Nuclear fallout is the residual radioactive material propelled into the upper atmosphere following a nuclear explosion. This fallout consists of fission products mixed with atoms that have become radioactive by neutron capture during the explosion.

There are two types of fallout: a small amount of fission products with long half-lives and a large amount of radioactive particulate matter produced from dust and sand with shorter half-lives. Some examples of the radioisotopes produced from a thermonuclear explosion, in order of their increasing half-lives, are shown in Table 12.4. The half-lives range from a few tens of microseconds to greater than 10^{19} years. The longest-lived radionuclide is ^{209}Bi, which is known as quasi-stable due to its extremely long lifetime. The radioisotopes with very short half-lives decay rapidly after formation. However, the longer-lived radioisotopes such as ^{90}Sr, ^{137}Cs, ^{3}H, ^{14}C, ^{129}I, and ^{239}Pu have all been observed in wet and dry deposition. While some of these radioisotopes are in gas form, most of them are released as metal oxides, which form fine-mode aerosols. Most of the heavy metal oxides are not highly water soluble, but species such as ^{137}Cs, an alkali metal, form highly soluble hydroxides.

Over 500 above-ground nuclear weapon tests were conducted, with the largest amount occurring between 1961 and 1962 by the United States and the former Soviet Union. The global measurements of radioactive fallout during this peak period caused health concerns and the two countries agreed to a partial test ban treaty, eliminating above-ground testing in 1963. The exposures to fallout from the two bombs dropped on Japan, along with the accidental exposure of inhabitants of the Marshall Islands during testing in 1954, caused radiation sickness among the populations. Exposure to lower doses of radioactive fallout, especially ^{90}Sr, became a global concern. Strontium-90 is an alkali earth element with chemistry similar to calcium. When ingested or inhaled, this element is carried through the blood stream to the bones. Thus, the bioaccumulation of fallout-derived ^{90}Sr was suspected to be a cause of bone cancer. There was also apprehension that small children could be receiving higher doses per body weight in

Table 12.4 The major radioisotopes produced from nuclear explosions including the daughters produced by radioactive decay, the type of decay process (α or β^-), and the half-life of the parent radioisotope.

Radioisotope	Half-life	Decay process	Daughter
^{213}Po	3.7 μs	α	^{209}Pb
^{217}At	32 ms	α	^{213}Bi
^{221}Fr	4.8 m	α	^{217}At
^{213}Bi	46 m	β^-	^{213}Po
^{209}Pb	3.3 h	β^-	^{209}Bi
^{239}Np	2.3 d	β^-	^{239}Pu
^{90}Y	2.7 d	β	^{90}Zr (stable)
^{237}U	6.8 d	β^-	^{237}Np
^{131}I	8.0 d	β^-	^{131}Xe (stable)
^{225}Ac	10 d	α	^{221}Fr
^{225}Ra	14.9 d	β^-	^{225}Ac
^{233}Pa	27 d	β^-	^{233}U
^{3}H	12.3 yr	β^-	^{3}He (stable)
^{90}Sr	29 yr	β^-	^{90}Y
^{137}Cs	30 yr	β^-	^{137}Ba (stable)
^{241}Am	432 yr	α	^{237}Np
^{14}C	5.7×10^3 yr	β^-	^{14}N (stable)
^{229}Th	7.3×10^3 yr	α	^{225}Ra
^{239}Pu	2.4×10^4 yr	α	^{235}U
^{233}U	1.6×10^5 yr	α	^{229}Th
^{99}Tc	2.2×10^5 yr	β^-	^{99}Ru (stable)
^{36}Cl	3.1×10^5 yr	β^-	^{36}Ar
^{237}Np	2.1×10^6 yr	α	^{233}Pa
^{129}I	1.6×10^7 yr	β^-	^{129}Xe (stable)
^{235}U	7.0×10^8 yr	α	^{231}Th
^{209}Bi	1.9×10^{19} yr	α	^{205}Tl (stable)

milk products that were contaminated with the ^{90}Sr. All of these potential negative impacts led to the United Nations ban on above-ground nuclear weapon testing.

The above-ground nuclear weapon testing released a measurable amount of anthropogenically produced radioisotopes in the atmosphere and geosphere. Since the timing of these releases is well known, the radioactive isotopes have been used as tracers for transport and depositional processes in the atmosphere and geosphere. Stratospheric transport from the equator to the polar regions was confirmed by long-range measurements of fallout from nuclear tests that were conducted near the Equator in the Pacific Ocean. The above-ground tests also produced excess amounts of the naturally produced ^{36}Cl and ^{14}C. Measurements of ^{36}Cl, ^{137}Cs, and ^{239}Pu have been used to confirm sedimentation rates in geology and oceanography. Excess amounts of ^{14}C, often called "bomb carbon," have been used in tree ring dating to identify the 1961–1962 peak years of nuclear testing. In addition, measurements of ^{239}Pu and excess ^{36}Cl from above-ground tests have been used to calibrate the dates obtained

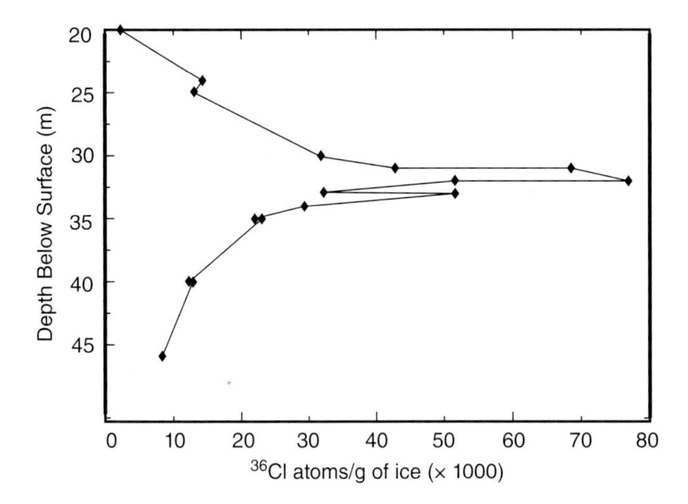

Figure 12.12 Measurements of ^{36}Cl in the Upper Fremont Glacier ice showing the increase in ^{36}Cl concentration from the large number of above-ground nuclear tests that occurred from 1960 to 1962. Source: Adapted from USGS, Wikimedia Commons.

from stable isotope ratios in ice cores, as discussed in Section 11.1. An example of the use of excess ^{36}Cl to identify the 1961–1962 time period in ice cores from the Upper Fremont Glacier located in Wyoming is shown in Figure 12.12. In areas where the rate of ice formation is known, measurements of the radioisotope tracers in ice cores are found to be temporally consistent with dates determined from stable isotope ratios. This is important to validate the climate measurements obtained from the ice cores.

12.6 Nuclear Power

12.6.1 Harnessing Nuclear Energy

While the testing of nuclear weapons labeled the environment with anthropogenic radioisotopes, the peaceful use of nuclear energy was also being pursued. It was quickly realized that a nuclear fission chain reaction could be used to produce a sustained source of energy, as demonstrated by the group led by Fermi at the University of Chicago. The initial reactors built by the Fermi group were constructed to produce ^{239}Pu for weapons. However, Enrico Fermi proposed that, since the ^{239}Pu was also fissionable, the production of ^{239}Pu from the ^{238}U in the ^{235}U nuclear reactors could produce much more fissionable nuclear fuel than could be obtained by using ^{235}U alone. The additional use of ^{239}Pu would increase the available nuclear fuel and extend the nuclear energy resources. The idea of also using the ^{239}Pu produced during the fission of ^{235}U as additional fuel was called a "breeder reactor," which was introduced in Section 11.7. In general, a breeder reactor is a nuclear reactor that generates more fissile material than it consumes. At the time, the known uranium deposits were very limited and the use of ^{239}Pu produced from thermal neutron capture by ^{238}U in a ^{235}U reactor would essentially make use of all of the available uranium as a nuclear fuel.

The first breeder reactor design was developed at ANL and the Experimental Breeder Reactor I (EBR-I) was constructed and successfully demonstrated at a site in Idaho, known as ANL-West, in December 1951, as shown in Figure 12.13(A). It became one of the world's first electricity-generating nuclear power plants when it produced sufficient electricity to illuminate

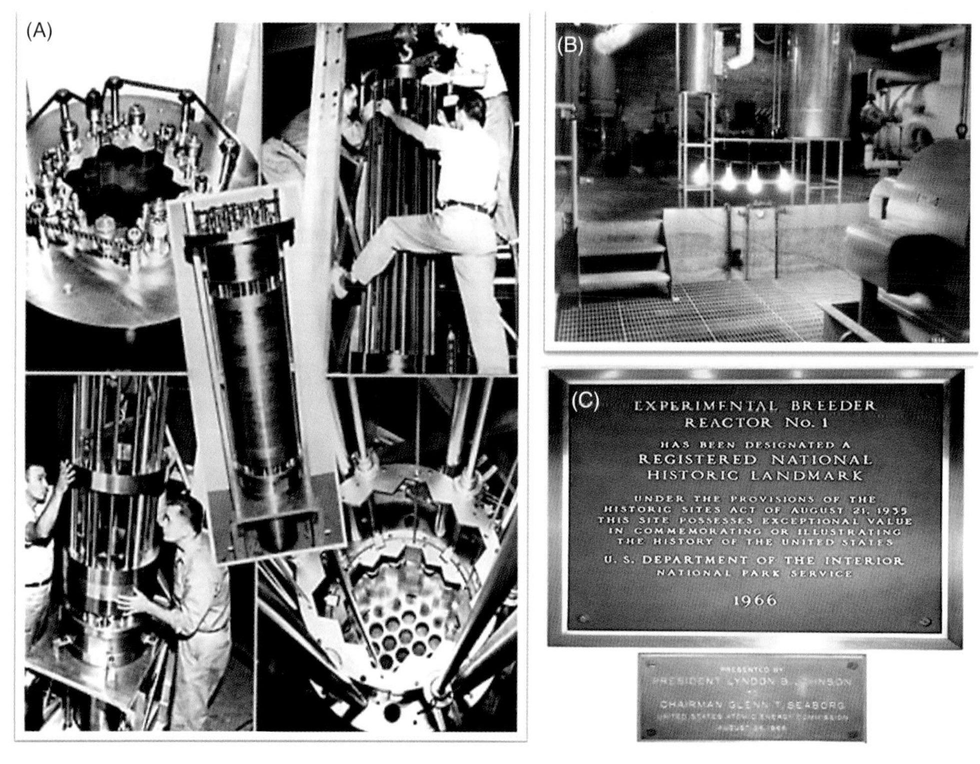

Figure 12.13 (A) Construction of the EBR-I. (B) The first power generated by nuclear fission in December 1951. (C) The plaque designating the EBR-1 a Registered National Historic Landmark. Source: A and B, U.S. DOE; C, ANL West, Wikimedia Commons.

four 200 W light bulbs (Figure 12.13(B)). The EBR-I eventually produced about 200 kW of electricity, generating enough electricity to power its building. It demonstrated that the breeder concept was feasible by producing more fissile ^{239}Pu fuel during its operation than the ^{235}U it consumed. The EBR-I site is now a museum and was designated as a Registered National Historical Landmark by Lyndon Johnson in 1966 (Figure 12.13(C)). The second-generation breeder reactor (EBR-II) is a fast neutron reactor, which uses high-energy neutrons carrying energies of 5 MeV or greater. It produced 20 MW of electricity using a conventional steam turbine generator. The EBR-II operated for 30 years before being decommissioned. During its lifetime it generated over two billion kilowatt hours of electricity, providing the majority of the electricity and heat for the surrounding ANL-West facilities.

The EBR-II was engineered for safe automatic shutdown in the case of a power failure. The nuclear fuel is submerged under molten sodium, which readily conducts heat from the fuel. In addition, the fuel rods are placed in a physical environment that allows them to expand if they become overheated. Since a sustained nuclear reaction depends on a high density of fuel, expansion of the fuel rods lowers the density of the fuel, stopping the nuclear reaction and allowing the excess heat to dissipate without damage to the plant. The expansion of the fuel during a heating event causes the system to shut down automatically, even without human operator intervention. Tests of this automatic shutdown were conducted in 1984 when the power to the coolant pumps was shut off with the power plant operating at full capacity. The plant successfully powered down and the large pool of molten sodium surrounding the nuclear reactor fuel rods allowed the heat to be released safely without damage. This nuclear plant design, and others

Table 12.5 The fissile radioisotopes generated in a thermal neutron reactor, the parent radioisotopes that produce them, and the fission processes in which they can be used (thermal neutron or fast neutron).

Daughter	Parents[a)]	Fission process
^{238}Pu	^{242}Cm (α)	fast
^{239}Pu	^{238}U, ^{238}Pu, ^{243}Cm (α)	thermal
^{240}Pu	^{239}Pu, ^{244}Cm (α)	fast
^{241}Pu	^{240}Pu	thermal
^{242}Pu	^{241}Pu, ^{242}Am (EC)	fast
^{243}Pu	^{242}Pu	—
^{241}Am	^{241}Pu (β^-)	fast
^{242}Am	^{241}Am	—
^{243}Am	^{242}Am, ^{243}Pu	fast
^{244}Am	^{243}Am	—
^{242}Cm	^{242}Am (β^-)	—
^{243}Cm	^{242}Cm	thermal
^{244}Cm	^{243}Cm, ^{244}Am (β^-)	fast
^{245}Cm	^{244}Cm	thermal

a) Decay processes are neutron capture unless otherwise stated in parentheses.

that followed, improved plant safety by addressing the flaws in the initial design of the reactor and fuel rod casings. Fuel rods are now designed to be able to expand when heated, lowering the density of the fuel and thus automatically stopping the chain reaction before damage to the plant can occur.

Nuclear fission reactor designs have evolved to make the most efficient use of the neutrons produced from a fission reaction. The breeder reactor concept accomplishes this by using the neutrons to produce more fuel than it consumes. This begins with the capture of a neutron by ^{238}U to form the fissile ^{239}Pu, followed by a number of decay processes that produce 13 other radioisotopes of Pu, Am, and Cm, as listed in Table 12.5. The thermal neutron fissile isotopes produced in this cycle are ^{239}Pu, ^{241}Pu, ^{243}Cm, and ^{245}Cm. The fast neutron fissile isotopes are ^{238}U, ^{238}Pu, ^{240}Pu, ^{242}Pu, ^{241}Am, ^{243}Am, and ^{244}Cm. All of these radioisotopes can be used as nuclear fuel. Radioisotopes produced in ^{238}U decay that are too short lived to be used as nuclear fuel are ^{243}Pu (5 hours), ^{242}Am (16 hours), ^{244}Am (10 hours), and ^{242}Cm (163 days). However, they rapidly form other radioisotopes that are fissionable. As with ^{239}Pu, all the fissile radioisotopes produced from ^{238}U decay can be chemically separated from the spent ^{235}U fuel for further use as fissionable nuclear fuels. The use of all of the fissile radioisotopes produced by thermal ^{235}U reactors will extend the ^{238}U resource, and holds promise for long-term electrical energy generation. It will also reduce the amount of radioactive waste materials produced in a nuclear power plant by recycling them as fuel.

Nuclear fusion has also been pursued as a source of energy. Nuclear fusion is expected to have several theoretical advantages over fission as a source of power. These include reduced radioactivity, little nuclear waste, ample fuel supplies, and increased safety. However, harnessing nuclear fusion has proven more difficult than harnessing nuclear fission. A controlled fusion reaction is extremely difficult to produce in a practical and economical manner. The nuclear fusion process requires a highly confined environment, with temperature and pressure high enough to create a plasma of millions of degrees in which the fusion can occur. A major problem with controlled nuclear fusion is the very high temperatures and pressures that must be reached. Attempts to produce a controlled fusion reaction between ^{2}H and ^{3}H, as shown in Figure 12.11, have included the use of high-intensity lasers to produce the required temperatures and pressures and electric or magnetic fields to contain the resulting plasma. This work has been ongoing since the 1960s. Advances in plasma production, high-flux neutron sources, and magnetic and electric containment designs have been made, but no design has successfully produced more fusion power output than the electrical power input.

There are also many issues with materials that must be resolved before the continuous operation of a commercial fusion power plant can become a reality. Because of the high neutron fluxes from fusion, a nuclear fusion power plant would have to deal with long-term neutron activation of the containment materials, which would lead to embrittlement and cracking. These same materials will also be exposed to very strong magnetic fields and high temperatures. Because of the high neutron flux and the high-energy α particles produced, it is also likely that the materials used in the fusion reactor will become radioactive. It has been estimated that the nuclear waste produced from a fusion reactor during its lifetime would be about the same as a fission reactor. However, the radioactivity would likely be from short-lived radioisotopes. Thus, the radioactivity from the waste would decay much faster than the long-lived fission reactor products. After about 500 years, the radiation from the waste from a fusion reactor is estimated to be about the same as the natural radioactivity that is found concentrated in coal fly ash. While the hope for a reliable and affordable fusion reactor continues, the development of a commercial fusion reactor still has considerable obstacles to overcome before it becomes feasible.

12.6.2 Uranium Production

Once the potential for uranium use in nuclear power was recognized, the demand and exploration for uranium ore was expanded. The top 10 uranium-producing countries during 2007–2016 are shown in Figure 12.14 (World Nuclear Organization, 2017). The largest producer of uranium is Kazakhstan, which is located in central Asia, producing 39% of the world's

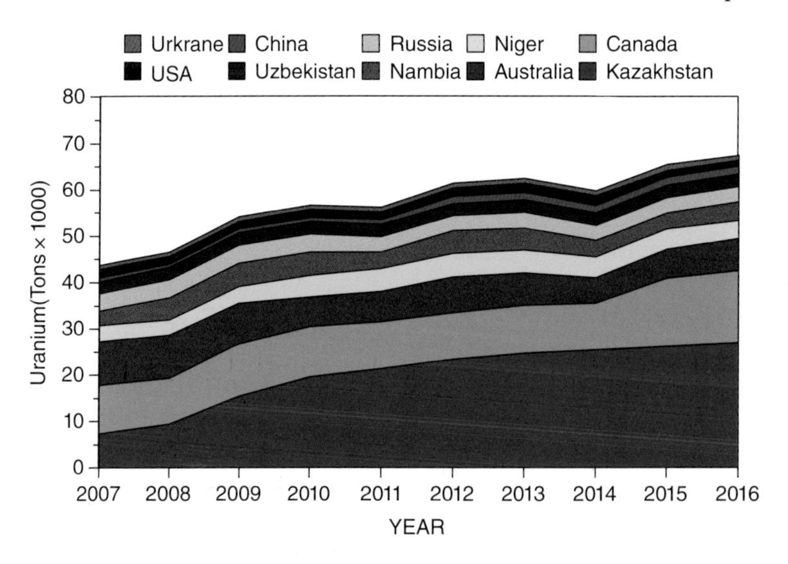

Figure 12.14 The amount of uranium produced (tons) by the top 10 uranium-producing countries from 2007 to 2016.

uranium in 2017. The second largest producer is Canada at 22%, followed by Australia at 10%. The world terrestrial uranium reserves are estimated to be about 44.1 million tons. Australia has the largest known uranium deposits, which are estimated to be close to 1.8 million tons and Kazakhstan has the second largest uranium deposits at 0.8 million tons. This large amount of available uranium has led to the development of nuclear power plants that do not make use of the breeder reactor approach in most countries, including the United States. Uranium is processed and used as fuel rods until they are spent, and then the fuel rods are stored as nuclear waste instead of being recycled as fuel. The known terrestrial uranium reserves will last about 100 years at current use levels. However, not included in this inventory is sea water, which contains about 3–4 ppb of uranium. The recent advances in specific uranium-absorbing materials using complexing agents in polymer fibers may allow uranium to be mined from sea water. If this becomes possible on a large scale, it would extend the uranium reserves from an estimated 100 to 1000 years. Widespread use of breeder reactors would extend the life of the ^{238}U nuclear fuel to a much longer time frame, while at the same time drastically reducing the amount of nuclear waste.

Originally, uranium prospecting was accomplished in ground searches for the presence of uranium ore deposits by using simple portable Geiger counters. Now, airborne gamma counting instruments are the primary tool used in searching for uranium ore deposits. Mining of uranium ore is carried out by four methods. Two of these methods are those used for coal extraction: open pit mines, which represent about 20% of uranium mining, and underground mining, which represents another 26.2%. Once mined, the uranium ore is crushed and concentrated by leaching with H_2SO_4. When the resulting solution is dried, it produces a concentrated powder known as "yellowcake." The composition of yellowcake depends on the leaching and precipitating conditions. The compounds that have been identified in yellowcake include uranyl hydroxide ($UO_2(OH)_2$), uranyl sulfate (UO_2SO_4), sodium para-uranate ($Na_2U_2O_7{\cdot}6H_2O$), tri-uranium octoxide (U_3O_8), uranium dioxide (UO_2), and uranium trioxide (UO_3). The yellow color of yellowcake is due to the presence of the U(VI) oxidation state. The oxidation states of uranium are vividly colored, as shown in Figure 12.15. The most common oxidation states are U(III) (brown–red), U(IV) (green), and U(VI) (yellow). The U(VI) oxidation state is present in all compounds found in yellowcake except for UO_2, which has an oxidation state of IV (green) and U_3O_8 with an oxidation state of V (colorless).

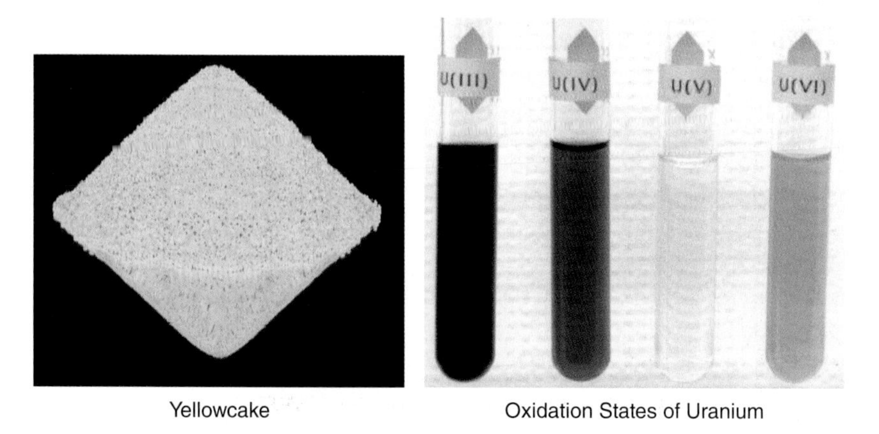

Figure 12.15 The concentrated yellowcake powder obtained from uranium ore mining operations compared to the colors of uranium oxidation states U(III) to U(VI). Source: left, Argonne National Laboratory; right, Los Alamos National Laboratory, Wikimedia Commons.

The other two methods of uranium mining are heap leaching and in-situ leaching, which involve chemical leaching of the uranium from the ore deposits at the mine site. Heap leaching is used on uranium oxide ores located at or near the surface. It was used more often in the past when surface uranium oxide ore deposits were more plentiful. It currently accounts for about 2% of uranium mining. In heap leaching, the uranium is removed from the ore by building a pile of uranium ore on top of a thick plastic liner supported by sand or clay. The pile is sprayed with dilute H_2SO_4, for 30–90 days. The runoff from the pile containing the uranium oxides is collected and processed in a uranium recovery plant to produce the yellowcake. During heap leaching, the piles present a hazard due to release of dust, formation of Radon gas, and leaching of liquids into surface waters. In-situ leaching is similar to natural gas fracking and accounts for about 45% of uranium mining. An oxidant, typically H_2O_2, is injected into the ore body directly to form the uranium oxides, followed by injection of H_2SO_4 solution to solubilize and extract the uranium oxides. The solution is then pumped to the surface and processed. This method requires a fairly porous ore body to be successful. It has the advantages of no mine tailings or surface disruption of the environment, which are involved in all other mining operations. The disadvantage is the possibility of direct contamination of unconfined groundwater aquifers.

After the uranium ore has been concentrated into yellowcake it is transported to two types of processing plants, which make three types of nuclear fuel. The type of fuel required depends on the type of fission process used in the nuclear plant. Some nuclear power plants do not require enrichment of the ^{235}U because they use a moderator that allows operation at the natural abundance. In this case, the yellowcake is formed into the appropriate shape for the fuel rods without enrichment. For many of the thermal nuclear reactor designs, the fuel has to be enriched in the ^{235}U from the natural abundance of 0.73% to about 3%. This is accomplished by converting all of the uranium ore to UF_6. The UF_6 is a gas at 57 °C and can be enriched by either using gaseous diffusion or centrifugation. After enrichment, the $UF_6(g)$ is converted back to $UO_2(s)$ and processed into a pelleted form using a sintering furnace. The solid pellets are initially formed into cylinders and then machined into uniform sizes for stacking into sealed non-corrosive metal alloy tubes.

12.6.3 Nuclear Plant Designs

There are many different successfully operated nuclear power plant designs based on the two main types of nuclear reactors: thermal neutron reactors and fast neutron reactors. Most fast neutron reactors use liquid metals, typically molten sodium, as coolant due to their low neutron cross-sections. Since sodium has safety issues, some designs have replaced the molten sodium coolant with helium gas coolant. Fast neutron reactors do not use a neutron moderator and require that the fuel be enriched to 20% or greater. A number of fast neutron reactors use a uranium blanket to capture the neutrons from the reactor core for production of ^{239}Pu, which is then used as a mixed oxide fuel (MOX). Since the fast neutron reactors are lighter and smaller than the thermal neutron reactors, they are typically used as power sources for military ships and submarines. The United States chose not to pursue fast breeder reactors for electricity generation in 1977 because the enriched ^{239}Pu from the reactors could be used in nuclear weapons. Also, further development of fast neutron reactor designs has not continued in any significant manner worldwide because of the high cost of fuel enrichment. France, however, has worked to obtain a closed nuclear fuel cycle program by using fuel reprocessing plants to produce enriched MOX nuclear fuels for the fast neutron reactors. This type of fast breeder reactor only accounts for about 0.5% of the current reactors in the world.

Thermal neutron reactor designs typically differ in the type of neutron moderator and coolant used in the reactor. Neutron moderators absorb kinetic energy from the neutrons so that the thermal neutron capture by the ^{235}U proceeds at a higher rate. Very pure graphite was originally used as the neutron moderator in the first thermal neutron reactor. Although many graphite reactors were built for use in nuclear power plants, most of them have been decommissioned due to the potential for fire accidents such as the one that occurred at Chernobyl. There are still a few in use today in Russia, with improved safety procedures added. Other moderators that have been used include water (both H_2O and D_2O), light molten salts (LiF and BeF_2), as well as organics in the form of biphenyl or terphenyl. Thermal nuclear reactors are often classified according to the coolant used. These include boiling water reactors (BWR), pressurized water reactors (PWRs), pressurized heavy water reactors (PHWR), molten salt reactors (MSR), and gas-cooled reactors (GCR). The PWR and BWR designs use H_2O as both coolant and neutron moderator. The PHWR design uses D_2O, also called heavy water, as both coolant and neutron moderator. The gases normally used as coolants in the GCR include He, CO_2, and N_2. The only graphite-moderated reactors still in operation use H_2O as the coolant. They are called light water graphite reactors (LWGR).

The different types of thermal neutron power plants operating in the world as of 2014 are shown in Figure 12.16 (IAEA, 2015). The most widely used thermal neutron power plant design is the PWR at 63% of the total number of reactors operating worldwide. The least widely used are the LWGRs and the GCRs. All the water-cooled reactor designs (PWR, BWR, and PHWR) operate on the same basic principle. The primary water coolant is pumped to the reactor core, where it is heated by the energy released by the fission reaction. The heated water then flows to a heat exchanger, known as a steam generator, where it transfers the thermal energy to a secondary water-cooling system which generates steam. The primary water coolant does not come in contact with the secondary system to prevent any possible transfer of radioactivity. The steam produced in the steam generator then flows to steam turbines which operate electric generators to generate the electricity. In contrast to a BWR, the pressure in the primary coolant loop of both the PWR and the PHWR prevents the water from boiling in the reactor.

A nuclear power plant design shown in Figure 12.17 was developed by Canadian nuclear scientists. This reactor, called the Canadian Deuterium Uranium reactor (CANDU), is a variant of the PHWR design using D_2O as both coolant and moderator. Since the deuterium already contains an extra neutron, it has a lower neutron cross-section than H_2O. This gives the reactor a better neutron economy than those that use H_2O coolant, and allows it to be operated using the natural abundance of ^{235}U (0.7%), thus eliminating the cost of fuel enrichment. The main issues with nuclear reactors are the need to store the nuclear waste once the fuel is depleted of ^{235}U. This requires the waste to be transported and placed into a repository for long-term storage. Since the CANDU system can operate on unenriched ^{235}U, it can use depleted uranium fuel rods or MOX fuels. This could extend the nuclear fuel output as well as reduce the potential

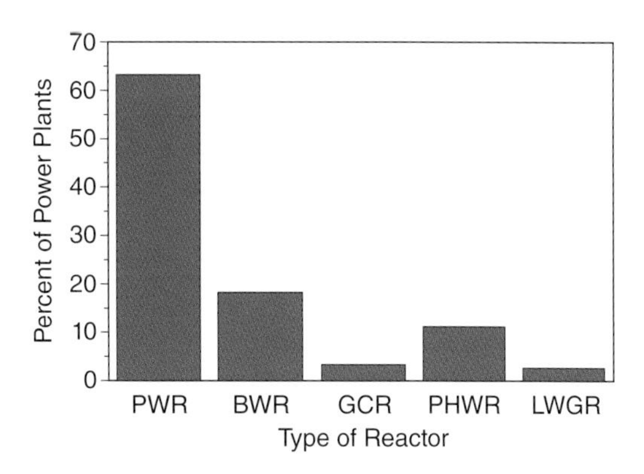

Figure 12.16 The percentage of the thermal neutron nuclear power plants by type operating in the world as of 2014.

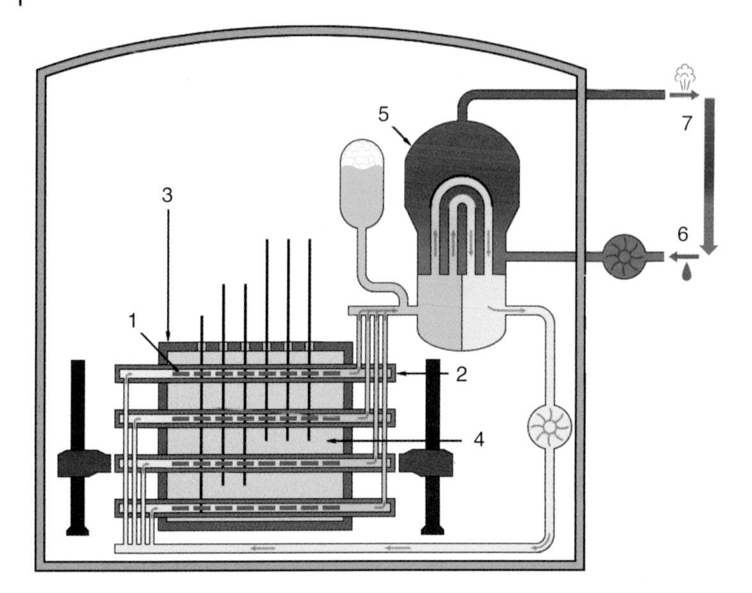

Figure 12.17 The CANDU nuclear reactor design showing the uranium fuel rods (1), fuel rod tubes (2), the large calandria (3), containing the D_2O moderator (4). The D_2O heated by the fission reaction enters a steam generator (5), where it heats water from a secondary cooling loop (6) to generate steam (7), which powers a steam generator. Source: Adapted from Inductiveload, Wikimedia Commons.

proliferation of ^{239}Pu by using it as fuel. The CANDU reactor design can also be used to "burn" nuclear waste actinides, reducing the need for waste storage. The process of nuclear burning is the introduction of a series of nuclear reactions, including neutron capture, fission, and other decay processes, which convert radioactive species to stable elements.

While most nuclear reactor designs place the fuel rods, moderator, and coolant in a very large pressurized container, the CANDU system uses much smaller fuel rods (about 10 cm diameter) placed in individual tubes. Each of the small fuel tubes is placed into a second tube filled with CO_2, which acts to insulate the tubes from each other. The fuel tubes are placed into an unpressurized container, called a calandria, which contains D_2O. The fission reaction heats pressurized D_2O in a primary cooling loop, shown in Figure 12.17 in yellow. A steam generator transfers the heat to a secondary cooling loop, shown in Figure 12.17 in blue and red. The steam powers a steam turbine which operates an electric generator. The exhaust steam from the turbines is then cooled, condensed, and returned as feed water to the steam generator.

Another useful aspect of the CANDU reactor design is that the control rods can be replaced over time with materials that form neutron capture products useful in other applications, such as nuclear medicine. For example, a few of the stainless-steel control rods in the CANDU reactor can be replaced with rods made of cobalt. The ^{59}Co undergoes neutron capture to form ^{60}Co ($t_{1/2}$ = 5.3 years). The ^{60}Co undergoes β^- decay accompanied by the emission of two γ-rays with energies of 1.17 and 1.33 MeV. Once irradiated, the cobalt rods are removed from the reactor and placed into shielded containers for use as γ-ray sources. Gamma rays are used in sterilization of medical equipment, in cancer therapy, and in food irradiation to kill any pathogens. High-energy γ-rays are also used in industrial radiography to determine flaws in materials.

12.6.4 Nuclear Waste

Since most of the nuclear reactors use PWR or BWR designs that require enriched uranium for fuel, the fuel rods must be replaced periodically once the fissionable material is depleted.

Figure 12.18 Dry cask storage vaults for above-ground spent fuel rod storage in the United States. Source: Nuclear Regulatory Commission, Wikimedia Commons.

The fuel rods are typically used for a period of about 6 years before they are considered to be spent fuel that can no longer support the nuclear chain reaction. The spent fuel rods, containing both short-lived and long-lived radioisotopes, are placed onto special racks and submersed into the bottom of a 12-m deep water pool known as a spent fuel pool. The water acts to cool the fuel rods as well as a radiation barrier. The rods are kept in the pools for 10–20 years to allow the short-lived radioisotopes to decay. They are then either sent to a reprocessing plant to separate the longer-lived fissile isotopes, or dried and placed into a long-term storage facility. Since no permanent long-term storage facility has currently been identified in the United States for spent fuel rods, dry cask storage is being used as an interim storage solution. In dry cask storage the spent fuel rods are placed into steel containers where they are surrounded by an inert gas. The steel containers, or casks, are sealed and placed into concrete, steel, or other material to provide additional radiation shielding. The dry casks are then placed in long term storage in concrete vaults such as those shown in Figure 12.18.

Other methods of long-term storage include underground storage in a waste facility or deep bore-hole injection. In addition, a number of nations actively disposed of nuclear radioactive waste in ocean sites from 1954 to 1993. Thankfully, this means of disposal has currently been banned by international agreement. Underground storage sites have been designed in the United States for the long-term storage of spent nuclear waste in Nevada and New Mexico. The New Mexico site is called the "Waste Isolation Pilot Plant" and is currently storing transuranic wastes in a geological salt bed at a depth of 655 m. The Nevada site is the "Yucca Mountain Nuclear Waste Repository," which was designed to handle 70 000 tons of high-level waste including spent nuclear plant fuel as well as nuclear wastes from other activities such as nuclear medicine. This proposed site is currently not being pursued due to state and regional opposition. Other countries have considered similar underground storage facilities. Finland is close to operating the first large spent nuclear fuel repository at a nuclear reactor site called Onkalo. This facility will handle all of the Finnish nuclear waste by placing it into a series of spiral tunnels that are over 500 m deep.

Since the amount of energy released in nuclear fission is about 10^8 times that obtained from combustion, the amount of spent fuel rods from a nuclear power plant is actually fairly small compared to the amount of waste from fossil fuel-powered power plants. For example, the Yankee Rowe nuclear power plant, located in the state of Massachusetts, was operated from 1960 to 1992 and generated 44 billion kilowatt hours of electricity. The spent fuel from its 32 years of operation is contained in 16 dry cask vaults similar to those shown in Figure 12.18. For comparison, a coal-fired power plant would need to combust 20 million tons of coal to produce the same amount of energy, with the associated release of CO_2, air pollutants, and fly ash. It would also have to store tons of fly ash and bottom ash containing metals and natural radioisotopes in storage ponds, as discussed in Section 10.6.1.

Considering that there are over 450 nuclear power plants in operation with an additional 150 that have been decommissioned, the overall safety record of nuclear power plants is actually quite good. There have been a few notable nuclear accidents over the years. These include the accident at the Three Mile Island Nuclear Power plant in Pennsylvania, the Chernobyl Reactor accident in the Ukraine, and the Fukushima Daiichi Nuclear accident in Japan. The Three Mile Island Nuclear Reactor number 2 (TMI-2) was a PWR design. It suffered a coolant handling problem on March 28, 1979, which led to a partial meltdown of the reactor. The problem began with a check valve in the secondary coolant system that was stuck open while indicator lights on the reactor control panel did not adequately alert operators to the situation. In addition, valves to auxiliary water pumps had been shut down for routine maintenance of the secondary cooling system, in violation of NRC rules. A cascade of effects happened quickly, resulting in the release of coolant from the reactor into the containment building and another building next to TMI-2. There was a small release of radioactive gases into the environment from the accident, which was primarily in the form of the radioactive noble gas ^{135}Xe. The reactor itself was closed off and isolated, and the cleanup was concluded in December 1993. This accident led to many improvements in the operating safety procedures but also led to a slowdown of the growth of the nuclear power industry in the United States. There was no evidence that this accident caused any significant radiation exposure to the local region.

The most significant nuclear accident to date occurred in Chernobyl in the Ukraine on April 26, 1986. The incident began during a systems test of the LWGR reactor when there was a sudden power surge. The power spike led to rupture of the reactor vessel and a series of steam explosions, which exposed the graphite moderator to air, causing it to ignite. The resulting fire sent long plumes of highly radioactive material into the atmosphere over an extensive geographical area. The plumes drifted over large parts of the western Soviet Union and Europe, with about 60% of the fallout landing in Belarus. The accident required the evacuation of 300 000 residents in the area and resulted in 30 deaths. This accident brought attention to the design flaws and potential fire hazard associated with the LWGR design used by the Soviet Union. This accident also led to revisions in reactor designs and improvements in nuclear reactor safety.

The third major nuclear accident happened at the nuclear power plant in Fukushima, Japan on March 11, 2011. A 9.0 undersea earthquake occurred off the coast of Tōhoku, with the epicenter approximately 43 miles east of the Oshika Peninsula at an underwater depth of approximately 18 miles. Often referred to in Japan as the Great East Japan Earthquake, it was the most powerful earthquake ever recorded in Japan. The earthquake was accompanied by a 15-m high tsunami which flowed over the 6 miles of protective sea wall and flooded the Fukushima BWR nuclear power plant. Sea water flooded the basements and disabled back-up power generators used for coolant pumping for the six nuclear reactors at the site. Back-up generators located at a higher location were also flooded and failed. Batteries were used for 8-hour emergency operation while attempting to bring other back-up power to the plant. However, the earthquake and tsunami had caused sufficient damage to roads and the surrounding area that delivery of back-up power was hopelessly delayed. On March 12 the water pumps stopped and the reactors began to overheat. A zirconium alloy, called zircaloy, which was used in the reactor fuel assembly, reacted exothermally with water to form hydrogen gas. This caused two hydrogen gas explosions on March 12 and March 14 in two of the nuclear reactors, severely damaging the reactor buildings. Damage to one of these nuclear reactor buildings led to another explosion in a third reactor through a connected vent. The reactor containment vessels were vented in order to reduce gas pressure and prevent the discharge of coolant water into the sea. This released radioactive isotopes into the environment. Most of the radioactive releases were confined to the local area, but traces of radioactivity from the radioactive gases were observed globally.

Since the accident there have been no deaths related to short-term exposure of radioactivity. However, a 2013 WHO report predicts that, for populations living in the most affected areas and exposed as infants, there is a 70% higher risk of developing thyroid cancer for girls, a 7% higher risk of leukemia in males, and a 6% higher risk of breast cancer in females (WHO, 2013). Preliminary estimates of radiation doses outside the most affected areas indicate that the risks are low and no observable increases in cancer above natural variation are anticipated. For comparison, in the Chernobyl accident only 0.1% of the 110 000 cleanup workers surveyed have so far developed leukemia and not all cases can be directly attributed to the accident. However, 167 Fukushima plant workers received radiation doses that slightly elevated their risk of developing cancer.

Follow-up examination of the accident in Fukushima concluded that locating the nuclear power plant close to the ocean was a major factor in the seawater flooding, which was the primary cause of the accident. The choice of location for a nuclear power plant in areas that are known to be seismically active needs to be considered carefully. Nuclear power plants and storage facilities need to prepare for worst-case scenarios or use other energy options in these seismically active regions to maintain safety. Despite these three accidents, nuclear electric power generation still has the lowest industrial risk factors when the entire lifecycle is taken into account. For example, it has been estimated that coal mining, processing, and combustion is more hazardous than nuclear power generation by a factor of one million on a per petawatt hour basis. All nuclear power plants are required to operate and collect air samples just outside the plant to ensure that no radioactivity is released during normal operation. The primary radioactive elements found in these samples are the naturally produced ^{210}Pb and ^{7}Be, which are attached to atmospheric aerosols.

12.7 Radioisotopes in the Environment

All isotopes of an element behave chemically in the same way. Thus, radioisotopes will undergo the same chemical reactions in the environment as the stable isotopes of the same element. This includes oxidation and reduction reactions in air and water as well as organic complexation reactions in aqueous media and in wet aerosols. In fact, radioisotopes are often used as a means of determining the mechanism of a chemical reaction of an element by following the easily measured radioactivity from a radioisotope of the same element. The chemistry, as well as physical properties such as solubility, of a radioisotope are defined by the known chemistry of the stable isotope of an element and its compounds. This same chemistry can be used for predicting the chemical transport and transformation of the radioisotope in the atmosphere, hydrosphere, biosphere, and geosphere. When considering the amounts of radioactivity in the environment, the natural sources of radioactivity are by far the most significant, with anthropogenically produced radioisotopes being a very small component. Thus, most of the ionizing radiation to which we are exposed is due to primordial and cosmogenic radioisotopes in our

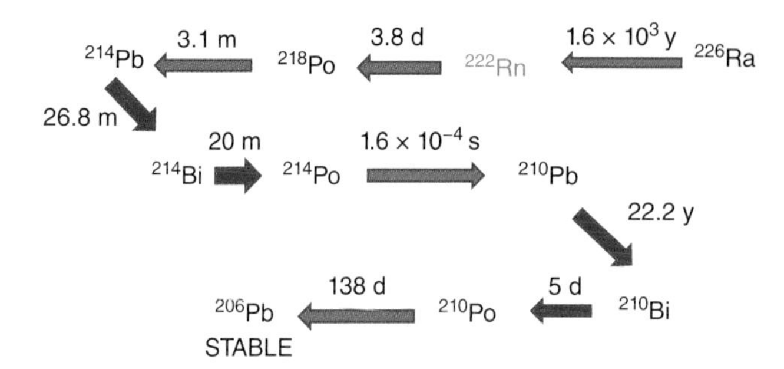

Figure 12.19 The decay series of ^{226}Ra producing ^{222}Rn gas (green) and its solid daughters by α decay (blue) and β^- decay (red).

environment. The most common primordial radioisotopes are ^{235}U, ^{238}U, ^{232}Th, ^{226}Ra, ^{222}Rn, and ^{40}K and the most common cosmogenic radioisotopes are ^{14}C, ^{3}H, and ^{7}Be.

A significant amount of the radioisotopes in the environment are produced from the ^{238}U decay series shown in Table 12.1. Of particular importance is the naturally occurring ^{226}Ra, which is produced from the decay of ^{238}U found primarily in rocks, ground water, and surface waters. The decay of ^{226}Ra, shown in Figure 12.19, forms the gas-phase radioisotope ^{222}Rn. The ^{222}Rn has 3.8 days to transport through the rocks, ground water, or surface waters where it is formed to reach the atmosphere before it decays to ^{218}Po. Upon decay, it forms radioisotopes that will diffuse to existing surfaces and then decay rapidly to ^{210}Pb ($t_{1/2}$ = 22.2 years). Some of the ^{222}Rn decays before it can reach the surface, resulting in the ^{210}Pb being deposited in the soil and upper sediments. However, most of the ^{222}Rn reaches the lower atmosphere where the ^{210}Pb is attached to atmospheric aerosols, as discussed in Section 6.3. Wet and dry deposition of the aerosols results in an increase of the ^{210}Pb levels found on soil surfaces. The ^{210}Pb exists as a lead oxide, which is not soluble in water. Thus, after deposition it remains in the soil and reaches secular equilibrium with the energetic α emitter ^{210}Po. A radioisotope reaches secular equilibrium when its production rate is equal to its decay rate. A number of plants, in particular tobacco, are known to uptake the ^{210}Po and its precursors from the soil and the air. Smoking tobacco concentrates these natural radioisotopes onto particles, which are inhaled. It is estimated that smokers inhale about 1.5×10^{-3} Bq of ^{210}Po per cigarette smoked. The exposure to ^{210}Po α emission in tobacco smoke has been proposed as a major mechanism for the high rate of cancer in the lungs, nasopharynx, and bronchia of smokers.

The ^{238}U, which produces ^{222}Rn and its daughters, is not uniformly distributed in the terrestrial environment. There are many regions that have higher concentrations of uranium in soil and rocks. In these areas, ^{222}Rn gas can move up from the ground into buildings through openings in floors or walls, as shown in Figure 12.20. In this manner, it can accumulate in buildings that have low outside air exchange rates. Exposure to ^{222}Rn gas in homes is the second leading cause of lung cancer after smoking. Because of this, the U.S. EPA has set standards for ^{222}Rn concentrations in buildings and in homes of 0.15 Bq l^{-1} of air. The air exchange rate for a building is a controlling factor in indoor air quality. The lower the air exchange rate, the higher the concentrations of indoor air pollutants such as ^{222}Rn gas. However, it is also a controlling factor for energy consumption. The higher the air exchange rate, the more energy is required to maintain a stable indoor temperature. The air exchange rate of a building typically depends on weather conditions, type of mechanical ventilation, surrounding terrain, and building characteristics such as tightness of the building. The national average for Radon in outdoor air is 0.015 Bq l^{-1}, while the national average for indoor air is 0.05 Bq l^{-1}.

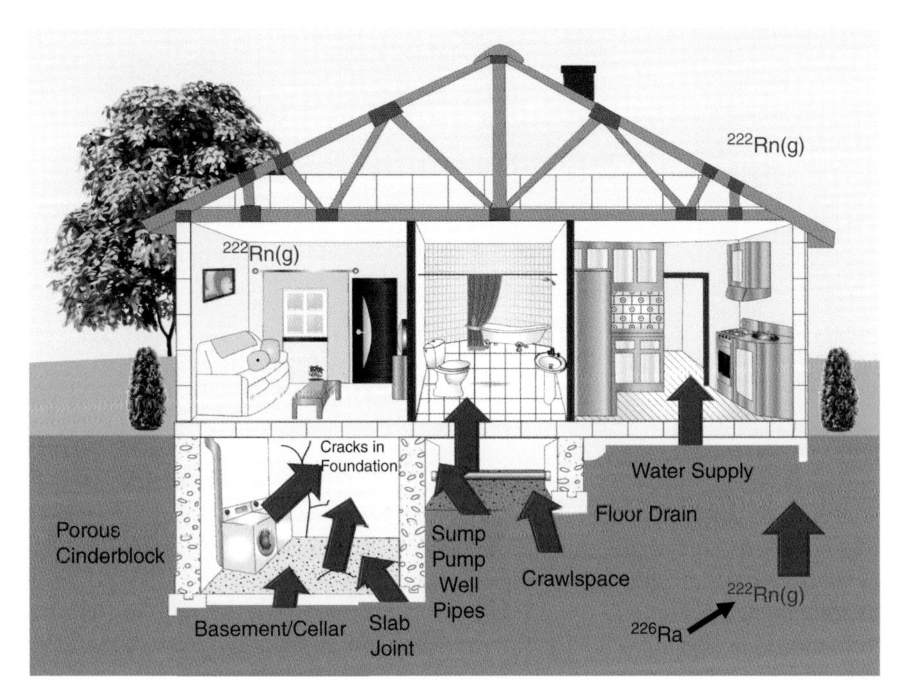

Figure 12.20 The sources of ^{222}Rn from ^{226}Ra decay in soils surrounding a home. Source: Adapted from U.S. EPA.

In the United States, it is commonly required that buildings or homes placed on the market for sale be tested for indoor levels of ^{222}Rn throughout the building. If the levels are found to be above the 0.15 Bq l^{-1} standard, building remediation is required to lower the levels before the real-estate transaction can be completed. This is usually accomplished by simply installing outside air exchange circulation fans in the areas where the ^{222}Rn levels are in exceedance of the standard. The measurement of ^{222}Rn levels in a home is actually quite simple. Since the nonpolar ^{222}Rn gas is soluble in nonpolar solvents, the sample is collected by allowing the ^{222}Rn to diffuse into a heavy oil held in a can that is opened and exposed to the air. After a fixed amount of time, the can containing the oil is sealed and the ^{222}Rn concentration in the oil is determined by α counting of the ^{222}Rn decay, as discussed in Section 9.2.5.

The levels of ^{222}Rn and its daughters can also build up in the air inside underground mining operations, especially uranium mines. This has caused miners to suffer high inhalation exposures of ^{222}Rn and its daughters that are attached to aerosol particles. In some areas, where thorium concentrations are high, exposure to ^{220}Rn can also be a concern. However, ^{220}Rn has a fairly short half-life of 55.6 s, which limits the distance it can diffuse through the rock into the air. Most older underground mining operations had low air exchange rates and the levels of ^{222}Rn in many mines, particularly coal mines, reached unacceptably high levels. The high exposure levels were linked to lung cancers in miners and led to safety regulations requiring that the air in mine shafts be exchanged with fresh air on a fairly short time scale. This is accomplished by additional air shafts and the use of large circulation fans to exchange mine air with surface air.

Many coals are found to be naturally enriched in ^{238}U and its daughters. Marie and Pierre Curie first isolated the radioactive elements radium and polonium from peat, which is a precursor to coal, because it was enriched in natural radioactivity. This enrichment is due to the complexation of the ^{238}U by natural humic and fulvic acids and peat tannins. After the coal is

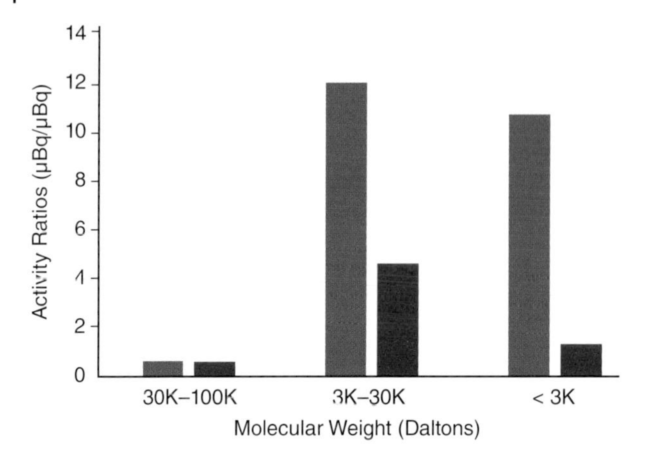

Figure 12.21 The ^{228}Th/^{232}Th (blue) and ^{230}Th/^{232}Th (red) activity ratios associated with natural humic materials in the size ranges of 30 000–100 000 Da, 3000–30 000 Da, and <3000 Da from Volo Bog in northern Illinois.

combusted, the remaining fly ash typically contains a factor of 100 higher levels of ^{238}U and ^{234}Th than nuclear power plant waste per kilowatt hour of electricity produced. Most of this natural radioactivity is found in the fly ash and bottom ash that is stored in ponds near the coal-fired power plants. This can be a major source of radioisotopes in surface waters when the water and sludge in the ponds is released, as discussed in Section 8.6.1 (Figure 8.20).

The chemical binding of radioisotopes to humic and fulvic acids is very strong under natural conditions in fresh, salt, and ground waters. These natural large organic polyelectrolytes are important in the aqueous transport of radioisotopes as well as toxic metals released from mining activities. The humic and fulvic acids have been found to have especially strong binding capabilities with the naturally occurring thorium isotopes (Gaffney et al., 1992). This is evident in measurements of thorium isotopes associated with natural organics in a peat bog. The natural organics in the bog water were separated into molecular size ranges of 100 000–30 000 Da, 30 000–3000 Da, and <3000 Da by ultrafiltration techniques as discussed in Section 9.3. The relative amounts of the ^{228}Th/^{232}Th and ^{230}Th/^{232}Th activity ratios in the different molecular size ranges are shown in Figure 12.21. The smaller size ranges contain more thorium radioisotopes than the larger size range. In addition, the ^{228}Th and ^{232}Th isotopes are not equally distributed in the two smaller size ranges. Although the chemistry of the two isotopes is the same, their sources are different. The ^{230}Th is produced from the decay of ^{234}U in the ^{238}U decay series, while the ^{228}Th is produced from the decay of ^{228}Ac in the ^{232}Th decay series. Since the chemistry of the parent radioisotopes is different, it is likely that the organic complexation occurred somewhere in the decay series before the thorium isotopes were formed.

The three thorium radioisotopes have lifetimes of 1.9 years for ^{228}Th, 7.5×10^4 years for ^{230}Th, and 1.4×10^{10} years for ^{232}Th. If there was a rapid exchange of thorium between the binding sites of the natural organics, the activity ratios would be the same for all size ranges. Since bogs have no inflow or outflow, the bog water is recharged only by local precipitation and the only source of natural organics is the bog peat tannins, as discussed in Section 8.5.4. The very different thorium activities observed for the three size ranges show that the smaller sizes bind the thorium more effectively and the thorium isotopes are not being exchanged between the binding sites of the organics in the smaller size ranges. This same binding profile has also been observed with uranium isotopes. The fact that humic substances can strongly complex radioisotopes has very important consequences with regard to their transport in soils and natural waters. These stable

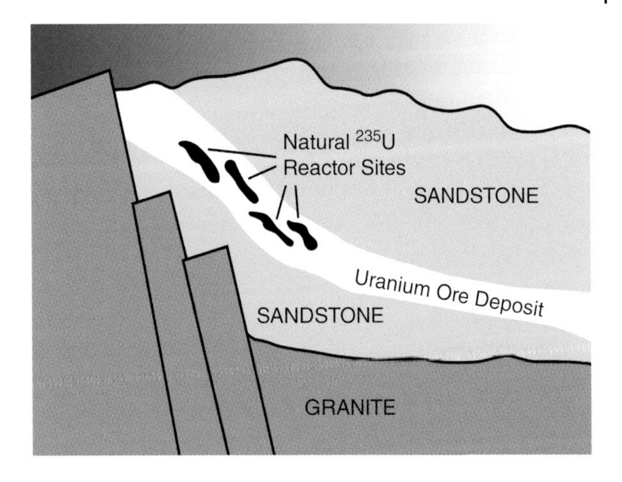

Figure 12.22 The geological formation of the Oklo natural ^{235}U fission reactor located in Gabon, Africa. Source: Adapted from MesserWoland, Wikimedia Commons.

complexes, once formed, can lead to enhanced solubilities and transport of radioisotopes as colloids that would not be predicted by the simple inorganic chemistry of the radioisotopes.

Before 1945 it was believed that there were no natural sources of the transuranic elements present in the environment and that they were all due to anthropogenic activities involving either nuclear weapons testing or the nuclear fuel cycle. In 1972, a naturally occurring uranium nuclear reactor was discovered in Oklo, Gabon Africa by Francis Perrin. The site, shown in Figure 12.22, was a large uranium deposit that was being mined for UO_2. Analysis of the uranium found that the ^{235}U levels were depleted from the natural 0.73% to about 0.6%. Further investigation indicated that other ore deposits nearby were depleted to as much as 0.4%. Examination of neodymium isotopes found that the ^{142}Nd was depleted from the natural 27% to 6%, accompanied by enhanced levels of ^{143}Nd. The ruthenium isotope ^{99}Ru was also enriched from 12.7% to as much as 30%. This excess ^{99}Ru could only have been produced from the β^- decay of ^{99}Tc. Both the ^{143}Nd and ^{99}Ru abundances could only have been produced from the natural nuclear fission of ^{235}U. Careful chemical isotopic characterization identified a number of ore bodies that had undergone ^{235}U fission about 1.7 billion years ago at the Oklo site.

The ^{235}U natural abundance at that time would have been 3.1%. This higher level of ^{235}U in the deposit allowed fission to occur when ground water, acting as a neutron moderator, moved into the ore bodies through the porous sandstone. After a period of time the heat released from the fission reaction would have converted the water to steam. The steam would have moved out of the ore body causing loss of the neutron moderator, which would cause the fission process to shut down and allow the reactor sites to cool. After cooling, the ground water could return and begin the cycle again. The ^{239}Pu would have been formed in the fission process as shown in Figure 12.9. However, since ^{239}Pu has a half-life of 2.41×10^4 years, it has all decayed away after the Oklo natural reactors ceased to operate when the ^{235}U levels were depleted. Only the longer-lived ^{143}Nd and ^{99}Ru isotopes remained behind to tell the story.

12.8 Radiation Exposure

While the activity of a radioisotope is given in Bq s^{-1}, the amount of radiation present is measured in roentgens (R), named in honor of William Röntgen, discover of the X-ray. A roentgen is the amount of ionizing radiation producing one electrostatic unit of positive or negative ionic charge in one cubic centimeter of air under standard conditions. It is equal to 2.59×10^{-4} coulombs per kilogram (C kg^{-1}). While the roentgen gives the amount of ionizing radiation in air and is a measure of radiation exposure, the important quantity is how much of the radiation is absorbed by a material. This is called the radiation dose or absorbed dose, which is the amount of energy absorbed by a medium per unit mass as a result of exposure to ionizing radiation. The rad was the original unit for a radiation dose, which is equal to 100 erg g^{-1}. The modern unit of absorbed radiation dose is the gray (Gy), which was named in honor of Louis Gray, who was one of the first to study the effects of different types of ionizing radiation on human tissue. The Gy is equal to an absorbed dose of 1 J kg^{-1}, which is equal to 100 rad.

The Gy does not necessarily account for the damage potential of the different types of radiation. An equivalent dose is a radiation dose that represents the health effects of low levels of ionizing radiation on the human body. It is derived from the absorbed dose but also takes into account the biological effectiveness of the radiation, which is dependent on the type of radiation. An equivalent dose is measured in Sieverts (Sv), named after Rolf Sievert, who studied the biological effects of different types of radiation exposure. The Sv is equal to the mean absorbed dose in Gy multiplied by a weighting factor which is dependent on the type of ionizing radiation and its energy. Another commonly used unit for the equivalent dose for humans is the rem, which is short for "roentgen equivalent man." The rem uses the dose in rad units, instead of grays, multiplied by a scaling factor. Since the Gy is equal to 100 rad, the rem is equal to 1×10^{-2} Sv.

There are three ways to minimize exposure to any radioactive source. These are limiting exposure time, reducing the distance from the source, and use of proper shielding materials. Limiting exposure time to the radiation directly lowers the dose. Since radiation is given in units of Bq s^{-1}, less exposure time (s) is equivalent to less radiation. Decreasing the distance from the source of the radiation also lowers the effective radiation exposure. The intensity of the ionizing radiation (I_r) from any radioactive source is inversely proportional to the square of the distance (d) from the source:

$$I_r = \frac{1}{d^2}$$

This is due to the fact that the ionizing radiation is emitted isotropically, radiating with the same intensity in all directions from the source. In the nuclear power industry, as well as in medical and research facilities, increasing the distance between the radioactive source and the exposed individual is often accomplished by making use of robotic arms, which allow the material to be handled from a distance.

The use of shielding to limit radiation exposure depends on the type of radiation. Shielding of α and β^- particles is accomplished by halting their forward trajectory, while shielding of γ-rays requires materials that can absorb the high-energy electromagnetic radiation. Since low-energy α particles are easily stopped by as little as a thin sheet of paper, simple protective clothing acts as an effective shield when handling an α emitter. The smaller, more energetic β^- particles are a bit more difficult to stop and require shielding with simple materials such as plate glass or Plexiglas. Since γ-rays are essentially very energetic X-rays, shielding from γ sources is accomplished by using a dense heavy metal, such as lead or tungsten, in the form of metal sheets or bricks, which are typically placed inside a dense steel or concrete container. The lead and tungsten materials, along with the steel and concrete, have high atom densities so that interactions with the γ-rays are higher, resulting in their absorption. In nuclear power plants and processing facilities, the buildings and work areas are usually made from heavy steel and concrete along with lead liners. This prevents the γ-rays from penetrating the building walls.

Nuclear physicists, chemists, and engineers working with or around nuclear materials wear film radiation badges to routinely monitor their exposure to ionizing radiation and ensure that exposure is limited. In the United States, adult occupational exposures to ionizing radiation are monitored and limited to 10 mSv per year of age, with any annual exposure not to exceed 50 mSv. Workers must maintain strict safety procedures to ensure that they stay below the exposure limits. The guiding principle for radiation safety for radiation workers is the ALARA principle, which stands for "as low as reasonably achievable." It means making every reasonable effort to maintain exposures to ionizing radiation as far below the dose limits as practical. The ALARA principle is defined by the U.S. Nuclear Regulatory Commission as:

> ... making every reasonable effort to maintain exposures to ionizing radiation as far below the dose limits as practical, consistent with the purpose for which the licensed activity is undertaken, taking into account the state of technology, the economics of improvements in relation to state of technology, the economics of improvements in relation to benefits to the public health and safety, and other societal and socioeconomic considerations, and in relation to utilization of nuclear energy and licensed materials in the public interest.
>
> (USNRC, 2018)

The current annual background exposure for the average U.S. citizen is approximately 6 mSv, which is 60% of the occupational limits. Natural exposures include γ-rays from the Sun, commonly referred to as cosmic rays. Populations living at high altitudes and aircraft crews will have higher exposures to these γ-rays than populations living near sea level. Airlines typically limit the number of working days for flight crews and pilots to make sure that they are under occupational radiation exposure limits. This type of radiation is a particular problem for astronauts and occupants of the International Space Station. Special care has to be taken to ensure that space vehicles and the Space Station have enough shielding to protect the individuals from the Sun's ionizing radiation.

The average annual radiation dose worldwide has been estimated to be about 3.0 mSv yr^{-1}. The major sources of the ionization radiation, shown in Figure 12.23, include ^{222}Rn inhalation, food and water, radiation from the ground, cosmic rays, medical X-rays and other procedures, nuclear testing, occupational exposure, and nuclear energy including fallout from the Chernobyl accident and releases from the nuclear power fuel cycle (UNSCEAR, 2010). The inhalation of ^{222}Rn and its daughters accounts for about one-third of the annual dose. In addition, the exposure due to all of the anthropogenic sources, with the exception of medical X-rays, is orders of magnitude lower in dose. Indeed, advances in medicine – including nuclear medicine

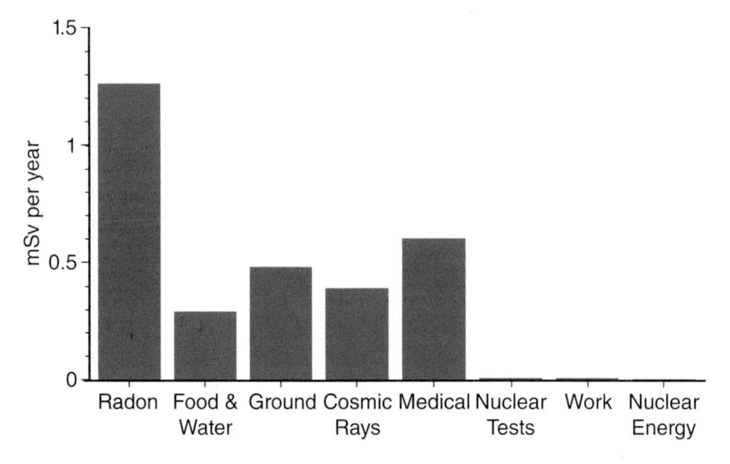

Figure 12.23 The world average annual ionizing radiation doses (mSv per year) from Radon inhalation, food and water, radiation from the ground, cosmic radiation, medical including X-rays, nuclear testing, occupational exposure, and nuclear energy including fallout from the Chernobyl accident and releases from the nuclear power fuel cycle.

and tomography – have increased the average U.S. citizen's exposure to about 6.0 mSv yr^{-1} from the global average of 3.0 mSv yr^{-1}.

The natural radiation being emitted from the ground varies by location across the globe, as would be expected since the concentrations of uranium and thorium in the soil, rocks, and water vary widely over the world. The natural radioactivity in food and water comes primarily from ^{14}C and ^{40}K radioisotopes. The average human contains about 20 mg of ^{40}K and 25 ng of ^{14}C, and both of these radioisotopes are constantly being renewed through consumption of food and water. The two radioisotopes each account for about 4×10^3 Bq of radioactivity in the human body. Although both radioisotopes are β^- emitters, the β^- from ^{40}K is about 10 times more energetic than the β^- emission from ^{14}C. Thus, the internal dose from ^{40}K is responsible for about 0.12 mSv yr^{-1}, while ^{14}C only about 0.012 mSv yr^{-1}. Uranium and thorium and their daughters account for an additional 0.17 mSv from food and water, with a total of about 0.3 mSv yr^{-1}. Some foods, such as bananas, are particularly enriched in potassium content, and therefore contain more ^{40}K. Due to its water solubility, ^{40}K is responsible for about 12 Bq l^{-1} of radioactivity in the oceans, which is 99% of the radioactivity in sea water.

12.9 Applications of Radioisotopes

Radioisotopes are commonly used in chemistry to determine reaction mechanisms. They can also be used to follow equilibrium reactions, such as the solubility of a solid in water. For example, the solubility of AgCl(*s*) in water to give $Ag^+(aq)$ and $Cl^-(aq)$ can be shown to be dynamic by using an AgCl compound that is labeled with a radioisotope of silver. A solid AgCl compound containing the radioisotope $^{111}Ag^+$ is added to a saturated solution of AgCl (1.3×10^{-5} M). This radioisotope has a 7.45-day half-life and undergoes β^- decay with the emission of a 342-keV γ-ray. After several hours have passed, the activity of $^{111}Ag^+$ in the solution is determined by β^- detection. Since the original solution is saturated, no direct dissolution of the $^{111}AgCl$ can occur. Thus, the presence of $^{111}Ag^+$ in solution can only occur by the dynamic equilibrium exchange of Ag^+ ions with the solution and the solid:

$$^{111}AgCl(s) + Ag^+(aq) \rightleftarrows {}^{111}Ag^+(aq) + AgCl(s)$$

Similarly, the equilibrium between $Fe^{2+}(aq)$ and $Fe^{3+}(aq)$ can be observed by using the radioisotope ^{55}Fe. The ^{55}Fe has a half-life of 2.7 years and undergoes EC to produce ^{55}Mn with the emission of a 5.9-keV X-ray, which can easily be detected. The $^{55}Fe^{3+}$ is added to a solution of Fe^{2+}. After a period of time, the Fe^{3+} and Fe^{2+} are separated from the solution by ion exchange and the amount of the radioisotope is determined in each oxidation state. The measurement of radioactivity as $^{55}Fe^{2+}$ demonstrates that the following oxidation exchange reaction has occurred:

$$^{55}Fe^{2+}(aq) + Fe^{3+}(aq) \rightleftarrows {}^{55}Fe^{3+}(aq) + Fe^{2+}(aq)$$

There are many other examples where a radioisotope has been used to show that a reaction occurs. In many cases where the concentration of the radioisotope is measured as a function of time, this method can also be used to quantitatively determine reaction rates and equilibrium constants very accurately.

Very small amounts of radioisotopes can easily be measured from the characteristic energies of the different types of radiation. In many cases the detection limits for the radioisotopes can approach the natural background levels. For this reason, radioisotope tracers can be used in biology and medicine without exposing workers or test subjects to unnatural levels of radioactivity. Imaging methods have been developed that use specific radioisotopes for the

$$^{238}U + n \rightarrow {}^{137}I + {}^{99}Y + 2n$$

$$^{99}Y \xrightarrow{1.5\,s} {}^{99}Zr + \beta^- \xrightarrow{2.1\,s} {}^{99}Nb \xrightarrow{15\,s} {}^{99}Mo + \beta^-$$

$$^{99}Mo \xrightarrow{66\,h} {}^{99m}Tc + \beta^- \xrightarrow{6\,h} {}^{99}Tc + \gamma \xrightarrow{2.1\times10^5\,y} {}^{99}Ru + \beta^-$$

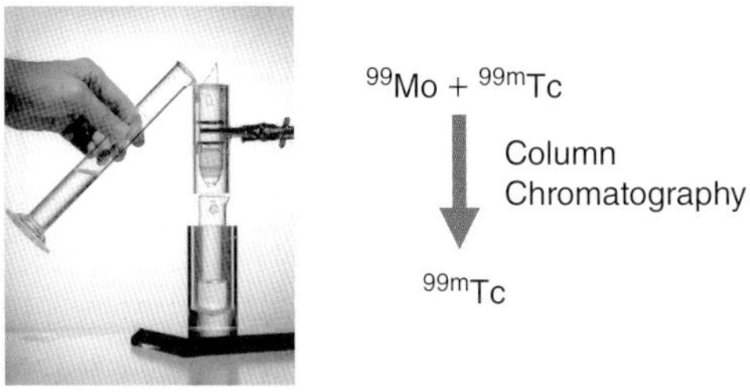

$^{99}Mo + {}^{99m}Tc$

Column Chromatography

^{99m}Tc

Figure 12.24 The ^{238}U decay chain, which produces the metastable ^{99m}Tc. Separation of ^{99m}Tc from the parent ^{99}Mo is accomplished rapidly by column chromatography (left). Source: Photograph by Brookhaven National Laboratory, Wikimedia Commons.

determination of organ functions and the identification of tumors. Cancerous tumors can be identified due to their increased uptake of the radioisotope. Radioisotopes are also used in treating cancer by releasing ionizing radiation into the cancerous tissue, causing cell death and preventing the cancerous cells from reproducing. However, once a radioisotope is used the material has to be treated as radioactive waste and disposed of in an appropriate manner.

One medical isotope that is commonly used in imaging techniques is technetium. This synthetic element was first discovered in early neutron bombardment of ^{238}U. It is produced from the ^{238}U fission product ^{99}Y as shown in Figure 12.24. The ^{99}Y undergoes a number of fast β^- decays to produce ^{99}Mo, the precursor of the metastable radioisotope ^{99m}Tc. The ^{99m}Tc has a 6-hour half-life and undergoes a nuclear isomerization to produce the ground-state species ^{99}Tc ($t_{1/2} = 2.1 \times 10^5$ years) with the emission of a 142.5-keV γ-ray. Chemical separation of the ^{99}Mo ($t_{1/2} = 66$ hours) from the ^{99m}Tc daughter is accomplished by simple column chromatography as shown in Figure 12.24 (left). After separation, the ^{99m}Tc is complexed with different binding agents to produce radiopharmaceuticals that are used for the imaging of specific organs. There are currently more than fifty ^{99m}Tc radiopharmaceuticals that allow imaging of the human brain, heart, lungs, liver, thyroid gland, kidneys, skeletal structure, gall bladder, tumors, and blood, which allows mapping of the circulation system. Today, this synthetic radioisotope is one of the most commonly used radiotracers in nuclear medicine.

The amount of a parent radioisotope to the amount of a stable isotope daughter can be used to determine the age of environmental materials by:

$$\Delta t = \lambda^{-1} \times \ln\left((N_P + N_D)/N_P\right)$$

where Δt is the age of the material, N_D is the number of atoms of the daughter isotope in a sample of the material, N_P is the number of atoms of the parent isotope in the sample, and λ is the radioisotope decay rate, which is described in Section 12.2. The common methods of radioisotope age dating are shown in Table 12.6. Uranium–thorium dating uses the amounts of the parent ^{234}U and daughter ^{230}Th ($t_{1/2} = 7.54 \times 10^4$ years) to determine the age of calcium carbonate materials such as coral, shells, or cave formations in the range of 1000–350 000 years. Potassium–argon dating uses the amounts of the parent ^{40}K and the stable daughter ^{40}Ar for the measurement of the age of volcanic rocks. The primordial radioisotope ^{40}K is present in molten lava and the decay product ^{40}Ar gas is able to escape as long as the rock is molten. However, it cannot escape after the rock solidifies. Thus, the ratio of the stable ^{40}Ar to the radioisotope parent ^{40}K can give the age of the rocks since they solidified. This method is most applicable

Table 12.6 Radioisotopes used to determine the age of dating environmental materials.

Radioisotope	Dating method	Half-life (yr)	Material
$^{234}U \rightarrow ^{230}Th$	uranium–thorium	2.5×10^5	calcium carbonate
$^{40}K \rightarrow ^{40}Ar$	potassium–argon	1.25×10^9	volcanic rocks
$^{147}Sm \rightarrow ^{143}Nd$	samarium–neodymium	1.06×10^{11}	meteorites
$^{235}U \rightarrow ^{207}Pb$	uranium–lead	7.038×10^8	zircon minerals
$^{238}U \rightarrow ^{206}Pb$	uranium–lead	4.47×10^9	zircon minerals
^{14}C	radiocarbon	5.73×10^3	organic materials
^{36}Cl	chlorine-36	3.01×10^5	ground water
^{210}Pb	lead-210	22.2	sediments, ice cores

for dating minerals and rocks more than 100 000 years old. The samarium–neodymium dating method uses the amounts of the parent ^{147}Sm to the daughter ^{143}Nd ($t_{1/2} = 1.06 \times 10^{11}$ years) to determine the age of rocks and meteorites with ages of about 2.5×10^9 years. The uranium–lead method of dating relies on two decay processes: ^{238}U to ^{206}Pb ($t_{1/2} = 4.47 \times 10^9$ years) and ^{235}U to ^{207}Pb ($t_{1/2} = 7.1 \times 10^8$ years). It is commonly used to date the mineral zircon ($ZrSiO_4$). Zircon can include U and Th in its crystal structure as it solidifies, but it cannot include Pb. Thus, the ^{206}Pb and ^{207}Pb content of a zircon mineral can only be from the radioactive decay of ^{238}U and ^{235}U after the formation of the mineral. Thus, the amounts of ^{238}U to ^{206}Pb or ^{235}U to ^{207}Pb can be used to determine the age of the mineral since it solidified.

Radiocarbon dating, also referred to as carbon dating, was discussed in Section 6.4 as a method of differentiating particulates emitted during the combustion of fossil fuel from the particulates emitted during the combustion of a biofuel. The ^{14}C ($t_{1/2}$ = 5700 years) decays by β^- emission to ^{14}N and the amount of ^{14}C in an organic material can determine the age of the material since the original organism died. The results are reported in fraction of modern carbon, which is defined as the ratio between the $^{14}C/^{12}C$ in the sample and the $^{14}C/^{12}C$ in a sample of modern carbon. Radiocarbon dating is accurate for ages from about 600 to 700 years up to about 57 000 years. The cosmogenic radioisotope ^{36}Cl ($t_{1/2} = 3.01 \times 10^5$ years) decays by β^- emission to ^{36}Ar. It is primarily used for dating old ground waters. After wet and dry deposition from the atmosphere, the ^{36}Cl is carried into ground water through recharge processes. It is not affected by geochemical reactions and so travels with the ground water until it reaches a confined aquifer. The age of isolated groundwater aquifers can be determined by comparing the concentrations of ^{36}Cl in the aquifer to that in normal recharge water. Chlorine-36 dating is generally applicable to water in the age range of 1×10^5 to 1×10^6 years. Measurements of ^{210}Pb ($t_{1/2}$ = 22.2 years) in sediment cores can be used to determine the sedimentation rates in lakes, oceans, and other water bodies and from this, the age of sediment from a particular depth in the sediment column can be estimated. The ^{210}Pb is produced in the ^{226}Ra decay chain shown in Figure 12.19. The ^{210}Pb, which is attached to atmospheric aerosols, is deposited into surface waters by wet and dry deposition and eventually becomes attached to sediment particles. The measurements of ^{210}Pb as a function of depth in the sediment have been used to determine the age of the sediment and the rate of sedimentation. This method can also be used in the same manner to confirm ice core dates.

A practical application of radioactivity is food irradiation. Food irradiation with γ-rays is a cold pasteurization method that is used as a means of preserving foods without the need to add chemicals. It was found at the Massachusetts Institute of Technology in the early 1900s that exposure to γ-rays could effectively kill bacteria present in food. The first patent for a food irradiation device was filed in France in 1906. During World War II further research into food irradiation was carried out by MIT to enhance the lifetimes of food used by soldiers in the field, where refrigeration was not available. The first commercial food irradiation facility was built and operated in Germany in 1958. It is currently used in over 60 countries to extend the shelf-life of food products. The isotope used for food irradiation is ^{60}Co ($t_{1/2}$ = 5.27 years), which decays by β^- emission to the stable isotope ^{60}Ni with the emission of two γ-rays of 1.17 and 1.33 MeV. Gamma irradiation is used to kill insects and pests in grain, bacteria such as *Escherichia coli* and salmonella in beef and chicken, trichinosis in pork products, and microbes in sea foods, spices, and herbs. It has also been used to slow down ripening and sprouting in fruits and vegetables. This is an effective way of increasing the shelf-life of the food products.

In the United States, food irradiation is treated as a food additive. Hence, the control and approval of food irradiation is covered by the U.S. Food and Drug Administration (FDA). Food irradiation has been approved by the U.S. FDA for use in packaged meat, poultry, pork products, lobsters, crabs, shrimp, and shellfish, along with spices, lettuce, and spinach products. Shipping of fruits and vegetable products from one region to another normally requires that products be placed in quarantine to ensure that they are not infested with pests or bacteria. Food irradiation eliminates that need and allows for the fruits, nuts, vegetables, and spices to be shipped directly from country to country. Food irradiation doses are carefully limited so that the food does not undergo any radiation damage. In addition, the doses are high enough to ensure that the bacteria or pests are killed, so that no radiation-resistant bacteria or pests can develop.

Food irradiation could virtually eliminate most food-borne diseases if implemented globally. Despite this potential, there remain concerns by the uninformed that food irradiation changes the food or that the food retains the radiation. It is impossible for γ-ray irradiation to produce radioactivity in the food product. In addition, the exposure to the γ-ray sources does not cause any more loss in vitamin or food value than do other means of food processing, such as chilling, freezing, drying, or heating. Some other means of food irradiation are X-rays and energetic electron beams. However, the production of X-rays is less efficient and more expensive than the production of γ-rays and energetic electron beams have much lower penetration distances than γ-rays, making them useful only for specific products.

Radioisotopes are used as very stable sources of ionizing radiation in a wide range of instruments. They range from ^{63}Ni used as a source of electrons in the electron capture detector, discussed in Section 7.2, to the use of ^{241}Am as a source of α particles in smoke detectors. The ^{241}Am in smoke detectors is in the chemical form of $^{241}AmO_2$ with a half-life of 432 years. It emits α particles with a minimal amount of γ radiation. A new smoke detector contains about 0.3 μg of ^{241}Am. The sealed radioactive ^{241}Am source, shown in Figure 12.25, is used to ionize the smoke particles in order to detect their presence. Note that the source label indicates an amount of radioactivity of 1 μCi or 37 kBq.

Some plutonium isotopes have been used for portable power generation due to the heat generated by the energetic decay of α particles. In particular, ^{238}Pu ($t_{1/2}$ = 87.7 years) has been used as a lightweight thermoelectric generation system in satellites. The ^{238}Pu is an α emitter with a very low γ and neutron emission. Because α particles have a low penetration, a source of ^{238}Pu is readily shielded by simple plastic or thin metal casing. A block of $^{238}PuO_2$ produces a large amount of thermal energy due to the self-absorption of the α particles by the ^{238}PuO metal oxide. This self-absorption generates about 570 W of power and will still produce about 300 W of power after 30 years of operation. The ^{238}Pu can act as a power source for approximately one human lifetime. The ^{238}PuO has been used as a power source by NASA since it was originally used in lunar experiment packages in 1969 as part of the Apollo program. Since then it has been used in both Voyager space probes, along with the Cassini, Galileo, and New Horizons probes,

Figure 12.25 The inside of a typical household smoke detector showing the labeled ^{241}Am α source used to produce ionized smoke particles. Source: MD111, Wikimedia Commons.

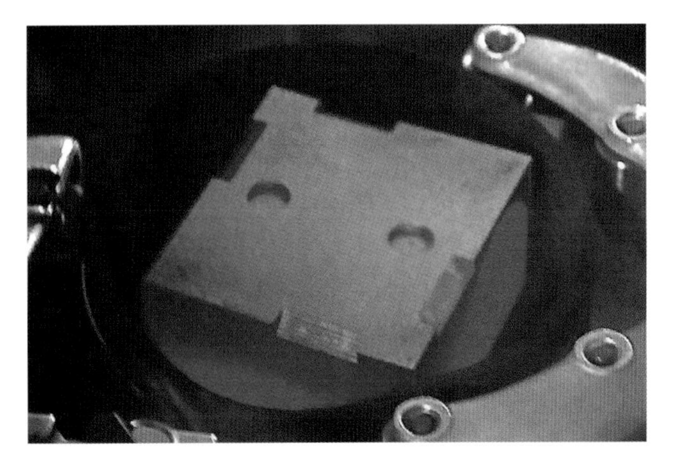

Figure 12.26 The $^{238}PuO_2$ thermoelectric heat generator used in the Curiosity Mars Rover. Source: Idaho National Laboratory, Wikimedia Commons.

and in the Curiosity Mars Rover shown in Figure 12.26. A small amount of $^{238}PuO_2$ has been used successfully as a power source in about 100 pacemakers, minimizing the need for surgical battery changes. This application has not been pursued after the development of lithium primary cells due to the fact that all plutonium species are toxic heavy metals.

References

Gaffney, J.S., Marley, N.A., and Orlandini, K.A. (1992). Evidence for thorium isotopic disequilibria in natural waters due to organic complexation: Geochemical implications. *Environmental Science & Technology* 26: 1248–1250.

IAEA 2015 *Nuclear power plants in the world*. Available online at https://www-pub.iaea.org/MTCD/Publications/PDF/rds2-35web-85937611.pdf (accessed May 26, 2018).

National Nuclear Data Center n.d. *Information extracted from the Chart of Nuclides database.* Available online at http://www.nndc.bnl.gov/chart (accessed December 1, 2018).
UNSCEAR 2010 *Sources and effects of ionizing radiation.* Available online at http://www.unscear.org/unscear/en/publications/2008_1.html (accessed May 30, 2018).
USNRC 2018 *ALARA*. Available online at https://www.nrc.gov/reading-rm/basic-ref/glossary/alara.html (accessed December 11, 2018).
WHO 2013 *Health risk assessment from the nuclear accident after the 2011 Great East Japan earthquake and tsunami, based on a preliminary dose estimation.* World Health Organization, Geneva.
World Nuclear Organization 2017 *World uranium mining production.* Available online at http://www.world-nuclear.org/information-library/nuclear-fuel-cycle/mining-of-uranium/world-uranium-mining-production.aspx (accessed December 6, 2018).

Study Problems

12.1 (a) Who discovered X rays?
(b) Who discovered radioactivity?

12.2 (a) What is radioactivity?
(b) In what unit is radioactivity measured?
(c) What is the symbol for the unit?

12.3 What are the two elements that were discovered by Pierre and Marie Curie?

12.4 How are radioisotopes different from isotopes?

12.5 What are the five major radioactive decay processes?

12.6 What are the two decay processes that can lead to the loss of a proton and the gain of a neutron in the nucleus?

12.7 (a) What is the decay process that leads to the emission of an energetic helium nucleus?
(b) How does this change the nucleus?

12.8 (a) What decay process leads to the emission of an energetic electron from the nucleus?
(b) How is the nucleus changed?

12.9 (a) What decay process leads to the emission of energetic photons?
(b) How is the nucleus changed?

12.10 (a) What are the three forms of radioactivity?
(b) What is their order of penetration?

12.11 What is the driving force behind nuclear decay?

12.12 (a) What type of kinetics does radioisotope decay follow?
(b) How is the activity of a radioisotope related to the decay rate?

12.13 (a) What charge does an α particle have?
(b) A β^- particle?
(c) A γ-ray?

12.14 How is the lifetime of a radioisotope related to the decay rate?

12.15 (a) What is the half-life of a radioisotope?
(b) What is the half-life equal to?

12.16 (a) What happens when a positron interacts with an electron?
(b) What is produced from this process?

12.17 What are the three sources of radioisotopes in the environment?

12.18 The primordial radioisotope ^{238}U has a half-life of 4.5 billion years and the Earth is about 4.5 billion years old. What fraction of the initial ^{238}U remains on the planet today?

12.19 (a) What radioisotope is responsible for the production of ^{40}Ar?
(b) By what radioactive decay process does this occur?

12.20 (a) What are the two common terms for ^{222}Rn and ^{220}Rn gases?
(b) What decay chains produce them?
(c) Which of the two radioisotopes has the longer half-life?

12.21 (a) How is the cosmogenic radionuclide ^{14}C produced?
(b) What is its decay process?

12.22 When ^{14}N atoms were bombarded with α particles, what were the products?

12.23 What is the nuclear process that occurs when a large radioisotope forms two smaller nuclei after being bombarded by a neutron?

12.24 What was the first known fissile radioisotope?

12.25 (a) If the concentration of ^{235}U is high enough, what type of reaction can occur?
(b) What happens if a nuclear moderator is present?
(c) What happens if a moderator is not present?

12.26 (a) What was the second fissile radioisotope discovered?
(b) How was it formed?

12.27 What radioisotopes were used in the two nuclear bombs used by the United States during World War II?

12.28 What nuclear reaction was the basis of a fusion bomb?

12.29 (a) What are the two types of fallout?
(b) Most of the long-lived radioisotopes are deposited in what form?
(c) What radioisotope was a major concern globally from radioactive fallout?

12.30 A significant amount of ^{137}Cs was injected into the atmosphere in 1960. This radioisotope has a half-life of 30 years. In 2020, what fraction of the original amount of ^{137}Cs will remain?

12.31 (a) What is a breeder reactor?
(b) What fuel was used for the first breeder reactor?
(c) What fuel does it produce?

12.32 (a) What was the first material used to moderate the neutron flux in a ^{235}U nuclear reactor?
(b) What hazard does this moderator present when used in a nuclear power reactor?

12.33 (a) What new nuclear reactor design allows for the nuclear chain reaction to cease if there is a power or coolant failure?
(b) Why does this result in a chain reaction shut down?

12.34 What is the biggest problem with developing a controlled fusion reaction?

12.35 (a) What oxide produces the yellow color in yellowcake?
(b) what uranium oxidation state causes this?

12.36 What are the two types of nuclear reactors used in nuclear power plants?

12.37 (a) Most of the nuclear reactors in operation today use what as a coolant?
(b) What type of reactors are these?
(c) What type of reactor is the CANDU?

12.38 (a) What are the three major sites where a nuclear reactor accident has occurred?
(b) Which of these was the most significant?

12.39 (a) What radioisotope is responsible for most of the natural inhaled radioactivity?
(b) What is its source?

12.40 (a) What radioisotope is concentrated in tobacco smoke?
(b) What is its decay process?

12.41 What two radioisotopes are responsible for most of the natural radioactivity exposure in foods?

12.42 What is the major anthropogenic source of radiation exposure?

12.43 What isotope of uranium was found to be depleted in the Oklo ore that led to this discovery?

12.44 (a) What is the modern unit for radiation dose?
(b) How does this new unit relate to the rad?

12.45 (a) What is an equivalent dose?
(b) What are two units for the equivalent dose?
(c) How are they related?

12.46 What are the three ways to minimize exposure to ionizing radiation?

12.47 What does rem stand for?

12.48 (a) What is the acronym that represents the guiding principle for radiation safety for radiation workers?
(b) What does the acronym stand for?

12.49 (a) What is the major source of exposure to ionizing radiation worldwide?
(b) What two radioisotopes are the primary sources of radioactivity in food and water?

12.50 (a) What synthetic radioisotope is one of the most commonly used radiotracers in medicine?
(b) What radioisotope is commonly used as a γ-ray source for food irradiation?
(c) How is it produced?

12.51 (a) What radioisotope is commonly used as an α source in household smoke detectors?
(b) What radioactive oxide has been used as a power source by NASA in satellites and in the Curiosity Mars Rover?

12.52 (a) What radioisotope is used to date old ground waters?
(b) What radioisotope is used to determine sedimentation rates in lakes and oceans?

13

Future Energy Sources and Sustainability

Most of the past work in environmental chemistry was focused on solving problems after they were recognized. In other words, environmental problems were dealt with reactively and not proactively. For example, photochemical air pollution occurred for many years in the region surrounding Los Angeles, CA before it was recognized that it was caused primarily by the emissions of NO_x and VOCs from gasoline-powered vehicles. Once this was recognized, measures were enacted to reduce the emissions, such as the requirement that all new vehicles be equipped with catalytic converters to reduce VOCs. While this led to the reduction of urban levels of tropospheric ozone by slowing the conversion of NO to NO_2, the transport of the NO_x emissions downwind resulted in a general increase in regional tropospheric ozone. Although the combustion of fossil fuels allowed us to expand our technologies by providing us with a large source of energy, this energy is not inexhaustible and we now know that the combustion of any carbonaceous fuel leads to environmental pollution problems on urban, regional, and global scales. Due to the huge amounts of energy now required for transportation and electrical power in most developed countries, alternatives to fossil fuel and biomass combustion will be needed in the near future to reduce air and water pollution and to slow climate change.

The continued combustion of fossil fuels also depletes an important natural resource that is needed for other industries, such as organic chemical and pharmaceutical production, which use petroleum as a feedstock for syntheses. The petrochemical industry produces numerous products, including polymers and fibers used in clothing, packaging, and construction materials. In addition, petrochemicals are used for producing some pharmaceuticals. The alternative to the petrochemical feedstocks is biomass feedstocks. However, the amounts of available biomass feedstocks will also be limited due to limits on other resources such as land area, nutrients, and water. There will eventually be trade-offs between food production and biomass production for use in biopolymers in clothing, packaging, and construction materials, as well as for use as an alternative fuel.

Fossil fuel combustion is also the major source of the increasing CO_2 levels that, along with other related greenhouse species, are changing our global climate at an ever increasing rate. Even if all fossil fuel CO_2 emissions were lowered to zero tomorrow, the elevated levels of CO_2 already present in the atmosphere would take several decades to fall to acceptable levels. Coping with the impacts of climate change will require significant adaptation, especially in those regions most affected. In addition, the increasing population levels will lead to increases in energy, water, and food demand, and ultimately to shortages in nonrenewable resources. These are the greatest challenges facing environmental chemists and other scientists working to find long-term solutions that support a sustainable world economy. Sustainable development is defined by the UN World Commission on Environment and Development as development that meets the needs of the present without compromising the ability of future generations to meet their own needs. Thus, achieving sustainability is reaching a state where our energy, water, and

Chemistry of Environmental Systems: Fundamental Principles and Analytical Methods, First Edition.
Jeffrey S. Gaffney and Nancy A. Marley.

food resources meet the demands of today's societies without causing harm to our environment or depleting them for future generations. Reaching this sustainable state in light of increasing populations will require a number of changes in our sources of energy, management of water resources, and development of reliable and economic methods for recycling products. Any proposed changes in chemical and energy sources need to be evaluated for potential environmental impacts, availability, and costs before they are implemented. This will require lifecycle analyses of both chemical and energy alternatives.

13.1 The Need for Non-Fossil Energy Sources

It is now clear that the combustion of fossil fuels as well as other carbonaceous fuels leads to the emission of CO_2 and other air pollutants (Gaffney and Marley, 2009). The IPCC is continuing to examine the current trends in increases of radiatively important greenhouse gases, particularly CO_2. In this analysis, the IPCC has made a number of recommendations for slowing down the increases in greenhouse gas emissions and halting the rise of CO_2 in the troposphere. Of particular importance is the recognition that the use of renewable energy sources, along with nuclear energy, will be needed to reduce and ultimately eliminate fossil fuel combustion and be able to mitigate climate change (IPCC, 2012).

The fossil fuel reserves are large but not limitless. As discussed in Section 10.2.1, it took hundreds of millions of years for the natural systems to produce the existing fossil fuel deposits. Although the global demand for energy is continually depleting the coal, oil, and gas reserves, currently the amount of commercially available fossil fuels has been able to keep up with demand. The production of fuels from these unrenewable resources is expected to reach a peak in the not so distant future, after which the availability will begin to decrease. The timing of this peak in fossil fuel availability will depend on the source location and recovery technologies available. Thus, the fuel availability can be extended by locating any remaining fossil reserves and enhancing extraction methods. However, the general consensus is that the peak in oil and gas is expected to occur around 2020, at which time demand will be higher than supply and costs will rise substantially. Peak coal production is expected to occur by 2050, however its use is likely to be limited due to the higher environmental impacts associated with its combustion. In any case, within the coming decades the amount of fossil fuels available for energy use is expected to decline. The fact is that fossil fuels are unsustainable and are presently causing significant damage to air and water quality, as well as being major contributors to climate change. In order to maintain a stable and sustainable energy economy, it is clear that the global community needs to move away from fossil fuels and toward other energy alternatives that do not release greenhouse gases or have major impacts on air and water quality.

It is also important to recognize that significant amounts of fossil fuels are required in the petrochemical industry, particularly in the production of polymers that are used in plastics, textiles, and building materials. The combustion of fossil fuels, and oil and gas in particular, is depleting these important chemical resources at a rapid rate. During the refining of crude oil and natural gas, petrochemical refineries separate three major organic petrochemical feedstocks, as shown in Figure 13.1. These are light alkanes, light alkenes, and light aromatics. The light alkanes are produced from the processing of raw natural gas and include methane, ethane, propane, and the butanes (*n*-butane and isobutane). The light aromatics, known as BTX, are produced from the refining of crude oil and include benzene, toluene, and the xylenes (*o*-, *m*-, and *p*-xylene). The light alkenes are produced from steam cracking of both gas and liquid feeds from natural gas processing and oil refining. Steam cracking is a process that breaks down the large saturated hydrocarbons into smaller hydrocarbons. It is the principal

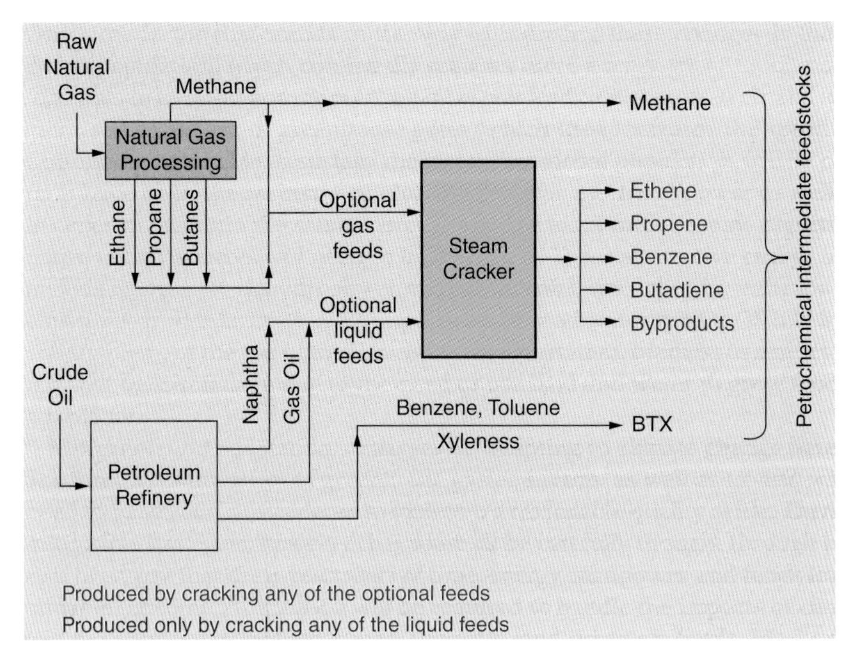

Figure 13.1 Natural gas and petroleum refinery processing of raw natural gas and crude oil into key petrochemical compounds used for numerous products. Source: Mbeychok, Wikimedia Commons.

petrochemical method for producing the lighter alkenes, which include ethene, propene, butene, and butadiene. Ethene and propene are two of the more important petrochemical feedstocks. The global production of ethene is on the order of 120 MT yr^{-1}, while the production of propene is about 70 MT yr^{-1}. The production of the light aromatics is similar to that of propene (70 MT yr^{-1}). These numbers can be compared to the approximately 4400 MT yr^{-1} of crude petroleum produced globally. In the United States about 2% of each barrel of crude petroleum is used as a petrochemical feedstock. The other 98% is primarily used for fuels that are combusted to produce electricity and power motor vehicles and diesel trucks.

The light alkanes are typically chlorinated for use as solvents or brominated for use as flame-retardant additives in polymers and fire extinguishers. The light alkenes are the building blocks for a wide range of materials, such as numerous types of polymers and oligomers that are used to produce fibers, plastics, resins, lubricants, gels, elastomers, solvents, detergents, and adhesives. Butene and butadiene are used in making synthetic rubber. The light aromatics are used in the production of polymers and fibers. Benzene is used in the production of dyes and synthetic detergents. Both benzene and toluene are used for the production of isocyanates, methylene diphenyl di-isocyanate (MDI) and toluene di-isocyanate (TDI), which are used in the manufacture of polyurethane. The xylenes are used to produce plastics and synthetic fibers.

The chemicals produced from ethene, propene, and benzene are shown in Figures 13.2 and 13.3 as examples of how widely these feedstocks influence the chemical industry and our lives. The primary products from ethene are polyethylene, ethanol, ethylene oxide, vinyl acetate, and 1,2-dichloroethane. Ethylene oxide is further used to produce ethylene glycol, used as an engine coolant and in the formation of polyesters. Also produced from ethylene oxide are glycol ethers and ethoxylates. The 1,2-dichloroethane is used to produce tetrachloroethylene, trichloroethylene, and vinyl chloride, which is used to produce polyvinylchloride (PVC). The primary products from propene are isopropyl alcohol, acrylonitrile, polypropylene, polypropylene oxide, acrylic acid, and allyl chloride. The secondary products are polyol,

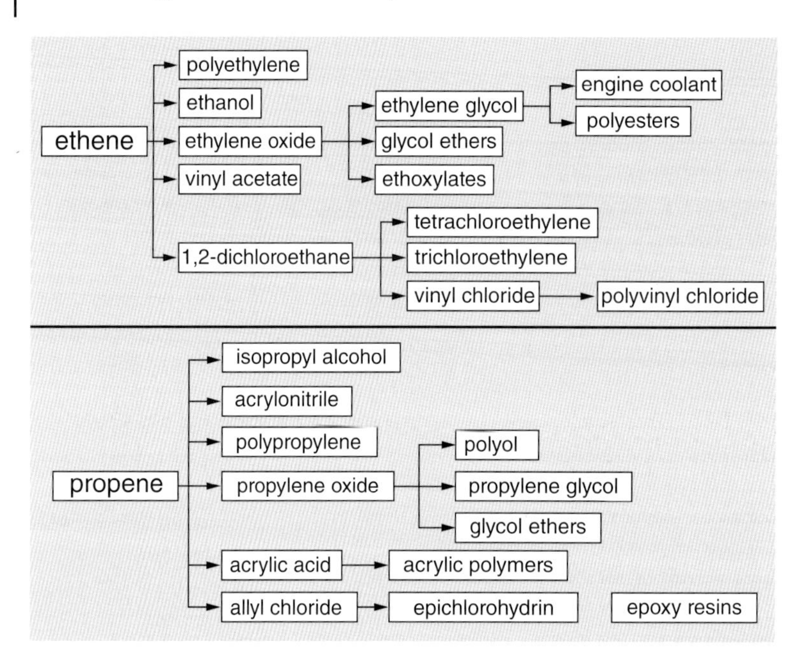

Figure 13.2 The chemicals synthesized from ethene and propene produced from petrochemicals. Source: Adapted from Mbeychok, Wikimedia Commons.

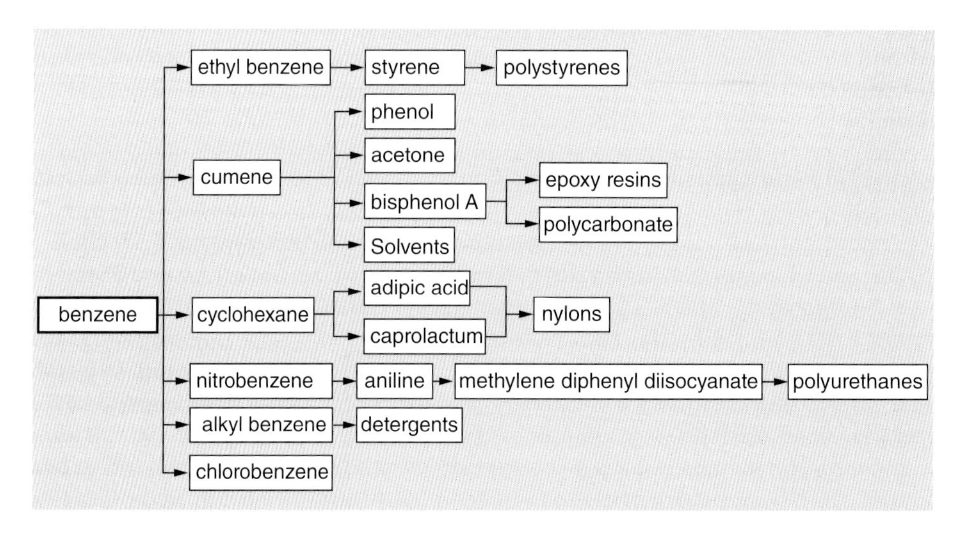

Figure 13.3 The chemicals synthesized from benzene produced from petrochemicals. Source: Adapted from Mbeychok, Wikimedia Commons.

propylene glycol, glycol ethers, acrylic polymers, and allyl chloride used in epoxy resins. The major products obtained from benzene, shown in Figure 13.3, are ethylbenzene used to produce styrene and polystyrenes, bisphenol A used to produce polycarbonates and epoxy resins, adipic acid and caprolactam used in the production of nylons, and aniline used in the production of polyurethanes. Also produced are cumene, phenol, acetone, cyclohexane, nitrobenzene, chlorobenzene, and allyl benzene, which is used in the production of detergents.

To ensure that the global production of chemical products is maintained, the conservation of the petroleum reserves used as chemical feedstocks needs to become a priority. If the combustion of fossil fuels was to be stopped, the amount of time that the petroleum reserves could continually be used by the petrochemical industry would be extended by a factor of 50 or more in the United States alone, and likely longer globally. Ceasing fossil fuel combustion would also extend the time required to find renewable chemical alternatives for all of the petrochemical products that are used globally. Thus, petroleum is not only an energy source but it is also an important source of daily use products. Once the petroleum reserves are depleted, these products will also face shortages unless we have developed alternative chemical feedstocks. This further supports the need for the development of alternative energy sources, not only to minimize environmental impacts of fossil fuel combustion but also to maintain this resource for the production of petrochemical products.

13.2 Alternative Energy Sources

The switch from fossil energy sources to cleaner alternative energy sources is not an easy process. New infrastructure must be put in place in order to be able to transition from one energy source to another, and this takes time and planning. Coal is still the major source of electrical energy globally. It is also the main source of fossil CO_2 emissions to the atmosphere. Natural gas power plants, which produce less CO_2 per unit of electrical energy, have begun to replace coal-fired power plants as a short-term solution to reduce CO_2 emissions and improve air and water quality. However, the combustion of natural gas still releases substantial amounts of NO_x into the troposphere as well as CH_4, which is a more potent greenhouse gas than CO_2. Furthermore, the atmospheric oxidation of CH_4 produces CO_2. Natural gas resources are also limited and are not sustainable, so the combustion of CH_4 is not a long-term solution for the production of electrical energy.

One possibility for extending the time frame for developing sustainable electric power is the building of nuclear fission power plants, including the current ^{235}U-fueled plants along with breeder reactors. Current nuclear power plants were designed for a 40-year lifetime. The lifetimes of nuclear power plants can likely be extended to 80–100 years with improvements in materials, designs, and maintenance procedures. If uranium can be extracted economically from sea water, the amount of available nuclear fuel could last for at least a couple of centuries and could be extended further by using breeder technologies. This would extend the effective peak time for nuclear energy sufficiently long that it can be considered as a sustainable energy source. The main drawback for nuclear power has always been the need to store or process nuclear waste. This problem can be greatly reduced if fuel rods are recycled in breeder reactors. The implementation of breeder reactors will require the construction of nuclear reactors and nuclear processing facilities on the same site to reduce the potential environmental impacts and lower overall costs. Although there have been a number of studies that show nuclear power has the lowest lifecycle greenhouse gas emissions and the lowest air and water quality impacts, the number of new nuclear power plants being constructed remains low and is essentially zero in the United States. Nevertheless, nuclear power remains one of the most reliable approaches to developing a sustainable energy economy with minimum impacts on the environment in most long-term planning studies.

Renewable energy options include wind, hydropower, tidal, geothermal, solar, woody biomass combustion, landfill gas, bioethanol, and biodiesel. Wind, hydropower, geothermal, tidal, and solar energy are all dependent on the availability and reliability of their source, which is region specific. The other renewable sources of energy are all tied to the combustion of biomass.

The total renewable energy sources worldwide account for about 20% of the current energy production, nuclear energy accounts for about 3% and the rest (77%) is provided from fossil fuels. The most used of the renewable energy sources worldwide is traditional biomass, which is wood burning for heat and cooking. The second most used is hydropower, followed by bioethanol, solar, and biodiesel.

13.2.1 Wind Power

Man has used wind power for centuries to power sail boats, windmills, and other devices. Modern wind power used to generate electricity relies on sustained winds whose energy is tapped by using the air movement to turn turbine blades, which then turn electric power generators. Wind power has the advantage of not releasing any greenhouse gases into the environment, other than the CO_2 released during the production of the materials used in the construction of the wind turbine blades and electric generator parts. A wind turbine, shown in Figure 13.4, is a device that converts the wind's kinetic energy into electrical energy. The turbine blades can be turned into the direction of the wind and the angle of the blades can be adjusted to the plane of the wind for maximum energy extraction. There is a limit on the amount of kinetic energy that can be converted into electrical energy using a wind turbine due to the laws of conservation of mass and energy. It occurs because the air velocity decreases as it turns the blades, causing the velocity of the air behind the turbine to also decrease. This is called the Betz limit after the German physicist Albert Betz. Betz showed that the maximum wind energy that can be extracted using a wind turbine is equal to 59.3% of the kinetic energy of the wind. In practice, large wind-extracting blades can capture up to 80% of the Betz limit, or about 47% of the wind's kinetic energy.

The shape and dimensions of the wind turbine blades are determined by the aerodynamics required to efficiently extract the kinetic energy and by the strength required to resist the forces applied to the blade. Longer wind turbine blades rotate at slower speeds with higher torque, which is an advantage over smaller blades that rotate faster and can be a hazard for birds, bats, and insects flying through the area. Originally the wind turbine blades were made of steel and were quite heavy, which limited their length. Modern wind turbines are now constructed of fiberglass and carbon fiber materials that are lighter and stronger than steel blades. These long, lightweight blades typically have lifetimes of about 20 years. New materials such as carbon nanotubes are being investigated for use in turbine blades that may further increase blade strength

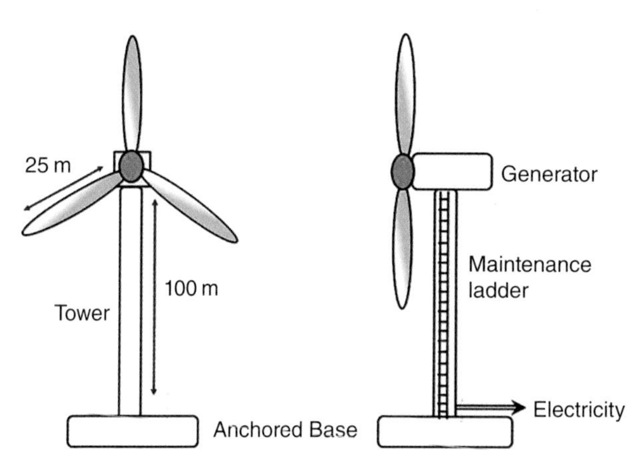

Figure 13.4 The twisted-blade propeller wind turbine design typically used in wind farms to obtain electrical energy from the kinetic energy of the wind.

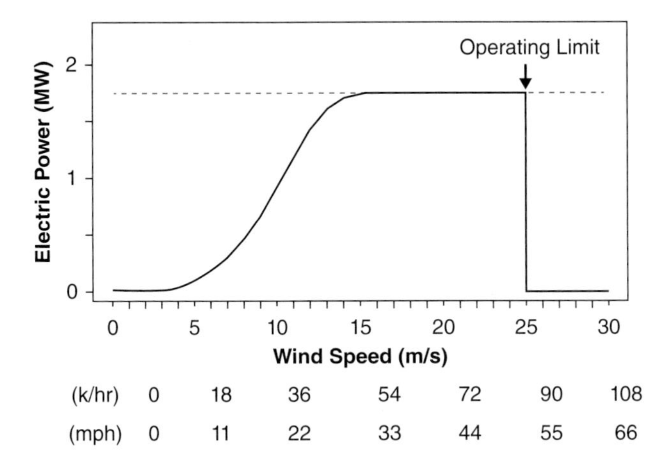

Figure 13.5 The typical power output for a wind turbine for different wind speeds (m s^{-1}) compared to the maximum power output (red dashed line). Source: Adapted from isjc99, Wikimedia Commons.

and decrease blade weight, while increasing the lifetime of the blades up to a century. In addition to the blade design, the design of a complete wind power system must also address the design of the rotor hub, gear housing, tower structure, generator, controls, and foundation. Wind power systems also make extensive use of sophisticated computer modeling and simulation tools to determine the best orientation of the blades to the wind, placement of the turbines, and times of operation.

A wind turbine is designed to produce power over a range of wind speeds. Wind turbines require minimum wind speeds in the range of 3–3.5 m s^{-1} (13 km h^{-1} or 8 mph) to be able to generate any power. The optimal maximum power range is between 14 and 25 m s^{-1} (50–90 km h^{-1} or 31–56 mph). The typical power output for a wind turbine under different sustained wind speeds is shown in Figure 13.5 compared to the maximum possible power output (red line), which occurs at the highest wind speed allowed by the turbine (25 m s^{-1}). Since the turbine can be damaged at higher wind speeds, a brake is installed to stop the blades from turning during times where the wind speed exceeds the allowed maximum in order to protect the turbine from damage. The actual power output is at the maximum level up to wind speeds of about 15 m s^{-1} and begins to decrease as wind speeds decrease. It reaches the lower limit (0 power output) at wind speeds of about 3 m s^{-1}.

In general, wind power has advantages over most other power sources in that it does not require water for operation and generates no greenhouse gases. Its two major drawbacks are that power is generated only when the wind speed is >3 m s^{-1} and that the wind speed should be >8 m s^{-1} to be able to extract a reasonable amount of power. One method of transforming wind turbines into a more reliable power source is to store power in batteries when the wind turbines are producing surplus power. The stored power from the batteries can then be used during time periods when there is insufficient wind to generate power. Another option is to use wind turbines to supplement other power sources such as natural gas or nuclear power in order to lower the demand put on these other sources during times of peak use. This is generally the method of choice in most regions where wind power is available.

The annual average wind speed patterns at 50 m height are shown globally in Figure 13.6 from 1983 to 1993. The areas that would most likely make use of wind power have minimum average wind speeds of 6 m s^{-1} and above (Figure 13.6, orange–red). Many of these areas are located in the open oceans. The places most likely to be considered for location of multiple wind turbines, called wind farms, are in the mid-latitude coastal areas where sea or lake breezes

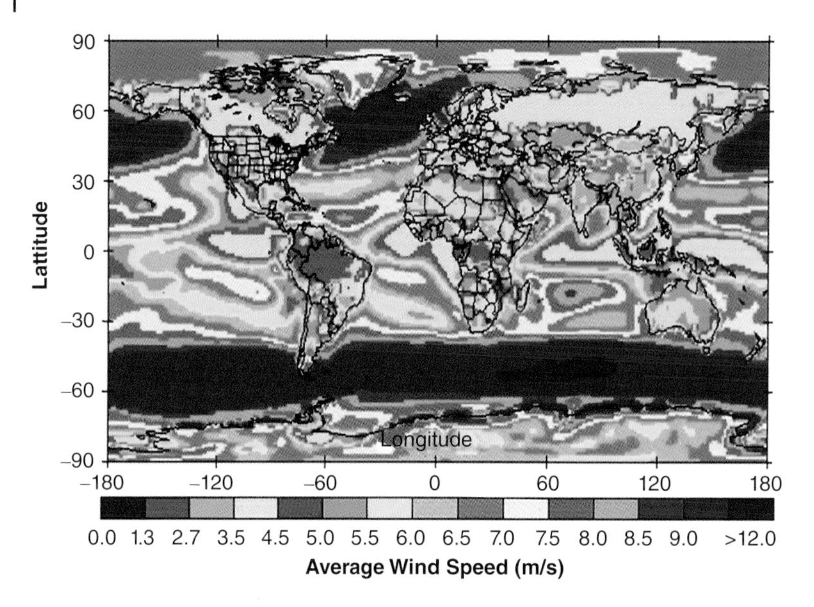

Figure 13.6 Average wind speed (m s^{-1}) globally at 50 m height for the period July 1983 to June 1993. Source: Adapted from NASA, Wikimedia Commons.

occur on a regular basis. Other areas include inland deserts such as the Mojave in the United States and the Sahara in Africa. A wind farm is a group of wind turbines in the same location used for the production of electric power. A large wind farm may consist of several hundred wind turbines distributed over an extended area. For example, the Gansu Wind Farm located in China has several thousand turbines with a capacity of over 6000 MW. There are currently 23 large wind farms (>450 MW capacity) around the world with 10 located in the United States, 9 in China, and 1 each in India, Romania, Egypt, and Scotland. Establishing the exact location of each turbine in a wind farm is important because of the effects of terrain on wind speed. A location difference of about 30 m could potentially double the energy output of the turbine. Wind farms in hilly or mountainous regions tend to be located on ridges at least 3 km inland from the nearest shoreline. This is to make use of the wind acceleration as it approaches the ridge. Wind farms can also be located offshore. Europe is the leader in offshore wind farms, with the first installed by Denmark in 1991. There are 39 wind farms in waters off Belgium, Denmark, Finland, Germany, Ireland, the Netherlands, Norway, Sweden, and the United Kingdom, with a combined operating capacity of 2396 MW as of 2010. The European Wind Energy Association has set a goal of wind farms totaling a capacity of 40 GW by 2020 and 150 GW by 2030.

There are currently considerable areas within the United States committed to wind farming. Large wind farms in the United States include four in California (3240 MW), four in Texas (2764 MW), one in Oregon (845 MW), and one in Indiana (600 MW). Conditions contributing to a successful wind farm location include wind conditions, access to electric transmission, physical access, and local electric prices. One of the main considerations is that wind turbine systems be located near currently accessible electrical transmission lines in order to lower the overall costs of adding the power generated by the wind farm into the power grid. The areas that are suitable for wind farming in the contiguous 48 states, along with accessible power transmission lines, are shown in Figure 13.7. Four of the large wind farms in the United States are located in Texas. The wind power in Texas is generally fair to marginal (Figure 13.7, brown to orange), with areas in the panhandle classified as good (Figure 13.7, pink). However, the transmission lines are quite dense in eastern Texas and all of the four large wind farms are located in central

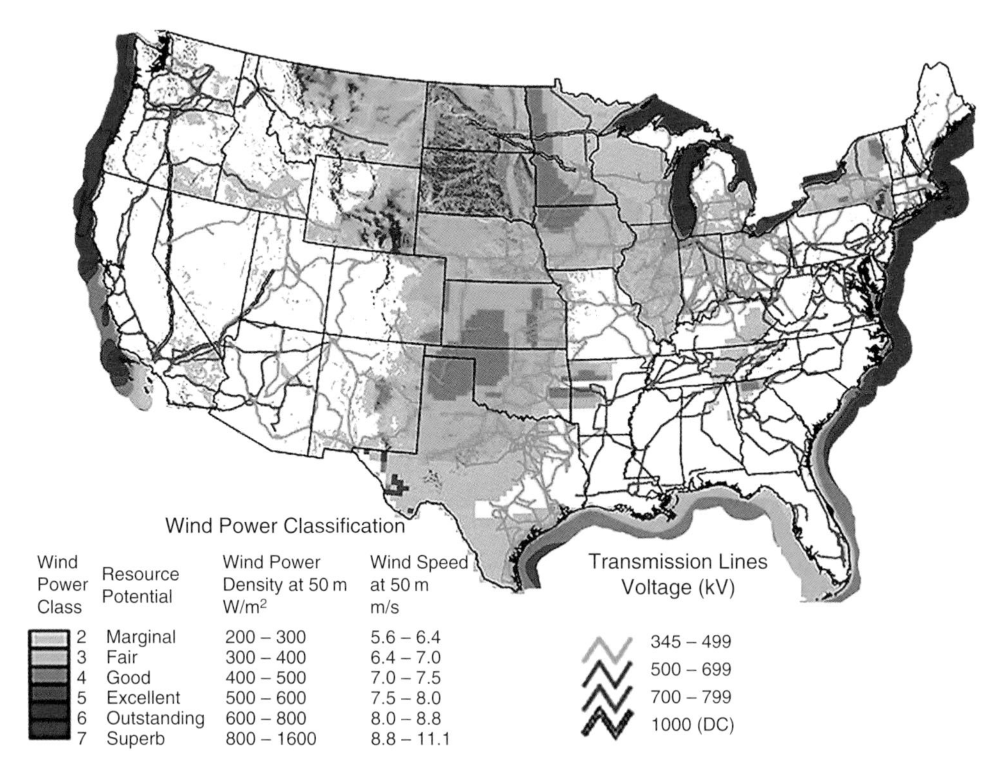

Figure 13.7 The classification of wind power in the United States compared to the location of electrical grid transmission lines. Source: Adapted from National Energy Renewable Laboratory, Wikimedia Commons.

Texas not far from the cluster of transmission lines to the east. The four large wind farms in California are also located near the high-voltage transmission lines in western California.

The areas with the highest sustained winds, rated outstanding and superb in Figure 13.7, are off the Pacific and northern Atlantic coasts, as well as in the Great Lakes. The United States has lagged behind Europe in the development of offshore wind farms. Offshore wind farms are currently in the early stages of development in the United States. The first commercial offshore wind farm, Block Island off the coast of Rhode Island, began operation in 2016. Offshore wind farms have the drawback of requiring special transmission lines to connect the generated electricity to the power grid. Scotland is an excellent example of how transmission lines impact wind power generation. About 50% of Scotland's total electrical energy is supplied from wind power. Currently, Scotland has over 5300 MW of wind power generation capacity. Of that, only 200 MW is accessed from offshore wind turbines due to the lack of offshore transmission lines. Scotland is currently planning to further increase their wind power in the future by increasing offshore transmission lines.

The major drawback for increasing global-scale wind power is the large amount of continental land mass areas where the winds are not sufficiently strong and/or regular enough to allow wind power to be useful and reliable. In regions where winds are of sufficient strength and regularity, the only major drawback is esthetics, since the addition of the large number of wind turbines required for a wind farm changes the landscape. One long-term concern is that as climate change continues the wind patterns may shift, which could result in wind farms losing the wind speed and regularity needed to operate the turbines, or producing wind speeds above the turbine operating limit.

13.2.2 Hydropower

Hydropower generation is currently the largest renewable energy source, accounting for 70% of renewable global electric power generation. Hydropower operates on the same principles as wind power. It converts the kinetic energy of a flowing fluid into electrical energy by using the movement of water to turn turbine blades, which then turn electric power generators. The generation of hydropower requires placing large turbines into a flowing body of water that is constantly being regenerated. This is commonly accomplished by placing a dam into a river system. The dam then directs the flow of water through the hydroelectric power generating system, as shown in Figure 13.8. Most of the kinetic energy is produced from the potential energy of the dammed water in the reservoir, which depends on the volume of the water and the difference in height between the reservoir and the outflow. This height difference is called the head, or hydraulic head, which is usually measured as a liquid surface elevation, expressed in units of length. A large pipe, called a penstock, delivers the water from the reservoir to the water turbine, which turns an electrical generator. The intake to the penstock is usually blocked by a screen to prevent fish from being carried into the turbines.

A number of large dams have been built around the world for the production of hydroelectric power. They account for about 17% of the total electricity generated globally. The Aswan Dam in Egypt is one of the largest and best known of the hydroelectric power stations. Along with generating electricity, the Aswan Dam serves to control annual flooding events on the Nile and to contain water for use in crop irrigation during times of drought. Hydroelectric power requires a significant capital investment to build the dams, and the construction process releases greenhouse gases from combustion-powered equipment. Carbon dioxide is also released from the production of concrete used to make the walls or barriers that hold the water. In addition, greenhouse gases can be generated when the land is slowly converted from dry land into wetlands. Dams typically flood large areas of land when put in place, changing land areas that were previously dry land into water or wetland environments. As discussed in Section 11.2.2, the anaerobic bacteria that live in wetlands are a primary source of the greenhouse

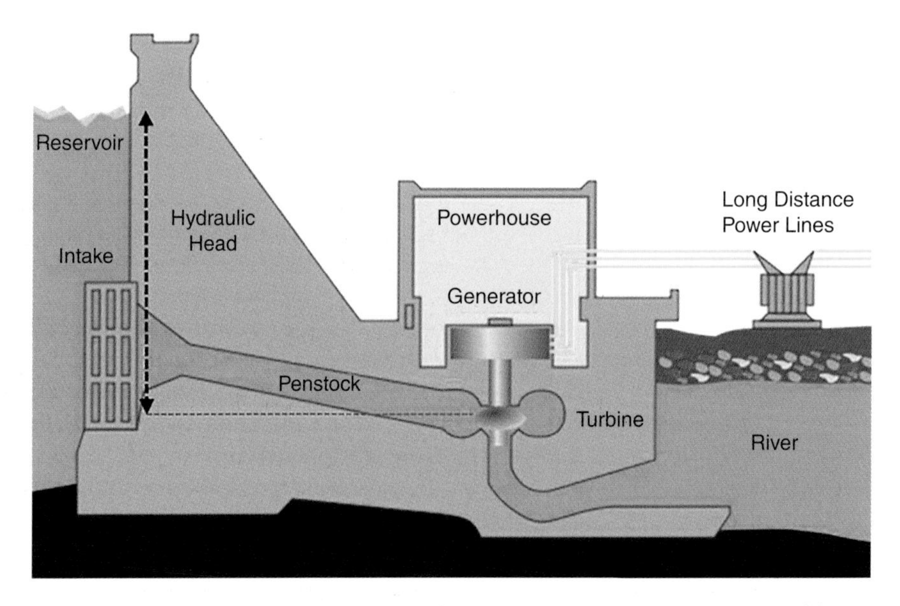

Figure 13.8 A typical hydroelectric dam used to generate electricity. Source: Adapted from Tennessee Valley Authority, Wikimedia Commons.

gas CH_4. While greenhouse gas emissions from dams in temperate and colder climates release negligible amounts of greenhouse gases, dams in tropical regions release measurable amounts, most notably CH_4. Overall, hydroelectric power plants release significantly less amounts of greenhouse gases than fossil fuel energy power plants during the lifecycle of the dam.

The major drawback from hydroelectric power generation is that the physical placement of a large dam causes significant changes to the surrounding environment and ecological systems. Dams require a continuous major source of water, usually a river. Control of the river flow at the dam changes the volume of the downstream water supply, which is usually needed for agricultural irrigation and drinking water. The downstream water flow reduction also affects the amounts of sediments that would naturally be carried by the river onto the surrounding land during flooding events. The sediment deposition replaces soil nutrients and can counter subsidence in areas of subsurface mining or oil and gas extraction. Dams also present an unnatural blockage of rivers, which can have a significant impact on fish that spawn upstream. Most notably the salmon fishing industry has been significantly impacted by dams in North America. Attempts to solve this problem have led to the removal of dams which have had significant impacts on fish populations. One such hydroelectric facility was located on the Elwha River in the state of Washington. Before the dam was built, about 400 000 salmon were observed to return to spawn in the river each year. After the dam became operational, the number of returning salmon was reduced to less than 3000. The dam was removed after its useful lifetime to allow the salmon fishing industry a chance to recover in this area.

While hydroelectric power is attractive due to its fairly low cost and long-term use as a renewable power source, it has a significant impact on the local ecological systems. Also, most rivers that are suitable for producing hydroelectric power already have dams, so hydroelectric power from river dams is not seen as an area for potential major increase in the future. Climate change may also have significant impacts on existing dams if changes in rainfall distributions significantly affect water availability. Dams are also susceptible to earthquakes and older dams require maintenance to ensure that they do not cause catastrophic flooding during severe weather, heavy rainfall, or seismic events.

Another form of hydroelectric power generation is tidal power, which converts the kinetic energy of tidal flow into electricity using turbines and electrical generators. The moving tidal water powers the water turbines in much the same way as water flowing through a dam penstock, except that the direction of flow changes. Tidal generators can be built into the structures of existing bridges or they can be entirely submersed. Land constrictions, such as straits or inlets, at some sites create higher water velocities during tidal flow. This higher kinetic energy in the faster water flow in these areas can produce more power when captured with the use of turbines. The world's first large-scale tidal power plant, the Rance Tidal Power Station, was built in France in 1966. It has 24 turbines to generate a capacity of 240 MW. Currently, the largest tidal power station is the Sihwa Lake Tidal Power Station in South Korea, which has 10 turbines with a total capacity of 254 MW.

Since tides are determined by the location of the Moon relative to the Earth and the Earth's rotation, they are easily predicted and are more reliable than winds. This makes tidal power attractive as a renewable energy source. In addition, extracting the kinetic energy in tidal flow is much more efficient than can be done in wind flow due to the higher density of water. A tidal current of 10 mph would yield an energy output equal to or greater than a 90 mph wind speed for the same size turbine. Currently, tidal power stations have been limited to regions where there are sufficient differences in the height of the tides to be able to extract energy efficiently. Alternatively, the incoming tide can be captured in a large basin by use of a tidal barrage. A tidal barrage is a structure similar to a dam but instead of damming the water on one side, it allows water to flow into a basin during high tide and allows it to flow out of the basin during low tide.

As the tidal water is released back into the ocean it passes through turbines to generate the electricity. This is accomplished by measuring the water flow and controlling the sluice gates in the tidal barrage at specific times of the tidal cycle. The kinetic energy produced by the use of a tidal barrage is based on the potential energy created by the difference in height between high and low tides, similar to the hydraulic head of a traditional dam. When the sea level rises and the tide begins to come into the basin, the temporary increase in water volume behind the barrage creates the large amount of potential energy. With the receding tide, this energy is then converted into mechanical energy as the water is released through large turbines. The mechanical energy is converted to electrical power by electrical generators.

The main drawback of tidal power for the production of electrical energy is the potential damage to fish and marine mammals by the turbines. In addition, the use of tidal barrages has the same impacts as traditional dams: they change the local ecological system due to the large-scale change in the environment caused by the formation of the tidal basin. The costs for construction of tidal power plants are high and corrosion of the turbines by the sea water is also a concern. In addition to shortening the life of the turbines, corrosion has the potential to release metal pollutants into the marine water. Tidal power is limited to coastal regions or islands and would require installation of transmission lines for the electricity to be carried significant distances from the tidal power source.

13.2.3 Geothermal Energy

Geothermal energy is another renewable and sustainable energy source that has been used successfully to generate electricity. Its energy originates from the formation of the planet and from radioactive decay in the interior of the planet. The geothermal gradient drives a continuous conduction of thermal energy from the core of the planet to the surface. In areas where the Earth's crust is thin, such as at tectonic plate boundaries, molten mantle rock can rise into the crust, heating the rock and water to temperatures as high as 370 °C. In areas where fissures occur in the crust, the heated water can be released to the surface as a hot spring or geyser. Geothermal hot water springs have long been used for bathing and space heating since the time of the Romans. Iceland currently uses hot spring water for space heating on a large scale. Iceland is a leader in the use of geothermal energy, which accounts for 66% of its total primary energy use.

Geothermal power plants were traditionally built on the edges of tectonic plates, where the geothermal resources are available near the surface. Surface geothermal energy comes in two forms: vapor dominated or liquid dominated. Rocks in vapor-dominated reservoirs are at temperatures from 240 to 300 °C, which can produce superheated steam. Liquid-dominated reservoirs are more common, with temperatures greater than 200 °C. They are typically found near young volcanoes surrounding the Pacific Ocean. Until recently, geothermal power systems have exploited these natural vapor-dominated and liquid-dominated hot deposits. However, most of the geothermal energy on the planet is located in deep dry rock. In addition, since the efficiency of energy extraction is dependent on temperature, large geothermal power plants require the higher temperatures of the deep hot rock resources to efficiently extract sufficient geothermal energy.

Improvements in drilling and extraction technologies have allowed for geothermal systems to be developed that take advantage of the higher temperatures in deep geothermal rock reservoirs. These deep resources are tapped by drilling into the crust to where the rock temperature is sufficient to be able to convert water into steam. Extraction of this deep geothermal energy is accomplished by enhanced geothermal systems (EGS) like the one shown in Figure 13.9. The EGS plants actively inject water into deep wells to be heated by the hot rocks and pumped back out. Two holes are drilled into the rock, one to inject water into the rock and a second to

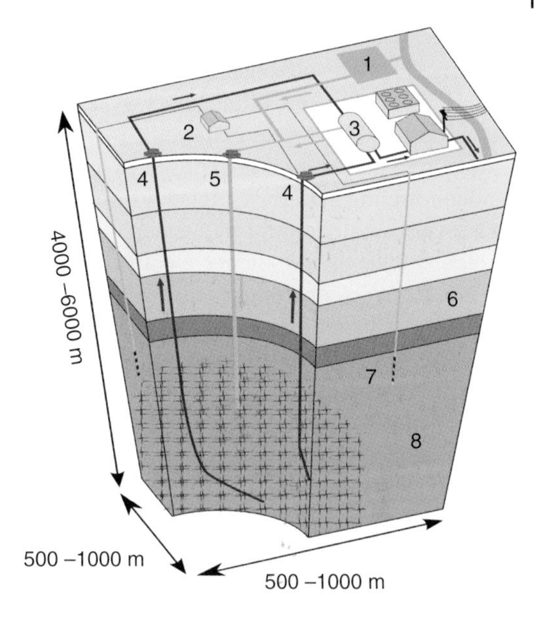

Figure 13.9 An enhanced geothermal system showing: water reservoir (1), pumping station (2), electric turbine (3), hot rock boreholes (4), water injection borehole (5), porous sediments (6), observation well (7), and hot bedrock (9). Source: Adapted from Geothermie-Prinzip, Wikimedia Commons.

extract the steam or hot water. The water is injected under high pressure in order to expand the existing rock fissures and enable the water to freely flow in and out. The injected water, which travels through fractures in the rock, is heated by the hot rock until it is forced out of the second borehole as steam or very hot water. The heat energy captured by the water is converted to electricity using a steam turbine. This water injection technique was adapted from the oil and gas extraction techniques discussed in Section 10.2.3. The differences are that the geologic formations are deeper and no toxic chemicals are used in the process, which reduces the possibility of environmental contamination. Good locations for this type of geothermal power are those over deposits of deep granite rock that is covered by 3–5 km (1.9–3.1 miles) of insulating sediments, which slow heat loss. An EGS plant is expected to have an economical lifetime of 20–30 years.

The main environmental impacts from geothermal power plants are the release of greenhouse gases that are carried with the hot water from the subsurface hot rock source. Since the hot water is a good solvent for minerals in the subsurface, it can contain toxic metals such as arsenic and mercury, which are a potential risk if released into nearby water supplies. However, recycling the extracted water back into the hot rock reservoir limits this potential impact. The other important impact from EGS power is the potential for causing seismic events from the deep subsurface injection of pressurized water into the hot rock. In most applications the seismic concerns have been minimal. However, there have been cases where hydraulic fracturing of the rock caused significant seismic activity to the point where operation of the geothermal project was ceased. The main economic drawback for geothermal power is the significant cost of drilling deep enough to reach the subsurface hot bedrock. These costs limit the number of regions to those where the overburden sediments are not so thick as to make drilling economically impractical.

13.2.4 Solar Power

Solar power converts the incoming energy from the Sun into electricity. This can be done by using photovoltaic (PV) devices, which convert sunlight to electricity directly, or by using a solar energy collector, which uses the sunlight to heat a working fluid that powers an electrical generator. The PV devices, commonly known as solar cells, operate by the photovoltaic effect, which is the process that generates a voltage or an electric current in a PV cell when it is exposed to sunlight. The photovoltaic effect is an application of the photoelectric effect, which was discussed in Section 3.1. In general, it is the emission of electrons or other free charge carriers from a material when it is exposed to light. The solar cells are constructed from two layers of semiconducting materials, typically doped highly purified silicon. The first silicon layer is doped with a very small amount of a group 15 element such as arsenic or antimony. Since the group 15 elements have five valence electrons, their addition into the silicon crystal, which has four valence electrons, introduces one more electron than is needed for bonding. This type of semiconductor – which has an excess amount of electrons in the crystal lattice – is known as an n-type semiconductor. The second silicon layer is doped with a very small amount of a group 13 element such as gallium. Since the group 13 elements have three valence electrons, their addition to the silicon crystal introduces one less electron than is needed for bonding. This type of semiconductor – which has a deficiency of electrons in the crystal lattice – is known as a p-type semiconductor. The crystal lattice in an electron-deficient p-type semiconductor contains electron holes, giving it an excess positive charge. These two types of semiconductors are brought together to create a p–n junction.

When sunlight hits the solar cell it is absorbed by the semiconductor materials. The free electrons in the n-type material become mobile and travel through the p–n junction toward the n-type semiconductor, creating an electric current. Once the electric current is generated, it is collected by metal plates on either side of the solar cell and transferred to wires where it can flow as electricity. Typically, many solar cells are connected together to form a larger device called a solar panel, shown in Figure 13.10. The solar panel produces direct current (DC), which fluctuates with the intensity of the sunlight. The DC current is converted to an alternating current (AC) by a power inverter. Solar panels can be placed on residential roofs to supplement the electrical power used in a home. A typical residential solar panel has approximately 60 solar cells combined to produce from 220 to <400 W of power. Storage batteries are often added to store any excess energy that is not required as it is produced. This allows for operation at night and at other times when sunlight is limited. Many residential solar panel systems are connected to the power grid whenever possible, which makes the storage of excess energy optional.

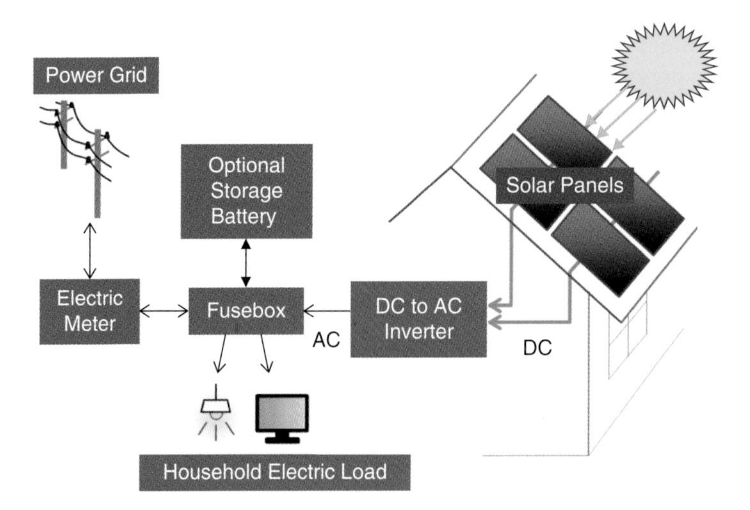

Figure 13.10 A typical residential solar panel system used for electrical power. Source: Adapted from S-kei, Wikimedia Commons.

The solar panels can be linked together in solar arrays, which are used to generate electricity in a commercial PV power plant, sometimes called a solar park. Most solar parks use ground-mounted solar arrays, which can be at a fixed tilt or use a single-axis or dual-axis solar tracker. Solar trackers automatically orient the solar panels toward the Sun. This minimizes the angle of incidence between the incoming sunlight and the panels, and increases the amount of energy produced. The solar panels in a solar park must be spaced out to reduce shading between them as they reorient toward the Sun during the day. Thus, solar parks that use solar trackers require more land area than solar parks that use fixed-tilt mounts with the same number of panels. However, the increase in energy output can be as much as 30% over fixed panels in locations with high levels of direct sunlight. While solar tracking improves the overall energy efficiency, it also increases the installation and maintenance costs.

Most commercial solar panels used in solar arrays have a 15–20% efficiency for converting the incoming solar radiation to DC electricity. The land area required for a specific power output varies depending on the location, the efficiency of the solar modules, the slope of the site, and the type of mounting used. Fixed-tilt solar arrays using typical modules of about 15% efficiency on horizontal sites need about 2.5 acres/MW in the tropics and over 5 acres/MW in the mid-latitudes. The required area is about 10% higher for a single-axis solar tracker and about 20% higher for a dual-axis solar tracker because of the longer shadow the array casts when tilted at a steeper angle. Figure 13.11 shows a large-scale commercial solar park located in Spain. The solar arrays are placed on poles with dual-axis solar trackers. The shadows cast by each solar array, while in steep tilt, determine the distance between each array and thus the total amount

Figure 13.11 A solar farm near Lerida, Spain, which uses a pole mount with dual-axis trackers. Source: Chixoy, Wikimedia Commons.

of land required by the solar park. The best locations for solar parks are sites where there is no other valuable land use.

The energy efficiency of a solar park is a function of the climate, the equipment used, and the system configuration. The primary energy input is the global light irradiance in the plane of the solar arrays, which is a combination of the direct and the diffuse solar radiation. A major factor in the energy output of the system is the conversion efficiency of the solar panels. The solar panels in the solar arrays are susceptible to collecting dust and debris which essentially block the solar light incident on the surface. They can also be damaged by dust storms, hail events, and slow corrosion of the surfaces by oxidants and acids in the air and precipitation. The deposition of debris or damage to the surface of the solar panels effectively reduces the intensity of the light absorbed by the semiconductor material and so reduces the efficiency of the PV devices. This requires that the panels be cleaned or replaced when necessary.

Another method used for collecting solar energy to produce electricity is by using a solar energy collector. Solar energy collectors use light-collecting optic systems, which concentrate the sunlight to a focal point where the energy can be converted into heat. A number of collector designs have been used to track the Sun and focus the sunlight. In all of these systems a working fluid is heated by the concentrated sunlight. The heated working fluid is then used for power generation in conventional steam-driven turbines. Solar power plants using this technology are called concentrated solar power (CSP) stations. CSP stations generate solar power by using mirrors or lenses to concentrate a large area of sunlight onto a small area. For example, two concentrated solar power tower stations located in Spain (PS10 and PS20) are shown in Figure 13.12. These two power stations produce 11 MW (PS10) and 20 MW (PS20) of power. The PS10 has 625 large collecting mirrors called heliostats, each with a surface of 120 m^2. The heliostats track the Sun and focus the light onto a tower where the heat is converted into

Figure 13.12 The PS10 (left) and PS20 (right) concentrated solar power stations located in Andalusia, Spain. Source: Koza 1983, Wikimedia Commons.

electricity by driving a steam turbine. PS20 has 1255 heliostat collectors, each with an area of 120 m^2, and produces about 48 000 MW yr^{-1}. Spain and the United States remain the global leaders in CSP stations, producing 2.3 and 1.7 GW of solar power, respectively. Areas where there is strong sunlight throughout most of the year are most suitable for the location of CSP stations. These areas include deserts and semi-arid regions, where cloudiness throughout the year is minimal.

A significant amount of energy is required to produce the materials used in both solar PV systems and CSP stations. The production of high-purity silicon used in the semiconductors for photovoltaic devices typically involves the conversion of SiO_2 to silicon by the reaction with high-purity carbon at temperatures greater than 2000 °C:

$$SiO_2 + 2C \rightarrow Si + 2CO$$

However, this reaction can also produce SiC as a side product:

$$SiO_2 + 3C \rightarrow SiC + 2CO$$

In order to minimize production of the side product SiC, the reaction is run with excess SiO_2 which can also react with the SiC to produce Si:

$$2SiC + SiO_2 \rightarrow 3Si + 2CO$$

The silicon derived from this high-temperature reaction requires further purification to produce the high-purity polycrystalline form of silicon used in semiconductors and PVs. This chemical purification process, called the Siemens process, involves the vaporization of volatile silicon compounds followed by thermal decomposition and vapor deposition. The volatile silicon compounds used are typically silane (SiH_4), which is decomposed to yield the purified silicon and H_2 gas:

$$SiH_4 \rightarrow Si + 2H_2$$

The decomposition of SiH_4 is then followed by chemical vapor deposition of the purified Si at low pressures (0.19–0.98 Torr). This is an energy-intensive process that leads to significant amounts of greenhouse gas emissions. Lifecycle analyses of the greenhouse gas emissions from solar PV plants, which include the process of forming purified silicon, estimate about 40 g CO_2 per kilowatt hour. By comparison, the greenhouse gas emissions from CSP stations are about 25 g CO_2 per kilowatt hour. Alternative materials are being explored that may reduce the greenhouse gas emissions from the production of PVs, including organic PVs and thin-film materials. Currently nuclear, hydropower, and wind power have lower lifecycle greenhouse gas emission estimates than solar power. Water usage for PV systems is very low, while the CSP systems require significant amounts of water for cooling as well as for the working fluid. In addition, long-term corrosion of both mirrors and PV devices could cause some local water quality problems, especially from cadmium and lead, which are used in the construction of solar cells. While rooftop solar systems do not have a significant land footprint, solar parks typically require a large land commitment. The large CSP facilities have land footprints slightly smaller than coal-fired power plants, which require large areas for storage of both coal and collected fly ash.

13.2.5 Biomass

The potential of the different energy sources derived from biomass, including woody biomass, bioethanol, and biodiesel, have been discussed in detail in Section 10.4. These renewable energy sources have numerous problems with regard to their impacts on the environment.

As all biomass fuels involve combustion, air pollution is still a major issue with their use. In addition to the NO_x and VOCs associated with the combustion of fossil fuels, the combustion of biomass fuels also emits other species that are related to air pollution health effects. This is primarily due to the biomass fuels containing oxygenated hydrocarbons. The liquid biomass fuels contain ethers, esters, or alcohols, while the woody biomass contains cellulose, which is a polysaccharide $(C_6H_{10}O_5)_n$. In all cases, the oxygenated hydrocarbons can crack to emit aldehydes and other species which contribute to eye irritation and respiratory stress. The liquid biomass fuels will also require use of land that could otherwise be used for growing food crops. There will always be a land use trade-off between food and fuel with the use of biomass fuels. In addition, biomass crops are water intensive and use fertilizers and pesticides that add to pollution of surface waters. Thus, the use of liquid biomass fuels has significant impact on both air and water quality. They also have higher greenhouse gas emissions than other alternative energy sources such as wind, hydroelectric, tidal, geothermal, nuclear, and solar power.

One possibility, which is currently being considered, is the use of natural woody biomass as an energy source for biomass electrical power plants. This woody biomass can be collected from areas where it is otherwise likely to provide fuel for wildfires in susceptible regions. With climate change causing natural forest underbrush to grow rapidly in wet springs and then dry out during dry summers, wildfires are expected to increase. In many areas this underbrush is cleared by controlled burns to remove it in areas susceptible to wildfires. Instead, this underbrush could be gathered to be used as fuel to produce biomass electrical power. This could reduce the potential for wildfires while creating jobs and adding to the total electrical capacity. The emissions from the biomass electrical plants could readily be controlled by post-combustion collection of pollutants to reduce atmospheric impacts from the fuel. This use of natural woody biomass for stationary electrical energy could be a viable alternative to the damage and air pollution created by uncontrolled wildfires or controlled burns.

13.2.6 Hydrogen

Hydrogen has also been proposed as another possible renewable energy fuel that could replace fossil fuels. Hydrogen can be produced from the electrolysis of water or from steam reforming of natural gas. Steam reforming combines water vapor and methane at temperatures between 700 and 1100 °C to yield H_2 gas as one of the products, with about 70% production efficiency:

$$CH_4(g) + H_2O(g) \rightleftarrows CO(g) + 3H_2(g)$$

The CO produced from the steam reformation can undergo further reaction with H_2O by the water gas shift reaction to also yield H_2:

$$CO(g) + H_2O(g) \rightleftarrows CO_2(g) + H_2(g)$$

Currently, steam reforming of natural gas is the major industrial source of H_2. However, this methodology requires a lot of thermal energy and produces the greenhouse gas CO_2.

Hydrogen can also be produced by the electrolysis of water, which requires a source of electrical power. If the source of the electrical power is another renewable energy source with low greenhouse gas emissions, the production of H_2 by electrolysis can be viewed as the conversion of electricity into H_2 without an increase in greenhouse gas emissions. However, if the electricity is obtained from fossil fuel-fired power plants, then the H_2 produced is accompanied by an increase in the emission of greenhouse gases and thus has no advantage over steam reforming.

Once obtained, the H_2 can be used to produce energy by combustion or by electrochemistry. If the H_2 is burned in pure O_2, the only product is water. However, the combustion of H_2 in air will produce NO_x under the high-temperature conditions of a hydrogen flame. The NO_x will

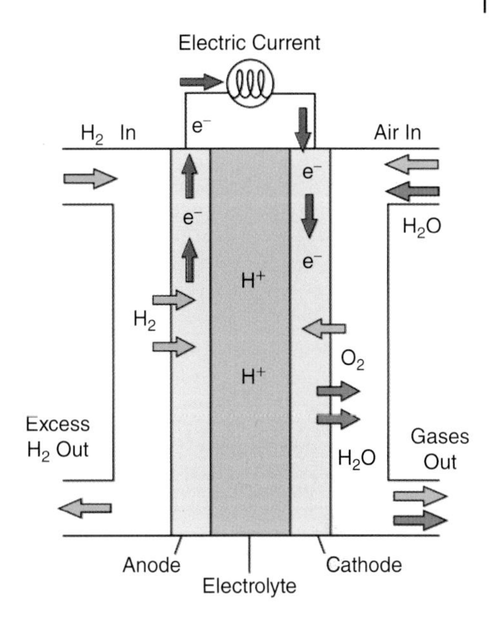

Figure 13.13 The operation of a hydrogen fuel cell. Source: Adapted from R. Dervisoglu, Wikimedia Commons.

contribute to air pollution, specifically the production of ozone and other products, as discussed in Section 5.4. The second option is to use the H_2 to produce energy in electrochemical fuel cells. A fuel cell is an electrochemical cell that converts chemical energy into electricity through an electrochemical reaction of H_2 with an oxidizing agent, typically O_2. The H_2 fuel cells consist of an anode, a cathode, and an electrolyte that allows positively charged hydrogen ions to move between the two sides of the fuel cell, as shown in Figure 13.13. The H_2 is oxidized at the anode to produce H^+ and electrons. The H^+ then flows from the anode to the cathode through the electrolyte, while electrons flow from the anode to the cathode through an external circuit, which produces electricity. At the cathode the H^+ ions react with O_2 in air to form H_2O, which is emitted from the fuel cell as water vapor.

The fuel cell effectively converts the H_2 into electricity. Every step in this process – from converting electricity into H_2 by steam reforming or electrolysis and then reconverting the H_2 back to electricity in a fuel cell – leads to loss of energy. Thus, the use of H_2 as a fuel is not an overall energy-efficient process. There are also safety issues in handling H_2 as a fuel, since it is a potentially explosive gas. In addition, most applications require the transport of the H_2 in heavy compressed gas cylinders at high pressures. The use of hydrogen fuel cells has been proposed as a source of power to drive electrically powered motor vehicles, since fuel cells are more energy efficient than fossil fuel combustion. However, the use of this type of technology has been limited due to the difficulties in the amounts of H_2 required and the hazards involved with the transport and storage of the H_2 fuel. The storage of compressed H_2 also adds a significant amount of weight to the vehicle, reducing its efficiency and requiring more fuel to travel the same distance. Although an alternative method of storage of the H_2 as a metal hydride has been explored, at this time the use of lighter lithium ion storage batteries is a safer and more reliable alternative to H_2 fuel cells as an energy source for motor vehicles.

While H_2 fuel cells are an emission-free method of producing electricity, lifecycle analysis has estimated that the CO_2 emissions from a fuel cell-powered vehicle are only about 10% less than that expected from a gasoline-powered vehicle when the production of the H_2 fuel is included (Dhanushkodi et al., 2008). In addition, if the use of H_2 as a fuel was to increase, it is very likely that its atmospheric concentrations would rise due to accidental or routine releases. Currently, the background levels of H_2 in the troposphere are at about 0.5 ppm, primarily from natural sources. The tropospheric lifetime of H_2 is similar to that of methane. The increased H_2 concentration in the troposphere would thus lead to the enhanced production of water vapor in the stratosphere similar that attributed to methane, as discussed in Section 4.5. The use of H_2 as a fuel would therefore have impacts on stratospheric ozone, increasing ozone depletion.

13.3 Sustainability

Sustainability is the process of maintaining change in a balanced fashion. Specifically, it means avoiding the depletion of natural resources in order to maintain an ecological balance. From an ecological perspective, sustainability involves meeting the current and long-term needs for any species survival, which includes food, water, reproduction, and a stable environment in which to prosper. When any species over-uses its resources, it is considered to have reached its carrying capacity. The carrying capacity of a biological species is the maximum population size that the environment can sustain indefinitely. When the carrying capacity is exceeded, a loss of population occurs until the population reaches a new equilibrium with the available resources. From an economical and societal perspective, sustainability involves the use of resources, direction of investments, development of technology, and change in institutions, which can meet both current and future human needs. Sustainability in the context of the human population deals with complex issues that include social values, economics, and the environment, as well as relationships with energy, water, and material resources.

The 2005 World Summit on Social Development identified the three goals of sustainability as economic development, social development, and environmental protection (UNGA, 2005). These goals have been expressed as three overlapping ellipses, shown in Figure 13.14, to indicate that the goals are interdependent – where none can exist without the others. The economy is dependent on the size of the society, and the size of both the economy and the society is constrained by environmental limits. The development of a sustainable economy and society relies on balancing local and global efforts to meet basic human needs without destroying or degrading the natural environment or depleting the natural resources. While the Earth is large in size and has significant resources, these resources are not limitless. In addition, our ability to be able to increase energy efficiency and use of the available resources is limited by chemical and physical laws.

Obtaining sustainability over the long term in the face of a changing environment caused by climate change will require rapid implementation of mitigation and adaptation strategies.

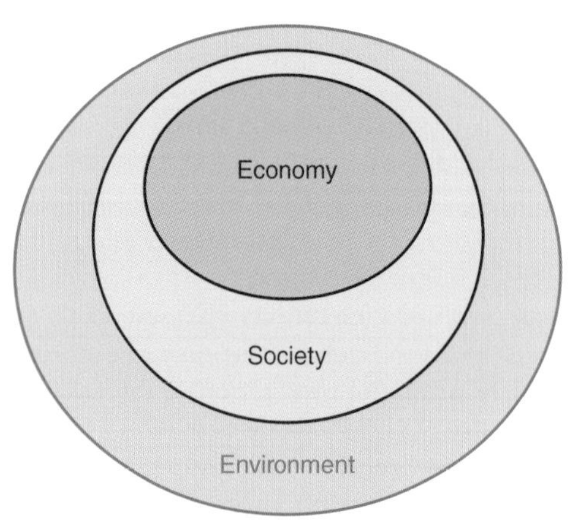

Figure 13.14 The interdependent relationship between the economy, society, and the environment, where the size of both the economy and society are constrained by environment limits. Source: KTucker via Wikimedia Commons.

One large factor that stands in the way of initiating these changes is the rapidly increasing global population, which continually requires more energy, more food, more water, and more materials to be sustained. Population increases lead to increases in air and water pollution, and increased emissions of greenhouse gases, which then increases the speed of climate change. Continuing to be able to sustain the increasing global population will be difficult, since fossil fuels have been the mainstay of global energy for electrical power as well as transportation. In order to maintain the same energy use as the fossil fuels become depleted, the use of alternative energy sources will need to be increased. These alternative energy sources are likely to include nuclear, wind, hydropower, and geothermal, since they have the lowest greenhouse gas emissions as well as the lowest impacts on air and water quality. While liquid biomass fuels can supplement the use of fossil fuels for transportation, biomass as a renewable energy source may not be sustainable due to the need to use land and water to grow food for the increasing population.

Mitigation and adaptation strategies for adapting to climate change have been discussed in Section 11.7. Mitigation of greenhouse gas emissions, as well as air and water pollutants, will need to be seriously considered to maintain a reasonable quality of life. There are many possible mitigation strategies, however they must all be carefully thought through before implementation to ensure that the investments of time, energy, manpower, and funds lead to a safe and sustainable outcome. Adaptation will be required to handle the impacts of climate change, which will include severe weather, floods, droughts, and rising sea levels. Most adaptation strategies will be easier to implement in countries with robust economies and politically stable societies. Coastal regions will be impacted globally and only those areas that can build the infrastructure to handle the predicted sea-level rise will be able to maintain a stable society. Island nations and low-lying coastal nations that do not have strong economies to build and support the necessary infrastructure to adapt to sea-level rise will lose land mass unless support is given by the more prosperous countries. In some cases, entire island nations may be lost. Both mitigation and adaptation will require global cooperation from all nations to be able to reach economic, social, and environmental sustainability for the world population.

Green chemistry, which is focused on the design of products and processes to minimize the use and generation of hazardous substances, has a role to play in reaching sustainability. The 12 principles established by the American Chemical Society, which are used to guide the practice of green chemistry in the development of chemical products and processes, are listed in Table 13.1. These principles focus on minimizing chemical waste, preventing chemical pollution of the environment, and minimizing the use and production of hazardous chemicals. Attempts are being made to quantify the "greenness" of a chemical process, while including other variables such as chemical yield, the cost of reactants, energy use, and simplicity of product synthesis and purification. One example of the use of green chemistry is the replacement of commonly used solvents with others derived from renewable resources such as biomass. Solvents are consumed in large quantities in many chemical syntheses, as well as for cleaning and degreasing. Traditional solvents are often toxic or environmentally harmful, such as chlorinated compounds. Green solvents, in contrast, are generally derived from renewable resources and biodegrade to safe products. A methodology has been developed in France by the chemical company Sanofi Chemie that uses solvent classification cards, which contain key information about materials, to assist organic chemists in finding suitable green replacement solvents (Prat et al., 2013). A number of the commonly used solvents, such as benzene, toluene, chloroform, and carbon tetrachloride, can be toxic to workers and are environmental pollutants. They are being replaced with solvents that have minimal impact on the environment and biodegrade rapidly if released into waste water streams. Some alternative solvents that are derived from biomass are being evaluated with regard to their availability, suitability, hazards, and costs, as

Table 13.1 The 12 principles of green chemistry.

Principle	Description
Waste prevention	Chemical waste prevention is easier than waste clean-up
Atom economy	Chemical synthetic methods should maximize material in the product
Less hazardous reagents	The least hazardous chemical reagents should be used to produce the best chemical product to safeguard workers and the environment
Safer chemical design	Design of chemicals should enhance their effectiveness while lowering any environmental impact
Use smallest amounts	All auxiliary materials used should be the smallest amounts needed and as nonhazardous as possible
Design for energy efficiency	Chemical syntheses and processes should use minimal energy
Use of renewable feedstocks	When possible, renewable chemical feedstocks should be used instead of non-renewables
Reduce steps	Unnecessary chemical steps should be avoided. The extra steps cost energy and produce waste
Use catalysts	Catalysts that can produce the product in small amounts repeatedly are preferred to reactants that are consumed
Design for degradation	Chemical products should not cause pollution and after use should degrade to non-harmful products
Real-time analysis	Analytical chemical methods should be developed to be real time for process monitoring and control to prevent hazardous materials from forming
Safer reactants	Chemical reactants should be chosen to minimize risks of explosions, fires, and accidental releases

part of a renewed effort in the chemical and pharmaceutical industries to become sustainable (Byrne et al., 2016).

Both the European Union and the United States have put in place regulations that require chemical companies to provide information regarding the safety of the chemicals that they produce. In 2007, the European Union established a chemical use program called "the regulation, evaluation, authorization, and restriction of chemicals" (REACH). Any chemical that is produced or imported in amounts over 1 T yr^{-1} is required to have data collected and submitted to the European Chemicals Agency regarding its safety and environmental hazards. Any chemicals that are likely to be carcinogens or have significant health risks must also be registered and authorized. The only exceptions to this are polymers and non-isolated chemical intermediates that occur in chemical syntheses. Similarly, the United States passed the Toxic Substances Control Act (TSCA) in 1976 and the Pollution Prevention Act in 1990, which are focused on preventing environmental impacts before they occur. The state of California has been proactive in this area by starting a green chemistry initiative in 2008 and establishing the Department of Toxic Substances Control, which prioritizes chemicals according to their hazards and promotes the replacement of harmful chemicals with safer ones following the principles of green chemistry.

Reaching a state of sustainability with depleting resources and an increasing population whose activities have impacts on the environment will first require a clear recognition of the problems before solutions can be formalized. Global climate change is occurring. The rate of this change – and how rapidly global communities can adapt to it – will depend on global societies working together on both short and long-term solutions. As technology advances, we will likely have many of the tools needed to accomplish these tasks, but it will require society to accept the need for change and to continue to initiate sustainability methods, such as those begun in green chemistry, if sustainability is to be reached in a timely manner.

13.4 Long-Term Planning

The yearly average global temperature anomalies that have been measured from 1880 to 2017 are shown in Figure 13.15 (black line) along with a 5-year moving average (Figure 13.15, red line). The global temperature is the average of the surface air temperatures and ocean surface temperatures across the globe. The atmospheric measurements are combined from the air above land and above the ocean surface collected by ships, buoys, and more recently satellites. Four major data sets are used to study the global temperature: HadCRUT4, produced jointly by the UK Met Office Hadley Centre and the University of East Anglia's Climatic Research Unit; GISTEMP, produced by the NASA Goddard Institute for Space Sciences (GISS); MLOST, produced by the National Oceanic and Atmospheric Administration (NOAA); and a fourth unnamed data set produced by the Japan Meteorological Agency (JMA).

A global temperature anomaly is the variation of the global temperature at a point in time from a baseline temperature determined over a 30-year average. This 30-year average was originally determined for 1951 to 1980 and after the turn of the century was moved to 1981 to 2010. The baseline temperature is set to zero and the variations are measured as positive for temperatures warmer than the baseline value and negative for temperatures cooler than the baseline value. Temperature anomalies are used instead of the absolute temperature because they tend to be highly correlated over larger distances (1000 km) and so are more typical of temperature changes over large areas, while the absolute temperatures vary significantly over even short distances. The reasonably reliable near-surface temperature records, which have at least semi-global coverage, are generally considered to begin around 1850. Temperature records earlier than this are sparse and not well standardized.

The baseline value used for the measurements in Figure 13.15 was determined for the period from 1951 to 1980. The negative temperature anomalies prior to 1970 in Figure 13.15 have mostly been attributed to higher loadings of atmospheric aerosols during this period, with some influences from changes in ocean circulation patterns such as La Niña. The temperature anomalies have all been positive and steadily increasing since 1975. The global average surface temperatures show a warming of 0.85 °C, in the period 1880 to 2012, based on the four major data sets. This is generally a trend of 0.064 ± 0.015 °C per decade. However, most of the observed warming occurred from 1900 to 1940 and after 1970. In addition, the warming trend is higher over land than over ocean, higher for the Arctic regions, and higher since 1970 than from 1880 to 1970 (IPCC, 2013).

Figure 13.16 shows the average temperature anomalies predicted by eight of the major models used to estimate future temperature anomalies from a baseline value determined from 1981 to 2010. It is assumed in the models that no specific policy measures are made to control future greenhouse gas emissions. The global temperature anomaly measured in 2017 from Figure 13.15 is approximately +0.95 °C and is expected to reach +1.0 °C by 2020. This is higher than the predicted value of about +0.4 °C in 2020. Although the difference in the measured and predicted temperature anomalies is due in part to the difference in baseline values used, the predicted value of +1 °C was not to be reached until the period between 2030 and 2050, depending on the model used. The increased warming by 2100 is predicted to be from +2.2 to +4.7 °C with an average between the models of +3.4 °C. The warming trend from 1900 to

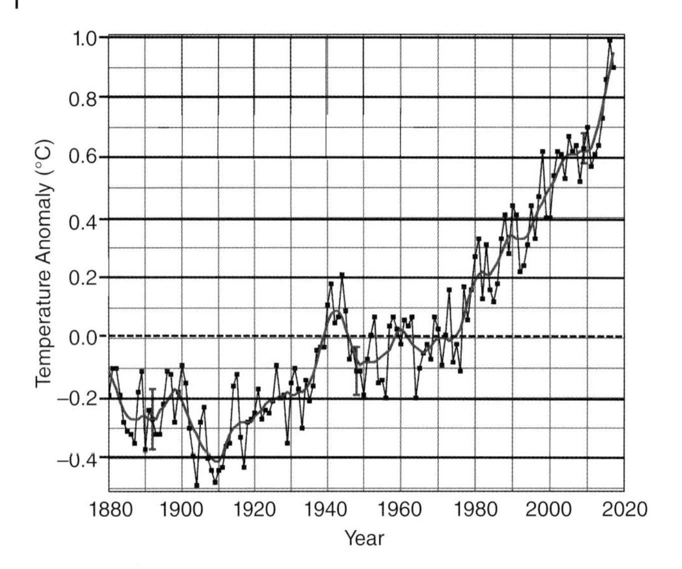

Figure 13.15 Measured global temperature anomalies (black) along with a 5-year moving average (red). Source: Adapted from NASA, Wikimedia Commons.

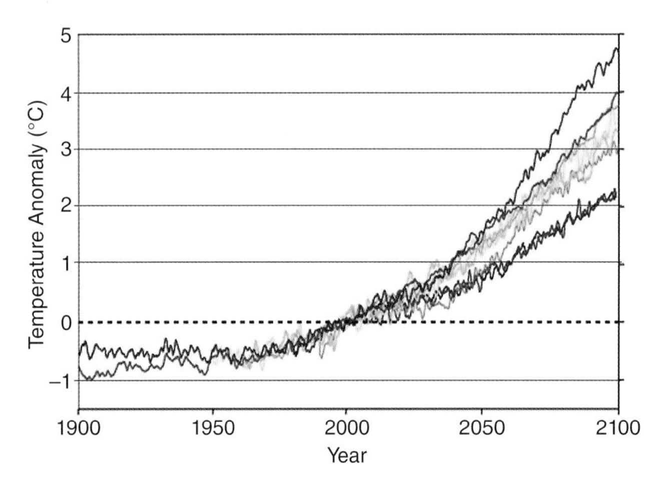

Figure 13.16 The global temperature anomaly predicted by eight major climate models. Source: Adapted from Robert A. Rohde, Wikimedia Commons.

2018 was measured as 1.25 °C (Figure 13.15) and the predicted value over the same time period is 1.0 °C (Figure 13.16). Thus, the general agreement between the model predictions and the measured values during the time period from 1900 to 2018 shows that the model predictions of increased warming from 2020 to 2100 need to be taken seriously.

Since the average mean global warming projections are based on global average temperatures, they do not imply that they should be applied to all regions of the globe. The projected change in annual mean surface air temperature from the late twentieth century to the middle twenty-first century from the NOAA Geophysical Fluid Dynamics Laboratory model (Figure 13.16, turquoise blue line) for different regions of the globe is shown in Figure 13.17. The projections assume that no specific policy is implemented to control greenhouse gas emissions. The estimates for the surface temperatures obtained from this model clearly show that the

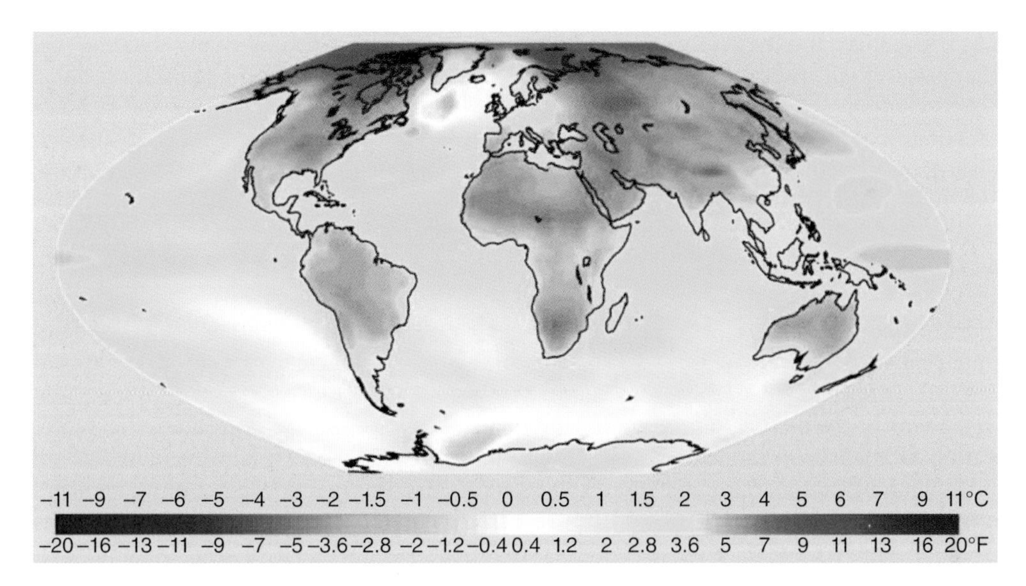

Figure 13.17 The projected changes in annual mean surface air temperature from 2050 to 2150 by the NOAA Geophysical Fluid Dynamics Laboratory model, assuming no limitations on greenhouse gas emissions. Source: NOAA Geophysical Fluid Dynamics Laboratory, Wikimedia Commons.

Arctic and Northern latitudes will be the most severely impacted regions, with temperatures increasing by 5–6 °C. In addition, land temperatures will increase more than ocean temperatures by about 3–4 °C. The results clearly indicate that adaptation strategies must be implemented prior to the mid-twenty-first century, especially in the most severely impacted regions of the globe, in order to keep pace with climate change. These projections of increasing temperatures will also be linked to changes in weather patterns, increases in severe weather, and increases in wildfires.

The amount of global mean sea-level rise will depend on how warm the polar regions become and how much of the ice melts, particularly the land ice in regions such as Greenland and Antarctica. Satellite radar measurements have reported a rise in global mean sea level of 7.5 cm (3.0 inches) from 1993 to 2017. This represents a trend of about 30 cm (12 inches) per century. The contributing factors to this rise are thermal expansion of the oceans (42%), melting of temperate glaciers (21%), and melting of land ice in Greenland and Antarctica (23%). The latest IPCC estimate of sea-level rise is 90 cm (2 ft) by 2099 (IPCC, 2013). In addition, as with temperature anomalies, sea-level rise will not be equal across the globe. Coastal subsidence, gravitational effects of changing ice masses, and variation in thermal expansion all influence regional sea-level rise. Sea level may actually fall in areas near current and former glaciers and ice sheets. The Atlantic Ocean is predicted to warm faster than the Pacific Ocean. This could cause a sea-level rise three to four times the global average on the U.S. east coast and Europe, as well as impacts on the Atlantic Meridional Overturning Circulation (AMOC) discussed in Section 11.5.

Alternative energy sources such as wind, hydropower, solar, tidal, and even biomass, will be impacted by these effects of climate change. The expected changes in rainfall, sea-level rise, and storm frequency need to be considered as renewable energy infrastructure is established. For example, stronger and more frequent storms, such as hurricanes, are projected by global climate models. Energy infrastructure that is placed in areas where high winds and flooding are expected to occur more frequently needs to be able to withstand this type of event without failure or further damage to the environment. The establishment of renewable energy sources such as wind, solar, and hydropower also needs to consider how potential changes in regional weather patterns will impact the choice of location and the ultimate performance of these alternate energy sources. Changes in cloud cover will have direct effects on solar power and changes in precipitation distribution patterns will affect river flows that are important for hydropower. Increasing temperatures, frequent droughts, and flooding will have impacts on agriculture, which will affect the sustainability of biomass fuels. Changing temperature patterns will also increase energy demand in areas where increases in summer temperatures and decreases in winter temperatures affect air conditioning and heating requirements. In order to consider the effects of climate change on the establishment of alternative fuels, our ability to predict weather and climate patterns needs to continue to be improved. A more thorough understanding of regional changes will require continued development of the RCMs in addition to critical field measurements for model validation.

While many of the projected impacts from climate change cannot be prevented, the implementation of adaptation strategies can help to minimize their effects. However, adaptation requires planning and a continued re-examination of current greenhouse gas emissions and energy use as they relate to anticipated future climate conditions. The main obstacle to the rapid establishment of adaptation strategies is that most societal and economic systems typically focus on short-term problems and not on long-term issues. This approach is one of reaction to the short-term problem as discussed in Sections 1.4 and 1.5. Nonetheless, proactive efforts toward the predicted impacts of climate change will be much more cost effective than short-term reactive efforts, as has been seen in regions such as the Netherlands, where efforts have been made to invest in adaptive strategies in advance of expected problems.

Lifecycle analysis, as discussed in Section 11.7, will be a useful tool required in making informed decisions with regard to adaptation strategies with the least short-term and long-term impacts. These lifecycle assessments should be applied to energy sources, as well as to commonly used products, chemicals, and materials. They are aimed at determining the best source of energy or the best product for a given application with the least impact on the environment. There are a number of approaches that have been applied to lifecycle assessments. These include "cradle to grave," "cradle to gate," and "cradle to cradle." The "cradle to grave" approach includes all energy use and environmental impacts, beginning with the point that the raw materials are obtained and ending with the final waste products and their disposal. The "cradle to gate" approach includes every impact from the extraction of the raw materials to the point that the product leaves the production facility. The "cradle to cradle" approach begins with the extraction of raw materials and ends with recycling the product for reuse as the same product, a different product, or raw materials. In most cases, the "cradle to grave" approach will give the most complete evaluation of the environmental impacts of a product or energy source.

The different lifecycle assessments have evolved due to the increasing potential for climate change impacts if greenhouse gas emissions are not reduced. This approach has led to the concept of the carbon footprint. A carbon footprint is the total amount of greenhouse gases emitted into the troposphere by an individual, an organization, an event, or a product, reported as an equivalent amount of CO_2. The determination of a carbon footprint is not as easy as it may first appear. In many cases, the exact amount of the carbon footprint cannot be determined due to lack of information about the complex interactions involved in the cycling of greenhouse gas species in the environment. In addition, greenhouse gas emissions can come from both direct and indirect sources. For example, the carbon footprint of a city or town is a combination of the greenhouse gas emissions from the energy (transportation and electrical) used by the population, the emissions from the transportation of goods to and from the population, and the emissions during the production of those goods. The greenhouse gas emissions from the energy used by the population are direct emissions, while the emissions from the production and transportation of goods to the population are indirect emissions. Both direct and indirect greenhouse gas emissions need to be considered to accurately determine the carbon footprint. Similar approaches have been made to determine a water footprint and a land footprint corresponding to the amounts of water and land required to support the individual, organization, event, or product.

There are always trade-offs when attempting to reduce the carbon footprint of an individual, organization, event, or product. For example, the largest component of an individual's carbon footprint is energy use. The use of air conditioning in summer and heating in winter can be reduced by adding insulation and reducing the air exchange in a building. This reduces the greenhouse gas emissions associated with the energy required for indoor temperature control. However, indoor air pollution in homes will increase due to the reduced air recirculation, as discussed in Section 12.7. The carbon footprint of a product can be reduced by preparing the product for reuse. However, care must be taken to ensure that the items being reused do not present a health risk. For example, plastic bottles are not reused because it is difficult to pasteurize them after use. Glass bottles can easily be pasteurized and reused, however there is an energy cost associated with pasteurization which adds to the product's carbon footprint. For sustainability purposes, the conservation, reuse, and recycling of materials requires that the materials and processes involved result in a reduction in the carbon, water, and land footprints without the addition of environmental or health risks.

While conservation, recycling, and reuse of materials can all help to lower the carbon, water, and land footprints, the major source of the carbon footprint of any individual, organization, or product is energy. The minimum, maximum, and average estimates of CO_2 equivalent greenhouse gas emissions from different energy sources are listed in Table 13.2, as determined from lifecycle analysis (IPCC, 2014). Both nuclear and wind power have the lowest CO_2 equivalent greenhouse gas emissions, with an average of 11–12 g kWh^{-1}. The high maximum value listed for hydropower (2200 g kWh^{-1}) arises from estimates of changes in land use as well as the greenhouse gases produced from the creation of the manmade reservoir. Both solar and geothermal have the highest emissions of all the non-combustion-related energy sources, with PV solar

Table 13.2 The average, minimum, and maximum estimates of the CO_2 equivalent greenhouse gas emissions in g CO_2 kWh^{-1} from the currently available energy sources as determined from lifecycle analysis (IPCC, 2014).

Energy source	Average	Minimum	Maximum
Coal, pulverized	820	740	910
Gas, combined cycle	490	410	650
Biomass	230	130	420
Biomass, co-firing with coal	740	620	890
Solar PV, utility scale	48	18	180
Solar PV, rooftop	41	26	60
Solar, CSP	27	8.8	63
Geothermal	38	6.0	79
Hydropower	24	1.0	2200
Wind, offshore	12	8.0	35
Wind, onshore	11	7.0	56
Nuclear	12	3.7	110

being the highest at 41–48 g kWh^{-1}. The majority of energy lifecycle assessments agree that nuclear power has the most promise for lowering the carbon footprint of energy production. However, nuclear power will require considerable education and security measures before it can move forward at a pace that will make a difference in the predicted impacts of climate change. This will require planning and acceptance by the global society for it to occur in a timely manner.

Estimates of the CO_2 equivalent greenhouse gas emissions from motor vehicles are based on standard driving cycles, which differ from region to region. For example, in the United States the estimated emissions are 200 g CO_2 per kilometer for passenger cars and 280 g CO_2 per kilometer for trucks, based on the USEPA standard driving cycle. This is compared to the European estimates of 127 g CO_2 per kilometer for passenger cars and 175 g CO_2 per kilometer for light commercial vehicles, using the European driving cycle. The major difference between these two driving cycles is that the USEPA driving cycle requires more stops and starts than the European driving cycle in order to represent U.S. urban driving conditions. The comparison of the CO_2 equivalent emissions between these two driving cycles demonstrates that driving conditions can have a significant effect on the emissions per kilometer. Fuel conservation methods that impose lower speed limits or help to maintain constant road speeds with minimal stopping and starting can reduce CO_2 equivalent greenhouse gas emissions from motor vehicles. However, to be effective these measures require changes in behavior patterns, which can conflict with economic and social habits. They would also require significant changes in infrastructure, including vehicles, roads, and traffic control devices.

In addition to long-term planning, the global community will need to begin to work together to build the necessary infrastructure required to avoid or minimize the impact of climate change. While the problems can seem overwhelming, if environmental scientists work together with societal leaders the global community may be able to limit the magnitude of the inevitable effects of climate change. In fact, the technologies are currently available and can be implemented now that would reduce the emission of greenhouse gases. The main obstacle to the implementation is economic and societal reluctance to changing the current patterns of energy use. In some cases, this includes a denial that anthropogenic activities are the cause of climate change, despite overwhelming scientific evidence to the contrary (IPCC, 2013).

In any case, the nonrenewable fossil energy sources, which are currently the main global energy fuels for electricity and transportation, will continue to be depleted in the coming decades. The inevitable shortage of fossil fuels will drive prices up and ultimately cause societies to seek alternatives. The recognition of this should encourage us to begin to adopt cleaner energy sources, renewable and environmentally safe products, and improved agricultural practices, while maintaining environmental quality. This is required to assure that future generations will be able to continue to maintain a reasonable quality of life. Climate change, with all of its predicted impacts, is happening and is predicted to worsen if we make no changes in our energy policies today. These changes will require global cooperation and planning to find and implement long-term sustainable solutions. It is preferable that the recognition of the impacts of fossil fuel combustion on the environment and the implementation of alternative energy sources occurs in time for us to slow the progress of global warming and climate change, instead of waiting for the worst to occur before we act.

References

Byrne, F.P., Saimeng, J., Paggiola, G. et al. (2016). Tools and techniques for solvent selection: Green solvent selection guides. *Sustainable Chemical Processes* 4 (7): https://doi.org/10.1186/s40508-016-0051-z.

Dhanushkodi, S.R., Mahinpey, N., Srinivasan, A., and Wilson, M. (2008). Life cycle analysis of fuel cell technology. *Journal of Environmental Informatics* 11: 36–44.

Gaffney, J.S. and Marley, N.A. (2009). The impacts of combustion emissions on air quality and climate: From coal to biofuels and beyond. *Atmospheric Environment* 43: 23–36.

IPCC 2012 *Renewable energy sources and climate change mitigation*. Available online at https://www.ipcc.ch/site/assets/uploads/2018/03/SRREN_Full_Report-1.pdf (accessed January 7, 2019).

IPCC 2013 *Climate change 2013: The physical science basis*. Available online at https://www.ipcc.ch/assessment-report/ar5 (accessed January 1, 2019).

IPCC 2014 *IPCC Working Group III – Mitigation of Climate Change, Annex III: Technology – specific cost and performance parameters*. Available online at https://www.ipcc.ch/pdf/assessment-report/ar5/wg3/ipcc_wg3_ar5_annex-iii.pdf (accessed November 17, 2018).

Prat, D., Pardigon, O., Flemming, H.W. et al. (2013). Sanofi's solvent selection guide: A step towards more sustainable processes. *Organic Process Research & Development* 17: 1517–1525.

UNGA 2005 *2005 World summit outcome*. Available online at http://data.unaids.org/topics/universalaccess/worldsummitoutcome_resolution_24oct2005_en.pdf (accessed December 29, 2018).

Study Problems

13.1 (a) What is sustainability?
(b) What is sustainable development?

13.2 What are the three major organic petrochemical feedstocks for the petrochemical industry?

13.3 What is the process that converts gas and liquid feeds from gas and oil refining into light alkenes?

13.4 In the United States refineries, what percentage of each barrel of crude petroleum is used as a petrochemical feedstock?

13.5 What are the two most important petrochemical feedstocks?

13.6 What effect would stopping the combustion of crude oil-derived fuels have on the petrochemical industry in the United States?

13.7 (a) What is the most used renewable energy source worldwide?
(b) What is the second most used renewable energy source?

13.8 (a) What is the minimum wind speed required by a wind turbine to be able to generate power?
(b) What is the optimum wind speed?

13.9 (a) What areas around the globe are best for wind power?
(b) What areas in the United States?

13.10 What percentage of electricity globally comes from hydropower?

13.11 (a) What is the major drawback to hydroelectric power?
(b) What type of fishing industry has been most affected by hydroelectric power plants in North America?

13.12 (a) What country is the leader in geothermal energy use?
(b) What percentage of its energy is from geothermal?

13.13 (a) What are the two methods for converting solar energy into electricity?
(b) How does a solar energy collector function?

13.14 (a) At what temperatures is SiO_2 converted to high-purity Si?
(b) What is the reaction for this process?
(c) What reaction conditions are used to minimize the formation of the side product SiC?

13.15 (a) What are the CO_2 equivalent greenhouse gas emissions for a photovoltaic solar power plant according to lifecycle analysis?
(b) What are the CO_2 equivalent greenhouse gas emissions for a concentrated solar power station?

13.16 Why are biomass fuels not as attractive an alternative to fossil fuel resources when compared to other renewable energy possibilities?

13.17 (a) What are the two methods of producing hydrogen for fuel?
(b) Which is the major industrial source of hydrogen?
(c) Once produced, what are the two methods used to convert hydrogen to energy?

13.18 What is the reaction that occurs at the cathode of a H_2 fuel cell?

13.19 If the use of H_2 as a fuel was to increase, how would this affect the stratosphere?

13.20 (a) What is carrying capacity?
(b) What happens when a species reaches its carrying capacity?

13.21 (a) What are the three goals of sustainability?
(b) How are these three goals related?

13.22 What will be required to be able to develop sustainability over the long term in the face of a changing environment and climate change?

13.23 (a) What is the focus of green chemistry?
(b) What is the general focus of the 12 principles of green chemistry?

13.24 (a) What is the global temperature?
(b) What is a global temperature anomaly?
(c) Why are temperature anomalies used to determine global temperature changes instead of absolute temperatures?

13.25 Most of the observed warming between 1880 and 2012 occurred in what two periods?

13.26 Which will be the regions with the greatest warming in the mid-twenty-first century?

13.27 (a) What is the largest contributing factor to sea-level rise?
(b) What is the second largest factor?

13.28 (a) What is the latest IPCC estimate of sea-level rise by 2099?
(b) What areas are expected to be most impacted by sea-level rise?
(c) Why?

13.29 What is included in the following lifecycle approaches:
(a) Cradle to grave?
(b) Cradle to gate?
(c) Cradle to cradle?

13.30 What is a carbon footprint?

13.31 According to lifecycle analysis:
(a) What two alternative energy sources have the lowest CO_2 equivalent greenhouse gas emissions?
(b) Which two alternative energy sources have the highest CO_2 equivalent greenhouse gas emissions of all the non-combustion-related energy sources?

Appendix A

Answers to Study Problems

Chapter 1

1.1 Environmental chemistry is traditionally defined as the study of the sources, reactions, transport, effects, and fates of chemical species in water, soil, and air environments, and the effect of human activity on these.

1.2 The four major environmental systems are the atmosphere, hydrosphere, geosphere, and biosphere.

1.3 (a) Green chemistry is the development of chemical processes that use smaller amounts of safer chemicals with less energy in order to lower their environmental impacts.
(b) Environmental chemistry is focused on understanding the chemical reactions and processes that control the environmental systems and how these are impacted by the addition of anthropogenic chemicals.

1.4 The areas of chemistry important to the study of environmental chemistry include analytical chemistry, physical chemistry, organic chemistry, inorganic chemistry, and simple biochemistry.

1.5 (a) Anthropogenic pollution is defined as the introduction of harmful substances or the creation of harmful impacts in the environment that are tied directly to man's activities, including agriculture, industry, and energy production and use.
(b) Anthropogenic pollution can be described as being intentional or nonintentional, with intentional pollution in many cases being tied to warfare.

1.6 The recognition of occupational exposure problems led to the establishment of the Occupational and Safety Administration (OSHA).

1.7 Dichlorodifluoromethane (CCl_2F_2), trademarked as "Freon," was developed for use in refrigeration units as an alternative to ammonia, propane, sulfur dioxide, and chloroform.

1.8 Freon acts as a stratospheric ozone-depleting agent and is also a potent greenhouse gas.

1.9 The time period before the Industrial Revolution can be considered an era of "the environment vs man," since our technology had not advanced to a point where we could

Chemistry of Environmental Systems: Fundamental Principles and Analytical Methods, First Edition.
Jeffrey S. Gaffney and Nancy A. Marley.

control or harness our environment and we lived at the mercy of environmental conditions. The period after the Industrial Revolution can be considered an era of "man vs the environment," where we begin to control the impacts of the environment on society.

1.10 (a) The most recent geological era, beginning about 8000 BCE, is considered by many geologists to be the Anthropocene.
(b) This is because the presence of *Homo sapiens* has had such a large impact on the Earth since that time.

1.11 It currently takes about 12–13 years for the Earth's human population to increase by a billion.

1.12 Cities with more than 10 million inhabitants are called megacities.

1.13 The trends in increasing human population and fossil fuel use are directly correlated.

1.14 Fossil fuels are in limited amount and are a non-renewable source of energy.

1.15 1,1″-(2,2,2-trichloroethane-1,1-diyl)bis(4-chlorobenzene) or dichlorodiphenyltrichloroethane is the chemical name of the pesticide DDT.

1.16 The use of DDT was banned because it was found to bioaccumulate in animals and was causing reproductive problems in bird populations.

1.17 The book written by Rachel Carson that described the problems associated with the use of the pesticide DDT was called *Silent Spring*.

1.18 The two alkyl lead compounds that were used as additives in gasoline to reduce engine knocking were tetramethyllead and tetraethyllead.

1.19 The chemical formula of cinnabar is HgS.

1.20 The lowest level of the atmosphere is the troposphere.

1.21 The atmosphere is the most mobile of the four environmental systems.

1.22 The goal of environmental chemistry is the development of a sustainable, sound, and safe Earth system for future generations.

Chapter 2

2.1 (a) Argon.
(b) It is sourced from the Earth's surface and geosphere from the radioactive decay of the long-lived ^{40}K that yields ^{40}Ar, which is the dominant isotope of argon in the Earth's atmosphere.

2.2 (a) Nitrogen and oxygen (N_2 and O_2).
(b) Oxygen is the more reactive gas and is sourced from the biosphere, being produced from photosynthesis.

2.3 Methane (CH_4) and carbon dioxide (CO_2).

2.4 (a) 7.5 people would be a part per billion.
(b) 7500 people would be a part per million.

2.5 The troposphere, the tropopause, and the stratosphere.

2.6 The Earth's surface is warmed by the Sun, causing it to be heated. That heat is transferred to the nearby air, making it less dense. As you go up in altitude away from the surface of the Earth, the pressure drops and the air is colder. Cold air sinks and warm air rises, so cold above warm leads to mixing until you reach the tropopause where there is constant temperature.

2.7 (a) Carbon dioxide, CO_2, but note that on Earth water is very important also in maintaining the Earth's temperature.
(b) Mars is colder because it has a much lower atmospheric pressure so there are fewer molecules of CO_2 to trap the heat.

2.8 The South Pole is completely covered by snow and ice that has a very high albedo. It is also at a high altitude of about 2800 m in elevation, which makes it colder than the North Pole which is closer to sea level. Both poles are very cold, so there is little water vapor in the air and it is very dry.

2.9 (a) A temperature inversion is a region of the atmosphere where there is warm air above cold air.
(b) This is not normal in the troposphere. It is a stable situation and the cold air below needs to be heated or moved to go to the normal situation in the troposphere where cold air is above warm. This typically happens at night or in the early morning, especially in valleys where the terrain can make them fairly common.

2.10 Tall stacks allow for the pollutants to be released at height, above thermal inversions which might occur in the valley regions. If the emissions were released at ground layer into cold air masses that are capped by warm air (an inversion which is stable), the concentrations of pollutants would be higher due to lack of mixing.

2.11 Hadley, Ferrel, and Polar Cells.

2.12 (a) Water vapor.
(b) Carbon dioxide (CO_2).
(c) Carbon dioxide.

2.13 Keeling curves in honor of the late Dr. David Keeling of the Scripps Institute of Oceanography.

2.14 Milankovitch cycles.

2.15 Oxygen, O_2 and ozone, O_3. They can both absorb ultraviolet radiation that leads to breaking and forming of molecular bonds, leading to heating the stratosphere.

2.16 Clouds can act to reflect incoming solar radiation and shade the planet. Dissolved components in the clouds and liquid water can absorb outgoing IR radiation, which can lead to greenhouse effects.

2.17 The deserts are dry regions (arid). The relative humidity (water vapor content) is much lower, so the outgoing longwave radiation going from the Earth's surface at night into space is not absorbed by water as it is in grasslands and forests that are more humid (i.e. water vapor content is higher).

2.18 There is more terrestrial land mass in the Northern Hemisphere than the Southern Hemisphere. The terrestrial biomass is also larger, and agricultural plants and deciduous forests that are seasonal in their photosynthetic uptake of carbon dioxide cause the observed changes in carbon dioxide levels in the Northern Hemisphere. Lower levels are due to uptake of CO_2 by plants during spring and summer times.

2.19 Thermohaline circulation.

2.20 The soot would act to change the surface albedo from a strongly reflecting to a strongly absorbing albedo. This could increase heating of the surface and melt ice from Greenland that would cause warmer freshwater flow into the oceans, causing a decrease in salinity and density that could affect the surface water circulations.

Chapter 3

3.1 Electromagnetic wave theory best describes the behavior of light with regard to reflection and refraction.

3.2 Snell's Law describes the change in direction of an electromagnetic wave when traveling from one medium to another medium of different density.

3.3 The Fresnel equation describes the ratio of the reflected light intensity to the intensity of incident light for a perpendicular beam of unpolarized light.

3.4 Constructive and destructive interference.

3.5 (a) $2.997\,924\,58 \times 10^8$ m s^{-1}.
(b) $2.997\,924\,58 \times 10^5$ km s^{-1}.

3.6 (a) As an electromagnetic wave crosses a boundary between different media, its velocity changes.
(b) Its frequency remains the same.

3.7 The long-wavelength light (red) is refracted less than the short-wavelength light (violet).

3.8 Only light absorbed by a molecule can cause a photochemical reaction.

3.9 Light absorption by a molecule is a one-photon process and the sum of the primary process quantum yields Φ must add up to one.

3.10 The inclusion of secondary processes results in an apparent quantum yield that is greater than one.

3.11 Beer's Law relates the absorbance of light by an absorbing species to the concentration.

3.12
1. Thermochemical reactions involve the absorption or emission of heat, while photochemical reactions involve the absorption of light.
2. Thermal reactions involve reactants in their ground electronic states, while photo chemical reactions involve reactants in their electronically excited states.

3.13 An einstein is a mole of photons.

3.14 (a) 109 kcal mol^{-1}.
(b) 142 kcal mol^{-1}.

3.15 (a) The shortest wavelength of light entering the stratosphere is 200 nm.
(b) The shortest wavelength entering the troposphere is 300 nm.

3.16 (a) Third order.
(b) First order.
(c) Third order.
(d) Second order.

3.17 (a) Second order.
(b) First order.
(c) First order.
(d) First order.

3.18 Collision of the product with the third body removes the excess energy stabilizing the product.

3.19 (a) Third-body reactions are termolecular as all three species are involved in the transition state.
(b) In cases where the atmospheric pressure is constant and the concentration of M does not change, the reaction is considered to be second order overall.

3.20 (a) $k = Ae^{-Ea/RT}$.
(b) The parameter A is called the frequency factor, which is related to the number of collisions with the correct orientation for reaction.

3.21 The half-life of a first-order reaction is a constant and does not depend on the initial reactant concentration.

3.22 (a) The three possible photochemical pathways for an electronically excited molecule are dissociation, ionization, and molecular rearrangement.
(b) The most important photochemical reaction in the lower atmosphere is dissociation.

3.23 (a) The physical deactivation pathways that involve the emission of light are fluorescence and phosphorescence.
(b) The difference between fluorescence and phosphorescence is that the electronic transition in fluorescence takes place between two electronic energy levels with the same electron spin, called singlet states. The electronic transition in phosphorescence involves two electronic energy levels of different electron spins, a singlet state and a triplet state.

3.24 (a) Collisional deactivation and internal conversion deactivate an electronically excited molecule by the emission of heat.
(b) Collisional deactivation involves the transfer of the excitation energy to other molecules in the system through energetic collisions. Internal conversion transfers the excitation energy to vibrational modes of the same molecule.

Chapter 4

4.1 The stratosphere lies between the tropopause and the stratopause.

4.2 About 20% of the mass of the atmosphere is in the stratosphere.

4.3 (a) The top of the stratosphere is approximately 50 km.
(b) The bottom of the stratosphere is about 18 km over the equator.
(c) 10–13 km over middle latitudes.
(d) About 8 km over the polar regions.

4.4 Vertical mixing is slow in the stratosphere because warmer air lies above colder air. Vertical mixing is faster in the troposphere because warmer air lies below colder air. The warmer air rises and the colder air sinks.

4.5 The increase in temperature with altitude in the stratosphere is due to the series of exothermic chemical reactions, primarily the dissociation of O_2 to O atoms and the reaction of O with O_2 to form O_3.

4.6 (a) Oxygen and ozone absorb most of the UVB and UVC radiation in the stratosphere.
(b) Oxygen primarily absorbs radiation in the UVC.
(c) Ozone primarily absorbs in the UVB.

4.7 (a) Ozone does not absorb at 325 nm so it is an estimate of the solar intensity at the top of the atmosphere (I_o). Ozone does absorb at 305 nm so it is the solar intensity after absorption (I). The total column ozone is determined by Beer's Law and the ratio I/I_o.
(b) Ozone is measured in Dobson units, defined as the thickness (in units of 10 μm) of a layer of pure ozone at STP that would give the same absorbance as that measured in the total atmospheric column.

4.8 Oxygen atoms are produced in the stratosphere by the photolysis of molecular oxygen.

4.9 The stratospheric ozone formation cycle is called the Chapman Cycle.

4.10 (a) The ozone formation cycle overpredicts the amount of ozone observed in the stratosphere.
(b) It only addresses the formation of ozone and it does not address the natural chemical cycles that are responsible for the loss of ozone.

4.11 The HO_x Cycle and the NO_x Cycle are natural ozone loss mechanisms.

4.12 $O(^1D)$ initiates both the HO_x and NO_x Cycles.

4.13 (a) $O(^3P)$ is the term symbol of the oxygen atom ground state.
(b) $O(^1D)$ is the term symbol of the first excited state of the oxygen atom.

4.14 (a) OH reacts with O_3 in the HO_x Cycle.
(b) $HO_2 + O \rightarrow OH + O_2$.

4.15 (a) NO reacts with O_3 in the NO_x Cycle.
(b) $NO_2 + O \rightarrow NO + O_2$.

4.16 $OH + NO_2 \rightarrow HNO_3$.

4.17 The net ozone depletion reaction is $O_3 + O \rightarrow 2O_2$.

4.18 The photolysis of the CFCs in the upper stratosphere to release Cl atoms.

4.19 $Cl + O_3 \rightarrow ClO + O_2$.

4.20 (a) $ClONO_2$.
(b) $ClO + NO_2 \rightarrow ClONO_2$.

4.21 (a) Ozone concentrations are lower at the equator because stratospheric transport is from the equator to the poles.
(b) The Brewer–Dobson circulation.

4.22 (a) Ozone depletion was first observed over Antarctica.
(b) The South Polar vortex isolates the air mass over Antarctica and polar stratospheric clouds enhance spring ozone-depleting reactions.

4.23 (a) The South Polar area over Antarctica is colder than the North Polar region.
(b) Polar stratospheric clouds sequester the Cl reservoir species HCl, $ClONO_2$, and HOCl, which can rapidly photolyze with the spring sunrise to produce two Cl free radicals. They also sequester any available HNO_3, a reservoir species for NO_2. This prevents it from taking part in the NO_x termination reaction, resulting in more ozone-depletion cycles for one Cl atom.

4.24 The CFCs were used in refrigeration, aerosol can propellants, and solvents.

4.25 The HCFCs have shorter lifetimes in the troposphere because they can react with OH by abstraction.

4.26 The global agreement to cease using the CFCs was called the "Montreal Protocol on Substances that Deplete the Ozone Layer" or simply the "Montreal Protocol."

4.27 (a) The layer of sulfate aerosols in the lower stratosphere is called the Junge layer.
(b) Two major sources of the sulfate in aerosols are COS transported from the troposphere and SO_2 from volcanic eruptions.

4.28 (a) CFC-11 is $CFCl_3$.
(b) CFC-12 is CF_2Cl_2.

4.29 (a) $C_2F_3Cl_3$ is CFC-113.
(b) $C_2F_4Cl_2$ is CFC-114.

4.30 (a) The ozone-depleting species from the photolysis of halon is Br.
(b) $Br + O_3 \rightarrow BrO + O_2$.

4.31 The reaction releases both Br and Cl atoms back into the catalytic cycles by the reaction $BrO + ClO \rightarrow Br + Cl + O_2$.

4.32 Because the photolysis reaction $ClOOCl + h\nu \rightarrow 2Cl + O_2$ occurs rapidly during daylight hours.

4.33 $Cl + H_2O \rightarrow HCl + OH$.

4.34 (a) CH_4 and N_2O are increasing in the troposphere due to changing agricultural practices, primarily rice production. Organic fertilizer usage leads to methane, and nitrate fertilizers lead to nitrous oxide.
(b) Increasing CH_4 affects the HO_x Cycle and stratospheric water levels and N_2O affects the NO_x Cycle, both of which can enhance stratospheric ozone depletion.

4.35 (a) The equivalent effective stratospheric chlorine (EESC) provides an estimate of the total effective amount of ozone-depleting halogens in the stratosphere.
(b) It is calculated from the emission amounts of CFCs and other important halogenated compounds in the troposphere, their ozone depletion potentials (ODPs), and their transport times from the troposphere to the stratosphere.

4.36 The decline in the EESCs since 1992 is primarily due to the reduction in emissions of CFCs and other ozone-depleting species into the troposphere due to the Montreal Protocol.

4.37 (a) The stratosphere is expected to get cooler.
(b) Colder temperatures would increase the formation of polar stratospheric clouds.

4.38 (a) Water freezes and precipitates out in the upper troposphere and so does not normally reach the stratosphere.
(b) The major sources of water vapor in the stratosphere are the oxidation of methane and hydrogen gases, which are transported from the troposphere.

4.39 (a) The types of stratospheric clouds are Ia, Ib, Ic, and II.
(b) Type II polar stratospheric clouds are not found in the Arctic.

4.40 (a) All type I polar stratospheric clouds contain nitric acid.
(b) Type II polar stratospheric clouds contain water ice.

4.41 (a) Methane is produced primarily by anaerobic bacteria.
(b) Nitrous oxide is produced primarily by anaerobic bacteria.

Chapter 5

5.1 The chemistry of the troposphere is essentially oxidative chemistry.

5.2 (a) OH is the most important oxidant in the daytime troposphere.
(b) NO_3 is the most important oxidant at night.

5.3 (a) The lowest layer of the troposphere is the planetary boundary layer (PBL).
(b) The behavior of the PBL is directly influenced by its contact with the planetary surface.

5.4 Our air is considered to be polluted when the natural background concentrations are exceeded by a significant amount by anthropogenic emissions, to the extent that they can produce harmful effects to the biosphere.

5.5 The major source of OH in the troposphere is the reaction of $O(^1D)$ with H_2O.

5.6 London.

5.7 (a) From the words smoke and fog.
(b) The combination of fly ash and wet acidic aerosols was considered to be a mixture of smoke and fog.

5.8 Coal burning in cities was always characterized by SO_2 and heavy particulate emissions.

5.9 This new type of smog is called "photochemical smog."

5.10 Arie Haagen-Smit found that the NO_x and VOC emissions from motor vehicles were causing the problems in Los Angeles.

5.11 (a) In 1970 the Clean Air Act established the U.S. Environmental Protection Agency.
(b) The EPA was given the responsibility for establishing regulations to protect the air and water quality.

5.12 (a) Six criteria of air pollutants were identified.
(b) They were NO_2, SO_2, ground-level ozone, CO, particulate matter (PM), and lead.

5.13 Primary pollutants are emitted directly from the source into the troposphere. Secondary pollutants are produced from chemical reactions of primary pollutants in the troposphere.

5.14 (a) NO_2, SO_2, CO, PM, and lead are primary pollutants.
(b) Ozone is a secondary pollutant.

5.15 (a) Pollutants that have been identified as posing a health hazard but are not listed as criteria pollutants are called hazardous air pollutants by the EPA and are listed as air toxics.
(b) There were originally 190 air toxics.
(c) Five have been removed from the list.
(d) None have been added to the list, but *N*-propylbromide has been under consideration for addition to the list since December 28, 2016.

5.16 O atoms are formed in the troposphere by the photolysis of NO_2.

5.17 (a) A photostationary state is a steady state reached by a photochemical reaction in which the rates of formation and reaction of a species are equal.
(b) The photostationary state predicts that ozone will only form if $[NO_2] > [NO]$.

5.18 OH reacts with alkanes by abstraction.

5.19 $CH_4 < RCH_3 < R_2CH_2 < CH_2O$.

5.20 (a) Alkyl peroxy radicals (RO_2).
(b) NO_2 and an alkoxy radical (RO).

5.21 Methane is the slowest to react with OH.

5.22 (a) A six-member cyclic intermediate including the oxygen atom and the hydrogen on the fifth carbon.
(b) The larger alkoxy radicals can internally isomerize by intramolecular hydrogen abstraction, transferring the hydrogen from the fifth carbon to the oxygen atom.
(c) A hydroxyl ketone or a hydroxy aldehyde.

5.23 (a) Predicted $k_{OH} = 5.8 \times 10^{-12}$ cm^3 molecule^{-1} s^{-1}.
(b) Measured $k_{OH} = 5.1 \times 10^{-12}$ cm^3 molecule^{-2} s^{-1}, so predicted value is slightly higher than measured value.

5.24 (a) Predicted $k_{OH} = 3.9 \times 10^{-12}$ cm^3 molecule^{-1} s^{-1}.
(b) Measured $k_{OH} = 4.0 \times 10^{-12}$ cm^3 molecule^{-1} s^{-1}, so predicted value is slightly lower than measured value.

5.25 Addition.

5.26 OH addition is faster.

5.27 A hydroxyperoxy radical.

5.28 (a) The NO_3 radical is formed from the reaction of NO_2 with O_3.
(b) It is only important at night since it photolyzes rapidly in sunlight to give O and NO_2.

5.29 (a) The NO_3 radical can react with alkenes by addition.
(b) The NO_3/alkene reaction cycle forms organonitrates.

5.30 (a) The NO_3 radical can also react with NO_2 to form N_2O_5.
(b) The formation of N_2O_5 is favored under high NO_2 conditions at night.

5.31 (a) The reaction of ozone with VOCs is only important with unsaturated hydrocarbons.
(b) Ozone reacts with unsaturated hydrocarbons by addition.

5.32 (a) The addition of ozone to the alkene forms an intermediate adduct called a molozonide.
(b) The final product of the decomposition of the molozonide is a diradical called the Criegee biradical.

5.33 (a) The reaction of OH with CH_3CHO, followed by the addition of O_2, forms the peroxyacetyl radical.
(b) The peroxyacetyl radical reacts with NO_2 to form peroxyacetyl nitrate (PAN).
(c) This equilibrium reaction favors the formation of PAN at colder temperatures.

5.34 (a) There are four higher analogs of PAN normally observed in the troposphere.
(b) The general name for this group of compounds is peroxyacyl nitrates, $RCOO_2NO_2$.
(c) They are in equilibrium with the peroxyacyl radicals and NO_2.

5.35 (a) The most important source of NO_x on a global level is anthropogenic combustion of fossil fuels, primarily in motor vehicles.
(b) The most important source of VOCs is biogenic, mostly from terrestrial plant emissions.

5.36 (a) Methane is the most stable hydrocarbon in the troposphere.
(b) Because it is slow to react with OH.

5.37 HNO_3 is formed by reaction of OH with NO_2.

5.38 (a) H_2SO_4 is produced by the reaction of H_2O_2 with SO_2 in wet aerosols and in clouds.
(b) The H_2O_2 is primarily produced from the reaction of HO_2 with itself in areas where NO is low. The SO_2 is produced primarily from the combustion of coal.

5.39 Methyl iodide is released from the oceans.

5.40 (a) The reaction of N_2O_5 with NaCl in the sea-salt aerosol forms $ClNO_2$.
(b) $ClNO_2$ photolyzes to give NO_2 and Cl atoms.

5.41 (a) Cl reacts with alkanes by abstraction, leading to peroxy radicals and HCl.
(b) Cl atoms react by addition to alkenes, leading to a chloroperoxy radical.
(c) Reactions with Cl would be most important near polluted coastal cities, because (i) Cl abstraction produces water-soluble HCl, so Cl abstraction tends to be limited to areas near the source of Cl atoms and (ii) the formation of Cl from N_2O_5 requires a significant level of NO_2.

5.42 (a) The rate of reaction = $k[HO_2]^2$.
(b) The reaction of HO_2 with NO to produce OH and NO_2 would compete in a polluted environment.

Chapter 6

6.1 2 nm to 100 μm.

6.2 (a) Primary aerosols are those that are emitted into the atmosphere directly from a source.
(b) Secondary aerosols are those that are formed in the atmosphere from chemical reactions of gaseous precursors.

6.3 (a) Primary.
(b) Primary.
(c) Primary.
(d) Secondary.
(e) Secondary.

6.4 (a) Total suspended particulate (TSP).
(b) TSP measured the mass concentration of particulate matter in air.

6.5 (a) Ultrafine or Aitken mode.
(b) Fine or accumulation mode.
(c) Coarse mode.

6.6 Cloud condensation nuclei (CCN).

6.7 (a) Ultrafine or Aitken mode.
(b) Fine or accumulation mode.
(c) Coarse mode.

6.8 Ultrafine and fine modes.

6.9 Black carbonaceous soot or black carbon (BC) aerosols.

6.10 Fly ash.

6.11 Hydrogen peroxide (H_2O_2).

6.12 OH radical.

6.13 NH_4NO_3.

6.14 (a) Sulfate (SO_4^{2-}) and nitrate (NO_3^-).
(b) SO_2 and NO_x.

6.15 (a) CO_2.
(b) H_2CO_3, HCO_3^-, and CO_3^{2-}.

6.16 $[NO_3^-(aq)] = 2 \times 10^{-4}$ M.

6.17 (a) atm M^{-1}.
(b) M atm^{-1}.

6.18 (a) Hygroscopicity is the ability of a molecule to take up water.
(b) Compounds that have a high hygroscopicity will have larger values for Henry's Law constants (in units of M atm^{-1}).

6.19 Polar compound molecules will have larger values for Henry's Law constants (in M atm^{-1}) than nonpolar compounds.

6.20 (a) Diffusion.
(b) Gravitational settling.

6.21 Wet deposition and dry deposition.

6.22 Diffusion and gravitational settling.

6.23 Fine mode.

6.24 (a) 1–10 days.
(b) 25–40 days.
(c) 100–200 days.

6.25 (a) C-3 plants.
(b) C-4 plants.

6.26 (a) C-4 plants have more ^{13}C than ^{12}C.
(b) The C-3 cycle preferentially uses CO_2 containing the lighter carbon isotope (^{12}C), so the C-3 cycle will be depleted in CO_2 containing the heavier carbon isotopes ^{14}C and ^{13}C.

6.27 $^{13}C/^{12}C$ isotopic ratio, and ^{14}C content.

6.28 Corn and sugar cane.

6.29 Above-ground nuclear testing produced atmospheric ^{14}C, called "bomb" carbon.

6.30 1. Oxidative biochemical reactions which produce hydrogen peroxide.
2. Macrophages to seek out and ingest the hydrophobic material.

6.31 (a) The interaction or cooperation of two or more organizations, substances, or other agents to produce a combined effect greater than the sum of their separate effects.
(b) The exposure to soot aerosols and SO_2. The respiratory system produces H_2O_2 deep in the respiratory system to oxidize the soot. This H_2O_2 can then react with the co-inhaled SO_2 gas to produce H_2SO_4 deep in the lung, causing damage.

6.32 A *cis*-diol epoxide of benzo(a)pyrene.

6.33 Light absorption by carbonaceous aerosols and NO_2 gas.

6.34 (a) Light absorption and light scattering.
(b) Light absorption causes heating of the atmosphere, while light scattering causes cooling of the atmosphere.

6.35 (a) The complex refractive index.
(b) $m(\lambda) = n(\lambda) - ik(\lambda)$, where n is the aerosol refractive index and k is the aerosol absorption index.

6.36 Externally mixed, heterogeneous internally mixed, and homogeneous internally mixed.

6.37 (a) Homogeneous internally mixed.
(b) Externally mixed.

6.38 (a) Geometric, Rayleigh, and Mie.
(b) Mie.
(c) Geometric.
(d) Rayleigh.

6.39 Rayleigh scattering.

6.40 Fine-mode aerosols.

6.41 The indirect effect.

6.42 The semi-direct effect.

6.43 (a) Clouds.
(b) Haze.

6.44 Geometric optics.

6.45 Visible and infrared radiation.

6.46 NO_3^- and NO_2^-.

Chapter 7

7.1 (a) Precision is the degree to which repeated measurements give the same value whether or not that value is the true value.
(b) Reproducibility.
(c) Random errors cause low precision.

7.2 (a) The accuracy of a measurement is how close the result can come to the true value.
(b) Systematic errors cause low accuracy.

7.3 (a) Sensitivity is the ratio of the change in measurement signal to the change in concentration or amount of species being measured.
(b) A detection limit is the lowest amount of a chemical species that can be detected by the method.

7.4 (a) A Federal Reference Method.
(b) FRMs must be used to measure the criteria pollutants or for the calibration of the Federal Equivalent Methods.

7.5 (a) Impingers are especially designed bubble tubes used for collecting airborne chemicals in a liquid medium.
(b) They are used to concentrate the sample in order to obtain sufficient sample for measurement.

7.6 (a) Acids and oxidants.
(b) Teflon.
(c) An inert polyperfluorocarbon.

7.7 (a) Oxidation and acid-catalyzed reactions.
(b) Both positive and negative errors.

7.8 Nylon filters.

7.9 Active sampling uses an air sampling pump to actively pull the air through the sample collection device. Passive sampling does not require active air movement from a pump.

7.10 (a) Titration with KI.
(b) In an ozonesonde.

7.11 (a) Ozone chemiluminescent reaction with ethene.
(b) Ozone chemiluminescent reaction with NO.
(c) Long-path UV absorption.

7.12 DOAS uses a polynomial fit to subtract the broad background from the narrow absorption bands of ozone instead of comparing two closely related wavelengths.

7.13 (a) A low-pressure mercury lamp.
(b) Aromatic hydrocarbons as well as elemental mercury.
(c) Interferences are removed using a sample blank created by passing the sample air through an ozone scrubber.

7.14 MnO_2-coated wire mesh.

7.15 (a) Ozone measurements as a function of altitude.
(b) A radiosonde.

7.16 (a) They are calculated indirectly from emission rates and atmospheric concentrations of compounds that are known to be removed from the atmosphere primarily by reaction with OH.
(b) DOAS and laser-induced fluorescence (LIF).

7.17 (a) Reaction of $H_2O_2(aq)$ with *p*-hydroxyphenylacetic acid (PHPAA) in the presence of a peroxidase enzyme to form a PHPAA dimer.
(b) The PHPAA dimer is detected by fluorescence at 405 nm after excitation at 320 nm.
(c) Organic peroxides and peracids.
(d) By HPLC.

7.18 (a) Nondispersive infrared spectroscopy (NDIR).
(b) IR bandpass filter.
(c) By a rotating fan blade, called a chopper.
(d) All signal components that are not pulsed are removed from the signal.

7.19 CO, CO_2, SO_2, and N_2O.

7.20 (a) A colorimetric method based on the reaction with *p*-rosaniline.
(b) Heavy metals and NO_x.
(c) The addition of EDTA for heavy metals and sulfamic acid for NO_x.

7.21 (a) SO_2 and NO_2.
(b) Using a spectral mask.
(c) The Sun.

7.22 A colorimetric method based on the conversion of NO_2 to NO_2^- in basic solution followed by reaction with sulfanilamide to form an azo dye.

7.23 (a) The chemiluminescent reaction of NO with O_3.
(b) A molybdenum-coated wire mesh heated to 300–400 °C.

7.24 Ozone chemiluminescent reaction with NO using a heated catalyst to convert NO_2 to NO.

7.25 (a) The sum of all the atmospheric nitrogen-containing compounds that the NO_x Box responds to.
(b) NO_x, HNO_3, HONO, N_2O_5, HO_2NO_2, organic nitrates, and peroxyacyl nitrates.

7.26 A red cut-off transmission filter.

7.27 (a) GC/ECD.
(b) The use of a radioactive source and 15-minute analysis times.
(c) O_2 and water.

7.28 Chemiluminescent reaction with luminol.

7.29 Peracetic acid and HNO_3.

7.30 HNO_3 is removed from the air sample by the use of a nylon filter and measured by comparing the signal from the NO_x channel with and without a nylon filter in place.

7.31 GC–ECD or GC–MS with negative ion detection.

7.32 (a) CF_4 and C_2F_6.
(b) They are primarily produced from the electrochemical production of aluminum.

7.33 (a) Flame ionization detector.
(b) The air is pulled through a short nonpolar column, which separates methane from the rest of the hydrocarbons.

7.34 (a) High-resolution capillary GC.
(b) A positive ion mass spectrometer (MS) or an FID.
(c) The ECD.

7.35 (a) Reaction with an aldehyde or ketone with 2,4-dinitrophenylhydrazine (DNPH).
(b) Reverse-phase HPLC with a UV absorption detector at 370 nm.

7.36 (a) The chemiluminescent reaction of alkenes with O_3.
(b) Isoprene.

7.37 Proton transfer mass spectrometry (PTMS).

7.38 The number of chlorine and bromine atoms in the molecule.

7.39 Quadrupole mass spectrometry and time-of-flight mass spectrometry.

7.40 (a) Accelerator mass spectrometry.
(b) Combustion to CO_2 followed by reduction to graphite.

7.41 HPLC–MS or HPLC with fluorescence detection.

7.42 Total organic carbon (OC) and total elemental carbon (EC).

7.43 (a) Nephelometers.
(b) Aerosol particle number concentrations, aerosol size ranges, and fine aerosol scattering.

7.44 The aethalometer.

7.45 The particle soot absorption spectrometer (PSAP).

7.46 The production of sound waves in a sample produced by the absorption of pulsing light.

7.47 ICP–MS after acid extraction.

7.48 (a) Filter media that is low in inorganic content, such as Teflon.
(b) Quartz fiber filters, which do not use organic binders.

7.49 A synthetic mixture of ultrahigh purity O_2 (22% v/v) and N_2 (78% v/v).

7.50 Baseline measurements are an accurate measurement of species concentrations over time, beginning before any change occurs.

7.51 They can be used as a standard for measuring future concentrations to help identify any problems as they occur and can help to determine the effectiveness of any solution that is applied to an identified problem.

Chapter 8

8.1 Gas, liquid, and solid.

8.2 Hydrogen bonding.

8.3 Water is a very good solvent and always has chemicals and minerals dissolved in it.

8.4 Water is a very good solvent and the oceans continually collect dissolved salts.

8.5 Evaporation from ocean surfaces.

8.6 Solid, as ice frozen in the polar ice caps, glaciers, and permanent snow.

8.7 Groundwater reservoirs.

8.8 Lakes.

8.9 Endorheic basins.

8.10 Temperature and salinity.

8.11 (a) Na^+ and Cl^-.
(b) Saline solution.

8.12 (a) CDOM for colored dissolved organic matter.
(b) Mouth of rivers.
(c) Main source is decomposed terrestrial organic matter that is carried in the rivers to the oceans.

8.13 Tannins.

8.14 Humic and fulvic acids.

8.15 Lacustrine or lentic water systems.

8.16 (a) Plankton blooms.
(b) Nutrients are brought up during upwelling from the bottom to the surface.

8.17 Coriolis effect.

8.18 Higher at the surface.

8.19 The water condition is called anoxic.

8.20 Lake surface waters are in equilibrium with CO_2, H_2CO_3, and HCO_3^-, while ocean waters also contain significant amounts of CO_3^{2-} which act to buffer the pH from 5.3 to 8.2.

8.21 Blue green algae.

8.22 Ammonium, NH_4^+ and nitrate, NO_3^-.

8.23 No.

8.24 $(NH_4)_2SO_4$.

8.25 It has increased the concentrations of H_2CO_3 and HCO_3^-, causing a decrease in pH in surface ocean waters.

8.26 Persistent organic pollutants.

8.27 DDE.

8.28 Humic and fulvic acids.

8.29 Humic and fulvic acids.

8.30 Coal combustion, smelters, natural erosion, orchard pesticide use, electronics, and glass-production waste.

8.31 Chloroform, bromoform, di-halogenated, and trihalomethanes, along with mono-, di-, and trichloroacetic acids.

8.32 Aerobic and anaerobic bacteria.

8.33 Garbage patches.

8.34 Methoxychlor.

8.35 (a) Clay.
(b) Silt.
(c) Sand and gravel.

8.36 Global distillation.

8.37 0–8 °C.

8.38 Udic.

8.39 (a) Days to years.
(b) Centuries to millennia.

8.40 Equilibrium solubility product or K_{sp}.

Chapter 9

9.1 (a) A grab sample is a single sample taken just below the water surface at a specific time or over as short a period as is feasible.
(b) Two samplers commonly used for obtaining water samples at depths of tens to about 100 feet are the Kemmerer sampler and the Van Dorn sampler.

9.2 (a) Measurements that should be taken in the field at the time the sample is obtained are temperature, pH, DO, turbidity, sample depth, and total depth.
(b) Turbidity, sample depth, and total depth are measured in situ.

9.3 (a) DO is dissolved oxygen concentration.
(b) BOD is biochemical oxygen demand.
(c) COD is chemical oxygen demand.

9.4 (a) Water transparency or turbidity is measured in the field with a Secchi disk.
(b) Results are reported as Secchi depth (m^{-1}).

9.5 (a) Composite sampling consists of a collection of individual discrete samples taken at regular intervals over a period of time, usually 24 hours.
(b) The rosette sampler.
(c) The rosette sampler can measure pH, DO, conductivity, and turbidity continuously.

9.6 (a) Turbidity can be measured directly using a turbidity meter or turbidity sensor.
(b) They measure the amount of light scattered by the particles in the water.
(c) They operate on the same principle as a nephelometer.

9.7 (a) The most common glass electrode is the pH electrode.
(b) The most common one measured by a crystal membrane electrode is fluoride.

9.8 (a) pH standards of 2 and 4 are used for samples of $pH < 7$.
(b) pH standards of 4 and 10 are used for samples of $pH > 7$.

9.9 (a) The DO membrane permeable electrode and the DO optical luminescent detection.
(b) The Winkler titration method.

9.10 (a) The determination of BOD involves the measurement of the initial DO and the DO after a 5-day incubation.
(b) BOD is reported in $mg\,l^{-1}$ DO.

9.11 (a) The total BOD of waste water is composed of a carbonaceous biochemical oxygen demand (CBOD) and a nitrogenous biochemical oxygen demand.
(b) 2-Chloro-6-(trichloromethyl) pyridine is added to the sample bottle to inhibit the nitrifying bacteria in the sample.

9.12 (a) Conductivity is the measure of the ability of a material to pass an electrical current.
(b) Conductivity is the reciprocal of resistivity.

9.13 (a) The original standard method for the measurement of NH_3 and NH_4^+ was a spectroscopic method.
(b) The reagent used was Nessler's reagent.
(c) The chemical composition of Nessler's reagent is potassium tetraiodomercurate(II) ($K_2(HgI_4)$) in an alkaline solution.

9.14 (a) An alkaline solution of phenol (C_6H_5OH) and the hypochlorite ion (ClO^-) with NH_3 forms the highly colored dye indophenol blue.
(b) The common name for the reagent is Bertholet's reagent.

9.15 The determination of total organic nitrogen is total Kjeldahl nitrogen (TKN).

9.16 (a) The concentration of NO_3^- in water is determined by reaction with brucine sulfate.
(b) The concentration is determined by absorption at 410 nm.
(c) Temperature and time must be carefully controlled during the reaction.

9.17 (a) NO_3^- in water can also be determined by reduction to NO_2^-.
(b) Reduction of the NO_3^- is accomplished by passing the filtered water sample through a column containing granulated copper-coated cadmium metal.
(c) NO_3^- concentration is determined by absorption at 540 nm after conversion to a colored product.

9.18 NO_3^- in water is measured by reacting it with 4-aminobenzesulfonamide called sulfanilamide and *N*-(1-naphthyl)-ethylenediamine to form a colored diazo dye.

9.19 (a) PO_4^{3-} is determined by reaction with ammonium molybdate ($(NH_4)_2MoO_4$) and antimony potassium tartrate ($K_2Sb_2(C_4H_2O_6)_2$) in acidic solution.
(b) A phosphomolybdate complex is formed.
(c) The ascorbic acid reduces the phosphomolybdate complex to molybdenum blue.

9.20 (a) $Ba^{2+}(aq) + SO_4^{2-}(aq) \rightarrow BaSO_4(s)$.
(b) The excess Ba^{2+} is determined by complexation with methylthymol blue under basic conditions and the colored complex is determined by absorbance.

9.21 (a) The measurement of free CN^- is accomplished by distilling an acidified sample and trapping the HCN gas in an NaOH solution.
(b) The final reagent in the CN^- determination is barbituric acid.

9.22 (a) Ca^{2+} exchanges with the Mg^{2+}, releasing it into the solution.
(b) The free Mg^{2+} is measured by the addition of the organic dye calmagite.
(c) Hardness is reported as $mg\,l^{-1}$ $CaCO_3$.

9.23 (a) The measurements of DOC and TOC are based on the spectroscopic determination of CO_2 generated from organic carbon by high-temperature combustion or low-temperature chemical oxidation.
(b) For the measurement of DOC the sample is first filtered through a 0.45-μm filter to remove any organic particulate matter.
(c) The results are reported in $mg\,l^{-1}$ C.

9.24 (a) The three main spectroscopic methods used for the analysis of metals and some metalloids in water are based on atomic absorption spectroscopy, atomic emission spectroscopy, and inductively coupled plasma mass spectroscopy (ICP–MS).
(b) ICP–MS has the lowest detection limits.

9.25 (a) The two line source lamps used in AAS are hollow cathode lamps and electrodeless discharge lamps.
(b) The free analyte atoms are created in a graphite furnace.
(c) Background correction is applied by using a continuum deuterium source or a Zeeman magnetic field.

9.26 (a) An inductively coupled plasma torch is used to create a population of free analyte atoms in atomic emission spectroscopy.
(b) The line emissions can be separated by a scanning monochromator or by a polychromator.

9.27 (a) The two methods used for mercury determination in water are cold vapor atomic absorption or atomic fluorescence spectroscopy.
(b) $SnCl_2$ is used as the reducing agent to convert Hg^{2+} to Hg^0.

9.28 (a) The detector in ion chromatography is a conductance cell.
(b) A suppressor reduces the background signal and lowers the detection limits in ion chromatography.
(c) The suppressor removes any background ions in the eluent and replaces them with nonionic species.

9.29 (a) The measurement of purgeable volatile organic compounds in water is accomplished by gas chromatography mass spectroscopy.
(b) Non-volatile organic compounds can be determined in water by a liquid–solid extraction followed by GC–MS detection.
(c) Organohalogens, pesticides, and insecticides are measured by GC–ECD.
(d) Carbonyl compounds are measured in water by reverse-phase HPLC–UV after dinitrophenylhydrazine (DNPH) derivatization.

9.30 Reverse-phase HPLC uses a hydrophobic stationary phase on the HPLC column instead of the normal hydrophilic stationary phase and uses a polar mobile phase instead of the normal hydrophobic solvent mobile phase.

9.31 (a) A GC is coupled to a mass spectrometer by a jet separator.
(b) An HPLC is coupled to the mass spectrometer by an electrospray ionization (ESI) interface.

9.32 (a) An excess amount of the strong oxidant potassium dichromate is used for the determination of COD in water.
(b) The main interference for the determination of COD in water is Cl^-.
(c) Mercuric sulfate.

9.33 (a) The major contributor to radioactivity in most water systems is from the decay of ^{226}Ra and its daughters.
(b) The ^{226}Ra concentration in water is determined by measuring the amount of the radiochemical decay product $^{222}Rn(g)$ from a sulfate precipitate.
(c) The ^{222}Rn gas is purged into a scintillation cell and counted for alpha activity.

9.34 Hollow-fiber ultrafiltration uses cross-flow filtration to minimize filter polarization while flat membrane ultrafiltration uses vertical filtration and stirs the sample above the filter to minimize filter polarization.

9.35 (a) An alpha particle is a helium atom nucleus with a +2 charge.
(b) A beta particle is an electron or positron.

9.36 Chemical separation methods are likely to lead to structural alterations that will change the chemical behavior of the colloids.

9.37 (a) Distillation and deionization.
(b) The production of DI water is less energy intensive and much faster than distillation.
(c) Ultrapure water can be produced by ultrafiltration methods along with inert gas purging to remove the dissolved gases.

9.38 Carbon dioxide and oxygen can contaminate and change water chemistry when samples are exposed to air.

9.39 The resistivity of ultrapure water at 25 °C is 18.18 MΩ.

9.40 (a) Laboratory dust can add trace metals and solids to blanks, standards, and samples if left open to the atmosphere for any length of time.
(b) Clean rooms are designed to prevent this type of contamination.

9.41 (a) Instrument calibration drift would result in consistently high or low standard results.
(b) Improperly cleaned glassware would result in randomly high blank values.
(c) Contaminated reagents would result in systematically high blank values.

Chapter 10

10.1 Relating to or derived from living matter.

10.2 Aldehydes and ketones.

10.3 Oxygen.

10.4 Anoxic or very low oxygen content.

10.5 Humin.

10.6 (a) Variation in temperature.
(b) A geothermal gradient.

10.7 (a) Kerogen.
(b) Catagenesis.

10.8 (a) Phytoplankton and zooplankton.
(b) Plants.
(c) Algae.

10.9 (a) Gas.
(b) Oil.
(c) Coal.

10.10 (a) Diagenesis.
(b) Coalification.

10.11 Lignite, sub-bituminous, bituminous, anthracite.

10.12 (a) Phenoxy (—OH), methoxy (—OCH_3), and carboxylic (—COOH) groups decrease.
(b) Aromatic ring structures increase.

10.13 (a) Strip mining, contour mining, and mountain top mining.
(b) The removal of top soil and overburden.

10.14 (a) Primary, secondary, and enhanced recovery.
(b) Primary recovery used the pressure difference between the source rock formation and the well bore, secondary recovery injects fluids to create pressure. Enhanced recovery designed to alter the properties of the oil itself.

10.15 Fracking.

10.16 Sour gas wells.

10.17 (a) United States.
(b) China.

10.18 (a) Cryogenic separation.
(b) Bubbling through an aqueous solution of a basic amine.

10.19 (a) Fischer–Tropsch reaction.
(b) Coal gasification.
(c) A variety of straight chain alkanes with the formula C_nH_{2n+2}.

10.20 (a) Fractional distillation.
(b) Light distillates, middle distillates, and heavy distillates.

10.21 *t*-Butyl mercaptan (TBM) and dimethyl sulfide (DMS).

10.22 (a) First generation and second generation.
(b) First generation are produced from food crops, while second generation are not.
(c) First generation.
(d) Second generation.

10.23 (a) Glycolysis.
(b) Transesterification.

10.24 The United States and Brazil.

10.25 Wood waste and wood byproducts.

10.26 (a) Fuel oil.
(b) Pelleted wood.
(c) Wood varies in energy content according to its density and moisture content.

10.27 (a) E25.
(b) E100.
(c) B5.

10.28 Organic alcohols, HCl, sand, and bactericides.

10.29 (a) Deepwater Horizon.
(b) Exxon Valdez.

10.30 (a) Oxygen and water.
(b) They act as sacrificial anodes for cathodic protection.

10.31 Fertilizers and pesticides used in producing crops.

10.32 (a) Natural seeps.
(b) Produced waters from fracking and secondary oil recovery.
(c) Tanker spills.
(d) Large vessel discharges.

10.33 (a) An area of hypoxia in the bottom water layers.
(b) Fertilizer discharge from the Mississippi River causes the phytoplankton to greatly increase in number.

10.34 (a) Bituminous.
(b) It has the second highest fixed carbon content.

10.35 (a) Coal.
(b) H_2SO_4.
(c) It can either add to the $PM_{2.5}$ levels or be taken up into clouds, leading to the formation of acid rain.

10.36 (a) Nitrogen in the fuel and from the high-temperature thermal reaction of N_2 and O_2 in the combustion air.
(b) From the nitrogen in the coal.
(c) From the high-temperature reaction of N_2 and O_2 in the combustion air.
(d) NO.

10.37 CO_2, CH_4, and N_2O.

10.38 Diesel particulate filter systems (DPFs).

10.39 Aromatics, branched alkanes, and oxygenates (primarily ethanol).

10.40 (a) Ethanol.
(b) Because of the hydrogen bonding between the ethanol molecules.
(c) E10 is higher than gasoline.
(d) When 10% ethanol is blended into gasoline, the gasoline molecules disrupt the ethanol hydrogen bonds allowing both the ethanol and the gasoline to easily evaporate, raising the vapor pressure of the blend.

10.41 Ethene and acetaldehyde.

10.42 The particulates from biodiesel have a larger number of smaller particles with higher benzene and soluble organic content than that of petroleum diesel.

10.43 (a) Bunker C.
(b) SO_2, NO_x, and soot.
(c) Soot.
(d) They strongly absorb radiation, warming the air, and if deposited on ice or snow they will decrease the albedo, increasing the melting in these regions.

10.44 Isobutene and formaldehyde.

Chapter 11

11.1 (a) Climate is defined as the average condition of the weather at a given region over a long period of time, as exhibited by temperature, wind velocity, and precipitation.
(b) 20, 100, and 500-year time frames.

11.2 Weather is the state of the atmosphere at a particular place and time, while climate is the average of weather over a long period of time.

11.3 (a) The five different climate groups are tropical, dry, temperate, continental, and polar.
(b) Each of the groups is subdivided according to seasonal precipitation types and level of heat.

11.4 (a) The average global monthly surface daytime temperature range on Earth is −25 to 45 °C.
(b) The mean annual surface temperature for the Earth in 2010 was 14.86 °C.
(c) This was 0.96 °C above the mean temperature for the twentieth century.

11.5 Global climate change is now occurring on a much shorter time scale than past climate change, which occurred gradually over thousands of years.

11.6 (a) $^{18}O/^{16}O$ and $^{2}H/^{1}H$ stable isotopic ratios to determine the variation in mean ocean temperature over time.
(b) Water molecules containing the heavier isotopes (^{2}H and ^{18}O) evaporate at a higher temperature than water molecules containing the lighter isotopes. The increase in evaporation of the heavier elements at warmer temperatures results in an enrichment of the heavier elements in frozen precipitation, which forms the ice sheets.

11.7 (a) If the ocean temperature decreases, $^{18}O/^{16}O$ and $^{2}H/^{1}H$ will decrease.
(b) If the ocean temperature increases, $^{18}O/^{16}O$ and $^{2}H/^{1}H$ will increase.

11.8 (a) $^{18}O/^{16}O$ is used to determine ocean temperature change from sediment cores.
(b) Macroscopic and microscopic shells found in the sediments.

11.9 (a) There are higher values of $^{18}O/^{16}O$ and $^{2}H/^{1}H$ in the shells when the ocean water is colder and lower values of $^{18}O/^{16}O$ and $^{2}H/^{1}H$ in the sediments when the ocean water is warmer.
(b) Since the shells are formed from minerals in the ocean water, the isotopic composition of the shells is directly proportional to the isotopic composition of the ocean water. The isotopic composition of the ice cores is directly proportional to the water evaporation rates.

11.10 (a) The ice core data is useful from the present back in time to about 800 000 years.
(b) The sediment data is useful to about 200 000 million years.

11.11 (a) The maximum CO_2 concentration during the past 400 000 years was 300 ppm.
(b) The CO_2 level reached 310 ppm in 1950.
(c) The level of CO_2 in 2018 was 410 ppm.

11.12 (a) The climate during 1650–1850 was known as the Little Ice Age.
(b) It is believed to have been caused by decreases in the solar radiation.

11.13 (a) The atmospheric temperature increases since the Little Ice Age have generally been restricted to the troposphere. In addition, atmospheric modeling, which includes current solar radiation levels, cannot explain the observed increases in mean global temperature without considering the increasing levels of greenhouse gases.
(b) The major cause is increases in greenhouse gases, particularly CO_2.

11.14 (a) The greenhouse gases in the troposphere are H_2O, CO_2, CH_4, N_2O, tropospheric O_3, CFCs, and HCFCs.
(b) Water vapor is the strongest greenhouse gas.
(c) It can be removed naturally from the atmosphere by condensation to form clouds followed by precipitation.

11.15 The three strongest greenhouse gases after water vapor are CO_2, CH_4, and N_2O.

11.16 (a) Atmospheric concentrations of CH_4 have increased by 1.16 ppm.
(b) CO_2 levels have increased by 130 ppm.
(c) N_2O levels have increased by 0.06 ppm.

11.17 (a) The CFCs and HCFCs have no natural sources.
(b) The levels of CFCs are decreasing due to their banning by the Montreal Protocol.
(c) The HCFCs are being used as replacements for the CFCs.

11.18 (a) The GWP of a greenhouse gas is the measure of how much IR energy will be absorbed by 1 ton of the gas in the atmosphere over a given period of time relative to 1 ton of CO_2.
(b) The magnitude of the GWP depends on the strength of the IR absorption bands, their spectral location, and the atmospheric lifetime of the greenhouse gas.

11.19 (a) The natural CO_2 emissions include plant respiration, microbial respiration and decomposition, and air–sea gas exchange.
(b) The natural CO_2 sinks are photosynthetic uptake and ocean uptake.
(c) Anthropogenic CO_2 emissions include the combustion of fossil fuels, cement production, and land use changes such as deforestation.

11.20 (a) The major source of CH_4 is anaerobic bacteria.
(b) The major anthropogenic sources of CH_4 are rice cultivation, animal farming, agricultural burning, and natural gas mining and transportation.
(c) The most important sink for CH_4 is reaction with OH radical.

11.21 (a) The major source of N_2O is due to anaerobic bacterial decomposition of organic materials in wetlands.
(b) The anthropogenic sources of N_2O are nitrate fertilizers and agricultural burning.

11.22 The major sink for N_2O is mixing into the stratosphere.

11.23 (a) The PFCs are produced as a byproduct of the production of aluminum by electrochemical reduction of Al_2O_3.
(b) SF_6 is industrially produced as an insulating gas in high-voltage operations.

11.24 (a) Climate forcing considers all of the influences that change the amount of energy absorbed or lost, which cause temperatures to rise or fall.
(b) The increases in greenhouse gases result in atmospheric warming and are positive climate forcings.
(c) Backscattering of incoming radiation by clouds, which results in atmospheric cooling, is a negative climate forcing.

11.25 (a) The aerosols that absorb incoming solar radiation the strongest are BC and OC.
(b) This results in a positive climate forcing.
(c) Wind-blown dust, fly ash, nitrate, and sulfate scatter incoming radiation.
(d) This results in a negative climate forcing.

11.26 The two factors contributing most to the uncertainty in determining climate forcing are the short-term aerosol and cloud forcing values.

11.27 (a) The atmospheric global-scale computer models are abbreviated AGCM.
(b) The oceanic global-scale models are abbreviated OGCM.
(c) The combined atmosphere/ocean global-scale climate models are GCM.

11.28 (a) GCMs are large-scale models with a grid size of 300 km × 300 km; larger RCMs are regional-scale models with a grid size of 10 km × 10 km.
(b) The RCMs use boundary conditions calculated by GCMs on short time scales.

11.29 (a) The four RCPs are RCP2.6, RCP4.5, RCP6, and RCP8.5.
(b) They are named for the possible changes in radiative forcing in W m^{-2} by the year 2100.

11.30 (a) The IPCC is the Intergovernmental Panel on Climate Change.
(b) The latest assessment report is AR5.

11.31 Even if CO_2 emissions were stopped today, many of the surface temperature changes and impacts on the terrestrial systems will continue due to the increased overburden of CO_2 relative to pre-1850 levels.

11.32 The mean global temperature is projected to increase in the late twenty-first century by 0.3–4.8 °C and the mean seal level is projected to rise by 0.17–0.82 m under all RCPs.

11.33 (a) The increase in plant growth during wetter springs and increased amounts of dry biomass is projected to increase large-scale wildfires.
(b) Wildfires produce NO_x, CO, VOCs, and aerosol emissions that add to air pollution problems on local and regional scales.

11.34 The surface currents of the thermohaline circulation in the northern Atlantic Ocean are known as the Atlantic Meridional Overturning Circulation (AMOC).

11.35 (a) Climate change mitigation is the effort being made to reduce or prevent the emission of greenhouse gases, which are the major climate forcing agents.
(b) Climate change adaptation strategies are actions taken to help communities and ecosystems cope with the impacts of changing climate.

11.36 Alternative energy sources to fossil fuel combustion include wind, solar, hydro, woody biomass combustion, geothermal, landfill gas, biofuels, and nuclear power.

11.37 (a) When comparing the various energy alternatives, it is important to examine the complete lifecycle of the energy-producing system.
(b) The complete lifecycle includes the fuels consumed in the production of the materials used in the construction of the power plant, routine and long-term maintenance, handling of any waste materials, and decommissioning of the plant.

11.38 (a) Nuclear power, hydroelectric, and wind have the lowest CO_2 emissions during their lifecycle.
(b) Nuclear power is the most reliable.

11.39 A breeder reactor uses uranium as a nuclear fuel. The uranium generates plutonium as it is consumed. The plutonium byproduct is then separated from the spent uranium and used in the reactor to generate energy.

11.40 (a) Carbon sequestration involves the capture of CO_2 from power plant flue gases for the purpose of placing it into long-term storage.
(b) The most effective method of CO_2 storage is to convert the CO_2 to carbonate by reaction with a metallic oxide base, followed by storage in a geological formation.

11.41 (a) Geoengineering, or climate engineering, is the deliberate and large-scale intervention in the Earth's climate system with the aim of mitigating the adverse effects of climate change.
(b) The two categories are the removal of greenhouse gases from the atmosphere and the reduction of incoming solar radiation.

11.42 The reduction of incoming solar radiation by geoengineering projects would have a direct impact on solar power.

Chapter 12

12.1 (a) Wilhelm Röngten discovered X-rays.
(b) Antoine Henri Becquerel discovered radioactivity.

12.2 (a) Radioactivity is the spontaneous emission of high-energy particles, γ-rays, or both, caused by the disintegration of atomic nuclei.
(b) Radioactivity is measured in becquerels.
(c) Bq.

12.3 Radium and polonium were discovered by Pierre and Marie Curie.

12.4 Radioisotopes are not stable and undergo radioactive decay processes.

12.5 The five radioactive decay processes are alpha decay, β^- decay, β^+ decay (positron emission), gamma ray emission, and electron capture.

12.6 Electron capture (EC) and positron decay (β^+) lead to the loss of a proton and the gain of a neutron.

12.7 (a) Alpha decay leads to the emission of a helium nucleus.
(b) The nucleus loses two protons and two neutrons.

12.8 (a) β^- decay leads to the emission of an energetic electron from the nucleus.
(b) The nucleus has one less neutron and one more proton.

12.9 (a) Gamma ray emission leads to the emission of energetic photons.
(b) The nucleus falls from an excited state to the ground state.

12.10 (a) alpha particles, beta particles, and gamma rays.
(b) $\alpha < \beta < \gamma$.

12.11 The driving force behind the nuclear decay process is the number of neutrons relative to the number of protons in the nucleus of an isotope.

12.12 (a) Radioisotope decay follows first-order kinetics.
(b) The activity of a radioisotope ($-dN/dt$) is equal to the amount of the radioisotope (N) times the decay rate (λ).

12.13 (a) An α particle has a +2 charge.
(b) A β^- particle has a −1 charge.
(c) A γ-ray has no charge.

12.14 The average lifetime (τ) of a radioisotope is equal to $1/\lambda$.

12.15 (a) The half-life ($t_{1/2}$) of a radioisotope is the amount of time that it takes for half of the parent radioisotope to decay ($N_0/2$).
(b) The half-life of a radioisotope is equal to a constant (0.693/decay rate).

12.16 (a) When a positron and an electron interact, they undergo annihilation.
(b) It produces two gamma rays emitted at 180° from each other with equal energies.

12.17 The three sources of radioisotopes in the environment are primordial, cosmogenic, and anthropogenic.

12.18 About half of the initial amount of primordial ^{238}U is still present on the planet today.

12.19 (a) ^{40}K.
(b) Electron capture.

12.20 (a) ^{222}Rn is called Radon and ^{220}Rn is called Thoron.
(b) Radon is produced from the ^{238}U decay chain and Thoron is produced from the ^{232}Th decay chain.
(c) ^{222}Rn has the longer half-life.

12.21 (a) Carbon-14 is produced by a neutron colliding with ^{14}N.
(b) It undergoes β^- decay to reform ^{14}N.

12.22 $^{14}N + \alpha \rightarrow {}^{17}O + \text{proton}$.

12.23 Nuclear fission occurs when a large radioisotope is bombarded by a neutron, forming two smaller nuclei.

12.24 ^{235}U is the first known fissile radioisotope.

12.25 (a) If the concentration of ^{235}U is high enough, a nuclear chain reaction can occur.
(b) If a neutron moderator is present, the nuclear reaction will occur at a steady pace.
(c) If a moderator is not present, energy is released explosively and a nuclear explosion occurs.

12.26 (a) ^{239}Pu was the second fissile radioisotope discovered.
(b) It was discovered by neutron bombardment of ^{238}U.

12.27 ^{235}U and ^{239}Pu were used in the nuclear bombs used by the United States during World War II.

12.28 The nuclear fusion bomb was based on the combination of two isotopes of hydrogen (deuterium and tritium) to form helium, with the release of about 18 MeV of energy and a neutron.

12.29 (a) The two types of fallout are a small amount of fission products with long half-lives and a large amount of radioactive particulate matter produced from dust and sand with shorter half-lives.
(b) Most of the long-lived radioisotopes are deposited as fine-mode heavy metal oxide aerosols.
(c) Exposure to ^{90}Sr was a major concern.

12.30 25% of the ^{137}Cs will remain in 2020 (60 years is two half-lives).

12.31 (a) A breeder reactor is a nuclear reactor that generates more fissile material than it consumes.
(b) The fuel used in the first breeder reactor was ^{235}U.
(c) It produced ^{239}Pu.

12.32 (a) Very high purity graphite was the first material used as a neutron moderator.
(b) It can catch fire should the coolant overheat.

12.33 (a) Most designs place the fuel rods in a physical environment that allows the nuclear fuel to expand if it becomes overheated.
(b) Since a sustained nuclear reaction depends on high-density fuel, expansion of the fuel lowers the density of the fuel, stopping the nuclear reaction.

12.34 A major problem with controlled nuclear fusion is the very high temperatures and pressures that must be reached.

12.35 (a) The yellow color in yellowcake is from UO_3.
(b) The oxidation state of uranium is U(VI).

12.36 The two types of nuclear reactors used in nuclear power plants are thermal neutron reactors and fast neutron reactors.

12.37 (a) Most of the reactors use water, either H_2O or D_2O, as a coolant.
(b) These are the pressurized water reactor (PWR), the boiling water reactor (BWR), and the pressurized heavy water reactor (PHWR).
(c) The CANDU is a pressurized heavy water reactor (PHWR).

12.38 (a) The three major nuclear accident sites are Three-Mile island, PA, USA; Chernobyl, Ukraine; Fukushima, Japan.
(b) The most significant nuclear accident to date occurred in Chernobyl in the Ukraine.

12.39 (a) The radioisotope responsible for most of the natural inhaled radioactivity is ^{222}Rn.
(b) ^{222}Rn is produced in the ground from the ^{238}U decay chain as a decay product of ^{226}Ra.

12.40 (a) ^{210}Po is concentrated in tobacco smoke.
(b) It decays by α emission.

12.41 The two major radioisotopes in food are ^{14}C and ^{40}K.

12.42 The major source of anthropogenic radiation is medical exposure to X-rays, nuclear imaging, and use of nuclear pharmaceuticals.

12.43 ^{235}U was found to be depleted in the Oklo site.

12.44 (a) The modern unit of absorbed radiation dose is the gray (Gy).
(b) The Gy is equal to 100 rad.

12.45 (a) An equivalent dose is a dose quantity representing the health effects of low levels of ionizing radiation on the human body.
(b) An equivalent dose is measured in Sieverts (Sv) or rems.
(c) The rem is equal to 1×10^{-2} Sv.

12.46 Three ways to minimize exposure to a radioactive source are limiting exposure time, reducing the distance from the source, and the use of shielding materials.

12.47 The rem is short for "roentgen equivalent man."

12.48 (a) The guiding principle for radiation safety for radiation workers is the ALARA principle.
(b) ALARA stands for "as low as reasonably achievable."

12.49 (a) The major source of exposure to ionization radiation is radon inhalation.
(b) Natural radioactivity in food and water comes primarily from ^{14}C and ^{40}K.

12.50 (a) ^{99m}Tc is one of the most commonly used radiotracers in medicine.
(b) ^{60}Co is used for food irradiation.
(c) It is produced by neutron bombardment of ^{59}Co.

12.51 (a) ^{241}Am is commonly used as an alpha source in household smoke detectors.
(b) $^{238}PuO_2$ has been used as a power source by NASA.

12.52 (a) ^{36}Cl is used for dating old ground waters.
(b) ^{210}Pb is used to determine the sedimentation rates in lakes and oceans.

Chapter 13

13.1 (a) Sustainability is the condition where our energy, water, and food resources meet the demands of today's society without causing harm to our environment or depleting it for future generations.
(b) Sustainable development is development that meets the needs of the present without compromising the ability of future generations to meet their own needs.

13.2 The three major organic petrochemical feedstocks are light alkanes, light alkenes, and light aromatics.

13.3 The light alkenes are produced by steam cracking.

13.4 About 2% of each barrel of crude petroleum is used as a petrochemical feedstock.

13.5 Ethene and propene are two of the more important petrochemical feedstocks.

13.6 It would extend the petroleum reserves used by the petrochemical industry by a factor of 50 or more.

13.7 (a) The most used of the renewable energy sources worldwide is traditional biomass, which is wood burning for heat and cooking.
(b) The second most used is hydropower.

13.8 (a) Wind turbines require minimum wind speeds in the range of 3–3.5 m s^{-1} (13 km h^{-1} or 8 mph) to be able to generate power.
(b) The optimal wind speed is between 14 and 25 m s^{-1} (31–56 mph).

13.9 (a) The areas that would be best for wind power are in the mid-latitude coastal areas.
(b) In the United States the best areas are off the Pacific and northern Atlantic coasts, as well as in the Great Lakes.

13.10 Hydropower accounts for 17% of global electric power generation.

13.11 (a) The major drawback to hydroelectric power is that the physical placement of a large dam causes significant changes to the surrounding environment and ecological systems.
(b) The salmon fishing industry has been significantly impacted by dams in North America.

13.12 (a) Iceland is the leader in the use of geothermal energy.
(b) Geothermal accounts for 66% of its total energy use.

13.13 (a) Solar energy is converted into electricity by using photovoltaic devices or a solar thermal collector.
(b) Solar energy collectors use light-collecting optic systems that act to concentrate the sunlight to a focal point where the energy can be converted into heat.

13.14 (a) The conversion of SiO_2 to silicon occurs at >2000 °C.
(b) $SiO_2 + 2C \rightarrow Si + 2CO$.
(c) The reaction is run with excess SiO_2 to minimize the formation of SiC.

13.15 (a) CO_2 equivalent greenhouse gas emissions from a solar PV plant are 40 g CO_2 per kilowatt hour.
(b) CO_2 equivalent greenhouse gas emissions from a CSP station are about 25 g CO_2 per kilowatt hour.

13.16 All biomass fuels involve combustion, which contributes to air pollution and has higher greenhouse gas emissions than other alternative renewable energy possibilities.

13.17 (a) Hydrogen can be produced from the electrolysis of water or from steam reforming of natural gas.
(b) Steam reforming of natural gas is the major industrial source of H_2.
(c) H_2 can be used to produce energy by combustion or by electrochemistry.

13.18 At the cathode of a H_2 fuel cell, the H^+ ions react with O_2 in air to form H_2O.

13.19 Increased H_2 concentration in the troposphere would thus lead to the enhanced production of water vapor in the stratosphere, which would increase ozone depletion.

13.20 (a) Carrying capacity is the maximum population size of a species that the environment can sustain.
(b) When a species reaches its carrying capacity, loss of population occurs until the population reaches a new equilibrium with the available resources.

13.21 (a) The three goals of sustainability are economic development, social development, and environmental protection.
(b) The economy is dependent on the size of the society, and the size of both the economy and the society is constrained by environmental limits.

13.22 Obtaining sustainability over the long term in the face of a changing environment caused by climate change will require rapid implementation of mitigation and adaptation strategies.

13.23 (a) Green chemistry is focused on the design of products and processes to minimize the use and generation of hazardous substances.
(b) The 12 principles of green chemistry focus on minimizing chemical waste, preventing chemical pollution of the environment, and minimizing the use and production of hazardous chemicals.

13.24 (a) The global temperature is the average of the surface air temperatures and ocean surface temperatures across the globe.
(b) A global temperature anomaly is the variation of the global temperature at a point in time from a baseline temperature determined over a 30-year average.
(c) Temperature anomalies are used because they tend to be highly correlated over larger distances, while the absolute temperatures vary significantly over even short distances.

13.25 Most of the observed warming occurred from 1900 to 1940 and after 1970.

13.26 The Arctic and Northern latitudes will be the most severely impacted regions.

13.27
(a) The largest contributing factor to sea-level rise is thermal expansion of the oceans, which accounts for 42%.
(b) The second largest contributing factor is the melting of ice in Greenland and Antarctica at 23%.

13.28
(a) The latest IPCC estimate of sea-level rise is 90 cm (2 ft) by 2099.
(b) Sea-level rise could be three to four times the global average on the U.S. east coast and in Europe.
(c) The Atlantic Ocean is predicted to warm faster than the Pacific Ocean.

13.29
(a) Cradle to grave includes all energy use and environmental impacts, beginning with the point at which the raw materials are obtained and ending with the final waste products and their disposal.
(b) Cradle to gate includes every impact from the extraction of the raw materials to the point when the product leaves the production facility.
(c) Cradle to cradle begins with the extraction of raw materials and ends with recycling the product for reuse as the same product, a different product, or raw materials.

13.30 A carbon footprint is the total amount of greenhouse species emitted into the troposphere by an individual, an organization, an event, or a product reported as an equivalent amount of CO_2.

13.31
(a) Both nuclear and wind power have the lowest CO_2 equivalent greenhouse gas emissions.
(b) Both solar and geothermal have the highest emissions of all the non-combustion-related energy sources.

Appendix B

List of U.S. EPA Hazardous Air Pollutants – Air Toxics

Chemical name	Health concern(s)	CAS number
Acetaldehyde	Lachrymator, irritant, possible carcinogen	75070
Acetamide	Suspected carcinogen	60355
Acetonitrile	Lachrymator, toxic, metabolized to cyanide	75058
Acetophenone	Lachrymator, irritant	98862
2-Acetylaminofluorene	Carcinogen	53963
Acrolein	Lachrymator, irritant	107028
Acrylamide	Irritant, possible carcinogen	79061
Acrylic acid	Lachrymator, irritant, corrosive agent	79107
Acrylonitrile	Lachrymator, irritant	107131
Allyl chloride	Lachrymator, toxic, possible carcinogen	107051
4-Aminobiphenyl	Carcinogen	92671
Aniline	Toxic, possible carcinogen	62533
o-Anisidine	Toxic, metabolized to cyanide, carcinogen	90040
Asbestos	Mesothelioma agent, asbestosis	1332214
Benzene	Carcinogen	71432
Benzidine	Carcinogen (bladder and pancreatic cancer)	92875
Benzotrichloride	Lachrymator, irritant, toxic, possible carcinogen	98077
Benzyl chloride	Lachrymator, irritant, toxic, possible carcinogen	100447
Biphenyl	Lachrymator, irritant, toxic	92524
Bis(2-ethylhexyl)phthalate (DEHP)	Endocrine disruptor, possible carcinogen	117817
Bis(chloromethyl)ether	Carcinogen	542881
Bromoform	Toxic, carcinogen, ozone depleting substance	75252
1,3-Butadiene	Lachrymator, carcinogen, teratogen	106990
Calcium cyanamide	Lachrymator, irritant	156627
Caprolactam	Irritant – *Removed from list in June 1996*	105602
Captan	Pesticide, fungicide, possible carcinogen	133062
Carbaryl	Insecticide, toxic, cholinesterase inhibitor	63252
Carbon disulfide	Lachrymator, irritant, toxic	75150

Chemistry of Environmental Systems: Fundamental Principles and Analytical Methods, First Edition.
Jeffrey S. Gaffney and Nancy A. Marley.

Chemical name	Health concern(s)	CAS number
Carbon tetrachloride	Toxic, liver cancer, ozone-depleting substance	56235
Carbonyl sulfide	Toxic	463581
Catechol	Lachrymator, irritant, toxic	120809
Chloramben	Lachrymator, irritant, possible carcinogen	133904
Chlordane	Insecticide, carcinogen	57749
Chlorine	Lachrymator, irritant, toxic	7782505
Chloroacetic acid	Irritant, toxic, possible carcinogen	79118
2-Chloroacetophenone	Lachrymator, irritant, toxic	532274
Chlorobenzene	Lachrymator, irritant, toxic	108907
Chlorobenzilate	Pesticide, toxic, possible carcinogen	510156
Chloroform	Toxic, ozone-depleting substance	67663
Chloromethyl methyl ether	Irritant, carcinogen	107302
Chloroprene	Lachrymator, irritant, possible carcinogen	126998
Cresols/cresylic acid, isomers and mixture	Lachrymator, irritant, toxic	1319773
o-Cresol	Lachrymator, irritant, toxic	95487
m-Cresol	Lachrymator, irritant, toxic	108394
p-Cresol	Lachrymator, irritant, toxic	106445
Cumene	Irritant, toxic	98828
2,4-D, salts and esters	Herbicide, irritant, toxic, possible carcinogen	94757
DDE	DDT byproduct, toxic	3547044
Diazomethane	Highly toxic, possible carcinogen	334883
Dibenzofurans	Toxic (aquatic organisms)	132649
1,2-Dibromo-3-chloropropane	Fumigant, toxic, carcinogen	96128
Dibutylphthalate	Toxic (aquatic organisms)	84742
1,4-Dichlorobenzene	Lachrymator, toxic (aquatic organisms)	106467
3,3-Dichlorobenzidene	Carcinogen	91941
Dichloroethyl ether (bis(2-chloroethyl)ether)	Highly toxic, carcinogen	111444
1,3-Dichloropropene	Pesticide, toxic, possible carcinogen	542756
Dichlorvos	Insecticide, highly toxic	62737
Diethanolamine	Lachrymator, irritant	111422
N,*N*-Dimethylaniline	Lachrymator, irritant, toxic	121697
Diethyl sulfate	Highly toxic, carcinogen	64675
3,3-Dimethoxybenzidine	Toxic, carcinogen	119904
Dimethyl aminoazobenzene	Possible carcinogen	60117
3,3′-Dimethyl benzidine	Irritant, possible carcinogen	119937
Dimethyl carbamoyl chloride	Lachrymator, irritant, toxic, carcinogen	79447
Dimethyl formamide	Lachrymator, irritant, teratogen	68122
1,1-Dimethyl hydrazine	Lachrymator, irritant, toxic, possible carcinogen	57147

Chemical name	Health concern(s)	CAS number
Dimethyl phthalate	Lachrymator, irritant, toxic	131113
Dimethyl sulfate	Toxic, carcinogen	77781
4,6-Dinitro-*o*-cresol and salts	Lachrymator, irritant, toxic	534521
2,4-Dinitrophenol	Highly toxic, possible carcinogen	51285
2,4-Dinitrotoluene	Highly toxic, carcinogen	121142
1,4-Dioxane (1,4-diethyleneoxide)	Lachrymator, irritant, possible carcinogen	123911
1,2-Diphenylhydrazine	Lachrymator, irritant, toxic, carcinogen	122667
Epichlorohydrin (l-chloro-2,3-epoxypropane)	Irritant, carcinogen	106898
1,2-Epoxybutane	Lachrymator, irritant, carcinogen	106887
Ethyl acrylate	Lachrymator, irritant, toxic, carcinogen	140885
Ethyl benzene	Toxic, possible carcinogen	100414
Ethyl carbamate (urethane)	Highly toxic, possible carcinogen	51796
Ethyl chloride	Toxic (aquatic organisms), possible carcinogen	75003
Ethylene dibromide (dibromoethane)	Toxic (aquatic organisms), carcinogen	106934
Ethylene dichloride (1,2-dichloroethane)	Lachrymator, irritant, toxic, carcinogen	107062
Ethylene glycol	Toxic, teratogen	107211
Ethylene imine (aziridine)	Toxic, possible carcinogen	151564
Ethylene oxide	Lachrymator, irritant, toxic, carcinogen	75218
Ethylene thiourea	Carcinogen	96457
Ethylidene dichloride (1,1-dichloroethane)	Possible carcinogen	75343
Formaldehyde	Lachrymator, irritant, toxic, carcinogen	50000
Heptachlor	Insecticide, toxic, irritant, possible carcinogen	76448
Hexachlorobenzene	Fungicide, carcinogen, toxic (aquatic organisms)	118741
Hexachlorobutadiene	Toxic, possible carcinogen	87683
Hexachlorocyclopentadiene	Toxic, carcinogen	77474
Hexachloroethane	Teratogen	67721
Hexamethylene-1,6-diisocyanate	Lachrymator, irritant	822060
Hexamethylphosphoramide	Toxic, carcinogen	680319
Hexane	Irritant, toxic (aquatic organisms)	110543
Hydrazine	Lachrymator, irritant, toxic (aquatic organisms), carcinogen	302012
Hydrochloric acid	Lachrymator, irritant, toxic, corrosive	7647010
Hydrogen fluoride (hydrofluoric acid)	Lachrymator, irritant, toxic, corrosive	7664393
Hydrogen sulfide	Toxic (aquatic organisms) – *Delisted December 1991*	7783064
Hydroquinone	Lachrymator, irritant, toxic (aquatic organisms), possible carcinogen	123319
Isophorone	Lachrymator, irritant, possible carcinogen	78591

Chemical name	Health concern(s)	CAS number
Lindane (all isomers)	Insecticide, toxic, possible carcinogen	58899
Maleic anhydride	Lachrymator, irritant	108316
Methanol	Toxic	67561
Methoxychlor	Insecticide, highly toxic	72435
Methyl bromide	Ozone-depleting substance, lachrymator, toxic, irritant	74839
Methyl chloride	Ozone-depleting substance, lachrymator, toxic, irritant	74873
Methyl chloroform	Ozone-depleting substance, lachrymator, toxic, irritant	71556
Methyl ethyl ketone	Lachrymator, irritant – *Delisted December 2005*	78933
Methyl hydrazine	Lachrymator, irritant, toxic (aquatic organisms), possible carcinogen	60344
Methyl iodide	Lachrymator, irritant, toxic, possible carcinogen	74884
Methyl isobutyl ketone (hexone)	Lachrymator, irritant, toxic	108101
Methyl isocyanate	Lachrymator, irritant, toxic, teratogen	624839
Methyl methacrylate	Lachrymator, irritant	80626
Methyl tert butyl ether	Irritant, possible carcinogen	1634044
4,4-Methylene bis(2-chloroaniline)	Carcinogen	101144
Methylene chloride	Lachrymator, irritant, toxic, possible carcinogen	75092
Methylene diphenyl diisocyanate (MDI)	Lachrymator, toxic – *Under review for delisting*	101688
4,4′-Methylenedianiline	Toxic, possible carcinogen	101779
Naphthalene	Irritant, toxic (aquatic organisms), possible carcinogen	91203
Nitrobenzene	Toxic (aquatic organisms), possible carcinogen	98953
4-Nitrobiphenyl	Lachrymator, irritant	92933
4-Nitrophenol	Lachrymator, irritant, toxic	100027
2-Nitropropane	Possible carcinogen	79469
N-Nitroso-*N*-methylurea	Carcinogen, mutagen, teratogen	684935
N-Nitrosodimethylamine	Highly toxic (aquatic organisms), possible carcinogen	62759
N-Nitrosomorpholine	Possible carcinogen	59892
Parathion	Insecticide, highly toxic (aquatic organisms)	56382
Pentachloronitrobenzene (quintobenzene)	Fungicide, toxic	82688
Pentachlorophenol	Pesticide, lachrymator, irritant, toxic, possible carcinogen	87865
Phenol	Lachrymator, irritant, toxic	108952
p-Phenylenediamine	Irritant, toxic, possible carcinogen	106503
Phosgene	Highly toxic, corrosive	75445
Phosphine	Highly toxic, corrosive	7803512
Phosphorus	Irritant, highly toxic	7723140
Phthalic anhydride	Lachrymator, irritant, toxic	85449
Polychlorinated biphenyls (aroclors)	Insecticides, highly toxic (aquatic organisms)	1336363
1,3-Propane sultone	Toxic, carcinogen, mutagen, teratogen	1120714
beta-Propiolactone	Carcinogen	57578

Chemical name	Health concern(s)	CAS number
Propionaldehyde	Lachrymator, irritant	123386
Propoxur (Baygon)	Insecticide, toxic	114261
Propylene dichloride (1,2-dichloropropane)	Toxic, possible carcinogen	78875
Propylene oxide	Possible carcinogen	75569
1,2-Propylenimine (2-methyl aziridine)	Lachrymator, irritant, toxic	75558
Quinoline	Irritant, toxic	91225
Quinone	Lachrymator, irritant, toxic	106514
Styrene	Lachrymator, irritant, toxic	100425
Styrene oxide	Lachrymator, irritant, toxic, possible carcinogen	96093
2,3,7,8-Tetrachlorodibenzo-*p*-dioxin	Toxic, possible carcinogen	1746016
1,1,2,2-Tetrachloroethane	Ozone-depleting substance, toxic	79345
Tetrachloroethylene	Toxic (aquatic systems), possible carcinogen	127184
Titanium tetrachloride	Reacts violently with water to produce HCl, corrosive	7550450
Toluene	Irritant, toxic, possible carcinogen	108883
2,4-Toluene diamine	Lachrymator, irritant, toxic, possible carcinogen	95807
2,4-Toluene diisocyanate	Highly toxic, lachrymator, irritant, possible carcinogen	584849
o-Toluidine	Highly toxic, possible carcinogen	95534
Toxaphene (chlorinated camphene)	Insecticide, highly toxic, possible carcinogen	8001352
1,2,4-Trichlorobenzene	Toxic, possible teratogen	120821
1,1,2-Trichloroethane	Toxic, possible carcinogen	79005
Trichloroethylene	Carcinogen	79016
2,4,5-Trichlorophenol	Lachrymator, irritant	95954
2,4,6-Trichlorophenol	Irritant, possible carcinogen	88062
Triethylamine	Lachrymator, irritant, toxic	121448
Trifluralin	Herbicide, toxic, possible carcinogen	1582098
2,2,4-Trimethylpentane	Irritant, toxic (aquatic organisms)	540841
Vinyl acetate	Lachrymator, irritant	108054
Vinyl bromide	Lachrymator, irritant, toxic, possible carcinogen	593602
Vinyl chloride	Lachrymator, irritant, toxic, carcinogen	75014
Vinylidene chloride (1,1-dichloroethylene)	Toxic, possible carcinogen	75354
Xylenes (isomers and mixture)	Lachrymator, irritant, toxic	1330207
o-Xylene	Lachrymator, irritant, toxic	95476
m-Xylene	Lachrymator, irritant, toxic	108383
p-Xylene	Lachrymator, irritant, toxic	106423
Antimony compounds[a)]	Lachrymators, irritants	—
Arsenic compounds[a)] (inorganic, including arsine)	Toxic, carcinogens	—
Beryllium compounds[a)]	Irritants, toxic (berylliosis), possible carcinogens	—

Chemical name	Health concern(s)	CAS number
Cadmium compounds[a)]	Irritants, toxic, possible carcinogens	—
Chromium compounds[a)]	Toxic, carcinogens	—
Cobalt compounds[a)]	Irritants	—
Coke oven emissions[a)]	Lachrymators, irritants, carcinogens	—
Cyanide compounds[b)]	Lachrymators, irritants, highly toxic	—
Glycol ethers[c)]	Toxic	—
Lead compounds[a)]	Toxic, possible carcinogens	—
Manganese compounds[a)]	Irritant, toxic – manganism	—
Mercury compounds[a)]	Toxic – central nervous system disorders	—
Fine mineral fibers[d)]	Irritants – pulmonary, possible carcinogens	—
Nickel compounds	Irritants, possible carcinogens	—
Polycyclic organic matter[e)]	Lachrymators, irritants, carcinogens	—
Radionuclides (including Radon)[f)]	Lung irritants, carcinogens	—
Selenium compounds[a)]	Toxic, possible carcinogens	—

a) Unless otherwise specified, includes any unique chemical substance that contains the named type of chemical compound (i.e. antimony, arsenic, lead, etc.) as per its formula.
b) Chemical structure is XCN, where X = H or any other group which can dissociate (e.g. NaCN, KCN, or $Ca(CN)_2$).
c) Includes mono-substituted and di-substituted ethylene glycol, including diethylene glycol and triethylene glycol structures. $R—(OCH_2CH_2)_n—OR'$, where $n = 1, 2, 3$, R = alkyl or aryl groups, R′ = R or H. Also includes groups which yield glycol ethers with the structure $R—(OCH_2CH)_n—OH$ when they undergo reactions. Excluded from this category are all glycol-containing polymers.
d) Includes mineral fiber emissions of average diameter 1 μm or less from facilities that are involved in the manufacturing or processing of glass, rock, slag fibers, or other mineral fiber materials.
e) Includes organic compounds with two or more benzene rings, having boiling point equal to or higher than 100 °C.
f) Any element that undergoes radioactive decay processes (e.g. uranium, radium).

Appendix C

Henry's Law Constants (H_x) for Selected Inorganic and Organic Compounds

Compound type	Compound name	Chemical formula	$H_x = [X]/P_x$ ($M\ atm^{-1}$)	$H_x = P/[X]$ ($atm\ M^{-1}$)
Inorganic	Oxygen	O_2	1.3×10^{-3}	7.7×10^{2}
	Nitrogen	N_2	6.1×10^{-4}	1.6×10^{3}
	Carbon monoxide	CO	9.5×10^{-4}	1.1×10^{3}
	Carbon dioxide	CO_2	3.4×10^{-2}	2.9×10^{1}
	Ozone	O_3	1.0×10^{-2}	1.0×10^{2}
	Sulfur dioxide	SO_2	1.3	7.7×10^{-1}
	Hydrogen sulfide	H_2S	8.7×10^{-2}	1.1×10^{-1}
	Hydrogen peroxide	H_2O_2	1.0×10^{5}	1.0×10^{-5}
	Nitric oxide	NO	2.0×10^{-3}	5.0×10^{2}
	Nitrogen dioxide	NO_2	1.0×10^{-2}	1.0×10^{2}
	Nitrous oxide	N_2O	2.5×10^{-2}	4.0×10^{1}
	Nitric acid	HNO_3	2.0×10^{5}	5.0×10^{-6}
	Pernitric acid	HO_2NO_2	4.0×10^{3}	2.5×10^{-4}
	Ammonia	NH_3	6.0×10^{1}	1.7×10^{-2}
	Nitrous acid	$HONO$	5.0×10^{1}	2.0×10^{-2}
	Hydrogen cyanide	HCN	1.2×10^{1}	8.3×10^{-2}
	Dinitrogen tetroxide	N_2O_4	1.5	6.7×10^{-1}
	Dintrogen pentoxide	N_2O_5	∞	—
	Nitrogen trifluoride	NF_3	7.9×10^{-4}	1.3×10^{3}
	Hydrogen fluoride	HF	1.5×10^{4}	6.7×10^{-5}
	Hydrogen chloride	HCl	2.5×10^{3}	4.0×10^{-4}
	Hypochlorous acid	$HOCl$	9.3×10^{2}	1.1×10^{-3}
	Chlorine nitrate	$ClNO_3$	∞	—
	Chlorine	Cl_2	9.5×10^{-2}	1.1×10^{1}
	Dichlorine monoxide	Cl_2O	1.7×10^{1}	5.9×10^{-2}
	Chloramine	$ClNH_2$	9.4×10^{1}	1.1×10^{-2}
	Dichloramine	Cl_2NH	2.9×10^{1}	3.4×10^{-2}
	Nitrogen trichloride	NCl_3	1.0×10^{-1}	1.0×10^{1}
	Hydrogen bromide	HBr	7.2×10^{-1}	1.4

Chemistry of Environmental Systems: Fundamental Principles and Analytical Methods, First Edition.
Jeffrey S. Gaffney and Nancy A. Marley.

Compound type	Compound name	Chemical formula	$H_x = [X]/P_x$ (M atm^{-1})	$H_x = P/[X]$ (atm M^{-1})
	Bromine nitrate	$BrONO_2$	∞	—
	Bromine	Br_2	1.8	5.6×10^{-1}
	Bromine chloride	BrCl	9.4×10^{-1}	1.1
	Iodine	I_2	3.0	3.3×10^{-1}
	Iodine chloride	ICl	1.1×10^{2}	9.1×10^{-3}
	Iodine bromide	IBr	2.4×10^{1}	4.2×10^{-2}
	Sulfur trioxide	SO_3	∞	—
	Sulfuric acid	H_2SO_4	1.5×10^{14}	6.7×10^{-15}
	Mercury	Hg^0	9.3×10^{-2}	1.1×10^{1}
	Radon	Rn	9.3×10^{-3}	1.1×10^{2}
	Arsine	AsH_3	8.9×10^{-3}	1.1×10^{-2}
	Carbonyl sulfide	OCS	2.2×10^{-2}	4.5×10^{1}
	Carbon disulfide	CS_2	5.5×10^{-2}	1.8×10^{1}
Alkanes	Methane	CH_4	1.3×10^{-3}	7.7×10^{2}
	Ethane	C_2H_6	2.0×10^{-3}	5.0×10^{2}
	Propane	C_3H_8	1.5×10^{-3}	6.7×10^{2}
	Butane	C_4H_{10}	1.1×10^{-3}	9.1×10^{2}
	Pentane	C_5H_{12}	8.0×10^{-4}	1.3×10^{3}
	Hexane	C_6H_{14}	8.0×10^{-4}	1.3×10^{3}
	Heptane	C_7H_{16}	4.0×10^{-4}	2.5×10^{3}
	Octane	C_8H_{18}	3.0×10^{-4}	3.3×10^{3}
	Nonane	C_9H_{20}	2.0×10^{-4}	5.0×10^{3}
	Decane	$C_{10}H_{22}$	1.8×10^{-4}	5.6×10^{3}
	Hexadecance (cetane)	$C_{16}H_{34}$	4.3×10^{-3}	2.3×10^{2}
Alkenes	Ethene	C_2H_4	4.8×10^{-3}	2.1×10^{2}
	Propene	C_3H_6	4.8×10^{-3}	2.1×10^{2}
	1-Butene	C_4H_8	4.0×10^{-3}	2.5×10^{2}
	2-Butene (cis or trans)	C_4H_8	4.4×10^{-3}	2.3×10^{2}
	1-Pentene	C_5H_{10}	2.5×10^{-3}	4.0×10^{2}
	2-Pentene (cis or trans)	C_5H_{10}	4.4×10^{-3}	2.3×10^{2}
	1-Hexene	C_6H_{12}	2.4×10^{-3}	4.2×10^{2}
	1-Heptene	C_7H_{14}	2.5×10^{-3}	4.0×10^{2}
	1,3-Butadiene	C_4H_6	1.4×10^{-2}	7.1×10^{1}
	2-Methylbuta-1,3-diene (isoprene)	C_5H_8	1.3×10^{-2}	7.7×10^{1}
	Pinene	$C_{10}H_{16}$	4.9×10^{-2}	2.0×10^{1}
Alkynes	Ethyne (acetylene)	$HC{\equiv}CH$	4.0×10^{-2}	2.5×10^{1}
	Propyne	$CH_3C{\equiv}CH$	9.4×10^{-2}	1.1×10^{1}
	Butyne	$CH_3CH_2C{\equiv}CH$	5.4×10^{-2}	1.9×10^{1}
	Pentyne	$CH_3CH_2CH_2C{\equiv}CH$	4.0×10^{-2}	2.5×10^{1}

Compound type	Compound name	Chemical formula	$H_x = [X]/P_x$ ($M\ atm^{-1}$)	$H_x = P/[X]$ ($atm\ M^{-1}$)
Aromatics	Benzene	C_6H_6	1.8×10^{-1}	5.6
	Methylbenzene (toluene)	C_7H_8	1.6×10^{-1}	6.3
	1,2-Dimethylbenzene (*o*-xylene)	C_8H_{10}	2.0×10^{-1}	5.0
	1,3-Dimethylbenzene (*m*-xylene)	C_8H_{10}	1.5×10^{-1}	6.7
	1,4-Dimethylbenzene (*p*-xylene)	C_8H_{10}	1.3×10^{-1}	7.7
	1,3,5-Trimethlbenzene (mesitylene)	C_9H_{12}	1.4×10^{-1}	7.2
PAHs	Naphthalene	$C_{10}H_8$	2.1	4.8×10^{-1}
	Anthracene	$C_{14}H_{10}$	3.5×10^{1}	2.9×10^{-2}
	Pyrene	$C_{16}H_{10}$	8.8×10^{1}	1.1×10^{-2}
	Benzo(a)pyrene (BaP)	$C_{20}H_{12}$	1.6×10^{-1}	6.3
Alcohols	Methanol	CH_3OH	2.2×10^{2}	4.5×10^{-3}
	Ethanol	C_2H_5OH	2.0×10^{2}	5.0×10^{-3}
	1-Propanol	C_3H_7OH	1.5×10^{2}	6.7×10^{-3}
	2-Propanol (isopropanol)	C_3H_7OH	1.3×10^{2}	7.7×10^{-3}
	1-Butanol	C_4H_9OH	1.3×10^{2}	7.7×10^{-3}
	2-Butanol	C_4H_9OH	1.0×10^{2}	1.0×10^{-2}
	1-Octanol	$C_8H_{17}OH$	4.1×10^{1}	2.4×10^{-2}
	Hydroxybenzene (phenol)	C_6H_5OH	1.9×10^{3}	5.3×10^{-4}
	Benzyl alcohol	$C_6H_5CH_2OH$	9.0×10^{3}	1.1×10^{-4}
	1-Hydroxy-2-methylbenzene (*o*-cresol)	$CH_3C_6H_4OH$	8.3×10^{2}	1.2×10^{-3}
	1-Hydroxy-3-methylbenzene (*m*-cresol)	$CH_3C_6H_4OH$	6.3×10^{2}	1.6×10^{-3}
	1-Hydroxy-4-methylbenzene (*p*-cresol)	$CH_3C_6H_4OH$	1.0×10^{3}	1.0×10^{-3}
Polyols	1,2-Ethanediol	CH_2OHCH_2OH	4.0×10^{6}	2.5×10^{-7}
	1,3-Propanediol	$CH_2OHCH_2CH_2OH$	9.2×10^{5}	1.1×10^{-6}
	1,2,3-Butanetriol	$C_4H_7(OH)_3$	3.0×10^{11}	3.3×10^{-12}
	1,2,3,4-Tetrahydroxybutane	$C_4H_6(OH)_4$	2.0×10^{16}	5.0×10^{-17}
	1,2,3,4,5-Pentahydroxypentane	$C_5H_7(OH)_5$	9.0×10^{20}	1.1×10^{-21}
	1,2-Dihydroxybenzene	$C_6H_4(OH)_2$	4.6×10^{3}	2.2×10^{-4}
	1,3-Dihydroxybenzene	$C_6H_4(OH)_2$	8.3×10^{6}	1.2×10^{-7}
	1,4-Dihydroxybenzene (hydroquinone)	$C_6H_4(OH)_2$	2.5×10^{7}	4.0×10^{-8}
Peroxides	Methyl hydroperoxide	CH_3OOH	3.1×10^{2}	3.2×10^{-3}
	Ethyl hydroperoxide	CH_3CH_2OOH	3.4×10^{2}	2.9×10^{-3}
	Hydroxymethyl hydroperoxide	$HOCH_2OOH$	1.6×10^{6}	6.3×10^{-7}

Compound type	Compound name	Chemical formula	$H_x = [X]/P_x$ ($M\ atm^{-1}$)	$H_x = P/[X]$ ($atm\ M^{-1}$)
Aldehydes	Methanal (formaldehyde)	H_2CO	6.3×10^3	1.6×10^{-4}
	Ethanal (acetaldehyde)	CH_3CHO	1.5×10^1	6.7×10^{-2}
	Propanal (propionaldehyde)	CH_3CH_2CHO	1.3×10^1	7.7×10^{-2}
	Propenal (acrolein)	$CH_2{=}CHCHO$	8.2	1.2×10^{-1}
	2-Methylpropenal (methacrolein)	$CH_2{=}C(CH_3)CHO$	5.0	2.0×10^{-1}
	Benzaldehyde	C_6H_5CHO	3.6×10^1	2.8×10^{-2}
	3-Hydroxybenzaldehyde	HOC_6H_4CHO	4.0×10^5	2.5×10^{-6}
	Ethanedial (glyoxal)	OHCCHO	3.6×10^5	2.8×10^{-6}
Ketones	Propanone (acetone)	CH_3COCH_3	3.0×10^1	3.3×10^{-2}
	3-Pentanone	$CH_3CH_2COCH_2CH_3$	2.0×10^1	5.0×10^{-2}
	2-Octanone	$CH_3COC_6H_{13}$	5.4	1.9×10^{-1}
	1-Phenylethanone (acetophenone)	$C_6H_5COCH_3$	1.0×10^2	1.0×10^{-2}
	2,3-Butadione	$CH_3COCOCH_3$	7.4×10^1	1.4×10^{-2}
Organic acids[a)]	Methanoic acid (formic acid)	HCOOH	6.0×10^3	1.7×10^{-4}
	Ethanoic acid (acetic acid)	CH_3COOH	1.0×10^4	1.0×10^{-4}
	Propanoic acid	CH_3CH_2COOH	6.0×10^3	1.7×10^{-4}
	2-Methyl-2-propenoic acid (methacrylic acid)	$CH_2{=}C(CH_3)COOH$	2.6×10^3	3.8×10^{-4}
	Benzoic acid	C_6H_5COOH	1.4×10^4	7.1×10^{-5}
	Ethanedioc acid (oxalic acid)	HOOCCOOH	5.0×10^8	2.0×10^{-9}
	Hexanedioc acid (adipic acid)	$HOOCC_4H_8COOH$	1.8×10^7	5.6×10^{-8}
	Ethanoic peroxyacid (peracetic acid)	CH_3CO_3H	8.4×10^2	1.2×10^{-3}
Esters	Methyl methanoate (methyl formate)	$HCOOCH_3$	4.1	2.4×10^{-1}
	Methyl ethanoate (methyl acetate)	CH_3COOCH_3	7.8	1.3×10^{-1}
	Methyl benzoate	$C_6H_5COOCH_3$	5.6×10^1	1.8×10^{-2}
	Ethyl ethanoate (ethyl acetate)	$CH_3COOCH_2CH_3$	5.9	1.7×10^{-2}
Ethers	Dimethyl ether	CH_3OCH_3	1.0	1.0
	Diethyl ether	$CH_3CH_2OCH_2CH_3$	1.2	8.3×10^{-1}
	Methoxybenzene (anisole)	$C_6H_5OCH_3$	2.4×10^{-1}	4.2
	Tetrahydrofuran (THF)	C_4H_8O	1.8×10^1	5.6×10^{-2}
Epoxide	1,2-Epoxypropane	C_3H_6O	5.3	1.9×10^{-1}
Multi-functional	Propanonal (methyl glyoxal)	CH_3COCHO	3.2×10^4	3.1×10^{-5}
	2-Hydroxyethanal (hydroxyacetaldehyde)	$HOCH_2CHO$	4.1×10^4	2.4×10^{-5}
	2-Oxoproanoic acid (pyruvic acid)	$CH_3COCOOH$	3.1×10^5	3.2×10^{-6}

Compound type	Compound name	Chemical formula	$H_x = [X]/P_x$ ($M\ atm^{-1}$)	$H_x = P/[X]$ ($atm\ M^{-1}$)
	1-Hydroxy-2-methoxybenzene (guiacol)	$HOC_6H_4OCH_3$	9.1×10^{2}	1.1×10^{-3}
	Hydroxybutanedioc acid (malic acid)	$HOOCCH_2CHOHCOOH$	2.0×10^{13}	5.0×10^{-14}
	2-Hydroxy-1,2,3-propanetricarboxylic acid (citric acid)	$HOOCH_2COHCOOHCH_2COOH$	3.0×10^{18}	3.3×10^{-19}
	2-Oxopentanedioic acid (α-keto glutaric acid)	$HOOC(CH_2)_2COCOOH$	1.0×10^{9}	1.0×10^{-9}
	2-Hydroxypropanoic acid (lactic acid)	$CH_3CHOHCOOH$	7.0×10^{7}	1.4×10^{-8}
	2,3-Dihydroxybutanedioic acid (tartaric acid)	$HOOCCHOHCHOHCOOH$	1.0×10^{18}	1.0×10^{-18}
	2,3-Dihydroxypropanal (glyceraldehyde)	$CH_2OHCHOHCHO$	2.0×10^{10}	5.0×10^{-11}
Amines	Methylamine	CH_3NH_2	9.0×10^{1}	1.1×10^{-2}
	Ethylamine	$CH_3CH_2NH_2$	8.1×10^{1}	1.2×10^{-2}
	1-Propylamine	$CH_3CH_2CH_2NH_2$	6.7×10^{1}	1.5×10^{-2}
	1-Butylamine	$CH_3CH_2CH_2CH_2NH_2$	5.8×10^{1}	1.7×10^{-2}
	1-Pentylamine	$C_5H_{11}NH_2$	4.1×10^{1}	2.4×10^{-2}
	1-Hexylamine	$C_6H_{13}NH_2$	3.7×10^{1}	2.7×10^{-2}
	Dimethylamine	$(CH_3)_2NH$	5.7×10^{1}	1.8×10^{-2}
	Diethylamine	$(CH_3CH_2)_2NH$	3.9×10^{1}	2.6×10^{-2}
	Dipropylamine	$(CH_3CH_2CH_2)_2NH$	1.9×10^{1}	5.3×10^{-2}
	Dibutylamine	$(CH_3CH_2CH_2CH_2)_2NH$	1.1×10^{1}	9.1×10^{-2}
	Trimethylamine	$(CH_3)_3N$	9.6	1.0×10^{-1}
	Triethylamine	$(CH_3CH_2)_3N$	6.7	1.5×10^{-1}
	1,2-Diaminoethane (ethylenediamine)	$H_2NCH_2CH_2NH_2$	5.9×10^{5}	1.7×10^{-6}
	Ethanolamine	$HOCH_2CH_2NH_2$	6.2×10^{6}	1.6×10^{-7}
Amino acids	Glutamic acid	$HOOCCH(NH_2)CH_2COOH$	1.0×10^{13}	1.0×10^{-13}
	Asparagine	$H_2NCOCH_2CH(NH_2)COOH$	1.0×10^{13}	1.0×10^{-13}
	Serine	$HOCH_2CH(NH_2)COOH$	4.0×10^{12}	2.5×10^{-13}
	Glutamine	$H_2NCOCH_2CH_2CH(NH_2)COOH$	1.0×10^{13}	1.0×10^{-13}
	Glycine	$CH_2(NH_2)COOH$	9.0×10^{7}	1.1×10^{-8}
	Arginine	$C_6H_{14}N_4O_2$	1.0×10^{17}	1.0×10^{-17}
	Alanine	$CH_3CH(NH_2)COOH$	6.0×10^{7}	1.7×10^{-8}
	Leucine	$(CH_3)_2CHCH_2CH(NH_2)COOH$	2.0×10^{7}	5.0×10^{-8}
N-Cyclics	Pyrrolidine (tetrahydropyrrole)	C_4H_8NH	4.2×10^{2}	2.4×10^{-3}
	N-methyl-pyrrolidine	$C_4H_8NCH_3$	3.3×10^{1}	3.0×10^{-2}
	Piperidine	$C_5H_{10}NH$	2.2×10^{2}	4.5×10^{-3}
	N-methyl-piperidine	$C_5H_{10}NCH_3$	2.9×10^{1}	3.4×10^{-2}
	Pyridine	C_5H_5N	1.1×10^{2}	9.1×10^{-3}

Compound type	Compound name	Chemical formula	$H_x = [X]/P_x$ ($M\ atm^{-1}$)	$H_x = P/[X]$ ($atm\ M^{-1}$)
	2-Methylpyridine	$C_5H_4NCH_3$	1.0×10^2	1.0×10^{-2}
	3-Methylpyridine	$C_5H_4NCH_3$	1.3×10^2	7.7×10^{-3}
	4-Methylpyridine	$C_5H_4NCH_3$	1.7×10^2	5.9×10^{-3}
	2-Methylpyrazine	$C_4N_2H_3CH_3$	4.5×10^2	2.2×10^{-3}
	Benzopyridine (quinoline)	C_9H_7N	3.7×10^3	2.7×10^{-4}
Organic nitrates	Methyl nitrate	CH_3ONO_2	2.0	5.0×10^{-1}
	Ethyl nitrate	$CH_3CH_2ONO_2$	1.6	6.3×10^{-1}
	1-Butyl nitrate	$CH_3CH_2CH_2CH_2ONO_2$	1.0	1.0
	2-Nitroxy ethanol	$HOCH_2CH_2ONO_2$	4.0×10^4	2.5×10^{-5}
	1-Nitrooxy-2-propanol	$CH_3CHOHCH_2ONO_2$	1.1×10^4	9.1×10^{-5}
	2-Nitrooxy-3-butanol	$C_4H_9O_4N$	1.0×10^4	1.0×10^{-4}
	Nitrooxyacetone	$CH_3COCH_2ONO_2$	1.0×10^3	1.0×10^{-3}
	1,3-Propanediol dinitrate	$O_2NOCH_2CH_2CH_2ONO_2$	1.3×10^2	7.7×10^{-3}
	Peroxyacetyl nitrate (PAN)	$CH_3COOONO_2$	4.0	2.5×10^{-1}
	Peroxypropionyl nitrate (PPN)	$CH_3CH_2COOONO_2$	2.9	3.4×10^{-1}
	Peroxybutyl nitrate (PBN)	$CH_3CH_2CH_2COOONO_2$	2.3	4.3×10^{-1}
	Peroxymethacryloyl nitrate (MPAN)	$CH_2{=}C(CH_3)COOONO_2$	1.7	5.9×10^{-1}
	Peroxyisobutyryl nitrate (PiBN)	$(CH_3)_2CHCOOONO_2$	1.0	1.0
Nitriles	Ethane nitrile (acetonitrile)	$CH_3C{\equiv}N$	5.3×10^1	1.9×10^{-2}
	Propane nitrile (propionitrile)	$CH_3CH_2C{\equiv}N$	2.7×10^1	3.7×10^{-2}
	Butane nitrile (butyronitrile)	$CH_3CH_2CH_2C{\equiv}N$	1.9×10^1	5.3×10^{-2}
	Benzene nitrile	$C_6H_5C{\equiv}N$	1.8	5.6×10^{-1}
	Ethane dinitrile (cyanogen)	$N{\equiv}CC{\equiv}N$	1.9×10^{-1}	5.3
	2-Propene nitrile (acrylonitrile)	$H_2C{=}CHC{\equiv}N$	1.1×10^1	9.1×10^{-2}
Nitro compounds	Nitromethane	CH_3NO_2	4.0×10^1	2.5×10^{-2}
	Nitroethane	$CH_3CH_2NO_2$	2.1×10^1	4.8×10^{-2}
	1-Nitropropane	$CH_3CH_2CH_2NO_2$	1.4×10^1	7.1×10^{-2}
	Nitrobenzene	$C_6H_5NO_2$	4.7×10^1	2.1×10^{-2}
	2-Nitrotoluene	$CH_3C_6H_4NO_2$	7.8	1.3×10^{-1}
	3-Nitrotoluene	$CH_3C_6H_4NO_2$	1.4×10^1	7.1×10^{-2}
	4-Nitrotoluene	$CH_3C_6H_4NO_2$	1.6×10^1	6.3×10^{-2}
	1-Methyl-2,4-dinitrobenzene (TNT)	$CH_3C_6H_3(NO_2)_2$	2.1×10^1	4.8×10^{-2}
	2-Nitrophenol	$HOC_6H_4NO_2$	7.0×10^1	1.4×10^{-2}
	3-Nitrophenol	$HOC_6H_4NO_2$	5.0×10^5	2.0×10^{-6}
	4-Nitrophenol	$HOC_6H_4NO_2$	2.6×10^6	3.8×10^{-7}
	3-Methyl-2-nitrophenol	$HOC_6H_3CH_3NO_2$	2.5×10^2	4.0×10^{-3}
	2,4-Dinitrophenol	$HOC_6H_3(NO_2)_2$	3.5×10^3	2.9×10^{-4}

Compound type	Compound name	Chemical formula	$H_x = [X]/P_x$ ($M\,atm^{-1}$)	$H_x = P/[X]$ ($atm\,M^{-1}$)
Fluoro compounds	Fluoromethane	CH_3F	5.9×10^{-2}	1.7×10^{1}
	Difluoromethane	CH_2F_2	8.7×10^{-2}	1.1×10^{1}
	Trifluoromethane	CHF_3	1.4×10^{-2}	7.1×10^{1}
	Tetrafluoromethane	CF_4	2.0×10^{-4}	5.0×10^{3}
	Fluoroethane	CH_3CH_2F	4.4×10^{-2}	2.3×10^{1}
	1,1-Difluoroethane	FCH_2CH_2F	5.4×10^{-2}	1.9×10^{1}
	Hexafluoroethane	C_2F_6	5.9×10^{-5}	1.7×10^{4}
	Formyl fluoride	FCHO	3.0	3.3×10^{-1}
	Trifluoro ethanoic acid (trifluoro acetic acid)	CF_3COOH	8.9×10^{3}	1.1×10^{-4}
	Fluorobenzene	C_6H_5F	1.6×10^{-1}	6.3
Chloro compounds	Chloromethane	CH_3Cl	2.9×10^{-2}	3.4×10^{1}
	Dichloromethane	CH_2Cl_2	3.1×10^{-1}	3.2
	Trichloromethane	$CHCl_3$	3.0×10^{-1}	3.3
	Tetrachloromethane	CCl_4	3.3×10^{-2}	3.0×10^{1}
	Chloroethane	CH_3CH_2Cl	6.9×10^{-2}	1.4×10^{1}
	1,1-Dichloroethane	CH_3CHCl_2	1.8×10^{-1}	5.6
	1,2-Dichloroethane	CH_2ClCH_2Cl	9.1×10^{-1}	1.1
	1,1,1-Trichloroethane	CH_3CCl_3	6.0×10^{-2}	1.7×10^{1}
	Chloroethene (vinyl chloride)	$CH_2{=}CHCl$	3.8×10^{-2}	2.6×10^{1}
	Trichloroethanal	CCl_3CHO	3.4×10^{5}	2.9×10^{-6}
	Trichloro ethanoic acid	CCl_3COOH	7.4×10^{4}	1.4×10^{-5}
	Chlorobenzene	C_6H_5Cl	2.5×10^{-1}	4.0
	Hexachlorobenzene	C_6Cl_6	5.9×10^{-1}	1.7
	1-Chloro-2-methyl benzene	$ClC_6H_4CH_3$	1.9	5.3×10^{-1}
	1-Chloronaphthalene	$C_{10}H_7Cl$	2.9×10^{-1}	3.4
	2-Hydroxychlorobenzene	HOC_6H_4Cl	1.2×10^{2}	8.3×10^{-3}
CFCs	Chlorofluoromethane	CH_2FCl	1.5×10^{-1}	6.7
	Chlorodifluoromethane	CHF_2Cl	3.7×10^{-2}	2.7×10^{1}
	Dichlorofluoromethane	$CHFCl_2$	1.9×10^{-1}	5.3
	Chlorotrifluoromethane	CF_3Cl	5.8×10^{-4}	1.7×10^{3}
	Dichlorodifluoromethane	CF_2Cl_2	2.5×10^{-3}	4.0×10^{2}
	Trichlorofluoromethane	$CFCl_3$	1.7×10^{-2}	5.9×10^{1}
	1,1,2,2-Tetrachloro-difluoroethane	$CFCl_2CFCl_2$	1.0×10^{-2}	1.0×10^{2}
	1,1,2-Trichlorotrifluoroethane	$CFCl_2CF_2Cl$	2.0×10^{-3}	5.0×10^{2}
	1,1-Dichlorotetrafluoroethane	$CFCl_2CF_3$	5.9×10^{-4}	1.7×10^{3}
	1,2-Dichlorotetrafluoroethane	CF_2ClCF_2Cl	8.2×10^{-4}	1.2×10^{3}
	Chloropentafluoroethane	CF_3CF_2Cl	3.2×10^{-4}	3.1×10^{3}

Compound type	Compound name	Chemical formula	$H_x = [X]/P_x$ ($M\ atm^{-1}$)	$H_x = P/[X]$ ($atm\ M^{-1}$)
	Dichlorotrifluoroethane	$C_2HF_3Cl_2$	2.9×10^{-2}	3.4×10^{1}
	1-Chloro-1,2,2,2-tetrafluoroethane	$CHClF_2CF_3$	1.1×10^{-2}	9.1×10^{1}
	2-Chloro-1,1,1-trifluoroethane	CH_2ClCF_3	3.7×10^{-2}	2.7×10^{1}
	1,1-Dichloro-1-fluoroethane	$CFCl_2CH_3$	7.9×10^{-3}	1.3×10^{2}
	1-Chloro-1,1-difluoroethane	CF_2ClCH_3	1.4×10^{-2}	7.1×10^{1}
	Chlorodifluoroethanoic acid	$CClF_2COOH$	2.5×10^{4}	5.0×10^{-5}
	Chlorodifluoroethanoic peroxy acid	$CClF_2COOOH$	3.0×10^{3}	3.3×10^{-4}
	Dichlorofluoroethanoic peroxy acid	$CCl_2FCOOOH$	3.0×10^{3}	3.3×10^{-4}
Bromo compounds	Bromomethane	CH_3Br	1.6×10^{-1}	6.3
	Dibromomethane	CH_2Br_2	1.1	9.1×10^{-1}
	Tribromomethane	$CHBr_3$	1.8	5.6×10^{-1}
	Bromoethane	CH_3CH_2Br	1.3×10^{-1}	7.7
	1,2-Bromoethane	CH_2BrCH_2Br	1.1	9.1×10^{-1}
	1-Bromopropane	$CH_3CH_2CH_2Br$	1.1×10^{-1}	9.1
	2-Bromopropane	$CH_3CHBrCH_3$	9.2×10^{-2}	1.1×10^{1}
	Bromoethanoic acid	$CH_2BrCOOH$	1.5×10^{5}	6.7×10^{-6}
	Dibromoethanoic acid	$CHBr_2COOH$	2.3×10^{5}	4.3×10^{-6}
	Tribromethanoic acid	CBr_3COOH	3.6×10^{5}	2.8×10^{-6}
	Bromobenzene	C_6H_6Br	5.4×10^{-1}	1.9
	4-Bromophenol	HOC_6H_4Br	7.0×10^{3}	1.4×10^{-4}
	Bromotrifluoromethane	CF_3Br	2.0×10^{-3}	5.0×10^{2}
	Bromodichloromethane	$CHCl_2Br$	4.6×10^{-1}	2.2
	Dibromochloromethane	$CHBr_2Cl$	1.2	8.3×10^{-1}
Iodo compounds	Iodomethane	CH_3I	1.4×10^{-1}	7.1×10^{-2}
	Diiodomethane	CH_2I_2	2.3	4.3×10^{-1}
	Triiodomethane	CHI_3	3.4×10^{-1}	2.9
	Iodoethane	C_2H_5I	1.4×10^{-1}	7.1
	Iodobenzene	C_6H_5I	7.8×10^{-1}	1.3
	Chloroiodomethane	CH_2ICl	8.9×10^{-1}	1.1
Sulfur compounds	Methanethiol	CH_3SH	2.6×10^{-1}	3.8
	Ethanethiol	CH_3CH_2SH	2.6×10^{-1}	3.8
	1-Propanethiol	$CH_3CH_2CH_2SH$	2.5×10^{-1}	4.0
	1-Butanethiol	$CH_3CH_2CH_2CH_2SH$	2.2×10^{-1}	4.5
	Thiophenol	C_6H_5SH	3.0	3.3×10^{-1}
	Dimethyl sulfide	$(CH_3)_2S$	9.6×10^{-1}	1.0

Compound type	Compound name	Chemical formula	$H_x = [X]/P_x$ (M atm^{-1})	$H_x = P/[X]$ (atm M^{-1})
	Thiophene	C_4H_4S	4.4×10^{-1}	2.3
	Dimethylsulfoxide	CH_3SOCH_3	5.0×10^{4}	2.0×10^{-5}
	Methanesulfonic acid	CH_3SO_3H	8.2×10^{11}	1.2×10^{-12}
	2,2′-Dichlorodiethylsulfide (mustard gas)	$ClCH_2CH_2SCH_2CH_2Cl$	3.0×10^{1}	3.3×10^{-2}
PCBs	2,2′-PCB	$C_{12}H_8Cl_2$	2.9	3.4×10^{-1}
	2,5-PCB	$C_{12}H_8Cl_2$	2.6	3.8×10^{-1}
	3,3′-PCB	$C_{12}H_8Cl_2$	4.2	2.4×10^{-1}
	3,4-PCB	$C_{12}H_8Cl_2$	4.8	2.1×10^{-1}
	4,4′-PCB	$C_{12}H_8Cl_2$	5.1	2.0×10^{-1}
	2,3,5-PCB	$C_{12}H_7Cl_3$	3.0	3.3×10^{-1}
	2,4,6-PCB	$C_{12}H_7Cl_3$	1.5	6.7×10^{-1}
	2,4,4′-PCB	$C_{12}H_7Cl_3$	3.6	2.8×10^{-1}
	2,2′,3,3′-PCB	$C_{12}H_6Cl_4$	4.9	2.0×10^{-1}
	2,2′,5,5′-PCB	$C_{12}H_6Cl_4$	2.9	3.4×10^{-1}
	2,2′,6,6′-PCB	$C_{12}H_6Cl_4$	1.8	5.6×10^{-1}
	2,2′,5,6-PCB	$C_{12}H_6Cl_4$	2.4	4.2×10^{-1}
	3,3′,4,4′-PCB	$C_{12}H_6Cl_4$	1.1×10^{1}	9.1×10^{-2}
	2,2′,4,5,5′-PCB	$C_{12}H_5Cl_5$	3.9	2.6×10^{-1}
	2,2′,4,6,6′-PCB	$C_{12}H_5Cl_5$	1.1	9.1×10^{-1}
	2,2′,3,3′,4,4′-PCB	$C_{12}H_4C_6$	3.3×10^{1}	3.0×10^{-2}
	2,2′,4,4′,5,5′-PCB	$C_{12}H_4Cl_6$	7.6	1.3×10^{-1}
	2,2′,4,4′,6,6′-PCB	$C_{12}H_4Cl_6$	1.3	7.7×10^{-1}
	Aroclor1221	PCB mixture	4.4	2.2×10^{-1}
	Aroclor1242	PCB mixture	2.4	4.2×10^{-1}
	Aroclor1248	PCB mixture	2.3	4.3×10^{-1}
	Aroclor1254	PCB mixture	3.0	3.3×10^{-1}
	Aroclor1260	PCB mixture	3.3	3.0×10^{-1}
	Aroclor1268	PCB mixture	2.5	4.0×10^{-1}
Pesticides	Hexachlorocyclopentadiene	C_5Cl_6	3.7×10^{-2}	2.7×10^{1}
	α-1,2,3,4,5,6-Hexachloro-cyclohexane	$C_6H_6Cl_6$	1.3×10^{2}	7.7×10^{-3}
	γ-1,2,3,4,5,6-Hexachlorocyclohexane (lindane)	$C_6H_6Cl_6$	2.8×10^{2}	3.6×10^{-3}
	Dodecachloropentacyclo-decane (mirex)	$C_{10}Cl_{12}$	1.2	8.3×10^{-1}
	Aldrin	$C_{12}H_8Cl_6$	3.6×10^{1}	2.8×10^{-2}
	Dieldrin	$C_{12}H_8OCl_6$	9.2×10^{1}	1.1×10^{-2}
	1,1,1-Trichloro-2,2-bis-(4chlorophenyl)ethane (DDT)	$C_{14}H_9C_{15}$	1.9×10^{1}	5.3×10^{-2}
	Molinate	$C_9H_{17}NOS$	1.7×10^{2}	5.9×10^{-3}

Compound type	Compound name	Chemical formula	$H_x = [X]/P_x$ ($M\ atm^{-1}$)	$H_x = P/[X]$ ($atm\ M^{-1}$)
	Parathion	$C_{10}H_{14}NO_5PS$	8.2×10^2	1.2×10^{-3}
	Malathion	$C_{10}H_{19}O_6PS_2$	2.7×10^3	3.7×10^{-4}
	Methylchlorpyrifos	$C_7H_7NO_3Cl_3PS$	3.3×10^2	3.0×10^{-3}
	Fenitrothion	$C_9H_{12}NO_5PS$	2.7×10^3	3.7×10^{-4}
	Dicapthon	$C_8H_9NO_5ClPS$	4.4×10^3	2.3×10^{-4}
	Ronnel	$C_8H_8O_3C_{13}PS$	4.8×10^1	2.1×10^{-2}
	Leptophos	$C_{13}H_{10}O_3BrCl_2P$	3.8×10^2	2.6×10^{-3}
Dioxins	2,7-Dichlorodibenzo[b,e][1,4]-dioxin (2,7 DiCDD)	$C_{12}H_6O_2Cl_2$	1.7×10^1	5.9×10^{-2}
	1,2,4-Trichlorodibenzo[b,e][1,4]-dioxin (1,2,4-TriCDD)	$C_{12}H_5O_2Cl_3$	2.8×10^1	3.6×10^{-2}
	1,2,3,4-Tetrachlorodibenzo[b,e][1,4]-dioxin (1,2,3,4-TCDD)	$C_{12}H_4O_2Cl_4$	5.1×10^1	2.0×10^{-2}
Radicals	Cyanide	CN	8.0×10^{-2}	1.3×10^1
	Hydroxyl	OH	3.0×10^1	3.3×10^{-1}
	Hydroperoxyl	HO_2	5.7×10^3	1.8×10^{-4}
	Nitrate	NO_3	1.8	5.6×10^{-1}
	Fluorine atom	F	2.1×10^{-2}	4.8×10^1
	Chlorine atom	Cl	1.5×10^{-2}	6.7×10^1
	Chlorine dioxide	ClO_2	1.0	1.0
	Bromine atom	Br	1.2	8.3×10^{-1}
	Iodine atom	I	8.0×10^{-2}	1.3×10^1

a) Note: For acids the appropriate K_a values were used to calculate the H_x values.

Source: Sander, R. (2015) Compilation of Henry's law constants (version 4.0) for water as solvent. *Atmospheric Chemistry and Physics*, vol. 15, pp. 4399–4981.

Appendix D

Organic Water Pollutants, their Chemical Structures, Sources, and Concentration Limits in U.S. Drinking Water

Organic pollutant	Structure[a)]	Major sources	Max. limit
Acrylamide	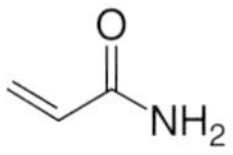 O, NH2	Used as flocking agent in wastewater treatment, plastics industry	Treatment limit of 0.05% water at 1 mg l^{-1} concentration
Alachlor	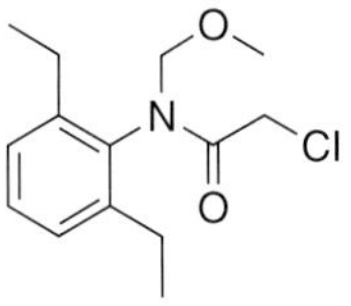 O, N, Cl, O	Herbicide, agricultural runoff	2 ppb
Atrazine	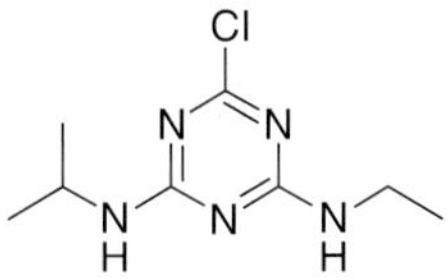 Cl, N, N, N, N H, N H	Herbicide, agricultural runoff	3 ppb
Benzene		Oil refinery runoff, leaking gasoline, and oil storage tanks	5 ppb
Benzo(a)pyrene (PAH)	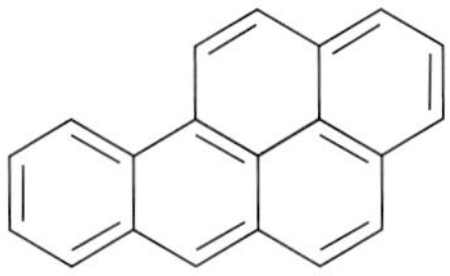	Leaching from water storage and pipe plastic linings	0.2 ppb

Chemistry of Environmental Systems: Fundamental Principles and Analytical Methods, First Edition.
Jeffrey S. Gaffney and Nancy A. Marley.

Organic pollutant	Structure[a)]	Major sources	Max. limit
Carbofuran	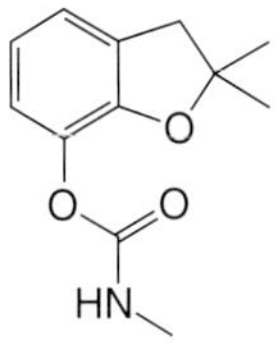	Pesticide; leaching from soils into water supplies; used for rice and alfalfa	40 ppb
Carbon tetrachloride	CCl_4	Solvent, industrial waste water; chemical plant waste release	5 ppb
Chlordane	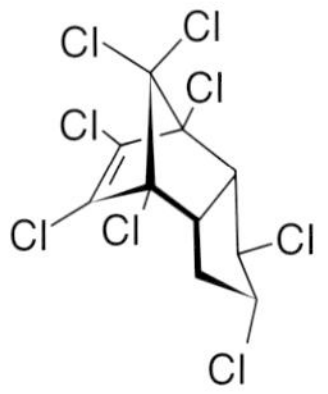	Banned pesticide used in homes for termites; leaching from past use in older homes into water supplies	2 ppb
Chlorobenzene	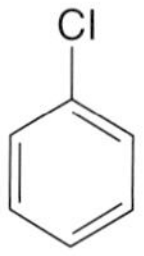	Solvent; wastewater discharge from chemical industries	100 ppb
2,4-Dichlorophenoxy acetic acid (2,4-D)	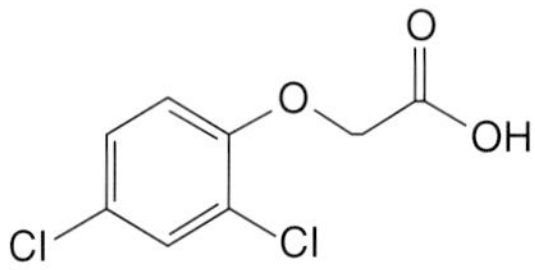	Herbicide; agricultural water runoff from crops and soils	70 ppb
Dalapon	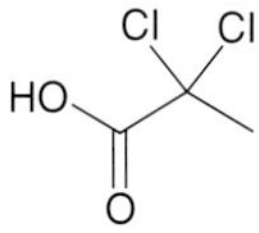	Herbicide; runoff from highway shoulder applications	200 ppb
1,2-Dibromo-3-chloropropane (DBCP)	$BrCH_2CHBrCH_2Cl$	Banned soil fumigant; runoff from fumigated agricultural fields	0.2 ppb

Organic pollutant	Structure[a)]	Major sources	Max. limit
o-Dichlorobenzene	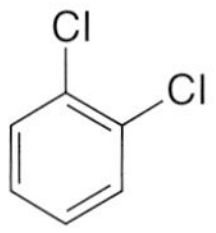 Cl Cl	Wastewater discharge from chemical industrial facilities	600 ppb
p-Dichlorobenzene	 Cl Cl	Wastewater discharge from chemical industrial facilities	75 ppb
1,2-Dichloroethane	$ClCH_2CH_2Cl$	Wastewater discharge from chemical industrial facilities	5 ppb
1,1-Dichlorethene	$Cl_2C{=}CH_2$	Wastewater discharge from chemical industrial facilities	7 ppb
cis-1,2-Dichlorethene	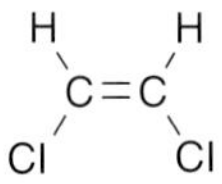 H H C=C Cl Cl	Wastewater discharge from chemical industrial facilities	70 ppb
trans-1,2-Dichloroethene	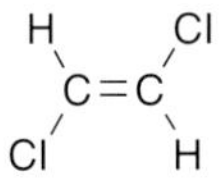 H Cl C=C Cl H	Wastewater discharge from chemical industrial facilities	100 ppb
Dichloromethane	CH_2Cl_2	Wastewater discharge from chemical and drug facilities	5 ppb
1,2-Dichloropropane	$ClCH_2ClCHCH_3$	Wastewater discharge from chemical industrial facilities	5 ppb

Organic pollutant	Structure[a)]	Major sources	Max. limit
Di-2-ethylhexyl adipate	O, O, O, O	Wastewater discharge from chemical industry	400 ppb
Di-2-ethylhexyl phthalate	CH_3, CH_3, O, O, O, O, CH_3, CH_3	Wastewater discharge from chemical, rubber, and plastic facilities	6 ppb
Dinoseb	OH, CH_3, O_2N, CH_3, NO_2	Herbicide; runoff from agricultural applications	7 ppb
Dioxin (2,3,7,8 TCDD)	Cl, O, Cl, Cl, O, Cl	Waste incineration and chemical industry discharge	0.03 ppt
Diquat dibromide	N⊕, ⊕N, 2 Br⊖	Herbicide; runoff from agricultural applications	20 ppb
Endothall	O, OH, O, OH, O	Herbicide; runoff from agricultural applications	100 ppb
Endrin	Cl, Cl, Cl, Cl, Cl, O, Cl	Banned pesticide; residual runoff from soils	2 ppb

Organic pollutant	Structure[a]	Major sources	Max. limit
Epichlorohydrin	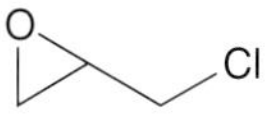	Water treatment chemical; chemical and polymer industry discharge	Treatment limit of 0.01% water at 20 mg l^{-1} concentration
Ethylbenzene	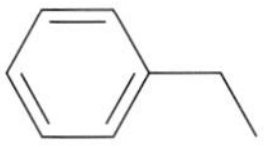	Petroleum refinery discharge	700 ppb
Ethylene dibromide (EDB)	$BrCH_2—CH_2Br$	Petroleum refinery discharge	0.05 ppb
Glyphosate	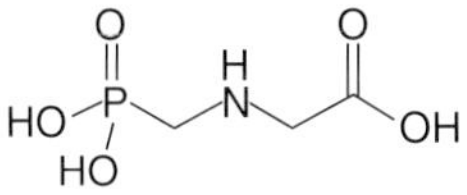	Herbicide; runoff from use	700 ppb
Heptachlor	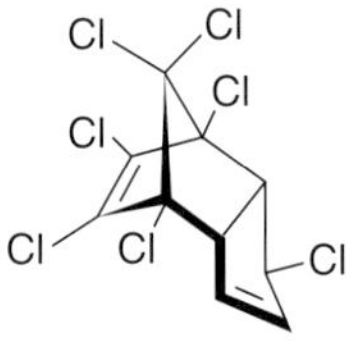	Banned pesticide used in homes for termites; leaching from past use in older homes into water supplies	0.4 ppb
Heptachlor epoxide	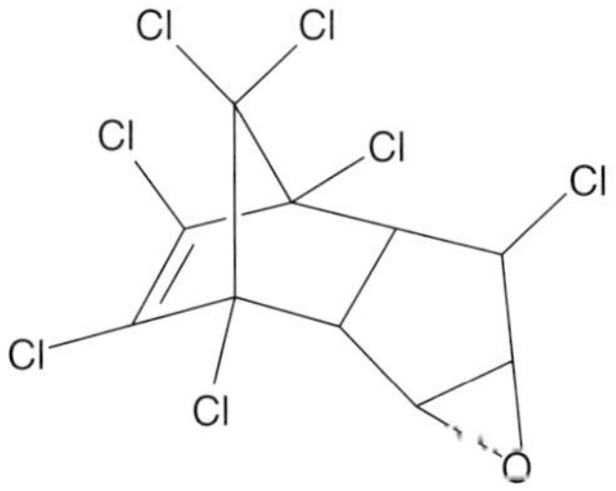	Oxidation product of heptachlor	0.2 ppb
Hexachlorobenzene	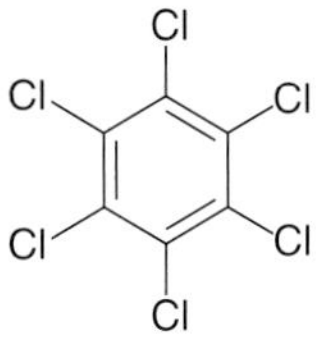	Metal refinery and agricultural chemical production waste water	1 ppb

Organic pollutant	Structure[a)]	Major sources	Max. limit
Hexachlorocyclo-pentadiene	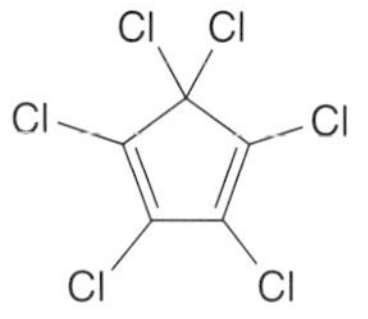	Chemical industry waste water	50 ppb
Lindane	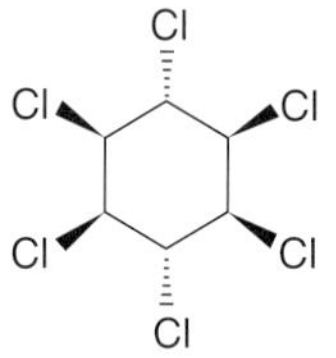	Insecticide; runoff from cattle and agriculture	0.2 ppb
Methoxychlor	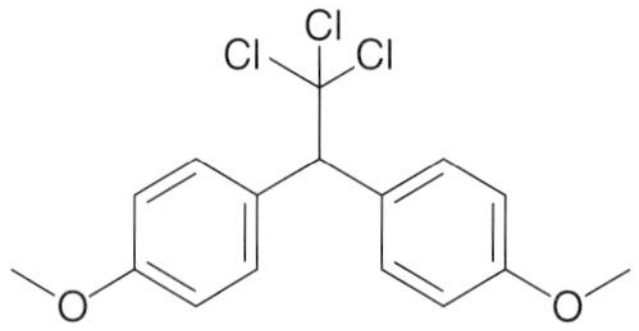	Insecticide; runoff from cattle and agriculture	40 ppb
Oxamyl or Vydate	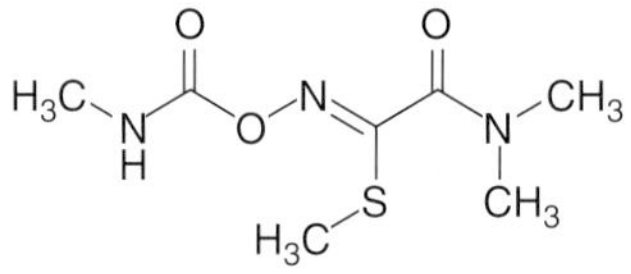	Insecticide; runoff from orchards and gardens	200 ppb
Polychlorinated biphenyls (PCBs)	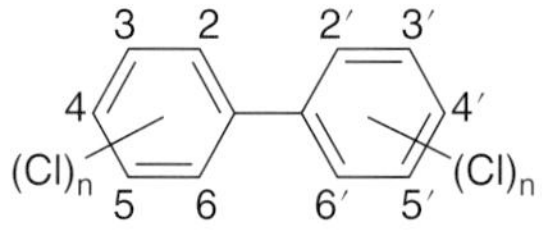	Waste chemical release, landfill runoff	0.5 ppb
Pentachlorophenol	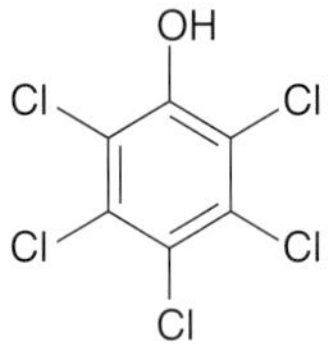	Waste water from wood preservation processing plants	1 ppb
Picloram	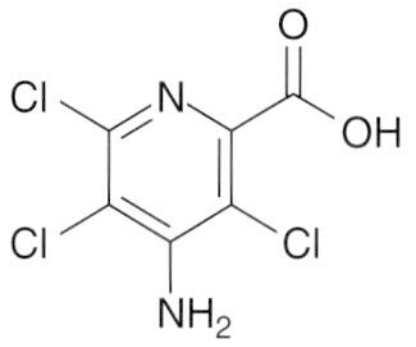	Herbicide; runoff from soils	500 ppb

Organic pollutant	Structure[a)]	Major sources	Max. limit
Simazine	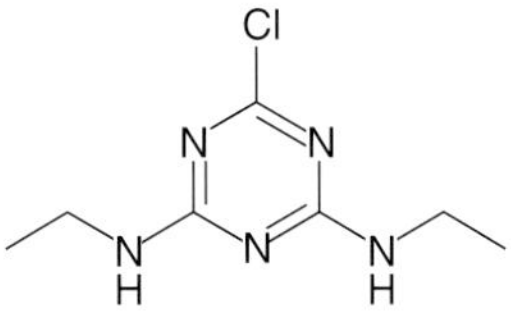	Herbicide; runoff from soils	4 ppb
Styrene	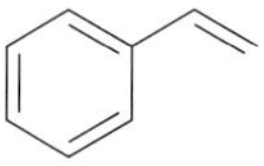	Discharge from polymer production, landfill runoff	100 ppb
Tetrachloroethene	$Cl_2C{=}CCl_2$	Dry-cleaning waste, chemical industry discharge	5 ppb
Toluene	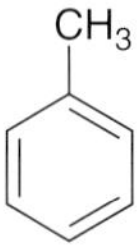	Petroleum refinery release	1 ppm
Toxaphene	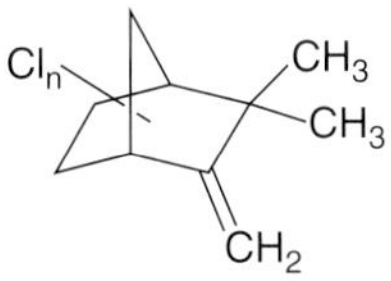	Banned insecticide; runoff and leaching from cattle and agricultural soils	3 ppb
2,4,5-TP or Silvex	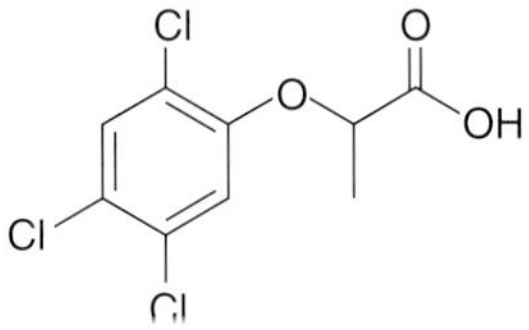	Banned herbicide; agricultural runoff	50 ppb
1,2,4-Trichlorobenzene	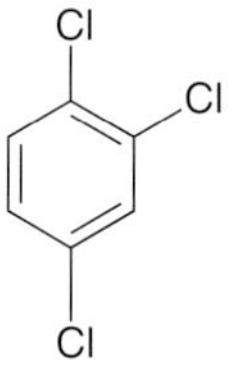	Discharge by textile industry	70 ppb

Organic pollutant	Structure[a)]	Major sources	Max. limit
1,1,1-Trichloroethane	$CCl_3—CH_3$	Discharge from industrial degreasing	200 ppb
1,1,2-Trichlorethane	$CCl_2H—CH_2Cl$	Chemical industrial release	5 ppb
Trichloroethene	$Cl_2C{=}CHCl$	Discharge from industrial degreasing	5 ppb
Vinyl chloride	$ClHC{=}CH_2$	PVC pipe monomer leaching, plastic factory discharge	2 ppb
Xylenes (total)	$CH_3—C_6H_4—CH_3$ (*o*, *m*, *p*)	Petroleum refinery discharge, chemical factory discharge	10 ppm

a) Structure from Wikimedia Commons.
Source: USEPA, *National Primary Drinking Water Regulations*, March 22, 2018. https://www.epa.gov/ground-water-and-drinking-water/national-primary-drinking-water-regulations#Organic (accessed June 26, 2018).

Appendix E

Chemicals Used in the Hydraulic Fracturing of Oil Shales for Natural Gas Extraction

Chemical	Property	Purpose
Hydrochloric acid	Acid	Dissolves rock, initiates cracks
Glutaraldehyde	Bactericide	Bacteria removal
Quaternary ammonium chloride	Bactericide	Bacteria removal
Tetrakis hydroxymethyl-phosphonium sulfate	Bactericide	Bacteria removal
Ammonium persulfate	Gel stabilizer	Delays gel breakdown time
Sodium chloride	Gel stabilizer	Delays gel breakdown time
Magnesium peroxide	Gel stabilizer	Delays gel breakdown time
Magnesium oxide	Gel stabilizer	Delays gel breakdown time
Calcium chloride	Gel stabilizer	Delays gel breakdown time
Choline chloride	Clay protector	Prevents clay from shifting or swelling
Tetramethyl ammonium chloride	Clay protector	Prevents clay from shifting or swelling
Sodium chloride	Clay protector	Prevents clay from shifting or swelling
Isopropanol	Freezing inhibitor	Slows corrosion, lowers freezing point
Methanol	Freezing inhibitor	Slows corrosion, lowers freezing point
Formic acid	Corrosion inhibitor	Keeps well piping from corroding
Acetaldehyde	Corrosion inhibitor	Keeps well piping from corroding
Light petroleum distillate	Fluid stabilizer	Maintains viscosity
Potassium metaborate	Fluid stabilizer	Maintains viscosity
Triethanolamine zirconate	Fluid stabilizer	Maintains viscosity
Sodium tetraborate	Fluid stabilizer	Maintains viscosity
Boric acid	Fluid stabilizer	Maintains viscosity
Zirconium complex	Fluid stabilizer	Maintains viscosity
Borate salts	Fluid stabilizer	Maintains viscosity
Ethylene glycol	Freezing inhibitor	Lowers freezing point, stabilizes gels
Methanol	Fluid stabilizer	Stabilized gels, lowers friction
Polyacrylamide	Friction reducer	Reduces water friction
Petroleum distillate	Friction reducer	Carrier for friction reducers
Guar gum	Fluid gelling	Allows sand suspension in water

Chemistry of Environmental Systems: Fundamental Principles and Analytical Methods, First Edition.
Jeffrey S. Gaffney and Nancy A. Marley.

Chemical	Property	Purpose
Polysaccharide blend	Fluid gelling	Allows sand suspension in water
Citric acid	Iron complexing	Prevents iron precipitation
Acetic acid	Iron complexing	Keeps iron from precipitating
Thioglycolic acid	Iron complexing	Keeps iron from precipitating
Sodium erythorbate	Iron complexing	Prevents iron precipitation
Lauryl sulfate	Emulsifier reducer	Prevention of emulsion formation
Sodium hydroxide	pH control	Maintain fluid effectiveness
Potassium hydroxide	pH control	Maintain fluid effectiveness
Sodium carbonate	pH control	Maintain fluid effectiveness
Potassium carbonate	pH control	Maintain fluid effectiveness
Copolymer of acrylamide and sodium acrylate	Scale reducer	Keeps well piping from building up scale
Sodium polycarboxylate	Scale reducer	Keeps well piping from building up scale
Phosphonic acid salt	Scale reducer	Keeps well piping from building up scale
Lauryl sulfate	Surfactant	Maintains viscosity, transport inorganics
Naphthalene	Surfactant carrier	Carrier fluid for surfactants
2-Butoxyethanol	Surfactant carrier	Fluid, surfactant stabilizer

Glossary

accuracy	How close the result of a measurement comes to the true value.
actinic flux	The spherically integrated photon flux incident at a point.
activation energy	The minimum energy needed for the chemical reaction to occur.
aerodynamic diameter	The diameter of a spherical particle with a density of $1\ g\ cm^{-3}$ that has the same settling velocity as an irregularly shaped particle.
algae bloom	A large unnatural growth of phytoplankton caused by a sudden increase in nutrients.
analyte	The species being measured.
analytical curve	A plot of measurement signal (y) as a function of the concentration of the chemical substance (x) that is being measured.
Anthropocene	The current geological period during which human activity has been the dominant influence on climate and the environment.
anthropogenic pollution	The introduction of harmful substances or the creation of harmful impacts in the environment that are directly tied to man's activities.
aphotic zone	Area of the ocean below the dysphotic zone where no sunlight can penetrate.
azeotrope	A mixture of liquids that has a constant boiling point because the vapor has the same composition as the liquid mixture.
baseline measurements	An accurate measurement of environmental species concentrations over time before any change occurs.
becquerel	The activity of a quantity of a radioisotope which results in the decay of one nucleus per second.
benthic zone	The bottom layer of a body of water.
biochemical oxygen demand (BOD)	A measure of the amount of oxygen consumed by bacteria during the decomposition of organic material present in a water sample.
biomolecular	A chemical reaction where the transition state involves two molecules.

Chemistry of Environmental Systems: Fundamental Principles and Analytical Methods, First Edition.
Jeffrey S. Gaffney and Nancy A. Marley.

blackbody	A theoretically ideal radiator and absorber of energy at all electromagnetic wavelengths.
bond dissociation energy	The minimum energy required to break a given bond in a molecule.
boundary layer	The height in the troposphere where atmospheric mixing is reduced due to cooling of the warmer air as it rises (approximately 1500 m).
breeder reactor	A nuclear reactor that generates more fissile material than it consumes.
Brewer–Dobson circulation	A large-cell stratospheric circulation consisting of an upwelling of air in the tropics, a horizontal movement at high altitudes toward the mid-latitudes, and a downwelling of air near the poles.
calibration curve	A method of determining the concentration of an analyte in an unknown sample from a plot of the method response of a set of standards (ordinate) to their concentrations (abscissa).
carbon footprint	The total amount of greenhouse gases emitted into the troposphere by an individual, an organization, an event, or a product, reported as an equivalent amount of CO_2.
carbon sequestration	The capture of CO_2 from power plant flue gases for the purpose of placing it into long-term storage.
carrying capacity	The maximum population size that the environment can sustain indefinitely.
catagenesis	A cracking process which results in the conversion of organic kerogens into hydrocarbons.
cathodic protection	A technique used to control the corrosion of a metal surface by connecting the metal to be protected to a more easily corroded "sacrificial metal" which act as the anode of an electrochemical cell.
cation exchange capacity (CEC)	The amount of positively charged ions that can be retained on a known mass of soil.
Chapman Cycle	The ozone–oxygen chemical reaction cycle that continually regenerates ozone in the stratosphere.
chemical actinometry	A method of determining the actinic flux by measuring the photolysis rate of a well-characterized species with known values for the absorption cross-section and the quantum yield.
chemical oxygen demand (COD)	A measure of the amount of oxygen required to oxidize both soluble and particulate organic matter in water.
chemiluminescence	The emission of light by an electronically excited molecule created as the product of a chemical reaction.
chemotroph	An organism that produces energy from electron capture instead of photon capture.
chlorofluorocarbons (CFCs)	Fully halogenated alkanes (methane, ethane, and propane) that contain chlorine, fluorine, and carbon.
circumpolar vortex	An upper-level, low-pressure area that rotates counterclockwise around the North Pole and clockwise around the South Pole.

climate	The average condition of the weather at a given region over a long period of time as exhibited by temperature, wind velocity, and precipitation.
climate forcings	The influences that change the amount of energy absorbed or lost, which cause temperatures to rise or fall.
cloud condensation nuclei (CCN)	Aerosols required as a substrate to allow for the condensation of water vapor to form clouds.
coal gasification	The process of producing syngas, a mixture of CO and H_2 from coal.
coalification	The process in which vitrinite becomes converted into coal of increasingly higher rank, with anthracite as the final product.
collisional deactivation	The transfer of the electronic excitation energy of a molecule to other molecules in the system through energetic collisions.
colloid	A heterogeneous mixture in which the particles are intermediate in size between those of a solution and a suspension and do not settle out upon standing.
composite sample	Samples obtained at one sample point over a specific time frame that are mixed together to represent an average of the water conditions over the period of time sampled.
conductivity	The measure of the ability of a material to pass an electrical current.
confined aquifer	A groundwater system that is covered by an impermeable layer, which prevents water from recharging the aquifer from the soil surface located directly above the aquifer.
Coriolis effect	The generation of a force on a rotating object which acts perpendicular to the direction of motion or the axis of rotation.
cosmogenic radioisotopes	Radioisotopes that are created by high-energy cosmic rays interacting with the nucleus of a naturally occurring stable atmospheric isotope, causing protons and neutrons to be expelled from the atom.
co-solvency	The increase in aqueous solubility of hydrocarbons due to the presence of other compounds in water that serve as a co-solvent.
crenon	The source of a river characterized by cooler temperatures, reduced oxygen content, low suspended materials, and slow-moving water.
Criegee biradical	A carbonyl oxide with two charge centers ($R_2C{\cdot}OO{\cdot}$).
criteria pollutants	Six air pollutants (NO_2, SO_2, ground-level O_3, CO, particulate matter, and lead), which have national air quality standards that define the allowable concentrations of these substances in ambient air.
decay series	The series of decay processes that converts a radioactive element into a series of different elements until it produces a stable isotope.
denuders	Gas collection devices that use a reactive solution or substrate to remove reactive gases before they reach the sample collection device or measurement system.

deposition velocity	The rate of deposition described as a flux, which is the amount of material moving from the air to a surface, including either the oceans or terrestrial surfaces.
detection limit	The lowest amount of a chemical species that can be detected by a measurement method.
detritus	Dead particulate organic material which includes the bodies or fragments of dead organisms.
diagenesis	The physical and chemical changes that occur during the conversion of organic sediment to sedimentary rock.
diffraction	The bending of light around the corners of an obstacle or surface opening, resulting in the separation of wavelengths.
diffusion denuder	An air-sampling device that collects the gas-phase species of interest on the walls of a flow-through tube reactor while the particulates pass through unaffected.
dissociation	A photochemical reaction where the electronic excitation results in breaking of the bonds between the atoms of the molecule.
Dobson unit (DU)	The thickness (in units of 10 μm) of a layer of pure ozone at STP that would give the same absorbance as that measured in the total atmospheric column.
Doppler effect	An apparent change in the frequency of an electromagnetic wave that occurs when the source is in motion relative to an observer.
dysphotic zone	The area of the ocean below the euphotic zone where sunlight intensity drops off significantly.
effluent	The outflow or discharge of liquid waste into a body of water.
einstein	The unit for 1 mol of photons.
Ekman spiral	A displacement of current direction by the Coriolis effect interacting with a surface wind over the ocean which results in a displacement of surface current at 45° to the right in the Northern Hemisphere and 45° to the left in the Southern Hemisphere, with successively deeper layers further displaced, resulting in an upward spiral motion of currents.
electromagnetic radiation	A form of energy that is produced by oscillating electric and magnetic waves.
energy transfer	The process of transferring energy from the excited state of 1 mol to a second molecule which is then raised to a higher-energy state.
enhanced oil recovery (EOR)	Oil recovery methods designed to alter the properties of the oil itself in order to increase oil recovery.
enthalpy of combustion	The enthalpy (heat) change that occurs when 1 mol of a compound is burned completely in oxygen, with all reactions and products in their standard states.
environmental chemistry	The study of chemistry in natural systems and how this chemistry changes when perturbed by anthropogenic activities and/or the release of chemicals into the environment that changes their natural background levels.
epilimnion	The topmost layer in a thermally stratified lake.

equivalent dose	A radiation dose quantity that represents the health effects of low levels of ionizing radiation on the human body.
equivalent effective stratospheric chlorine (EESC)	An estimate of the total effective amount of ozone-depleting halogens in the stratosphere calculated from the emission amounts in the troposphere, their ozone depletion potentials, and their transport times from the troposphere to the stratosphere.
euphotic zone	The top area of the ocean where photosynthesis occurs.
eutrophication	A state of excessive richness of nutrients in a lake or other body of water, which causes a dense growth of plant life resulting in a decrease in DO levels.
fallout	The residual radioactive material propelled into the upper atmosphere following a nuclear explosion.
flowback fluid	Murky, salty fluids which return to the surface from natural gas fracking wells once the well pressure is released.
fluorescence	The emission of light of longer wavelength after the absorption of a photon of shorter wavelength with a very short time interval.
fracking	The process of injecting liquids at high pressure into subterranean rocks in order to crack the source rock and release natural gas trapped as bubbles in the rock.
Fresnel's equation	An equation that describes the ratio of the reflected light intensity to the intensity of incident light for a perpendicular beam of unpolarized light.
fuel cell	An electrochemical cell that converts chemical energy into electricity through an electrochemical reaction of H_2 with an oxidizing agent, typically O_2.
gasoline gallon equivalent (GGE)	The amount of non-gasoline fuel that is required to equal the energy content of 1 gal of gasoline.
geoengineering	The deliberate and large-scale intervention in the Earth's climate system with the aim of mitigating the adverse effects of climate change.
geothermal gradient	The rate of increasing temperature with respect to increasing depth.
glycolysis	The enzymatic breakdown of glucose to produce pyruvate.
global distillation effect	A repeating cycle, which acts to carry persistent pollutants from warmer regions to colder regions.
global warming potential (GWP)	The measure of how much IR energy will be absorbed by 1 ton of a greenhouse gas in the atmosphere over a given period of time relative to 1 ton of CO_2.
grab sample	A single sample taken just below the water surface at a specific time or over as short a period as is feasible.
green chemistry	The development of chemical processes that use smaller amounts of safer chemicals with less energy in order to lower their environmental impacts.
greenhouse gases	Atmospheric gases that can absorb the IR radiation emitted from the Earth's surface, trapping it in the atmosphere and preventing it from exiting into space.

half-life	The amount of time required for the reactant concentration to fall to half of its original value.
hardness	The amount of Ca^{2+} and Mg^{2+} salts in water.
Henry's Law	A gas law which states that the amount of dissolved gas is proportional to its partial pressure in the gas phase, with a proportionality factor called the Henry's Law constant.
heterogeneous reaction	A chemical reaction in which the reactants are of two or more different phases, or where one or more reactants undergo chemical change at an interface.
humin	An inhomogeneous dark brown class of organic compounds that are insoluble in water at all pHs.
hydrochlorofluorocarbons (HCFCs)	Halogenated alkanes that contain chlorine, fluorine, and carbon along with one or more hydrogens.
hydrological cycle	The sequence of processes that control the circulation of water throughout the hydrosphere.
hydrosphere	The combined sources of water in all its forms found on, under, and above the surface of the Earth.
hygroscopicity	The ability of a molecule to take up water.
hypolimnion	The lower layer of water in a stratified lake.
hypoxia	An environmental phenomenon where the concentration of dissolved oxygen in the water column decreases to a level that can no longer support living aquatic organisms.
impingers	Specially designed bubble tubes used for collecting airborne chemicals in a liquid medium.
interference	A phenomenon where two waves of the same frequency superimpose to form a new wave with a higher or lower amplitude depending on their phase difference.
internal conversion	The transfer of the electronic excitation energy of a molecule to vibrational modes of the same molecule.
intersystem crossing	A radiationless process involving an electronic transition between two electronic states with different spin multiplicity.
ion selective electrode (ISE)	An electrode constructed with a permeable membrane that passes the ion of interest.
irradiance	The photon flux crossing a surface.
isomers	Compounds with the same molecular formula but different chemical structures.
kerogen	Organic material in rocks that cannot be taken up into solution by the use of normal organic solvents because of its high molecular weight.
kinetic isotope effect	The change in the rate of a chemical reaction due to a change in reactant isotopes leading to the favoring of one isotope over another in a chemical reaction.
lacustrine deposits	Sedimentary rock formations in the bottom of ancient lakes.
light absorption	The interaction of light with a particle or gas molecule that removes energy from the incident light and converts it to another form.

light scattering	The interaction of light with a particle or gas molecule that maintains the total amount of energy in the incident light but, in most cases, alters the direction in which the light travels.
limnetic zone	The sunlit surface waters of a lake or pond, away from the shore.
littoral zone	The area of a lake or pond close to the shore.
matrix effect	A situation where the sample matrix can contribute to the analyte signal.
methane clathrate	A solid similar to ice in which a large amount of CH_4 is trapped inside the crystal structure of water.
megacity	A large metropolitan area with a population of 10 million or greater.
mixing ratio	The ratio of the amount of one chemical species to the amount of the entire air sample.
molecularity	The number of reactants involved in the transition state of a chemical reaction.
Montreal Protocol (on Substances that Deplete the Ozone Layer)	An international treaty signed in 1987, designed to protect the ozone layer by phasing out the production of numerous substances that are responsible for ozone depletion.
Multi-Filter Rotating Shadow Band Radiometer (MFRSR)	An instrument used to measure the diffuse and total components of the solar irradiance at different wavelengths in the visible or UV spectral regions.
non-criteria pollutants	Air or water pollutants that have been identified as posing a health hazard but are not listed as criteria pollutants.
nuclear burning	The introduction of a series of nuclear reactions, including neutron capture, fission, and other decay processes, which convert radioactive species to stable elements.
octane rating	A standard measure of the performance of an engine using fuel.
oil equivalent volume (OEV)	The volume of fuel required to obtain the same energy as 1 m^3 of fuel oil.
orographic lift	The forcing of an air mass from a low altitude to a higher altitude due to rising terrain.
ozone-depleting potential (ODP)	The relative amount of degradation that a chemical species can cause to the stratospheric ozone layer.
ozonesonde	A lightweight, compact, and inexpensive balloon-borne instrument for measuring atmospheric ozone.
paleoclimatology	The study of climatic conditions, and their causes and effects, in the geologic past, using evidence found in glacial deposits, fossils, and sediments.
persistent organic pollutants (POPs)	Hazardous organic chemical compounds that are resistant to biodegradation and thus remain in the environment for a long time.
phagocytosis	The process by which a cell engulfs a solid particle to form an internal compartment known as a phagosome.
phosphorescence	The emission of light of longer wavelength after the absorption of a photon of shorter wavelength with a much longer time interval than occurs in fluorescence.

photoacoustic effect	The production of sound waves in a sample produced by the absorption of pulsing light.
photochemical reactions	Chemical reactions initiated by the absorption of energy in the form of light.
photochemistry	The study of chemical and physical reactions of molecules caused by the absorption of visible and/or ultraviolet light.
photoelectric effect	The emission of electrons or other free charge carriers from a material when it is exposed to light.
photostationary state	A steady state reached by a photochemical reaction in which the rates of formation and reaction of a species are equal.
photovoltaic effect	The generation of a voltage or an electric current in a photovoltaic cell when it is exposed to sunlight.
planetary albedo	The fraction of the solar radiation that is reflected by the different surfaces of the Earth's crust, clouds, and atmospheric aerosols.
planetary boundary layer (PBL)	The layer of air whose behavior is directly influenced by its contact with the planetary surface.
polar stratospheric clouds (PSCs)	Clouds composed of nitric acid and water that form in the lower stratosphere during the winter months at temperatures of 195 K.
potamon	The downstream portion of a river characterized by warmer temperatures, higher suspended particulates, lower oxygen levels, slow flow, and sandier bottoms.
precision	The degree to which repeated measurements give the same value whether or not that value is the true value.
primary pollutants	Air pollutants that are emitted directly from the source into the troposphere.
primordial radioisotopes	Radioisotopes that have existed in their current form since before the Earth was formed.
produced water	Water that occurs naturally in a shale reservoir which flows to the surface with the fracking fluids in a natural gas fracking well when the pressure is released.
profundal zone	The deep-water area of a pond or lake located below the range of effective light penetration.
quantum yield	The number of molecules undergoing a photochemical or photophysical process divided by the number of photons absorbed.
radiation dose	The amount of energy absorbed by a medium per unit mass as a result of exposure to ionizing radiation.
radiative forcing	The change in the Earth's energy balance when one of the variables is changed while all the others are held constant.
radioactivity	The spontaneous emission of high-energy particles, electromagnetic radiation, or both, caused by the disintegration of atomic nuclei.
rain shadow	An area where no rain falls on the leeward side of a mountain due to the prevailing winds having lost their moisture before crossing the mountain range.

random errors	Measurement errors that arise from unpredictable fluctuations and are not constant or of the same sign.
reflection	The change in direction of an electromagnetic wave at an interface between two media with different densities so that the wave returns to the medium from which it originated.
reformulated gasoline (RFG)	Gasoline containing oxygenated additives resulting in an oxygen content of 2% by weight.
refraction	The change in direction of an electromagnetic wave when traveling from one medium to another of different density.
refractive index (*n*)	The speed of light in a vacuum divided by the speed of light in the medium.
relative humidity	The percentage of the amount that is needed for saturation at the same air temperature.
reservoir species	Chemical compounds that act to sequester a more reactive species, thereby preventing it from taking part in chemical reactions. Under the right conditions, they can slowly react or photolyze to release the more reactive species.
resolution	The ability to detect small changes in the signal from a measurement method or the concentrations of the species being measured.
response time	The length of time that it takes to make a measurement after the sample has been introduced into the measurement system.
retention time	The time it takes for an analyte to pass through a chromatography column from the injection port to the detector.
retroreflector	A reflection device commonly used in open-path spectroscopy that reflects light back along the incident path.
rhithron	The upstream portion of a river characterized by cool temperatures, high oxygen levels, and fast, turbulent flow.
roentgen	The amount of ionizing radiation producing one electrostatic unit of positive or negative ionic charge in one cubic centimeter of air under standard conditions.
SATP (standard ambient temperature and pressure)	Atmospheric conditions of a pressure of 1 atm and a temperature of 298 K (25 °C).
scintillation	A flash of light produced in certain materials when they absorb ionizing radiation.
Secchi disk	An opaque disk, which can be either white or black and white, used to gauge the transparency of water by measuring the depth at which the disk ceases to be visible from the surface.
secondary pollutants	Air pollutants produced from chemical reactions of primary pollutants.
secular equilibrium	The situation where the production rate of a radioisotope is equal to its decay rate.
sensitivity	The ratio of the change in a measurement signal to the change in the concentration or amount of the species being measured.
settleable solids	Any solid particulates that do not remain suspended or dissolved in water not subject to motion.

smog	The combination of soot, fly ash, and wet acidic aerosols from coal and wood burning, considered to be a mixture of smoke and fog.
solution	A homogeneous mixture of one or more solutes dissolved in a solvent.
solvation	State where a solute is surrounded or complexed by solvent molecules, leading to the stabilization of the solute species in the solution.
standard addition	A standard is added directly to a sample and the result of the sample is compared to the result of the same sample with standard added to determine the presence of interferences.
steam cracking	A process that breaks down the large saturated hydrocarbons into smaller hydrocarbons.
steam reforming	The process of producing hydrogen from water vapor and methane at high temperatures.
Stoke's Law	A mathematical expression that describes the force of viscosity on a small sphere moving through a fluid.
Stokes number	A unitless number that characterizes the behavior of particles suspended in a flowing fluid, defined as the ratio of the characteristic stopping time of the particle to the characteristic flow time of the fluid around the surface of an object.
surfactants	Surface active compounds that contain at least two functional groups: one that is water soluble and another that is oil soluble.
suspension	A heterogeneous mixture in which some of the particles settle out of the mixture upon standing.
sustainability	A state where energy, water, and food resources meet the demands of today's societies without causing harm to the environment or depleting resources for future generations.
sustainable development	Development that meets the needs of the present without compromising the ability of future generations to meet their own needs.
synergism	The interaction of two or more substances to produce a combined effect greater than the sum of their separate effects.
systematic errors	Errors in a measurement method that are constant and always of the same sign.
termolecular	A chemical reaction where the transition state involves three molecules.
thermal inversion	An increase in temperature with height, which is the inverse of the normal condition.
thermocline	A steep temperature gradient in a body of water such as a lake that has a layer of water above and a layer of water below at different temperatures.
thermohaline circulation	The component of general oceanic circulation that is controlled by differences in temperature and salinity.
total solar irradiance (TSI)	The amount of solar radiative energy incident on the top of the Earth's atmosphere.

total suspended particulates (TSP)	The first regulatory measure of the mass concentration of particulate matter (PM) in air.
transesterification	A chemical reaction that exchanges an alkyl group of an ester with an alkyl group of an alcohol.
transpiration	The process of water movement through a plant, including evaporation from plant surfaces such as leaves, stems, and flowers.
unconfined aquifer	A groundwater system that is recharged by water percolation through the ground surface directly above the aquifer.
unimolecular	A chemical reaction where the transition state involves only one reactant molecule.
vacuum UV	The highest-energy ultraviolet radiation at wavelengths between 10 and 300 nm.
vitrinite	A coal-like material formed from the original deposition of woody plants.
weather	The state of the atmosphere at a particular place and time with respect to the meteorological conditions.
window region	The wavelength region where the major greenhouse species do not absorb.
zero air	A synthetic mixture of ultrahigh purity O_2 at 22% by volume and N_2 at 78% by volume.
zwitterion	An organic molecule that has one functional group with a positive charge and one functional group with a negative charge and the net charge of the entire molecule is zero.

Index

Chemistry of Environmental Systems: Fundamental Principles and Analytical Methods, First Edition.
Jeffrey S. Gaffney and Nancy A. Marley.

q

r

S